INTRODUCTION TO
SOLID STATE IONICS
PHENOMENOLOGY AND APPLICATIONS

INTRODUCTION TO
SOLID STATE IONICS
PHENOMENOLOGY AND APPLICATIONS

C. S. SUNANDANA

UNIVERSITY OF HYDERABAD
INDIA

CRC Press
Taylor & Francis Group
Boca Raton London New York

CRC Press is an imprint of the
Taylor & Francis Group, an **informa** business

A CHAPMAN & HALL BOOK

CRC Press
Taylor & Francis Group
6000 Broken Sound Parkway NW, Suite 300
Boca Raton, FL 33487-2742

First issued in hardback 2019

ISBN-13: 978-1-4822-2970-7 (hbk)

Library of Congress Cataloging-in-Publication Data

Sunandana, C. S., 1949- author.
    Introduction to solid state ionics : phenomenology and applications / C.S. Sunandana.
        pages cm
    Includes bibliographical references and index.
    ISBN 978-1-4822-2970-7
    1. Superionic conductors. 2. Solid state physics. I. Title.

QC717.S76 2016
530.4'16--dc23
2015021360

Visit the Taylor & Francis Web site at
http://www.taylorandfrancis.com

and the CRC Press Web site at
http://www.crcpress.com

Contents

Preface

Solid state ionics is the science and technology of superionic conductors, also called fast ion conductors or solid electrolytes. It is a modern science and technology that started growing rapidly in the 1990s when the Li-ion battery was first introduced to the global market by Sony Corporation. It is in the making and could be called the science and technology of the twenty-first century. Earlier Samar Basu demonstrated the ability of Li to be intercalated into graphite, laying the foundation for the Li-ion battery we know today [P.1].

Michael Faraday sowed the seeds of this science in the nineteenth century by giving us the basic vocabulary—"ion," "cathode," "anode"—and enunciating the laws of electrochemical equivalence and of electrolysis. "Faraday" is a unit of capacitance as important as the Coulomb, the unit of charge. He demonstrated that the semiconducting silver sulfide becomes a solid ionic conductor at not too high temperatures. Einstein chipped in with his random walk theory of diffusion while Nernst demonstrated the dramatic "Nernst glower." Einstein and Nernst are responsible for the fundamental connection between ion diffusion and ionic conductivity.

The advent of solid state ionics happened in the early twentieth century when Tubandt and Lorenz measured the electrical conductivity of silver halides and discovered superionic conductivity in silver iodide at a temperature of 420 K. This triggered a fundamental study of silver and copper ion conductors besides leading to the discovery of the room temperature superionic conductor Ag_4RbI_5. These studies integrated the concepts of solid state physics and electrochemistry into solid state ionics.

The science and technology of superionic conductors has grown to an extent that warrants a pedagogic study of this fascinating subject at the (under)graduate level. This book is based on this conviction. It is meant to be an introduction to the subject essentially inspired by the solid state physics as expounded in many textbooks beginning with C. Kittel.

This book essentially contains 10 chapters, 7 of which are devoted to phenomenology and 3 to applications.

The idea of writing this book was born in 2011 and a 6-month sabbatical sanctioned in 2012 by the University of Hyderabad helped me do some spade work. A kind invitation by Aastha Sharma on behalf of CRC Press in April 2013 made me send a formal proposal that led to the making of this book. I thank Stephanie Morkert for coordinating this project. Alexander Edwards, Richard Tressider, and Christine Selvan are thanked for very professional production support.

This book draws upon my research experience in solid state ionics for over three decades. A large number of masters and graduate students have taken part in this endeavor. I owe a debt of gratitude to all these people. It is heartening to find that my first PhD student, associate professor Dr. Palani Balaya, National University of Singapore, Singapore, is a leader in this area in his own right. My employer, the University of Hyderabad, has always given me unstinted support, currently in the form of an honorary professorship.

I am grateful to the publishers of books, journals, patents, and e-literature—too numerous to mention—for permission to cite works of other authors.

I am grateful to Professor Dinkar Sirdeshmukh, formerly of Kakatiya University, Warangal, India, for his encouragement. On a personal front, I am grateful to my parents, wife, Prabha, son, Dr. Sumohana Channappayya, daughter-in-law, Krupa, and granddaughter, Avani, for all the love and encouragement that made this work possible.

I would gratefully welcome feedback on this book, which can be addressed to sunandana@gmail.com.

Channappayya Shamanna Sunandana
Andhra Pradesh, India

REFERENCE

P.1   S. Basu, C. Zeller, P. J. Flanders, C. D. Fuerst, W. D. Johson, J. D. Fischer, Synthesis and properties of lithium-graphite intercalation compounds, *Mater. Sci. Eng.*, 38 (1979) 275.

Author

Channappayya Shamanna Sunandana studied at Sarvodaya High School, Tumkur, Karnataka, and Government College, Tumkur. He earned his BSc with distinction from the University of Mysore in 1968, majoring in physics and mathematics. Thereafter, he went to the Indian Institute of Technology Madras (IIT-M) from where he earned his MSc in physics in 1970 and PhD in physics in 1976, with a thesis on "ESR studies on some x-irradiated and doped sulphates," under the guidance of the late Professor C. Ramasastry. After a five year postdoctoral stint at the Materials Science Research Center (MSRC) in IIT-M, he joined the faculty of the School of Physics, University of Hyderabad (UoH), Hyderabad. He was professor at UoH from 1987 to 2014 and is currently honorary professor. He has conducted experimental research in condensed matter physics and physics of materials supervising 20 MPhil, 3 MTech, and 10 PhD projects. PhD students trained by him have occupied national and international academic positions and scientist positions. Dr. Sunandana has published more than 150 research articles in peer-reviewed journals, including more than 10 review articles and book chapters. Two of his articles appear on *Wikipedia*. He has coauthored three books, one published by Springer International Publishing and the other two by Lambert Academic Publishing, Germany. He has presented at conferences in Japan, Malaysia, and the United States.

Dr. Sunandana has taught several courses, including "Introduction to Solid State Physics" and "Probes of Condensed Matter and Physics of Materials" at the master's level, besides instructing undergraduate and postgraduate students in laboratory courses such as mechanics, solid state physics and microwave lab. He has been running a monthly poster called "Impertinent Questions" since 1991, featuring motivating science issues and original poetry. He has won many scholarships during his educational career. He is a fellow of the Andhra Pradesh Academy of Sciences (now Telangana State Academy of Sciences), foundation member of the Materials Research Society of India, and a life member of the Indian Physics Association.

1

What Is Solid State Ionics?

1.1 Perspectives

What is solid state ionics? To answer this question, we should know what an *ion* is. Hydrogen and oxygen ions rule this world, with a little help from carbon. The people of this terra firma stand firmly on essentially ionic structures made of Si^{4+} and $O^=$ aided a little by Ca^{++} and Mg^{++}. H^+ is a proton and $O^=$ is stable ion.

An ion is an atom or molecule in which the total number of electrons is not equal to the total number of protons, giving the atom a net positive or negative electrical charge and, more importantly, mobility and stability. Ions can be created by both chemical and physical means. In chemical terms, if a neutral atom loses one or more electrons, it has a net positive charge and is known as a cation. If an atom gains electrons, it has a net negative charge and is known as an anion. An ion consisting of a single atom is an atomic or monatomic ion; if it consists of two or more atoms, it is a molecular or polyatomic ion. In the case of physical ionization of a medium, such as a gas, what are known as "ion pairs" are created by ion impact, and each pair consists of a free electron and a positive ion.

The word *ion* comes from the Greek ἰόν (*ion*), meaning "going," the present participle of ἰέναι (*ienai*), meaning "to go." This term was introduced by English physicist and chemist Michael Faraday in 1834 for the then-unknown species that *goes* from one electrode to the other through an aqueous medium. Faraday knew that since metals dissolved into and entered a solution at one electrode, and new metal came forth from the solution at the other electrode, some kind of substance *moved through the solution in a current, conveying matter from one place to the other.* Faraday also introduced the words *anion* for a negatively charged ion and *cation* for a positively charged one. In Faraday's nomenclature, cations were named thus because they were *attracted to the cathode* in a galvanic device and anions were named due to their *attraction to the anode.* There are free ions and ions in liquids and solids. Figure 1.1 shows the energies of the alkali and halide ions both in the free state and in ionic crystals. This figure dramatically illustrates the effect of placing free ions in an ionic crystal on their energies. Note particularly the proximity of the energies of Cu^+, Ag^+ with those of the halide ions.

Ions come with positive or negative charges in different sizes. Size and mass are very important in solid state ionics (SSI) as they decide the forces experienced by them when they sit in a solid and move. Figure 1.2 shows ions in various sizes with diameters ranging between ~0.1 (Li^+) and ~0.4 (I^-) nm.

SSI is the science of the motion of ions in solids. Ions are the prime movers of SSI just as electrons are the current carriers in solid state electronics (SSE). While the fundamental particles, electrons, in solids are ubiquitous, general and nonspecific ions are atom specific even if they are basic constituents of all solids (metallic semiconducting and insulating). Just as electrons rule the science of the solid state (physics and chemistry), ions in solids particularly the "fast" or the "super" ones that are the subject of this book possess a science that embraces all the basic concepts of solid state physics but in a modified form. Structure and dynamics along with thermodynamics and statistical mechanics play a vital role in the science of ion motion. The emergence of nanoscience has revitalized the SSI setting challenging goals for future technologies. SSI is a physicochemical science at once dynamic in nature involving nonequilibrium processes at the iono-electronic level. According to

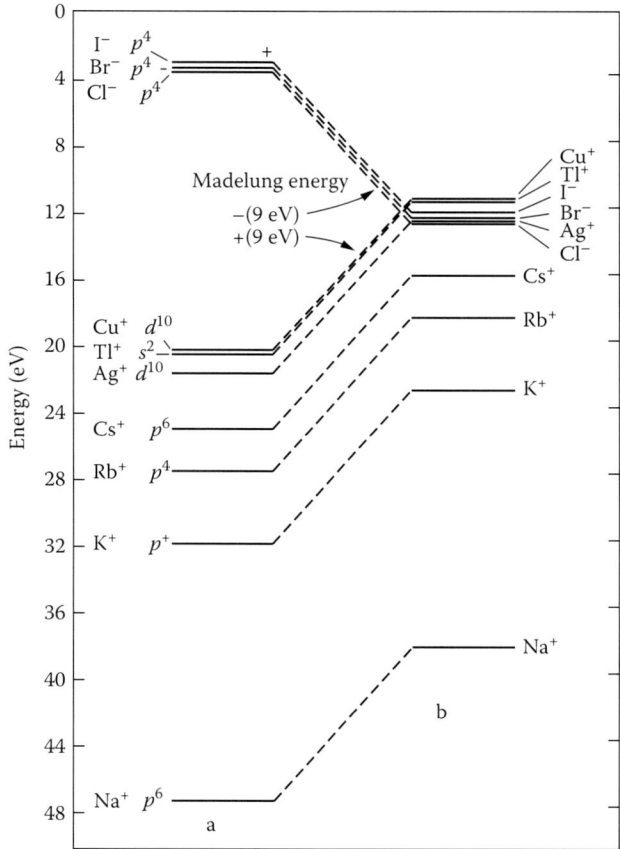

FIGURE 1.1   Energies of monovalent cations in *a*, free state and *b*, in crystals. Madelung energy is the electrostatic energy of an ionic crystal. (Adapted from Brown, F.C., *The Physics of Solids*, W.A. Benjamin, New York, 1967.)

Whittingham, the founding principal editor of the international journal *Solid State Ionics* launched in 1980, SSI is defined as follows:

> Solid state ionics is an interdisciplinary subject area that encompasses chemists, physicists and materials scientists and engineers. It incorporates the synthesis of materials, their characterization (including physical, thermodynamic, kinetic), potential applications and theory.

Trinity crystal structure (usually polymorphic), lattice vibrations, and electronic band structure are involved in a fundamental way in SSI—an emerging science with an exciting array of futuristic applications in daily life. Structures in SSI are framework structures supporting charged matter transport over macroscopic distances in a material or in a device. SSI is supportive of and complementary to solid state electronics as Figure 1.3 illustrates. Let us now focus on the theme of this book through a few perspectives beginning with the (un)common salt.

1.1.1 The (Un)common Salt

Sodium iodide, like sodium chloride, is a common salt used as an additive to form the "iodized" table salt. But silver iodide is an uncommon salt. So is lead fluoride, zirconia, and urania, and of course sodium beta alumina. Remember that lead is the vital component of the lead–acid battery where relativity is at work to contribute to the dc voltage the battery generates. bcc iodine ion framework in silver iodide

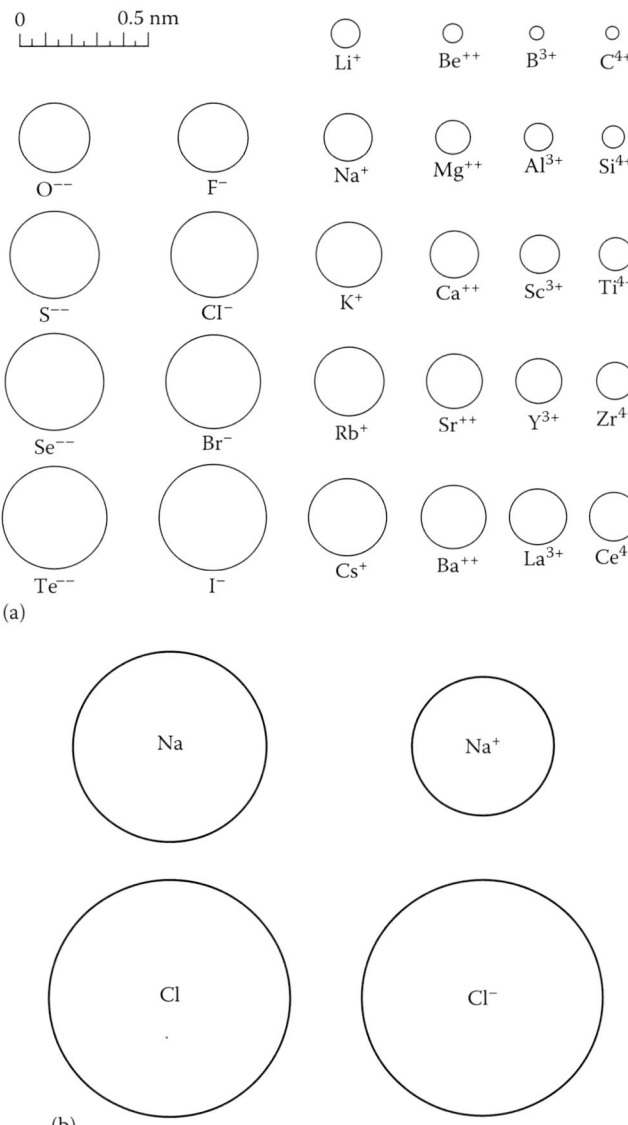

FIGURE 1.2 (a) Positive and negative ions that make up a large class of insulating and conducting ionic solids. Li⁺, Na⁺, O=, and F⁻ are prime movers of fast ion conductors. (b) Atom shrinks when a cation is formed while it is enlarged upon anion formation. This has significance for crystal structure formation and transformation. HI is unique among halides. It is the most covalent. Li⁺ in LiI is a rattler. NaI has no vacancy pairs (Schottky pairs). KI, RbI, and CsI team up with AgI to produce room temperature fast ionic or superionic conductors to be discussed in this book. (Adapted from Pauling, L., *The Nature of the Chemical Bond*, 3rd ed., Cornell University Press, Ithaca, NY, 1969; *General Chemistry*, Dover, New York, 1970.)

structure attained through a structural phase transition supports the rapid diffusive motion of Ag⁺ ions. Likewise, the very robust lead ion framework in the fluorite structure of lead fluoride supports the fast motion of F⁻ ions.

SSI helps in the understanding of such motion in a variety of corresponding specific crystal structures, which assists in the rational design of SSI materials and the development of microionic and nanoionic power devices. The chemical energy of an ionic solid has its origin in the storage and motion of its mobile ions. Its conversion to suitable useable forms, say, mechanical and electrical, is the theme of SSI technology that includes Li battery technology and fuel cell technology.

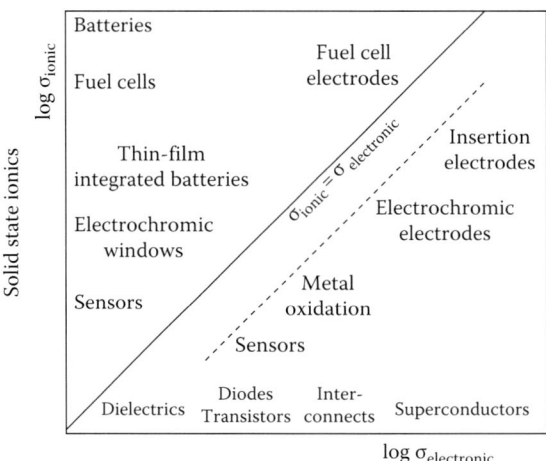

FIGURE 1.3   The interface formed by solid state ionics (characterized by ionic conductivity σ_{ionic} due to ion currents) and solid state electronics (characterized by electronic conductivity $\sigma_{electronic}$ due to electron currents). The devices based on the two disciplines are indicated. Note that the electrodes and sensors appear closest to the interface. (From Tuller, H.L., Highly conducting ceramics, in Buchanan, R.C. (ed.), *Ceramic Materials for Electronics*, 3rd edition, Marcel Dekker Inc., 2004.)

Let us now compare two iodides: NaI and AgI (Figure 1.4). NaI expands on heating while AgI contracts on heating! NaI is soluble in water while AgI is not. NaI has a rocksalt structure until it melts, while AgI is polymorphic and has a wurtzite-type hexagonal structure at room temperature.

NaI has a conductivity of 10^{-8} S/cm at 282°C (555 K), while for AgI the value is 1 S/cm at roughly half of that temperature, namely, 147°C. NaI melts only once at 661°C, whereas AgI melts twice: Ag^+ component at 147°C and the bcc iodine framework at 552°C. The density of NaI is 3.67 g/cm^3 while that of AgI is 5.675 g/cm^3. NaI exhibits soliton-type 1D excitations due to isolated Na^+ vacancies (there are no vacancy pairs!) at temperatures greater than or equal to 636 K. AgI is "plasmonic" at temperatures greater than 147°C—an unusual one-component plasma of Ag^+ ions forming a liquid-like disorder in high ion-conducting state of AgI.

Na^+ positions in the fcc lattice of NaI (Figure 1.4a) are identifiable by x-ray diffraction (XRD) right from room temperature up to its melting point; Ag^+ positions in hexagonal AgI (Figure 1.4b) are identifiable through XRD beyond 147°C (Figure 1.4c) when it transforms to the I-centered bcc phase. Like

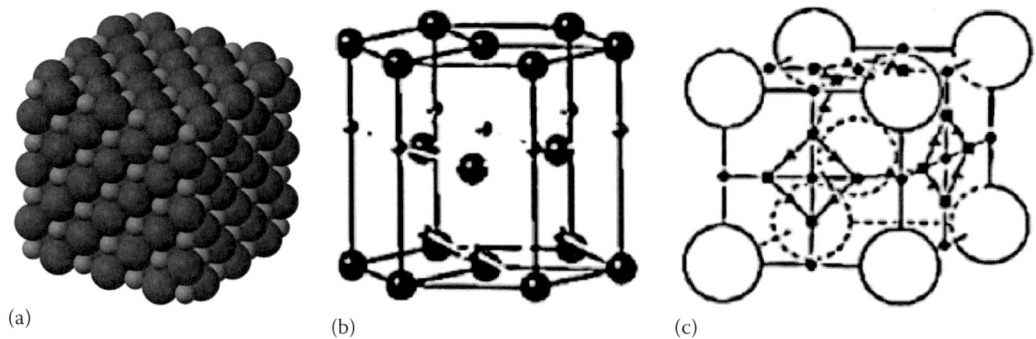

(a)   (b)   (c)

FIGURE 1.4   (a) Face-centered cubic NaI structure (black spheres: I$^-$, white spheres: Na$^+$). I–I distance = 0.647 nm is the lattice parameter identical to that of the AgI (zincblende) lattice(not shown). (b) Hexagonal β-AgI structure, a = 0.459 nm, c = 0.749 nm (big spheres: I$^-$, small spheres: Ag$^+$). (c) A model for α-AgI structure (big circles: I$^-$; ▲, ■, ● represent a total of 42 sites available for occupation by just two Ag^+ ions in the body-centered cubic lattice with a = 0.506 nm). Note that α-AgI has a higher density than β- or γ-AgI.

NaCl, NaI is a "common" salt used as an additive to form "iodized" table salt. But AgI is an uncommon salt like PbF_2 or ZrO_2 or sodium beta alumina. What is the origin of ionic conductivity? Section 1.2 would answer this question. Let us now look after a few SSI perspectives from human body processes.

1.1.2 Nature's Brain Power Source

Human brain needs power to function. This power has to be and indeed is generated by the human body itself and is perhaps the first electrochemical cell ever designed and fabricated. As one uses the brain, this cell gets "discharged," and like the cellphone battery, it has to be charged again. Sleep and rest help the body regenerate the "brain power." It is one of the most important and primary biological activities for sustainable, creative development of human civilization in the form of intellectual and technological development. It is thus necessary to see how it is done in the context of SSI because it could motivate us to seek a better solution to the problem of conversion of chemical energy to mechanical/electrical energy. It is curious to note that intense thinking—creative thinking—uses the natural cell power more and more and the back of the head is found to get warm. A similar process happens in a laptop computer. Both the consequences are due to the heat generated according to the second law of thermodynamics—the work–heat principle. This heat is irreversibly lost to the environment although at present efforts are on to utilize the waste heat.

From a biophysicist's perspective [1.1], the brain is a massively parallel analog electrochemical computer, implementing algorithms through biophysical processes at the ion-channel level. The fluid surrounding a neuron is like dilute seawater. It is abundant in Na^+ and Cl^- ions together with traces of Ca^{++} and Mg^{++}. The cytoplasm inside the cell is rich in K^+ and many charged organic molecules. Neuron membrane is a selective barrier, separating these two milieus. In the resting state, only K^+ can trickle through. As K^+ escapes, it creates a negative charge on the cytoplasm. Potential energy represented by this difference in polarity for each cell is its "resting potential." This can be a measurable ~1/10 V. Twenty such cells equal the electrical energy of a 2 V flashlight battery.

The cell membrane actually has myriads of channels [1.2]. Each channel (Figure 1.5a) has a molecular structure configured to let only one particular ion substance pass through its gate. In the resting state most of these gates are closed. A nerve is triggered to fire when a critical number of its excitationary synapses receive neurotransmitter signals that tip the balance of the cell's interior charge. The gates on the sodium channel are caused to open. Na^+ ions near these gates down the rush. Single positive charge left by the missing outermost electron is attracted to the negative ions inside the cell. This event always begins in the cell body. As soon as the in-rushing Na^+ makes the polarity more positive in that local region (italics), nearby Na gates further down the axon are likewise triggered to open (Figure 1.5b).

This phenomenon is called depolarization. It constitutes the electrical action potential. Most importantly, it is self-perpetuating. Regardless of how long the journey down the axon, in some cases longer than a meter, the signal strength is maintained! What if a damaged region of the cell interrupts the chemical chain reaction? The signal resumes its full power as it continues down the axon beyond the compromised region. The event last about a millisecond as the system quickly returns to equilibrium. This is the excitation phase that tells us why cells fire.

Conversely, many synapses are inhibitory in that they increase the cells' negative polarity and do not decrease, in part by opening Cl^- gates. Thus, cells do not fire.

How do nerves fire then? When does a given cell begin the action potential chemical reaction? These occur only when the cell's polarity changes sufficiently to open *enough* of the right kind of gates. Depolarizing it further will continue the process. The required threshold is a summation of impulses that occur due to either one or both of the following reasons: Either the sending neuron fires rapidly in succession or sufficiently large number of neurons are firing. However, the result is the same: When the charge inside the cell decreases to a certain characteristic value, Na^+ gates start to open when neuron depolarizes further. Once begun the sequence continues throughout the axon and all of its branches. Thereby, it transforms the receiving nerve cell into a transmitting one whose inherent electrical signal continues to all other cells connected to it. Thus, a signal is transmitted from one neuron to another. The path of electricity defines the circuit.

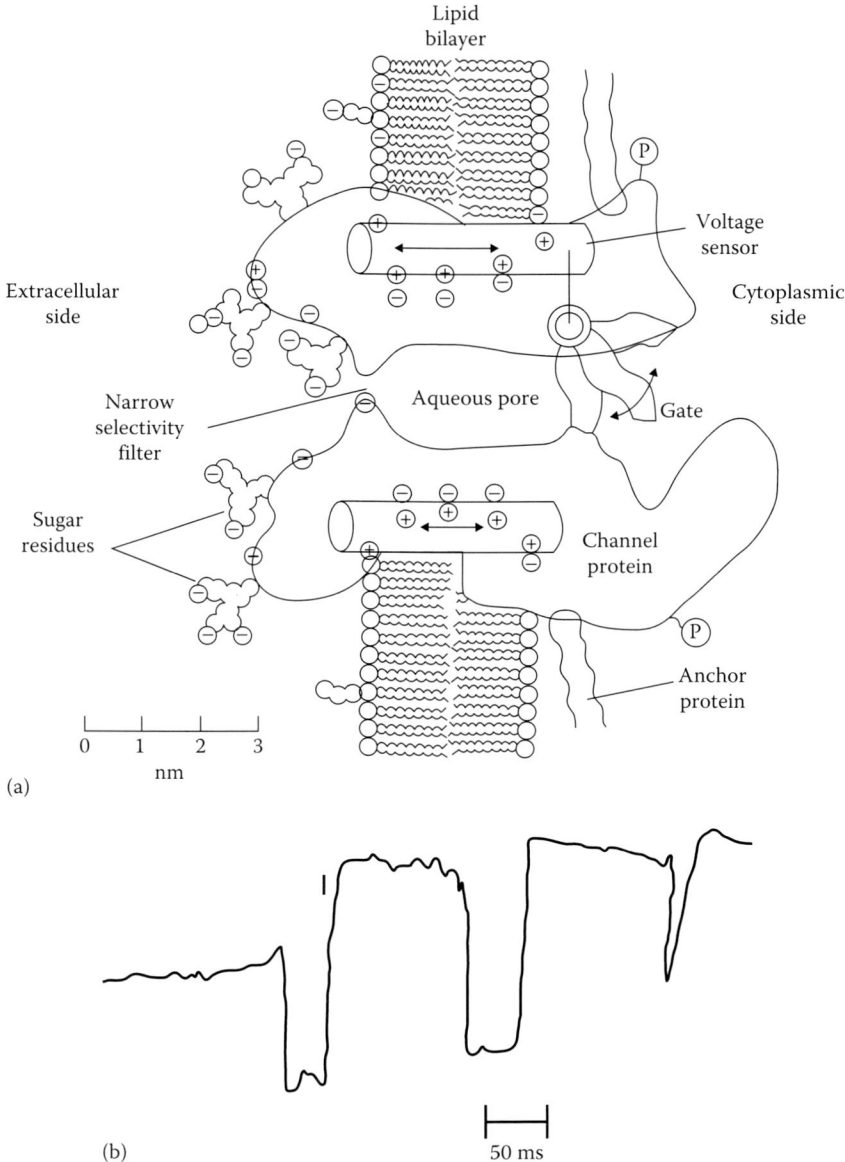

FIGURE 1.5 (a) Na-channel macromolecule in a membrane. (From Kobayashi M, *Solid State Ionics*, 174, 57, 2004.) (b) opening and closing of single ion channels in a membrane.

A threshold depolarization induces a phase transition of these glycol-protein macromolecules to a superionic phase in which the conductance of Na$^+$ ions is high. This hypothesis (1) explains the origin of transient inward current and other channel properties and (2) offers an approach to microscopic structure–function relationships [1.3].

Human body is a wonderfully wired communication system. Nerve conduction is a crucial function in this very dynamic and functional system involving the transport of Na$^+$, Cl$^-$, and also K$^+$ ions. Information is transmitted in the human body by electrical pulses in nerve fibers known as axons. An axon has a very high resistance of ~250 MΩ. While a pulse travels through a wire with nearly the speed of light, the velocity of a nerve pulse is about one-millionth of that speed. Two electrical potentials observed in nerves are (1) the resting potential and (2) the action potential. Nature has devised an amazingly efficient communication system using a "wire" considered a good insulator! The electrical properties of an axon can be understood by a poorly insulated electrical cable model, which allows current leaks to surroundings. More precisely, an axon is assumed to consist of a cylindrical membrane containing a conducting fluid called the axoplasm. The axoplasm resistivity ρ_a is related to the resistance R of a length of a wire l and radius r so that

$$R = \rho_a l / \pi r^2 \tag{1.1}$$

A thin membrane of length l and surface area has a nearly flat section, so it may be considered as a parallel plate capacitor with a capacitance C given by

$$C = C_m \left(2 \pi r l \right) \tag{1.2}$$

where C_m is the capacitance per length l of the axon.

The RC product is the time constant T of a series circuit that is related to the action potential velocity, which eventually constitutes an important evolutionary step in rapid transmission of large amounts of information. This is because in the so-called myelinated axon with interrupted myelin sheaths, the axon potential propagates faster than in an unmyelinated axon.

An axon thus resembles a poorly insulated electrical cable whose electrical properties are analogous to those of an RC network. A resting axon has a potential below that of the surrounding interstitial fluid called axoplasm whose electrical resistivity is a few ohm meter or a few kilo-ohm centimeter. The resting potential and the nonequilibrium concentrations of Na^+ and K^+ are maintained by the active Na–K pump mechanism that offsets passive ionic flows due to diffusion and electrical forces.

Weak electrical stimuli are sufficient to produce proportional axon responses. These responses are similar to those of an analog RC circuit and decreases rapidly with distance. A stimulus that raises the axon potential above a critical threshold triggers a current pulse and an accompanying action potential.

The network connection between nerve conduction and solid state ionics is intriguing, amazing, and instructive. The generation of the action potential and the regional differences observed throughout the heart result from the selective permeability of ion channels distributed on the cell membrane. The ion channels reduce the activation energy required for ion movement across the lipophilic cell membranes. During the action potential, the permeability of ion channels changes; each ion, for example, X, moves passively down its electrochemical gradients ($\Delta V = [V_m - V_X]$, where V_m is the membrane potential and V_X the reversal potential of ion X) to change the membrane potential of the cell. The electrochemical gradient determines whether an ion moves into the cell (depolarizing current for cations) or out of the cell (repolarizing current for cations). Homeostasis of the intracellular ion concentrations is maintained by active and coupled transport processes that are linked directly or indirectly to ATP hydrolysis.

Ion channels have two fundamental properties: ion permeation or the movement through the open channel and gating or the mechanism of opening and closing of ion channels. The selective permeability of ion channels to specific ions results in Na^+, K^+, and Ca^{2+} channels. Size, valency, and hydration energy determine selectivity. The selectivity ratio of the biologically important alkali cations is high. For example, the Na^+:K^+ selectivity of sodium channels is 10:1. Ion channels do not function as simple fluid-filled pores but provide multiple binding sites for ions as they traverse the membrane. Ions become dehydrated as they cross the membrane as ion-binding site interaction is favored over ion–water interaction. Like an enzyme–substrate interaction, the binding of the permeating ion is saturable. Most ion channels are singly occupied during permeation; certain K^+ channels may be multiply occupied. The equivalent circuit model of an ion channel is that of a resistor. The electrochemical potential, ΔV, is the driving force for ion movement across the cell membrane. Simple resistors have a linear relationship between ΔV and current I, namely, the Ohm's law,

$$I = \Delta V / R = \Delta V G, \tag{1.3}$$

where G is the channel electrical conductance (1/channel conductivity). Most ion channels have a nonlinear current–voltage relationship. For the same absolute value of ΔV, the magnitude of the current depends on the direction of ion movement into or out of the cells. This property is termed rectification and is an important property of K^+ channels; they pass little outward current at positive (depolarized) potentials. The molecular mechanism of rectification varies with ion channel type. Block by internal Mg^+ and polyvalent cations is the mechanism of the strong inward rectification demonstrated by many K^+ channels.

Gating is the mechanism of opening and closing of ion channels and is their second major property. Ion channels are also subclassified by their mechanism of gating: voltage-dependent, ligand-dependent, and mechanosensitive gating. Voltage-gated ion channels change their conductance in response to variations in membrane potential. Voltage-dependent gating is the most common mechanism of gating observed in ion channels. A majority of ion channels open in response to depolarization. The pacemaker current channel (I_f channel) opens in response to membrane hyperpolarization. The steepness of the voltage dependence of opening or activation varies between channels. Sodium channels increase their activation by $\approx$e-fold (2.73) for 4 mV of depolarization; in contrast, the K^+ channel activation increases e-fold for 5 mV of depolarization.

Ion channels have two mechanisms of closure. Certain channels like the Na^+ and Ca^{2+} channels enter a closed inactivated state during maintained depolarization. To regain their ability to open, the channel must undergo a recovery process at hyperpolarized potentials. The inactivated state may also be accessed from the closed state. Inactivation is the basis for refractoriness in cardiac muscle and is fundamental for the prevention of premature reexcitation. The multiple mechanisms of inactivation are discussed in the following texts. If the membrane potential is abruptly returned to its hyperpolarized (resting) value while the channel is open, it closes by deactivation, a reversal of the normal activation process.

Seawater and the seaworld ionics is important and interesting because life on earth first arose in water [1.4]. The principal characteristic of the water in the earth's oceans is their salinity that arises from (1) hydrothermal discharges and (2) weathering of continental crust materials. Primeval oceans must have contained significant concentrations of Cl^-, Na^+, Ca^{2+}, and Mg^{2+}, the first two being dominant anion and cation, respectively. Composition of seawater is defined by salinity which is the total amount of dissolved solids per kilogram of water. Typical concentrations of ions (in ppm) are Na^+ 10,500, Cl^- 19,000, SO_4 2,700, HCO_3^- 142, Mg^{2+} 1350, Ca^{2+} 410, K^+ 390, and Br^- 67 among others, totaling 34,487. There is thus an intimate connection between ocean ionics and mineralogy. The latter offers its own perspective as discussed later in this chapter.

Electric fish produce their electrical fields from a specialized structure called an electric organ made up of modified muscle or nerve cells, which became specialized for producing bioelectric fields stronger than those that normal nerves or muscles produce. Typically, this organ is located in the tail of the electric fish. The electrical output of the organ is called the *electric organ discharge* (EOD).

Fish with an EOD that is powerful enough to stun prey are called strongly electric fish. The amplitude of the signal can range from 10 to 600 V with a current of up to 1 A. The electric eel and electric rays are examples. Strongly electric marine fish deliver low-voltage, high-current electric discharges, while freshwater fish have high-voltage, low-current discharges. *This is because of the different conductance of salt and fresh water.* To maximize the power delivered to the surroundings, the impedances of the electric organ and the water must be matched. In salt water, a small voltage can drive a large current limited by the internal resistance of the electric organ. Hence, the electric organ consists of many electrolytes in parallel. In freshwater, the power is limited by the voltage needed to drive the current through the large resistance of the medium. Hence, these fish have numerous cells in series.

The extremely interesting electric ray [1.5] has two large electric organs on either side of its head, where current passes from the lower to the upper surface of the body. Composed of columns, each column consists of 140 to half a million gelatinous plates. In saltwater fish, these batteries are connected in parallel. In freshwater, the batteries are connected in series transmitting discharges of higher voltage. Why? Fresh water has higher resistivity than salt water (seawater), so that to be effective a higher voltage is required. With such a battery, an average electric ray can electrocute a fish delivering 50 A at 50 V—the inverter power sufficient to maintain an average home! Thus, we are again led to a discussion of human body ionics.

1.1.3 Electrolyte Balance in Human Body

Ion conduction in materials enables active functions including actuation, sensing, and transport, and these functions are as important to the functioning of the microionic devices as they are to the working of the human body.

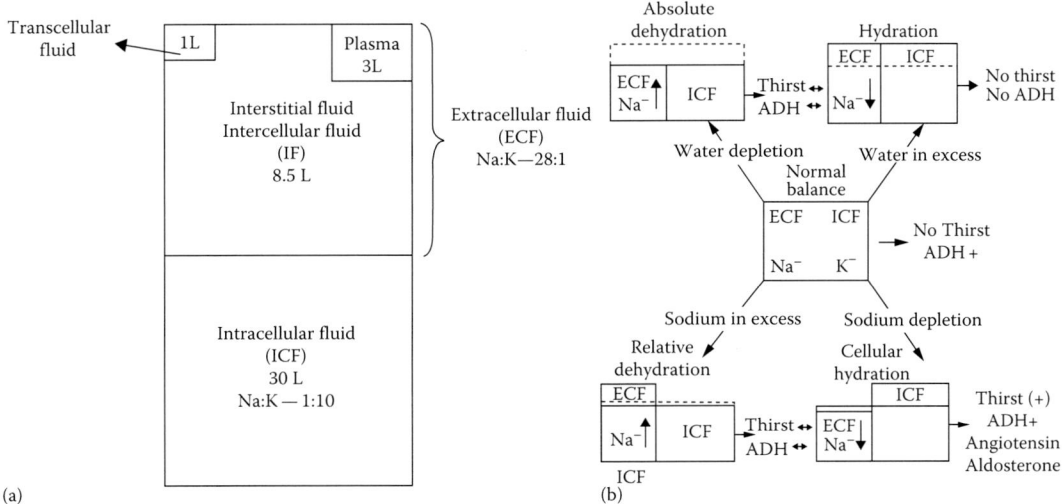

FIGURE 1.6 Fluids in human body support Na^+ and K^+ ion transport: (a) intracellular fluids and (b) extracellular fluids. (Adapted from Srilakshmi, B., *Nutrition Science*, New Age International Publishers, New Delhi, India, 2010, Chapter 20.)

Human body—like the earth—is made of nearly two-thirds of water besides bones and muscles. This water somewhat like the seawater has dissolved ions mainly Na^+, K^+, and Cl^-. The body needs a daily supply of sodium chloride for its sustenance via electrolyte balance. Indeed, it has a built-in reservoir of Na^+ and K^+ in the extracellular and intracellular fluids in the body, respectively (Figure 1.6). Extracellular fluid includes plasma and intestinal fluid besides transcellular fluids including aqueous humor and cerebrospinal fluid. From the SSI perspective, the main cation Na^+ present to the extent of ~1.8 g/kg fat-free body weight controls the body fluid osmolarity and thus the body fluid volume itself. Na^+ ions and the ratio of Na^+ to other ions are vital to this function.

Na^+ ions can permeate the cell membrane. Muscle contraction and nerve transmission require a temporary exchange of extracellular Na and intracellular K. An active mechanism of pump subsequently transfers the Na out of the cell. Na in the bone is essentially a reservoir forming part of the active labile Na pool in the body.

K^+ present to the extent of 250 g mostly in the intracellular fluid controls muscular activity. During muscle contraction, K from the muscle cells is lost into the extracellular fluid only to be returned to the muscle cell after contraction. Variations in K^+ concentration cause abnormalities in the conduction and activity of cardiac muscle. Thus, life depends on K^+ among other things. Within the cell, K^+ acts as a catalyst in many biological reactions including the release of energy and in glycogen and protein synthesis. Thus, Na^+ and K^+ are the chief liquid electrolytes in the body whose balance is maintained by several mechanisms [1.6]. Interestingly, the gastrointestinal tract constantly regulates electrolyte levels every day.

Water and electrolyte balance are complementary so that proper water and electrolyte (6 g/day of NaCl through food as prescribed by WHO) intake help maintain this balance.

1.1.4 Mineralogy and History

Mineralogy is an important branch of geology that has an important conenction to solid state physics and chemistry through crystallography and is thus relevant to SSI. Therefore, it provides an important motivation as demonstrated by the very recent discovery of fluorocronite which is nothing but lead fluoride which is an important F^- ion conductor [1.7]. Starting from the fluorite or calcium fluoride to fluorocronite, there have been many SSI-related minerals [1.8] and are illustrated in Figures 1.7 and 1.8.

Oxide minerals also have important roles in technology, for example, as semiconductors, thermal and electrical insulators, fuel cell components, substrates for thin films, photovoltaic materials, and as products of metal oxidation. Because of their technological applications, and their fundamental interest to geosciences, materials science, physics, and chemistry, diffusion properties of many oxide minerals have been studied intensively, using a wide range of experimental, analytical, and computational approaches. In many cases, particular attention has been devoted to deciphering the atomic-level mechanisms involved in diffusion. With the possible exceptions of metals and halides, oxides have probably been studied in more detail with regard to their point defect and diffusion properties than any other group of minerals. The oxide minerals considered in this chapter are relatively simple in terms of their structure and chemistry, but nonetheless exhibit quite complicated diffusion behavior in many cases. Due to the simplicity

(a)

(b)

FIGURE 1.7   Important minerals based on solid state ionic materials: (a) fluorite, (b) fluorocronite (see [1.7] and Table 1.1).

(Continued)

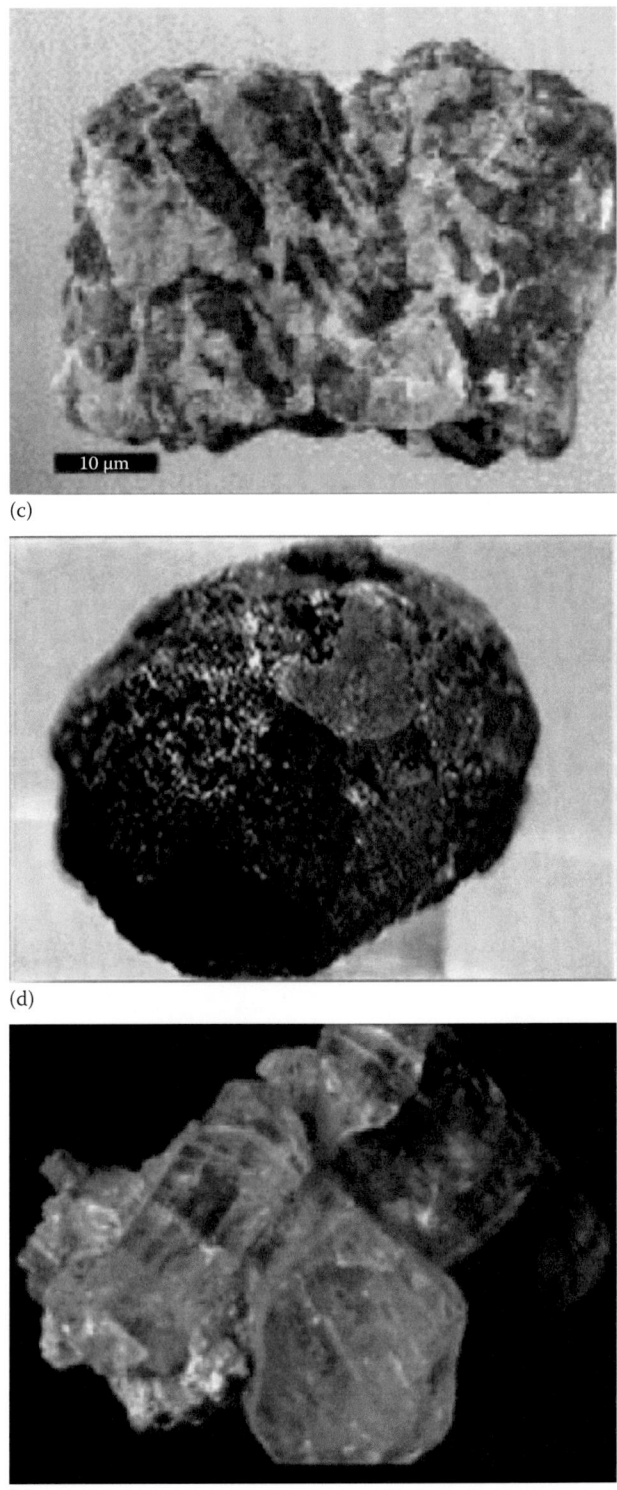

10 μm

(c)

(d)

(e)

FIGURE 1.7 (*Continued*) Important minerals based on solid state ionic materials: (c) tryphilite, (d) baddeleyite, (e) apatite $Ca_5(PO_4)_3X$, where X = F (fluorapatite), Cl (chlorapatite), OH (hydroxyapatite). (*Continued*)

(f)

(g)

FIGURE 1.7 (*Continued*) Important minerals based on solid state ionic materials: (f) perovskite, and (g) garnet (Photo courtesy of webmineral.com/data/Perovskite.shtml; www.minerals.net/mineral/garnet.aspx; Apatite: www. concretemender.com; Jeff Weissman: Photographic Guide to Mineral Species.)

of the minerals, and to the amount and quality of data available, the origin of many of these complicated behaviors is fairly well understood. In magnetite, for example, cation diffusion rates have a complex dependence on oxygen fugacity. This dependence is due to internal redox reactions and due to a transition from an interstitial diffusion mechanism at low oxygen fugacities (or pressures) to a vacancy mechanism under more oxidizing conditions. Oxygen and titanium diffusion rates in rutile also vary strongly with fO_2, due to internal reduction of titanium and the associated production of oxygen vacancies and titanium interstitials. Some cations in rutile exhibit strong diffusional anisotropy, which is thought to result from rapid diffusion along interstitial "channels" that extend along the c-direction in the rutile structure.

In periclase, trivalent cations diffuse rapidly compared to divalent cations, opposite to the trend observed in most silicate minerals. This behavior arises from the Coulombic attraction between trivalent cations and cation vacancies. Similar interactions between oxygen vacancies and cation vacancies appear to be responsible for lattice diffusion of oxygen in periclase.

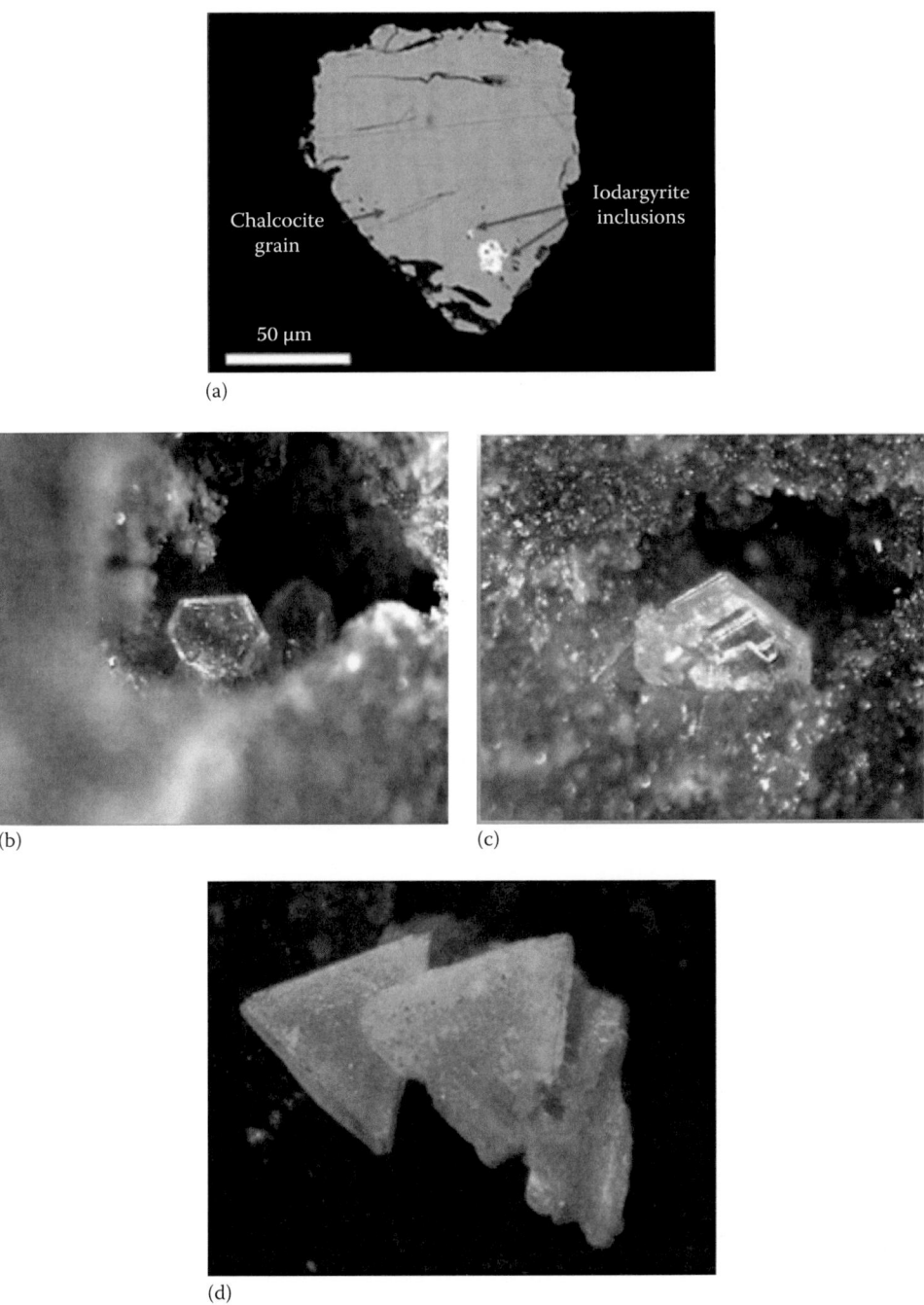

FIGURE 1.8 Minerals based on AgI and CuI. (a) Micron-size AgI (iodargyrite) inclusions in mineral chalcocite grain, (b) a natural crystal of AgI, (c) a macrocrystal of β-AgI showing hexagonal platelets stuck together, and (d) triangular prisms of mineral marshite (CuI). (Photo courtesy of www.mineralatlas.com.)

To appreciate the importance of AgI and also CuI, it is appropriate to mention that these two iodides occur as the rarest of rare minerals containing iodine. Known by the names iodargyrite (AgI) and marshite (CuI), they grow either as micron-sized inclusions in other minerals or by themselves as hexagonal platelets (beta AgI) or as triangular tablets (gamma CuI) (Figure 1.8).

SSI has developed over a period of three centuries now. Curiously, frogs were used by Galvani as "live solids" to demonstrate animal electricity! Volta's battery provided the impetus for researches by Michael Faraday who could be considered as the founder of SSI having recognized for the first time that lead fluoride and silver sulfide conduct electricity in the solid state. Equally important, Faraday gave us a vocabulary for solid state electrochemistry and enunciated the laws of electrolysis. Indeed the first research reports of compounds that possess exceptionally high ("liquid-like") values of ionic conductivity within the solid state were given by Faraday in the first half of the nineteenth century. In the case of the "fluoride of lead," the extraordinary nature of this behavior is clear from the original text [1.9]: "When a piece of that substance, which had been fused and cooled, was introduced into the circuit of a voltaic battery, it stopped the current. Being heated, it acquired conducting powers before it was visibly red hot in daylight…."

Table 1.1 gives a representative but not exhaustive timeline of SSI events since the late eighteenth century.

1.1.5 Batterivity

"Batterivity" is a word coined from two words "battery" and "relativity." Although the first is so common and practical and the second is so profound and theoretical, there is an important connection between the two discovered very recently providing a rare perspective discussed in the following text.

Cars are started even today using lead-acid batteries (Figure 1.9). More compact now than before, they generate energy using electrochemical reactions [1.10]. These reactions occur between lead compounds and sulfuric acid. Let us model the reactions. As electrons move at high speed around a lead nucleus, their energy levels change owing to the special theory of relativity. The change accounts for as much as 1.7–1.8 V of a standard 2.13 V lead-acid cell which is about 80% of the rated battery power. Car batteries would yield far less voltage were it not for the relativistic effects on the lead atoms according to simulations. The relativity perspective to an electrochemical reaction involving a liquid electrolyte is quite unique that it would amaze anyone starting an automobile every day, with the engine getting 80% of its power through relativity. To appreciate the uniqueness of the heavy lead atom note that tin–acid batteries do not work even though tin and lead are similar otherwise. Let us explore why.

Electrons typically orbit their atoms at speeds much less than the speed of light. Relativistic effects are thus largely negligible while describing atomic properties. Notable exceptions are to be found in the heaviest elements of the periodic table. The electrons of such elements must orbit at near-light speed because they have to counter the strong attraction of their large nuclei. According to the theory of special relativity, these high-energy electrons act in some ways as if they have greater mass so that their orbitals must shrink in size relative to the slower electrons. Thus, they maintain the same angular momentum. This contraction is most pronounced in the spherically symmetric s orbitals of heavy elements. Interesting consequences are the yellowish hue of gold and the liquid nature of mercury at room temperature. Although relativistic effects on the crystal structure of lead have been known for a long time, interesting chemical properties have been revealed only after a century and a half since the first battery was assembled. Exciting lead chemistry happens in the lead-acid battery based on electrochemical cells. These cells consist of two plates: one of lead metal and the other of lead dioxide (PbO_2) immersed in sulfuric acid. Lead releases electrons to become lead sulfate ($PbSO_4$). Lead oxide gains electrons and also becomes lead sulfate. The combination of these two reactions results in a voltage difference of 2.1 V between the two plates. Theoretical models of the lead-acid battery have been available but not the fundamental physical principles until 2011 [1.11]. To find the cell's voltage, we need to calculate the energy difference between the electron configurations of the reactants and the products: the initial and the final energies. To create H_2SO_4 electrolyte, start from SO_3 (gas) and use the experimental reaction

TABLE 1.1

Timeline of Electrochemical and Solid State Ionics Events

1791	Publication of Galvani's work on animal electricity; Living muscles as Leyden jars storing electricity
1800	Nicholas and Carlisle discovered electrolysis of water by means of a battery; Production of hydrogen
1801	Volta's demonstration of a working battery that won Napoleon's Gold Medal; Prime mover for researchers like Michael Faraday
1807	Davy discovered potassium by means of a 2000-element battery through electrolysis of potash, besides Na and Ce; role of reaction at electrodes decomposition of electrolytes identified
1807	Grotthus theory of electrolytes, movement of charges (separation of the charges on H and O in water molecule)
1826	Becquerel, polarizing effects of electrodes causing H evolution
1827	Georg Ohm discovered the law that the current (I) through a conductor between two points is directly proportional to the potential difference (V) across the two points. $I = V/R$. Resistance (R) is the proportionality constant independent of the current. General form of this law due to Kirchhoff, valid for ionic conduction, is $J = \sigma E$ is the current density at a given location in a resistive material, E is the electric field at that location, and σ is a material-dependent parameter called the conductivity. For reactive circuits, it is $V = IZ$, where Z is complex impedance, the basis of impedance spectroscopy in solid state ionics.
1833	Faraday developed electrochemical vocabulary (electrode, anion, anode, cation, cathode), linked mass of compound produced or consumed with the amount of charges passed thus discovering the laws of electrolysis, and found that lead fluoride and silver sulfide conduct electricity, laying foundations of solid state ionics
1836	Two-components battery—the Daniell cell
1837	Jacobi's galvanoplasty
1839	Grove discovered reversibility of electrolysis—basis of first fuel cell; 120 years later NASA used fuel cells on its space vehicles
1859	Plante's lead–acid battery—still in use. Its links with special theory of relativity of Pb discovered in 2011!
1868	Leclanche's saline battery based on Zn and MnO_2—still very useful
1874	Kohlrausch theory of the conductivity of electrolytes
1887	Arrhenius theory on acid–base reactions and ionic dissociation
1889	Nernst's discovery of thermodynamics of electrochemistry
1895	Bottger's H-electrode enables pH measurements
1899	First electric car, La Jamais Contente, reached record speeds of 100 km/h over a few kilometers
1899	Nernst suggested ZrO_2 as a solid conductor for oxygen ions
1902	Cotterel equation describing electrode kinetics with mass transport by diffusion
1905	Tafel's empirical law links the electrode overpotential to current
1906	Cremer's glass bulb pH electrode still in use
1914	Edison's Ni/Fe alkaline secondary battery; Tubandt and Lorenz discovered fast Ag+ ion conduction in alpha silver iodide
1923	Debye and Huckel published their seminal paper on "Theory of Electrolyte" in *Phys. Z.*
1925	Burt from Millikan's laboratory demonstrated Na^+ ion conduction through a glass bulb by an ingenious electrolysis—thermionic emission experiment using a 60 W metal filament electric bulb
1924–1930	Butler–Volmer charge transfer theory at an electrode
1943	Wagner explained conduction mechanism in CaO, Y_2O_3, and MgO-doped ZrO_2
1967	Discovery of β-alumina ($Na_2O \cdot 11Al_2O_3$), the two-dimensional Na^+ solid electrolyte by Kummer, Weber, and Yao; development of 2.08 V sodium–sulfur battery initiated
1973	Liang discovered Li^+ ion conductor LiI-35 mol% Al_2O_3—a composite electrolyte for Li-ion battery
1975	A Rahman did a pioneering molecular dynamics study of F– ion diffusion in CaF_2 and deduced the diffusion coefficient
1976	Goodenough, Hong, and Kafalas discovered NASICON—$Na_3Zr_2Si_2PO_{12}$, the three-dimensional Na^+ ion conductor
1980	*Solid State Ionics*—An international multidisciplinary journal launched by the North Holland Publishing Co., now called Elsevier

(Continued)

TABLE 1.1 (*Continued*)

Timeline of Electrochemical and Solid State Ionics Events

1981, 1983	Li-ion batteries developed using the "rocking chair" concept or intercalation and deintercalation of Li ions into anode/cathode during cell charge/discharge
1997	Padhi, Nanjundaswamy, and Goodenough discovered electrochemical activity in the tryphlite-/olivine-structured LiFePO4
2008	Colossal ionic conductivity discovered at the interfaces of epitaxial ZrO_2:Y_2O_3/$SrTiO_3$ heterostructures
2011	Fluorocronite, a natural analogue of PbF_2, discovered by Mills et al.
2011	The substantial Pb relativistic contribution to the voltage of the lead–acid (car battery) discovered and computed by Ahuja et al.

Source: Funke, K., *Sci. Technol. Adv. Mater.*, 14, 043502, 2013, for a recent motivating historical account.

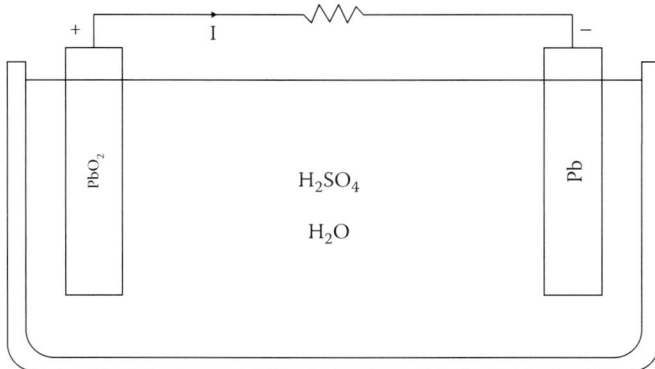

FIGURE 1.9 The car battery schematic showing cathode (PbO_2), anode (Pb), electrolyte (H_2SO_4, H_2O), and the external circuit delivering micro ionic power to load I^2 (load resistance), where I is the battery-generated current.

$$H_2O \text{ (liquid)} + SO_3 \text{ (gas)} \rightarrow H_2SO_4 \text{ (liquid)}$$

for which the energy is known. Subtract this energy from the total energy.

The average calculated standard voltage is 2.13 V as against the experimental value of 2.11 V. It emerges from calculations that ~1.7 V of this standard voltage comes from relativistic effects, not only from PbO but also from $PbSO_4$. Thus, about 10 of the 12 V in a car battery comes from relativistic effects. For details of calculations the reader is referred to Ref. [1.11].

Why are the inner electrons of lead atom relativistic? The energy of the 1s orbital is approximately proportional to the square of the nuclear charge Ze, where Z is the atomic number and e the electronic charge. For hydrogen atom, it is one-half of the rest energy ($E = m_0c^2$) of an electron multiplied by the fine structure constant squared: $(mc^2/2)(1370)^2$. For Pb with $Z = 82$, a first approximation is $(mc^2/2)(82/137)^2$ which is of the same order of magnitude as $mc^2/2$ making it relativistic. A better approximation for the energy is $(mc^2/2)(n - 1)^2/137^2$, where n is the principal quantum number. Subtracting 1 accounts for the shielding by the other 1s electron.

Any heavy atom conforming to *jj* coupling in atomic physics taking part in an electron transfer event is expected to show relativistic effects. The connection of the car battery to the theory of special relativity is one aspect of SSI, while there is another equally important aspect: principles and practice of solid state electrochemistry [1.12,1.13]. Interestingly, both these connections involve chemical reactions that take place at the two interfaces in the electrolytic cell: the cathode–electrolyte interface and the electrolyte–anode interface.

The battery charger and the mobile/laptop/EV battery represent a study in contrast. As much as the ac and the dc generators, even the human body is considered a natural battery. But as all organisms need food to survive, it was already clear even when the first human set foot on this planet that there is no free lunch!

Lattice (3D spatial) periodicity is at the root of all computable and measureable properties of solids. These are mainly structural, thermal, and electronic. Thus the treatment of these properties as highlighted by reciprocal lattice and Brillouin zone are similar. Symmetry reduces the problem of an infinite solid to that of the reciprocal lattice and Brillouin zone that holds just a few (usually <10) and not 10^{23} atoms!

Structural properties are studied by diffraction of x-rays, neutrons, and electrons, when they are "Bragg scattered" by the atoms of the structure. Vibrational (dynamic properties of the lattice) or the so-called lattice dynamics is conveniently studied by inelastic neutron scattering to experimentally obtain phonon dispersion relations of materials.

Heat capacity of solids involves harmonic vibrations while thermal expansion and thermal conductivity involve anharmonic vibrations. Mechanical properties such as elastic constants and elastic moduli such as compressibility or reciprocal bulk modulus are measured by ultrasonic pulse echo methods.

1.2 Ionic Solid State: Home of Defects and Disorders

Defects and disorders in a real finite ionic crystal—as compared to the ideal infinite insulating solid made from the attraction of positive and negative ions—develop at temperatures greater than absolute zero. Thus, the ionic solid is the natural home of defects and disorder as the solid begins its "journey" from absolute zero to its melting point (T_m) and even beyond occasionally [1.14]. SSI materials are encountered somewhere in between the extended range $0 < T < T_m$ (K). Figure 1.10 traces the journey taking AgBr as the example.

The ionic solid made of positively charged ions such as Na^+ and negatively charged ions such as F^- held together by predominantly electrostatic long-range Coulomb forces of attraction acts as an electronic and ionic insulator at absolute zero temperature. This is in sharp contrast to a pure metal whose electrical

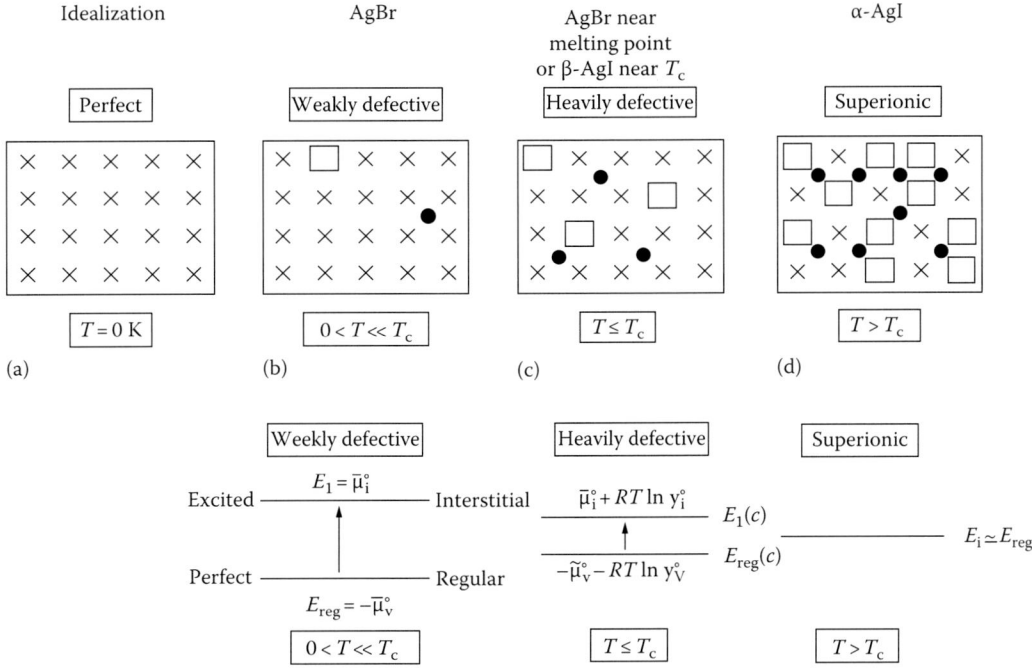

FIGURE 1.10 Top: Distinction between (a) an ionic insulator, (b) an ionic conductor with a small concentration of point defects, (c) a heavily defective (or disordered) ionic crystal, and (d) a fast ionic conductor. Bottom: A flat-band picture of an "ionic semiconductor" corresponding to the three situations (b), (c), and (d). The energy equivalents of regular (perfect) lattice position and an (excited) interstitial position are indicated. (From Kitttel, C. and Kroemer, K., *Thermal Physics*, 2nd ed., W.H. Freeman, New York, 1980.)

resistivity is zero at absolute zero. An ionic solid can only support polarization currents that may flow in solids with polarizable cations or anions. At a nonzero temperature ($T > 0$ K), however, crystal lattice vibrations set in, which can knock off ions in pairs (Na^+, F^-), say, in NaF.

Intuitively, ionic bond formation between metal M and nonmetal X may be looked upon as a redox or reduction–oxidation "reaction" leading to the formation of cation M^+ and anion X^-. This process involves the ionization potential of M and the electron affinity of X. Consider the formation of the ionic bond in LiF. Let a ^7Li atom and a ^9F atom be brought closer from $-\infty$ to $+\infty$, respectively. Li ($1s^2 2s^1$) and F ($1s^2 2s^2 2p^5$) have such a combination of electronic structure/configuration that the most electronegative or "electron hungry" F bonds with the most electropositive Li. Li^+ ($1s^2$) and F^- ($1s^2 2s^2 2p^6$) are formed by mutual donation and acceptance of the $2s^1$ electron of Li through which the F atom attains the inert gas (Ne) configuration obeying the octet rule. The Li^+ ion and the F^- ion are formed even before the quantum mechanical Li–F bond is born through the overlap of the respective wave functions. Note that the first ionization potential of Li is 5.39 eV and the electron affinity of F^- formation is only 0.74 eV. The binding energy of the LiF crystal is just the energy that has been gained by bringing the free atoms Li and F together to form the ionic crystal. Thermodynamics begins at the beginning, that is, even when the temperature of the solid slightly deviates from absolute zero temperature. Lattice begins to vibrate creating defects and disorder that are essential for the ions to conduct electricity.

Generally, ionic solids form from atomic and molecular ions through the balance of attractive and repulsive forces between ions of opposite charge. These forces are essentially electrostatic and classical in nature, although repulsion does contain a quantum mechanical component. Once the ionic solid is formed, it is already in an imperfect state! The imperfections arise from the thermodynamic necessity of minimizing the free energy of the solid. The ionic crystalline structure adopted by, say, NaI and AgI is an essentially "vibrating structure" at $T > 0$ K and these vibrations generate defects. In normal ionic crystals such as NaI, these defects (missing Na^+ and I^-) are small in number and their concentration increases only near their melting points. In fast ion conductors such as AgI—the subject of this book—the displaced Ag^+ ions and the associated Ag vacancy arise in substantial concentration at temperatures ~150°C much less than its melting point contributing to naturally enhanced ionic conductivity.

How does an ionic crystal such as NaI or AgI differ from a metallic crystal such as Na or Ag and a semiconductor crystal such as Si or GaAs?

1.2.1 Metals, Semiconductors, Insulators, and Fast Ion Conductors

Solids—whether crystalline or noncrystalline—are best divided into metals, semiconductors, and insulators on the basis of the property of electrical resistivity or electrical conductivity, an important theme of this book. This is because of the widest range of values, for example, 10^{32}, spanned by the best metals (electrical resistivity 10^{-10} Ω cm) and the best insulators (electrical resistivity 10^{22} Ω cm). SSI materials figure at the borderline between metals and semiconductors.

The band structure theory can explain this 10^{32} difference in electrical resistivity between conductors and insulators [1.15]. Simply put, fully occupied energy bands are ineffective for electron conduction. Electrons in electric fields cannot absorb energy unless interband transitions are involved.

Metals are important for SSI especially because they are anode materials in a solid state battery. Semiconductors such as Si and semimetals such as graphite too are important in this context.

Silver and copper—the metals in common use—are known for their exceptional current-carrying capacity. They have electrical conductivities of 0.588 and 0.621 MS/cm at room temperature. They are also good thermal conductors. Reactive alkali metals such as Li, the first solid in the periodic table, and Na—both mechanically soft solids—are also good conductors of electricity with electrical conductivities of 0.107 and 0.211 MS/cm, respectively. Moving across the periodic table, Si, the workhorse of the microelectronic/nanoelectronic world, is a semiconductor with an electrical conductivity of 1.56 mS/cm at 293 K. NaI and AgI are both nonmetallic solids in which ions carry current but not electrons simply because they have no free electrons at all. They are electronic insulators. So is diamond, which is an allotropic form of carbon with an electrical conductivity of ~10^{-13} S/cm at 293 K. The temperature dependences of electrical conductivity of these solids show variation characteristic of their metallic, semiconducting, or insulating character (Figure 1.11a, b).

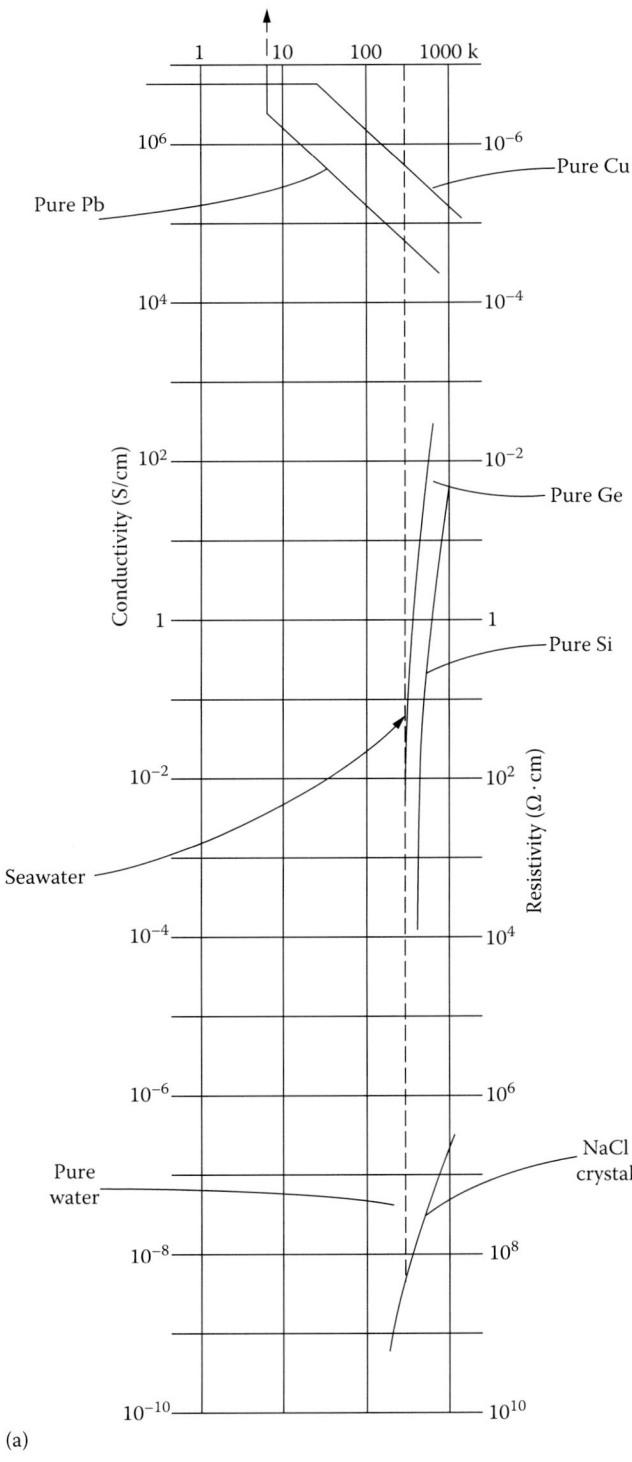

FIGURE 1.11 Distinction between metals, semiconductors, and insulators. (a) Conductivity or resistivity of lead, germanium and silicon, and NaCl crystal. In a metal, conductivity decreases with increasing temperature, while in a semiconductor or insulator, it increases with increasing temperature. Note also that a natural liquid ionic conductor like seawater conducts much better than NaCl and its conductivity is comparable to that of GE and Si which are electronic conductors. (Adapted from Purcell, E.M., *Electricity and Magnetism*, 2nd edn.,Tata McGraw Hill, New Delhi, India, 2008, p. 140.)   (*Continued*)

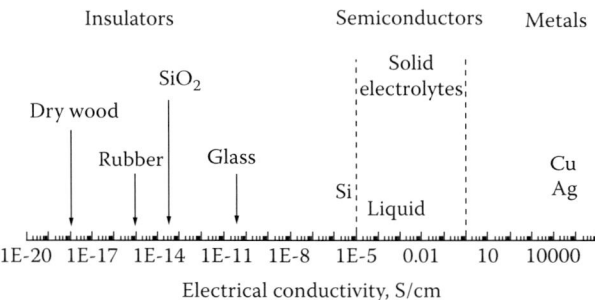

FIGURE 1.11 (*Continued*) Distinction between metals, semiconductors, and insulators. (b) Order-of-magnitude comparison of electrical conductivities of insulators, semiconductors, and metals. Dotted lines demarcate solid electrolytes from other solid types. Note the overlap of solid electrolytes and semiconductors. (After Thangadurai, V. and Weppner, W., *Ionics*, 12, 81, 2006.)

Figure 1.11b notes the overlap of solid electrolytes with semiconductors from the conductivity perspective.

Two facts emerge from these observations: In metals electrical conductivity decreases with increasing temperature, whereas in semiconductors the conductivity (whether electronic or ionic) increases with increasing temperature. A comparison between the temperature-dependent conductivities of NaI and AgI shows that while NaI conductivity exhibits a modest, monotonic increase, the conductivity of AgI shows a sudden step-like increase of three orders of magnitude already at 147°C, which is unusual for a normal ionic conductor such as NaCl. The Na ions in NaI or NaCl must be moving slowly compared to Ag in AgI. Thus, prima facie we could call AgI a fast ion conductor or a "super" ionic conductor in general SSI material as distinct from a normal or slow ion conductor. A following section introduces the electronic origin of these solids while the forthcoming chapters of this book discuss various families of fast ion conductors or SSI materials.

Ionic solids at room temperature generally have very low conductivities for the motion of ions through the crystal, less than 1 mS/cm. But several families of compounds based on H, Li, Cu, Ag, F, and O have been found to possess conductivities ~0.001–1 S/cm at ambient or moderately high temperatures. In the family MAg_4I_5 (M = K, Rb, NH_4), Ag^+ ions occupy only a fraction of the equivalent lattice sites and the conductivity proceeds by the hopping of a silver ion from one site to a nearby vacant site. The crystal structures possess parallel open channels to assist fast ion transport.

Energy bands of a crystal are occupied by the available electrons according to (1) the Pauli principle and (2) the Fermi–Dirac statistics. Insulators involve either fully occupied or fully unoccupied bands ineffective for conduction. Conductors are constituted by a partially filled band. Energy gap between occupied and unoccupied state is zero. Thus, it is possible to have a response to a steady electric field. Electrical conductivity of a metal usually decreases with increasing temperature, due to reduction in relaxation time elapsing between two successive collisions of the free carriers.

Semiconductors have fully occupied and fully unoccupied bands at $T = 0$ K. But the energy gap between occupied and unoccupied states is small, typically less than 1–2 eV. Upon varying T, a number of electrons occupy the conduction band and a number of holes are left in the valence band. Thermal excitation of carriers depends strongly on T (just as defect concentrations in ionic conductors) giving rise to a conductivity that highly increases with T.

1.2.2 Defects in Ionic Crystals

Three basic kinds of defects occur in ionic crystals: lattice vacancies, interstitials, and impurities. Vacancies and interstitials occur in all crystals with a probability that increases rapidly with temperature. However, unlike in metals, defects in ionic solids carry an effective charge that is transported across the solid over macroscopic distances under the influence of an externally applied electric field. Now the question is how a vacancy, say, Na^+ vacancy, in NaI is created.

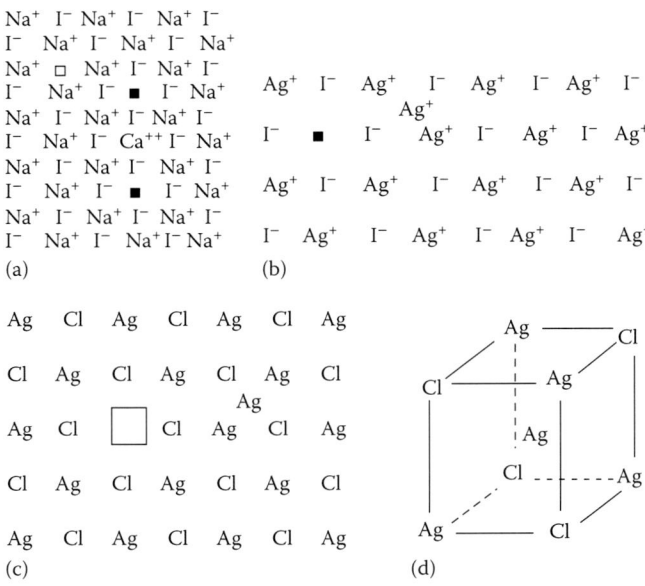

FIGURE 1.12 (a) Schottky defects in NaI. Doping with ions such as Ca^{++} creates extra Na^+ vacancies. Note that NaI and NaBr unlike NaCl do not contain cation and anion vacancy pairs. (b) Frenkel defects in AgI. (c) Frenkel defects in 2D and (d) 3D in AgCl. Note that unlike NaI and AgCl, AgI cannot be doped with Ca^{++}.

NaI has Schottky defects, that is, Na^+ and I^- vacancies, while AgI has Frenkel defects, that is, Ag^+ interstitials and Ag^+ vacancies (Figure 1.12). A Schottky defect pair in an ionic crystal such as NaI consists of a missing Na^+ ion and a missing I^- ion (Figure 1.12a). These are formed in the crystal out of thermodynamic necessity because at a temperature $T > 0$ K, the ions or atoms are in nonstop vibratory motion. There is a nonzero probability of the cation and the anion to break free of the solid and come to the surface of the crystal. This probability increases with increasing temperature so that at room temperature or 300 K there is a considerable number of such defect pairs. These defects facilitate ion migration in an ionic crystal under the influence of an external electric field. Diffusion of ions via defects gives rise to an ion current if the crystal is in an electrical circuit. Thus arises the ionic conductivity. The way diffusion takes place depends both on the crystal structure and on the nature of the defects present in it.

The Frenkel defect in an ionic crystal is in a sense an electron–hole pair in a semiconductor. An unusual one at that. Like in a semiconductor with electrons and holes both being mobile, the vacancy and the interstitial in an ionic crystal could both be mobile. But unlike in a semiconductor, the interstitial could be positive or negative and would always be more mobile than the vacancy. Some examples are Ag^+ in AgI and AgCl (Figure 1.12b, c) and F^- in CaF_2 or PbF_2.

Defects are the current carriers in ionic crystal and the basic physics of such conduction is introduced in Section 1.3.

1.2.3 Open Structures, Stuffed Structures, and Sublattices

Physical ideas of structure are intimately related to chemical ideas of bonding. Much of structural data of binary compounds AB understandable in terms of (1) electronegativity or the power of attraction of a metal ion for a nonmetal ion, (2) tendency for covalency or the formation of a partially covalent bond between a positive ion and a negative ion, and (3) atomic/ionic size qualitatively distinguishes between various types of close-packed crystal structures [1.15].

From the viewpoint of SSI materials, marginal/critical covalency in a polar covalent semiconductor such as AgI or CuBr or CuI favors four-coordinated cations rather than six-coordinated cations as in NaI.

Covalent or polar covalent semiconductors have directed bonds that favor open structures, while Coulomb interactions in ionic crystals favor closed structures, such that each cation is surrounded by anions and vice versa. Ionic liquids also feature this trend. While ionic conductors with thermodynamic point defects do not exhibit high ionic conductivity except near their melting temperatures, fast ion conductors with built-in structural disorder and associated defect–defect interactions exhibit ~<mS/cm ionic conductivities at temperatures only moderately higher than the room temperature.

Do not all solids have ions, be they metals or nonmetals? Not quite. Inert gas solids—helium, neon, argon, krypton, and xenon—are atomic in nature, although they can be ionized. Lithium (Li) and sodium (Na) or silver (Ag) and copper (Cu) have monovalent positive ions whose packing determines their structure. While the first two adopt body-centered cubic structures, the latter two possess face-centered cubic structures that are connected to form a diamond structure or silicon structure. The negative "ions" in these solids are the electrons. For example, a gram atom of Na metal consists of an Avogadro number of Na^+ ions and an equal number of electrons. Although for every Na^+ ion there is an electron, the metallic bond is not a proper ionic bond. Nor is it a covalent bond. Thus, only quantum mechanics can explain this bond. A unique characteristic of a proper ionic bond is that the electronic charge is proportional to the internuclear separation.

A strong covalent bond leads to a rigid open structure as in diamond or silicon structures. Metallic bond may be a special case of a covalent structure, but a metal cannot form enough metal–metal bonds to produce a rigid structure!

What are stuffed structures? These are framework structures such as silica that possess cavities. Into these cavities, one can "stuff" ions such as Li^+ and Na^+ to form $LiAlSiO_4$ and $NaAlSiO_4$ to realize SSI materials. Another structure that enables such "stuffing" is that of the garnet mineral discussed earlier.

A gram mole of NaI consists of an Avogadro number of Na^+ ions and an equal number of I^- ions. Even an elemental semiconductor such as Si (IV–IV) or SiC (IV–IV) or a compound semiconductor such as GaAs (III–V) is made up of ions with its valence electrons shared appropriately. The latter is not as free as in a metal but broken Si–Si bonds can release electrons that can carry current making Si or GaAs intrinsic semiconductors. Thus, it is fair to say ions rule the world with a little help from the associated electrons.

What kind of a bond exists between the Ag and I atoms in AgI? The tetrahedral Ag–I bond is found to be 70% ionic and 30% covalent but yet this much of a covalency is sufficient to produce an open structure with a semiconductor-like density. The weakness of the Ag–I bond renders it thermodynamically metastable while the covalency favors polymorphism similar to that of carbon. The Ag–I bond collapses at 147°C forcing Ag^+ ions to form a close-packed, denser-than-room temperature structure (either zincblende or wurtzite), which will be discussed in Chapter 5. The tetrahedral nature of the Ag–I bond persists even through the liquid state making it a unique bond in a solid state. The Na–I bond or the Li–I bond in NaI or LiI by comparison is a typical ionic bond in the two face-centered cubic structures.

All forms of solid matter are made up of ions except perhaps rare gas solids. Thus, a gram atom of sodium metal consists of an Avogadro number of Na^+ ions and an equal number of electrons, which could be viewed as the simplest form of negative ions (just as protons in a solid are perhaps the simplest positive ions). Likewise, a gram mole of electronic insulator NaCl has an Avogadro number of Na^+ ions and an equal number of Cl^- ions. Even an elemental semiconductor such as Si (IV–IV) or a compound semiconductor such as GaAs (III–V) is made up of ions (Si^{4+} or Ga^{3+} and As^{3-}) with their valence electrons shared appropriately in the rather open structures. The latter are not free as in a metal but only nearly free. But broken Si–Si bonds at temperatures greater than absolute zero release electrons that are responsible for the intrinsic conductivity of Si (and also GaAs). It is thus fair to say that ions rule the world from deep sea to the upper atmosphere and beyond with a little help from the associated electrons. Figure 1.13a illustrates the four important bonding types. Figure 1.13b gives an intuitive picture of the metallic, covalent semiconducting, and ionic insulating materials based on the relative fractions of metallicity and polarity depicted as Harrison's phase diagram [1.16].

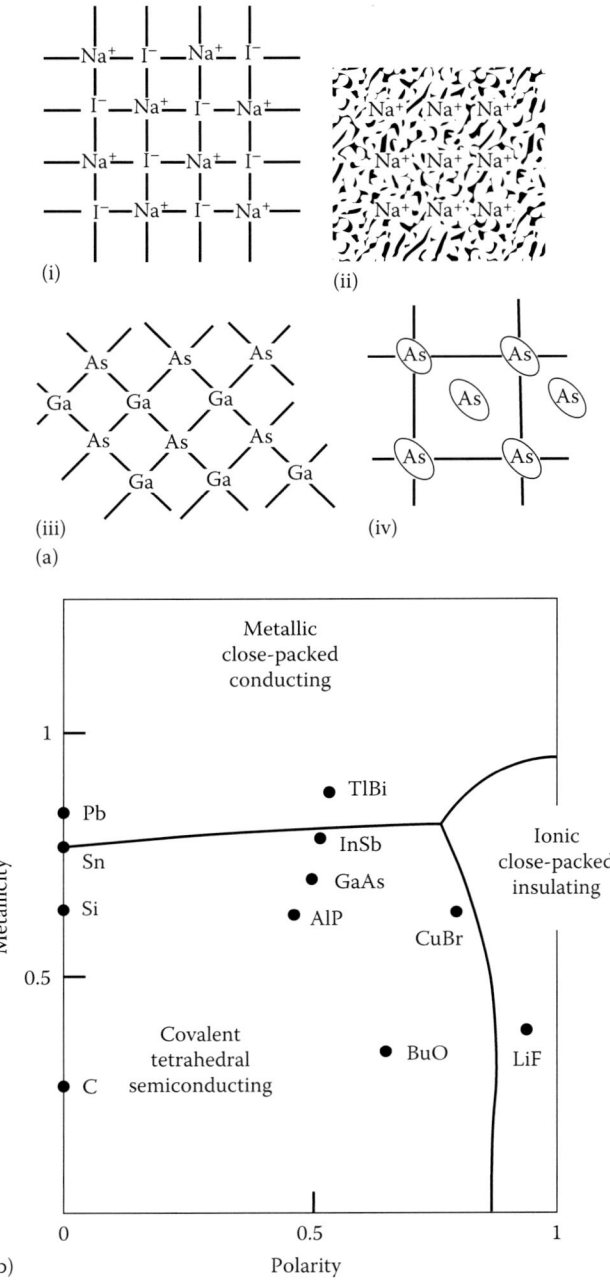

FIGURE 1.13 (a): (i) Ionic bond in NaI with electron transfer, energy 5 eV/atom or mol (typically); (ii) metallic bond in Na-free 3*s* electron gas, energy 3 eV/atom; (iii) covalent or electron pair bond in GaAs, energy 10 eV/atom; (iv) van der Waals bond (Ar) correlation of dipolar electron distributions, ~0 eV/atom. (From Beiser, A., *Perspectives of Modern Physics*, McGraw Hill, New York, 1969.) (b) Harrison's phase diagram separating metals, semiconductors, and insulators based on metallicity versus polarity. (From Harrison, W.A., *Electronic Structure and Properties of Solids—The Physics of the Chemical Bond*, Dover, New York, 1989.)

Normal ion conductors usually adopt close-packed crystal structures such as rocksalt and CsCl. These are dictated based on the criteria of Pauling: size, cation-to-anion radius ratio, electronegativity differences between anion and cation, and complete charge transfer from metal to nonmetal atom [1.17]. A feature of these structures is the monotonic variation of lattice parameters and defect densities upon changing thermodynamic constraints such as temperature and pressure. Furthermore, such structures cannot accept unlimited quantities of aliovalent impurities to create a sizeable concentration of extrinsic defects to give a high ionic conductivity. These features make them "textbook examples" and offer little insight into the problem and process of fast ion conduction. AgBr is an exception. Its thermodynamic (heat capacity) and conductivity behaviors near the melting point are characteristic of an incipient fast ion conductor.

While normal ionic solids and metals besides semiconductors are rigid structures with harmonically vibrating lattices (unit cells), fast ionic conductors are different in many respects:

1. The crystal lattice (see later) especially in the high conducting phase is naturally disordered. The way ionic conductivity develops with increasing disorder is a fascinating problem in itself. Considering the lattice to be made up of two sublattices, the cation sublattice and the anion sublattice, disorder develops in one of them either suddenly or gradually. The disordered sublattice is indescribable in terms of specific crystallographic positions or "basis." Thus, the Kittel definition [1.18] *lattice + basis = crystal structure* needs a considerable modification to have some relevance. Although the framework lattice is identified as I-centered bcc in α-silver iodide or as lead-centered fluorite in lead fluoride, the basis in the fast ionic phase is not at all well defined as far as the mobile ions are concerned. Indeed, XRD or even neutron diffraction of α-AgI is unable to determine the basis of Ag^+. Further, the I-centered bcc framework itself is a puzzle because it makes one feel that the phase transition from the wurtzite/zincblende phase at 300 K temperature to the bcc phase at 420 K has more to do with a restructuring of the I lattice! The Ag-ion sublattice just collapses at that temperature signaling the inability of the fragile Ag–I polar–covalent bond to maintain the open structure any longer. Thus for a "real" solid such as an SSI material, *real solid = perfect solid + defect/ disordered solid*.

2. The nature of the interatomic binding forces is different in fast ion conductors than in normal ones. In the former anharmonic lattice, vibrations play a crucial role in controlling physical properties. Their restoring forces are weaker in AgI than in NaI. Because of this feature, they tend to be mechanically soft.

3. The concept of Debye temperature is difficult to comprehend in fast ion conductors.

All these features make the fast ionic conductors neither Einstein solids with an exponential drop of heat capacity at low temperatures nor Debye solids with a heat capacity varying as the cube of absolute temperature. It is more appropriate to think of a combination of the two types to describe them. The nature of the motion of fast ions implies that the density of phonon states includes rather sharp maxima both at low energies and at high energies as in AgI. The usual Debye approximation does not seem to apply to the class of materials represented by AgI and PbF_2. The conductivity of these materials is also highly frequency dependent. Thus for a fast ion conductor, *framework sublattice + partial basis + disorder = fast ionic crystal structure*.

The existence of the sublattice disorder is a necessary condition for the appearance of fast ionic or superionic conduction. The cooperative features of ion transport in disordered systems are responsible for the *reduction* in the *overall activation energy* of ionic conduction from that of one-ion motion. The effects of ordering on the conductivity arise in three different ways: (1) ordering among conducting ions, (2) ordering of ions in the rigid sublattice, and (3) a cooperative excitation of ions from their normal sites to interstitial sites. In each case, a sharp decrease in the ionic conductivity is expected. The decrease in the ionic conductivity in the ordered state is mainly due to the *decrease in the correlation factor* rather than to the *increase in the activation energy* of ionic motion. Chapter 5 deals with this aspect.

1.2.4 Crystal Lattice, Reciprocal Lattice, and Brillouin Zones

All solids may not possess crystal structure or long-range order in space (e.g., glasses, gels, and poly-mers) but those which are crystalline help understand relatively easily the atomistic nature of matter. Crystals arise because of the attractive interactions between atoms and atomic or molecular ions, which fix them in 3D space. Furthermore, they may be characterized for their special properties including electrical, optical, magnetic, and chemical, which arise as a linear response (say, electrical conductivity) to an applied perturbation such as a sinusoidal electric field.

What do we mean by crystal structure? It is an arrangement of atoms or groups of atoms in a mathematical lattice such that the lattice arrangement looks the same when viewed from any two points separated by a translation vector. More importantly, as the solid contains an enormous number ($\sim 10^{23}$) of atoms, the concept of a crystal structure helps us to come up with a smallest possible cell or unit cell that may be translationally varied to build the macroscopic crystal. This unit cell for all practical purposes is the solid that possesses all the physical properties characteristic of the macroscopic solid including thermal (vibrational), elastic, ionic, and electronic properties.

A lattice net may be generated by a translational vector R_{uvw} and three basis vectors a_1, a_2, a_3 by means of the following relation:

$$R_{uvs} = ua_1 + va_2 + wa_3 \qquad (1.4)$$

This net is called a Bravais lattice. There are 14 unique Bravais lattices based on translational symmetry and superimposed rotational and rotation–reflection symmetry operations (called point-group symmetry operations). The more important SSI lattices (Figure 1.14) could be derived by a systematic "destruction" or deformation of the most symmetric solid, namely, the cube, thereby reducing its symmetry [1.18]. This helps to distinguish between seven crystal systems, namely, cubic, tetragonal, orthorhombic, hexagonal, rhombohedral, monoclinic, and triclinic in decreasing order of symmetry. Cubic is the most symmetric

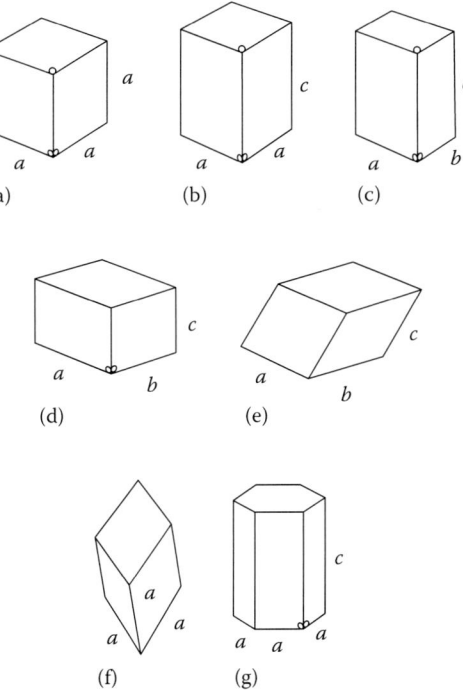

FIGURE 1.14 (a–g) Bravais lattices of interest in solid state ionics derived from the most symmetric cube. (After Kittel, C., *Introduction to Solid State Physics*, 7th ed., Wiley-New York, 2009.)

structure while triclinic is the least symmetric. Cubic and hexagonal structures are of interest in the context of SSI.

By "decorating" the lattice with a group of atoms called the basis, the crystal structure is generated. This is done by defining atom positions (x_{1j}, x_{2j}, x_{3j}, where j refers to all atoms in the unit cell) w.r.t. the basis vectors a_1, a_2, a_3:

$$r_j = x_{1j}a_1 + x_{2j}a_2 + x_{3j}a_3 \tag{1.5}$$

The basis of the LiI (NaCl type) structure is Li: (000) (½½0) (½0½) (0½½) and I: (00½) (½½½) (0½0) (½00) while that of the zincblende (ZnS) form of AgI is Ag: (000) and I: (¼¼¼). PbF_2 with fluorite (CaF_2) structure has the basis Pb: (000) (0½½) (½0½) (½½0) and F: (¼¼¼) (¼¾¾) (¾¼¾) (¾¾¼) (¾¼¼) (¼¾¼) (¼¼¾). Crystal structure determination involves the identification of the Bravais lattice type, unit cell dimensions, and most importantly the positions of atoms in the basis concerned. Point groups (32 in number) characterize the local symmetry of a given atom or ion in a crystal. Interestingly, LiI with predominantly ionic Li–I bond adopts the cubic closed-packed NaCl structure, while AgI with a polar covalent Ag–I bond adopts the open zincblende structure underlying the influence of chemical bonding on crystal structure and the resulting physical properties. If both basis atoms (Ag and I) are of identical chemical species, then the zincblende structure shown becomes the diamond structure. The other important structures are the non-centrosymmetric (i.e., lacking an inversion symmetry) wurtzite structure adopted by AgI, the spinel structure adopted by sodium beta alumina, and the perovskite structure adopted by $La_{2/3}TiO_3$. All these structures and others have a geological origin.

A complete set of rigid body motions that take a crystal to itself is called the space group. The crystallographic space groups in three dimensions are derived from a combination of the 32 crystallographic point groups with the 14 Bravais lattices. Each of these combinations belongs to one of the seven lattice systems mentioned earlier. This results in a space group being some combination of the translational symmetry of a unit cell including (1) lattice centering, (2) the point group symmetry operations of reflection, rotation, and improper rotation (also called rotoinversion), and (3) the screw axis and glide plane symmetry operations. The combination of all these symmetry operations results in a total of 230 different space groups describing all possible crystal symmetries and offer a description of the symmetry of the crystal. Of the aforementioned structures, LiI and PbF_2 share the space group $Fm3m$, while zincblende AgI has the space group $F\bar{4}3m$ and its wurtzite polymorph has the space group $P6_3mc$.

The way the spatiotemporal reciprocal space operates enables us to understand the world around us. From the naked eye observations of the colorful nature to XRD and to MRI scan of the brain, all involve the "reciprocals" of space and time.

Reciprocal lattice is thus a concept that shows how a periodic structure produces so many interesting phenomena in a solid state including the reflection of light by a silver mirror or a silicon surface or a diamond crystal and the optical transparency of a sodium chloride crystal. The reciprocal lattice is the essential link between the crystal diffraction experiment and the establishment of the complete structural identity of a crystalline solid in terms of the lattice parameters, basis, and electron density mapping. At the most fundamental level, it is necessary to know the conditions under which diffraction would occur by a crystal.

Consider the relevant incident radiation (x-rays or electrons or neutrons or even gamma rays) as a plane wave with a wavelength λ and wave vector $\mathbf{k}_0$ (or momentum)

$$\mathbf{k}_0 = 2\pi/\lambda s_0 \tag{1.6}$$

where s_0 is the unit vector along the propagation direction. In terms of the phase $\mathbf{k}_0 \cdot \mathbf{r}$, the incident wave is

$$\psi_0 = A\exp\left[i\left(\mathbf{k}_0 \cdot \mathbf{r} - \omega t\right)\right] \tag{1.7}$$

where ω is the angular frequency. Crystal diffraction produces a diffracted wave similar to (Equation 1.7) but with a wave vector $\mathbf{k}$. The von Laue condition for diffraction in terms of $\mathbf{k}$ reads

$$(1/2\pi)\left[\mathbf{k}\cdot(\mathbf{a}/a)-\mathbf{k}_0\cdot(\mathbf{a}/a)\right]=e/a$$

$$(1/2\pi)\left[\mathbf{k}\cdot(\mathbf{b}/b)-\mathbf{k}_0(\mathbf{b}/b)\right]=f/b \qquad (1.8)$$

$$(1/2\pi)\left[\mathbf{k}\cdot(\mathbf{c}/c)-\mathbf{k}_0(\mathbf{c}/c)\right]=g/c$$

where
   $\mathbf{a}$, $\mathbf{b}$, and $\mathbf{c}$ are the Bravais or direct lattice vectors
   a, b, and c are the crystallographic lattice parameters
   e, f, and g are integers such that $e = mh$, $f = mk$, $g = ml$ where in turn h, k, and l are integers
   m is a common integer

We can write the first terms on the LHS of (Equation 1.8) in terms of projections of $\mathbf{k}$ vectors on the different crystallographic axes.
   Note that the RHS of (Equation 1.8) has reciprocals of the lattice spacings a, b, and c which is justifiable since the LHS involves wave vectors. Now define a vector $\mathbf{a}^*$ so that

$$\mathbf{a}^*\cdot\mathbf{a}/a = 1/a, \quad \mathbf{a}^*\cdot\mathbf{b}/a = 0, \quad \mathbf{a}^*\cdot\mathbf{c}/a = 0 \qquad (1.9)$$

or

$$\mathbf{a}^*\cdot\mathbf{a} = 1, \quad \mathbf{a}^*\cdot\mathbf{b} = \mathbf{a}^*\cdot\mathbf{c} = 0 \qquad (1.10)$$

Similarly,

$$\mathbf{b}^*\cdot\mathbf{b} = 1, \quad \mathbf{b}^*\cdot\mathbf{a} = \mathbf{b}^*\cdot\mathbf{c} = 0 \qquad (1.11)$$

$$\mathbf{c}^*\cdot\mathbf{c} = 1, \quad \mathbf{c}^*\cdot\mathbf{a} = \mathbf{c}^*\cdot\mathbf{b} = 0 \qquad (1.12)$$

In terms of the starred vectors, Equation 1.8 reads

$$(1/2\pi)(\mathbf{k}-\mathbf{k_0}) = mh\mathbf{a}^* + mk\mathbf{b}^* + ml\mathbf{c}^* = \mathbf{r}_{hkl}^* \qquad (1.13)$$

where

$$\mathbf{r}_{hkl}^* = \mathbf{G}_{hkl} = m\left(h\mathbf{a}^* + k\mathbf{b}^* + l\mathbf{c}^*\right) \qquad (1.14)$$

$\mathbf{a}^*$, $\mathbf{b}^*$, and $\mathbf{c}^*$ are basis vectors of a reciprocal lattice and $\mathbf{r}_{hkl}^*$ is a lattice vector in reciprocal or wave vector space.
   By definition (1.9), $\mathbf{a}^*$ is perpendicular to the plane of the direct lattice vectors b and c. Similar remark holds for $\mathbf{b}^*$ and $\mathbf{c}^*$.

It can be shown that

$$\mathbf{a}^* = \mathbf{bxc}/\left[\mathbf{a}\cdot(\mathbf{bxc})\right], \quad \mathbf{b}^* = \mathbf{cxa}/\left[\mathbf{a}\cdot(\mathbf{bxc})\right], \quad \mathbf{c}^* = \mathbf{axb}/\left[\mathbf{a}\cdot(\mathbf{bxc})\right] \tag{1.15}$$

where the denominator is the unit cell volume.

The $\mathbf{r}_{hkl}^*$ or $\mathbf{G}_{hkl}$ of Equation 1.14 is normal to the *hkl* plane of the direct lattice. The *hkl* plane intersects the axes at *a/h*, *b/k*, *c/l*. Thus,

$$\left|\mathbf{r}_{hkl}^*\right| = \mathbf{G}_{hkl}m/d_{hkl} \tag{1.16}$$

d_{hkl}, the spacing between successive (*hkl*) planes, is

$$d_{hkl} = n\mathbf{a}/h = \mathbf{r}^*/\left|\mathbf{r}^*\right| = \mathbf{G}/\left|\mathbf{G}\right| = \mathbf{r}^*\cdot\mathbf{a}/h\left|\mathbf{r}^*\right| = m/\mathbf{r}^* = m/\mathbf{G} \tag{1.17}$$

The reciprocal lattice *substitutes points* for *planes* in the direct lattice.

Figure 1.15 illustrates the "transformation" of *hkl* planes to reciprocal lattice.

The von Laue condition (Equation 1.14) may be rewritten as

$$\mathbf{k} = \mathbf{k}_0 + \mathbf{K_n} \tag{1.18}$$

where

$$\mathbf{K_n} = 2\pi\mathbf{r}_{hkl}^*$$

The momentum conservation condition follows:

$$\hbar\mathbf{k} = \hbar\mathbf{k}_0 + \hbar\mathbf{K_n} \tag{1.19}$$

The incident photon with a momentum $\hbar k_0$ can be Bragg reflected to come out with momentum $\hbar k$ transferring a momentum of $\hbar K_n$ to the lattice. The momentum transfer is restricted to the discrete values of K_n only due to lattice periodicity.

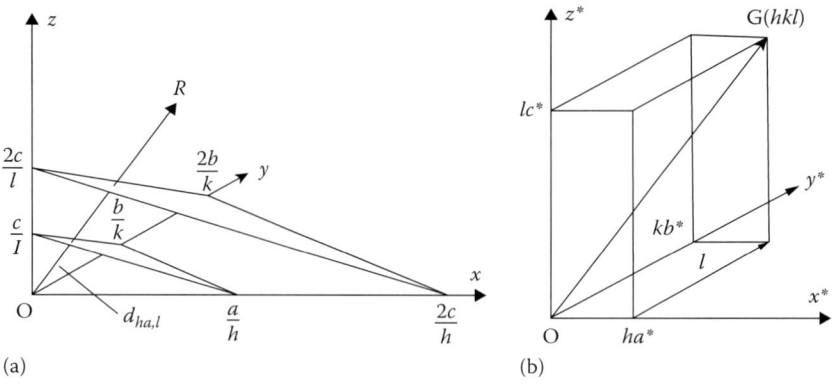

FIGURE 1.15 (a) Miller indices (*hkl*) planes in real space and (b) reciprocal lattice (r* or G(*hkl*)) with planes transformed to points. (From Fredricksson, H. and Akerlind, U., *Physics of Functional Materials*, Wiley, Chichester, U.K., 2008.)

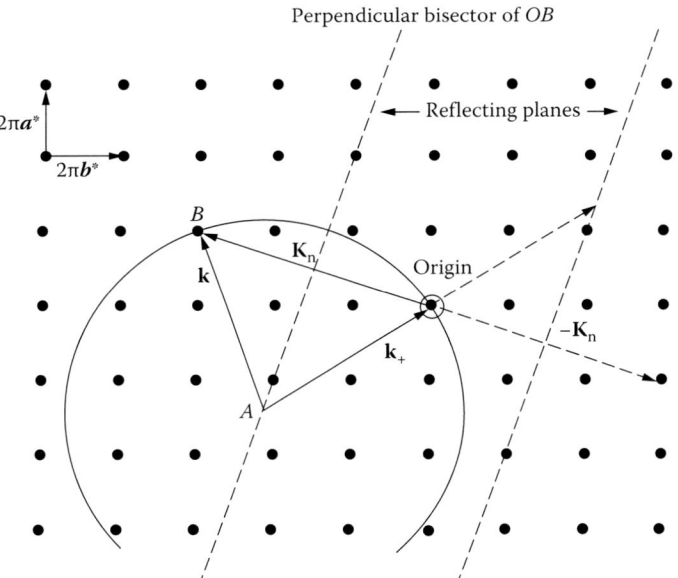

FIGURE 1.16 Construction of Ewald sphere. (Adapted from Brown, F.C., *The Physics of Solids*, W.A. Benjamin, New York, 1967.)

Brillouin zones are an essential ingredient of the theory of solid state and materials physics. They are used to describe and analyze the electron energy in the band energy structure of solids. The different Brillouin zones correspond to the primitive cells of different types that arise in the theory of electronic levels in a solid in a spatially periodic potential. The first Brillouin zone is the so-called Wigner–Seitz cell in the reciprocal lattice. It is thus a geometric construction to the Wigner–Seitz primitive cell in the wave vector or k-space ($k = 2\pi/\lambda$). The first Brillouin zone of the bcc lattice is just the fcc Wigner–Seitz cell and vice versa. A geometrical construction that illustrates both the concepts of reciprocal lattice and Brillouin zone is that of the Ewald sphere (Figure 1.16). The Ewald sphere is used in electron, neutron, and x-ray crystallographic analysis of materials; it demonstrates the relationship between (1) the wave vector (2π/wavelength) of the incident and diffracted x-ray beams and (2) the diffraction for a given reflection. How is it constructed?

Treating a crystal as a lattice of points of equal symmetry, what is the requirement for constructive interference in a diffraction experiment? The requirement is that in momentum space or reciprocal space, the values of momentum transfer where constructive interference occurs to cause crystal diffraction, by themselves form a lattice which is the reciprocal lattice. For example, the reciprocal lattice of a simple cubic real-space lattice is also a simple cubic structure. Another example, the reciprocal lattice of an fcc crystal real-space lattice is a bcc structure and vice versa. Ewald sphere helps to determine which lattice planes (represented by the grid points on the reciprocal lattice) will result in a diffracted signal for a given wavelength, λ, of incident radiation.

The incident plane wave falling on the crystal has a wave vector $\mathbf{k_0}$ whose length is $2\pi/\lambda$. The diffracted plane wave has a wave vector $\mathbf{k}$. If no energy is gained or lost in the diffraction process (it is elastic), then $\mathbf{k}$ has the same length as k_0. The difference between the wave vectors of diffracted and incident waves is defined as scattering vector $\Delta K = k - \mathbf{k_0}$ and $\mathbf{k}$ have the same length the scattering vector must lie on the surface of a sphere of radius $2\pi/\lambda$. This sphere is called the Ewald sphere.

The reciprocal lattice points are the values of momentum transfer. At these points, the Bragg diffraction condition is satisfied. For diffraction to occur, the scattering vector must be equal to a reciprocal lattice vector. Geometrically, this means that if the origin of reciprocal space is placed at the tip of $\mathbf{k_0}$, then diffraction will occur only for reciprocal lattice points that lie on the surface of the Ewald sphere. The concept of a sphere is extended to Debye theory of specific heats (Debye wave vector and Debye sphere) and to the electronic structure of metals (Fermi wave vector and Fermi sphere).

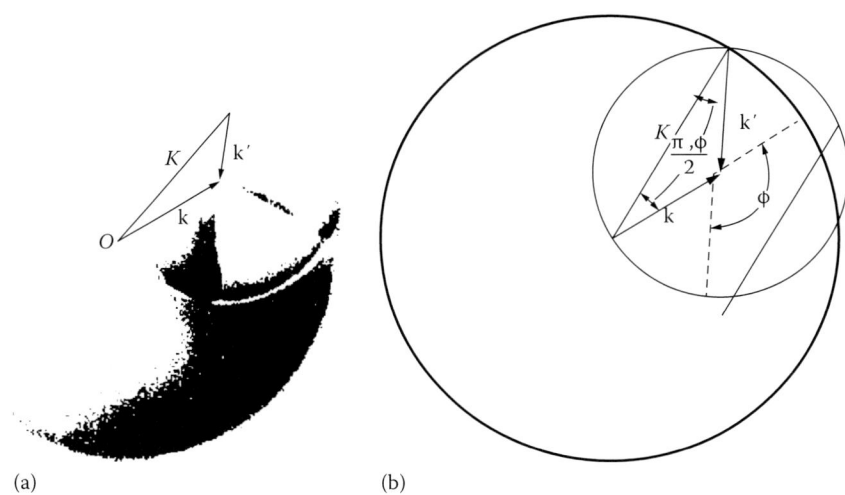

(a)   (b)

FIGURE 1.17   Ewald construction for the powder diffraction method. (a) The smaller sphere centered on the tip of the incident wave vector k with radius K. The origin is on its surface. The two spheres intersect in a circle. (b) Planar section of (a) containing the incident wave vector. (From Ashcroft, N.W. and Mermin, N.D., *Solid State Physics*, Holt Rinehart and Winston, New York, 1976.)

A slightly different Ewald construction introduces the concept of the Brillouin zone.

Consider that expanded reciprocal lattice vectors are $\mathbf{k_n} = 2\pi\mathbf{r^*} = 2\pi\mathbf{G}$ and $-\mathbf{k_n} = -2\pi\mathbf{r^*} = -2\pi\mathbf{G}$. Also another reflecting plane bisects the vector $-\mathbf{k_n}$ at $-\mathbf{k_n}/2$. Draw the incident wave vector $\mathbf{k_0}$ starting from the origin *O*. Then for diffraction to occur, $\mathbf{k_0}$ must terminate on the $-\mathbf{k_n}/2$ reflecting plane. Thus, the Brillouin zones may be constructed from the reciprocal lattice. The boundaries of these zones satisfy the conditions for crystal diffraction stated earlier.

X-ray powder diffraction is an important method of materials characterization. Thus, it is interesting to consider the Ewald sphere for powder diffraction (Figure 1.17) [1.19]. In Figure 1.17a, it is shown that Bragg reflections will occur for any wave vector k′ connecting any point on the circle of intersection to the tip of the vector k. The scattered rays thus lie on a cone that opens in the direction opposite to *k*. The triangle in Figure 1.17b is isosceles so that $K = 2k \sin \varphi/2$. Measure the angles φ at which Bragg reflections occur to know the lengths of all reciprocal lattice vectors shorter than $2k$. With this input and a few facts about macroscopic crystal symmetry and recognizing that the reciprocal lattice is a Bravais lattice, we can construct the reciprocal lattice.

The idea of Brillouin zone follows from the reciprocal lattice. Figure 1.18 displays the Brillouin zones in two cubic structures (bcc, fcc) and the hexagonal structure, which are adopted by a large number of SSI materials and will be discussed in Chapter 3.

Appendix 1A.1 at the end of this chapter contains an independent discussion of crystal symmetry in terms of point groups and space groups followed by a few problems.

1.2.5 Einstein Solids and Debye Solids

What are Einstein solids and Debye solids? Both are essentially vibrating solids. Structures that allow the existence and propagation of low-energy excitations are called Einstein solids. An ideal solid that is assumed to possess a continuous elastic energy spectrum of lattice vibrations is termed a Debye solid. As we shall see in Chapter 6, a vibrating lattice generates phonons that play a fundamental role in thermal properties and are related to SSI. While the crystal structure of atomistic solids discussed earlier is a static property, what is known about their thermal properties? Atoms in solids vibrate around their equilibrium positions because of their thermal energy.

Heat capacity of solids including SSI materials is an important perspective. The quantum theory treatment of specific heat capacity at constant volume (C_v) pioneered by Einstein was motivated by a sharp drop

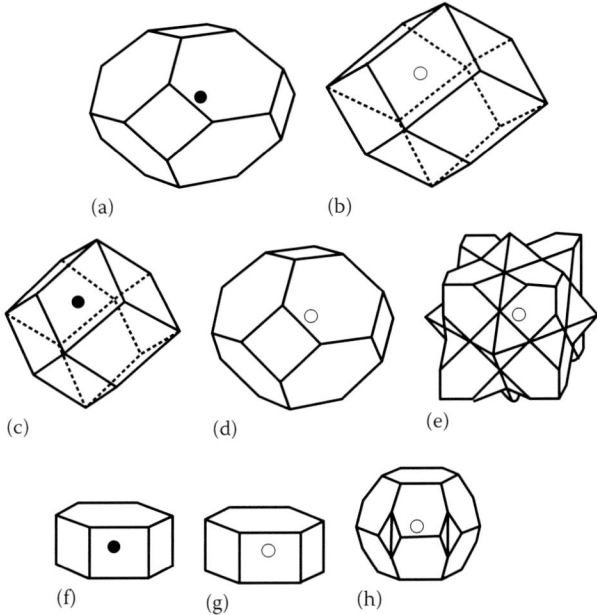

FIGURE 1.18 Primitive unit cells and Brillouin zones in bcc (a, b), fcc (c–e), and hcp (f–h) structures. The lattices are all monatomic. (From Fredricksson, H. and Akerlind, U., *Physics of Functional Materials*, Wiley, Chichester, U.K., 2008.)

in experimental data for diamond at temperatures lower than 1000 K, from the classical value of Petit and Dulong, that is, 5.9 cal/mol K. He explained it by assuming an ensemble of atomic harmonic oscillators of frequency $\nu_E = 3 \times$ number of atoms present. Einstein assumed for simplicity only one vibrational frequency on the average [1.21], now called the Einstein frequency ν_E. Thus we have a single oscillator Einstein model (Figure 1.19). An Einstein solid is one whose thermal property obeys the Einstein model. This model

1. Recognizes the atomistic nature of the solid state of matter; as defects in SSI materials are also atomistic, the model has a special role to play in such materials
2. Recognizes mechanical oscillations (vibrations) as a mechanism of energy exchange in solids and its connection to statistical thermodynamics
3. Motivates the development of anharmonic model for the lattice dynamics of solids
4. Quantizes the elastic energy of the solids leading to the concept of phonons (discussed in Chapter 6)

Let us discuss it further. Einstein proceeded to find the energy of these material oscillators as Planck did for radiation oscillators. Thus, the Einstein solid is a solid that vibrates with an Einstein frequency and possesses a heat capacity because of the vibrational free energy.

The average thermal energy of such a solid is

$$\langle E \rangle = \frac{(3R/N)\beta\nu_E}{\left[\exp\left(h\nu_E/k_BT\right)-1\right]} \tag{1.20}$$

where R is the gas constant/mol which in the modern context is $\beta = \hbar/k_B$.

Differentiating $\langle E \rangle$ w.r.t. T, the heat capacity of the solid at constant volume C_v is obtained as

$$C_v = \frac{d\langle E \rangle}{dT} + \frac{5.9\exp[h\nu_E/k_BT](h\nu_{E'}k_BT)}{[\exp(h\nu_E/k_BT)-1]^2} \tag{1.21}$$

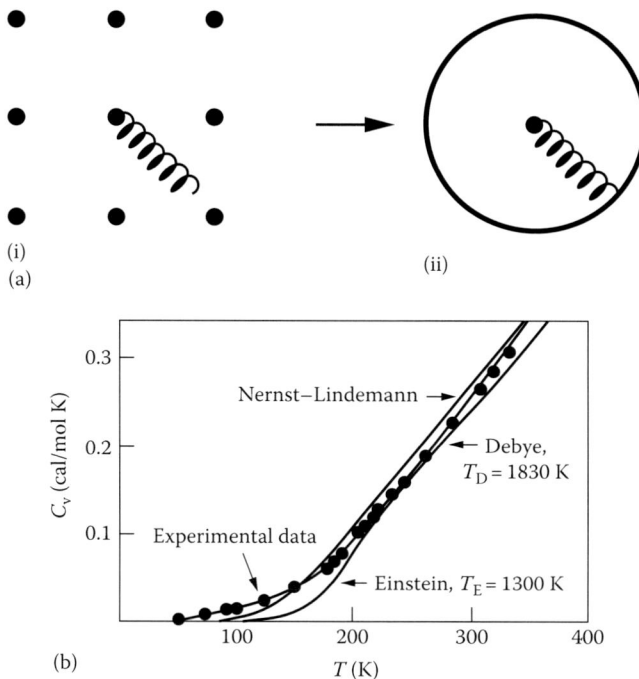

FIGURE 1.19   (a): (i) Einstein solid as a periodic array. Central atom connected to its eight neighbors through springs. (ii) To avoid summation peripheral atoms are replaced by a sphere (circle in 2D) smearing their masses uniformly over the surface. (b) Nernst–Lindemann fit to heat capacity data diamond in a modified Einstein model showing remarkable agreement with experiment. (From Cardona, M., Albert Einstein as the father of solid state physics, arxiv.org/pdf/physics/0508237, 2005.)

Equation 1.21 reduces to Petit–Dulong law as $T \gg h\nu_E/k_B$. At room temperature, the contribution of free electrons (in metals) to C_v is negligible.

Defects and disorders in a solid that are typical of SSI materials introduce interactions between lattice vibrations and result in anomalous heat capacity behavior characteristic of non-Debye solids. How does this arise will be discussed in Chapter 6.

The diamond specific heat fit to the Einstein model leaves a discrepancy at the lowest of frequency (Figure 1.19b). What is the reason for this discrepancy? Nernst and Lindemann examined the origin by using a two-Einstein frequency model to improve heat capacity data of diamond over single frequency fit obtained by Einstein. Nernst and Lindemann constrained one of the two frequencies to be half of the other to get better fit. These two frequencies correspond to the two averages of the acoustic and optic frequencies of diamond. The prediction of infrared frequency in diamond at 1330 cm^{-1} turned out to be close to the Raman mode in diamond! Einstein model is valid for low-energy optical modes that arise in SSI materials.

Note that in the diamond structure the tetrahedral symmetry and strong covalency are responsible for the high Debye temperature (~1000 K). In the weakly covalent IB–VII compounds particularly CuBr, CuI, and AgI which possess zincblende structures due to breaking of inversion symmetry and marginal covalency of the IB–VII bond, there arise low-energy optical modes that are important for the occurrence of fast Cu$^+$ and Ag$^+$ ion conduction and thus influence heat capacity below Debye temperature as it does in diamond. Chapter 6 would discuss this aspect in more detail. As would be discussed in Chapter 6, optical phonon modes that are independent of the wave vector are outside the purview of the Debye model but described quite well by the Einstein model [1.22]. Overall, an Einstein solid has an exponential temperature dependence of low-temperature specific heat capacity while a Debye solid, as would be discussed in detail in Chapter 6, has a heat capacity that varies as the cube of temperature.

After introducing the aspects of crystal structure and vibrations of atoms thereof, let us look at the basic aspects of electronic structure that essentially explain the wide range of electrical resistivity (Figure 1.11).

1.2.6 Electronic Structure

There are two ways to seek the origin of energy bands in solids. First is to look into what happens to the energy levels of isolated atoms as they are gradually brought closer and closer to form a solid. Second is to carefully consider and analyze the restrictions imposed by the periodicity of a crystal lattice on the motion of electrons, which is of course a more powerful approach to the theory of the solid state. We shall discuss the first approach and postpone the second approach until Chapter 7.

The electronic structure of a solid is really the electronic band structure that is a "crowded version" of the discrete energy level structure seen in Figure 1.1. The discussion of electronic structure begins from the Sommerfeld model which led to the model of a metallic solid as a "free electron Fermi gas" [1.18,1.19]. Two questions of relevance to SSI are as follows: (1) Can one envisage a free ion metal as an ideal fast ion or superionic conductor? (2) What is the basic (necessary but not sufficient) criterion for an electronic insulator which could be an ionic conductor? While question (1) would be answered in Chapter 7, a straightforward answer to question (2) is the following: A crystal can be an insulator only if the number of valence electrons in a primitive cell of the crystal is an even integer. This implies that if the number of such electrons is odd, then one would get a metallic crystal. Certain classes of SSI materials such as silver and copper halides could be considered as IB–VIII semiconductors to which nearly free electron model and tight-binding approximation are relevant as would be discussed in Chapter 7.

The electronic structure of a monatomic or diatomic solid such as Na, Si, NaI, and AgI is the way in which the constituent electrons are arranged among available energy "states." Figure 1.1 shows the free ion states and the states of the ion when placed in the Coulomb field of a solid. The electronic structure is the formation and relative arrangement of energy bands, which arise when $\sim 10^{23}$ atoms of Na or Si or 2×10^{23} atoms of NaI or AgI are brought together to form an equilibrium solid at absolute zero temperature. The nature of the interatomic forces and the chemical bonds as well as the periodicity of the crystal lattice all decide the energy band structure of a solid. At the most fundamental level, the number of valence electrons of a given atom provides the basic input for calculating the energy bands in a solid. The concepts of reciprocal lattice and Brillouin zone are involved in the representation of the electronic structure. Valence electrons in a metal such as sodium ($3s^1$) form a free electron Fermi gas whereas shared electrons in Si or AgI form a nearly free electron system. In NaI, there are virtually no free or shared electrons.

A fundamental model that serves to introduce the concept of electronic band structure is the Kronig–Penney model. Let us consider this model in the reciprocal space of the solid, a concept introduced earlier.

Let us formulate the band structure problem of a 1D solid, for instance a nanowire. The periodic potential of the lattice is modeled by delta functions. However, this problem is usually solved in direct space [1.18]. But it would be simple and insightful to solve it in the reciprocal space as follows [1.23].

For a lattice spacing of a units (Figure 1.20a), the reciprocal lattice is formed by $2\pi n/a$ ($n = 0, 1, 2, ..., \infty$) so that for $a \sim 6$ units, the reciprocal lattice would have atoms spaced more closely (Figure 1.20b).

Set up the 1D Schrodinger equation

$$\frac{-\hbar^2}{2m}\frac{d^2\psi}{dx^2} + U(x)\psi(x) = E\psi(x) \tag{1.22}$$

where the potential offered by the atoms $U(x)$ is periodic in x with a period a so that

$$U(x+a) = U(x) \tag{1.23}$$

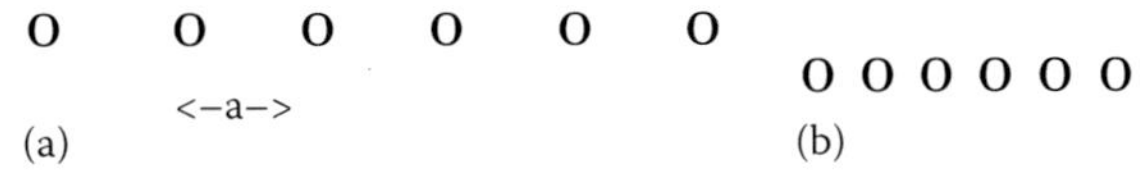

FIGURE 1.20 (a) 1D direct lattice and (b) 1D reciprocal lattice.

Expand the wavefunction $\psi_k(x)$ and the potential $U(x)$ in a Fourier series

$$\psi_k(x) = \sum_k C(k)\exp(ikx) \tag{1.24}$$

$$U(x) = \sum_{G_n} U(G_n)\exp(iG_n x) \tag{1.25}$$

where $C(k)$ and $U(G_n)$ are the appropriate Fourier coefficients.

Substitute in Equation 1.22 to get the central equation for band structure [1.23]

$$(\lambda_k - E)C(k) + \sum_{G_n} U(G_n)C(k - G_n) = 0 \tag{1.26}$$

where

$$\lambda_k = \frac{\hbar^2 k^2}{2m},$$

and

$$G_k = \pm\frac{2\pi n}{a} \quad n = 0,1,2,\dots,\infty \tag{1.27}$$

are the reciprocal lattice vectors for the 1D solid (Figure 1.22).

The Kronig–Penney model is described by the potential

$$(P/Ka)\sin Ka + \cos Ka = \cos ka \tag{1.28}$$

where P represents the strength of each delta function. The Fourier components are $\hbar^2 P/ma^2$. So, the central equation becomes

$$(\lambda_k - E)C(k) + (\hbar^2 P/ma^2)\sum_{G_n} C(k - G_n) \tag{1.29}$$

Let $f(k)$ denote the sum. Rearrange (Equation 1.29) to get

$$C(k) = \frac{-2Pf(k)a^{-2}}{(k^2 - 2mE/\hbar^2)} \tag{1.30}$$

Change k to $k - G_n$ and perform the sum over G_n.

The crucial point here is that

$$f(k - G_n) = \sum_{G_{n'}} C(k - G_n - G_{n'}) = \sum_{G_{n''}} (k - G_{n''}) = f(k) \tag{1.31}$$

Using this equality and setting $E = \hbar^2 K^2 / 2m$, we get the central equation for reciprocal space band structure

$$-a^2/2P = \sum_{G_n}\left[\left(k-G_n\right)^2 - K^2\right]^{-1} \tag{1.32}$$

The summation on the right side is done by splitting each term into partial fractions using the standard mathematical expression

$$\pi\cot x = \left(\pi/x\right) + 2\pi x \sum_{n=1}^{\infty}\left(x^2 - n^2\pi^2\right)^{-1} = \sum_{n=-\infty}^{\infty}\left(\frac{\left(n+x\right)}{\pi}\right)^{-1} \tag{1.33}$$

$$\text{RHS of } \left(1.32\right) = \frac{\left(a^2 K\right)\sin Ka}{\left(\cos Ka - \cos ka\right)} \tag{1.34}$$

Substitution in Equation 1.32 gives the basic equation of the Kronig–Penney model:

$$\cos ka = \cos Ka + \left(P/Ka\right)\sin Ka \tag{1.35}$$

This equation when solved numerically gives "energy bands" both allowed and forbidden. Thus, a consistent description of concepts in reciprocal space helps in a conceptual understanding of energy band. Note that the delta function description of potential used here is similar to the delta function representation of density of states in the Einstein model for the heat capacity of solids. It is also possible to work in the reciprocal space and derive the phonon dispersion relation of a 1D lattice of monatomic/diatomic crystal.

Solution of the Kittel's central equation for the energy at a Brillouin zone boundary is rather simple and motivating. It is the subject of a problem given at the end of this chapter. Chapter 6 considers in detail the electronic structure of SSI materials. Figure 1.21 displays the band structure in the Kronig–Penney model using 1D potential $U(x)$. Note that the allowed energy bands cannot cross each other. Furthermore, (1) for any allowed E, there are only two linearly independent solutions of the 1D Schrodinger equation

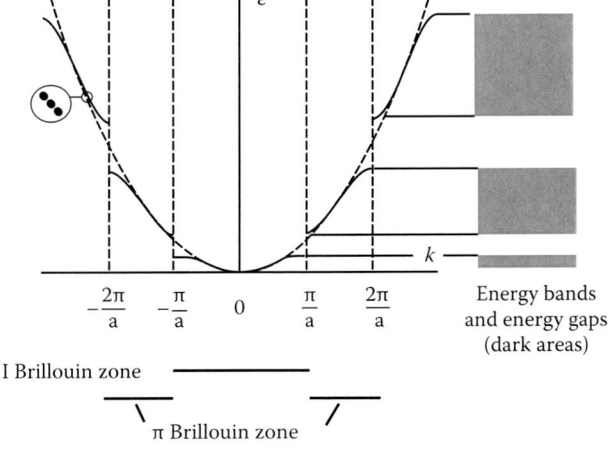

FIGURE 1.21 Electron energy band structure of a 1D crystal showing occupied bands and forbidden gaps. The three dots in the circle on the left indicate that the abrupt change in electron energy for certain k values occurs in three dimensions in an actual solid. (After Fredricksson, H. and Akerlind, U., *Physics of Functional Materials*, Wiley, Chichester, U.K., 2008.)

and (2) the feature $E(k) = E(-k)$ rules out any degeneracy at a given k. In any allowed energy band, the $E(k)$ versus k relation is a monotonic function of k for $0 \leq k \leq k/a$. Extremal energies occur only at $k = 0$ and π/a, where dE/dk generally vanishes.

1.3 Physics of Ionic Conduction

Monovalent cations such as H^+, Li^+, Cu^+, and Ag^+ or anions such as F^-, Cl^-, and $O^=$ can move in a special ionic solid and generate an ionic (not electronic) current when the solid forms a part of an electrical circuit. Size, position in the periodic table, and freedom of movement in a solid with appropriate structure offer a continuous chain of vacant sites. For facile ion motion, diffusion has to be exact, with activation energies ~0.1 eV so that ionic conductivities ~0.01–1.0 S/cm could be realized at or slightly above 300 K.

The problem of ion diffusion and ion mobility and thus ionic conductivity is fundamentally related to the crystal structure of the solid. The latter provides (1) mobile ions as part of a sublattice (Figure 1.22a) and (2) a large number of defects—vacancies of Schottky or Frenkel (Figure 1.22b) type—that provide easy pathways for long-range ion transport in the SSI materials.

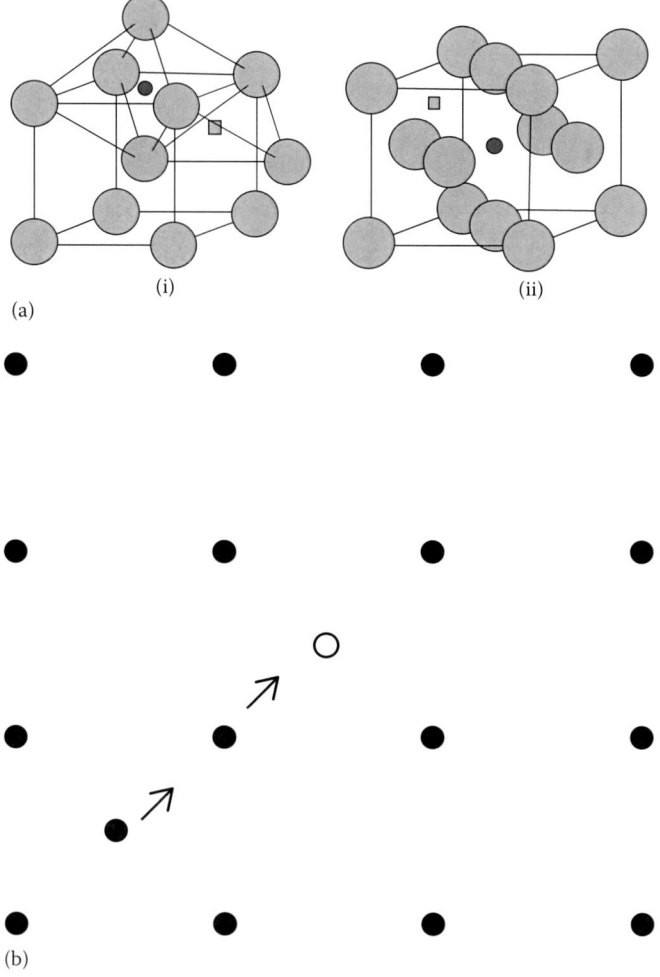

FIGURE 1.22 (a) Illustrating tetrahedral (squares) and octahedral (black circles) interstitials in (i) bcc lattice and (ii) fcc lattice for ion transport. (b) Cation interstitial jumps along a line (collinear interstitially jump) models ion conduction in AgBr.

1.3.1 Ion Diffusion

Movement of ions through a material medium—gas or liquid or solid not containing free electrons or holes—over macroscopic distances is ion diffusion. Ion diffusion of mobile ions in an ionic solid is responsible for ionic conduction and ionic conductivity. Preparation of oral rehydration solution that is a solution of sugar and sodium chloride in water involves molecular (sugar) and ion (Na^+) diffusion. Einstein determined the Avogadro number and the size of a sugar molecule from an analysis of diffusion data on sugar solutions. Kohlrausch measured the electrolytic conductivity of salt solutions. A dramatic demonstration of ion diffusion was carried out by Burt [1.24]. He observed that Na^+ ions in a molten $NaNO_3$ bath into which a 60 W electric bulb is immersed diffused through the glowing bulb and combined with the electrons released from thermionic emission from tungsten filament to form a Na film that was deposited on the cooler interior of the bulb.

A small piece of indium metal placed on a Si wafer and touched by a hot soldering rod melts the indium metal enabling an atom to diffuse through Si and providing donor atoms in the semiconductor.

Thus, diffusion can occur whenever there is a concentration gradient or chemical potential gradient between a solute and a "solvent." The result is the movement of ionic or atomic species from a region of high concentration to one of zero concentrations of diffusing species. Diffusion obeys Fick's law [1.8] while conductivity obeys Ohm's law. The ion diffusivity and ion mobility are connected through the Nernst–Einstein relation as would be seen later. These two parameters are ultimately related to the interionic potential and all the solid state properties that depend on this potential.

There is an important difference between neutral matter transport and charged matter transport in solids, even though both of them involve the existence of defects such as vacancies. The first process depends on size and pathway with no appreciable interaction of the diffusing species with the immediate surroundings. The second process of SSI is important in that the diffusing charge invariably involves the polarization of its immediate surroundings as it moves. If this ion diffusion is unusually fast then the corresponding ionic conductivity is a function of the frequency of the applied electric field and to a minor extent temperature. While diffusion in liquid media is isotropic, diffusion through solid media is anisotropic and is dependent on the crystal structure through which the ions move.

Classically speaking, a diffusing ion in a solid is said to traverse a barrier of height Δ or the activation energy at a rate

$$\nu = \nu_0 \exp\left(-\Delta / k_B T\right) \tag{1.36}$$

called the jump frequency. The ion makes ν_0 attempts (to jump) per second, the number of the order of a lattice phonon frequency. Thus, it is the frequency of oscillations of the ion within the cell associated with its lattice site.

In SSI materials, one species either cations such as Li^+ and Ag^+ or anions such as F^- and $O^=$ is more mobile than others. The less mobile ions form the "cage." ν_0 measures the vibration of the mobile ion within the cage. Elementary theory based on Equation 1.36 leads to the diffusion constant

$$D = \nu_0 a^2 = \nu_0 a^2 \exp\left(-\Delta / k_B T\right) \tag{1.37}$$

where a is the jump distance (Figure 1.23).

The diffusivity of an ionic species is a strongly temperature-dependent parameter. The exponential dependence is actually a dimensionless energy ratio defect migration energy/thermal energy. When the numerator is approximately equal to denominator, fast ion conduction occurs in a solid state ionic material, the ionic conductivity σ is related to D by the Einstein relation

$$\sigma - ne^2 D / k_B T \tag{1.38}$$

where n is the number density of diffusing ions.

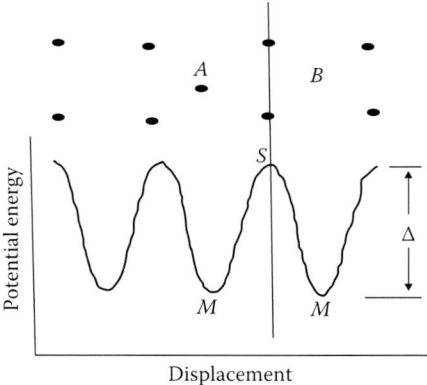

FIGURE 1.23 Illustrating interstitial ion diffusion in a Frenkel defect solid such as AgBr. Top: *A* is the interstitial. To occupy a place such as *B* by crossing the barrier (straight line). Bottom: Potential energy versus displacement curve. *M*, minimum; *S*, Saddle point. Ion jump over *S* ensures diffusion of defects.

The solid is considered as an *N*-body system with 3*N* vibrational degrees of freedom expressed as 3*N* normal coordinates. In this model, nu is computed from the rate at which the representative point of the system moves from one potential maximum to the next. In the configuration space (what is that?), the minima are separated by a saddle point (what is that?) in the 3*N*-dimensional potential space. It is this space over which the representative point must pass. In this approach, the activation energy measures the difference in potential between the bottom of the well and the saddle point. Let us now discuss conductivity basics.

1.3.2 Conductivity Basics: Ionic, Electronic, and Mixed Conductivities

Charged lattice defects in an ionic crystal (say, Ag^+ interstitials and Ag^+ vacancies in AgBr or AgI) move in response to an applied electric field resulting in an ion current density. Put simply, when a current is passed through an ionic crystal, electrolytic transport of matter occurs as found by Tubandt and Lorenz in silver iodide (Section 1.1.5). Ohm's law (Section 1.1.5) connects two quantities, current and the electric field, through ionic conductivity or electrolytic conductivity whose temperature dependence is opposite to that of metallic conductivity: ionic conductivity increases with increasing temperature which is also true of a semiconductor. We discuss the basics of ionic conductivity here:

$$\sigma = NZe\mu \qquad (1.39)$$

where
  N is the mobile ion concentration
  Ze is the ion charge
  μ is the ion velocity per unit applied field

The thermodynamics of point defects is central to this discussion.

While Schottky defects exist in normal ionic conductors such as sodium iodide (alkali halides), Frenkel defects arise in silver iodide and other fast ion conductors. Therefore, we discuss Frenkel defect conductivity basics. Utilizing thermodynamics and statistical mechanics, we can calculate the density of Frenkel defects. Thermodynamics deals with the number of cation interstitial-vacancy pairs in a crystal in equilibrium at temperature *T*. How are the defects formed? What is the rate at which equilibrium is attained? Such questions are beyond the realm of thermodynamics.

What is the number of defects in equilibrium at constant pressure? Let us write the expression for Gibbs free energy (and later minimize it)

$$G = W + PV - TS = H - TS \tag{1.40}$$

where

    H (enthalpy) is the internal energy
    W is the mechanical work
    P is the pressure
    V is the volume
    S is the entropy

Both H and S may depend on the defects present. If G_0 is the Gibbs free energy of the perfect crystal (crystal with zero defects) Equation 1.40 is rewritten as

$$G = G_0 + N_i g_i + N_v g_v - TS_{mix} \tag{1.41}$$

where

    N_i is the number of interstitials/unit volume
    N_v is the number of vacancies/unit volume
    g_i and g_v are the defect free energy contributions to G through interstitial and vacancy formation energies W_i and W_v

The entropy of mixing S_{mix} is calculated from the Boltzmann equation

$$S = k_B \log \pi \tag{1.42}$$

where

    k_B is the Boltzmann constant
    $\pi = w_i w_v$ is the thermodynamic probability, which is equal to the number of distinct ways of taking atoms from N lattice sites for form N_v vacancies

$$w_v = \frac{N!}{(N - N_v)! N_v!} \tag{1.43}$$

This is to be multiplied by the number of ways of placing these atoms on N' interstitial sites to form N_i interstitials

$$w_i = \frac{N'!}{N_i!(N' - N_i)!} \tag{1.44}$$

From Equations 1.42 and 1.43, the entropy of mixing is

$$S_{mix} = k_B \left[\frac{\ln N!}{(N - N_v)! N_v!} + \frac{\ln N'!}{N_i!(N' - N_i)!} \right] \tag{1.45}$$

Factorials in Equation 1.45 are large numbers, so we may use Stirling's approximation

$$\ln N! \approx N \ln N \tag{1.46}$$

Retaining only terms which depend on the disorder (Equation 1.41 with the help of Equation 1.45), we get

$$G \approx N_i g_i + N_v g_v - k_B T \left[N \ln N - (N - N_v) \ln(N - N_v) - N_v \ln N_v + N' \ln N' - (N' - N_i) \ln(N' - N_i) - N \ln N_i \right]$$

(1.47)

Now let us see why and how this G is minimized. A system with minimum G corresponds to an equilibrium configuration. G is minimized by making the ionic crystal electrically neutral (the principle of charge conservation):

$$-eN_v + eN_i = 0$$

(1.48)

In the range of temperatures in which thermal defects are produced (intrinsic region), $N_v = N_i$ already, but in general the variations δN_v and δN_i that arise as the crystal approaches equilibrium are equal. Thus, we require that in equilibrium

$$dG = \frac{\partial G}{\partial N_i \delta N_i} + \frac{\partial G}{\partial N_v \delta N_v} = 0$$

(1.49)

This is satisfied by setting the sum of partial derivatives equal to zero:

$$g_i + g_v - k_B T \left[\ln(N - N_v) - \ln N_v + \ln(N' - N_i) - \ln N_i \right] = 0$$

(1.50)

Rearrangement of Equation 1.49 gives

$$\left[\frac{N_i}{(N' - N_i)} \right] \left[\frac{N_v}{(N - N_v)} \right] = \exp \frac{-g}{k_B T}$$

(1.51)

where $g = g_i + g_v$, the free energy of formation of a Frenkel defect pair, which may be written as $g = h - T\Delta s$ or enthalpy minus the disorder-induced thermal entropy. The second term is negligible in normal ionic conductors but very important in fast ion conductors. The energy of formation of a pair is the most important experimentally determinable quantity, as is a pre-exponential factor which is $\sim 10^3 - 10^5$.

Equation 1.51 appears similar to a solubility product resulting from the application of the law of mass action. If the fractional concentrations of interstitials and vacancies are x_i and x_v, respectively, then

$$x_i x_v = K^{-1} = x_0^2$$

(1.52)

which is quite similar to the carrier concentration equation of an intrinsic semiconductor! K^{-1} is temperature dependent.

What is the expression for the ionic conductivity of a crystal with intrinsic Frenkel defects?

The movement of Frenkel defects gives ionic conductivity. If μ_i and μ_v are the defect velocities per unit electric field or mobilities of interstitials and vacancies, respectively, then an explicit form of Equation 1.38 is

$$\sigma = N_i e \mu_i + N_v e \mu_v$$

(1.53)

At high temperatures, let $N_i = N_v = n$, $n \ll N, N'$. Equation 1.51 reduces to

$$n = (NN')^{1/2} \exp \left(\frac{-w}{k_B T} \right)$$

(1.54)

How does ion mobility arise? The motion of the defect is essentially a jump from one interstitial site to another. The ion displaces in an applied field due to a drift velocity (similar to what happens in a metal or a semiconductor) superposed on a random thermal motion. The Nernst–Einstein equation connects ion mobility to an ion diffusion constant D_i:

$$\frac{\mu_i}{D_i} = \frac{e}{k_B T} \tag{1.55}$$

While Ohm's law relates current density to an applied electric field, Fick's first law relates current density to the concentration gradient

$$J_i = -D_i \frac{\partial n}{\partial x} \tag{1.56}$$

As discussed earlier, D has dimensions of cm^2/s. The important thing is to relate ion mobility to diffusion of an ion over a barrier of height Δ_i as the ion jumps over a length a through a random walk mechanism in three dimensions. Thus, the mobility of an interstitial is

$$\mu_i = \left(\frac{e a^2 v_0}{6 k_B T} \right) \exp\left(\frac{-\Delta_i}{k_B T} \right) \tag{1.57}$$

The conductivity is finally given by Equations 1.53 and 1.57 as

$$\sigma = (NN')^{1/2} \left[\frac{e^2 a^2 v_0}{6 k_B T} \right] \exp\left(\frac{-w}{2 k_B T} \right) \left\{ \exp\left(\frac{-\Delta_i}{k_B T} \right) + \left(\frac{-\Delta_v}{k_B T} \right) \right\} \tag{1.58}$$

Vacancies are less mobile than interstitials so that the second term in curly brackets may be neglected. Then σ assumes a simple form

$$\sigma T = A \exp\left(\frac{-(w/2 + \Delta_i)}{k_B T} \right) \tag{1.59}$$

where A is the pre-exponential factor $(NN')^{1/2}[e^2 a^2 v_0 / 6 k_B]$. The slope of the experimental plot of log σT versus $1/T$ called the Arrhenius plot (see Table 1.1) measured for pure samples of, say, AgBr gives the intrinsic defect energy exponent $w/2 + \Delta_i$ while the intercept gives A (Figure 1.24). The effect of doping with $CdBr_2$ conductivity of AgBr doped with less than 1 mol% $CdBr_2$ is seen as an enhancement by two orders of magnitude between 280 and 400 K. Doping is a process common to both semiconductors and ionic crystals, which introduces electrons or holes and excess cation or anion vacancies, respectively.

The tendency of some mobile ions such as $O^=$ in ZrO_2 to adopt lower or higher oxidation states either naturally or through doping of aliovalent (Ca^{++} or Y^{3+} in ZrO_2) results in a substantial "electronic" component to the total conductivity of the material. Thus, the ionic conductor becomes a mixed electronic–ionic conductor (MEIC). This prospect/aspect introduces an additional thermodynamic variable, namely, oxygen partial pressure in ZrO_2 and allied oxides, which makes conductivity to depend on concentration besides temperature and pressure. Nonstoichiometry arises as a consequence. Figure 1.25 illustrates the influence of oxygen "partial pressure" on temperature which demarcates three regions: (1) electronic conduction, (2) ionic conduction, and (3) electronic defect conduction in zirconia. An example of a cation conductor in which this happens is $Cu_{2-x}S$ where $1 \le x \le 2$. Nonstoichiometric compounds are quite common for d-, f-, and some p-block metals in combination with (1) soft anions such as S, Se, and H, Te (?) and (2) somewhat hard anions like oxygen. Halides, sulfates, and phosphates being hard anions,

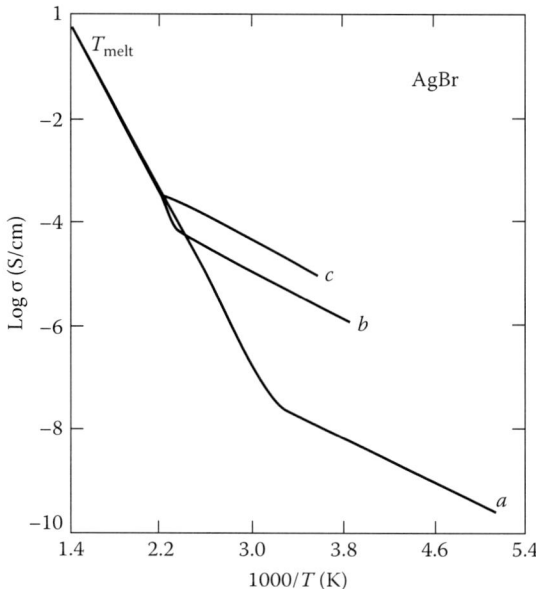

FIGURE 1.24   Experimental plots of log ionic conductivity versus reciprocal temperature for crystals of *a*, highly pure (1 ppm heavy metal impurity) AgBr; *b*, AgBr with 0.028 mol% $CdBr^2$; and *c*, AgBr with 0.12 mol% $CdBr_2$. Effect of Cd^{++} in enhancing conductivity via the formation of Ag^+ interstitials is clearly seen. (From Brown, F.C., *The Physics of Solids*, W.A. Benjamin, New York, 1967.)

nonstoichiometric compounds arise. Generally, two criteria favor the formation of such compounds: (1) crystal structure must allow changes in composition and (2) transition metals must have accessible oxidation states. Monoxides of the first-row transition metals NiO, CoO, MnO, and FeO meet these criteria and form layered structures that incorporate Li- such as $LiMnO_2$, $LiFeO_2$, $LiCoO_2$, and $LiNiO_2$. Chapter 2 elaborates on these aspects.

In principle, any material may possess nonzero electronic and ionic conductivities σ_{el} and σ_i. A *small* σ_{el} in a pure solid *ionic* conductor is a *necessary* condition for ion permeation through solid electrolyte, but application wise, this circumstance shortens the lifetime of a battery based on solid electrolyte. However, a *small* σ_i in an *electronic* conductor is a *necessary* condition for the permeation of ions through the electronic conductor which is advantageous for electrode materials development. It is necessary to balance σ_{el} and σ_i to realize enhanced performance in batteries. Thus, an important issue is to measure the partial conductivity based on changing the stoichiometry. Figure 1.22 shows a schematic plot.

MEICs exhibit both ionic and electronic (electron/hole) conductivities. In principle, any material may possess nonzero electronic and ionic σ's(σ_{el}, $\sigma_i \neq 0$). But MEICs are those materials in which σ_i, $\sigma_{el} \geq 10^{-5}$ *S/cm*. Of course there are processes in which minority carriers play an important role even if $\sigma_{el} < 10^{-5}$ *S/cm*. σ_{el} in a "pure" solid ionic conductor ($\sigma_i \gg \sigma_{el}$) is a necessary condition for the permeation of ions through the electronic conductor which is advantageous for the development of electrode materials. Chapter 4 discusses the phenomenological and experimental aspects of mixed conductors.

1.3.3 Structure–Defect–Ion Dynamics Nexus

A nexus apparently forms between three different aspects, namely, structure, defect scenario, and ion dynamics, and provides a unique motivating force to this book on solid state ionics. At the foremost level, the structure as an electron density map, the spectrum of atomic vibrations arising from structural peculiarities and deviations from perfection arising from the existence of a large concentration of atomistic defects and disorder and electronic band structure originating from the nature of the chemical bond are inextricably connected. The moving ion in an SSI material is almost as free as an electron in a metal it

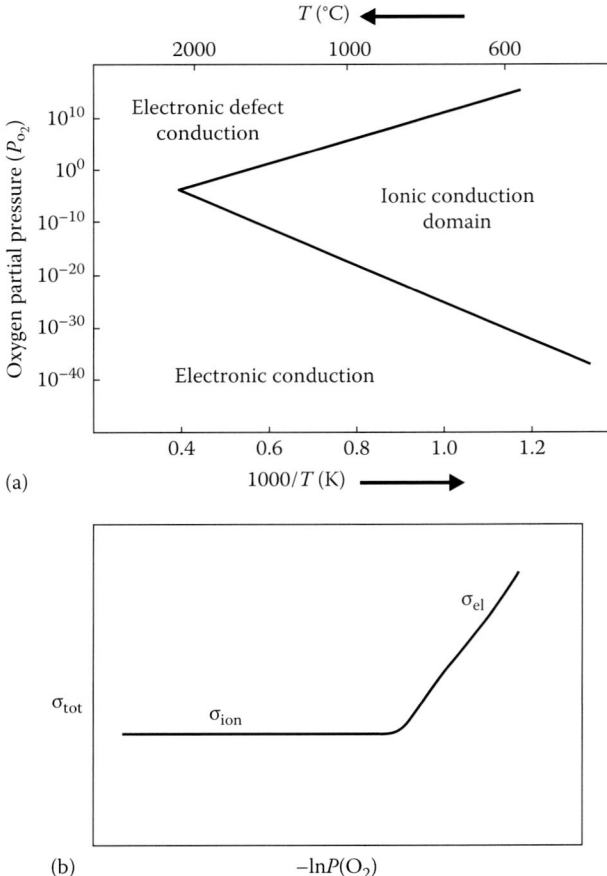

FIGURE 1.25 Characteristic features of mixed electronic–ionic conductors. (a) Oxygen ion conduction occurs in, say, zirconia over a domain sandwiched between a region of electronic defect conduction and electronic conduction. (b) σ_{tot} The total conductivity is the sum of ionic (σ_{ion}) electronic (σ_{el}) contributions. The latter is a sensitive function of oxygen partial pressure PO_2 which is an experimental independent variable. (From Maier, J., *Physical Chemistry of Ionic Materials*, Wiley, Chichester, U.K., 2004.)

possesses, in the sense that its motion does not disturb the solid considerably and its motion is aided by the bonding and unique structure. This is the phenomenological nexus. Next, the electrical conductivity of an SSI material made up of a pre-exponential factor and an energy exponent connect to their unique crystal structure and band structure, respectively, with defects and disorder playing a key role. Of course, more fundamentally, ion diffusion is directly connected to structure. These practical parameters link phenomenology to applications through solid state electrochemistry.

Let us go back to the picture of electron sharing in solids. If all valence electrons (say, $3s^1$ electrons in Na) are shared by all atoms of the crystal, then a metal results. Suppose four electrons are shared between four atoms as in C or Si, then semiconductor results. What if there are no electrons to share which means the "give and take" of electrons is essentially complete as in NaCl or NaI? An insulator is formed as an electronic insulator. However, defects—point defects such as Na^+ and Cl^- or I^- vacancies—ensure that a nonzero ionic conductivity is present. Among nonmetals, the largeness (>1–2 eV) or smallness (<1–2 eV) of the forbidden energy gap decides the semiconducting or insulating nature. Sharing of electrons with Si-like cores results in a semiconductor, whereas sharing of electrons with diamond-like cores results in an insulator.

There is thus a connection or a nexus between crystal structure, vibrations, and electronic structure in SSI materials such as AgI and PbF_2, which is missing in "common" salts such as NaI or NaCl. The chapter concludes with an outline of this book.

1.4 Scope of the Book

Monovalent cations H^+, Li^+, Cu^+, and Ag^+ or anions such as F^-, Cl^-, and $O^=$ can move in a rather special ionic solid and generate an ionic (not electronic) current when the solid forms a part of an electrical circuit. Size, position in the periodic table, and freedom of movement in a solid with appropriate structure offer a continuous chain of vacant sites. For facile ion motion, diffusion has to be exact, with activation energies ~0.1 eV so that ionic conductivities ~0.01–1.0 S/cm could be realized at or slightly above 300 K.

SSI is an applicable miniaturizable science that has made computers and portable and mobile communications possible through the solid state battery and is expected to revolutionize vehicular transport besides offering standby captive power to run homes and offices. Sensing and charge storage are emerging as versatile industrial applications.

Solid state physics and solid state electrochemistry besides materials chemistry make this happen. That is the story of solid state ionics, of which this book is an introduction. Crystal structure, phonon dispersion, and electronic band structure; defects, disorder, and nonstoichiometry; and nonequilibrium thermodynamics and statistical mechanics of iono-electron transport are the basic phenomenological threads that go into its fabric. Materials design and synthesis as well as characterization are naturally involved in an attempt to optimize diffusion coefficients and ionic conductivities. Special crystal structures attained through a phase transition are the rule, as are media such as polymers, gels, and ionic liquids that support high ionic conduction.

The three preceding sections have produced hopefully a broad-brush image of this book—something like a photographic negative that needs developing and printing. The chapters to follow essentially do that.

Chapter 2 is devoted to the thermodynamic and materials aspects of SSI materials. It begins with the study of phase diagrams and phase stability of ionic materials. Important binary and ternary systems of halides, oxides, and chalcogenides of important mobile cationic and anionic systems based on Li to Zr are considered. The aspects of nonstoichiometry and disorder are discussed. Classification of SSI materials is considered based on how the fast ion conducting state is achieved. This is followed by a discussion of important materials in the proton (H), Li, Na, noble metal, oxygen, and halogen families. Finally, the phase transitions in these materials are discussed as a paradigm of SSI. This chapter essentially sets the stage for the next chapter.

Chapter 3 considers the basic techniques of synthesis of SSI materials. It is essential to remember that these are not mere experimental methods to be routinely applied to make a given material. As these techniques have sound physicochemical basis, there is much scope for innovation and adaptability to particular situations. Thus, these techniques could lead to vastly improved or totally new ones. Discussed in this chapter are high-temperature solid state, controlled precipitation, sol–gel combustion, mechanochemical reaction, and hydrothermal and microwave synthetic procedures for bulk ionic materials; this is followed by physical and chemical vapor deposition techniques as well as ion-beam evaporation of single and multilayer thin films. The chapter concludes with a procedure for the fabrication of a Li-ion battery.

Chapter 4 is concerned with the natural next step, namely, the characterization of materials and devices. This includes the study of crystal structure through diffraction and microscopy, electrical characterization through diffusion and conductivity behavior, and thermal behavior including heat capacity, thermal expansion thermal conductivity, and thermoelectric coefficients, the last one being crucial for the development of thermoelectric materials. Characterization of SSI materials by impedance, dielectric, ultraviolet and visible absorption and photoacoustic spectroscopies, whose results connect to diffusion and conduction mechanisms and electronic structure, is discussed next. Complementary molecular spectroscopies such as infrared and Raman whose results focus on the role played by low-energy modes exhibited by SSI and magnetic resonance spectroscopies such as nuclear magnetic resonance (NMR), electron paramagnetic resonance (EPR), and Mossbauer or nuclear gamma resonance (NGR) that bear on the nuclear and electronic properties of these materials especially through structural phase transitions conclude the section on materials. Electrochemical characterization of the Li-ion battery, capacitor characteristics and those of fuel cells and sensors form the rest of the chapter.

Chapter 5 develops, following up on the paradigm of Section 2.5, a basic understanding of the phase transitions in SSI materials such as AgI and PbF_2 with the help of intuitive models and theories rooted in thermodynamics. Such an understanding is vital because the high conducting state is often attained through a first-order or second-order or a Faraday transition. Besides enriching the physics and chemistry of the superionic state, it would also help develop metastable fast ion conductors. This chapter discusses the so-called athermal phase transitions and order–disorder phase transitions in Li-transition metal oxides, which are relevant to the Li-ion battery applications.

Phonons that arise in a fundamental thermodynamic way in all solids through atomic vibrations of the crystal lattice form the basis of Chapter 6. SSI materials that contribute to defects and disorder also contribute to electrical and thermal properties besides electronic structure. Starting from the interatomic potential the phonon dispersion of typical materials is developed. Anharmonic vibrations that account for thermal expansion and thermal conductivity, and optical phonons in AgI-type compounds as well as Raman phonons in silver beta alumina, are discussed as typical case studies. The relevance of the Kramers problem is briefly discussed. Finally, the importance of soft modes to fast ion conduction is discussed taking doped zirconia as an example.

Chapter 7 deals with the electronic structure of SSI materials. Electronic structure basically arises from the nature of the "superionic" bond between a metal and a nonmetal. The bond fluctuations in a typical fast ion conductor AgI are also discussed as it relates to the phase transitions. Following up on Sections 1.2.1 and 1.2.6, the electronic structure is discussed in terms of bonding peculiarities. The formalism of density functionals is introduced, and the determination of electronic structure is outlined with specific examples. Electron relaxation in $LiFePO_4$ is considered in some detail as it relates to the magnetic phase transitions in that compound. Again as a follow-up of Section 1.3.3, the role of electronic structure in the ion dynamics of Li compounds is considered.

Chapter 8 is devoted to a comprehensive discussion of the most important application of SSI—solid state batteries. Following up on Sections 1.1.2 and 1.1.5, the basics of battery are discussed. The Li^+ ion physics of battery components is considered followed by a discussion of specific battery types including thin-film battery, Na^+ ion battery, and polymer battery. The next sections are devoted to advanced Li-air battery technology, and the battery for the electric vehicle. The chapter closes with observations on nanoscale battery materials for future work and also as a prelude to Chapter 10.

Chapter 9 is devoted to two applications: fuel cells and electrochemical sensors. Starting with how fuel chemical energy converts to electrical power, the basics of energy conversion process are delineated. Fuel cell elements and the physics of the solid oxide are discussed next, followed by a consideration of the design and evaluation of the solid oxide fuel cell and the polymer electrolyte membrane fuel cell. Supercapacitors are briefly discussed. The next two sections are devoted to sensor basics and typical sensors. The chapter concludes with an outlook.

Chapter 10 lays the foundations of the emerging area of nanoionics. The ionic nanosolid relative to the bulk solid is discussed for defects and conductivity, thermodynamics and electrostatics, and phase transitions. The physics for nanoscale devices is briefly outlined. Superionic superlattices, the prime movers for such devices as supercapacitors, are considered as they are novel material strategies. Thin-film devices are briefly discussed followed by some projections for the future.

All chapters carry summaries and some chapters a few problems. A few appendices on selected topics provide more detailed information.

1A.1 Appendix

1A.1.1 Crystal Symmetry Means Point Groups and Space Groups

What is the motivation to study crystal structures from the viewpoint of symmetries? It is because these are intimately connected to the experimental observations on materials—solid state ionic materials in the present context [1.25]. The immediate connection of crystal symmetry is to crystal diffraction. The sharp peaks that arise in x-ray or neutron diffraction is due to the symmetry of the concerned lattice. The intensities and widths of these peaks provide details, including dynamics of the ions when studied as a function of temperature. Another important connection of symmetry is to the allowed solutions of the wave equations for electrons in periodic crystals.

We shall briefly consider space groups and two subgroups thereof. What is a space group? Quite generally, it is a complete set of rigid body motions that take a crystal into itself. It is a group because it consists of a set of operations with a natural product along with the "doing nothing" identity element.

What are the rigid body motions of the lattice? Rotation, reflection, and inversion. All such operations can be described as the sum G of a translation a plus a rotation $R(n, \theta)$.

Two subgroups of the space group are of special interest:

1. The translation group made up of translation through all lattice vectors $n_1 a_1 + n_2 a_2 + L$, which leave the crystal invariant.

2. The *point* group or the operations leave the crystal invariant and *additionally* map some particular *point* onto itself. The space group is not simply the product of the point group and the translation group because there exist combinations of translation and reflection or rotation that leave a crystal invariant when used *together but not separately*. The 2D honeycomb lattice adopted by alumina which hosts Na^+ ions in Na-β alumina is an example [1.26]. Such a lattice is invariant when translated vertically by $a/(2\sqrt{3})$ and then reflected about a glide plane. Screw axes make a lattice invariant under a combination of translation plus a proper rotation is essentially a 3D operation. Note that while a space group defines the lattice, the point group does not.

The elements of the space group fixing a point of space are rotations, reflections, the identity element, and improper rotations.

The translations form a normal abelian subgroup of rank 3, called the Bravais lattice. There are 14 possible types of Bravais lattice. The quotient of the space group by the Bravais lattice is a finite group which is one of the 32 possible point groups. Translation is defined as the face moves from one point to another point.

The general formula for the action of an element of a space group is

$$y = Mx + D$$

where M is its matrix and D is its vector. The element transforms point x into point y.

In general, $D = D(\text{lattice}) + D(M)$, where $D(M)$ is a unique function of M that is zero for M being the identity. The matrices M form a point group that is a basis of the space group; the lattice must be symmetric under that point group.

A space group is denoted by a set of four symbols. The first describes the centering of the Bravais lattice (P, A, B, C, I, R, or F). The next three describe the most prominent symmetry operation visible when projected along one of the high symmetry directions of the crystal. These symbols are the same as used in point groups, with the addition of glide planes and screw axis, described earlier. By way of example, the space group of quartz is $P3_1 21$, showing that it exhibits primitive centering of the motif (i.e., once per unit cell), with a threefold screw axis and a twofold rotation axis. Note that it does not explicitly contain the crystal system, although this is unique to each space group (in the case of $P3_1 21$, it is trigonal).

There are 32 crystallographic point groups and 230 distinct crystallographic space groups. Table 1A.1 gives selected point groups and the corresponding space groups adopted by superionic conductors found in all the seven crystal systems. More comprehensive tables have been provided in Chapter 2.

There are 2 triclinic, 13 monoclinic, and 59 orthorhombic space groups. Let us learn some general properties of orthorhombic space groups, which are adopted by many practical superionic conductors such as NASICON and $LiFePO_4$. The orthorhombic space groups are best divided into three groups

TABLE 1A.1

Selected Superionic Point Groups and the Associated Space Group

Superionic Conductor	Crystal System	Point Group	Space Group	References
$Na_2FeP_2O_7$	Triclinic	1	$P1$	Kim et al. [1.31]
$Na_2MZr\ (P_2O_7)_2$ (M = Ni or Co)	Triclinic	1	P1	Byrappa et al. [1.32]
Cu_6PS_5Br	Monoclinic ($T < 166$ K)	m	C_c	Studenyak et al. [1.33]
Ag_2S	Monoclinic ($T < 452$ K)	$2/m$	$P2_1/c$	Hull [1.34]
Ag_2Se	Orthorhombic	222	$P2_12_12_1$	Hull [1.34]
γ-Li_3PO_4	Orthorhombic	mmm	$Pnma$	Hull [1.34]
$LiFePO_4$	Orthorhombic	mmm	$Pnma$	Hull [1.34]
$BaCeO_3, LaGaO_3$	Distorted perovskite	mmm	$Pnma$	Hull [1.34]
$LiCoO_2$	Rhombohedral	$3m$	$R3m$	
$Zr_3Y_4O_{12}$	Rhombohedral	-3	$R\overline{3}$	Hull [1.34]
β-AgI	Hexagonal wurtzite	C_{6v}	$P6_3mc$	Hull [1.34]
γ-AgI, γ-CuBr	Cubic zincblende	T_d	F-$43m$	Hull [1.34]
α-AgI	Cubic (bcc)	$m\overline{3}m$	$Im\overline{3}m$	Hull [1.34]
AgI	Cubic (fcc)	$m\overline{3}m$	$Fm\overline{3}m$	Hull [1.34]
β-PbF_2	Cubic(fluorite)	$m\overline{3}m$	$Fm\overline{3}m$	Hull [1.34]
$LiMn_2O_4$	Cubic(spinel)	$m\overline{3}m$	$Fd3m$	

based on crystal class: those that contain just 2 or 2_1 axes (class 222), those that contain a single 2 or 2_1 axes plus two mutually perpendicular planes (class $mm2$), and those that are centrosymmetric with both 2 or 2_1 axes and three mutually perpendicular planes (class mmm). Orthorhombic space groups belonging to the crystal class 222 are enantiomorphic, while those belonging to the crystal class $mm2$ are polar. The result of the mutually perpendicular axes and/or planes is to constrain all of the unit cell angles to 90°, that is, the unit cell axes are orthogonal to each other, but without constraint on their magnitude. (http://www.cryst.ehu.es/cgi-bin/cryst/programs/nph-wp-list?gnum=166&grha=rhombohedral.)

The space group symbol for the orthorhombic space groups is composed of the lattice centering followed by the symmetry with respect to the *a*, *b*, and *c* axes directions.

To illustrate some points concerning orthorhombic space groups, we will consider the space group *Pnma*, adopted by $LiMPO_4$ (M = Fe, Mn, Co, Ni), which is shown in Figure 1A.1.

The long space group symbol tells that this orthorhombic space group possesses three mutually perpendicular two-one screw axes. Note that none of the screw axes intersects in space. The origin for this space group has been chosen to coincide with a *point of inversion*. Note that the origin in this space group is equidistant from each of the 3 two-one screw axes, that is, at one-fourth of a unit cell length in *a*, *b*, and *c* from each screw axis.

Another important space group belongs to the hexagonal crystal system: $P6_3/mmc$ adopted by wurtzite-type β-AgI. To get an idea of the possible occupancies for Ag and I in this structure, Table 1A.2 presents the Wyckoff positions. A problem is set on the origin of wurtzite and zincblende structures adopted by AgI and CuBr, respectively.

The resolution of a crystal structure of solid stat ionic material determined by diffraction methods, to be described in Chapter 4, requires the knowledge of the following:

1. Its approximate composition
2. The unit cell dimensions
3. The symmetry
4. The diffracted intensities

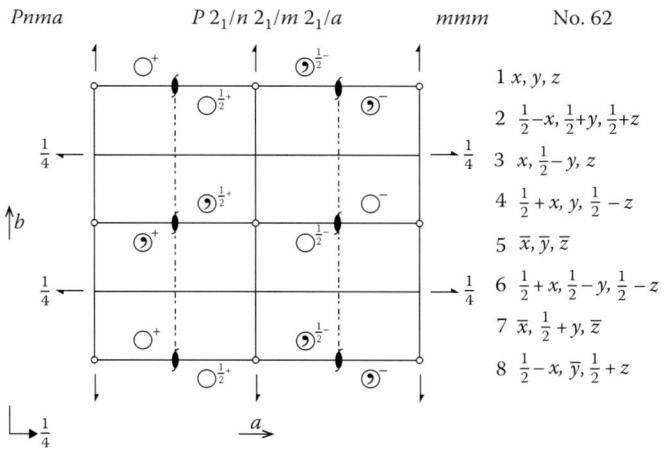

FIGURE 1A.1   The *ab*-plane projection of the orthorhombic lattice adopting the space group *Pnma*.

TABLE 1A.2

Wyckoff Positions for Occupation of Ag and I Atoms in β-AgI-Type Structure

Multiplicity	Wyckoff Letter	Site Symmetry	Coordinates			
24	*l*	1	(x, y, z)	$(-y, x - y, z)$	$(-x + y, -x, z)$	$(-x, -y, z + 1/2)$
			$(y, -x + y, z + 1/2)$	$(x - y, x, z + 1/2)$	$(y, x, -z)$	$(x - y, -y, -z)$
			$(-x, -x + y, -z)$	$(-y, -x, -z + 1/2)$	$(-x + y, y, -z + 1/2)$	$(x, x - y, -z + 1/2)$
			$(-x, -y, -z)$	$(y, -x + y, -z)$	$(x - y, x, -z)$	$(x, y, -z + 1/2)$
			$(-y, x - y, -z + 1/2)$	$(-x + y, -x, -z + 1/2)$	$(-y, -x, z)$	$(-x + y, y, z)$
			$(x, x - y, z)$	$(y, x, z + 1/2)$	$(x - y, -y, z + 1/2)$	$(-x, -x + y, z + 1/2)$
12	*k*	*m*	$(x, 2x, z)$	$(-2x, -x, z)$	$(x, -x, z)$	$(-x, -2x, z + 1/2)$
			$(2x, x, z + 1/2)$	$(-x, x, z + 1/2)$	$(2x, x, -z)$	$(-x, -2x, -z)$
			$(-x, x, -z)$	$(-2x, -x, -z + 1/2)$	$(x, 2x, -z + 1/2)$	$(x, -x, -z + 1/2)$
12	*j*	*m..*	$(x, y, 1/4)$	$(-y, x - y, 1/4)$	$(-x + y, -x, 1/4)$	$(-x, -y, 3/4)$
			$(y, -x + y, 3/4)$	$(x - y, x, 3/4)$	$(y, x, 3/4)$	$(x - y, -y, 3/4)$
			$(-x, -x + y, 3/4)$	$(-y, -x, 1/4)$	$(-x + y, y, 1/4)$	$(x, x - y, 1/4)$
12	*i*	*.2.*	$(x, 0, 0)$	$(0, x, 0)$	$(-x, -x, 0)$	$(-x, 0, 1/2)$
			$(0, -x, 1/2)$	$(x, x, 1/2)$	$(-x, 0, 0)$	$(0, -x, 0)$
			$(x, x, 0)$	$(x, 0, 1/2)$	$(0, x, 1/2)$	$(-x, -x, 1/2)$
6	*h*	*mm2*	$(x, 2x, 1/4)$	$(-2x, -x, 1/4)$	$(x, -x, 1/4)$	$(-x, -2x, 3/4)$
			$(2x, x, 3/4)$	$(-x, x, 3/4)$		
6	*g*	*.2/m.*	$(1/2, 0, 0)$	$(0, 1/2, 0)$	$(1/2, 1/2, 0)$	$(1/2, 0, 1/2)$
			$(0, 1/2, 1/2)$	$(1/2, 1/2, 1/2)$		
4	*f*	*3m.*	$(1/3, 2/3, z)$	$(2/3, 1/3, z + 1/2)$	$(2/3, 1/3, -z)$	$(1/3, 2/3, -z + 1/2)$
4	*e*	*3m.*	$(0, 0, z)$	$(0, 0, z + 1/2)$	$(0, 0, -z)$	$(0, 0, -z + 1/2)$
2	*d*	*−6m2*	$(1/3, 2/3, 3/4)$	$(2/3, 1/3, 1/4)$		
2	*c*	*−6m2*	$(1/3, 2/3, 1/4)$	$(2/3, 1/3, 3/4)$		
2	*b*	*−6m2*	$(0, 0, 1/4)$	$(0, 0, 3/4)$		
2	*a*	*−3m.*	$(0, 0, 0)$	$(0, 0, 1/2)$		

Source: After http://pd.chem.ucl.ac.uk/pdnn/symm3/sgortho.htm.

Problems for Appendix 1A.1

1A.1 T_d and O_h are two of the most important point groups that generate many superionic conductor space groups in the cubic family. Write down and illustrate the elements of T_d and O_h. How do the zincblende and the bcc structures arise from T_d and O_h point groups, respectively?

1A.2 (a) Given the basis for $LiCoO_2$ (Li 3a (000), Co 3b (000), O 6c (0 0 0.25)), construct two layers of the structure and identify the unit cell. (b) In $LiCoO_2$ as well as in Na-β alumina, the $O^=$ ions form 2D layers that construct the overall crystal structure. How does Co–O bonding and Al–O bonding in these two materials help stabilize the crystal structure that enables fast Li^+ and Na^+ motion, respectively? Consider how the point group symmetry (C_{6v}) generates the rhombohedral or trigonal space group of $LiCoO_2$ Rm.

1.A3 Two special symmetry operations screw axis and glide plane are involved in the generation of the space group. Visualize these operations by considering the simple screw and transflection. Transflection can be imagined as seeing yourself in a mirror even as the mirror is being moved parallel to itself. See illustrations in Figure 1A.3.

1A.4 (a) Wurtzite–zincblende coexistence in AgI. The tetrahedrally bonded β-AgI that is a topological insulator and a material with negative thermal expansion adopts both wurtzite (stable) and zincblende (metastable) structures at room temperature. The two structures coexist and the difference between them is subtle [1.18]. Illustrate with a sketch the stacking order A(I)A(Ag)B(I)B(Ag)A(I)A(Ag)B(I) B(Ag) that produces the "eclipsed" arrangement of zincblende (in contrast to the "staggered" arrangement of wurtzite). Bravais lattice is also hexagonal. The axis perpendicular to the hexagon is the c-axis. Identify the symmetry operations of this lattice. Obtain the elastic stiffness constant (c_{ij}) tensor. Set up the elastic wave equation. Solve it to find the longitudinal and transverse wave velocities. (b) The *hcp* lattice adopted by β-AgI among others contains a screw axis in the c-direction. Sketch a lattice and illustrate it. (*Hint*: The lattice remains invariant under translation along c by $c/2$ followed by rotation through 60°. Consider the Cartesian coordinates of a point on the c-axis.) The glide plane of the *hcp* is parallel to a plane containing both a- and c-axes. Where can such a plane be located so that translation along $c/2$ followed by reflection about the plane leaves the lattice unchanged or invariant (Figure 1A.4).

This lattice can be generated by the translation vector $r = n_1a_1 + n_2a_2 + n_3a_3$, where ns are integers and primitive translation vectors $a_1a_2a_3$ are given by $a_1 = c(001)$, $a_2 = (1/2)a[(1, 0, 0) + 3^{1/2}(0, 1, 0)]$, and $a_3 = (1/2)a[(-1, 0, 0) + 3^{1/2}(0, 1, 0)]$. Note the arrows at the top. The space group of this structure is $P6_3mc$. What is the angle between a_2 and a_3?

1A.5 The garnet structure has recently emerged as one supporting high Li^+ ion conductivity. Figure 1A.5 shows the original $Ca_3Fe_2Si_3O_{12}$ structure. Figure 1A.5a shows the complete unit cell (space group Ia-3d (230), the last cubic space group, $a = 1.15$–1.25 nm). Figure 1A.5b shows the arrangement of metal atoms in the garnet unit cell. $Li_5 La_3M_2O_{12}$ (M = Nb, Ta) adopts this structure. Also $Li_6BaLa_2Ta_2O_{12}$ has a conductivity of 40 μS/cm at 22°C [1.27].

Redraw these structures for the Li compounds and identify positions for Li^+ and possible paths for ion conduction. (*Hint*: Li^+ is tetrahedrally coordinated to oxygen like Si.)

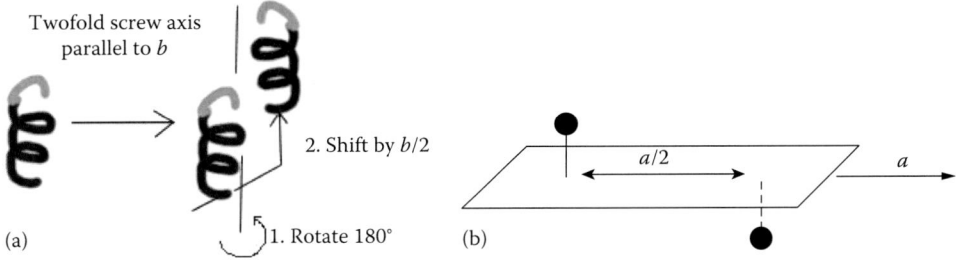

FIGURE 1A.3 Illustrating (a) screw axis and (b) glide plane. Using these concepts, try to understand the symmetry operations of *Pnma*.

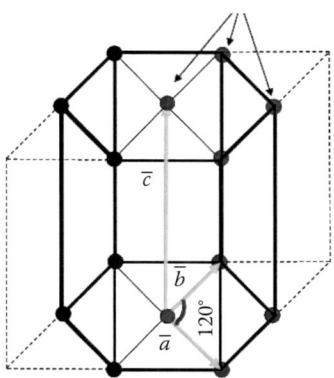

FIGURE 1A.4 Illustrating hexagonal structure; arrows indicate basis atoms. Lattice vectors a, b, c, and the *hexagonal angle* 120° are shown.

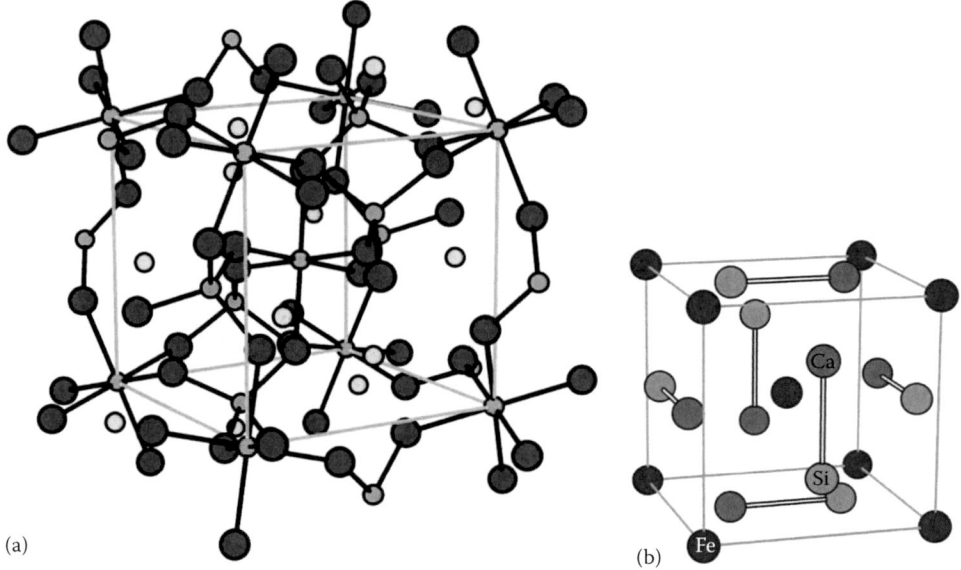

(a)

(b)

FIGURE 1A.5 In (a) dark gray circles are oxygen ions. In (b) rest of the ions are shown. Note that the composition of the Li compound us decided by the valences of Li, La and M. (From Garnet ($Ca_3Fe_2Si_3O_{12}$), https://chemistry.osu.edu/woodward/ch754/struct/Ca3fe2Si3.htm, modified April 22, 2002.)

Problems

1.1 Consider a set of points **R** constituting a Bravais lattice and a plane wave $e^{i\mathbf{k}\mathbf{r}}$, where **k** is the wave vector and **r** is the lattice vector. For what value of **k** will such a plane wave have the periodicity of the Bravais lattice?

1.2 Using a computer or by hand stack six rows of atoms are represented as circles. Call the first row "*A*." The second row "*B*" is formed by filling "spaces left between successive *A*s." The third row can be formed in two ways: (1) directly over A so that the fourth row becomes *B* and (2) by filling the third row *C* in the spaces between successive *B*s and the fourth row is again A. Thus generate two stacking sequences (1) *ABAB…* and (2) *ABCABC…*. Sketch the immediate neighborhood of a given atom in cases (1) and (2). These correspond to the hexagonal close-packed (hcp) and cubic close-packed (CCP) structures in two dimensions, the basis of the hexagonal and cubic crystal structures adopted by SSI materials.

1.3 (a) The x-ray powder diffraction pattern of the PbF_2 mineral fluorocronite taken using AgK_α x-rays (λ = 0.0559 nm for K_α radiation) gave five prominent reflections at Bragg angles (2θ) of 9°, 11°, 15°, 18°. and 19° with relative intensities 100, 46, 44, 42, and 21, respectively. Find the *d*-spacings and index the pattern on a cubic basis. Deduce the lattice parameter. Compare it with the β-PbF_2 value of *a* = 0.95397 nm. (b) Given the basis Pb (000) and F(¼¼¼), compute the structure factor and relative intensities and compare these with the aforementioned experimental values.

1.4 Trace the "symmetry breaking" operations involved in the "evolution" of the seven Bravais lattices shown in Figure 1.14 starting from the cubic lattice.

1.5 Using the basis of the fluorite lattice given in the text and the lattice constant of Problem 1.4, sketch the 3D unit cell of PbF_2. Locate the possible positions of the F^- Frenkel defect pair and highlight the F^- sublattice.

1.6 Find the reciprocal lattices corresponding to (a) bcc, (b) fcc, and (c) hexagonal direct lattices. Given that the three vectors **a**, **b**, **c** of Equation 1.14 are $\mathbf{a} = (a/2)(-\boldsymbol{x} + \boldsymbol{y} + \boldsymbol{z})$, $\mathbf{b} = (a/2)(\boldsymbol{x}-\boldsymbol{y} + \boldsymbol{z})$, $\mathbf{c} = (a/2)(\boldsymbol{x} + \boldsymbol{y}-\boldsymbol{z})$ for bcc; $\mathbf{a} = (a/2)(\boldsymbol{y} + \boldsymbol{z})$, $\mathbf{b} = (a/2)(\boldsymbol{x} + \boldsymbol{z})$, $\mathbf{c} = (a/2)(\boldsymbol{x} + \boldsymbol{y})$ for fcc, *a* being the corresponding lattice parameter; and $\mathbf{a} = 3^{1/3}(a/2)\boldsymbol{x} + (a/2)\boldsymbol{y}$, $\mathbf{b} = -3^{1/2}(a/2)\boldsymbol{x} + (a/2)\boldsymbol{y}$, $\mathbf{c} = c\boldsymbol{z}$, for hexagonal lattice *a*, *c* being lattice parameters. $\boldsymbol{x}, \boldsymbol{y}, \boldsymbol{z}$ are unit vectors.

1.7 Jump frequency of an ion in a solid state ionic material is a fundamental quantity that is related to ionic conductivity. Find an upper limit of the jump frequency and the maximum conductivity realizable. Take jump distance as 0.1 nm and temperature as 300°C.

1.8 Find the concentration of Frenkel defects in (a) AgCl at 300 K, (b) AgBr at (i) 27°C and (ii) 430°C (2° below melting temperature), (c) wurtzite AgI, and (d) zincblende AgI at 300 and 400 K. Compare the results to appreciate the difference between slow and fast ion conductors.

1.9 (a) Compute the axoplasm resistance of 1 cm long, 2 μm diameter segments of (i) unmyelinated and (ii) myelinated axons. Also compute their membrane capacitances and membrane leakage resistances. (b) A parameter λ called the space parameter indicates how far a current travels along the axon before most of it has leaked out through the membrane. It is given by the relation $\lambda = \sqrt{[R_m r/2\rho_a]}$ where *r* is the segment radius and ρ_a is the axon resistivity. Find λ for an axon of radius 0.2 μm if it is (i) myelinated and (ii) unmyelinated.

1.10 Examine the discontinuities in Figure 1.21. Discuss the origin of the forbidden gaps and of the first and second Brillouin zones in one dimension. Compare this to the 3D case illustrated in Figure 1.18d and e for the fcc structure and Figure 1.18f and g for the hexagonal structure.

REFERENCES

1.1 W. M. Grill, *Am. Sci.* 98(January–February 2010) 48.

1.2 M. Kobayashi, *Solids State Ionics* 174(2004) 57.

1.3 J. W. Kane, M. W. Sternheim, Nerve conduction, *Physics,* 3rd edn., Wiley, 1988, Chap. 18; A. Faller, M. Schuenke, *The Human Body: An Introduction to Structure and Function*, Thieme, New York, 2004.

1.4 F. Pirajno, *Hydrothermal Processes and Mineral Systems*, Springer, New York, 2009, Chap. 1.

1.5 P. A. Tipler G. P. Mosca. *Physics for Scientists and Engineers with Modern Physics*, 6th edn., W H Freeman, New York, 2007.

1.6 B. Srilakshmi, *Nutrition Science*, New Age International Publishers, New Delhi, India 2010, Chap. 20.

1.7 S. J. Mills, P. M. Kartashov, G. N. Gamyanin, P. S. Whitefield, A. Kern, H. Guerault, A. R. Kampf, M. Raudsepp, *Eur. J. Mineral.* 23(2011) 695.

1.8 M. Reich, *Mineralium. Deposita* 44(2009) 719.

1.9 M. Faraday, *Philos. Trans. R. Soc.* 90(1838) 507.

1.10 C. Kitttel, K. Kroemer, *Thermal Physics*, 2nd edn., W H Freeman, New York, 1980.

1.11 R. Ahuja, A. Blomqvist, P. Larsson, P, Pyykkö, P. Zaleski-Ejgierd, *Phys. Rev. Lett.* 106(2011) 018301.

1.12 H. Rickert, *Solid State Electrochemistry*, Springer-Verlag, Berlin, Germany, 1982.

1.13 J. Maier, *Physical Chemistry of Ionic Materials*, Wiley, Chichester, U.K., 2004.

1.14 C. S. Sunandana, P. Senthil Kumar, *Bull. Mater. Sci.* 27(2004) 1.

1.15 A. Beiser, *Perspectives of Modern Physics*, McGraw Hill, New York, 1969.

1.16 W. A. Harrison, *Electronic Structure and Properties of Solids—The Physics of the Chemical Bond*, Dover, New York, 1989.

1.17 L. Pauling, *The Nature of the Chemical Bond*, 3rd edn., Cornell University Press, Ithaca, NY, 1969; *General Chemistry*, Dover, New York, 1970.

1.18 C. Kittel, *Introduction to Solid State Physics*, 7th edn., Wiley-India, New York, 2009.

1.19 H. Fredricksson, U. Akerlind, *Physics of Functional Materials*, Wiley, Chichester, U.K., 2008.

1.20 N. W. Ashcroft, N. D. Mermin, *Solid State Physics*, Holt Rinehart and Winston, New York, 1976.

1.21 M. Cardona, Albert Einstein as the father of solid state physics, arxiv.org/pdf/physics/0508237, 2005, accessed August 17, 2015.

1.22 M. A. Omar, *Elementary Solid State Physics: Principles and Applications*, Pearson Education India, Delhi, 1999.

1.23 S. Singh, *Am. J. Phys.* 51(1983) 179.

1.24 R. C. Burt, *J. Opt. Soc. Am.* 11(1935) 87.

1.25 M. P. Marder, *Condensed Matter Physics*, 2nd edn., Wiley, Hoboken, NJ, 2010.

1.26 G. Collin, J. P. Boilot, R. Comes, *Phys. Rev. B* 34(1984) 5850.

1.27 V. Thangadurai, W. Weppner, *Ionics*, 12(2006) 81.

1.28 K. Kilner, Defects, mobile ions and disordered structures, SSI-17 tutorial Kilner.pdf, 2009, accessed August 18, 2015.

1.29 E. M. Purcell, *Electricity and Magnetism*, 2nd edn., Tata McGraw Hill, New Delhi, India, 2008, p. 140.

1.30 K. Funke, *Sci. Technol. Adv. Mater.* 14(2013) 043502.

1.31 H. Kim et al., *Adv. Funct. Mater.* 23(2013) 1147.

1.32 K. Byrappa et al., *Indian J. Phys.* 63A(1989) 321.

1.33 I. Studenyak et al., *Cond. Matter Phys.* 10(2007) 149.

1.34 S. Hull, *Rep. Prog. Phys.* 67(2004) 1233.

2

Solid State Ionic Materials

2.1 Preamble

This chapter deals with the structural diversity of solid state ionic materials (SSIMs) vis-a-vis their phase stability, often against phase transitions. Phase equilibria naturally involve thermodynamics. Among the numerous phases involving Li^+, Na^+, Ag^+, Cu^+, F^-, and $O^=$ with structures from triclinic to cubic, two parameters break symmetry: cooling from a high temperature and cation substitution especially in cubic structures such as perovskites. A thorough discussion of these aspects offers a bottom line: A thermodynamically stable defect-ridden, mobile ion path–optimized superionic conductor ready for application.

A perfect structure is the starting point for the "operating" "functional" reality of the SSIMs.

A 2D "structure" may be a quadrilateral or a square or a few figures in between.

A 3D structure may be a "quadrilateraloid" or a cube or many shapes in between.

The structure of a functional SSIM is similar to that of human lungs—into which air may go out or come in. Therefore, SSIMs are considered "ion breathing" structures that are themodynamically stable—often against possible phase transitions.

Phase diagrams are essential as a basis for understanding the relationships between stoichiometry and structure and properties of inorganic materials, especially complex materials that have variable composition. For materials based on solid–solution phases, it is necessary to know the solid–solution mechanisms in order to understand the detailed crystallography and properties of the phases. This is because, in many cases, properties vary dramatically with solid–solution composition. An overview of solid–solution equilibria and structures and properties of six ceramic systems is given in this chapter: lithium ion–conducting solid electrolytes based on Li_4SiO_4 solid solutions, Na^+ ion–conducting β-aluminas and the effect of Li/Mg additions, lanthanum zirconium tantalate ceramics, doped lithium niobate and lithium tantalate ferroelectrics, bismuth cuprate superconductors, and silicon aluminum oxynitride (SIALON) engineering ceramics. As well as providing fundamental data on the stoichiometry and stability of the phases concerned, it is shown how, in each case, valuable clues regarding the likely crystallography of the solid solutions can be obtained.

The concept of dynamic/thermodynamic/kinetic equilibrium is the basis of phase equilibrium and phase stability of solids. This perspective of understanding, exploring (experimentally), and computing phase diagrams in SSIMs is just at its early stage of investigation.

The use of a solid electrolyte immersed in a liquid electrolyte as in a lead–acid battery has an inherent advantage, as that of "dissolving" any gases evolved during "half-cell reactions." The all-solid state battery, however, has to operate in an environment according to the thermodynamics of the "partly open–partly closed" system, particularly in a multicomponent system involving a gaseous component. In such systems, the desired phase has to coexist in equilibrium with another phase, involving the evolution or consumption of oxygen at the electrode–electrolyte interface. Therefore, apart from the temperature and pressure, a new variable, the chemical potential, would be involved in the thermodynamic description of the system.

Therefore, there arises a need to investigate phase equilibria and phase stability and thus phase diagram of binary and multinary systems such as Li_2O–TiO_2, Na_2O–Al_2O3, CaO–ZrO_2, Li–Mn–O, Li–Fe–P–O, and Li–La–Zr–O in a search for new electrode/electrolyte materials and also in devising strategies for the synthesis and characterization of such materials. The importance of crystal structures with built-in 2D and 3D pathways for ion transport and thermodynamic control of defects, disorder, and defect pathways arise naturally. Structures with anion-based (I^-, S^-, PO_4^{3-}) and cation-based (Pb^{++}, La^{3+}, Zr^{4+}) frameworks have to be stable with respect to a range of temperatures.

Although it was Faraday who laid the foundations of the field of solid state ionic with silver sulfide and lead fluoride as Ag^+ and F^- conductors, to be joined much later by Nernst with zirconia, the oxygen conductors, and Tubandt by silver iodide, the Ag^+ conductor, the materials of the veteran lead–acid battery, the proton conductor H_2SO_4–H_2O, continue to serve for over a century until now as the prime component of UPS and as promising fuel cells for automobiles. The Li^+ conductor–LiI-based composite used as pacemaker initiated the search for the materials present in Li-ion batteries. Sodium–sulfur system chipped in as a candidate for high-voltage battery system initiated the search for ion power systems to run the cars of the near future.

Thus, a discussion of several solid state ionic systems based on different cations and anions puts into context several issues that form the basis of the future chapters.

2.2 Perspectives

SSIMs are binary ionic compounds made up of elements starting from H to Ag to Zr to U and N to O to Te to I. Thus, the well-known periodic table of elements may be modified to make it semiconductor/insulator-centric so that a new vision for synthesis properties and applications emerges. This modified Partial Table (Table 2.1) shown carries forward the picture of Harrison [2.1]. Electronic conductivity is the link between solid state ionics and semiconductor physics.

"Fast ions" whose rapid diffusion in certain special structures to give SSIMs are indicated earlier. Note the special position of H^+ and Li^+ along with He. Na^+, Cu^+, Ag^+, and Au^+ together form cationic conductors while F^- and $O^=$ form anionic conductors. Pb^{++} plays a special role in H^+-based "car batteries" and "UPS batteries." "Left wing" elements that begin with Sc and end with Pt are transition metals, a few of which have mixed valencies (Ti, V, Mn, Fe, Co, among others) that lead to a variety of structural frameworks along with O, S, P, and halogens. "Right wing" elements are the rare earths, which along with alkaline earths, Sr and Ba, give robust structures such as the perovskites that support $O^=$ or H^+ transport. It is noteworthy that the central portion of the modified Periodic Table contains elements with p covalency (Ge, Si, P, S, I), which through hybridization can yield diverse structures through phase equilibria studies often exhibiting phase transitions. This table, therefore, has inspired us to write this chapter, which connects the previous chapter and the succeeding chapters.

Indeed, the superionic conductor as a semiconductor exemplified by AgI, CuBr, CuI, and the many binary and ternary chalcogenides and chalcohalides of Ag and Cu has been considered from basic and applied viewpoints [2.2]. All these compounds incorporate the suitably hybridized weak covalent bond between the 3d or 4d electron of the metal and the p electron of the halogen/chalcogen. It is this bond that gives a Janus face to these compounds: the electronic subsystem along with the ionic subsystem results in superionic semiconductors—essentially mixed ionic–electronic conductors, which provides an inherent tendency to admit excess metal atoms over and above the ideal 2:1 metal: chalcogen as in Ag_2S gives rise to the possibility of defects (electronic and ionic), disorder, and the so-called nonstoichiometric condition [2.3]. At the atomic level, this makes possible a "switch" between the noble metal (Ag) and the semiconductor (Ag_2S), which will be discussed in Chapter 10. Layered oxides such as $LiCoO_2$ and $AgCrS_2$ are also possible due to covalent bonding. So let us now discuss why one should worry about SSIMs from the thermodynamics view point. Why one should focus on phase diagrams and phase stability.

There is a special need to understand SSIMs as an active, dynamic, mobile ion–based solid phase—a material made up of at least three elements. Examples are $LiMn_2O_4$ and $RbAg_4I_5$, which are cation conductors in which current is carried by Li^+ and Ag^+, respectively. This phase has a 3D translation degree of freedom for the mobile ion which is nearly free, more like an electron or a hole in a semiconductor. In view of the macroscopic nature of the ion transport, ion motion is intimately connected to the entire structure in which it exists. Thus, the thermodynamic conditions under which the phase exists become

TABLE 2.1

Modified Periodic Table—A Vision for Solid State Ionics

1	2	3	4	5	6	7	8	9	10	11	12	13	14	15	16	17	18
H^+																	He
Li^+	Be											B	C	N	$O^=$	F^-	Ne
Na^+	Mg^{++}											Al	Si	P	S	Cl	Ar
K	Ca	Sc	Ti	V	Cr	Mn	Fe	Co	Ni	Cu^+	Zn	Ga	Ge	As	Se	Br	Kr
Rb	Sr	Y	Zr	Nb	Mo		Ru	Rh	Pd	Ag^+	Cd	In	Sn	Sb	Te	I	Xe
Cs	Ba	Lu	Hf	Ta	W	Re		Ir	Pt	Au^+	Hg	Tl	Pb^{++}	Bi			

Lanthanides: La Ce Nd Sm Eu Gd

Actinides: Th U

important. Now the question that arises is under what pressure and temperature windows the phase is stable. That is, do the materials undergo any changes in the composition or structure if the temperature is varied under ambient pressure for instance?

Most of the SSIMs that are discussed in this chapter are usually synthesized by forming solid solutions of two or more binary compounds. A few of these materials undergo a phase change. A direct and convenient method of investigating these two aspects, namely, materials synthesis and phase transitions, is to construct the so-called equilibrium phase diagrams. The phase stability is dependent on the "equilibrium" state of these materials.

Equally important, SSIM is involved in all the three components of, say, a Li⁺ ion battery: the anode, the cathode, and the electrolyte. Each of these components supports one-way traffic in its own specialized way as the materials involved are robust enough—structurally and electrochemically—to *sustain* this traffic. They should also sustain the "forced" return traffic by restoring the lost "ion power" to the battery. In addition, note that the electrolyte is more often not a liquid but a desirable solid, as in the case of an all-solid state battery. Thus the roles of a solids state material as a "component" and "phase" should be considered in a thermodynamic sense, which make the discussion of phase diagrams and phase stability of these materials mandatory.

Figure 2.1 puts into perspective the intimate connection between phase stability, phase diagram (equilibrium), and phase transitions.

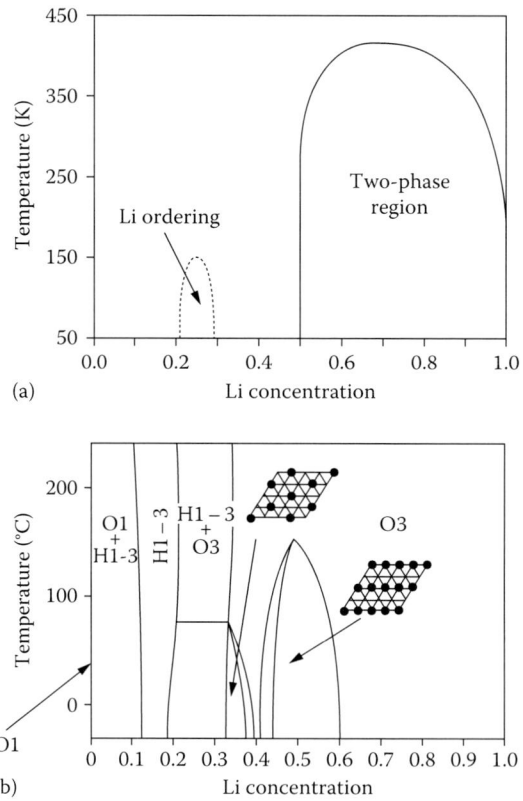

FIGURE 2.1   (a) Calculated phase diagram of "simulated" spinel Li_xCoO_2. Full lines: boundary of the region with two immiscible phases $0.5 < x < 1.0$; dashed line: boundary of a region where a second order, disordered to ordered phase transition takes place (centred on $x = 1/4$). Li⁺ occupy every second site while the other tetrahedral and octahedral sites are empty. (After Kraytsberg, A. et al., *Adv. Energy Mater.*, 2, 922, 2012.) (b) First-principles calculation of phase stability in Li_xCoO_2 as a function of x. The relevant structures are indicated within the figure. (After Van der Ven, A. et al., *Phys. Rev. B*, 58, 1998.)

2.3 Phase Diagrams and Phase Stability of Ionic Materials

2.3.1 Thermodynamics

Thermodynamics is the basis of macroscopic behavior of all materials or condensed matter systems and solid state ionics is no exception. Thermodynamics of SSIMs is apparently the initial point to discuss the intense variety of ionic materials that conduct electricity through fast ion motion/transport over macroscopic distances in the material/device. For a material to participate in the ion transport process, as in a micropower device, it has to be structurally robust and thermodynamically stable over a temperature range, say, $300 < T < 500$ K. This requirement calls for a systematic search for materials or a "scan" which is usually done by constructing a phase diagram—experimentally earlier but computationally now. Important question regarding phase stability are clarified through experiments on selected solid phases with temperature as an independent variable. Experimental phase diagrams for solid state ionic systems are based on the observation of (1) phase (co)existence, (2) reactions, and (3) transitions. Crystal chemistry has had a strong influence on the phase stability of SSIMs especially the oxides. Therefore, this chapter focuses on two aspects:

1. Gibbs phase rule, thermodynamics of phases as derived from phase diagrams, and the phase stability of fast ionic solid systems; the importance of pseudo-binary phase diagrams as unique to SSIM systems is emphasized.
2. A discussion on the major classes of SSIMs based on the mobile ion (e.g., Li ion) involved, from the view point of crystal structure; as more new materials are being discovered, this has become an active area of research and development.

The optimal material discovered through a phase diagram is often a metastable phase realized by effectively suppressing a high-temperature phase transition (to be discussed in detail in Chapter 5). Thus, the thermodynamic/structural phase transition emerges as an important paradigm of solid state ionics.

This chapter naturally brings in the question of materials synthesis which is discussed in Chapter 3.

Thermodynamics deals with three kinds of systems (Figure 2.2): (1) an isolated system in which *no* exchange of *energy or matter* with the bath or external surroundings is possible, (2) a closed system that may exchange *only* energy but not matter with surroundings, and (3) an open system that can exchange *both* matter and energy [2.5].

A solid state ionic system under normal atmospheric pressure and temperature is an open system. For a solid state ionic system of n components ($n > 2$), it is practical to look for a phase diagram of the "second kind." Thus, a ternary system $A–B–C$ can be represented by a pseudo-binary phase diagram just as the quaternary system $A–B–C–D$ can be specified by a pseudo-ternary phase diagram.

2.3.2 Phase Diagrams

Phase diagrams of SSIMs are quite unlike those of metallic alloys and even conventional ionics. They involve oxygen, fluorine, sulfur, phosphorus, and even nitrogen as "counter elements." In the case of

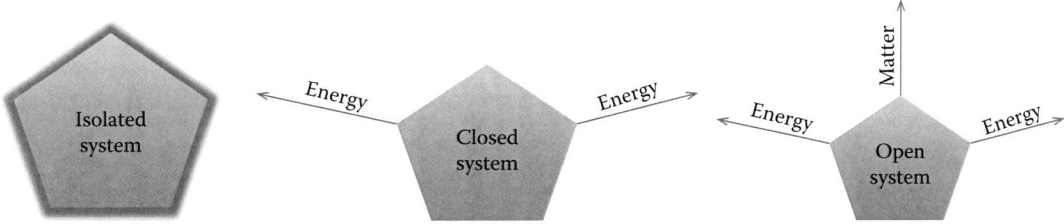

FIGURE 2.2 Three types of systems in thermodynamics. An SSIM is an open system unless it is held in a controlled, inert atmosphere. An open system allows both matter and energy exchange with surroundings.

oxygen, its partial pressure, that is, the fractional concentration of oxygen, in the chamber is used for the investigation of phase diagram.

Let us see what a phase diagram is.

It is a graphical summary of the range of temperatures and compositions over which certain phases or mixtures of phases of a material system exist under conditions of thermodynamic equilibrium. The seminal work of J Willard Gibbs provides the basis for interpreting and applying phase diagrams. This rule that is strictly applicable under conditions of thermodynamic equilibrium also has much significance in certain nonequilibrium situations.

What is the relationship between (1) the number of phases P present in a solid state ionic system at thermodynamic equilibrium and (2) the number of components (C) necessary to form these phases under given conditions of temperature, pressure, and composition?

The previous question could be answered based on the phase rule that Gibbs developed.

$$F = C - P + 2 \qquad (2.1)$$

where

P is the number of phases in equilibrium, physically distinct and mechanically separable
C is the number of components required to specify the system
F is the number of variables that describe the system, which is also called "degrees of freedom"

F is the real key variable because it includes the concentration of each phase which is of practical importance in materials synthesis and phase stability. F in fact is the number of independent intensive variables that define the state of a system. Let us consider a system having two chemically different substances. If the concentration of one substance gives a mole fraction of 0.3, then the concentration of the other should be 0.7, because the sum of the two substances is fixed to be 1.0. This "unity sum rule" applies irrespective of the number of components the final SSIMs to be synthesized has. This is "variance" and should not be confused with C.

Equation 2.1 is valid for any form of matter, but for condensed systems it is modified as

$$F + C - P + 1 \qquad (2.2)$$

which implies that one of the intensive variables is held constant.

Equation 2.2 should be applied to SSIMs after careful consideration of factors such as stoichiometry. In any case it depends on the experimentally obtainable phase diagrams such as Na–S and Li–S.

When a system is said to be in equilibrium with its surroundings, no useful energy enters into or passes out of the system. "Time" as a variable is not explicitly considered in such systems. Only in experiments such as thermal relaxation calorimetry, time is considered as a variable, when the system approaches equilibrium.

SSIM phase is essentially time invariant. This connects to the important issue of phase stability, which is essentially kinetic. An important analogy is the thermodynamic instability of the human body in the presence of oxygen! Here kinetic stability is achieved because the oxidation of the particular carbon composition involved exhibits slow kinetics at room temperature.

In the context of SSIM, zincblende-structured γ-AgI is a relevant example. If the wurtzite-structured β-AgI and γ-AgI are two phases in equilibrium at a fixed temperature, the application of Rule (2) gives $P = 2$ and $F = 0$. A reverse situation occurs for a one-component system at high pressure. Here $P - 1$, $C = 1$, $F = 0$.

Exactly what constitutes the phase of a solid state ionic material?

For a macroscopic system in equilibrium the basic thermodynamic relationship is

$$dU = TdS - PdV \qquad (2.3)$$

Equation 2.3 can be explained as follows:

d (Internal energy of a thermodynamic system) = Absolute temperature · d (Entropy) or Heat

− Pressure · d (volume) or Work

Gibbs [2.6] equation can be modified as

$$dG = -SdT + VdP + \gamma dA - \sum_i \mu_i dN_i \qquad (2.4)$$

where
  G is the Gibbs free energy
  μ_i is the chemical potential of component i of the multicomponent system
  A is the surface (interface) area
  γ is the surface (interface) energy

Equation 2.2 connects incremental changes in the internal energy of a system to heat and work (essentially a statement of the second law of thermodynamics). Equation 2.3 generalizes Equation 2.2.

In fact, Equation 2.3 shows that the number of molecules N_i of different components in the system could be varied. Thus, the treatment of various equilibria of interest to solid state ionics including chemical, phase, osmotic and surface becomes possible. Also one can examine the equilibrium condition of a solid and its surrounding medium. This motivating and easier-to-use equation could be used to treat phase equilibrium and phase diagrams. Based on statistics, it is valid only for materials being greater than a micrometer in size—where nanoscale ends and bulk begins. Parameter "size" of Equation 2.3 is a constant of the bulk. Note that surface/interface thermodynamics emphasizing the dependence of size on thermodynamic quantities is crucial to nanoscience and has to be treated separately [2.7,2.8].

The equilibrium phase diagram of a multinary (binary and higher) system links thermodynamics, solid state, and materials chemistry, and applications through phase stability, materials synthesis, and characterization. As a "roadmap," it is the spine of the "body" of this book. Realize that a proper discussion of and natural connection to the other chapters of this book is vital to a holistic understanding and application of solid state ionics. In fact this idea goes much beyond the theme of this book—to the evolution of life on earth as it exists today. And perhaps the evolution of the universe at large, although the tricky concept of time is involved here! The very concepts—phase, component, and degrees of freedom—are the elements that go to paint the mechanistic view of this world.

The simplest phase diagrams are those of single-component systems.

A single-component system such as I, S, and P has three possible phases: solid, liquid, and gas. Thus $C = 1$ and $P = 3$ which gives $F = 0$. For such a system to be at equilibrium, there exist no free intensive variables. Therefore, in a single-component system, the three-phase mixture can coexist only at one pressure and temperature, which is known as a triple point. Figure 2.3a and b illustrates this case for I and S, which are individually and together important components for the synthesis of Ag^+ ion–conducting solids such as AgI, Cu_2S, and Ag_3SI.

Elemental iodine, I_2, which forms dark gray crystals that appear metallic is a solid that is easily sublimed, and these vapors could be used to iodize thin films of copper and silver to create fast ion conducting films for measuring oxidation and in chemical sensor applications. The most notable feature of iodine's phase behavior is the very small difference (less than a degree) between the temperatures of its triple point **1** and melting point **2**.

The vapor pressure of iodine at room temperature is quite small—only about 0.3 torr (40 Pa). The fact that solid iodine has a strong odor and is surrounded by a purple vapor in a closed container is mainly a consequence of its strong ability to absorb green light.

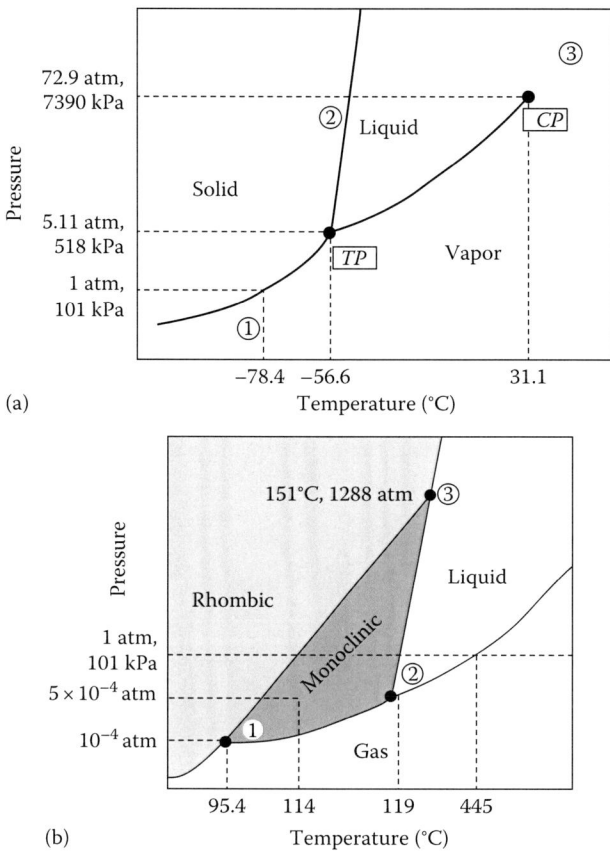

FIGURE 2.3 (a) *P–T* phase diagram of iodine. *TP* is triple point. *CP* is critical point. (b) *P–T* phase diagram of sulfur with three critical points. (After Lower, S., *Changes of State,* Chapter 7.5, http://www.chem1.com/acad/webtext/virtual-textbook.html.

Sulfur exhibits a very complicated phase behavior that remains a puzzle. Figure 2.3b is a simplified diagram. The tendency of S_8 molecules to break up into chains (especially in the liquid above 159°C) or to rearrange into rings of various sizes (S_6–S_{20}) makes the diagram not so simple. Even the vapor can contain a mixture of species S_2 through S_{10}. This input is very useful in planning SSIMs synthesis strategies.

The phase diagram of sulfur indicates two solid phases: *rhombic* and *monoclinic*—crystal structures in which the S_8 molecules arrange themselves. This gives rise to three triple points indicated by the numbers in the diagram. When rhombic sulfur (the stable low-temperature phase) is heated slowly, it changes to the monoclinic form at 114°C, which then melts at 119°. But when the monoclinic form is heated rapidly, there is no time period for the molecules to rearrange themselves, and hence the rhombic arrangement persists as a metastable phase, until it melts at 119°–120°. At very high pressures, more than one solid phase is formed. Li_2S and Na_2S are important sulfides.

Let us now look at the phase diagram of AgI which is the prime mover of a number of SSIMs that will be discussed later in this chapter. Figure 2.4a is the *P–T* diagram delineating the four solid phases wurtzite/zincblende, bcc, rocksalt, and "IV" besides the liquid phase (Figure 2.4b).

AgI–RbI and Ag_2S–AgI are important "binary" systems that contain the room temperature superionic conductor $RbAg_4I_5$ and the antiperovskite Ag_3SI. Their phase diagrams are shown in Figure 2.5a and b.

Let us now look at phase diagrams of Li-based solid state ionic systems. The motivation is to understand the relationships between stoichiometry (deviations from standard chemical formula), crystal

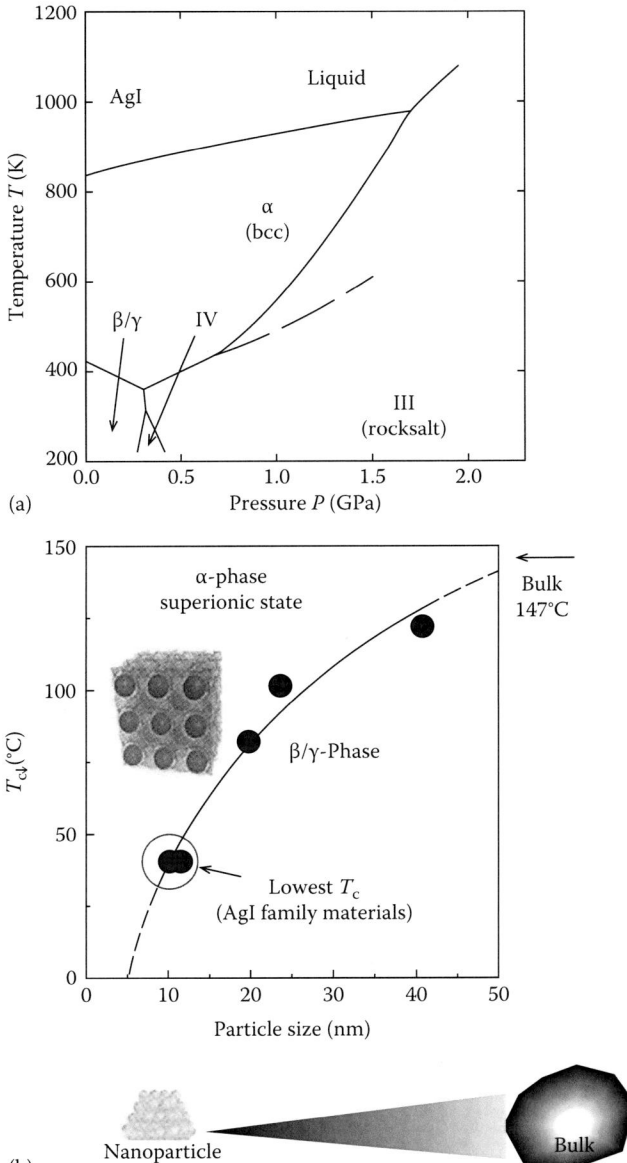

FIGURE 2.4   (a) *P–T* diagram of AgI. (Adapted from Hull, S., *Rep. Prog. Phys.*, 67, 1233, 2004.) Note the crystalline phase boundaries between β/γ (wurtzite. zincblende, α(bcc), rocksalt and phase IV besides liquid phase. Illustrates phase equilibrium of a typical single-component solid state ionic material. (b) AgI Wurtzite/zincblende phase transition temperature ($T_c\downarrow$) versus nanoparticle size showing drastic reduction in $T_c\downarrow$ (38°C) relative to bulk (147°C). (After Makiura, R. et al., *Nat. Mater.*, 8(6), 476, 2009.)

structure, and properties of inorganic materials, especially complex materials that have variable composition. In materials based on solid–solution phases, it is necessary to have information on solid–solution mechanisms in order to understand the detailed crystallography and properties of the phases. This is because, in many cases, properties vary dramatically with solid–solution composition. The concept of dynamic/thermodynamic/kinetic equilibrium is the basis of phase equilibrium and phase stability of solids. This perspective of understanding, exploring (experimentally), and computing phase diagrams in SSIMs is just at the early stage of its investigation.

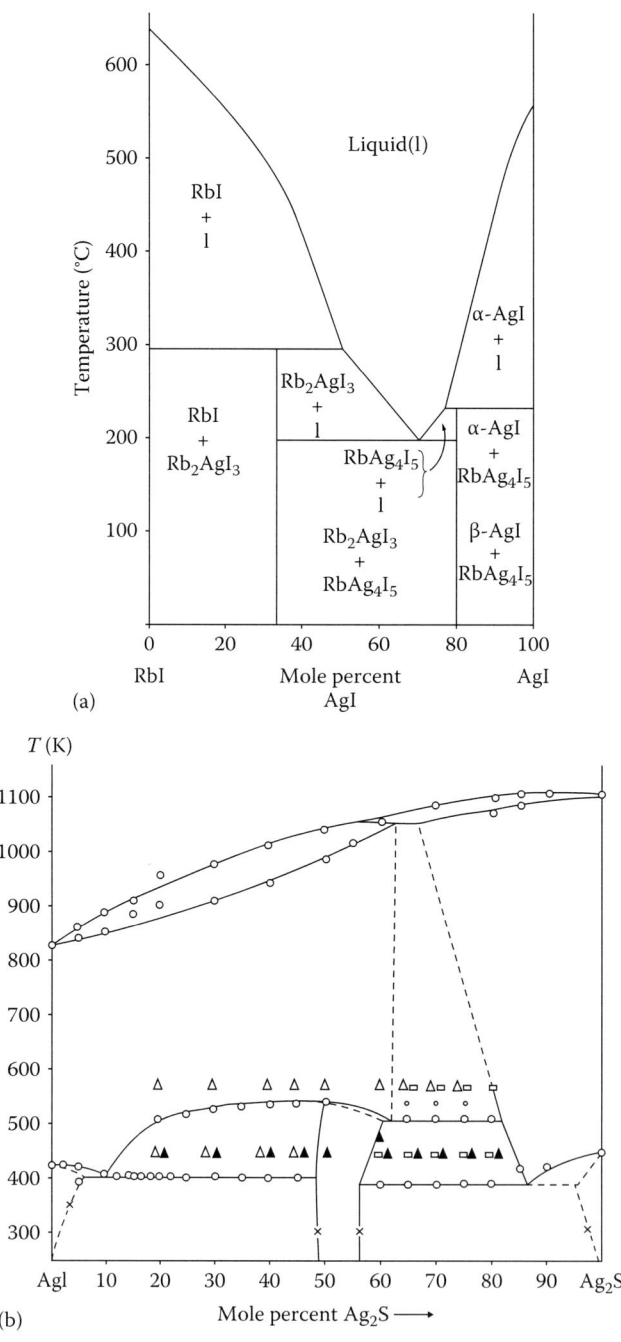

FIGURE 2.5 (a) Phase diagram of AgI–RbI due to Argue and Owens. RbAg$_4$I$_5$ occurs in the trapezium pointed by an arrow. Also note the solid solutions of RbAg$_4$I$_5$ with α-AgI and β-AgI besides with Rb$_2$AgI$_3$. (Adapted from Solid ionic conductor, US Patent 3,519,404, July 7, 1970.) (b) Ag$_2$S–AgI phase diagram. Antiperovskite Ag$_3$SI (▲) is an important phase besides α-AgI- and α-Ag$_2$S-based solid solutions. (After Dreisbach, B., *J. Solid State Chem.*, 60, 115, 1985.)

The use of a solid electrolyte immersed in a liquid electrolyte as in a lead–acid battery has an inherent advantage of "dissolving" any gases evolved during "half-cell reactions." The all-solid state battery, however, has to operate in an environment dictated by the thermodynamics of the "partly open–partly closed" system, particularly in a multicomponent system involving a gaseous component. In such systems, the desired phase has to coexist in equilibrium with another phase, often involving the evolution or consumption of oxygen at the electrode–electrolyte interface. This introduces oxygen concentration as an important thermodynamic variable (formally called chemical potential) apart from temperature and pressure in the thermodynamic description of the system.

Thus, phase diagrams involving Li are special because they often involve framework structures that support extraction and insertion of LI. Lithiation/delithiation of an electrode is thus a way of examining phase formation and phase stability not only of the electrode but also of the device of which it is a part.

To begin with, we look at the Li–C phase diagram that is most relevant both from the phenomenology of Li-intercalation or insertion into graphite carbon and from device application mainly to Li-ion battery. Figure 2.6a shows that graphite C can take between 52 and >99 atomic percent Li.

Figure 2.5b shows a carefully compiled phase diagram of Li_xC_6 showing different stages of "uptake" of Li by C. Intercalation takes place in "stages" when Li^+ occupies the weakly coupled interlayer space in the hexagonal graphite lattice. Also note that intercalation occurs at temperatures less than 70°C.

As a contrast to the simple diagram of Figure 2.6, let us consider the issue of phase diagram in an experimental perspective. Focus on the system Li_2O–TiO_2. What do the experiments tell us? This two-oxide system contains four stable phases: Li_4TiO_4, Li_2TiO_3, $Li_4Ti_5O_{12}$, and $Li_2Ti_3O_7$, besides one metastable phase, H. Li_2TiO_3 undergoes an order–disorder phase transition at 1215°C. High-temperature Li_2TiO_3 forms an extensive range of solid solution between ~44 and 66 mol% TiO_2. Low-temperature Li_2TiO_3 forms a more limited range of solid solution between ~47% and 51% TiO_2. The temperature of the order–disorder transition **decreases** to *either side* of the Li_2TiO_3 composition. The battery anode material—spinel phase $Li_4Ti_5O_{12}$—has an upper limit of stability at 1015°C ± 5°C, *above* which it decomposes to high Li_2TiO_3 solid solution and $Li_2Ti_3O_7$. $Li_2Ti_3O_7$ has a lower limit of stability at 957°C ± 20°C, *below* which it decomposes to $Li_4Ti_5O_{12}$ and rutile. During this decomposition of $Li_2Ti_3O_7$, phase H, a metastable phase of *unknown composition*, forms as an *intermediate*. $Li_2Ti_3O_7$ forms a *short range* of solid solutions between ~74% and 76% TiO_2. These and related results help construct a phase diagram for the system Li_2O–TiO_2 (Figure 2.7 [2.16]).

Zirconia stabilized by CaO, CeO_2, and Y_2O_3 is by far the most important SSIM working as a fuel cell electrode and an oxygen sensor at high temperatures (Chapter 9). Figure 2.8 shows the phase diagrams of ZrO_2–CaO/CeO_2/Y_2O_3. Note that Ca^{++}, Ce^{4+}, and Y^{3+} substitute Zr^{4+} and create oxygen vacancies to compensate for the charge deficit in the case of Ca and Y substitution. From this diagram, it is understood that the important inputs for materials synthesis are the optimal stabilizer composition and the temperature range over which the cubic phase is stable. The well-known 8YSZ corresponds to 8 mol% Y_2O_3.

Bulk form of samples is not the only type studied through determining phase diagram. Thin films, which are often device ready, are also the subject of such studies. An example is the Li–Mn–O system. Figure 2.9 illustrates such a diagram—a ternary triangle. As such systems have tremendous practical importance, it is important to know their thermodynamics.

Now let us consider solid state ionic system Li–Mn–O [2.21].

The phase rule connects the number of components ($C = 3$), the number of phases in equilibrium, physically distinct and mechanically separable (P), and the number of free thermodynamic variables F (i.e., temperature (T) and pressure (p)). Assume that two free thermodynamic variables, temperature and total pressure, can be chosen. Then, the *equilibrium* phase configuration it *that* configuration of *all* possible configurations which gives the *minimum* Gibbs free energy. (Note that one can also use equivalently the concept of Helmholtz free energy $F = U$(internal energy) $– TS$(entropy)). Basically, Gibbs free energy of a thermodynamic system is

$$G = \text{Internal energy } U + pV(\text{volume}) - TS(\text{entropy}) = \text{Enthalpy } H - TS \qquad (2.5)$$

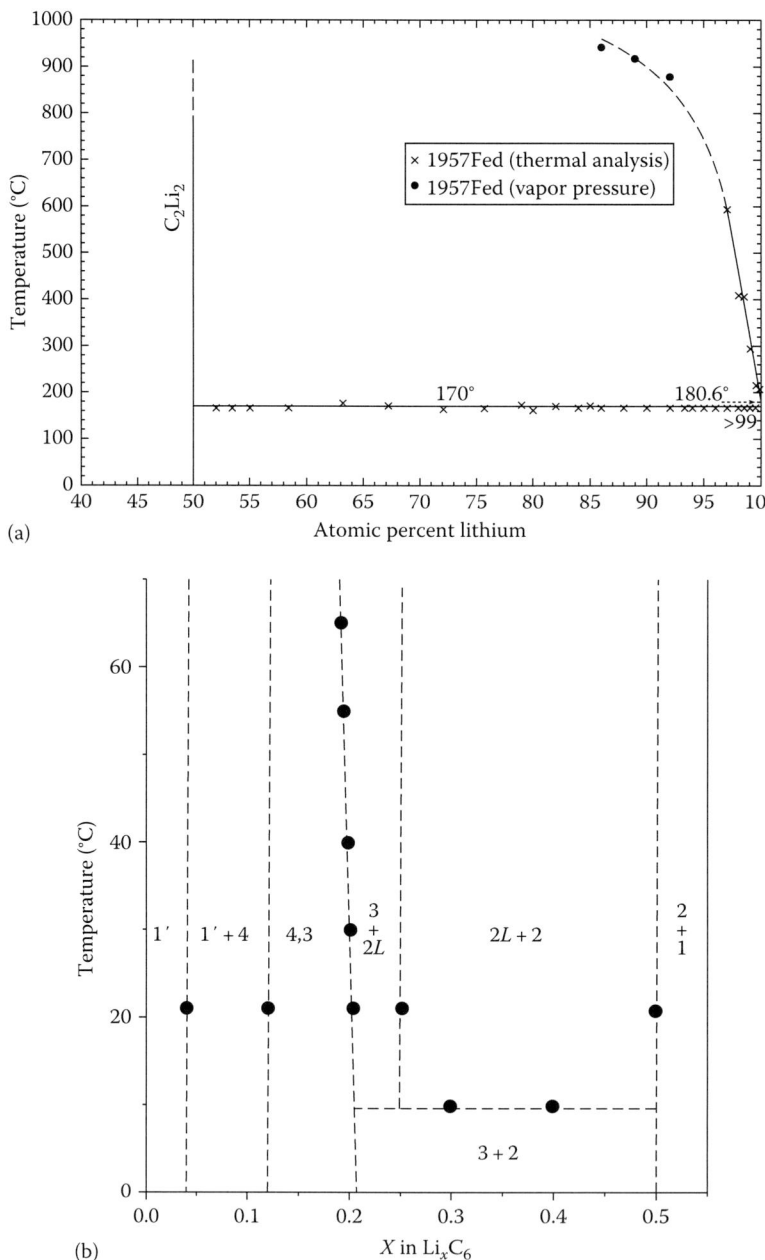

FIGURE 2.6 (a) The assessed C–Li phase diagram. (The assessment by J. Sangster is based on the thermal analysis and vapor pressure data of P.J. Fedorov; Mezn-Tszen, S. and Xuebao, H., 23(1957) 30 (in Chinese) (1957Fed referred to).) (b) Phase diagram of Li_xC_6 in the temperature range $273 < T < 343$ K. The numbers refer to the stages of intercalation of Li into C. The diagram is based on *in situ* XRD and dx/dV (at constant T) versus V data. (From Dahn, J.R., *Phys. Rev. B*, 44, 9170, 1991.)

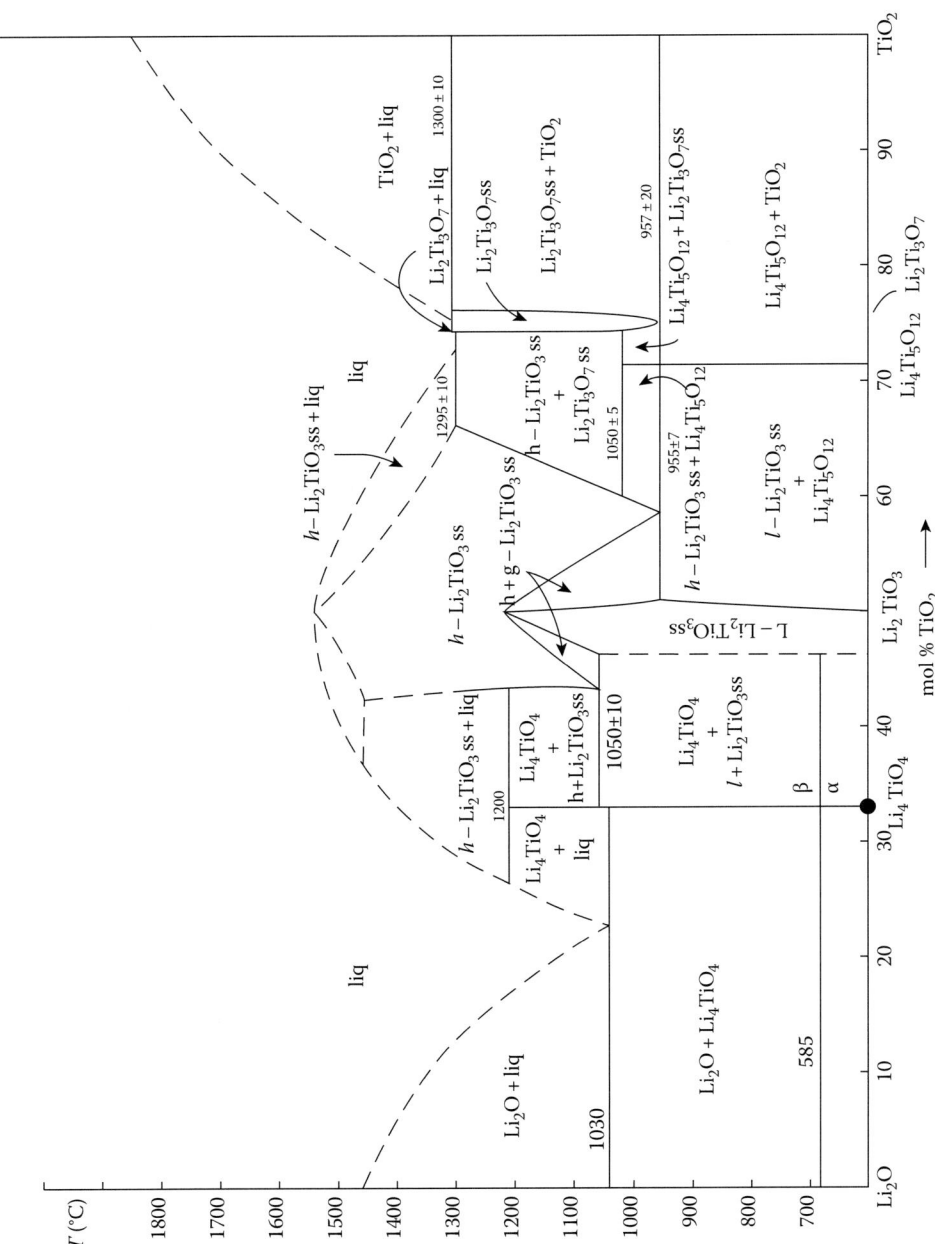

FIGURE 2.7 The comprehensive phase diagram of the Li$_2$O–TiO$_2$ system. Note the important phases on the concentration axis. More importantly, it illustrates many features of a typical phase diagram including the eutectic, the equilibrium line, and the peritectic. (After Izquierdo, G. and West, A.R., *Mater. Res. Bull.*, 15, 1655, 1980.)

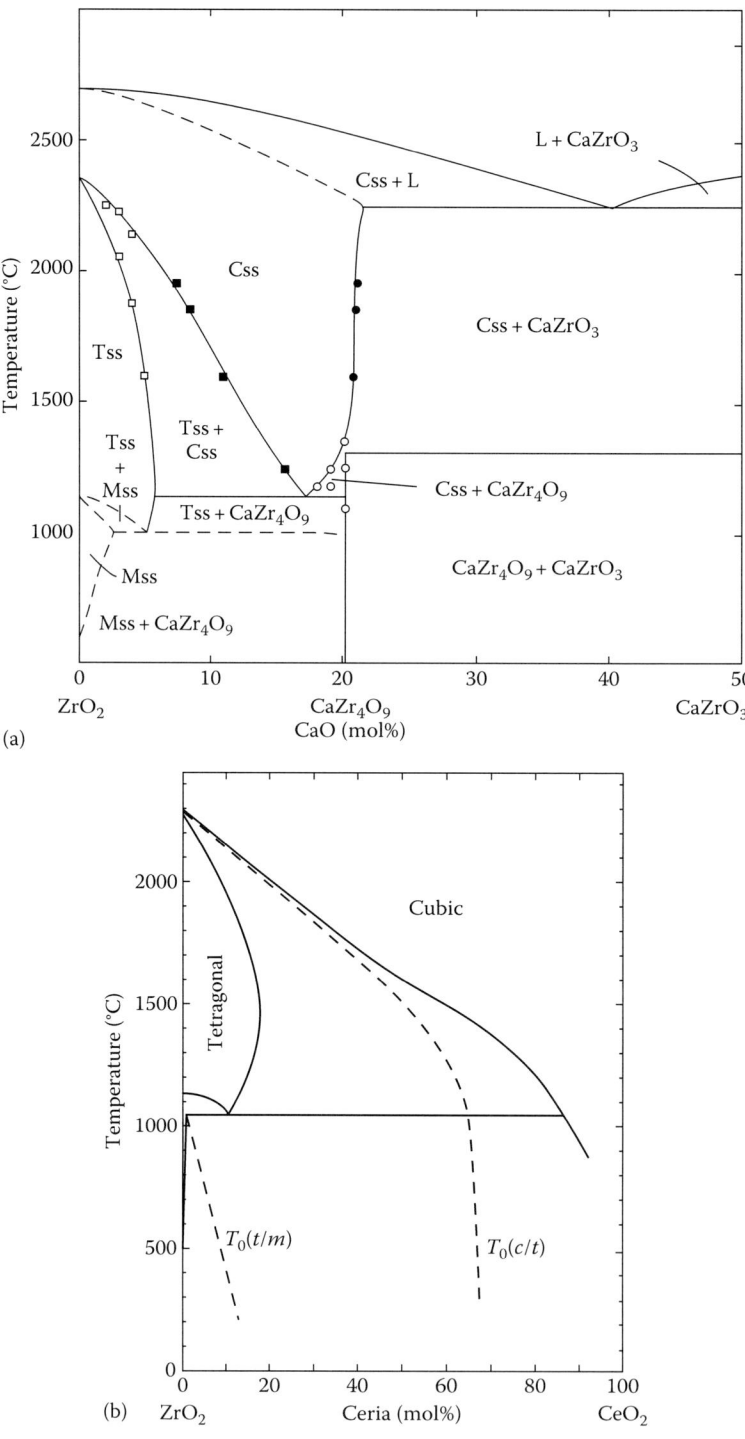

FIGURE 2.8 (a) Phase diagram of the system ZrO_2–CaO for CaO upto 50 mol%. Tss, Tetragonal solid solution; Mss, Monoclinic solid solution; Css, cubic solid solution. Note that the fast oxygen conductor CSZ or calcia-stabilized zirconia occurs as an exclusive cubic solid solution in the region marked "Css." (After Stubican, V.S. and Hellmann, J.R., in Heuer, A.H. and Hobbs, L.W. (eds.), *Advances in Ceramics*, Vol. 3, *Science and Technology of Zirconia*, American Ceramic Society, 1980, p. 25.) (b) Zirconia–ceria phase diagram. (From Yashima, M. et al., *SSI*, 86–88, 1131, 1996.) (*Continued*)

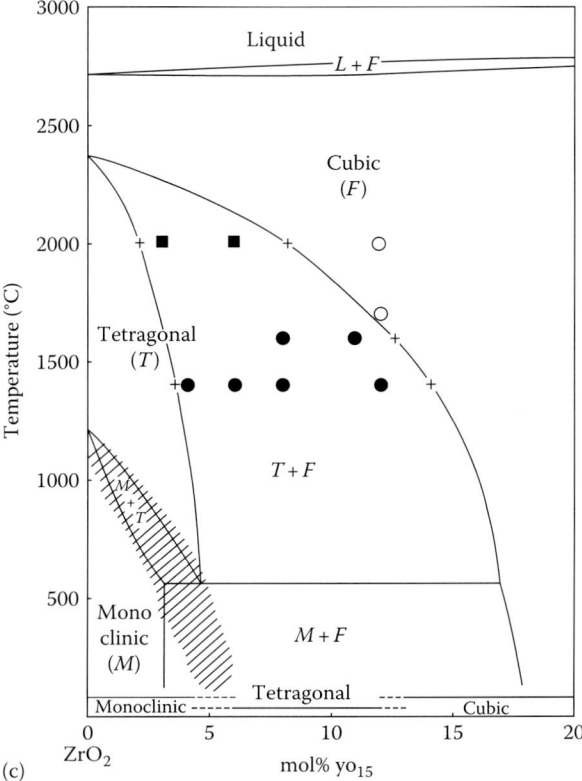

FIGURE 2.8 (Continued) (c) Y_2O_3–ZrO_2 system. Phases found at room temperature are monoclinic (Δ), tetragonal (□), cubic (O), monoclinic + tetragonal (■), and monoclinic + cubic (●). Hatched region corresponds to nonequilibrium monoclinic–tetragonal transition. (From Scott, H.G., *J. Mater. Sci.*, 10, 1527, 1975.)

In this context,

$$G(T,p,N) = N_1\mu_1(T,p) + N_1\mu_1(T,p) + N_1\mu_1(T,p) => \min \tag{2.6}$$

In Equation 2.6, μ_i is the chemical potential of the *i*th component which is the same for all co-existing phases and N_i is the number of moles of the *i*th component.

What is done in a practical situation? One takes measured amounts of Li (usually lithium carbonate) and Mn (as oxide) to perform a solid state reaction (to be discussed in Chapter 3). Note that a measured amount of the *third* component oxygen is *not* taken! What happens during the reaction? The oxygen stoichiometry of the sample equilibrates with the oxygen partial pressure of the gas phase at the given temperature. Thus, it is not meaningful to deal with oxygen as free adjustable component. Oxygen as a component must, therefore, be expressed by the *corresponding* chemical potential by performing a Legendre transformation onto G in Equation 2.2. Through such a transformation, the old free enthalpy G, a function of T, p, and the amounts of the three components are replaced by a new free enthalpy G_t and the amounts of only *two* components. G_t is a function of T, p, and oxygen partial pressure $p(O_2)$. Now the amount of the *third* component (oxygen) is replaced by its chemical potential μ_3. Therefore,

$$G(T,p,N_1,N_2,N_3) \Rightarrow G_t(T,p,\mu_3,N_1,N_2) = G(T,p,N_1,N_2,N_3) - N_3\mu_3 G_t(T,p,\mu_3,N_1,N_2)$$

$$= N_1\mu_1(T,p,\mu_3) + N_2\mu_2(T,p,\mu_3) \Rightarrow \min \tag{2.7}$$

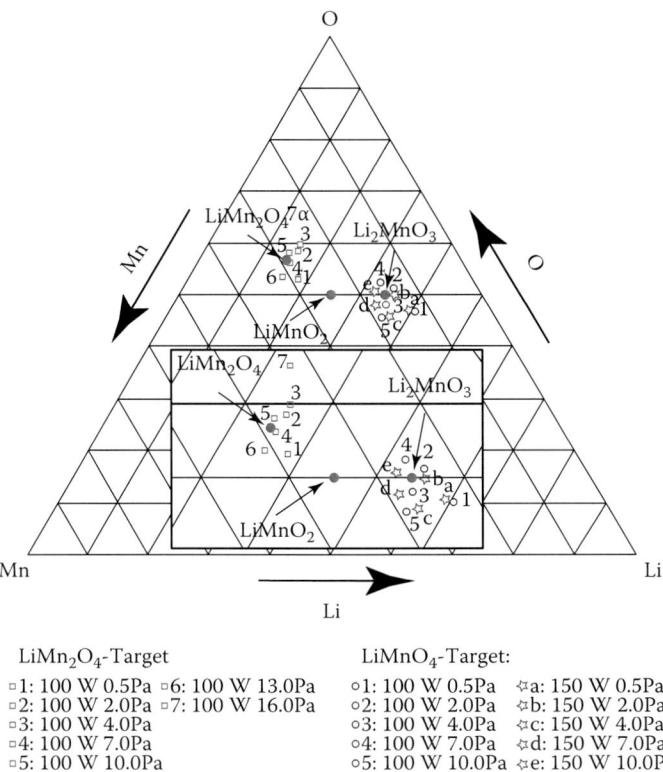

FIGURE 2.9 Typical ternary triangle: Chemical composition (a/o) of different radio frequencies of sputtered Li–Mn–O thin films using $LiMn_2O_4$ and $LiMnO_2$ targets (this technique is discussed in Chapter 3). Dot, one deposited film. Sputtering parameters listed later. Inset: details of area of interest where films were obtained. (After Fisher, J. et al., *Thin Solid Films*, 528, 217, 2013.)

$$\mu_3 = \left(\frac{1}{2}\right) RT \ln\left(p\left(O_2\right)\right) \tag{2.8}$$

where R is the gas constant.

The chemical potential of oxygen can be chosen freely by changing the oxygen partial pressure $p(O_2)$ of the gas phase. T and p can also be chosen freely so that we now have three free variables T, p, and $p(O_2)$. Gibbs phase rule says that two thermodynamic phases coexist in equilibrium. So, the phase relation of the system Li–Mn–O is considered a *binary* phase diagram, where x-axis is the relative cationic composition $x = Li/(Li + Mn)$ and y-axis is the temperature or the oxygen partial pressure.

Figure 2.9 is an example of the phase diagram of the second kind or a pseudo-binary phase diagram with cationic composition plotted along x-axis and temperature T along y-axis.

What is the difference between $p(O_2)$ and pH when independent variables of the phase diagram are concerned?

Viewed strictly as a ternary system, the same system Li–Mn–O may be represented in a triangle diagram (see [2.29b]).

The description of a closed n-component system at constant temperature and pressure requires $(n - 1)$ dimensional space in the form of a composition polyhedron. Thus, a closed four-component system requires 3D space and can be represented as a composition tetrahedron. In view of the large number of phases likely to arise in the typical solid state ionic system Li–Fe–P–O, we need a more amenable representation for easy analysis. More importantly, the relevant phase diagram is one in which the phase equilibria are described in an environment with controlled Li, Fe, P, and O_2 compositions and is not reflective of the environments of interest [2.22].

The Li–Fe–P–O_2 system serves as a model system for the construction of a phase diagram. To do so, we need to compare the relative thermodynamic stability of phases belonging to the system using a suitable free energy model.

Consider an isothermal, isobaric (i.e., constant T, constant p) closed Li–Fe–P–O_2 system. Gibbs free energy, a relevant thermodynamic potential, is expressed as a Legendre transform of the enthalpy H and internal energy U as follows:

$$G\left(T,p,N_{\text{Li}},N_{\text{Fe}},N_{O_2}\right) = H\left(T,p,N_{\text{Li}},N_{\text{Fe}},N_{O_2}\right) - TS\left(T,p,N_{\text{Li}},N_{\text{Fe}},N_{O_2}\right) \tag{2.9}$$

$$G\left(T,p,N_{\text{Li}},N_{\text{Fe}},N_{O_2}\right) = U\left(T,p,N_{\text{Li}},N_{\text{Fe}},N_{O_2}\right) + pV\left(T,p,N_{\text{Li}},N_{\text{Fe}},N_{O_2}\right) - TS\left(T,p,N_{\text{Li}},N_{\text{Fe}},N_{O_2}\right) \tag{2.10}$$

where T, S, p, V, and N_i are temperature, entropy, pressure, volume, and number of atoms or molecules of species I of the system, respectively. Our concern is the relative stability of condensed phases, for which pV is usually negligible. Thus at a temperature of 0 K at which phase diagram is to be calculated (note that experimental phase diagrams have a reference temperature of 300 K), G simplifies to just U. U has to be normalized with respect to the total number of particles of the system. To proceed further, recognize that the key variable in the synthesis of the SSIM $LiFePO_4$ (to be discussed in detail in Chapter 3) is the oxygen chemical potential. Note that Fe in $LiFePO_4$ is in divalent state or a reduced state of Fe atom, Fe^{3+} being the more common and more stable oxidation state. Obviously, $LiFePO_4$ is formed in a reducing environment. Therefore, we have an isothermal and isobaric system that is open with respect to oxygen and closed with respect to Li, Fe, and P. Oxygen grand potential is the appropriate thermodynamic potential to study phase equilibria with respect to an oxidizing/reducing environment. Now let us discuss what a grand potential is.

Grand potential is a quantity used in statistical mechanics, especially for irreversible processes in open systems. The grand potential is the characteristic state function for a grand canonical ensemble [2.23].

$$\Phi_G = U - TS - \mu N \tag{2.11}$$

where
  U is the internal energy
  T is the temperature of the system
  S is the entropy
  μ is the chemical potential
  N is the number of particles in the system

The change in the grand potential is given by

$$d\Phi_G = dU - TdS - SdT - \mu dN - Nd\mu = pdV - SdT + \mu dN \tag{2.12}$$

where
  p is the pressure
  V is the volume, using the fundamental thermodynamic relation (combined first and second thermodynamic laws);

$$dU = TdS - pdV + \mu dN \tag{2.13}$$

When the system is in thermodynamic equilibrium, Φ_G is minimum. This can be seen by considering that $d\Phi_G$ is zero if the volume is fixed and the temperature and chemical potential have stopped evolving.

As would be seen in Chapter 3, the synthesis of $LiFePO_4$ requires reducing environments, that is, either a high T processing or a low oxygen environment or even the presence of a carbon or hydrogen as reducing agent thus making the system open with respect to oxygen. Therefore, to study phase equilibria with respect to an oxidizing/reducing environment, the oxygen grand potential is defined as follows:

$$\Phi\left(T,p,N_{Li},N_{Fe},N_{O_2}\right) = G\left(T,p,N_{Li},N_{Fe},N_{O_2}\right) - \mu_{O_2}N_{O_2}\left(T,p,N_{Li},N_{Fe},N_{O_2}\right) \tag{2.14}$$

$$\Phi\left(T,p,N_{Li},N_{Fe},N_{O_2}\right) = U\left(T,p,N_{Li},N_{Fe},N_{O_2}\right) - TS\left(T,p,N_{Li},N_{Fe},N_{O_2}\right) - \mu_{O_2}N_{O_2}\left(T,p,N_{Li},N_{Fe},N_{O_2}\right) \tag{2.15}$$

where the term pV is negligible.

Normalize the potential with respect to the Li–Fe–P composition. Just retain the symbols U, S, and N_0 and drop the dependences so that

$$\Phi_n\left(T,p,x_{Li},x_{Fe},x_P,\mu_{O_2}\right) = \frac{\left(U - TS - \mu_{O_2}N_{O_2}\right)}{\left(N_{Li} + N_{Fe} + N_P\right)} \tag{2.16}$$

where $x_i = N/(N_{Li} + N_{Fe} + N_P)$ is the fraction of component I in the Li–Fe–P composition space.

A few assumptions may now be made which would help obtain a useful phase diagram. In the system being open with respect to oxygen, phase equilibria changes take place basically through reactions that involve the absorption or loss of oxygen gas. In such reactions, (1) the dominant entropy is that of oxygen gas and (2) the effect of T is sensed by changes in the oxygen chemical potential. The latter is related to T and p_{O_2} and is shown by the equations here:

$$\mu_{O2}\left(T,p_{O_2}\right) = \mu_{O_2}\left(T,p_O\right) + k_B T \ln\left(\frac{p_{O_2}}{p_O}\right) \tag{2.17}$$

$$\mu_{O_2}\left(T,p_{O_2}\right) = h_{O_2}\left(T,p_O\right) - T\left[s_{O_2}\left(T,p_O\right) - k_B \ln\left(\frac{p_{O_2}}{p_O}\right)\right] \tag{2.18}$$

where
  p_{O_2} is the partial pressure of oxygen
  p_O is the reference oxygen partial pressure (say 0.1 MPa)
  $s_{O_2}(T, p_O)$ is the entropy per molecule of oxygen at p_O and T
  $\mu_{O_2}(T, p_O)$ is the oxygen chemical potential at p_O and T
  $h_{O_2}(T, p_O)$ is the enthalpy per molecule of oxygen at p_O and T
  k_B is Boltzmann's constant

In Equation 2.17, consider TS, the entropy contribution to the grand potential of the condensed system. This term is negligible compared to the much larger entropy effect coming from $N_{O_2}S_{O_2}$ of μ_{O_2}. Therefore, Φ_n simplifies to

$$\Phi_n\left(\mu_{O_2},x_{Li},x_{Fe},x_P\right) = \frac{\left(U - \mu_{O_2}N_{O_2}\right)}{\left(N_{Li} + N_{Fe} + N_P\right)} \tag{2.19}$$

The assumption stated earlier enables the effect of T and p_{O_2} to be fully captured by the single variable the oxygen chemical potential μ_{O_2}. Note that according to Equation 2.19, a more negative value of μ_{O_2} corresponds to higher T or lower p_{O_2}. Thus by varying μ_{O_2}, the Li–Fe–P–O_2 or indeed any similar partly closed–partly open system can be constructed as *constant μ_{O_2} sections in the* $(\mu_{O_2}, x_{Li}, x_{Fe}, x_P)$ space.

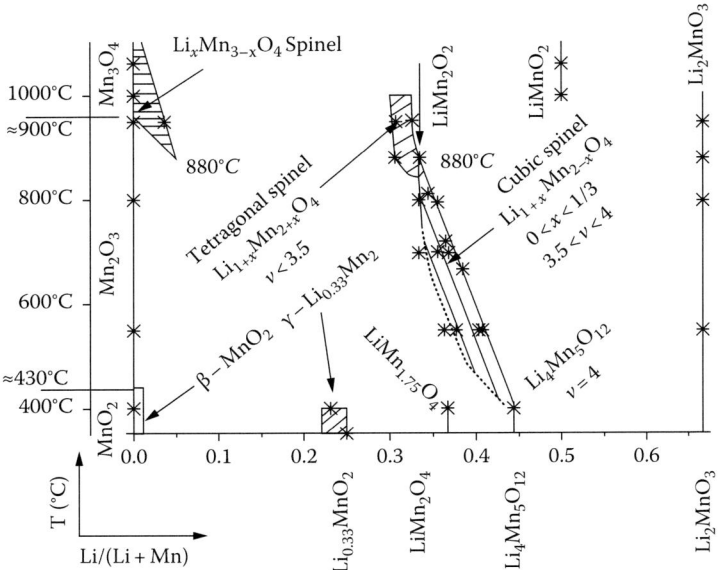

FIGURE 2.10 A pseudo-binary phase diagram of the Li–Mn–O system with cationic composition plotted along *x*-axis and temperature along *y*-axis. (From Poulsen, J.M. and Dahn, J.R., *Chem. Mater.*, 11, 3065, 1999.)

This is done (1) by taking the convex hull of Φ_n for all phases at a particular μ_{O_2} and (2) by projecting the stable nodes onto a 2D Li–Fe–P Gibbs triangle. Each isochemical potential phase diagram then represents phase equilibria at a particular oxidation environment. Of course each point in the phase diagram gives the phase or combination of phases with the lowest Φ_n.

This thermodynamic basis is implemented computationally using the methodology of density functional theory, which would be discussed in the context of electronic structure in Chapter 7.

Figure 2.10 shows the phase diagram (Gibbs triangle with P_2O_5, Li_2O, and Fe_2O_3 at the vertices) of the Li–Fe–P–O_2 system at oxygen chemical potential of −11.42 eV in a reducing environment just necessary for the *first* appearance of stable $LiFePO_4$ containing Fe^{2+}.

Another important system is Li–Mn–Co–Ni–O which is a candidate cathode for a Li-ion battery. It may be represented as a pseudo-ternary system as a triangle with $LiCoO_2$, $LiNiO_2$, and $LiMnO_2$ as the three vertices (Figure 2.11) and may be experimentally realized as a layered–layered composite [2.24].

The Li–P–S forms an important ternary system whose phase diagram is given in Figure 2.12. It is instructive to observe that the hatched region of the Li–P–S triangle accommodates many phases including Li_3PS_4 and $Li_4P_2S_6$.

After having gained a reasonable perspective on phase diagrams, let us consider the question of phase stability.

2.3.3 Phase Stability

The intuitive understanding of the stability of solid state ionic crystals is ultimately to be obtained in terms of visualizable interionic interactions. The Bragg hypothesis is particularly applicable to such materials: "If a crystal is composed of large ions and small ions, its structure will be determined essentially by the large ions and may approximate a close-packed arrangement of the large ions alone with the small ions tucked away in the interstices in such a way that each one is equidistant from four or six large ions" [2.26]. In some cases not all of the close-packed positions are occupied by ions leading to an open structure.

A basic question is why silver iodide, alumina, titania, zirconia, and lead fluoride among others do assume different structures under different conditions.

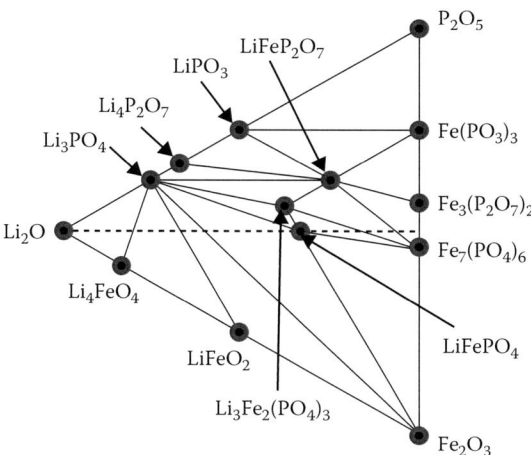

FIGURE 2.11 Calculated phase diagram—represented as the triangle Li_2O–Fe_2O_3–P_2O_5—under a reducing environment corresponding to an oxygen pressure $p_{O_2} = -11.52$ eV. This is just the pressure at which $LiFePO_4$ appears. (After Ong, S.P. et al., *Chem. Mater.*, 20, 1798, 2008.) Such a phase diagram would take a long time to determine experimentally requiring conditions that would be tough to meet.

It is because of the ion–ion repulsive (I^-, $O^=$, F^- in the examples cited earlier) potential energy that varies as the high inverse power (n) of the distance (r) between them: $\sim r^{-n}$. Thus, the equilibrium energy of an ionic crystal is

$$\Phi = \left(\frac{-z^2 e^2 A}{R} \right) \left(1 - \frac{1}{n} \right) \tag{2.20}$$

where

   R is the equilibrium distance between two adjacent ions in the crystal
   A is the Madelung constant characteristic of the structure

It is important to know how R changes from structure to structure for a given material. In this regard, Equation 2.20 helps in predicting the structure that is stable if Φ is sufficiently accurate. Modern methods of structure prediction would involve the use of density functional theory (to be discussed in Chapter 7).

From a thermodynamic viewpoint, the question of phase stability is answered from energetic considerations of a macroscopic system.

In any given system of two or more components, many possible structures (or phases) could be stable. These phases compete for (co)existence based on the respective values of the Gibbs free energy of each competing phase and the variation of this energy with pressure, temperature, composition, and other extensive parameters. The consequences of this competition are (1) the extent of solid solubility of phases, (2) the stability of phases and its temperature dependence, and (3) the variety of structures actually observed in phase diagrams as a subset of numerous possible structures.

The Gibbs free energy that is a measure of chemical energy can be written as

$$G = H - TS \tag{2.21}$$

where

   H is the enthalpy
   T is the absolute temperature
   S is the entropy

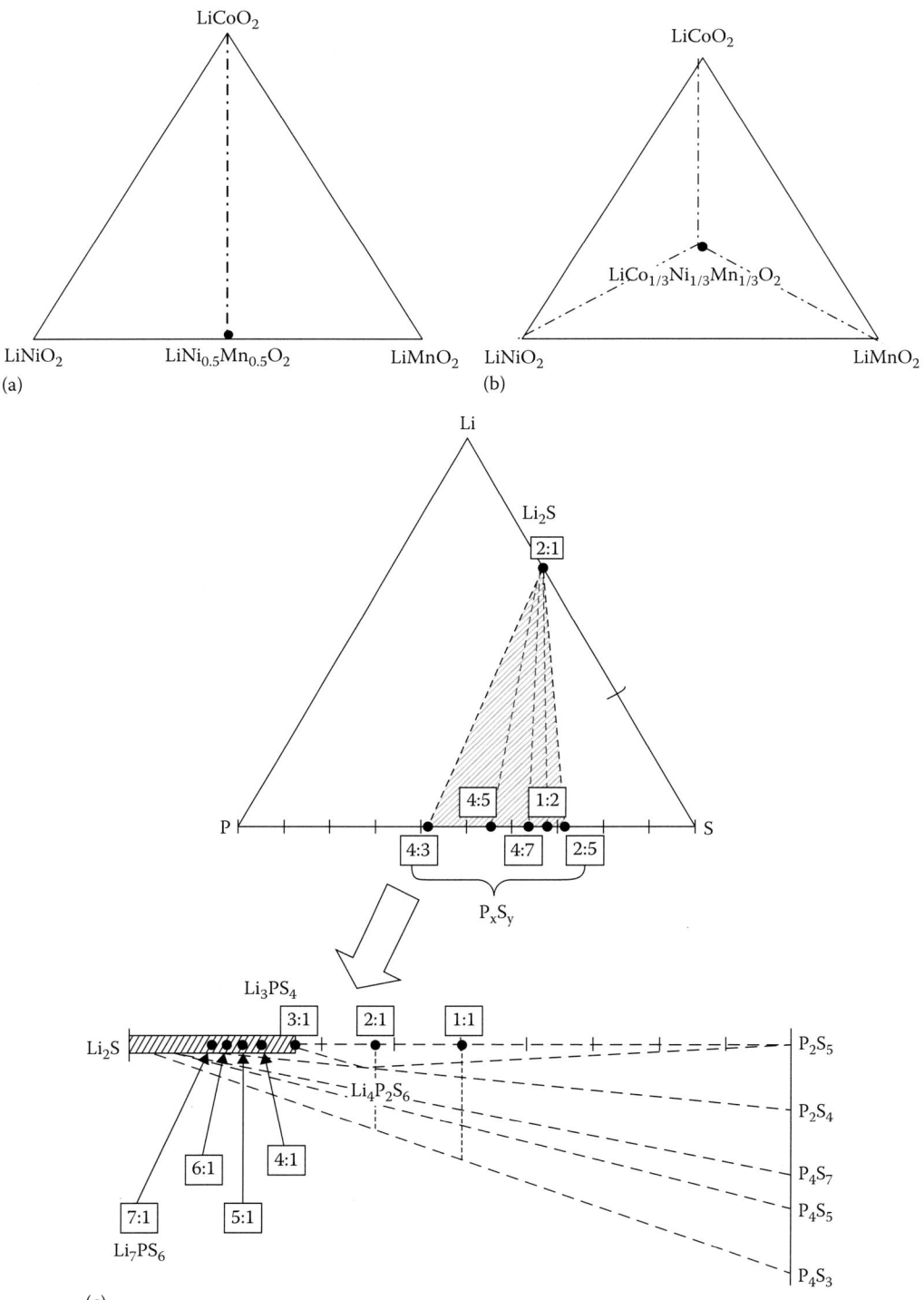

FIGURE 2.12 The Li–Mn–Co–Ni–O pseudo-ternary system. (a) LiNi$_{0.5}$Mn$_{0.5}$O$_2$ composition located on the base of the LiMO$_2$ triangle (midpoint). (b) LiCo$_{1/3}$Ni$_{1/3}$Mn$_{1/3}$O$_2$ composition located on the centroid of the LiMO$_2$ triangle. Phase diagram for the Li–P–S system. (After Murayama, M. et al., *Solid State Ionics*, 170, 173, 2004.)

H being the "heat content" connects the specific heat of the given phase.

$$H = U + PV \tag{2.22}$$

where
  U is the total "bonding" energy
  P and V are the pressure and volume

A system is in equilibrium when its Gibbs free energy does not change for infinitesimal changes in thermodynamic parameters, or, when

$$dG = 0 \tag{2.23}$$

which is a fundamental thermodynamic criterion for phase stability.

All chemical systems naturally tend toward states of minimum Gibbs free energy. Thus, G is a measure of relative chemical stability of a phase. For any phase, we can only determine volume, pressure P, temperature T, and chemical potential μ, but not G or H. We can only determine *changes* in G or H, as we change some other parameter of the system say P or V. For example, calorimetry measures changes in enthalpy ΔH for an endothermic or exothermic process such as melting or crystallization or structural phase transitions.

High-temperature oxide calorimetry was recently used to measure the enthalpy of formation of $x\mathrm{Ce}_{0.8}\mathrm{Y}_{0.2}\mathrm{O}_{1.9}$–$(1 - x)\mathrm{Zr}_{0.8}\mathrm{Y}_{0.2}\mathrm{O}_{1.9}$ solid solution system with different x content from 0 to 1. The calculated enthalpies of formation referred to the two oxides $\mathrm{Ce}_{0.8}\mathrm{Y}_{0.2}\mathrm{O}_{1.9}$ and $\mathrm{Zr}_{0.8}\mathrm{Y}_{0.2}\mathrm{O}_{1.9}$ show the following trends. As x changes the formation enthalpies increase for $0 < x < 0.4$, decrease for $0.6 < x < 0.4$ and decrease marginally for $0.6 < x < 1$. Comparing the result with other YSZ (yttria-stabilized zirconia) system and YDC (yttria-doped ceria), an important stabilizing mechanism was found: structure of oxygen vacancies near Zr^{4+} is energy favorable than those near Y^{3+}. This is because oxygen vacancies near Zr^{4+} will form seven-coordinate monoclinic zirconia that is a stable phase. While Ce^{4+} cannot attract the oxygen vacancies around Y^{3+}, it can only stabilize YDC by diluting the vacancy defect.

Comparison of this result with the structure and ionic properties of Zr and Ce showed that Zr^{4+} can stabilize the end member $\mathrm{Zr}_{0.8}\mathrm{Y}_{0.2}\mathrm{O}_{1.9}$ and cancel size mismatch effect by attracting oxygen vacancies from Y^{3+} to near the Zr^{4+} site, which is called "scavenging effect." The study also estimated that 20YDC–20YSZ (20% of yttria in YSZ and YDC) is only stabilized in a range of 0%–37.5% of YSZ based on the scavenging effect [2.27].

Let us now discuss the thermodynamic phase stability specifically for zirconia, which can exist in monoclinic, tetragonal, and cubic phases [2.28].

Taking the free energy of the insulating monoclinic phase at room temperature as reference, we can designate F_{tetra} and F_{cubic} as the free energies of high conducting tetragonal and cubic phases of ZrO_2. Thus, the energy difference $\Delta F = F_{\text{tetr}} - F_{\text{cubic}}$ may be used to discuss phase stability (Figure 2.13).

At a given temperature, the lowest value of the free energy determines the stable phase. The Helmholtz free energy F of a crystal is a sum of a ground-state energy and the free energy of lattice vibrations (see Chapter 6). The first term is usually calculated at $T = 0$ K. The second term is temperature dependent. In harmonic approximation (Chapter 1), it is calculated from the density of lattice vibrational or phonon states using the formula, neglecting thermal expansion,

$$F_{\text{harmonic}} = rk_{\text{B}}T \int_{0}^{\infty} d\omega\, g(\omega) \ln \left[2 \sinh \left(\frac{h\text{cut}\,\omega}{2k_{\text{B}}T} \right) \right] \tag{2.24}$$

where
  r is the number of degrees of freedom in a primitive unit cell
  ω is the phonon frequency
  $g(\omega)$ is the density of states that is the number of vibrational states per unit energy interval
  h cut is the Planck's constant
  k_{B} is the Boltzmann's constant

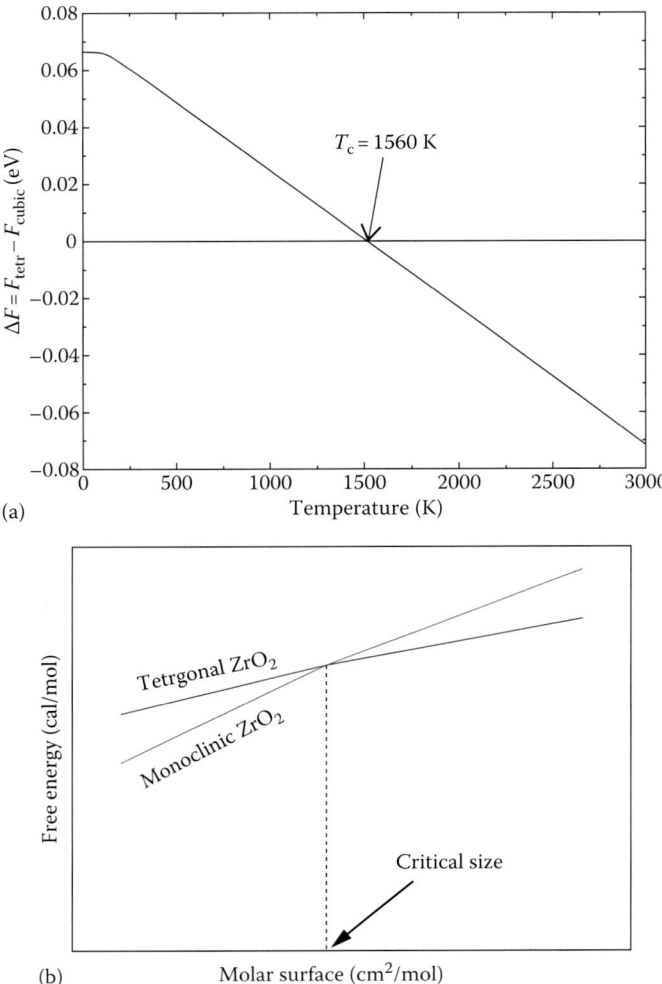

(a)

(b)

FIGURE 2.13 (a) Plot of the temperature dependence of calculated excess of free energies of the tetragonal phase of ZrO_2, $\Delta F = F_{tetr} - F_{cubic}$ with respect to the monoclinic phase taken as zero reference. ΔF changes sign from + to – at $T_c = 1560$ K which is close to the measured value of 1400 K. (After Sternik, M. et al., *J. Chem. Phys.*, 122, 064707, 2005.) (b) Schematic free energy diagram of monoclinic and tetragonal nanoparticles. (After Chen, L. et al., *J. Phys. Chem. C*, 115, 9370, 2011; Original paper Garvie, R.C., *J. Phys. Chem.*, 69, 1238, 1965.)

Gibbs theory that enables the calculation of phase diagrams is based on the establishment of a minimum free energy criterion. The calculation of phase boundaries of a given binary or ternary solid state ionic system is the mathematical problem of finding the minimum for the free energies of the phases concerned. This problem is relevant to the phase transition problem and would be discussed in detail in Chapter 5.

Important inputs of phase stability studies of solid state ionic systems are given in Table 2.2.

Phase stability is more directly expressed as the change in Gibbs free energy per oxygen molecule as a function of temperature. Therefore, oxygen partial pressure is a parameter. This is the so-called modified Ellingham diagram and is shown in Figure 2.4, drawn for $p_{O_2} = 0.1$ MPa. Note that ΔG and p_{O_2} are plotted on the left and right extremes of the vertical axis in the xy plane. It is modified to represent reactions in which a phase of interest is being reduced as in the case of $LiFePO_4$ rather than being oxidized. Thermal reduction occurs at a transition temperature that may be generally found at the intersection of $\Delta G(T)$ with the relevant p_{O_2} line. An Ellingham diagram is a graph that shows the temperature dependence

TABLE 2.2

Thermodynamic Parameters of Selected Precursors of Solid State Ionic Materials[a]

Materials	Standard Enthalpy of Formation (kJ/mol)	Standard Gibbs Energy of Formation (kJ/mol)	Standard Entropy at 298.15 K (J/mol K)	Heat Capacity at Constant Pressure at 298.15 K (J/mol K)
$LiBH_4$	−190.8	−125.0	75.9	82.6
LiH	$\Delta_f H^0$: −90.5 kJ/mol	$\Delta_f G^0$: −68.3	S^0: 20.0 J/mol	KCp: 27.9
Li_2SO_4	−1436.5	−1321.7	115.1	117.6
Li_2O	−597.9	−561.2	37.6	54.1
Li_2S				
LiF	−616.0	−587.7	35.7	41.6
LiCl	−408.6	−384.4	59.3	48.0
LiBr	−351.2	−342.0	74.3	
LiI	−270.4	−270.3	86.8	51.0
$NaBH_4$	−188.6	123.9	101.3	86.8
NaF	−576.6	−546.3	51.1	46.9
NaCl	−411.2	−384.4	59.3	48.0
NaBr	−361.1	−349.0	86.8	51.4
NaI	−287.8	−286.1	98.5	52.1
Na_2O	−414.2	−375.5	75.1	69.1
Na_2SO_4	−1387.1	−1270.2	149.6	128.2
Al_2O_3	−1675.7	−1582.3	50.9	79.0
AgCl	−127.0	−109.8	96.3	50.8
AgBr	−100.4	−96.9	107.1	52.4
CuCl	−137.2	−119.9	86.2	48.5
Cu_2S	−79.5	−86.2	120.9	76.3
CuS	−53.1	−53.6	66.5	47.8
CaO	−634.9	−603.3	28.1	42.0
CoO	−237.9	−214.2	53.0	55.2
Co_3O_4	−891.0	−774.0	102.5	123.4
Bi_2O_3	−573.9	−493.7	151.5	113.5
SiO_2 (α-quartz)	−910.7	−856.3	41.5	44.4
UO_2	−1085.0	−1031.8	77.0	63.6
PbO_2	−277.4	−217.3	68.6	64.6
ZrO_2				
CeO_2				
BaF_2	−1207.1	−1156.8	96.4	71.2
$BaCl_2$	−858.6	−810.4	123.7	95.1
$BaBr_2$	−757.3	−736.8	146.0	
CaF_2	−1228.0	−1175.6	68.567.0	
H_3PO_4	−1284.4	−1124.3	110.5	106.1
HgI_2	−105.4	−101.7	180.0	—
C (graphite)	0	0	5.7	8.5
C (diamond)	1.9	2.9	2.4	6.1
Liquids				
H_2O	−285.8	−237.1	700	75.3
H_2SO_4	−814.0	−690.0	156.9	138.9
H_3PO_4	−1271.7	−1123.6	150.8	145.0
Gases				
H_2	0	0	130.7	28.8
H	218.0	203.3	114.7	20.8
HI	26.5	1.7	206.6	29.2
H_2S	−20.6	−33.4	205.8	34.2
SO_2	−296.8	−300.1	248.2	39.9

Source: Adapted from Kondepudi, D. and Prigogine, I., *Modern Thermodynamics*, Wiley, Chichester, U.K., 2004.

[a] Crystalline solids unless stated otherwise.

of the stability for compounds. This analysis is usually used to evaluate the ease of reduction of metal oxides and sulfides that form the precursors of SSIMs.

The phase equilibria relevant to the synthesis of $LiFePO_4$ will be discussed in Chapter 3.

The phase stability of the Li–Mn–O system has recently been studied by Longo et al. [2.29b] from *ab initio* calculations. This study involves (1) the thermodynamic stability of Li and oxygen vacancies and (2) the electrochemical activation mechanisms of these cathode materials. The density functional theory calculations provide phase diagrams in both physical and chemical potential spaces and ranges of electrochemical activity are predicted. The calculated effects of p_H on the Li–Mn–O system phase stability elucidated the mechanism of Mn^{2+} formation from the spinel phase under acidic conditions.

2.4 Classification of Solid State Ionic Materials

SSIMs are by and large crystalline solids that are judged from the large number of mineral equivalents that are found to exist. Why are solids generally crystalline? Crystalline ordering is probably the simplest way for atoms to form a macroscopic solid. Basically, such an ordering results in low-energy arrangements of atoms. Why are such arrangements periodic? A profound question indeed! If an optimal neighborhood exists for each atom, then the lowest energy state for a large number of atoms gives the same neighborhood to every atom. SSIMs could have crystal structures that could possess many such "nearly lowest" energy states so that one could stabilize them in a given structure. Thus, structure is an important criterion for the classification of these materials. Equilibrium lattice structure could be a function of temperature and pressure so that the classification follows logically from what has been discussed in earlier sections of this chapter. Vibrational entropy and disorder often play an important role in the stabilization of various crystal structures.

Like superconductors (s-band, p-band, d-band and f-band, or beta-W, triple perovskite, etc.), superionic conductors are also classified based on structural components but more conveniently on the basis of cation or anion which is mobile. In this chapter, we consider the latter as the basis of classification and include important structures for illustration. SSIMs have defied/eluded classification essentially because of their complex "dynamic" structural varieties [2.10]. Why consider *mobile ion families*? Note that SSIMs tend to adopt a surfeit of cubic and allied structures. Furthermore, the structure -physicochemical property relationships exhibited by them is intriguing. Then there is the plurality of ions capable of fast ion diffusion and the interesting aspect of a particular ion tying up with a specific application. Li^+ ion for batteries and H^+ and $O^=$ for fuel cell among others. The latter aspect throws in the temperature range of phase stability and that of the operation of a device as a basis for classification just as the "phase transition types" (see Chapter 5) provides a basis for yet another classification. A basic point to note is all conducting ions (except perhaps H^+) are classical objects and any quantum mechanical effects such as quantum confinement occur at the nanoscale. A further point is electrons and holes in superconductors are common to all varieties—low- and high-temperature superconductors and also types I and II, but a fast ion conductor is ion specific. Although phase transitions occur in both superconductors and superionic conductors, a common basic interaction such as the electron–phonon interaction does not exist in the latter. And also the generalized concept of a superionic conducting transition temperature is missing, again due to the plurality of mobile ions.

Thus, it stands to reason that we consider a "taxonomy" of ion families. An important justification for this is the current assiduous search for new structures and materials containing mobile ions including Li^+, Na^+, Ag^+, and $O^=$ that offer device possibilities.

The relationship between thermodynamics and crystal structure is subtle. The structure adopted by two or three or four elements as in AgI and PbF_2, and ZrO_2, $LiMnO_2$, $LiFePO_4$ is such that it (1) provides for, at a moderate or high temperature and/or "partial pressure," a surfeit of defects, essentially vacancies or interstitials that through specific coordination (say, tetrahedral or octahedral usually in combination) and connectivity lead to paths/pathways for ion conduction and (2) possesses a sequence of phase transitions, over a range of temperatures, with the cubic structure arising at the highest of temperatures, providing an opportunity for stabilization, through doping and use of green-synthesis methods, of the high conducting phase at modest operating temperature. Thus phase stability as discussed earlier becomes important along with phase coexistence. This naturally connects to the materials innovation and synthesis challenges, the latter forming the basis for Chapter 3.

A discussion of mobile ion–based families focuses on the innovation of new structures involving a given mobile ion and also on materials and device development. Of course materials based on, say, Li^+ and $O^=$ may possess the same structure. But the discussion on mobile ion–based family provides an opportunity to compare and contrast the potential of a given mobile ion on the basis of its structural and allied properties.

Eight fundamental questions interrelate this chapter to almost all of the following chapters. They are as follows:

1. What atoms are involved and what are the electronic configuration of each of them?
2. What are the types of chemical bonds formed between a given pair of atoms?
3. What is the symmetry of the crystal?
4. How are the atoms arranged in the crystal structure?
5. Are there chains (1D) and layers (2D) that cause anisotropy?
6. Do such arrangements promote certain mechanisms for long-range ionic motions and/or atomic distortions?
7. How do these mechanisms give rise to the properties observed?
8. Which properties are crucial to device applications?

More details on the properties crucial to device applications are found in the book by Newnham [2.30].

The classification of SSIMs into six major families not only helps trace the materials revolution in solid state ionic but also helps us realize that several mobile ions may adopt the same structure (as in the β-alumina system) while a single mobile ion existing in different crystalline polymorphs exhibits differential ion transport properties and yet possesses the same interatomic forces as in β-AgI and α-AgI.

2.4.1 Li Family

Most simply put, lithium ($1s^2 2s^1$) lacks a third shell. Thus lithium, and the other second row elements of the periodic table, is much smaller. Because of its size, it has a high charge density and so shows *some covalent character* when involved in compounds. Comparatively, the charge density of sodium is not nearly as high as lithium and hence it forms *mostly ionic bonds*. This is probably a significant difference between lithium and any of the other group 1 elements.

Li^+ ion being the lightest and the smallest (apart from proton, see later) is also the fastest, whose transport is possible in a large number of compounds. The unusual nature of Li^+ transport is illustrated in Figure 2.14.

Thus, it would not be in contact with the surrounding anions. Consequently, the Li^+–I^- distance 0.302 nm in the crystal is ~10% greater than the radius SUM 0.276 nm, making the material softer than

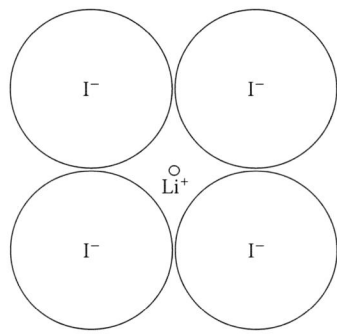

FIGURE 2.14 I^- ion contact in LiI leaves lot of room for tiny light Li^+ to execute "rattling" motion in the fcc lattice of LiI.

other alkali halides, and lowering its melting point, boiling point, and heats of fusion and vaporization. The bonding situation in LiI could also be responsible for its highest ionic conductivity and the ease of Li intercalation, which is important for its use in various applications. In contrast, consider NaI with a lower than expected melting point. Here, both anion–anion contact and cation–anion contact operate leading to a slight expansion of the crystal lattice.

LiI belongs to a small rare class of solids with cations that are *too small for their sites*. In such case, it is interesting to note that the cation moves off the center of the site and adopts a distorted octahedral coordination. These solids typically exhibit novel properties, such as ferroelectricity and piezoelectricity. $LiNbO_3$ is an example of ferroelectric superionic!

Small size and extreme lightness of Li^+ ion not just contribute to its fast diffusion in any Li-based crystal structure but also let it adopt unique structures. Indeed, many ternary and quaternary compounds exist in which a plenty of specially organized interstitial space is available with thermodynamically controlled disorder that favors fast ion conduction.

Pauling [2.26] has given a motivating discussion of the garnet structure with general formula $R^{++}_3Al_2Si_3O_{12}$ by applying coordination theory. In this theory, compounds of simple structure are those in which the number of essentially different kinds of anions is small. $R^{++}_3Al_2Si_3O_{12}$ is a simple orthosilicate containing Al and a divalent cation. Each oxygen would form the corner of a Si tetrahedron and one or two R^{++} polyhedra. Li-based minerals spodumene ($LiAlSi_4O_{10}$) and petalite ($LiAlSi_4O_{10}$) have oxygen ions common to a Si tetrahedron, an Al tetrahedron, and a Li tetrahedron! Thus when R^{++} is replaced by Li^+ ion, structures with unique coordinations, often favorable for solid state ionic activity, are obtained. The garnet structure $Li_7La_3Zr_2O_{12}$ stabilized by Murugan et al. [2.31] is a unique case to show how the adaptability of Li^+ in a familiar structure can result in a fast ion conductor.

The uniqueness of the Li^+ family of SSIMs is that it spans the entire gamut of crystal systems—from triclinic to cubic. Important and quantifiable deviations from ideal structures such as the perovskite structure are apparently a hallmark of these materials. It is intriguing that the perovskite structure is involved in both fast ion conducting and in high-temperature superconducting systems alike. This fact underscores the importance of thermodynamic systems with built-in entropy and metastability to these solid state phenomena. Redox capability is a feature common to these systems.

The ability of Li^+ to be extracted out of or inserted into a structural framework in a process called (de) intercalation is the most fundamental and unique of physicochemical phenomena. It is of course applicable to the design and development of electrochemical energy conversion and storage systems.

Lithium ion conductors can not only be crystalline but also composites with ceramics such as alumina, glasses, and polymeric as we shall discuss briefly following Table 2.3, which summarizes important data on the representative Li^+ ion conductors.

A few important structures selected from Table 2.3 are illustrated later: Among the most important is $LiCoO_2$, the Li-ion battery cathode material. Figure 2.15 shows the "unordered" and Li-ion ordered structures of this compound. The parent monoclinic structure and the hexagonal structure obtained by deintercalation and their crystallographic relationship are illustrated.

Figure 2.16 illustrates the crystallographic relationship between the monoclinic and hexagonal forms of $LiCoO_2$ both in terms of lattice vectors of the two phases (graphical) and as a transformation matrix between the vectors of the two phases.

Figure 2.17 illustrates the structure of $LiFeBO_3$.

Figure 2.18 gives the structures of the high-temperature tetragonal phase and the low-temperature monoclinic phase of the NASICON phase $LiZr_2(PO_4)_3$.

The ideal spinel structure adopted by the battery electrode material Li-deficient $Li_xMn_2O_4$ ($x < 2.0$) is shown in Figure 2.19.

The antipertovskite structure that is the inverse of the perovskite ABO_3 with the positions of cations and anions reversed is shown in Figure 2.20.

Intercalation of the well-known carbon superconductor C_{60} or fullerene with Li produces a superionic conductor at room temperature. Its structure and conductivity profile are presented in Figure 2.21.

The garnets form a solid solution with lithium distributed over a mixture of oxide tetrahedra and heavily distorted octahedral (Figure 2.22). Fast ion conduction is observed for all compositions and occurs exclusively via a network of edge-linked distorted oxide octahedra with the tetrahedrally coordinated

TABLE 2.3

Lithium Family

Li Metal	bcc	$a = 0.349$ nm	
$LiFe(SO_4)F$	$P\text{-}1$	$a = 0.51751$, $b = 0.54915$, $c = 0.72211$ nm, $\alpha = 106.5060$, $\beta = 107.1780$, $\gamma = 97.8650$, $V = 0.182441$ nm³	Tripathi et al. [2.73]
$LiFePO_4OH$ (natural mineral tavorite)	$P\text{-}1$	$a = 0.5350$, $b = 0.7291$, $c = 0.5117$ nm, $\alpha = 109.300$, $\beta = 97.75$, $\gamma = 106.470$, $V = 0.17478$ nm³	Roberts et al. [2.74]
$LiVPO_4F$	$P1$	$a = 0.5170$, $b = 0.5308$, $c = 0.7263$ nm, $\alpha = 107.590$, $\beta = 107.970$, $\gamma = 98.390$, $V = 0.17436$ nm³	Aleba Mba et al. [2.75]
$Li_7P_3S_{11}$	$P1$ triclinic	$a = 1.25009$ nm, $b = 0.603160$ nm, $c = 1.25303$ nm, $\alpha = 102.845(3)0$, $\beta = 113.2024(18)0$, $\gamma = 74.4670$	Yamane et al. [2.76]
$LiFeSO_4F$	$P\text{-}1$ triclinic	$a = 0.51751$ nm, $b = 0.54915$ nm, $c = 0.72211$ nm, $\alpha = 106.5060$, $\beta = 107.1780$, $\gamma = 97.8650$, $V = 0.182441$ nm³	Pasero et al. [2.77]
Li_2MnO_3	$(C2/c)$	$a = 0.49257$ nm, $b = 0.85250$ nm, $c = 0.96278$ nm, $\beta = 99.9760$	Tripathi et al. [2.73]
$FeSO_4F$ (framework)	$C2/c$	0.73637, 0.70753, 0.73117, 119.758, 0.328017 nm³	
Li_2SO_4	Monoclinic (300 K)	$a = 0.8239$, $b = 0.4954$, $c = 0.8474$ nm, $\beta = 107098'$	Kvist and Lunden [2.78]
	Cubic (fcc) (848 K)	$a = 0.707$ nm	
$LiZr_2(PO_4)_3$	High-temperature phase (tetragonal)	$a = 0.885$, $c = 2.24$ nm, $V/Z = 0.2514$ nm³	Sudreau et al. [2.79]
	Low-temperature phase (monoclinic)	$a = 1.5299$, $b = 0.8940$, $c = 0.8816$ nm, $\beta = 125.980$	
$LiFePO_4$	Orthorhombic $Pnmb$	$a = 0.9436$, $b = 0.5944$, $c = 0.4665$ nm, $V = 0.260807$ nm³	Sundarayya et al.[2.80]
$LiFeBO_3$	Monoclinic $C2/c$	$a = 0.5169$ nm, $b = 0.8924$ nm, $c = 1.0138$ nm, $\beta = 91.390$	Janssen et al. [2.81]
Li_2PS_4	$T = 906$ K $Pbcn$	$a = 0.86125$ nm, $b = 0.90211$ nm, $c = 0.84262$	Homma et al. [2.82]
Li_2PO_2N	$Cmc2$	$a = 0.90692$ nm, $b = 0.53999$ nm, $c = 0.46856$ nm	Senevirathne et al. [2.83]
$LiTi_2(PO_4)_3$		$a = 0.85135$ nm, $c = 2.08705$ nm, $c/a = 2.4514$	
$LiCoO_2$	298 K Hex $R\text{-}3m$	$a = 0.2815$ nm, $c = 1.405$ nm, $V = 0.0964$ nm³, $Z = 3$	Completely ordered rocksalt structure with alternate planes of Li and Co atoms Orman and Wiseman [2.84]
$LiNi_{1/3}Mn_{1/3}Co_{1/3}O_2$ (NMC)		$a = 0.2863$ nm, $c = 1.4244$ nm	Guo et al. [2.85]
$Li_xCo_{1-x}O$		$0.075 \leq x \leq 0.24\text{–}0.31$	
$Li_{0.2}Co_{0.8}O$		$a = 0.83488$ nm, average Co/Li–O bond length $= 0.2087$ nm	Wu et al. [2.86]

(Continued)

TABLE 2.3 (Continued)

Lithium Family

Li Metal	bcc	a = 0.349 nm	
$Li_{1/2}La_{1/2}TiO_3$	Cubic	$a_0 = 0.3871$ nm	Inaguma et al. [2.87]
$Li_2FeMn_3O_8$	Spinel $Fd3m$	a = 0.82508 nm	Kawai et al. [2.88]
$Li_4Ti_5O_{12}$	Spinel	a = 0.8355 + 0.02 nm	Izquierdo et al. [2.89]
LiH		a = 0.409 nm (NaCl type)	
LiC_6	(First-stage graphite intercalation compound) $P6/mmm$	a = 0.4305 nm, c = 0.3706 nm	
LiC_{12}	(Second-stage graphite intercalation compound) $P6/mmm$	a = 0.4288 nm, c = 0.7065 nm	Guerard and Herold [2.90]
Li_4C_{60}	Body-centered monoclinic ($I2/m$)	a = 0.9326 nm, b = 0.9048 nm, c = 1.50329 nm, β = 90.970	Ricco et al. [2.91]
LiI		a = 0.600 nm (NaCl type)	
Li_2S		a = 0.571 nm (fluorite)	
Li_2O		Antifluorite a = 0.462 nm	
Li_3N	$P6/mmm$ Hexagonal	a = 0.3648 nm, c = 0.3875 nm Z = 1	Rebenau and Schulz [2.92]
Li_3OCl	$Pm3m$ (with minor distortions)	a = 0.391 nm	
Li_3OBr	$Pm3m$ (with minor distortions)	a = 0.402 nm	Zhao and Daemen [2.93][a]
Li_3Sb			
$α-Li_3Sb$	Hexagonal	a = 0.4701 nm, c = 0.8039 nm	
$β-Li_3Sb$	Cubic	a = 0.6559 nm ($T > 6500°C$)	
Li_3Bi	Cubic	a = 0.6708 nm	
Li_3P	Hexagonal	a = 0.4264 nm, c = 0.2579 nm	Brauer et al. [2.94]
Li_6PS_5Cl	Cubic $F-43m$	a = 0.98397 nm	
Li_6PS_5Br	Cubic $F-43m$	0.99689	
Li_6PS_5I	Cubic $F-43m$	1.01490	Boulineau et al. [2.95]
LGPS $Li_{10}GeP_2S_{12}$	Tetragonal $P42/nmc$	a = 0.87177, c = 1.263452 nm	Kamaya et al. [2.96]
$Li_7La_3Zr_2O_{12}$	Cubic LLZO $Ia-3d$	a = 1.2975 nm	Geiger et al. [2.97]

[a] Antiperovskites can be structurally manipulated quite easily by chemical substitution, for example, by introducing large Br^- anions at the dodecahedral site to replace Cl^- anions. The use of mixed halogens (e.g., $Cl_{1-z}Br_z$) can push the tolerance factor: $t = (rA + rX)/[\sqrt{2}(rB + rX)]$ of the antiperovskite to vary from 0.85 for pure Li_3OCl to 0.91 for pure L_3OBr as the substitution goes from the chlorine end-member to the bromine end-member. A higher tolerance factor indicates that the antiperovskite structure approaches a less distorted.

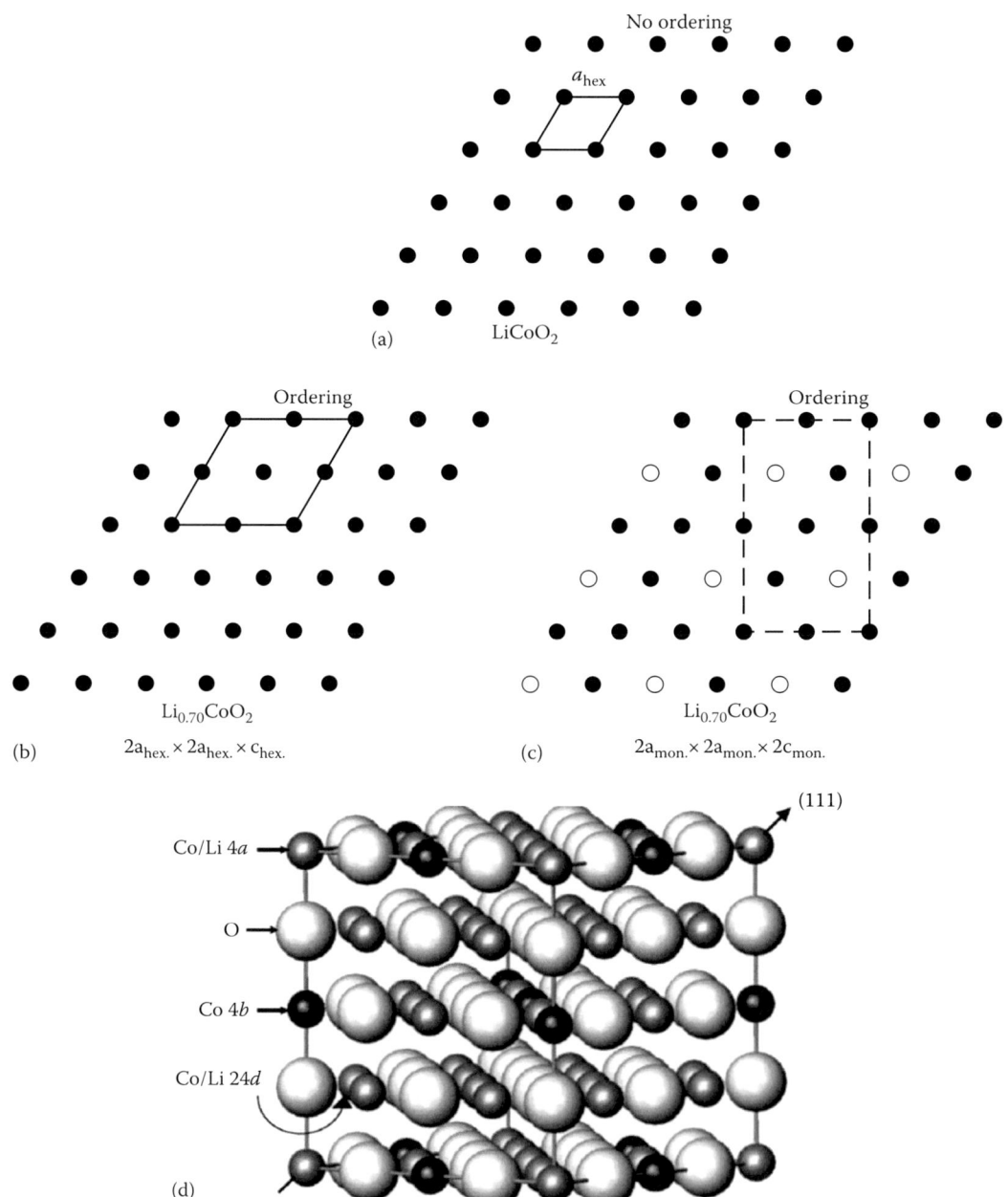

FIGURE 2.15 Schematics of (a) Li-ion positions in the LiCoO$_2$ parent sample, (b) proposed Li-ion positions in Li$_{0.7}$CoO$_2$ in a hexagonal cell, and (c) proposed lithium ion positions in Li$_{0.7}$CoO$_2$ in a monoclinic cell. The black dots represent lithium ions, the white dots represent vacancies and the gray dots represent a statistical distribution of lithium ions and vacancies. (After Clemencon, A., In-situ and ex-situ observations of Li deintercalation from LiCoO$_2$: AFM and TEM studies, MS thesis, MIT, Cambridge, MA, June 2005.) (d) Average crystal structure of Li$_{0.2}$Co$_{0.8}$O—a partially ordered rock-salt-like solid solution phase Li$_x$Co$_{1-x}$O (0.075 $\leq$ x $\leq$ 0.24–0.31. The cation stacking sequence along [111] consists of alternating planes of Co and Co/Li. Nanosized domains of this ordered phase appear alongside disordered regions. Depending upon Li content domain size increases from 2 to 8 nm. Compositions of ordered and disordered regions are Li and Co rich, respectively, leading to frozen-in incipient phase separation. This microstructure could be a precursor to precipitation of fully ordered, rhombohedral LiCoO$_2$. (After Wu, Y. et al., *Proc R. Soc. A*, 465, 1829, 2009.)

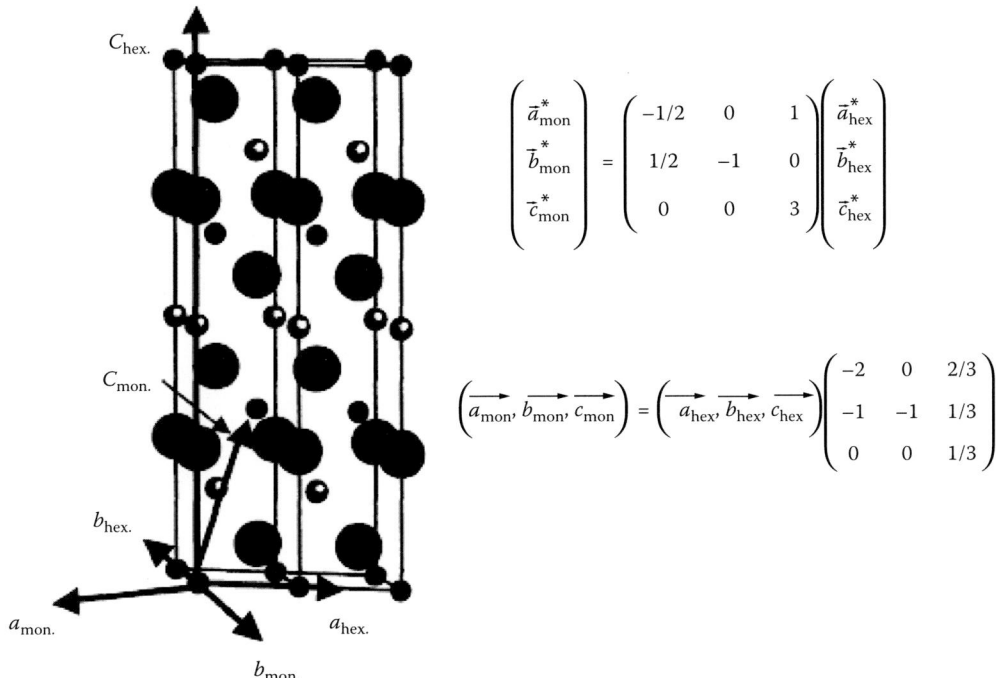

FIGURE 2.16 Monoclinic and hexagonal unit cells of LiCoO$_2$ and their crystallographic relationship. (After Clemencon, A., In-situ and ex-situ observations of Li deintercalation from LiCoO$_2$: AFM and TEM studies, MS thesis, MIT, Cambridge, MA, June 2005.)

site playing no part in the transport properties. The garnet structure can accommodate between three and seven Li$^+$ ions offering an opportunity to optimize Li-ion conductivity [2.38].

Next, we will look at Li$^+$-ion composites that gave probably one of the earliest Li$^+$-ion batteries.

2.4.1.1 Lithium-Ion Composites

Composite materials are heterogeneous mixtures of solid phases. The elaboration of composites offers a new degree of freedom in the search for advanced functional materials, because specific properties can to a certain degree be tailored by mixing appropriate phases [2.39]. In solid state ionics, two routes can lead to improved solid ionic conductors: (1) a search for new compounds and structures sustaining high levels of ionic conductivity as found from Table 2.3 and (2) a modification of existing compounds, by heterogeneous or homogeneous doping. This modification involves homogenous dissolution of a certain amount of aliovalent (a valency higher or lower than host cation) dopant in the bulk of the ionic conductor M$^+$ X$^-$ in order to increase the concentration of mobile charge carriers according to bulk defect equilibrium.

For example, additional metal vacancies can be created by doping with cations of higher valence, such as D^{2+}, in substitution of M$^+$:

$$DX_2 + 2M_M \rightarrow D^{\cdot}_M + V'_M + 2MX$$

V'_M is the vacancy created when D^{2+} settles down as D$^{\cdot}_M$ in the MX lattice.

Heterogeneous doping, however, involves *mixing with a second phase* with *very limited solid solubility* and the *formation of defect concentration profiles in the proximity of interfaces*. The deviations from local electrical neutrality (space charges) are a consequence of point defect equilibrium at interfaces. Apart from the improvement of the electrical properties, such as high conductivity and ionic transference

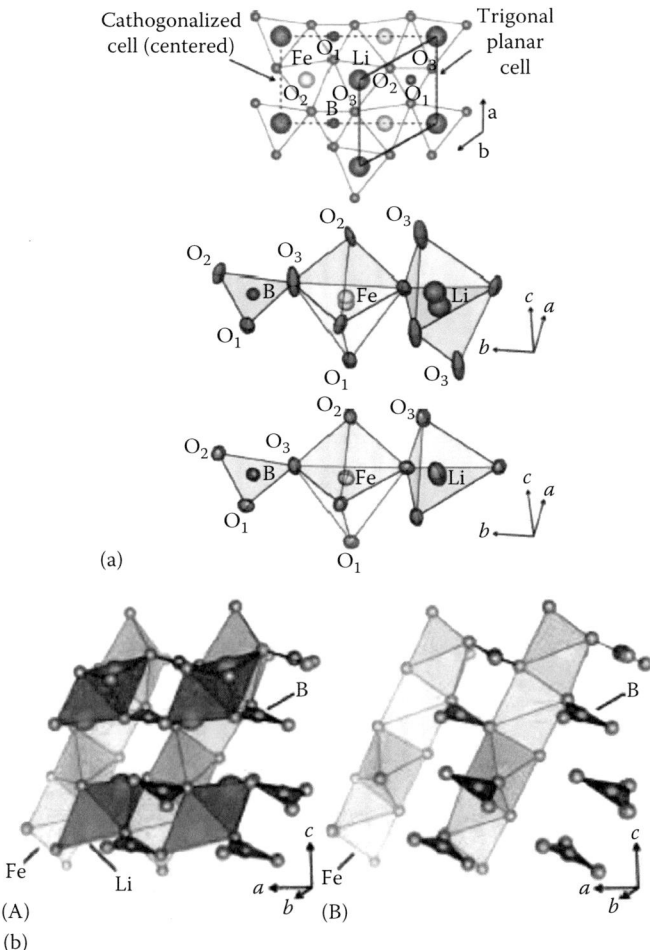

FIGURE 2.17 Model structure for $LiFeBO_3$ (a) Top: idealized 2D layer (space group $P3$, $a = 0.5066$ nm); Middle: Structure from literature. Note orthogonalized central cell with $a = 0.5066$, $c = 0.8774$ nm; Bottom: Anisotropic displacement of O ions. "Splitting" of Li and Fe sites and "elongation" of O ellipsoids displayed; (b) DFT optimized crystal structure for (A) average structure approximated to $C2/c$. (B) Structure after delithiation with unmodulated $C2/c$ symmetry. (From Janssen, Y. et al., *J. Am. Chem. Soc.*, 134, 12516, 2012.)

number, such composite materials can also develop better shock resistance or higher strength. Solid state ionics deals primarily with two-phase mixtures.

Liang discovered enhanced ionic conductivity in $LiI–Al_2O_3$ composites (Figure 2.23), LiI being hygroscopic, the importance of water in undried samples must be emphasized. Even higher increases, by two to three orders of magnitude, were found for a large number of $LiI–Al_2O_3$ compositions, whatever be the source of the starting materials and the fabrication route. At least two distinct Li^+ sites existing in these composites and their proportion depend on temperature and composition. At 50 vol% alumina, the population of bulk Li ions almost equals that in the interface regions—an aspect discussed in Chapter 10. This goes with the observed reduction of the ionic conductor grain size with increasing alumina concentration and reflects the very large effective interfacial regions in this material.

The composite $LiI–Al_2O_3$ solid electrolytes find use in commercial batteries: $Li/LiI–Al_2O_3/PbI_2/$ Pb and $Li/LiI–Al_2O_3/TiS_2/S$. In the ionic conductor analogue of semiconductor sensors, surface conductivity variations upon adsorption of gases are a result of chemical interactions of acid–base type. The transfer resistance, which plays a decisive role for response time and selectivity of electrochemical sensors at reduced temperature, is largely controlled by defect concentrations at the interface, so that

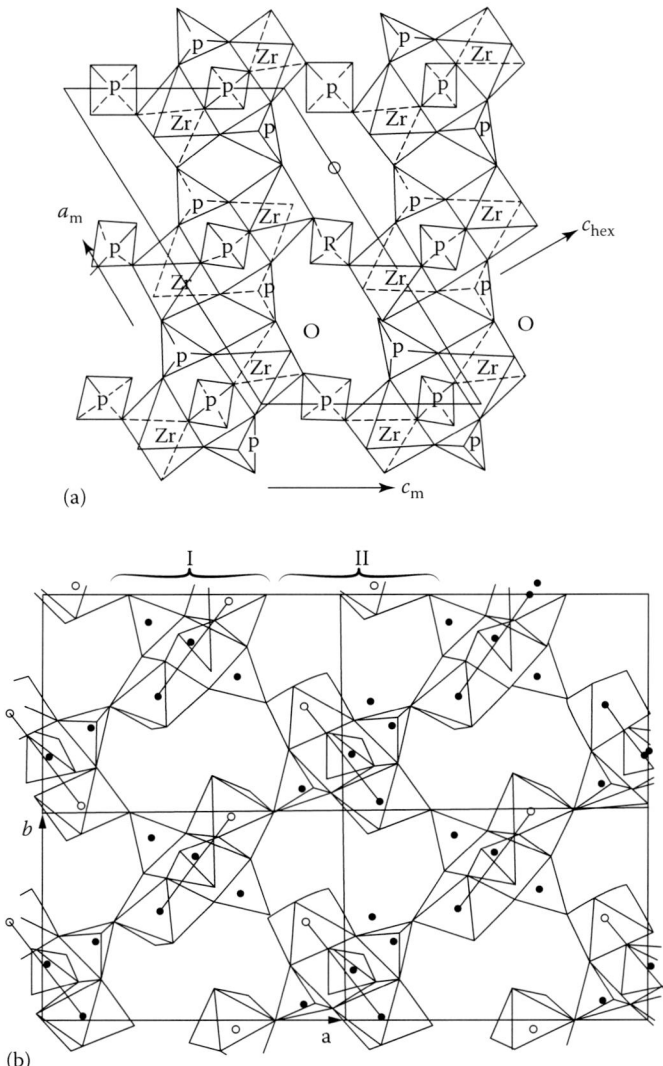

FIGURE 2.18 Structure of LiZr$_2$(PO$_4$)$_3$ (a) HT phase(tetragonal) projection of one-half unit cell along a axis of rhombo-hedral NASICON type structure. (b) LT phase (monoclinic) projection on (001) plane of beta ferric sulfate monoclinic type structure. (After Sudreau, F. et al., *JSSC*, 83, 78, 1989.)

heterogeneities, for example, in a composite material, can reduce the transfer resistance. See Chapter 9 for a detailed discussion.

2.4.1.2 Li-Based Glassy Polymers

A polymer such as polyethylene oxide is essentially a macromolecule and can exist in a noncrystalline state. Thus, one can visualize glassy polymers that are basically nonconducting.

By dissolving ionic salts, it is possible to induce ionic conductivity into insulating polymers. If the polymer is polyethylene oxide or polyvinyl alcohol, then the ionic salt could be LiClO$_4$ (for Li$^+$) or NaCl (for Na$^+$) or NH$_4$SCN (for H$^+$). The unique properties of polymer electrolytes include thin-film forming ability, easy processability, flexibility, lightweight, elasticity/plasticity, transparency, and reasonably high ionic conductivity. Easy pathways for ion diffusion are readily provided by the polymer host. Additionally, polymers maybe amorphous or crystalline, both types (simultaneously present usually)

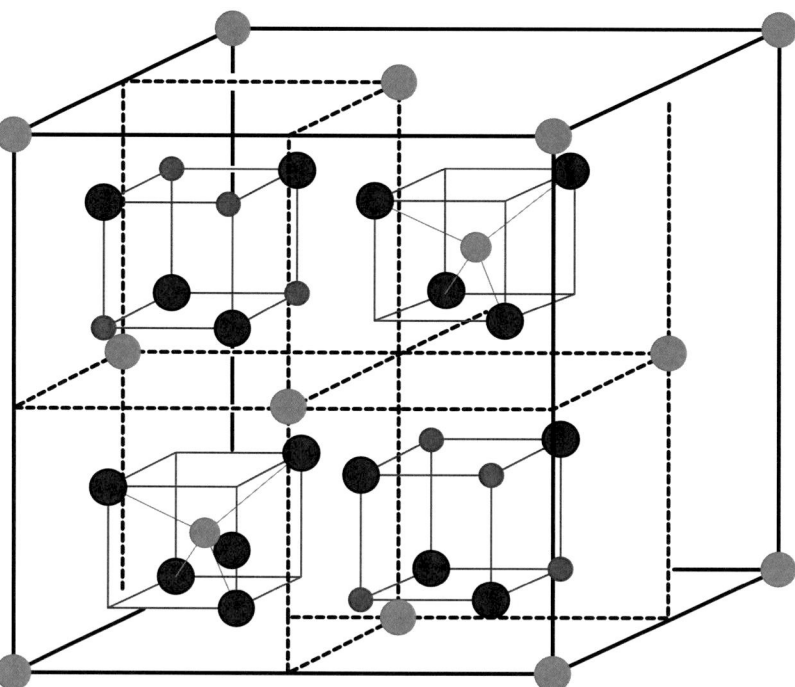

FIGURE 2.19 Ideal spinel structure with formula AB_2O_4 showing atoms only; the front half of the full-unit cell containing four alternate simple cubic and tetrahedral building blocks embedded in the full fcc type unit cell. There are four more such units at the back making up a total of eight units. Small light gray circles, A atoms; big dark gray circles, B atoms; big light gray circles, oxygens. This structure is adopted by $LiMn_2O_4$. Li^+ ions occupy only cube corners whereas Mn ions alternately occupy centers of tetrahedral and cube corners (sharing space with Li^+). Small size and high mobility of Li^+ ion makes the structure Li-deficient so that the formula is $Li_xMn_2O_4$.

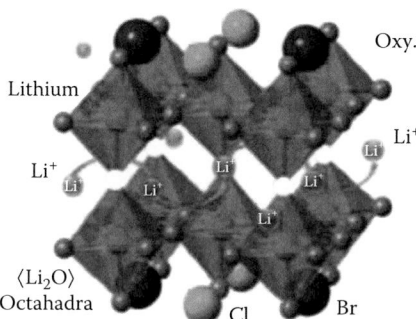

FIGURE 2.20 Li_3O (Cl, Br) antiperovskite structure. (After Zhao, Y. and Damen, L.L., *J. Am. Chem. Soc.*, 134, 15042, 2012.)

exhibiting ionic conductivity. However, biphasic nature influences both conductivity and crystallinity. Ions move in a dynamic environment created by the polymer chain in the amorphous phase when the polymer is held at temperatures above the glass transition temperature (T_g), usually T_g is greater than or of the order of ambient temperature. The short segments of the polymer chain execute crank-shaft like motion, thereby creating randomly, suitable coordination sites adjacent to ions so that these ions may hop. Segmental modes actually involve the motion of groups of atoms on the polymer chains and such motions are essentially slow relatively speaking. This unique type of motion limits hopping rate and maximum conductivity attainable to ~100 µS/cm at room temperature. These are essentially composite electrolytes and their conductivity is influenced by the thermally induced morphological changes in the polymer component such as glass transition and crystallization of uncrystallized regions [2.42].

It is interesting to compare polymers with glasses (to be discussed next). In polymer electrolytes, the ionic and segmental motions are strongly coupled to each other (Figure 2.24), while in inorganic glasses, the ionic motion is *decoupled* from that of the network.

High conductivity, solvent-free polymeric electrolytes are both intellectually challenging and technologically attractive and include the desirability of "single ion" conductors for alkali metal batteries [2.43].

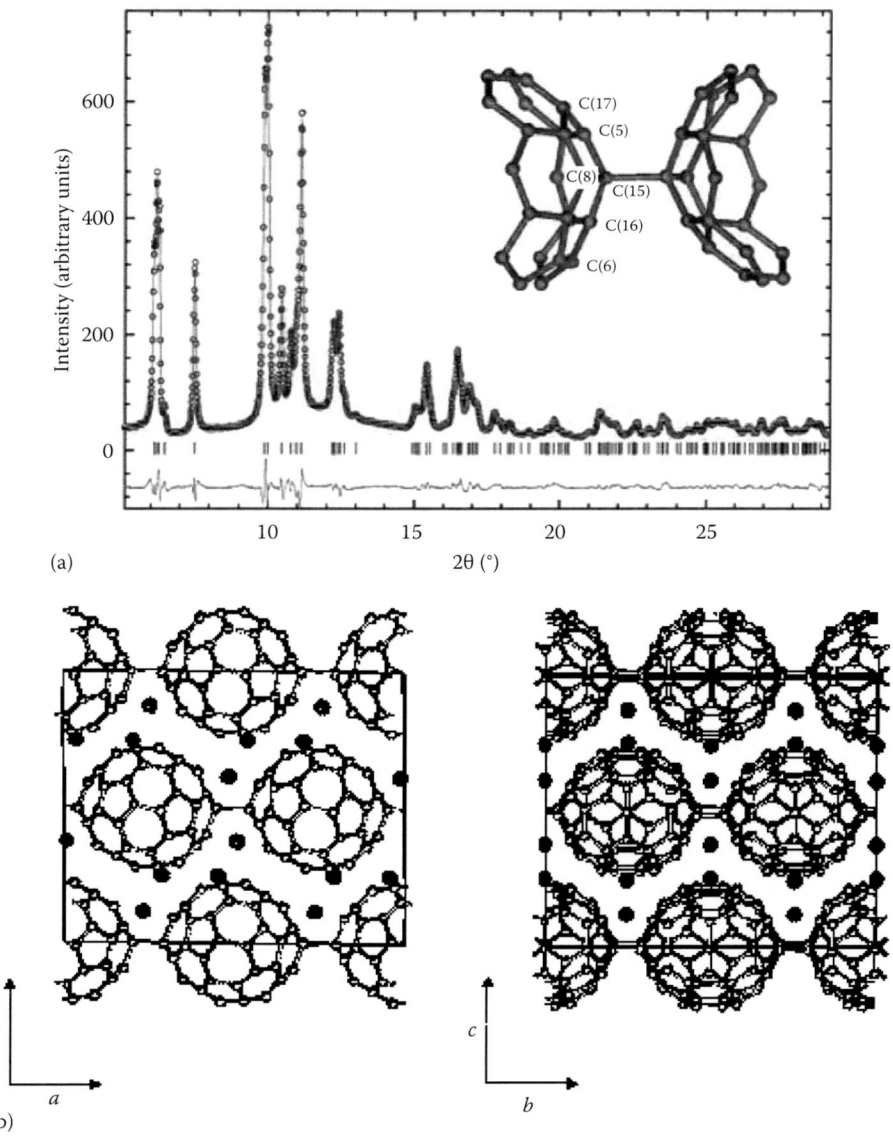

FIGURE 2.21 (a) Synchrotron x-ray diffraction profile for Li_4C_{60} at 295 K. O, observed; -, calculated. Lower solid line: difference profile, ticks: reflection positions. Inset: geometry of t_j frontier (C(5) and C(16)) and bridging (C(15)) carbon atoms on adjacent fullerenes along the direction of single C–C polymerization (*a* axis. (b) Crystal structure of polymerized Li_4C_{60} determined by synchrotron x-ray diffraction. Projections on the ac-(left) and bc-(right) basal planes are shown. Li^+ ions residing in pseudo-tetrahedral (close to the fullerene surface) and octahedral holes (intersitital space) are shown as closed circles). (After Margadonna, S. et al., *J. Am. Chem. Soc.*, 126, 15032, 2004.) *(Continued)*

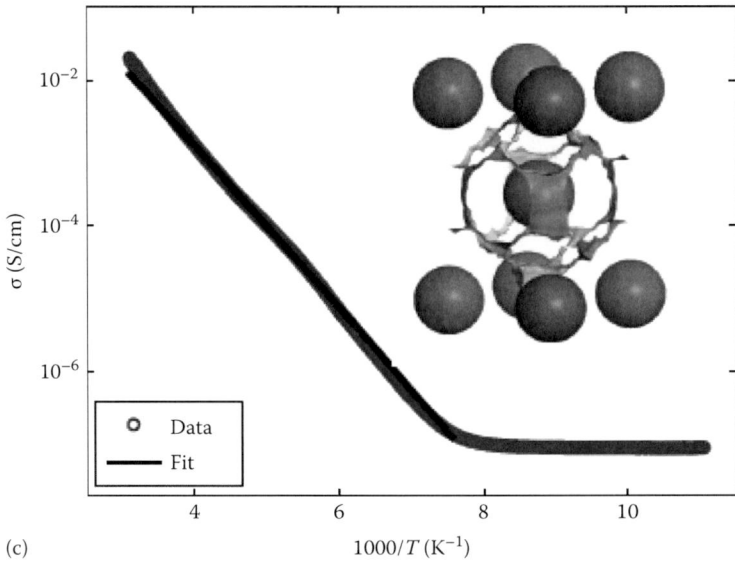

(c)

FIGURE 2.21 (*Continued*)   (c) Temperature dependence of DC conductivity of superionic Li_4C_{60} ($\sigma = 10$ mS/cm at 300 K activation energy for uncorrelated hopping is 200 meV) polymeric Li_4C_{60}. In the low-temperature plateau sample resistance is 20 MΩ. Insert shows possible paths for Li^+ ion diffusion in the fullerene lattice. Medium/light grey region is the volume for occupation of diffusing Li ions. (After Ricco, M. et al., *Phys. Rev. Lett.*, 102, 145901, 2009.)

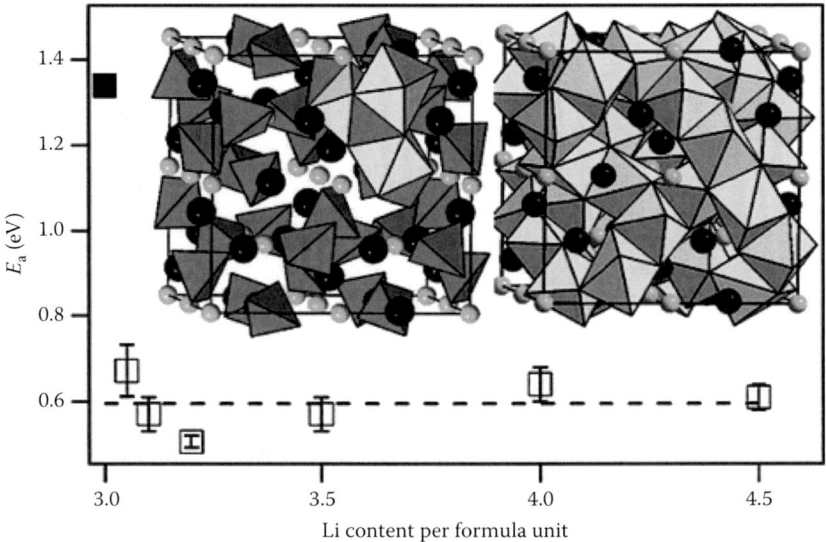

FIGURE 2.22   Polycrystalline compounds of the garnets $Li_{3+x}Nd_3Te_{2-x}Sb_xO_{12}$ series crystallize in the space group $Ia\bar{3}d$ with lattice parameters in the range 12.55576(12) Å for $x = 0.05$ to 12.6253(2) Å for $x = 1.5$. The lithium is distributed over a mixture of oxide tetrahedra and heavily distorted octahedra. Increasing the lithium content leads to vacancies in the tetrahedral position to an increased occupation of lithium in the octahedra. The latter exhibit considerable positional disorder: two lithium cations position within each octahedron. Fast Li^+ ion conduction occurs with a composition independent activation energy of ~0.6 eV exclusively via a network of edge-linked distorted oxide octahedra. (From O'Callaghan, M.P. et al., *Chem. Mater.*, 20, 2360, 2008.)

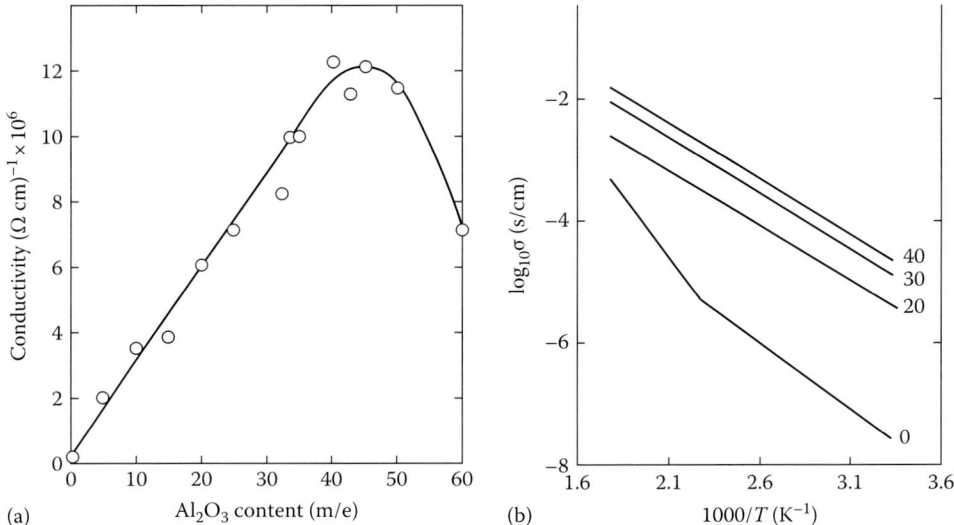

FIGURE 2.23 The LiI–Al$_2$O$_3$ composite showing a linear increase in Li$^+$ conductivity with Al$_2$O$_3$ content reaching a maximum at ~50 m/o alumina. (After Liang, C.C., *J. Electrochem. Soc.*, 120, 1289, 1973.) (b) Arrhenius plots of ionic conductivity of LiI–Al$_2$O$_3$ composites relative to LiI showing enhanced conductivity of composites. The enhancements are presumed to arise from highly conducting paths created along the interface between LiI and the filler (alumina) material. (After Poulsen, F.W. et al., *Solid State Ionics*, 9–10, 119, 1983.)

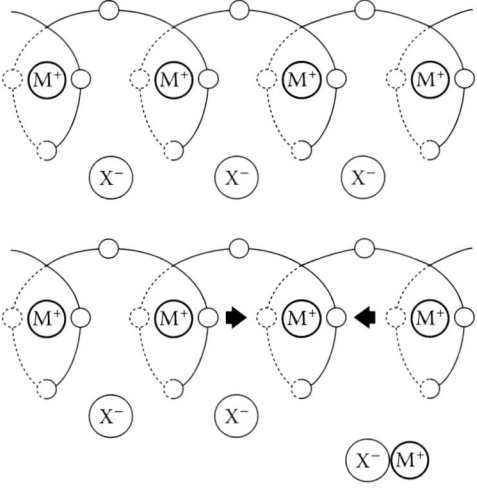

FIGURE 2.24 Motion of ion M$^+$ in a salt MX in polymer. (After Wright, P.V., *Electrochim. Acta*, 43, 1137, 1998.)

Since ion transport in polymer electrolytes (typically solutions of lithium salts in polyether systems) involves segmental motions of the host polymer, as described earlier, significant levels of ionic conductivity will exist only above the glass transition temperature, T_g. By contrast, in inorganic systems, ionic conduction is widespread in the glassy state, and in some "superionic" glasses reaches values of 10^{-2} S/cm at ambient temperature. Is decoupled ion transport possible in polymers? It is indeed possible in solvent-free glassy polymer electrolytes. So that instead of the usual "salt-in-polymer" one has "polymer-in-salt" electrolytes. An example is the side-group liquid crystal polymer electrolyte where mesogenic side groups impart a special "ultrastructure" on *both* the polymeric molecule and to a lesser extent on the

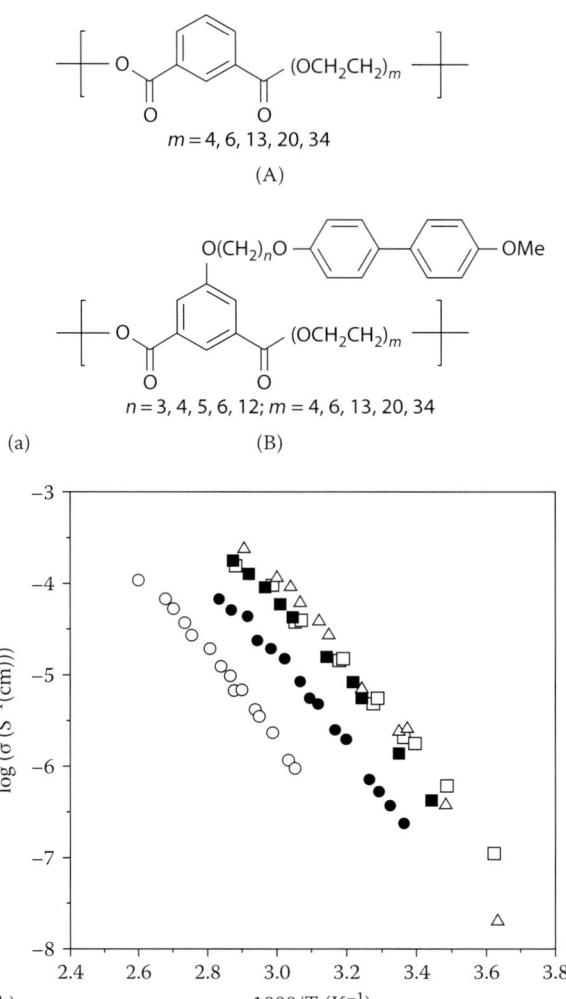

FIGURE 2.25 (a) Molecular structures of polymer precursors for LiClO$_4$-based solid polymer electrolytes: (A) the 0G*m* series, where 0 indicates the *absence* of pendent side chains and m is the number of ethylene oxide (EO) units per repeat unit; (B) the MeOC*n*G*m* series, where *n* is the number of CH2 spacers separating the mesogenic groups from the polymer backbone. (b) Arrhenius plots of ionic conductivities of the LiClO$_4$ complexes (AO:Li) (10:1) for polymers in the 0G*m* series, as in (a). Symbols: O, *m* = 4; ●, *m* = 6; □: *m* = 13; ■, *m* = 20; Δ, *m* = 34. (After Imrie, C.T. et al., *J. Phys. Chem. B*, 103, 4132, 1999.)

polyether backbone (Figure 2.25a). Figure 2.25b shows the ionic conductivity as a function of ethylene oxide groups (*m*) per repeat unit.

A brief discussion of Li-ion glasses and glass ceramics follows.

2.4.1.3 *Li-Ion Glasses and Glass Ceramics*

A glass combines the isotropy of a liquid and at the same time retains the rigidity of a solid. For glass formation, one needs flexible building blocks that can often come from the compatibility of the metal and nonmetal species forming the glass. It is generally obtained by rapid or slow quenching of the melts of the constituents.

Li$^+$ ion, chemically speaking, is a hard acid, according to the "Hard and Soft Acids Bases" theory of Pearson. Between O$^=$ and S$^=$, S$^=$ is a soft base with which Li$^+$ is more compatible [2.45]. Thus Li$_2$S–SiS$_2$

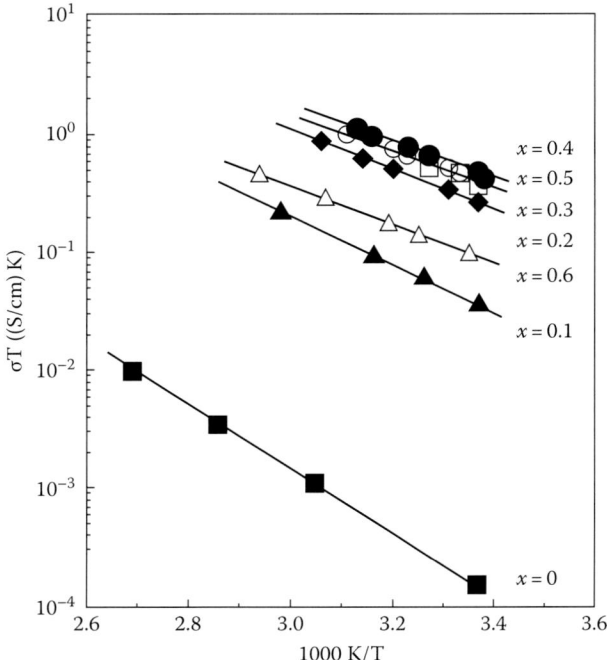

FIGURE 2.26 Arrhenius plot of the glass ceramic system $2[Li_{1+x}Ti_2Si_xP_{3-x}O_{12}]$–$AlPO_4$ showing how the Si content helps enhance the Li-ion conductivity.

glasses are ionically more conducting than Li_2O–SiO_2 glasses. Generally, glass formation helps achieve higher ionic conductivities than crystalline solids. A typical example of a stable Li^+ ion glass is $0.03Li_3PO_4$–$0.59Li_2S$–$0.38SiS_2$. This can be obtained at ambient pressure by quenching in liquid nitrogen. Its conductivity at room temperature is ~0.7 mS/cm [2.46].

Glass ceramics are obtained from glasses by careful heat treatment of quenched glasses. Fast Li^+ conducting glass ceramics in the pseudo-binary system $2[Li_{1+x}Ti_2Si_xP_{3-x}O_{12}]$-$AlPO_4$ may be made as follows: The glasses are made first by quenching the melts of required compositions from 1450°C, pouring them onto preheated stainless steel plates, and pressing them into 1–2 mm thick plates. After annealing in a furnace at 550°C, glass transition temperature (T_g) and crystallization temperature (T_x) are determined. A two-stage heat treatment produces glass ceramic: heating of annealed glass specimen at a $T > {\sim}T_g$ overnight and further heating to a $T > T_x$ and holding overnight. Figure 2.26 shows how Si content helps enhance conductivity of the glass ceramic.

Let us next discuss about the Na family.

2.4.2 Na Family

Faraday's early experiment on ion migration used a jelly containing sodium sulfate as the electrolyte in the following sandwich cell configuration:

+/wetted litmus paper A/unsaline jelly/wetted litmus paper B/jelly with sodium sulfate/wetted turmeric paper A/unsaline jelly/wetted turmeric paper B/−

When the contact was made [+ −], *wetted litmus paper A* and *wetted turmeric paper B changed color,* establishing an ionic current. No change was observed at *wetted litmus paper B* and *wetted turmeric paper A*. This is perhaps a pioneering demonstration of a gel-based sodium ion battery!

The importance of these Na compounds from application perspective is that while Li is abundant in nature to an extent of 20 mg/kg, Na has an abundance of 2.36×10^4 mg/kg. Na is only 1/20th as expensive as Li but available worldwide. In view of the role of Na^+ ion transport through membranes and ion channels in the human body processes, Na-based materials occupy a unique place among SSIM. The development of high-energy-density batteries for industrial applications involves the synthesis of better and newer Na-based materials. In this section, important material systems from Na–S to Na_xFePO_4 will be considered.

The question now arises: *Why sodium?* Open and layered structures based on Na^+ are better able to accommodate large Na^+ ions. Thus they are better intercalation compounds. In terms of phase stability, these structures also generally exhibit both Na and Li versions of the same compound. However, as illustrated by Na_2FePO_4F, the development of new structures and framework types *based specifically on sodium*—not variations on the lithium analogue—is more interesting for phenomenology and applications [2.47].

Let us look at the structures of a few Na-ion conductors.

Among the most important Na-ion conductors is sodium beta alumina whose structure is depicted in Figure 2.27. This is the structure in which Na^+ migration takes place in the oxygen ion planes located between two "spinel blocks."

The basic parent structure $NaZr_2(PO_4)_3$ (NZP) is a framework structure made up of corner-linked ZrO_6 octahedra. The latter are joined by PO_4 tetrahedra, each of which corner shares *four* of the octahedra. A 3D system of channels running through the structure is thus created. In this structure, two types of vacant sites arise: type I, a single distorted octahedral site occupied by Na^+ in NZP and three larger type II sites *vacant in NZP, resulting* in an incredibly stable and versatile structure. Indeed it has been adopted by hundreds of compounds; in NASICON or NA SuperIonic CONductor, $Na_3Zr_2(PO_4)$ $(SiO_4)_2$, three out of four vacant sites are occupied by Na^+ enabling a correlated motion as ions diffuse through the channels. This family has many members including $L_{1.5}Fe_{0.5}Ti_{1.5}(PO_4)_3$ and $Li_3V_2(PO_4)_3$ (Figure 2.28).

The structure of a related compound $Na_3V(PO_3)_3N$ is shown in Figure 2.29.

An important electrode material for Na-ion battery is $Na_{0.44}MnO_2$ whose structure is illustrated in Figure 2.30.

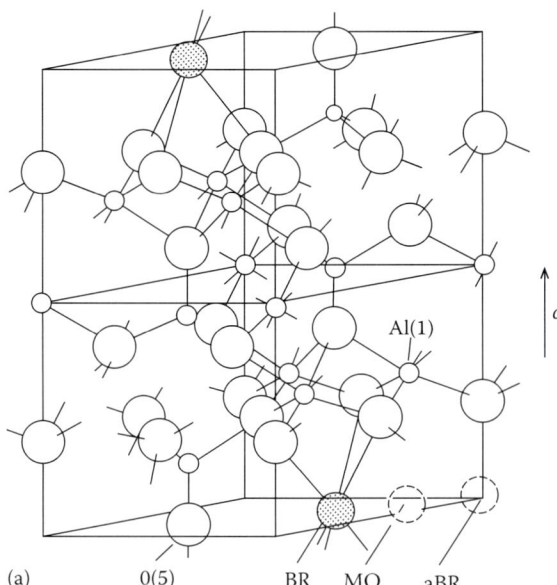

(a)   0(5)   BR   MO   aBR

FIGURE 2.27 Sodium beta alumina ($NaAl_{11}O_{17}$) structure (a) 3D hexagonal unit cell showing Beevers–Ross (BR), anti Beevers–Ross (ABR) and mid-oxygen (MO) sites for Na-ion migration   *(Continued)*

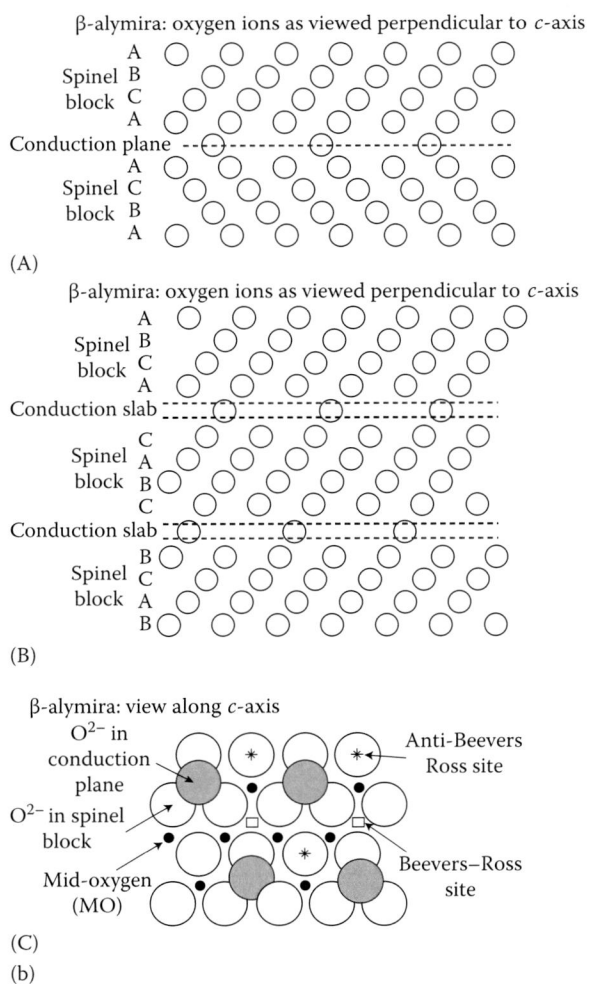

(A)

(B)

(C)

(b)

FIGURE 2.27 (*Continued*)   Sodium beta alumina (NaAl$_{11}$O$_{17}$) structure (b) 2D illustration of the structure showing the two variants of the basic structure-β′ (A) and β″ alumina (B), and, clear locations of the BR, ABR and MO sites (C).

Intercalation of the well-known fullerene (C$_{60}$) with Na (as with Li discussed earlier) produces Na$^+$ ion–conducting Na$_4$C$_{60}$ which is isostructural with the Li compound whose structure is shown in Figure 2.31.

Table 2.4 presents the crystallographic data on many sodium ion conductors—a family that continues to grow so that the list is only representative and not exhaustive.

2.4.3 Noble Metal Family

The unique noble metal family of Cu, Ag, and Au ion conductors has not only provided gut-level physics and chemistry but also motivated a search for applications. The question of ion dynamics in SSIM, particularly the dynamics of defect formation, has been successfully addressed using just one family of silver halides. Indeed, the so-called α-AgI is a fast ion conductor par excellence while AgCl and AgBr have helped focus on the thermodynamics of Frenkel defect formation and the mechanisms of ion conduction.

In this section, we shall list selected well-known and recent noble metal ion conductors and discuss salient features of their crystal structures. As mentioned in Table 1.1, the discovery of unusually high (~1 S/cm) ionic conductivity even at a rather low temperature of 420 K initiated the field of SSI. Indeed, silver and copper halides along with AuI have provided the basis for the synthesis and

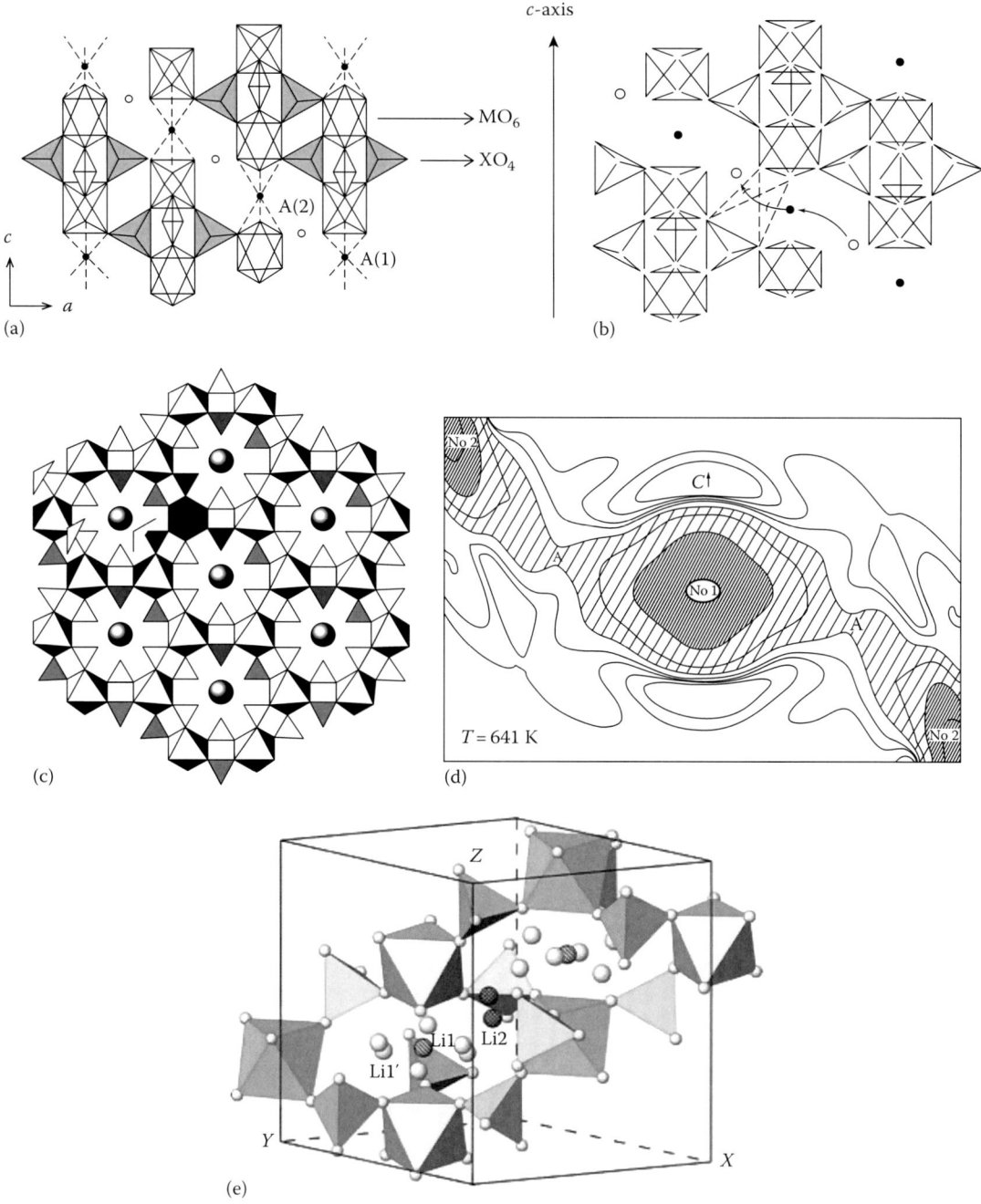

FIGURE 2.28 The structure of Nasicon ($Na_{1+x}Zr_2P_{3-x}Si_xO_{12}$) showing: (a) The A1 (type 1) and A2 sites (type 2). (From Goodenough, J.B. et al., *Mat. Res. Bull* 11, 203, 1976.) (b) Conduction pathway, and (c) Hexagonal array of the [$A_2(XO_4)$] groups in the plane (001). (From Anantharamulu, N. et al., *J. Mater. Sci.*, 46, 2821, 2011.) This is a framework structure with suitable tunnel size for Na^+ migration in three dimensions. (d) A conductivity channel obtained from a temperature-dependent x-ray experiment. (From Maier, J., *Physical Chemistry of Ionic Materials: Ions and Electrons in Solids*, Wiley, Chichester, U.K., 2004, Figure 6.10.) (e) Perspective view of a portion of the neutron diffraction derived crystal structure of $L_{1.5}Fe_{0.5}Ti_{1.5}(PO_4)_3$ at 298 K (space group $R\bar{3}c$, $Z = 6$), showing coordination polyhedra and O atoms (small spheres). Li^+ ions are emphasized as large spheres within the cavities M1 (hatched Li_1 and open Li_1') and M2 (cross-hatched Li_2). This shows the structural similarities shared by the Na- and Li-ion-conducting compounds. *(Continued)*

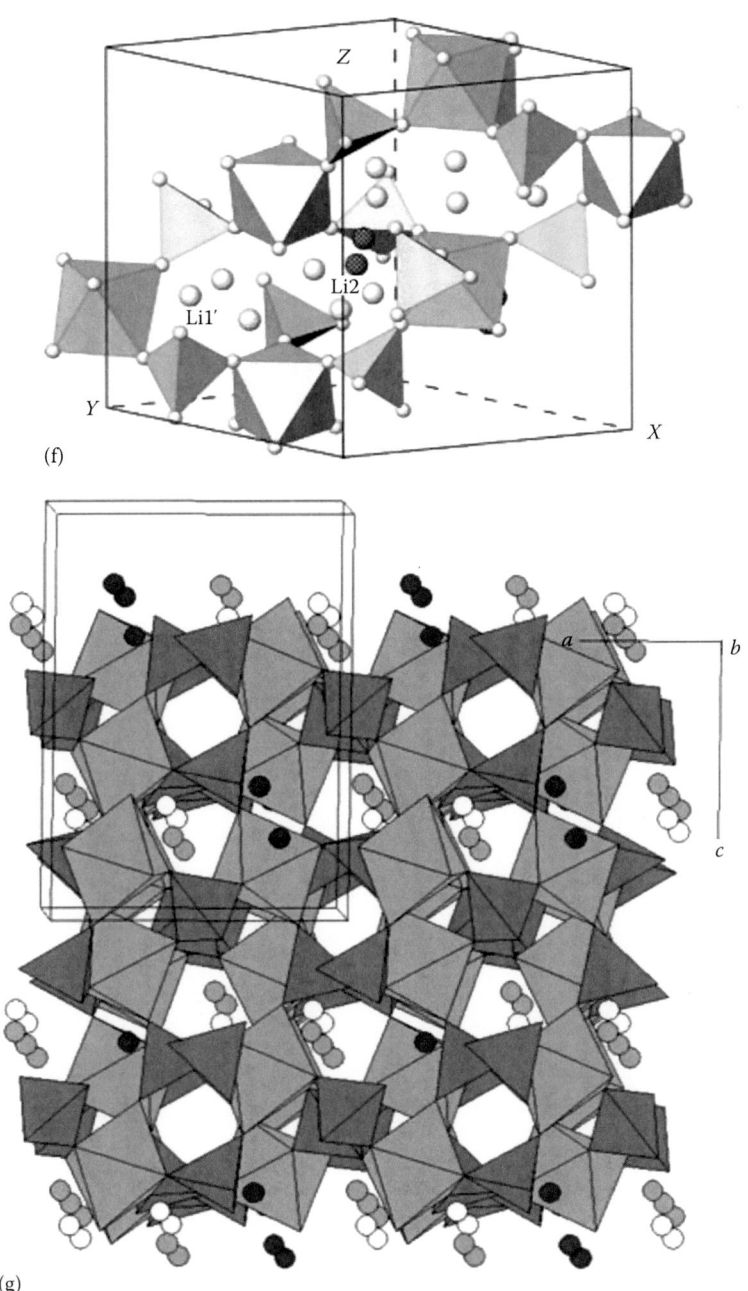

FIGURE 2.28 (*Continued*) (f) Perspective view of a portion of the crystal structure of $L_{1.5}Fe_{0.5}Ti_{1.5}(PO_4)_3$ at 673 K, emphasizing the Li^+ ions as large spheres within the cavities M1 (open Li_1') and M2 (cross-hatched Li_2). (After Catti, M. et al., *J. Mater. Chem.*, 14, 2004, 835.) (g) Schematic of NASICON-type $Li_3V_2(PO_4)_3$ structure, again highlighting the structural similarity of Na and Li compounds where only the ion size makes a difference in this case. (After Gover, R.K.B. and Slater, P.R., *Annu. Rep. Prog. Chem., Sect. A*, 100, 525, 2004.)

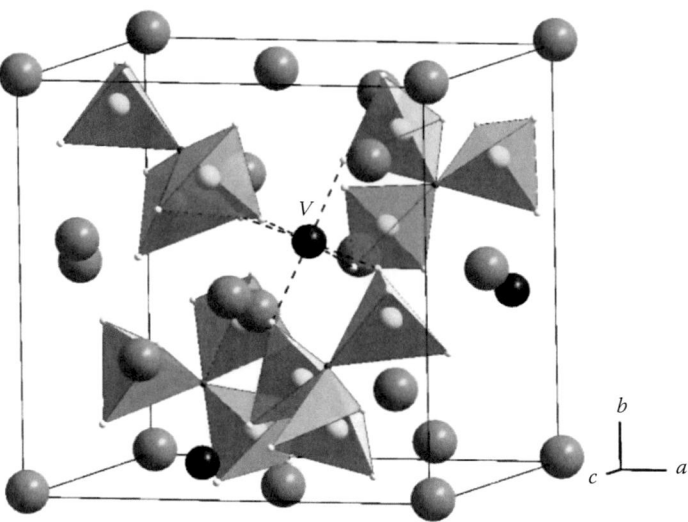

FIGURE 2.29   The crystal structure of $Na_3V(PO_3)_3N$. PO_4 tetrahedra are shown with gray shading; Na atoms are gray spheres and V atoms are black spheres. (From Kim, M. and Kim, S.J., *Acta Cryst. E*, 69, i34, 2013.)

investigation of a large number of ternary and quaternary (and even higher) fast ion conductors enriching the knowledge and paving way for device applications (Table 2.5). Perovskite oxides are tailor-made for solid oxide fuel cells [2.56].

2.4.4 Oxygen Family

The dual role of oxygen family in solid state ionics is noteworthy: oxygen conductors and proton conductors. The largest category of metal oxide crystals is the perovskites with general formula ABO_3, where A is a cation of larger size than B. Ideally, these lattices consist of a cubic close-packed arrangement of both oxide anions and the larger cation (Figure 2.32). The *smaller* cation occupies the octahedral hole at position (½½½). Physical properties of perovskites are not limited to superionic conductors indicated in Table 2.6.

In practice, most perovskite lattices have distorted cubic unit cells. The degree of structural distortion d is

$$d = \frac{\left(r_A + r_O\right)}{\sqrt{\left(r_B + r_O\right)}}$$

As $d \rightarrow 0$ the perovskite will become more perfectly cubic. For $d \leq 0.8$, however, ionic radius of the A site will be smaller than the ideal thus resulting in BO_6 octahedra becoming tilted to occupy the available volume. Stable perovskites occur for d values in the range $0.78 \leq d \leq 1.05$. Bond covalency helps realize stable structures outside this range.

In ABX_3 compounds (A, B metal atoms, X could be oxygen) that adopt perovskite structures, two different metal sites could be substituted with lower valence metal cations. An example is $LaGaO_3$ (Figure 2.33) in which La(3+) could be doped with Sr(2+) and Ga(3+) with Mg(2+) to effect dramatic enhancement of ionic conductivity at lower temperatures than it is possible with stabilized zirconia (Figure 2.34). In the same family, one finds Sr-doped $LaMnO_3$ and $LaCrO_3$ which can conduct both ions and electrons like stabilized zirconia.

Figures 2.35 and 2.36 show the structures of $LaBaGaO_4$ isomorphous with the orthorhombic β-K_2SO_4 structure and apatite-type structure.

Table 2.7 provides the structural characteristics of the oxygen family of SSIMs.

Next, we will discuss about halogen family, especially F-based compounds.

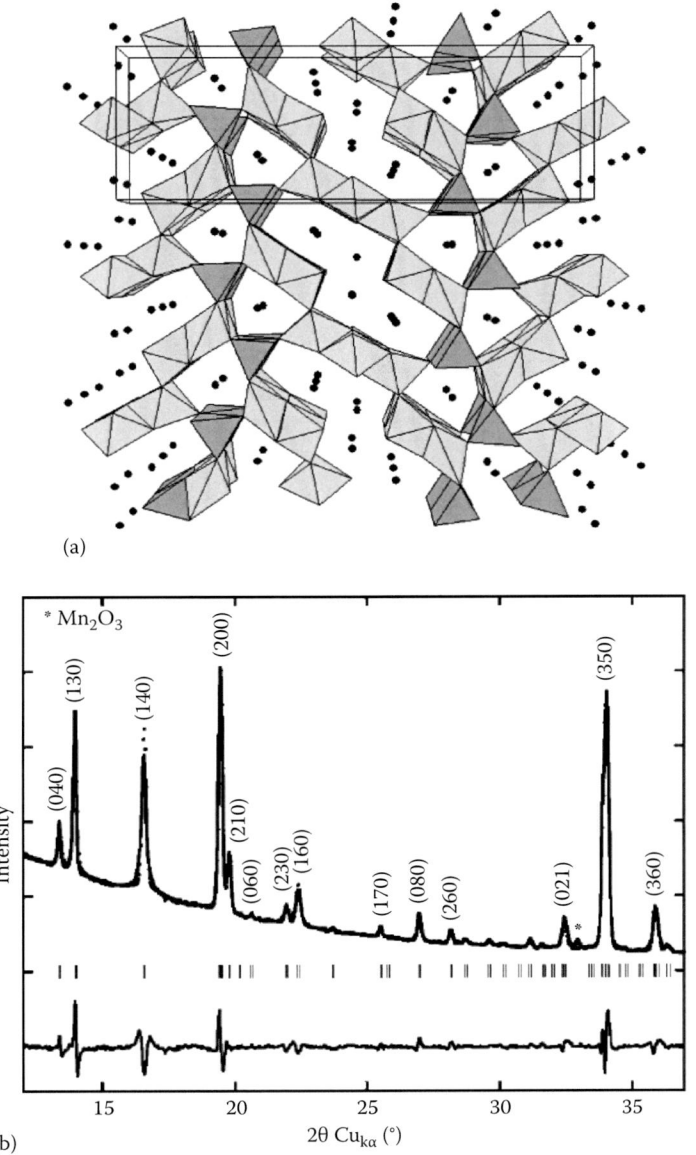

(a)

(b)

FIGURE 2.30   (a) $Na_{0.44}MnO_2$ is isostructural while $Na_4Mn_4Ti_5O_{18}$ crystallizes in an orthorhombic lattice (*Pbam* space group). The figure shows this structure looking down the *c*-axis. MnO_6 octahedra (yellow) and MnO_5 square pyramids (gray) link to form tunnels in which Na^+ ions (black spheres) are located. (b) XRD patterns of $Na_{0.44}MnO_2$—experimental and calculated. High-temperature solid state reaction was used to synthesize the material. (From Saint, J.A. et al., *Chem. Mater.*, 20, 3404, 2008.)

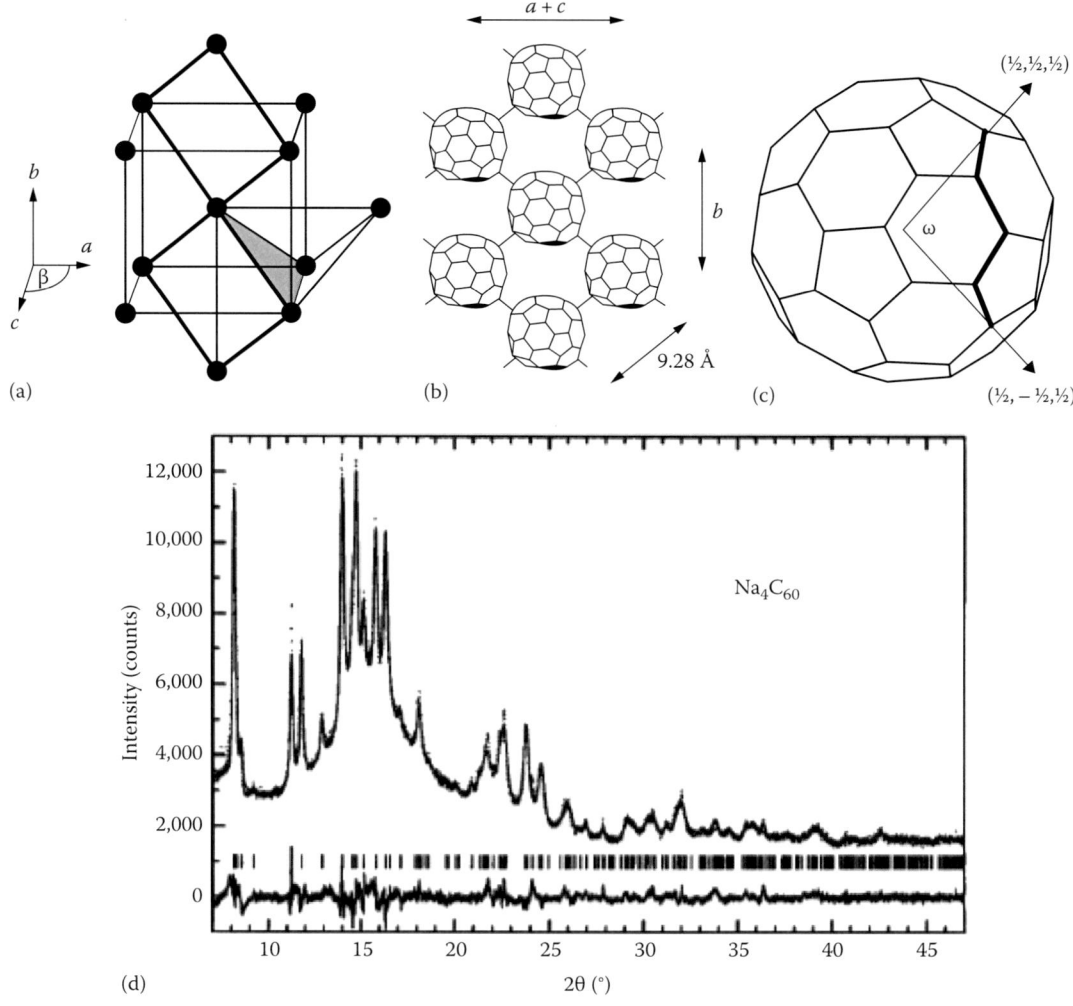

FIGURE 2.31 Structure of Na-intercalated fullerene, Na_4C_{60}. (a) Body-centered monoclinic unit cell: balls are fullerene molecules, bold lines are polymer planes, (b) polymer plane of C60 showing the correct orientation and approximate distortion of the molecule. Orientational disorder in $I2/m$ is described by a twofold rotation around b, (c) bonding atoms of the undistorted C60 molecule lying at ends of a zigzag line of three 56 and one 66 bonds. The atoms close an $\omega = 78.8°$ central angle which is within $0.5°$ the intraplanar angle determined by the lattice. (d) X-ray diffraction pattern of Na_4C_{60}. X rays of wavelength 1.14893 Å were selected by a double Si (111) mono chromator, and the diffracted beam was analyzed by a reflection from a Ge (111) crystal before the NaI scintillation detector. +: Observed data points, solid lines: calculated and difference plots, ticks: allowed reflection positions. (After Oszlanyi, G. et al., *Phys. Rev. Lett.*, 78, 4438, 1997.)

2.4.5 Halogen Family

Among the four elements of the halogen family, F, Cl, Br, and I, only F possesses, as in PbF_2 and allied fluorides notably CaF_2, SrF_2, and BaF_2, superionic conductivity at temperatures greater than 700 K. $SrCl_2$ also qualifies for a halide ion conductor (Cl^-). All these compounds crystallize in the fluorite structure shown in Figure 2.18. LaF_3 possesses tysonite structure with three sublattices for F^- ion.

In PbF_2, a new kind of phase transition was discovered by Faraday in which the highly ion-conducting state is quite unlike in AgI. It appears to be continuously spread over a temperature range of $\sim >100$ K.

Figure 2.36 shows the thermodynamic and ohmic nature of the Faraday transition in PbF_2. A detailed discussion of this transition appears in Chapter 5.

TABLE 2.4

Sodium Family

Na metal	bcc	$a = 0.423$ nm	
NaFeSO$_4$F	P21/c	$a = 0.66739$, $b = 0.86989$, $c = 0.71869$ nm, $\beta = 113.5250$, $V = 0.382567$ nm^3	Tripathi et al. [2.73]
FeSO$_4$F	(framework) C2/c	0.73637, 0.70753, 0.73117, 119.758, 0.328017 nm^3	
α-NaMnO$_2$	O$_3$ JT distorted monoclinic C2/m	$a = 0.5662$, $b = 0.2860$, $c = 0.5799$ nm, $\beta = 113.10$	Jansen and Hoppe [2.98]
Na$_3$Zr$_2$Si$_2$PO$_{12}$	(NASICON) C2/c	$a = 1.5616$ nm, $b = 0.9030$ nm, $c = 0.9211$ nm, $\beta = 123.850$, $V = 1.0789$ nm^3	
Na$_3$V$_2$(PO$_4$)$_3$	C2/c	$a = 1.5112$, $b = 0.8727$, $c = 0.8824$ nm, $\beta = 124.540$, $V = 0.958$	Lalere et al. [2.99]
Na$_2$Ti$_3$O$_7$	Monoclinic, P21/m	$a = 0.8571$, $b = 0.3804$, $c = 0.9135$ nm, $\beta = 101.570$	Andresson and Wadsley [2.100]
Na$_{0.44}$MnO$_2$	Orthorhombic, *Pbam*	$a = 0.9078$, $b = 2.644$, $c = 0.2827$ nm	Sauvage et al. [2.101]
Na$_{0.7}$CoO$_2$	Hexagonal *P63/mmc*	$a = 0.2833$ nm, $c = 1.0880$ nm	Bhide and Hariharan [2.102]
β-NaMnO$_2$	Double stacked sheet structure		
Na-β-alumina, Na$_2$O$_{0.11}$Al$_2$O$_3$	*P63/mmc*	$a = 0.559$, $c = 2.253$ nm	
Na-β″ alumina, 3Na$_2$O$_{0.16}$Al$_2$O$_3$	R3m (rhombohedral)	$a = 0.559$, $c = 3.39$ nm	Brook [2.103]
Na$_2$Co$_2$TeO$_6$	Pink	$a = 0.52727$, $c = 1.12301$, $c/a = 2.130$, $V = 0.2704$ nm^3	
Na$_2$Co$_2$TeO$_6$	Light pastel	0.52889, 1.12149, 2.120, 0.2717	
Na$_2$Ni$_2$TeO$_6$		0.52074, 1.11558, 0.26198	
Na$_2$Zn$_2$TeO$_6$		0.52796, 11.12941, 0.27264	
Na$_3$V(PO$_4$)$_3$N	(298 K) cubic P213	$a = 0.944783$ nm, $Z = 4$, $V = 0.843$ nm^3	Kim and Kim [2.104]
Na$_{0.7}$FePO$_4$	*Pnma*	$a = 1.02886$, $b = 0.60822$, $c = 0.49372$ nm $V = 0.30895$ nm^3	Moreau et al. [2.105]
NaFePO$_4$	*Pnma*	$a = 1.04063$, $b = 0.62187$, $c = 0.49469$ nm, $V = 0.32014$ nm^3	
Na$_{0.5}$Bi$_{0.5}$TiO$_3$	R3c (*RT*)	$aH = 0.54887$ nm, $cH = 1.35048$ nm, $V = 0.35233$ nm^3 with antiphase a-a-a- oxygen tilt system $\omega = 8.240$ with parallel cation displacements	Jones and Thomas [2.106]
Na$_4$C$_{60}$	Body-centered monoclinic (*I2/m*)	$a = 1.1235$ nm, $b = 1.1719$ nm, $c = 1.0276$ nm, $\beta = 96.160$	Oszlanyi et al. [2.107]

TABLE 2.5

Noble Metal Family

Au	Cubic(fcc)	0.408 nm	
AuI	Tetragonal	$P42/ncm$, $a = b = 0.435$, $c = 1.373$ nm, 8.25 g/cm³	Jagodzinski [2.108] Teicher and Weil [2.109]
Ag	Cubic (fcc)	0.409 nm	
AgF	fcc	0.4936 nm	
AgCl	fcc	0.55401	
AgBr	fcc	0.57745	
AgI	Zincblende	0.647 nm	
AgI	Wurtzite	$a = 0.4599$ nm, $c = 0.7520$ nm (296 K)	Shapiro and Reidinger [2.110]
Ag_2MI_4 (M = Cd, Zn, Pb, Hg)			
KAg_4I_5	$T = 584$ K, $P4132$, $Z = 4$	$a = 1.087195$ nm	
$RbAg_4I_5$	298 K, $P4132$, $Z = 4$	$a = 1.123934(3)$ nm, describe this SG also hexagonal P63mmc	
α-Ag_2S	Monoclinic, $P21/n$	$a = 0.423$, $b = 0.691$, $c = 0.787$ nm, $\beta = 99035'$	
Ag_2S	Tetragonal	$a = b = 0.690$, $c = 0.477$ nm	
β-Ag_2S	($T > 453$ K), bcc sulfur sublattice	$a = 0.488$ nm	
γ-Ag_2S	$T = 873$ K, fcc sulfur sublattice	$a = 0.6267$ nm	
Ag_2Se			
α-Ag_2Se	Orthorhombic $P212121$	$a = 0.4333$, $b = 0.7062$, $c = 0.7764$ nm	
β-Ag_2Se	$T = 443$ K, Cu_2O type $Im3m$ bcc anionic lattice	$a = 0.499$ nm	
Ag_2Te			
B phase	<423 K, monoclinic, $P21/c$	$a = 0.8185$ nm, $b = 0.8934$ nm, $c = 0.8418$ nm, and $\beta = 113.70$	Samal and Pradeep [2.111]
α phase	423–1075 K, fcc $Fm\text{-}3m$	$a = 0.657$ nm (8 Ag⁺ ions in a unit cell have 24 Ag⁺ sites available for hopping)	Chandra [2.112]
γ phase	1075–1223 K bcc $Im\text{-}3m$		
Ag_2CdI_4			
ε-phase	Tetragonal($I\bar{4}$, $I\bar{4}2m$)	$a = 0.635$ nm, $c = 1.270$ nm	Velgosh et al. [2.113]
β-phase	Hexagonal	$a = 0.4578$ nm, $c = 0.7529$ nm	Brightwell et al. [2.114]
Ag_4PbI_6	($T = 1900°C$) cubic(fcc)	$a = 0.6335$ nm	
Ag_2HgI_4	Hexagonal	$a = 0.629$ nm, $c = 1.255$ nm	Sukeshini and Hariharan [2.115]

(Continued)

TABLE 2.5 (*Continued*)

Noble Metal Family

Au	Cubic(fcc)	0.408 nm	
Ag_3SBr			
γ-phase	Orthorhombic (*Cmcm*) (118 K)	$a \sim b = 0.9423$ nm, $c = 0.9711$ nm	Sakuma and Hoshino [2.116]
$Ag_3SBr_{0.5}I_{0.5}$			
β-phase	(Room temperature) cubic	$a = 0.4842$ nm	Sakuma et al. [2.117]
Γ-phase	(99 K) Orthorhombic	$a \sim b = c = 0.9637$ nm	
Ag_3SI			
α-phase	($T > 519$ K)Cubic $Im3m$		
β-phase	($519 < T < 157$ K) $Pm\bar{3}m$		
γ-phase	($T < 157$ K) Rhombohedral $R3$		
$Ag_4Sn_3S_8$	$P4132$	with $a = 1.080898$ nm and $Z = 4$	Hull et al. [2.118]
$Ag_7Fe_3(P_2O_7)_4$	Monoclinic $P21/c$	$a = 0.9532$, $b = 0.8421$, $c = 2.8021$ nm, $\beta = 93.2250$, $V = 2.245$ nm^3	
$Ag_7Fe_3(As_2O_7)_4$	Monoclinic $C2/c$	$a = 0.95561$, $b = 0.84417$, $c = 2.8226$ nm, $\beta = 93.4650$, $V = 2.272$ nm^3	Quarez et al. [2.119]
Cu	Cubic fcc	$a = 0.361$ nm	
$CuCl$	Zincblende	0.541 nm	
$CuBr$	Zincblende	0.569	
CuI	Zincblende	0.604	
Cu_2S			
γ-Cu_2S	Orthorhombic $Ab2m$ 96 formula units(1)	$a = 1.11881$, $b = 2.7323$, $c = 1.3491$ nm	
β-Cu_2S	Hexagonal $P63/mmc$ 4 formula units, ($376.5 < T < 708$ K)	$a = 0.3961$ nm, $c = 0.6722$ nm	
α-Cu_2S	Cubic ($T > 708$ K) 4 formula units	$a = 0.5725$ nm	
Cu_2Se	fcc $T = 443$ K	$a = 0.585$nm	
Cu_2Se	$T = 383$ K, distorted cuprite type (tetragonal) $I4/mmm$	$a = 1.040$, $c = 0.393$ nm	
$Cu_{1.9}Se$	Cubic $Fd3m$	$a = 0.574$ nm	
KCu_4I_5	584 K, space group $P4\ 132$, $Z = 4$	$a = 1.123934$ nm	Hull et al. [2.120]
Cu_2HgI_4	Hexagonal	$a = 0.607$ nm, $c = 1.215$ nm	Sukeshini and Hariharan [2.115]
Cu_6PS_5Br	$F\bar{4}3m$ (*RT*)	$a = 0.9728$ nm density $= 4.68$ g/cm^3	Kuhs et al. [2.121]
Cu_7SiS_5I	$F\bar{4}3m$	$a = 0.99517$ nm	Studenyak et al. [2.122]
Cu_7GeS_5I	$F\bar{4}3m$	$a = 1.000483$ nm	Studenyak et al. [2.122]
$RbCu_4Cl_3I_2$	$P4332$	$a = 1.0032 + 0.03$ nm	Geller et al. [2.123]

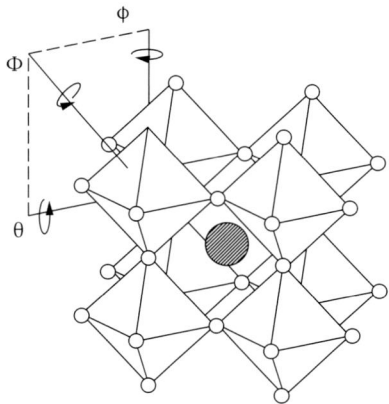

FIGURE 2.32   An ideal cubic perovskite structure with space group *Pm3m*. Distortions indicated by operations φ, θ, and Φ convert it to an orthorhombic perovskite with space group *Pbnm*.

TABLE 2.6

Properties Exhibited by the Perovskite Structure and Typical Compounds

Properties	Typical Compounds Exhibiting the Property
Ferroelectricity	$BaTiO_3$, $PbTiO_3$
Piezoelectricity	$Pb(Zr, Ti)O_3$, $(Bi, Na)TiO_3$
Electrical conductivity	ReO_3, $SrFeO_3$, $LaCoO_3$, $LaNiO_3$, $LaCrO_3$
Superconductivity	$La_{0.9}Sr_{0.1}CuO_3$, $YBa_2Cu_3O_7$, $HgBa_2Ca_2Cu_2O_8$
Ionic conductivity	$La(Ca)AlO_3$, $CaTiO_3$, $La(Sr)Ga(Mg)O_3$, $La(Sr)Ga(Mg)O_3$, $BaZrO_3$, $SrZrO_3$, $BaCeO_3$
Magnetic property	$LaMnO_3$, $LaFeO_3$, La_2NiMnO_6
Catalytic property	$LaCoO_3$, $LaMnO_3$, $BaCuO_3$
Electrode action	$La_{0.6}Sr_{0.4}CoO_3$, $La_{0.8}Ca_{0.2}MnO_3$

Source: Ishihara, T. (ed.), *Perovskite Oxides for Solid Oxide Fuel Cells*, Springer, New York, 2009.

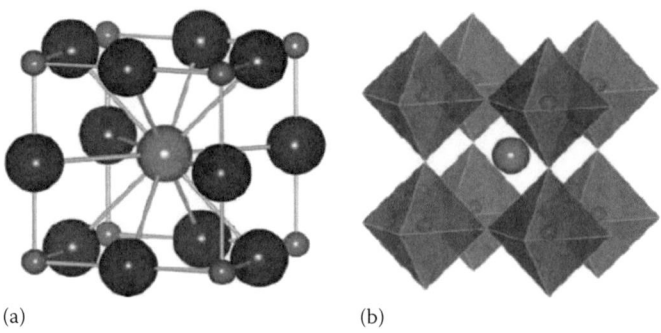

(a)   (b)

FIGURE 2.33 $LaGaO_3$ (a) La-centered perovskite unit cell, (b) corner-shared GaO_6 octahedra with La centered on 12-coordinate sites. (From Malavasi et al., *CSR*, 39, 4370, 2010.)

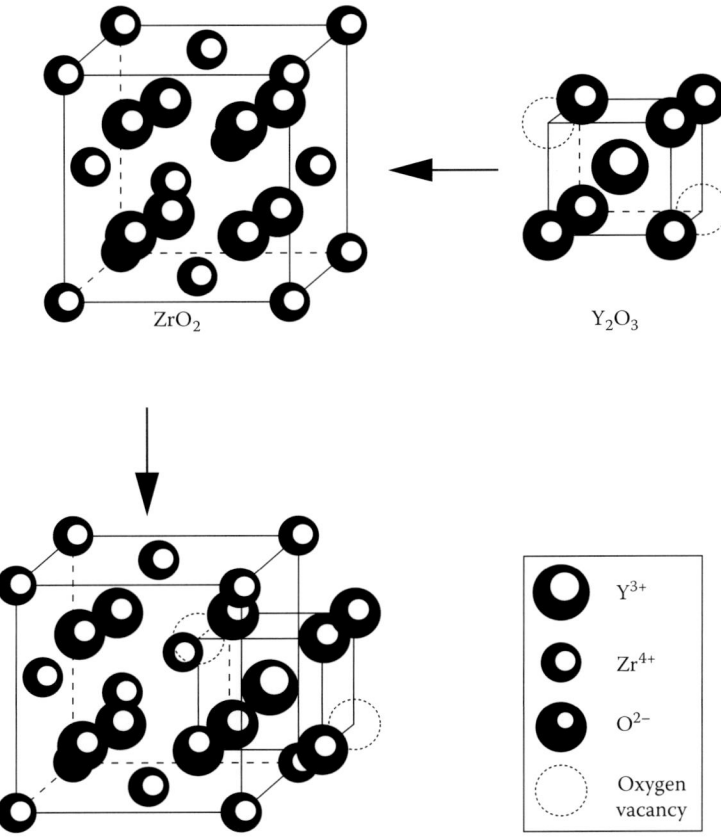

FIGURE 2.34   Structure of yttria (Y_2O_3)-stabilized zirconia (YSZ).

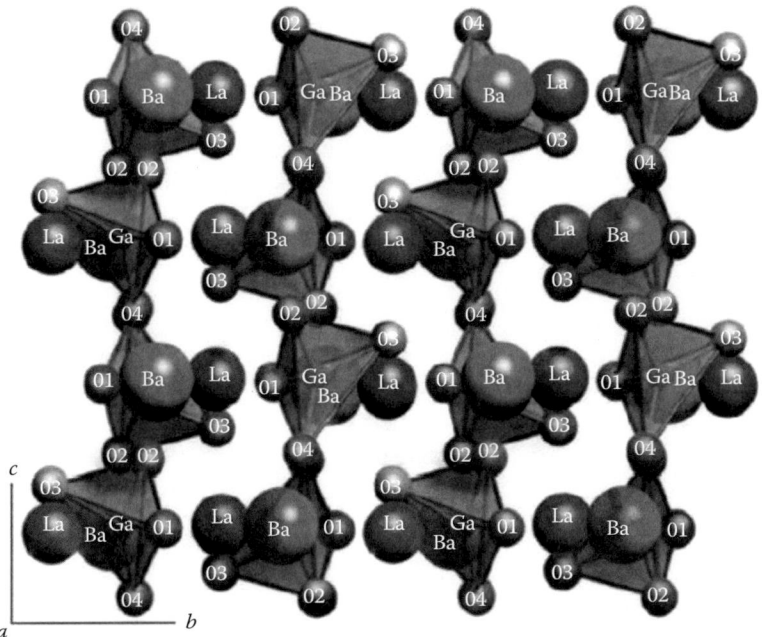

FIGURE 2.35   Structure of $LaBaGaO_4$ (β-K_2SO_4 –type). (From Kendrick, E. et al., *Nat. Mater.*, 6, 871, 2009.)

This important discovery of Faraday focused on the structure–property relationship, that is, the energy barrier for ion hopping in the fluorite structure is relatively small. This is relevant to oxide ion conductors discussed earlier especially doped zirconia.

Fluoride ion conductors meet several criteria characteristic of high mobility: low entropy of fusion, vacancies in the anionic sublattice due to nonstoichiometry, high cation polarizability, and presence of cations having different oxidation states. Therefore, they deserve to be classified as an important class of superionic conductors (Table 2.8).

Just as PbF_2, SnF_2 is also an F^- ion conductor but at much lower temperatures, possibly due to the high polarizability of Sn^{++} and the weak coordination of F^- ion. SnF_2 is monoclinic at room temperature ((α-SnF_2) phase). It transforms to γ-SnF_2 (tetragonal structure) over 413–453 K which is stable up to

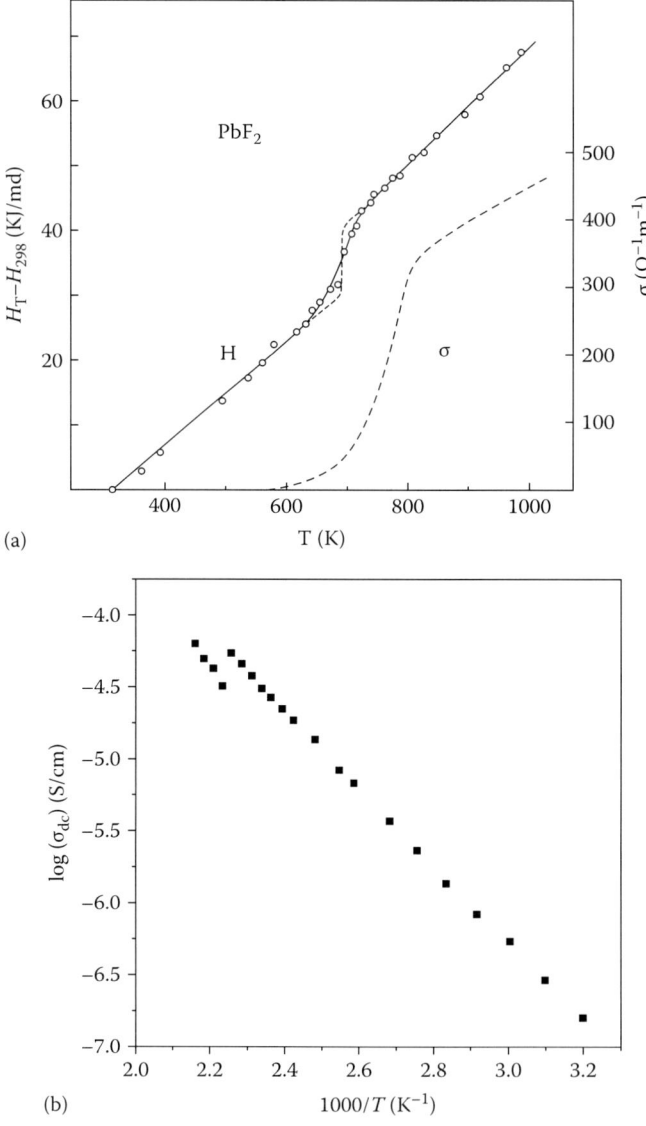

(a)

(b)

FIGURE 2.36 (a) Depicting the variations in the heat content and ionic conductivity across the "Faraday transition" in PbF_2. (Adapted from Funke, K., *Sci. Technol. Adv. Mater.*, 4, 043502, 2013.) (b) Arrhenius plot of DC conductivity of SnF_2 showing the α-SnF_2 to the high-conducting γ-SnF_2 phase at 443 K. Note the change in slope from 0.53 eV ($T < 4443$ K) to 0.76 eV ($T > 443$ K). (After Patro, L.N. and Hariharan, K., *Mater. Chem. Phys.*, 116, 81, 2009.)   (*Continued*)

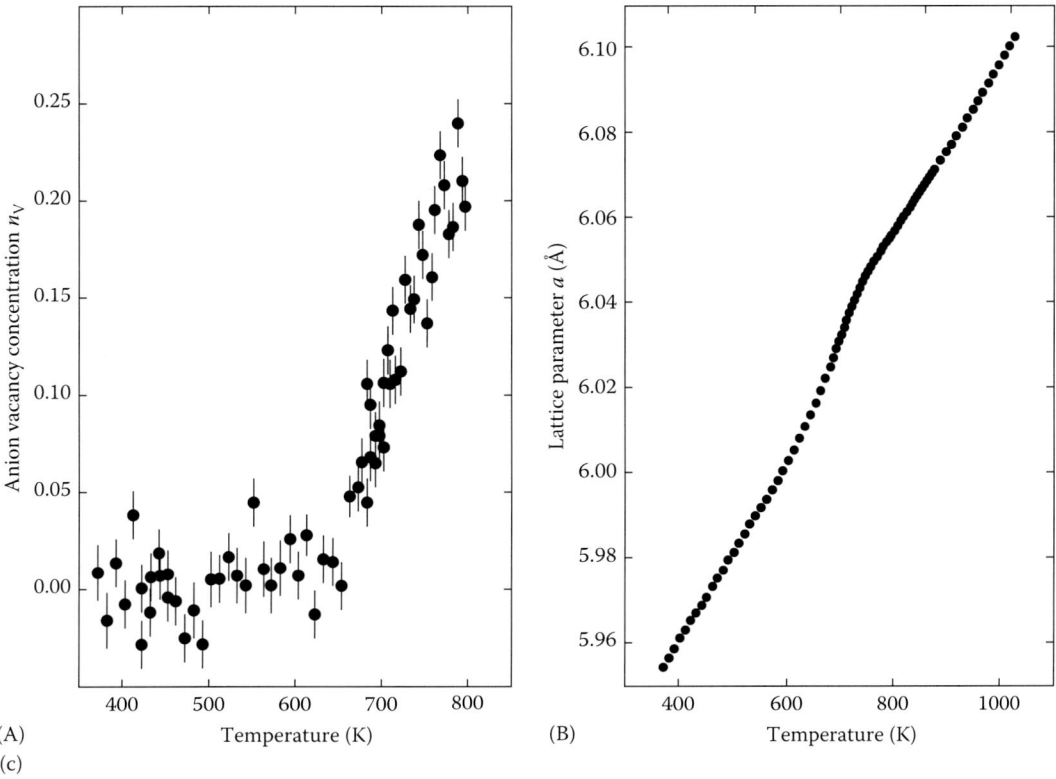

FIGURE 2.36 (Continued) (c) The temperature dependence of (A) anion vacancy concentration n_V and (B) lattice parameter of β-PbF_2. Note the similarity of (a) and c (B). (After Hull, S., *Rep. Prog. Phys.*, 67, 1233, 2004.)

the melting temperature (488 K). However, upon cooling, γ-SnF_2 transforms to β-SnF_2 (orthorhombic structure) through a second-order phase transition [2.60].

We conclude this section by providing a collection of Arrhenius plots of ionic conductivity several ternary fluorides and also LuF_3 in Figure 2.37.

Besides the mechanism of phase transition, the origin of high F^- conductivity in terms of structural elements in the F^- sublattice is a main question in F^- conductors.

We shall discuss nanoscale fluorides in Chapter 10.

2.4.6 Proton Family

Proton is a unique "ion." It is the cation of "protium," the most common isotope of hydrogen. Intermediate in size between an electron and Li^+, it is a "nonconformist ion." Proton family of SSIMs is, therefore, a special one. It is involved in the time-honored lead–acid battery and the contemporary fuel cell. Proton mobility in water is abnormally high, and thus, the limiting conductivity of water at room temperature is about seven times larger than Na^+ conduction. The thermal stability of proton-conducting materials is directly connected to the nature of the proton species size, diffusion rate within the host framework pathway, and the strength of the H bond [2.63]. Proton conduction is a fundamental problem involving hydrogen bonding and long-range proton transport, in which a two-step reaction mechanism, reorientation and transfer, is involved. Proton transfer mechanism in water is the paradigm. Water is a liquid with solid-like properties as far as proton transport takes place! But if very high proton conductivity is the solid state needs solids with liquid-like properties!! Proton charge follows the center of symmetry of the hydrogen-bond pattern. Reorientation is bond-breaking and forming or making a weakly bound outer part of the complex. Proton transfer occurs in the central part of the complex. The two processes are

TABLE 2.7

Oxygen Family

Eight YSZ	(Yttria-Stabilized Zirconia)		
	Monoclinic ($P21/c$)	$a = 0.51507$ nm, $b = 0.52028$ nm, $c = 0.53156$ nm, $\beta = 99.1960$	
	Cubic ($Fm3m$)	$a = 0.5128$ nm	ICDD data cited in Dercz et al. [2.124]
CeO_2	0.541 nm (fluorite)		
UO_2	0.547 nm		
$SrCo_{0.95}Mo_{0.05}O_{3-\delta}$	$P4/mmm$ 3D perovskite	$a = 0.385516$, $c = 0.774204$ nm, $V = 0.11506$ nm^3	Ahuadero et al. [2.125]
$LaGaO_3$	Orthothombic (room temp.), $Pbnm$	$a = 0.55245$, $b = 0.54922$, $c = 0.77740$ nm	Howard and Kennedy [2.126]
$La_{0.9}Sr_{0.1}Ga_{0.8}Mg_{0.2}O_{2.85}$		$a = 0.7816$ nm, $b = 0.5539$ nm, $c = 0.5515$ nm, $\beta = 90.060$, $V = 0.3581$ nm^3	Slater et al. [2.127]
$La_{0.9}Sr_{0.1}Ga_{0.8}Mn_{0.2}O_{2.92}$		$a = 0.5518$ nm, 1.3335 nm, $V = 0.3516$ nm^3	Thangadurai et al. [2.128]
$La_{0.8}Sr_{0.2}Ga_{0.8}Mg_{0.2}O_{2.8}$	Orthorhombic (Imma)	$a = 0.54976$, $b = 0.77922$, $c = 0.55333$ nm	Biswal and Biswas [2.129]
β-$La_2Mo_2O_9$	Cubic (890 K), $P213$	$a = 0.72014$ nm; $Z = 2$	Goutenoire et al. [2.130]
$Ce_{0.8}Gd_{0.2}O_{1.9}$		$a = 0.54317$nm	Cheng et al. [2.131]
$Ce_{1-x}Sm_xO_{2-\delta}$			
$x = 0$		$a = 0.54020$ nm	
$x = 0.1$		$a = 0.54273$	
$x = 0.2$		$a = 0.54305$	
$x = 0.3$		$a = 0.54390$	Wang et al. [2.132]
$La_{10}W_2O_{21}$	Cubic, $F\bar{4}3m$, $Z = 4$	$a = 1.11793$ nm	Chambrier et al. [2.133]
Apatite-type			
$M_{10}(XO_4)_6O_{2+y}$	(M = rare earth or alkaline earth; X = p-block element P, Si, Ge; y = extent of nonstoichiometry)		
SG	$P63/m$	a ~0.97–0.99 nm, c ~0.7 nm	
Si-apatite			
$La_{10}(SiO_4)_6O_3$, similar to $La_{4.67}(SiO_4)_3O$	$P63/m$ hexagonal	$a = b = 0.955$ nm, $c = 0.74$ nm	Nakayama et al. [2.134]
Ge apatite			
$La_{9.60}(GeO_4)_6O_{2.4}$	$P63/m$ hexagonal	$a = 0.99374$ nm, $c = 0.72835$ nm, $V = 0.62290$ nm^3	Leon-Reina et al. [2.135]
$La_{1.54}Sr_{0.46}Ga_3O_{7.27}$	$P421m$, tetragonal	$a_0 = 0.8056$, $c_0 = 0.5333$ nm	Liu et al. [2.136]
$La_{1.64}Ca_{0.36}Ga_3O_{7.32}$	$Cmm2$	a ~ 1.1416, b ~ 1.1226, c ~ 0.5248 nm	[2.137]
$Ca_{12}Al_{14}O_{33}$	$I4-3d$, cubic	$a = 1.199$ nm	Bartl et al. [2.138]
$(Bi_2O_3)_{0.75}(Gd_2O_3)_{0.25}$	Cubic ($Fm3m$)	$a = 0.5512$ nm (123 K), 0.5521 nm (292 K)	Ito et al. [2.139]
γ-$Bi_4V_2O_{11}$	$T = 885$ K, $I4/mmm$, tetragonal	$a = 0.3988$, $c = 1.542$ nm	Abraham et al. [2.140]
Bi_3NbO_7	$Fm-3m$, cubic	$a = 0.54788$ nm	Castro et al. [2.141]
$Na_{0.5}Bi_{0.5}TiO_3$	$R3c$ (RT)	$a_H = 0.54887$ nm, $c_H = 1.35048$ nm, $V = 0.35233$ nm^3 with antiphase a–a–a– oxygen tilt system $\omega = 8.240$ with parallel cation displacements	Jones and Thomas [2.106]

TABLE 2.8

Halogen Family

		a (nm)	$T\sigma_{Saturation}$ (K)[a]	$T_{melting}$ (K)	Conductivity (S/cm)	Transition Temp. (K)/ Temperature Range of Superionic Phase	
PbF_2	Fluorite	0.5940[b]	703	1095	5	~873	
CaF_2	Fluorite	0.5451[c]	1423	1691	0.0001–45	713–1373	
SrF_2	Fluorite	0.578[c]	1453	1673	0.00014–1.4	853–1253	
BaF_2	Fluorite	0.6184[c]	1233	1593	0.0001–25	703–1353	
$SrCl_2$	Fluorite	0.69744[d]	993	1148	0.2	1000	
LaF_3	Tysonite	$a = 0.7186,^e c = 0.7352$–1766	0.003–1.2	403–943		$T = 288$ K	
$PbSnF_4^f$	(300 K) $P4/nmm$	$a = 0.42262$ nm, $c = 1.14407$ nm					Kanno et al. [2.142]
α-SnF_2	Monoclinic	$a = 1.33520$ nm, $b = 0.49099$ nm, $c = 1.37888$ nm, $\beta = 1090$, $V = 0.854$ nm^3				$T_t = 413$–453 K (tetragonal) $T_m = 488$ K	Denes et al. [2.143]

a   Temperature at which conductivity saturates/.
b   link.springer.com/content/pdf/10.1007%2F10681727_905.pdf.
c   http://ocw.mit.edu/courses/materials-science-and-engineering/3-014-materials-laboratory-fall-2006/labs/w3_a1.pdf.
d   Wyckoff [2.144].
e   Korczak and Mikolajczak [2.145].
f   Neutron diffraction/.

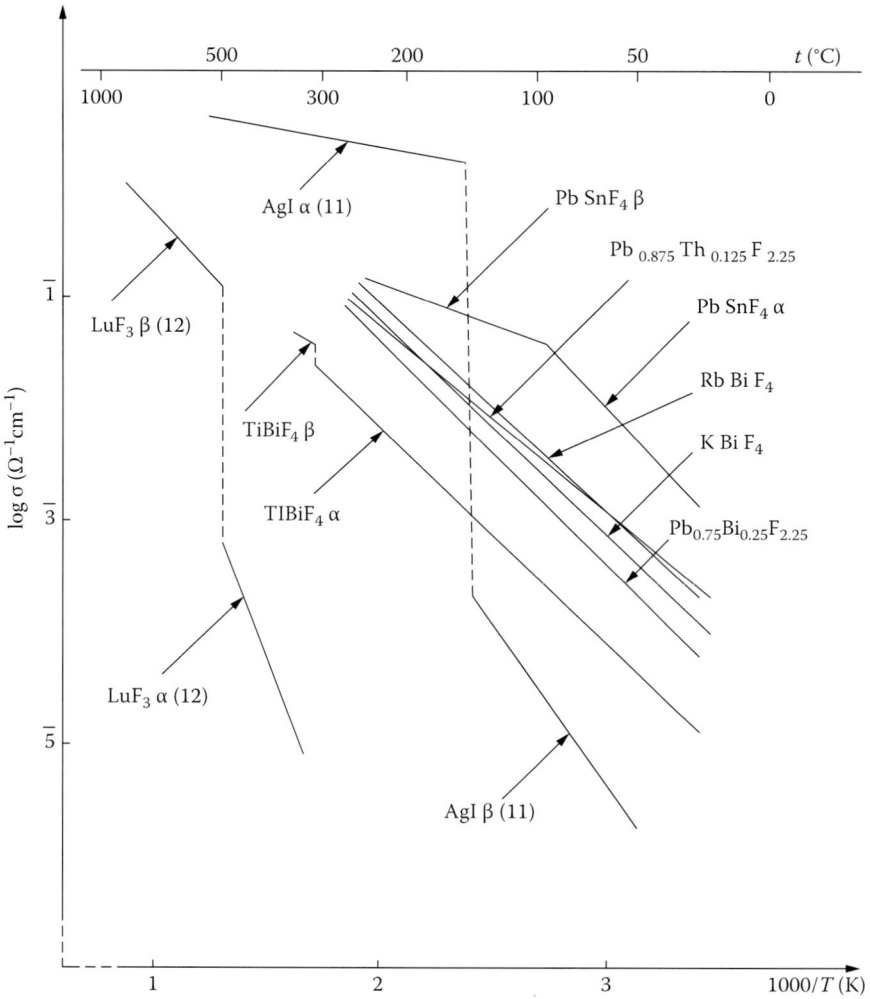

FIGURE 2.37 Log σ versus 1000/T (K) plots for LuF₃ and several ternary fluorides: KBiF₄, RbBiF₄, α- and β-TlBiF₄, α- and β-PbSnF₄, Pb₀.₇₅Bi₀.₂₅F₂.₂₅, Pb₀.₈₇₅Th₀.₁₂₅F₂.₂₅, and α- and β-LuF₃. The plot of α- and β-AgI is shown for comparison. (From Lucat, C. et al., *J. Solid State Chem.*, 29, 373, 1979.)

strongly coupled. The process is discussed in terms of the so-called Zundel cation $H_5O_2^+$ which is formed from a proton and two water molecules, the Eigen cation $H_9O_4^+$ and a hydronium ion (H_3O^+), besides three water molecules. According to the Grotthuss mechanism in which "hydron hopping" is visualized, these three ion species play a vital role. The transitions from zundel cation to eigen cation and back to zundel cation take ~10^{-13} s to occur [2.64].

Proton conductors contain H_2O and decompose readily on heating. The proton family includes

1. Hydrates such as hydrogen uranyl phosphate (HUP), $HUO_2PO_4.4H_2O$ with a conductivity of 40 mS/cm at 298 K (crystal structure is illustrated in Figure 2.39)

2. Hydronium β-alumina, $H_3O^+Al_{11}O_{17}$, usually made by ion exchange $Na^+\rightarrow\leftarrow H_3O^+$ from Na β-alumina

3. Materials that ionize protons easily: $Sb_2O_5.nH_2O$

4. Nonstoichiometric $BaCeO_{3-\delta}$—with mixed valence Ce^{3+} and Ce^{4+}—a perovskite that absorbs a small amount of water and is strongly bonded. H2O molecules dissociate and protons are attracted in the region of Ce^{3+}; possesses conductivity ~1 mS/cm at 873 K [2.63].

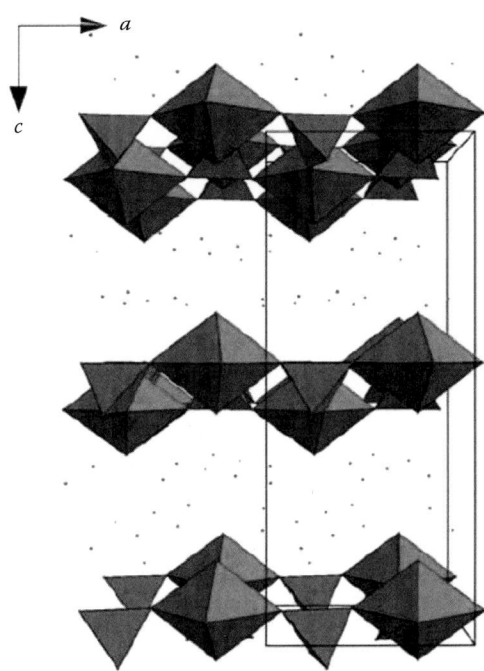

FIGURE 2.38 Crystal structure of H UO_2 PO_4 $4H_2O$ (HUP) viewed along the *b*-axis. (From Morosin, B.B., *Acta Cryst.*, 34, 3732, 1978.) Infinite sheets of UO_2PO_4 are separated by a two-level water molecule layer shown as dots. O atoms of this layer form squares and are hydrogen bonded. One of the four water molecules is replaced by a H_3O^+ ion which could be ion exchanged with Na^+, Ag^+ among others to yield good ionic conductors.

TABLE 2.9

Proton Family

$HSbO_3$ $0.5H_2O$	*Fd-3m* (neutron diffraction $\lambda = 0.15480$ nm)	$a = 1.0362$ nm	Slade et al. [2.146]
HUO_2PO_4 $4H_2O$	*P4/ncc* tetragonal	$a = 0.6995$, $c = 1.7491$ nm, $Z = 4$, $V = 0.85584$ nm³	Morosin [2.147]
$Rb_3H(SeO_4)_2$	$T = 455$ K *R-3m* hexagonal	$a = b = 0.61181$ nm, $c = 2.2674$ nm	Magome et al. [2.148]
$BaCeO_3$	*Pmcn*	$a = 0.8791$, $b = 0.6252$, $c = 0.6227$ nm Solid state reaction 1400°C–1500°C	Melekh et al. [2.149]
$BaCeO_3$ (neutron diffraction)		$a = 0.8780$, $b = 0.5232$, $c = 0.6218$ nm (annealed under 100% O_2)	
$BaCe_{0.85}Y_{0.15}O_{2.975}$ (ditto)	*R-3c*	0.6221, 1.5179, ditto	
$BaCe_{0.85}Y_{0.15}O_{2.975}$	*I2/m*	0.6237, 0.8758, 0.6220, $\beta = 90.70°$ (annealed under 4% $H_2 + N_2$)	
$BaCe_{0.75}Y_{0.25}O_{2.875}$	*I2/m*	0.6243, 0.8723, 0.6239, $\beta = 91.34°$ ditto	Tekeuchi et al. [2.150]
$La_{0.8}Ba_{1.2}GaO_{3.9}$	Orthorhombic $P2_12_12_1$	$a = 1.00665$ nm, $b = 0.73417$ nm, $c = 0.59433$ nm	Jalarvo et al. [2.151]

Representative proton conductors could be (1) amorphous xerogels prepared from hydrolysis-polycondensation reactions, (2) semicrystalline polyaniline polymers (PANI), and (3) "anhydrous" protonated perovskites. To these we could add (4) inorganic compounds such as KHF_2 and $KHSO_4$, and CsH_2PO_4 and (5) mineral acids such as H_2XO_4 (X = S, Se), $HClO_4$, H_5IO_6, and carboxylic acids such as formic acid and acetic acid [2.66]. Table 2.9 lists proton conductors.

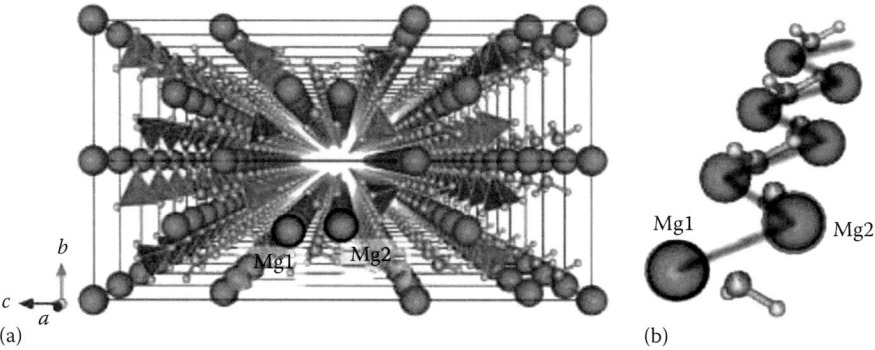

(a)   (b)

FIGURE 2.39   (a) Crystal structure of $Mg(BH_4)(NH_2)$ obtained by density functional theory. Atomic sizes are depicted by sphere radii. (b) Mg zigzag structure from (a). (From Higashi, S. et al., *Chem. Commun.*, 50, 1320, 2014.)

2.4.7 Mg Family

A recent addition to the pantheon of SSIMs is the small group of Mg-ion conductors. Mg^{++} has a radius of 0.078 nm, the same as that of Li^+ and much less than that of F^-. Magnesium metal has many useful properties, including high volumetric capacity and negative reduction potential. Unlike Na and Li, which react vigorously with water, Mg when in contact with water is stable at room temperature. Its natural abundance makes Mg metal anodes promising future energy storage systems. Besides high energy density and rate capability, safety, a most important aspect of a commercial battery system, is the demand that Mg batteries have the potential to meet.

Thus, we shall include a short discussion on Mg-ion conductors. $Mg(BH_4)(NH_2)$ material exhibited a high ionic conductivity of 10 μS/cm at 150°C, and its electrochemical window was estimated to be approximately 3 V using cyclic voltammetry. The theoretical crystal structure of this Mg^{++} conductors is shown in Figure 2.39 [2.68].

Next, we will consider phase transitions as a paradigm of solid state ionics.

2.5 Phase Transition as a Paradigm of Solid State Ionics

An SSIM is a unique phase of matter: solid to look at but fluid-like in macroscopic ion transport behavior. The electron density map referred to earlier indicates that the crystal structure adopted by the mobile ion sublattice does not have fixed locations for the mobile ions at all! Location of such ions in the usual way becomes very difficult. The question naturally arises: How does such a state come about? The fluid state of alkali halides such as NaI is attained through a solid–liquid phase transition at a well-defined melting temperature. In a gel type of soft solid electrolyte, the gel state is attained from a colloidal liquid state at a gel temperature T_{gel} such that $T \geq T_{gel}$. Thus, a phase transition is apparently the norm for the very existence of a solid state phase of matter.

It is at once a dynamic phase as also perhaps a quasi-equilibrium phase. Most fundamentally, it is a dissipative system.

What is a paradigm? It is a typical example of something, say, a model. Synonymous with a pattern, it could be a standard template, prototype, or archetype.

Fast ion conduction is itself a paradigm for structure–property relationship, the property being ionic conductivity σ: $\sigma = nq\mu$. The carrier mobility μ is the critical factor that determines conductivity. Note that this formula is valid for electron and electron–hole conductivity in a metal and a semiconductor, respectively, besides being applicable to mixed ionic–electronic conductors. μ depends on the geometry of the anion array in a conductor where the mobile ions are positive such as Ag^+ and the structure of the solid.

What factors are intrinsic to the structural framework? These include but are not limited to the following:

1. "Free" volume (or excluded volume)
2. "Bottleneck" size (limits free ion passage)
3. Lattice disorder

Phase stability is essentially dynamic. It is electrochemical in fact, that too in a device in operation. Phase transition is usually the route to the useable SSIM. So the three words "diagram," stability," and "transition" are intricately connected. To produce a "paradigm shift" in solid state ionics. So phase transition itself becomes a paradigm. We shall encounter phase transitions often and the entire Chapter 5 is devoted to the theory of phase transitions.

2.6 Summary

In this chapter, the thermodynamic aspects of SSIMs have been discussed with a focus on phase diagrams and phase stability. This is followed by a classification of SSIMs families based on mobile ion. Important crystal structures are discussed, following up on Chapter 1. Finally, the phase transitions in these materials are briefly discussed as a paradigm of SSI. A detailed discussion is found in the following chapters, including Chapter 5 where phenomenology is discussed.

Problems

2.1 In a typical binary system $(A_2O)_x-(BO_2)_{1-x}$, where A could be an alkali atom and B could be a glass former atom as one varies x, soon a point is reached where the mixture possesses the lowest melting point. This is the eutectic point. Look at the following phase diagram (From Balaya. P. and C.S. Sunandana, *J. Noncryst. Solids*, 162, 253, 1993.) and identify the eutectic temperature and composition. Also discuss the significance of other features of the diagram—the Li_2O-TeO_2 system.

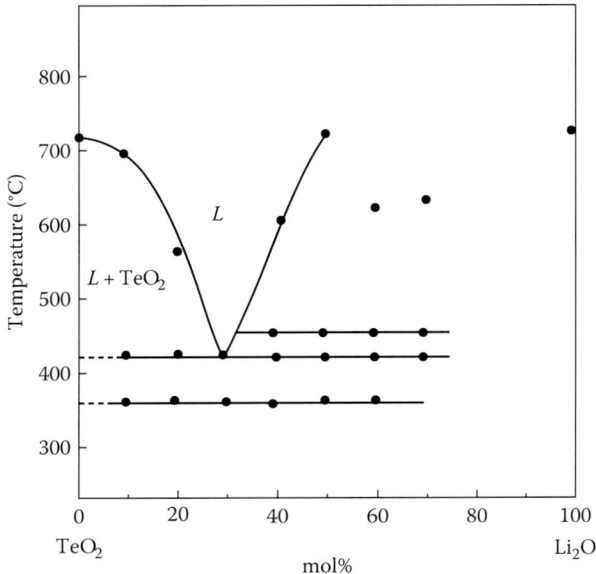

2.2 The phase diagram of the system CuBr–PbBr$_2$ shown later has many illustrative features of a typical pseudo-binary phase diagram. It was determined by Takahashi et al. [2.70] by DTA or differential thermal analysis technique. In this technique, a predecessor of the more sophisticated Differential scanning calorimetry (DSC) to be described in Chapter 4, 0.5 g of test sample was sealed under vacuum in a Vycor tube. A similar tube holds alfa alumina as the standard sample. The two samples placed side by side in a controlled furnace were heated at 2°C/min and transition temperatures were measured and plotted versus mol% lead bromide. From the diagram (From Takahashi, T. et al., *J. Solid State Chem.*, 21, 37, 1977.), determine the eutectic composition and temperature. What do the horizontal and vertical lines indicate?

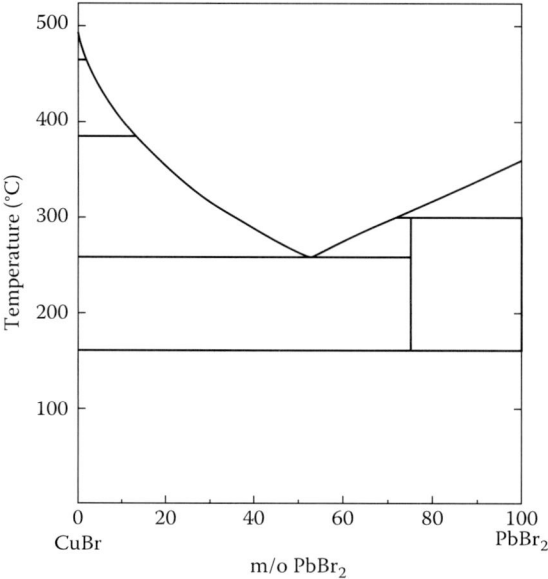

2.3 Shown here is the phase diagram of the Li$_3$PO$_4$–LiPO$_3$ system. (From Osterfeld R.K., *J. Inorg. Nucl. Chem.*, 30, 3174, 1968.) Li$_4$P$_2$O$_7$ melts incongruently while Li$_3$PO$_4$ phase persists on cooling from 1200°C [2.71].

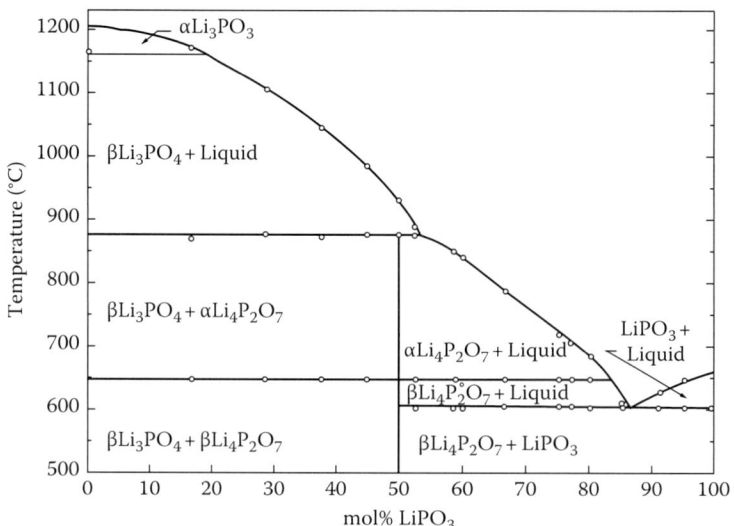

Based on the diagram, suggest a way of synthesizing α-Li_3PO_4 and $LiPO_3$.

2.4 The following is the phase diagram of system $FePO_4$–$LiFePO_4$. (From Dodd, J.L. et al. *Electrochem. Solid State Lett.*, 9, A151, 2006.)

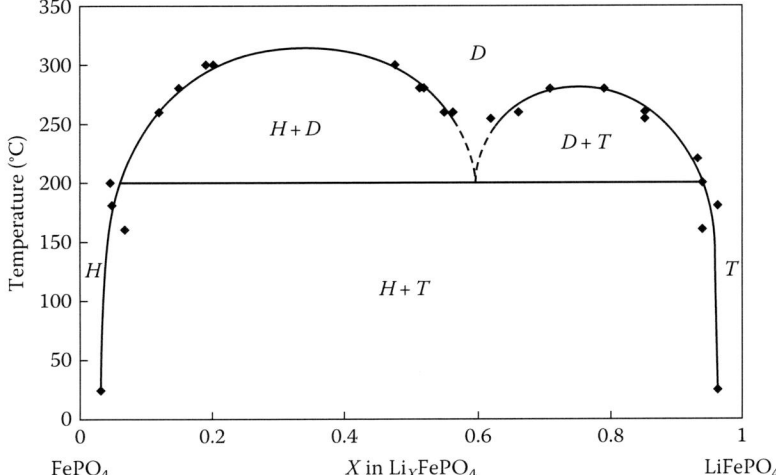

Here *H* means heterosite, *D* means disorder, and *T* means tryphlite—the three structural varieties of $LiFePO_4$. What can you say about these phases by looking at the phase diagram?

2.5 Phase diagrams often help discover new fast ion conductors. An example is system $CuBr$–$PbBr_2$. Using differential thermal analysis and x-ray diffraction analysis (to be described in Chapter 4), a new compound $CuPb_3Br_7$ has been found in this system. This compound is stable between its incongruent melting point (300°C) and 160°C, which below 160°C disproportionates to $CuBr$ and $PbBr_2$. It has a relatively high ionic conductivity of 3×10^{-2} $(\Omega$ cm$)^{-1}$ at 200°C and a low activation energy for the conduction of 22 kJ/mol. The transport number measurement (see Chapter 4) by Tubandt's method reveals that copper ions are the only charge carriers. Discuss the thermodynamics of this system. Also focus on the phase stability of the new phase [2.70].

2.6 Display the binary phase diagram as $p(O_2)$–T–x diagram [2.20]. Consider a reaction involving thermal decomposition of a nonstoichiometric oxide. Introduce a reaction constant K. Use the Gibbs–Helmholtz equation $RT \ln K = \Delta G$. Derive a relation between $\ln p(O_2)$ and reciprocal temperature ($1/T$).

2.7 High-temperature crystal structures as well as metastable structures of superionic conductors are all either cubic or hexagonal. In fcc, bcc, and hcp structures, there arise as a result of close packing of atoms concerned octahedral and tetrahedral "holes" or interstitial spaces which can be occupied by mobile ions. The table here gives the coordinates of these holes for an fcc structure.

Octahedral holes	Tetrahedral holes
½00, 0½0, 1½0, 1½0, ½10,	¼¼¼, ¼¾¼, ¾¾¼, ¾¼¼,
00½, 01½, 11½, 10½,	¼¼¾, ¼¾¾, ¾¾¾, ¾¼¾ (8)
½01, 0½1, 1½1, ½11,	
½½½ (13)	

Using an fcc lattice sketch in 3D, locate these positions. Also identify the possible "pathways" through which ions can move through the structure.

(a) Sketch bcc structure and locate octahedral and tetrahedral hole coordinates. (Hint: 18 octa-hedral and 24 tetrahedral holes arise.)

(b) Sketch hcp structure and identify octahedral and tetrahedral holes. (Hint: 2 octahedral and 10 tetrahedral holes arise.)

Which of these structures are most favorable for fast ion conduction?

2.8 AgBr–CuBr phase diagram. Consider the rich phase diagram shown here essentially due to the superionic component AgBr. Note that AgBr does not have any phase transition right up to its melting point, although its ionic conductivity increases just below its melting point.

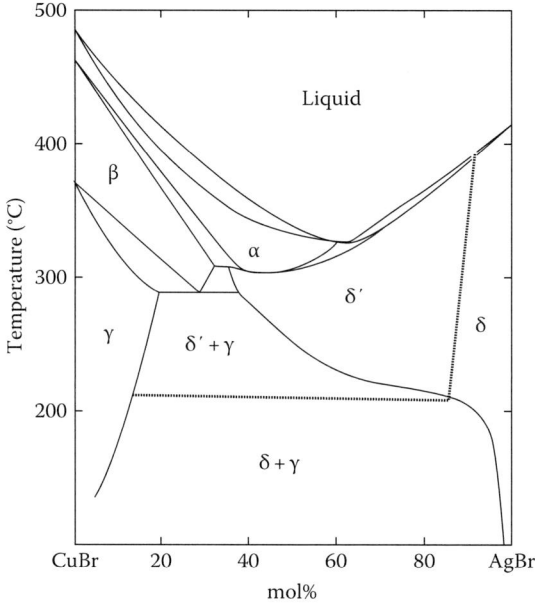

Classify the phases in this diagram into (1) single-phase and (2) two-phase systems. Discuss their thermodynamics and phase stability. (From Saito, M. and Tamaki, S., *Solid State Ionics*, 60, 1237, 1993.)

REFERENCES

2.1 W. A. Harrison, *Electronic Structure and the Properties of Solids: The Physics of the Chemical Bond*, Dover, New York, 1989.

2.2 Yu. Ya. Gurevich, A. K. Ivanov-Shitz, Semiconductor properties of superionic materials, in *Semiconductors and Semimetals*, Vol. 26, R. K. Willardson, A. C. Beer (Eds.), Academic Press, New York, 1988, Chap. 4.

2.3 H. Rickert, *Electrochemistry of Solids-An Introduction,* Springer-Verlag, Berlin, 1982.

2.4a A. Kraytsberg, Y. Ein-Eli, *Adv. Energy Mater.* 2(2012) 922.

2.4b A. Van der Ven et al., *Phys. Rev. B* 58(1998) 2975.

2.5 D. Kondepudi and I. Prigogine, *Modern Thermodynamics*, Wiley, Chichester, U.K., 2004.

2.6 J. W. Gibbs, *Trans. Conn. Acad.* 3(1878) 343.

2.7 A. I. Rsanov, *Surf. Sci. Rep.* 23(1996) 173.

2.8 J. Wang, *Curr. Nanosci.* 4(2)(2008) 179.

2.9 S. Lower, UC Davis Hypertext project, *Changes of State*, 2013, Chapter 7.5, http://chemwiki.ucdavis.edu/Wikitexts/8Simon_Fraser_Chem1%3A_Lower/06._States_of_Matter/Changes_of_State.

2.10 S. Hull, *Rep. Prog. Phys.* 67(2004) 1233.

2.11 R. Makiura et al., *Nat. Mater.* 8(6)(2009) 476–480.

2.12 G. R. Argue, B. B. Owens, Solid ionic conductor, US Patent 3,519,404, July 7, 1970.

2.13   B. Dreisbach, *J. Solid State Chem.* 60(1985) 115.

2.14   S. Mezn-Tszen H, *Xuebao*, 23(1957) 30 [in Chinese].

2.15   J. R. Dahn, *Phys. Rev. B* 44(1991) 9170.

2.16   G. Izquierdo, A. R. West, *Mater. Res. Bull.* 15(1980) 1655.

2.17   V. S. Stubican J. R. Hellmann, in A. H. Heuer, L. W. Hobbs (Eds.), *Advances in Ceramics*, Vol. 3, Science and Technology of Zirconia, American Ceramic Society, Columbus, OH, 1980, p. 25.

2.18   M. Yashima et al., *Solid State Ionics* 86–88(1996) 1131.

2.19   H. G. Scott, *J. Mater. Sci.* 10(1975) 1527.

2.20   J. Fisher et al., *Thin Solid Films* 528(2013) 217.

2.21   J. M. Paulsen and J. R. Dahn, *Chem. Mater.* 11(1999) 3065.

2.22   S. P. Ong, L. Wang, B. Kang, G. Ceder, *Chem. Mater.* 20(2008) 1798.

2.23   C. Kittel, H. Kroemer, *Thermal Physics*, 2nd Edn., W H Freeman, New York, 1980.

2.24   E. McCalla et al., *Chem. Mater.* 25(2013) 912.

2.25   M. Murayama et al., *Solid State Ionics* 170(2004) 173.

2.26   L. Pauling, *JACS* 51(1929) 1010.

2.27   W. Chen, A. Navrotsky, *J. Am. Ceram. Soc.* 90(2007) 584.

2.28   M. Sternik, K. Parlinski, *J. Chem. Phys.* 122(2005) 064707.

2.29a L. Chen et al., *J. Phys. Chem. C* 115(2011) 9370. Original paper: R. C. Garvie, *J. Phys. Chem.* 69(1965) 1238.

2.29b R. C. Longo et al., *Phys. Chem. Chem. Phys.* 16(2014) 11218.

2.30   R. E. Newnham, *Properties of Materials*, OUP, Oxford, U.K., 2005.

2.31   R. Murugan, V. Thangadurai, W. Weppner, *Angew. Chem. Int. Ed.* 46(2007) 7778.

2.32   A. Clemencon, In-situ and ex-situ observations of Li deintercalation from $LiCoO_2$: AFM & TEM studies, MS thesis, MIT, Cambridge, MA, June 2005.

2.33   Y. Wu et al., *Proc. R. Soc. A* 465(2009) 1829.

2.34   F. Sudreau et al., *JSSC* 83(1989) 78.

2.35   Y. Zhao, L. L. Damen, *J. Am. Chem. Soc.* 134(2012) 15042.

2.36   S. Margadonna et al., *J. Am. Chem. Soc.* 126(2004) 15032.

2.37   M. Ricco et al., *Phys. Rev. Lett.* 102(2009) 145901.

2.38   M. P. O'Callaghan et al., *Chem. Mater.* 20(2008) 2360.

2.39   P. Knauth, *J. Electroceramics* 5(2000) 111.

2.40   C. C. Liang, *J. Electrochem. Soc.* 120(1973) 1289.

2.41   F. W. Poulsen et al., *Solid State Ionics* 9–10(1983) 119.

2.42   C. S. Sunandana, P. Senthil Kumar, *Bull. Mater. Sci.* 27(2004) 1 and references therein.

2.43   P. V. Wright, *Electrochim. Acta* 43(1998) 1137, 1217.

2.44   C. T. Imrie, M. D. Ingram, G. S. McHattie, *J. Phys. Chem. B* 103(1999) 4132.

2.45   C. S. Sunandana, *J. Phys. Chem. Solids* 58(1998) 1359.

2.46   S. Kondo, K. Takada, Y. Yamamura, *Solid State Ionics* 53–56(1992) 1183.

2.47   B. L. Ellis, L. M. Nazar, *Curr. Opin. Solid State Mater. Sci.* 16(2012) 168.

2.48   N. Anantharamulu et al., *J. Mater. Sci.* 46(2011) 2821.

2.49   J. B. Goodenough, H. Y. P. Hong, J. A. Kafalas, *Mater. Res. Bull.* 11(1976) 203.

2.50   J. Maier, *Physical Chemistry of Ionic Materials: Ions and Electrons in Solids*, Wiley, Chichester, U.K., 2004, Fig. 6.10.

2.51   M. Catti et al., J. *Mater. Chem.* 14(2004) 835.

2.52   R. K. B. Gover, P. R. Slater, *Annu. Rep. Prog. Chem., Sect. A* 100(2004) 525.

2.53   M. Kim, S. J. Kim, *Acta Cryst. E* 69(2013) i34.

2.54   J. A. Saint, M. M. Doeff, J. Wilcox, *Chem. Mater.* 20(2008) 3404.

2.55   G. Oszlanyi et al., *Phys. Rev. Lett.* 78(1997) 4438.

2.56   T. Ishihara (Ed.), *Perovskite Oxides for Solid Oxide Fuel Cells*, Springer, New York, 2009.

2.57   L. Malavasi et al., *Chem. Soc. Rev.* 39(2010) 4370.

2.58   E. Kendrick et al., *Nat. Mater.* 6(2009) 871.

2.59   K. Funke, *Sci. Tech. Adv. Mater.* 4(2013) 043502.

2.60   L. N. Patro, K. Hariharan, *Mater. Chem. Phys.* 116(2009) 81.

2.61   C. Lucat et al., *J. Solid State Chem.* 29(1979) 373.

2.62   Ph. Colomban, *Fuel Cells* 13(2013) 6.

2.63   K. D. Kreur, *Solid State Ionics* 136–137(2000) 149.

2.64   A. R. *West Solid State Chemistry & Its Applications*, 2nd edn.–Student Edition, John Wiley & Sons, Chichester, U.K., 2014.

2.65   P. Knauth, M. L. Di Vona, *Solid State Proton Conductors*, John Wiley & Sons, Chichester, U.K., 2012.

2.66   B. B. Morosin, *Acta Cryst.* 34(1978) 3732.

2.67   S. Higashi et al., *Chem. Commun.* 50(2014) 1320.

2.68   C. S. Sunandana, T. Kumaraswami, *J. Non-Cryst. Solids* 85(1986) 247.

2.69   T. Takahashi et al., *J. Solid State Chem.* 21(1977) 37.

2.70   R. K. Osterheld, *J. Inorg. Nucl. Chem.* 30(1968) 3174.

2.71   M. Saito, S. Tamaki, *Solid State Ionics* 60(1993) 237.

2.72   R. Tripathi et al., *Angew. Chem. Int. Ed.* 49(2010) 8738.

2.73   A. C. Roberts et al., *Powder Diffract.* 3(1988) 93.

2.74   J. M. Aleba Mba et al., *Chem. Mater.* 24(2012) 122.

2.75   H. Yamane et al., *SSI* 178(2007) 1163.

2.76   D. Pasero et al., *Chem. Mater.* 17(2005) 345.

2.77   A. Kvist, A. Lunden, *Z. Naturforsch.* 20A(1965) 235.

2.78   F. Sudreau et al., *J. Solid State Chem.* 83(1989) 78.

2.79   Y. Sundarayya et al., *Mater. Res. Bull.* 42(2007) 1942.

2.80   Y. Janssen et al., *JACS.* 134(2012) 12516.

2.81   K. Homma et al., *JPSJ.* 79(Suppl A)(2010) 90–93.

2.82   K. Senevirathne et al., *SSI.* 233(21)(2013) 95–101.

2.83   H. J. Orman, P. J. Wiseman, *Acta Cryst. C* 40(1984) 12.

2.84   Z. P. Guo et al., *J. New Mater. Electrochem. Syst.* 6(2003) 263.

2.85   Y. Wu et al., *Proc. R. Soc. A* 465(2009) 1829.

2.86   Y. Inaguma et al., *SSC.* 86(1993) 689.

2.87   S. Kawai et al., *CM.* 10(1998) 3266.

2.88   A. D. Izquierdo et al., *MRB.* 15(1980) 1615.

2.89   D. Guerard, A. Herold, *Carbon* 13(1975) 337.

2.90   M. Ricco et al., *Phys. Rev. Lett.* 102(2009) 145901.

2.91   A. Rebenau, H. Schulz, *J. Less Common Met.* 50(1976) 155.

2.92   Y. Zhao, L. L. Daemen, *JACS.* 134(2012) 15042.

2.93   G. Brauer, E. Zintl, *Z. Physikal. Chem. B* 37 (1937) cited by H. J. Beister, K. Syassen, *Z. Naturforsch.* 45(b)(1990) 1388.

2.94   S. Boulineau et al., *SSI.* 1(2012) 221.

2.95   N. Kamaya et al., *Nat. Mater.* 10(2012) 682.

2.96   J. Geiger et al., *IC.* 50(2011) 1089.

2.97   M. Jansen, R. Hoppe, *ZAAC.* 163(1973) 399.

2.98   F. Lalere et al., *J. Power Sources* 247(2014) 975.

2.99   S. Andresson, A. D. Wadsley, *Acta Cryst.* 14(1961) 1245.

2.100 F. Sauvage et al., *Inorg. Chem.* 46(2007) 3289.

2.101 A. Bhide, K. Hariharan, *SSI.* 192(2011) 360.

2.102 R. J. Brook (Ed.), *Concise Encyclopedia of Advanced Ceramic Materials*, Pergamon Press, New York, 2012.

2.103 M. Kim, S. J. Kim, *Acta Cryst. E* 69(2013) i34.

2.104 P. Moreau et al., *Chem. Mater.* 22(2010) 4126.

2.105 G. O. Jones, P. A. Thomas, *Acta Cryst. B* 58(2002) 168.

2.106 G. Oszlanyi et al., *Phys. Rev. Lett.* 78(1997) 4438.

2.107 H. Jagodzinski, *Z. Kristallogr.* 112(1959) 80–87. en.Wikipedia.org/Wiki/Gold_monoiodide

2.108 M. Teicher, R. Weil, *PRB.* 18(1978) 7134.

2.109 S. M. Shapiro, F. Reidinger, in M. B. Salamon (Ed.), *Physics of Superionic Conductors*, Springer Verlag, Berlin, Germany, 1979, p. 45.

2.110 A. K. Samal, T. Pradeep, *Nanoscale* 3(2011) 4840.

2.111 S. Chandra, *Superionic Solids*, North Holland, Amsterdam, the Netherlands, 1981.

2.112 S. Velgosh et al., *Solid State Ionics* 188(2011) 31.

2.113 J. W. Brightwell et al., *Solid State Ionics* 15(1985).

2.114 A. M. Sukeshini, K. Hariharan, *J. Solid State Chem.* 101(1992) 87.

2.115 T. Sakuma, S. Hoshino, *J. Phys. Soc. Jpn.* 49(1980) 678.

2.116 T. Sakuma et al., *J. Phys. Soc. Jpn.* 51(1982) 2628.

2.117 S. Hull et al., *JPCM.* 17(2005) 1067.

2.118 E. Quarez et al., *New J. Chem.* 33(2009) 998.

2.119 S. Hull et al., *J. Solid State Chem.* 165(2002) 363.

2.120 W. F. Kuhs et al., *Acta Cryst. B* 34(1978) 64.

2.121 I. P. Studenyak et al., *J. Cryst. Growth* 30(2007) 326.

2.122 S. Geller et al., *Phys. Rev. B* 19(1979) 5396.

2.123 G. Dercz et al., *J. Achiev. Mater. Manuf. Eng.* 31(2008) 408.

2.124 A. Ahuadero et al., *Chem. Mater.* 24(2012) 2655.

2.125 C. J. Howard, B. J. Kennedy, *J. Phys. Cond. Matter* 11(1999) 3229.

2.126 P. R. Slater et al., *J. Solid State Chem.* 139(1998) 135.

2.127 V. Thangadurai et al., *Chem. Commun.* (1998) 2647–2648.

2.128 R. C. Biswal, K. Biswas, *J. Eur. Ceram. Sci.* 33(2013) 3053.

2.129 F. Goutenoire et al., *Chem. Mater.* 12(2000) 2575.

2.130 J. Cheng et al., *J. Chil. Chem. Soc.* 54(2009) 445.

2.131 S. F. Wang et al., *J. Mater. Res. Technol.* 2(2013) 141.

2.132 M.-H. Chambrier et al., *Inorg. Chem.* 53(2014) 147.

2.133 S. Nakayama et al., *J. Mater. Chem.* 5(1995) 1801.

2.134 Leon-Reina et al., *Chem. Mater.* 15(2003) 2099.

2.135 B. Liu et al., *Solid State Ionics* 191(2011) 168.

2.136 M.-R. Li et al., *Angew. Chem.-Ger. Edit.* 122(2010) 2412. https://research-archive.liv.ac.uk/4477/1/li_melilite_deposited.pdf.

2.137 H. B. Bartl, T. Scheller, *Neues Jahrb. Mineral. Monatsch.* 35(1970) 547.

2.138 Y. Ito et al., *Solid State Ionics* 79(1995) 81.

2.139 F. Abraham et al., *Solid State Ionics* 40/41(1990) 934.

2.140 A. Castro et al., *Mater. Res. Bull.* 33(1998) 31.

2.141 R. Kanno et al., *Solid State Ionics* 70–71(1994) 253.

2.142 G. Denes et al., *J. Solid State Chem.* 30(1979) 335.

2.143 R. W. G. Wyckoff, *Cryst. Struc.* 1.

2.144 W. Korczak, P. Mikolajczak, *J. Cryst. Growth* 61(1983) 601.

2.145 R. C. T. Slade et al., *Solid State Ionics* 92(1996) 171.

2.146 B. Morosin, *Acta Cryst.* 34(1978) 3732.

2.147 E. Magome et al., *Ferroelectrics* 378(2009) 157.

2.148 B. T. Melekh et al., *Solid State Ionics* 97(1997) 465.

2.149 K. Tekeuchi et al., *Solid State Ionics* 138(2000) 63.

2.150 N. Jalarvo et al., *Chem. Mater.* 25(2013) 2741.

3

Materials Synthesis

3.1 Preamble

Phase diagrams and phase stability discussed in the last chapter are a useful guide to materials synthesis in solid state ionics to be discussed in this chapter.

The existence of a well-defined melting point is necessary for the growth of single crystals that are needed only in special circumstances such as single-crystal neutron diffraction studies. For most purposes, the focus is on polycrystalline ceramics that are obtainable from a variety of techniques discussed here. The existence of a eutectic point (one or more cusps seen in the temperature composition diagrams) in the phase diagram of, say, Li_2O–P_2O_5 or Li_2S–P_2S_5 helps in the synthesis of glasses and glass ceramics. The latter-type materials important for applications are made through careful heat treatments of relevant multinary glass compositions over a restricted temperature range between the so-called glass transition temperature and the crystallization temperatures. Pressure–temperature behavior of aqueous solutions of precursor materials helps in the synthesis of solid state ionic materials by hydrothermal synthesis. The solubility of organometallic precursors in (non)aqueous solvents enables powder and thin-film synthesis by solvothermal synthesis often assisted by relatively low-temperature (~200°C–300°C) combustion. Water insolubility with very poor dissolution kinetics of silver and copper halides (and chalcogenides) helps temperature-controlled precipitation synthesis enabling not only nanoscale synthesis of individual materials but isomorphous substitution and solid solution synthesis as in AgI–CuI system. This could be extended to making ternary/quaternary compounds. Peritectic points (defined by a convex region in the phase diagram) as seen in RbI–AgI and Li_2O–TiO_2 binaries with narrow regions of solid–liquid phase coexistence help in the synthesis of poly/nano/monocrystalline phases for basic physical property measurements and applications.

The general mechanical softness of solid state ionic materials such as LiI and $LiCoO_2$, and Ag and Cu chalcogenides, among others, enables the application of the mechanochemical (reaction) methods to make device-ready micro/nanopowders of composites (LiI·Al_2O_3) and compounds. Even commercial powders (AgI or Cu_2S) may be converted to nanopowder by dry milling. AgI–CuI phase diagram has been generated by agate mortar grinding of Ag, Cu, and I [3.1]. This advantage essentially stems from their water insolubility and rapid nucleation rates and the weakly covalent metal–nonmetal bond in their semiconductor-like open crystal structures.

A physical vapor of a material or a chemical vapor—both thermodynamically stable phases—assists in the deposition of thin-film solid state ionic materials. Finally, an intercalation or insertion of Li atoms into matrices such as graphene, graphite, or TiS_2 is an important chemical "synthesis" technique to produce electrodes for Li-ion batteries. Intercalation and deintercalation are dynamic processes that occur during charging and discharging of a solid state rechargeable battery.

Materials synthesis in solid state ionics is an aspect that is at once evolving. And as technology advances, the challenge of making materials can only get more sophisticated. In addition to the liquid electrolyte–solid electrode compatibility, it is the goal of solid electrolyte synthesis to fabricate all solid state microionic and higher power devices with the eventual aim of realizing electric heavy vehicles for the highway and captive high-power installations.

Building on the edifice of the previous two chapters, we consider in this chapter the basic principles and practice of solid state ionic materials synthesis. This issue is to be considered in the larger perspective of ceramic materials synthesis that is a part of the discipline of materials science and technology.

It is important to realize that the challenge of optimized materials synthesis brings several techniques into consideration. Often more than one technique needs to be harnessed in a hybrid manner to obtain the desired solid state ionic material in the form specific to the target application. To emphasize the importance of innovative approaches to materials innovation, we cite the use of a combinatorial robot to determine the phase diagram of the Li–Co–Mn–O system for the synthesis in a pure oxygen atmosphere. As many as 400 compositions are made on the robot [3.2]. The coexistence regions between the spinel and layered structures are quite complicated, but the results help clarify a great deal of data known so far.

3.2 Bulk Solid State Ionic Materials

3.2.1 High-Temperature Solid State Synthesis

This type of materials synthesis requires the use of high temperature (~900°C–1000°C) primarily because the reaction is diffusion limited. The temperature for the reaction is chosen to be >2/3 of the melting point of the lowest melting reactant. Nucleation of the desired phase is the key step. High temperatures usually lead to a thermodynamically stable product. Kinetic control of the reaction is difficult and multiple grinding–sintering cycles (at least two) may be necessary to obtain a usable microstructure. A typical pathway for a solid state reaction is Grinding $\rightarrow$ Decomposition $\rightarrow$ Cooling #1 $\rightarrow$ Sintering $\rightarrow$ Cooling #2 $\rightarrow$ Final Product. Air or inert atmosphere may be used accordingly as the desired product has to be in an oxidized or reduced state for one of the reactants. As an example for the latter possibility, consider the synthesis of $LiFePO_4$. This compound may be synthesized using lithium carbonate (Li_2CO_3), ferrous oxalate ($FeC_2O_4 \cdot 2H_2O$), and ammonium dihydrogen phosphate ($NH_4H_2PO_4$) taken in stoichiometric ratio. These reactants are weighed out, thoroughly mixed, and grinded in an agate mortar and pestle for a few hours to obtain a fine powder. This powder is kept in a furnace at 550°C for 12 h in a reducing atmosphere (because Fe in $LiFePO_4$ must be preserved in the Fe^{2+} state). Nitrogen gas with 10% hydrogen may be used for this purpose. Varying time and temperature helps optimize particle size. The color of the resulting product indicates that the desired product is obtained. In this case, light-brown particles are obtained ensuring the absence of Fe^{3+} impurities in the mixture. The powders thus obtained are subjected to characterization by x-ray diffraction (XRD), microscopy, IR, and in this case Mössbauer spectroscopy. These are considered in the next chapter.

3.2.2 Temperature-Controlled Precipitation as a Strategy for SSI Materials Synthesis

Water insolubility or at least sparing solubility is the hallmark of solid state ionic materials, particularly the oxides, chalcogenides, and a few important halides such as Li_2O, $LiCoO_2$, ZrO_2, Li_2S, AgI, CuBr, and CuI.

AgI and CuI possess the same structure at ambient temperature and pressure, namely, zincblende (although zincblende AgI is metastable). Solid solubility is an important criterion for optimizing the conductivity of an SSI material. Knowledge of the location of the solute ions in a solid solution is crucial in distinguishing the substitution of metal ion A from metal ion B in a solid solution $A_xB_{1-x}X_y$ where X could be a halide or a chalcogenide or an oxide.

These features help realize a material property change in a single component system. For example, the system AgI–CuI has an extended solid solubility, although the radius of Ag^+/radius of Cu^+ = 0.126 nm/0.096 nm = 1.31 rather than 1. Add to this the differential water solubility between AgI and CuI, which could be exploited on the synthesis of solid solutions of AgI and CuI, through a controlled precipitation method at temperatures ~100°C [3.3].

A flowchart (Figure 3.1) illustrates the method to synthesize Ag_yCu_xI ($x + y = 1$).

Wet-chemical-chelating reaction processing may be used to synthesize a series of single β-phase nano-$Ag_{1-x}Cu_xI$ ($x = 0$–0.5) solid solution powders. Citric acid as complexing agent takes part in the process of chemical reaction and the chemical reactions are indicated in Figure 3.2a. A series of single β-phase nano-$Ag_{1-x}Cu_xI$ ($x = 0$–0.5) solid solution powders were synthesized by this wet-chemical-chelating reaction processing and citric acid used as complexing agent. The $Ag_{1-x}Cu_xI$ powders

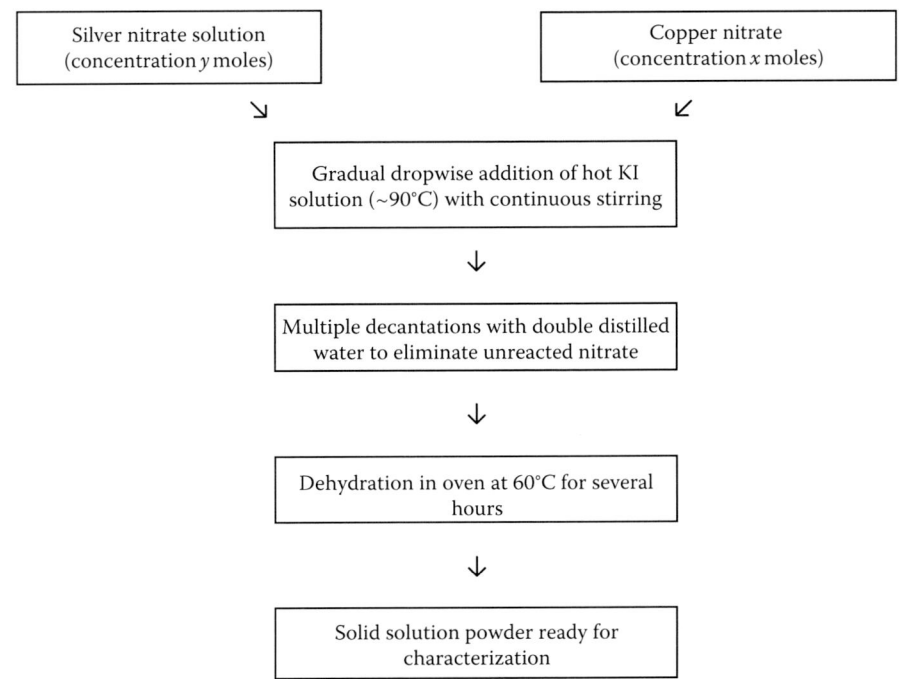

```
┌─────────────────────────┐          ┌─────────────────────────┐
│   Silver nitrate solution│          │     Copper nitrate       │
│  (concentration y moles) │          │  (concentration x moles) │
└─────────────────────────┘          └─────────────────────────┘
            ↘                                    ↙
        ┌──────────────────────────────────────────┐
        │   Gradual dropwise addition of hot KI      │
        │  solution (~90°C) with continuous stirring │
        └──────────────────────────────────────────┘
                            ↓
        ┌──────────────────────────────────────────┐
        │ Multiple decantations with double distilled│
        │   water to eliminate unreacted nitrate     │
        └──────────────────────────────────────────┘
                            ↓
        ┌──────────────────────────────────────────┐
        │   Dehydration in oven at 60°C for several  │
        │                  hours                     │
        └──────────────────────────────────────────┘
                            ↓
        ┌──────────────────────────────────────────┐
        │     Solid solution powder ready for        │
        │              characterization              │
        └──────────────────────────────────────────┘
```

FIGURE 3.1 Flowchart for temperature-controlled precipitation synthesis.

determined by XRD and transmission electron microscopy (TEM) demonstrated that the crystalline size and lattice parameter of the $Ag_{1-x}Cu_xI$ powders decrease with an increase in the amount of CuI substitution. The copper in the lattice of $Ag_{1-x}Cu_xI$ can effectively prevent the crystalline growth of the $Ag_{1-x}Cu_xI$ powders and the citrate used in the $Ag_{1-x}Cu_xI$ powders synthesis process can accelerate single β-phase crystalline structure formation. The lattice parameters have been ascertained by the results of XRD. Crystallite sizes, which decrease with copper iodide concentration increasing, have been demonstrated by TEM (Figure 3.2b), XRD pattern evolution (Figure 3.2c), and lattice contraction (Figure 3.2d).

Both undoped and Al-doped $LiMn_2O_4$ powders, the latter with the composition $LiMn_{2-y}Al_yO_4$ ($0 \leq y \leq 0.3$), may be conveniently prepared by the succinic acid–assisted wet chemistry. In this procedure, the dicarboxylic acid $C_4H_6O_4$ with a molecular weight 118.09 acts as a chelating agent. Acetates of high purity Li, Mn, and Al taken in stoichiometric proportions are dissolved in a minimum amount of methanol or distilled water. An equal volume of the complexing agent, namely, a 1 mol/L aqueous solution of succinic acid, is added to the aforementioned methanolic/aqueous solution, a solution with pH in the range between 3 and 4, by a careful adjustment of the concentration of the complexing solution. Homogenous, finely dispersed precipitates are obtained because of the poor solubility of Mn and Li succinates. The carboxylic (COOH) groups on the succinic acid are likely to form chemical bonds with the metal ions, and these mixtures develop into an extremely viscous paste upon slow evaporation of methanol. Li and transition metal cations are reasonably expected to be trapped within the paste, which ensures molecular-level mixing. This procedure carried out at ambient temperature amazingly eliminates the need for long-range diffusion during subsequent formation of the lithium manganate spinel.

The precursor mass is obtained by a drying of the paste at 120°C. In a final step, the precursor is made to decompose in air at ~300°C. The decomposition reaction is highly exothermic owing to the combustion of organic species present in the precursor mass. This exothermicity enhances the oxidation reaction and the onset of the spinel phase. The end product is a blue-black mass of the $LiMn_{2-y}Al_yO_4$ of the Al-doped spinel (Figure 3.3).

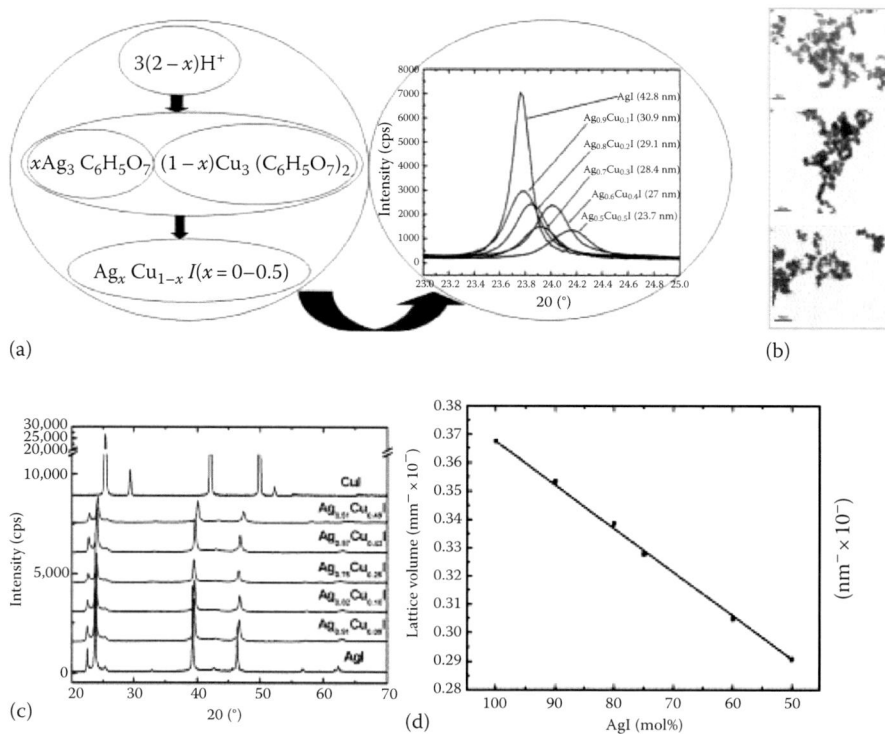

FIGURE 3.2 (a) Reaction scheme for wet chemical synthesis of nanoscale β-$Ag_xCu_{1-x}I$ solid solution powders and the corresponding progressive development of x-ray diffraction peak. (b) Typical transmission electron micrograph of nanopowder. (c) Evolution of XRD patterns of AgI with an increase in Cu content. (d) Linear decrease in lattice volume of AgI with an increase in Cu content. (From Liu, X. et al., *Mater. Res. Bull.*, 46, 910, 2011.)

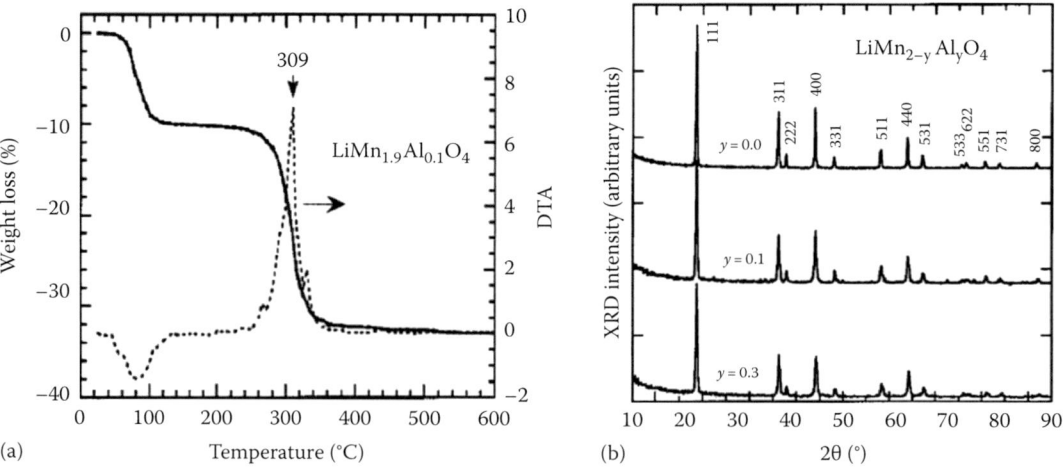

FIGURE 3.3 (a) TG (—) and DTA (···) monitoring of the precursor (acetates of Li, Mn, and Al) decomposition in $LiMn_{1.9}Al_{0.1}O_4$ synthesis. The sharp, heat-evolving, or exothermic peak at 390°C indicates onset of decomposition and product formation. (b) Phase identification by XRD. (After Julien, C. et al., *J. Mater. Chem.*, 11, 1837, 2001.)

3.2.3 Sol–Gel Combustion or Solvothermal Synthesis

Sol–gel synthesis is an offshoot of colloidal and interfacial chemistry and physics. A common example of sol–gel synthesis in the kitchen is the making of yoghurt from milk. Milk is a colloid and yoghurt is the gel. Sol–gel conversion is an unusual type of phase transition from a fluid like sol to a soft-solid like gel with built-in rigidity.

The science and technology of sol–gel evolved as an attempt to produce better ceramics through chemistry.

The sol–gel synthesis technique is essentially a method for producing solid materials from small molecules. The method is readily used for the fabrication of metal oxides of Si and Ti among others.

The sol–gel process (a) converts monomers (building blocks of polymers) into a colloidal solution or sol and (b) acts on the precursor for an integrated network or gel, of either discrete particles or network polymers.

The sol (or solution) gradually evolves toward the formation of a gel-like diphasic system (liquid phase + solid phase). Morphologies of target materials range from discrete particles to continuous polymer network.

For colloids, the volume fraction of particles or particle density is so low that an appreciable amount of fluid needs to be removed initially. For gel-like properties to be recognized, (a) allow sedimentation to occur, pour off the remaining liquid, or (b) use centrifugation and accelerate the phase separation.

Remove remaining liquid (solvent) by drying. Drying causes shrinkage and densification. The rate of solvent removal is determined by the distribution of porous regions (porosity) within the gel.

The ultimate microstructure of the target component is strongly influenced by the changes imposed upon the structural template during this crucial processing step.

The firing process, which involves thermal treatment, favors further polycondensation. It enhances mechanical properties and structural stability. Final sintering, densification, and grain growth are the processes leading to the final product.

In the sol–gel synthesis, one begins with a mixture of alkoxide precursors dissolved in an alcohol, preferably an alcohol based on the same alkyl group as the alkoxide. Commonly used alkyl groups are C_2H_5, C_4H_9, and C_6H_{13}, among others. Addition of water to the solution starts the process of hydrolysis. Microparticles arising from the condensation reactions between the hydrated metal ions constitute the sol—a stable suspension of small colloidal particles—so small that their weight is negligible compared with the fluctuating forces provided by the liquid. Thus, sedimentation does not occur. In the gel phase, the hydrated alkoxides condense into a polymer whose backbone is made up of alternating metal and oxide ions. The resulting network percolates throughout the volume of the liquid resulting in a rigid soft solid.

The biggest advantage of the sol–gel technology is that densification is often achieved at a temperature much lower than in, say, high-temperature solid state synthesis. One could deposit a film, make a monolithic ceramic or glass, or synthesize powders as indicated in the following:

	Precursor Sol ↓	
Deposition on a Substrate to Form a Film	Casting into a Suitable Container	Synthesize Powders
Dip/spin coating	Monolithic ceramics, glasses, fibers, membranes, aerogels	Microspheres, nanospheres

A sol consists of 10–100 nm diameter solid colloidal particles dispersed in a liquid. Brownian (really "unsettling") motion keeps colloidal sols stable against settling due to their extremely small sizes. Surface charge may aid in maintaining a dispersion of the particles. The term "sol–gel" describes processes that involve sols (note that sols are different from solutions) of inorganic colloids suspended in liquid media. Precursor liquids are sols or mixtures of sols and solutions. The "sol–gel" technology

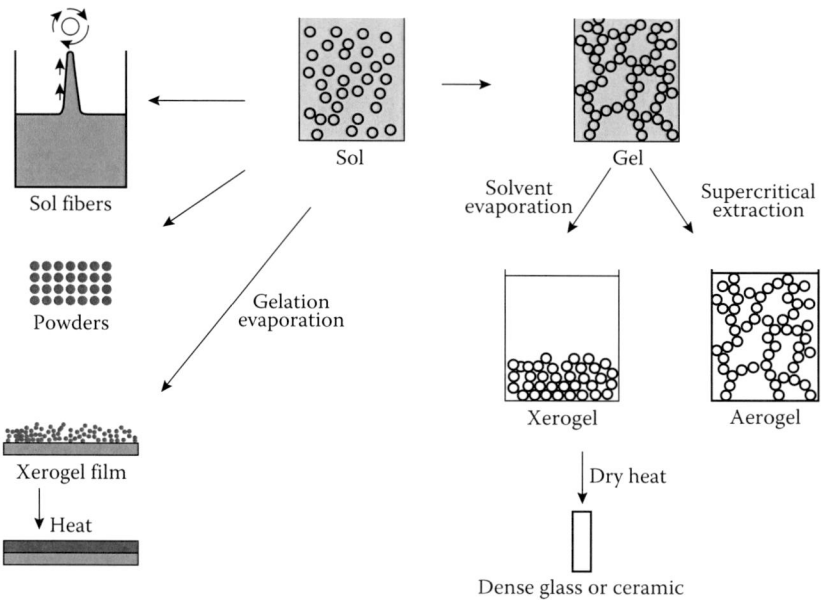

FIGURE 3.4 Processing options in sol–gel synthesis. (From Schubert, U. and Husing, N., *Synthesis of Inorganic Materials*, Wiley VCH, Weinheim, Germany, 2005.)

generally uses organometallic compounds, such as alkoxides of transition metals or of Si. These can be partly hydrolyzed and polymerized into a uniform gel.

There are two main approaches to sol–gel-based materials synthesis:

1. Colloidal suspension preparation via precipitation from a solution prepared by a metal salt
2. Preparation of a polymeric gel usually from the hydrolysis of an organometallic compound in an organic solvent

Both (1) and (2) involve hydrolysis of precursor while a condensation–polymerization reaction occurs simultaneously. However, they differ in (a) the rate of hydrolysis and (b) solvent used during hydrolysis.

Rapid hydrolysis occurs in (1). A schematic is shown in Figure 3.4.

The sol–gel process for materials synthesis—glass or ceramic or nanostructured film or nanosphere—is a versatile bottom-up process. A sol or colloidal solution is made from a suitable precursor and medium and converted into a gel from which either a glass or a nanoceramic is realized. The process is a classic demonstration of the evolution of the Si–O bond. Making SiO_2 glass by this method involves (a) hydrolysis, (b) gelation, (c) aging, and (d) drying, each of which is a complex kinetic process that needs to be carefully controlled and optimized. Specific Si–O bonds are involved in the sol–gel precursors leading to the overall reaction:

$$Si(OR)_4 + H_2O \rightarrow SiO_2 + ROH \tag{3.1}$$

in which the Si–O bonds are formed in aqueous/nonaqueous media depending on the choice of the precursor [3.6].

We give four examples of sol–gel applications to the synthesis of two Li-ion, a Na-ion and an O-ion, conductors in the following:

1. Synthesis of $Li_{1.4}Ti_{1.6}Al_{0.4}(PO_4)_3$. The flowchart illustrates the method (Figure 3.5).
2. Non-aqueous sol combustion synthesis of $LiFePO_4$. The flowchart delineates the method (Figure 3.6).

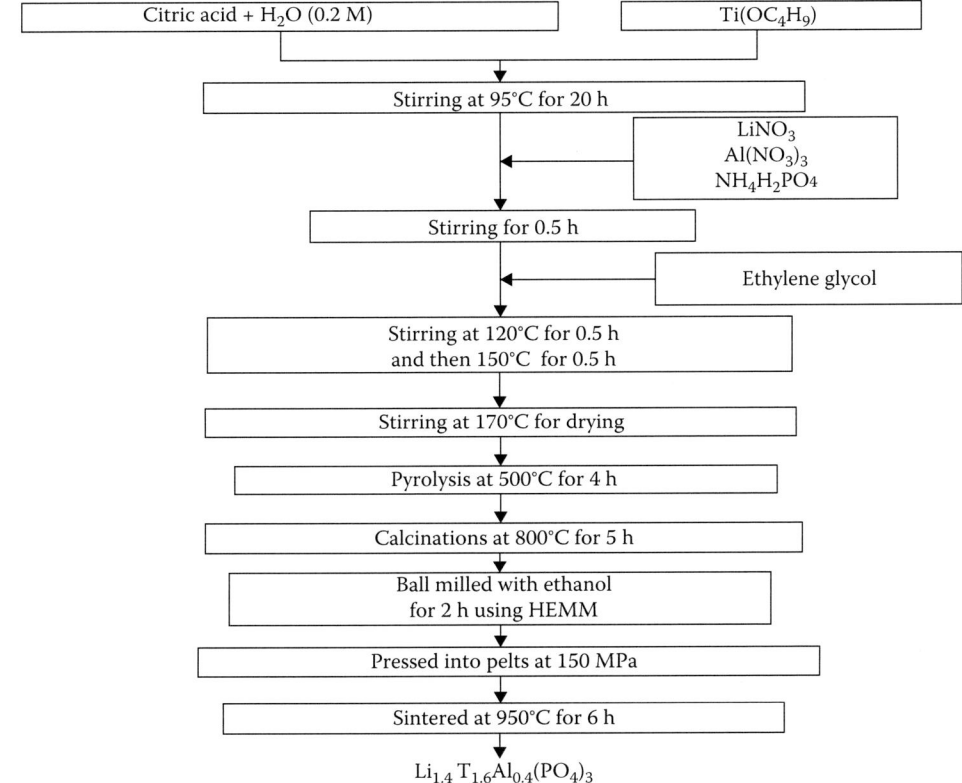

FIGURE 3.5 Flowchart for the citric-acid-based sol–gel synthesis of $Li_{1.4}Ti_{1.6}Al_{0.4}(PO_4)_3$. (After Takahashi, K. et al., *J. Electrochem. Soc.*, 159, A342, 2012.)

3. Aqueous solution–based synthesis of NASICON. The flowchart (Figure 3.7) describes the method step by step.
4. Gd-doped ceria is an important fuel cell material. Figure 3.8 shows the flowchart to make $Ce_{0.8}Gd_{0.2}O_{1.9}$ and the XRD pattern. A brief description follows.

Cerium nitrate ($Ce(NO_3)_3 \cdot 6H_2O$) was dissolved in deionized water, and the stoichiometric amount of gadolinia was dissolved in nitrate solution. They were mixed together with citric acid solution (the mole ratio of cation to citric acid was 1/1.5). The pH value of the system was adjusted to 7–8 with ammonia solution under continuous stirring at 40°C, and a homogeneous sol was formed. The sol was then heated at 120°C for about 1 h when a white alveolate precursor was obtained. The precursor was then calcined at different temperatures (650°C, 750°C, and 850°C) to get the final composition powders. The obtained powders were pressed into pellets under a pressure of about 200 MPa and then sintered at different temperatures (1000°C–1300°C) in air for 4 h and used for characterization.

Next, we discuss the mechanochemical synthesis.

3.2.4 Mechanochemical Reaction Synthesis

Although known since antiquity, the pioneering work of Michael Faraday in 1820 firmly established mechanochemical synthesis within the realm of materials research [3.6,3.11]. However, a fundamental understanding of many aspects of mortar and pestle or (sealed) ball-mill-assisted mechanochemical processes remains sparse. Only recently an experimental setup has been devised to monitor mechanochemical processes *in situ* yielding direct insights into mechanistic and kinetic aspects of solid state reactions that are promoted by grinding such as the synthesis of metal-organic frameworks [3.12,3.13].

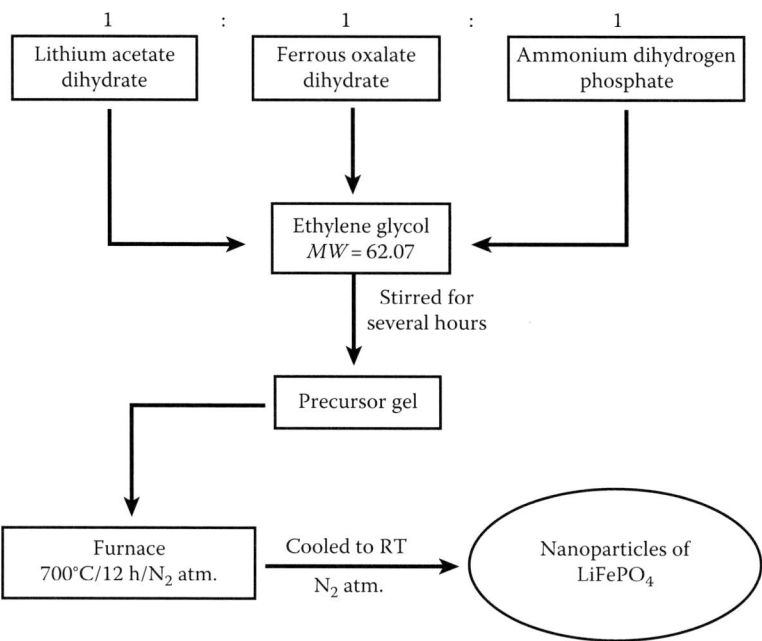

FIGURE 3.6 Flowchart for nonaqueous sol combustion synthesis of LiFePO$_4$. (From Sundarayya, Y. et al., *Mater. Res. Bull.*, 42, 2942, 2007.)

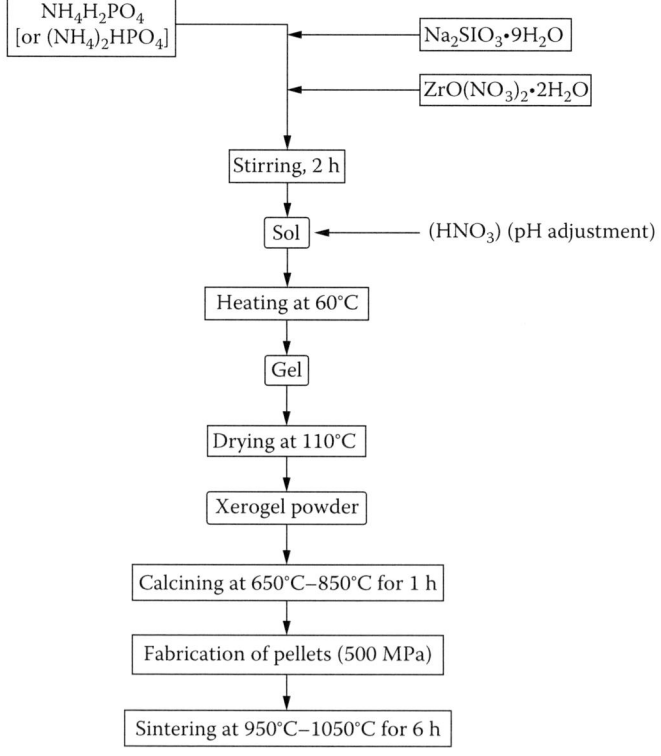

FIGURE 3.7 Flowchart for aqueous-solution-based sol–gel synthesis of NASICON. (After Shimizu, Y. et al., *J. Mater. Chem.*, 7, 1487, 1997).

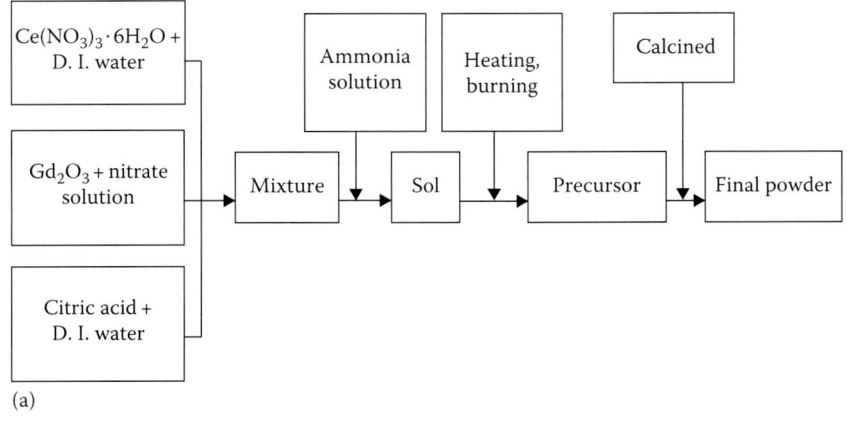

(a)

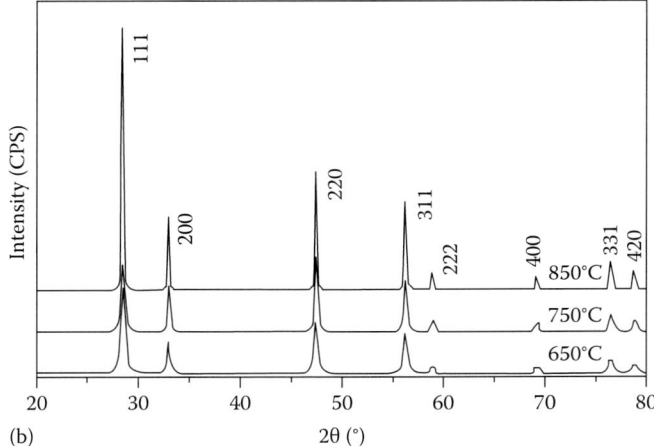

(b)

FIGURE 3.8 (a) Flowchart for sol combustion synthesis of Gd-doped CeO_2. (From Cheng, J. et al., *J. Chilean Chem. Soc.*, 54, 445, 2009.) (b) X-ray diffractograms of the evolution of $Ce_{0.8}Gd_{0.2}O_{1.9}$.

The process of grinding itself and the attendant repeated fracture with the high "local" temperatures involved effectively catalyze low-temperature reactions. Figure 3.9 visualizes the action of mechanochemical effects to yield the required product.

Essentially, a "dry" processing synthesis technology involves manual grinding or ball milling of two or more components of the material intended to be synthesized, exploits the effects of non-hydrostatic mechanical stress and plastic strain on the chemical processes, and thereby induces change in the energy and entropy, as well as change in structure and chemical composition of molecular crystals and other aggregates of matter.

Mechanochemical reaction (MCR) starts at the surface and proceeds to completion in three steps: shearing, plastic deformation, and readily occurring local chemical reaction assisted plausibly by diffusion. When one of the components is a sublimating solid such as iodine, the sheared surface of say a grain of copper or silver in the powder is bathed in the confined vapors of iodine generated by grinding of flakes taken in an agate mortar.

The reactions would be local. Bond formation *ab initio* could take place. How does a single nascent Ag–I or Cu–I bond lead to a cluster—tetrahedral AgI_4 or CuI_4? Propagate laterally and lead to a monolayer of AgI or CuI? Propagate across the interface possibly by percolation/diffusion to make a nanocrystalline compound? This is the crux of soft MCR of noble metal halides. What about the hard MCR of transition metal oxides?

Hard MCR is nothing but an MCR followed by "soft annealing." Here, MCR provides a mechanically activated mixture. Soft annealing ensures complete transformation of reactants to the desired material at a significantly lower temperature than that used in conventional high-temperature solid state synthesis.

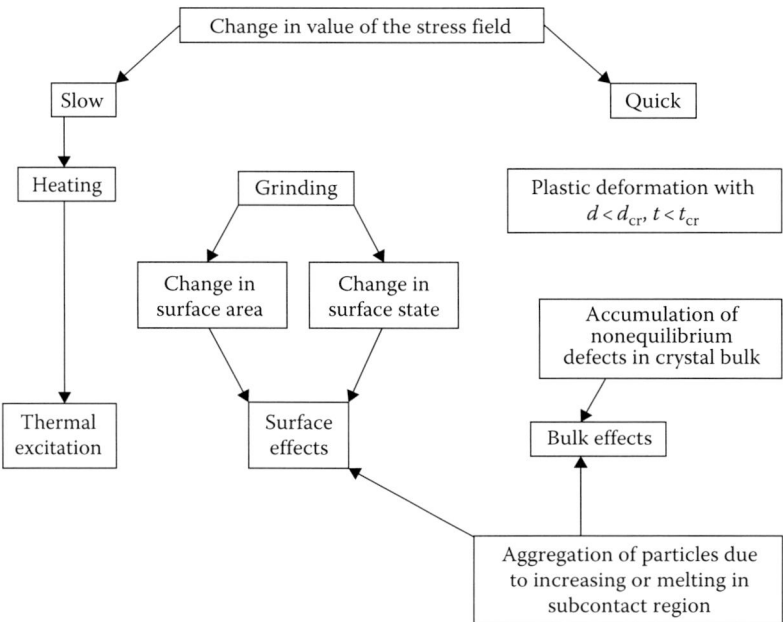

FIGURE 3.9 How mechanochemical effects work largely through surface and also through bulk effects. (From Boldyrev, V.V., *J. Chim. Phys.*, 83, 821, 1986.)

In hard MCR, mere increase in milling time only leads to incomplete conversion of starting materials to the desired compound/material. Soft annealing is thus a necessary preparative step.

Important events that lead to the final product could be

1. Activation of intramolecular bonds
2. Plastic deformation
3. Unusual mass transport processes
4. Intimate mixing of atomic species

A new (nano)crystalline or polymorphic or amorphous chemical system is the product.

Some thermodynamics, some kinetics, and some atomic/molecular physics/chemistry would be at work. Together they constitute a product technology based on mechanical activation and thermal promotion of reactant species.

Two questions arise as follows:

1. What are the basic factors that limit the yield of the simple MCR?
2. What does the mild annealing actually accomplish to take the reaction from the half-way house to the destination?

The IUPAC Compendium of Chemical Technology defines MCR as follows: a "chemical reaction that is induced by mechanical energy." Heinicke defines mechanochemistry thus: Mechanochemistry is a branch of chemistry that is concerned with chemical and physicochemical transformations of substances in all states of aggregation produced by the effect of mechanical energy. Structural disordering, structural relaxation, and structural mobility simultaneously affect the reactivity of solids.

There are three stages during high-energy ball milling (and also mortar and pestle grinding in favorable cases, Figure 3.10.

In region (a) where interaction of particles being milled is negligible, the degree of dispersion varies linearly with grinding time. The dispersion soon hits a saturation stage called aggregate stage (b) where van der Waals forces (0.04–4 kJ/mol) arise between grains. This is quickly followed by stage (c) where increase

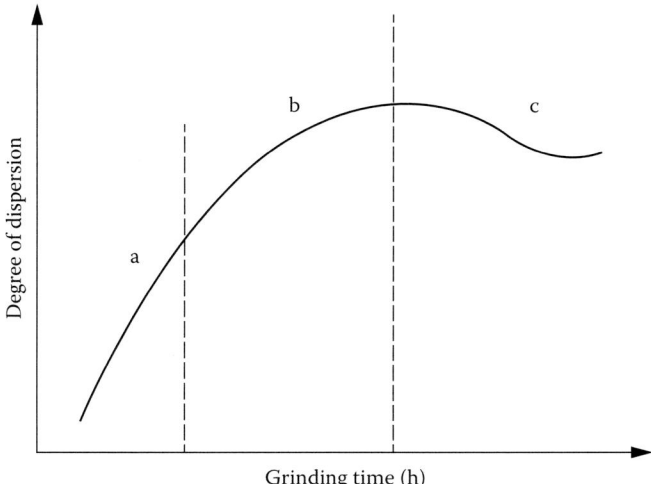

FIGURE 3.10   Three stages of high-energy ball milling. (From Balaz, P. et al., *Chem. Soc. Rev.*, 42, 7571, 2013.)

in dispersion ceases and drops stopping altogether eventually. This is the agglomeration stage where chemical bonds with energies 40–400 kJ/mol are formed and MCR causes changes in crystal structure.

The role of defects generated during MCR is fundamental to the process evolution. These include point defects, dislocations, grain boundaries, amorphous regions, and two-phase regions among others.

By way of illustration, we discuss (a) attrition in an agate mortar to produce $Ag_{1-x}Cu_xI$ solid solutions ($0 \leq x \leq 1$) and (b) ball milling to produce $LiMn_2O_4$ spinel and $LiCoO_2$ layered oxide.

(a) *$Ag_{1-x}Cu_xI$ solid solutions ($0 \leq x \leq 1$)*

Nanoscale crystallites of Ag-rich ($Ag_{1-x}Cu_xI$, x = 0.05, 0.10, 0.15, and 0.25), Cu-rich ($Cu_{1-y}Ag_yI$, y = 0.05, 0.10, 0.15, and 0.25), and intermediate $Ag_{1-x}Cu_xI$ (x = 0.50) solid solutions and end-members AgI and CuI with sizes in the range of 46–13 nm were synthesized by attrition at ambient temperature in a soft MCR of Ag, Cu, and I using an agate mortar and pestle. Monophasic γ-AgI (zincblend, a = 638 pm) with disordered Ag^+ sublattice and the crystallite size of about ~31 nm was realized in the case of $Ag_{0.75}Cu_{0.25}I$ (x = 0.25) composition. Lattice parameter decreases linearly from 649 to 604 pm with increasing Cu concentration in the AgI–CuI system validating Vegard's law. Smallest size (~13 nm) agglomerated nanocrystals were realized in the Cu-rich composition $Cu_{0.75}Ag_{0.25}I$ (a = 615 pm), while unagglomerated uniform-sized (~17 nm) and spherical shape nanocrystallites of $Ag_{0.50}Cu_{0.50}I$ (a = 626 pm) with maximum strain were synthesized for sensor applications using MCR (Figure 3.11). Differential scanning calorimetry study shows the systematic changes in the phase transition temperature with Cu substitution. Ag-rich composition possesses less enthalpy (ΔH [x or Cu = 0.05, 0.10, 0.15, 0.25] = 6.0, 6.11, 6.6, 6.3 in kJ/mol) and entropy (ΔS [y or Ag = 0.05, 0.10, 0.15, 0.25] = 14.15, 14.1, 15.03, 13.6 in J/mol K) when compared to undoped AgI (ΔH = 9.63 kJ/mol, ΔS = 22.8 J/mol K) implying greater thermal stability of γ-phase due to Cu-strengthened Ag–I bond. Enhanced entropy (ΔS = 8.17 J/mol K) in $Cu_{0.75}Ag_{0.25}I$ (Cu-rich) solid solutions relative to CuI (ΔS = 1.0 J/mol K) indicates Ag-induced cation disorder. Fifteen percent of Ag-doped CuI ($Cu_{0.85}Ag_{0.15}I$) nanocrystals apparently behave like microscopic p–n junctions with currents in the range of 10^{-6}–10^{-8} A characterized by a nonlinear *I–V* curve.

(b) *$LiMn_2O_4$*

The mechanochemical method may be used for the synthesis of highly dispersed stoichiometric and nonstoichiometric $Li_xMn_2O_4$ spinel either directly during mechanical activation at room temperature or by preliminary activation followed by relatively low-temperature heat treatment. One could start from different manganese (MnO_2, Mn_2O_3, MnO) and lithium (LiOH, $LiOH \cdot H_2O$, Li_2CO_3) compounds. The oxidation state of manganese has a great influence on the kinetics of MCR. On the other hand, different crystal structure and mechanical properties of initial lithium compounds result

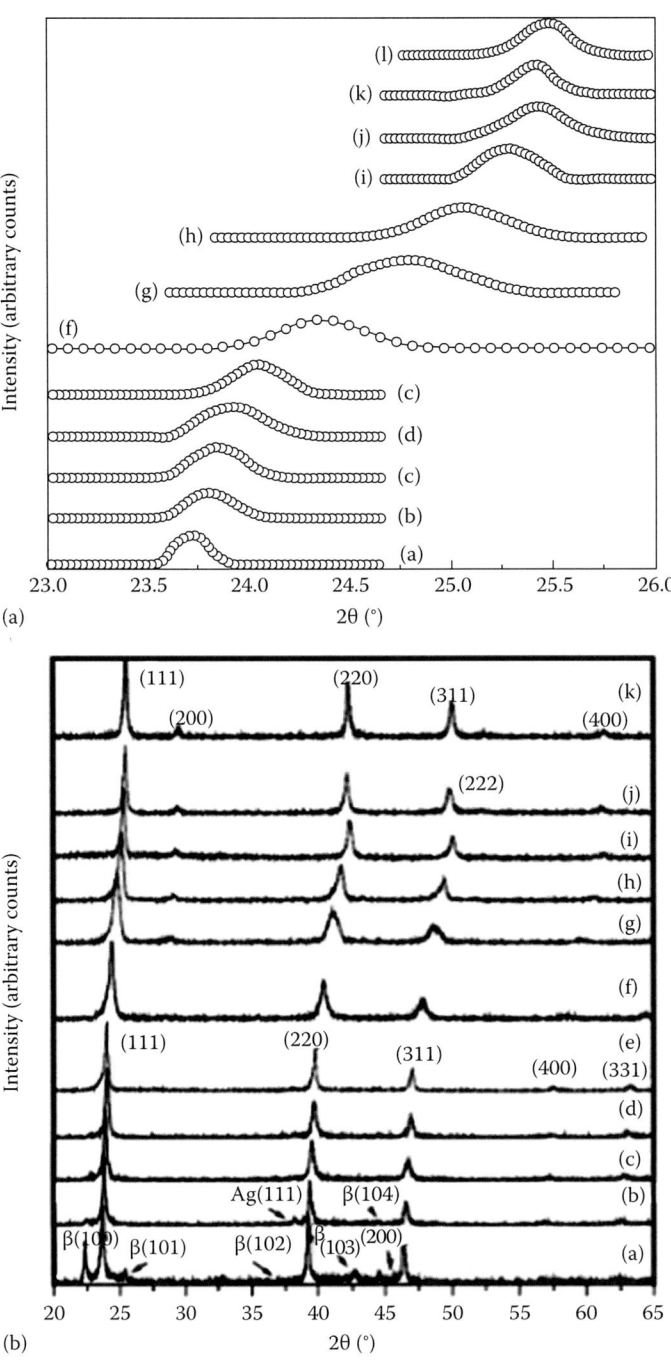

FIGURE 3.11   (a) Dramatic evolution of the nanoscale AgI–CuI solid solution phase upon systematic increase of Cu fraction in the dry Ag–Cu–I mixture precursor upon attrition at room temperature in an agate mortar and pestle. Shown is the (111) Bragg peak in the XRD pattern of the zincblende phase. Notice progressive broadening upon progressive increase of Cu indicating the crucial role of Cu as noticed in the wet synthesis discussed earlier. (b) Evolution of the entire zincblende XRD pattern in the zincblende AgI–CuI system.   *(Continued)*

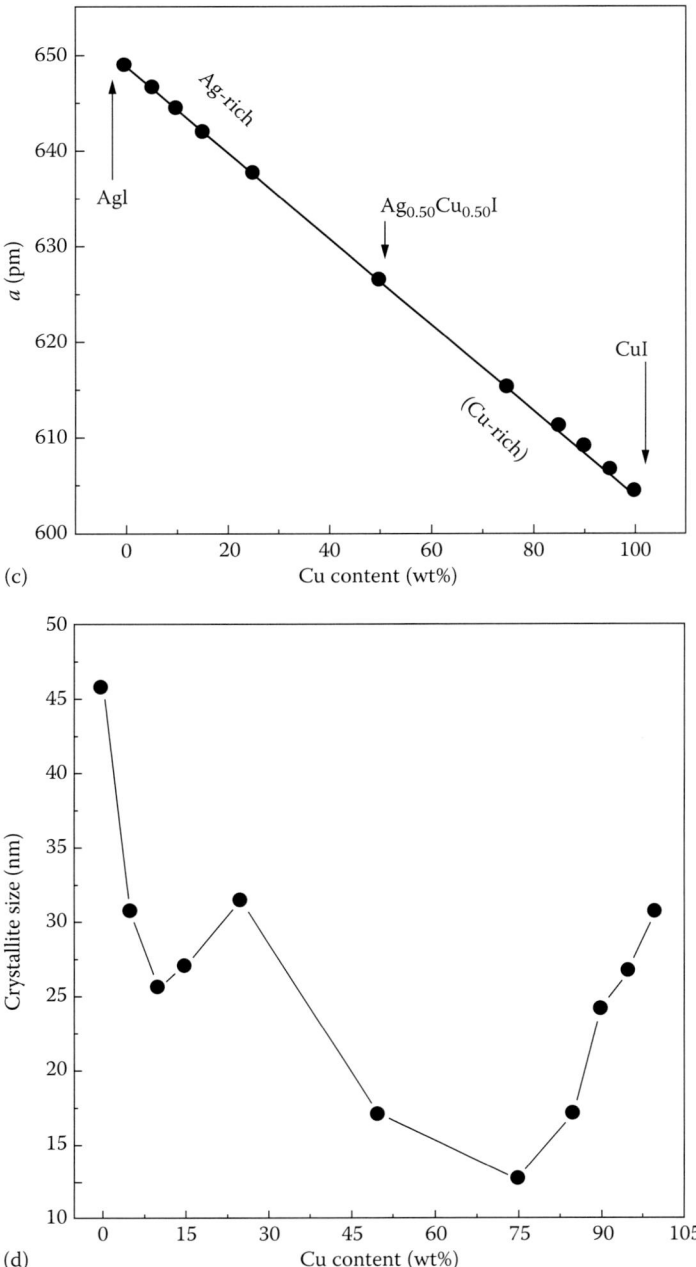

FIGURE 3.11 (*Continued*) (c) Plot of lattice parameter (*a*) versus Cu composition in the AgI–CuI system of solid solutions showing smooth linear decrease from AgI to CuI. (d) Crystallite size changes in AgI–CuI solid solutions made by mechanochemical reaction. Smallest crystallite size occurs for $Cu_{0.75}Ag_{0.25}I$ but unagglomerated crystallites of 17 nm are seen in $Ag_{0.50}Cu_{0.50}I$. (From Bharathi Mohan, D. and Sunandana, C.S., *J. Phys. Chem. Solids*, 65, 1669, 2004.)

in different mechanisms of mechanochemical action on the activated mixtures. Temperature and lithium content strongly affect the composition and lattice constant of the final products [3.16].

During milling or grinding, externally applied mechanical energy is transferred to the particles of reactants causing many changes including deformation, friction, fracture, amorphization, and quenching. Subsequently, new product phase(s) is/are formed at the interfaces of the reactants. With diffusion of atoms of the reactant phases, the product phase grows further and the chemical reaction is realized. The principal advantages of this process are the simplicity, effective mixing accompanying a break-off of the chemical bonds and their recombination, and decreased energy expenditure.

The method has been initially applied to the preparation of $LiCoO_2$ cathode material to mix thoroughly the salts or oxides of Co and Li followed by calcination. The preliminary step considerably shortens the high-temperature ($\geq 800°C$) calcination time.

Table 3.1 gives details of raw materials and processes involved in making cathode materials. Table 3.2 lists the types mechanochemical methods involved in the synthesis of graphite and alloy-based anode materials [3.17].

It is clear from this discussion that mechanochemical technique is indeed an efficient method of synthesis that allows to obtain highly dispersed compounds, either directly during mechanical activation at room temperature or by preliminary activation followed by heat treatment at relatively low temperatures. Mechanochemical synthesis involves mechanical grinding of the dry reactants (essentially solvent free) without heating apart from any heating that results from the conversion of the mechanical energy of grinding into heat.

An interesting recent novel variant is the solvent-assisted mechanochemistry [3.18]. Here, a small quantity of liquid is deliberately added to the reaction mixture. The added liquid can dramatically accelerate, and even enable, MCRs between solids. Interesting links may be seen between the major techniques of synthesis from Figure 3.12.

3.2.5 Hydrothermal Synthesis

Hydrothermal processing can be defined as any heterogeneous reaction in the presence of aqueous solvents or mineralizers under high pressure and temperature conditions to dissolve and recrystallize (recover) materials that are relatively insoluble under ordinary conditions.

TABLE 3.1

Cathode Materials Made by Using Different Precursors

Raw Materials	Process	Features of Structure	Electrochemical Performance
1. $LiCoO_2$			
$LiOH + Co(OH)2$	Milling/10 h	HT-$LiCoO_2$	—
$LiOH + Co(OH)_2$	Milling/40 h	HT-$LiCoO_2$ + spinel Co_3O_4	—
$LiOH·H_2O + Co(OH)2$	Ball milling + 600°C/2 h	$LiCoO_2$	—
$LiOH + Co(OH)_2(Co(OH)_2)$	Milling + 600°C/4 h	Disordered, dispersed HT-$LiCoO_2$	Medium
$Li_2O + CoO$	Milling	Spinel $LiCoO_2$	Poor
$LiOH·H_2O + Co(OH)_2$	Ball milling + 850°C/24 h	HT-$LiCoO_2$	Good
HT-$LiCoO_2$	Ball milling	$LiCoO_2$ + disordered Li_xM_{1-x}	Poor
2. $LiMn_2O_4$			
$LiOH(LiOH·H_2O, Li_2CO_3)$ + MnO_2	Milling	Nanocrystalline $LiMn_2O_4$	4 V, good
$LiOH + Mn_2O_3$ (MnO)	Milling	No $LiMn_2O_4$	—
$Li_2O + MnO_2$	Grinding	Disordered nano spinel $LiMn_2O_4$ with strain or defects	3 V, 4 V plateaus, good cycling
$LiMn_2O_4$ from solid state reaction	Ball milling	Nanodomains, strains, defects	3 V, good cycling

TABLE 3.2

Types of Mechanochemical Methods for the Synthesis of Anode Materials

Factors of Mechano- Chemical Methods	Features of the Factors	Changes of Structure Parameters						Electrochemical Performance
		Particle	Surface Area	Surface Structure	Radicals	Microstructure	Crystal Structure	
1. Graphite								
Jet and turbo milling	Soft, mild	Cut into smaller	Increase		Localized spins, larger g-value		Little change in d_{002}, decrease in L_c	Decrease in irreversible capacity, increase in Coulomb efficiency under optimal condition
Ball and colloid milling		Nanometer, agglomerates	Increase		Localized spins, smaller g-value		Increase or little change in d_{002}, and decrease in content of hexagonal phase	Increase in reversible capacity and coulomb efficiency, improved cycling
Impact/shock-type mechanical milling/grinding	Drastic, strong	Torn into pieces of disordered carbon	Increase	O-terminated carbon		Large ratio of disorder such as vacancies, microcavities, and voids	Decrease of L_a and L_c	Increased reversible Capacity mainly associated with the charge slope >1.0 V, increase in irreversible capacity, capacity fading
Shear-type grinding	Shear force		No evident change	O-terminated carbon		Less disorder or disorganization such as vacancies, microcavities	Little change in crystal size and d_{002}	Increase in reversible and irreversible capacity
Reactive milling	Graphite + lithium		Increase	Lithiated surface				Spontaneous formation of surface passivating film, high reversible capacity, lower hysteresis
Reactive atmosphere	Such as O_2, CO_2, H_2O	Slight increase		Oxygen-containing groups		Disorder present	Little change of d_{002} and crystal size	Loose surface passivating film, exfoliation of graphite, capacity fading, slight increase in reversible capacity

(Continued)

TABLE 3.2 (Continued)

Types of Mechanochemical Methods for the Synthesis of Anode Materials

Precursors	Composition	Structure	Electrochemical Performance	
			Reversible Capacity	Cycling
2. Alloy-type				
Cu + Sn	Cu_6Sn_5	Hexagonal structure	200 mAh/g	Good
Mg + Sn	Mg_2Sn	Cubic + orthorhombic phase	250–300 mAh/g	Good
		Cubic + orthorhombic phase	460 mAh/g	Good
Ni + Sn	Ni_3Sn_4	Nanocrystalline	125–200 mAh/g	Good
Graphite + Sn	Composite such as $C_{0.9}Sn_{0.1}$, $C_{0.8}Sn_{0.2}$	Amorphous graphite + nanocrystalline Sn	400–600 mAh/g	Good
Sn + Mn + C	$SnMn_3C$	Perovskite, nanoparticles	150 mAh/g	Good
Si + Ag	SiAg	Nanosized Si in Ag matrix	ca. 280 mAh/g	Good
Si + TiN		Nanocrystalline or amorphous Si dispersed in TiN	ca 300 mAh/g	Good
Ni (Fe) + Si	NiSi (FeSi)	Mixture of NiSi (FeSi) and Si	600–1000 mAh/g	Poor
Graphite + Si	Composite such as $C_{0.8}Si_{0.2}$		1039 mAh/g	Satisfactory
$Fe_{80}Si_{20}$ + graphite	Composite		ca. 600 mAh/g	Good
β-Zn_4Sb_3	ZnSb + unknown structure		Increased to 566 mAh/g	Poor
Co + Sb	$CoSb_3$	Fine powder, <100 nm	586	Poor
Zn_4Sb_3 + graphite	Composite		581	Good

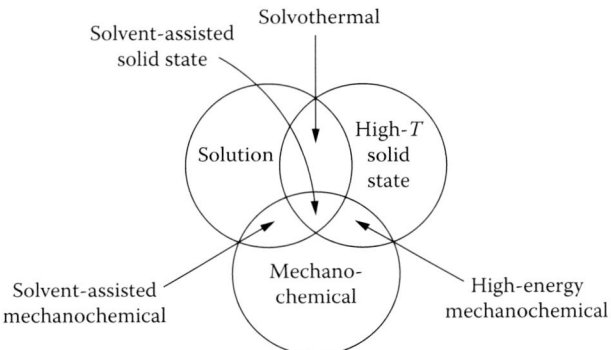

FIGURE 3.12 Mutual overlap of solvent-assisted and solution-based materials synthesis techniques with mechanochemical and high-temperature solid state techniques. Note the comprehensive overlap of the mechanochemical method. (From Bowmaker, G.A., *Chem. Commun.*, 49, 334, 2013.)

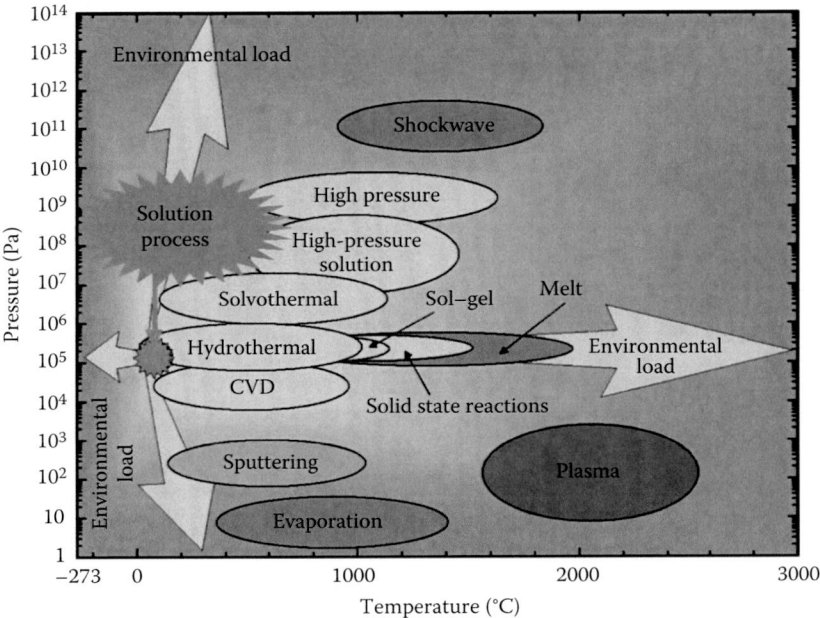

FIGURE 3.13 Hydrothermal synthesis in perspective. (From Byrappa, K. and Adshiri, T., *Prog. Cryst. Growth Char. Mater.*, 53, 117, 2007.)

"Hydrothermal" reaction could be any heterogeneous chemical reaction in the presence of a solvent—aqueous or nonaqueous—above room temperature and at pressures greater than 1 atm. in a closed system.

"Solvothermal" is used to mean any chemical reaction in the presence of a nonaqueous solvent or solvent in supercritical or near-supercritical conditions.

Figure 3.13 compares this method with other synthesis techniques in the form of a pressure versus temperature diagram.

It is important to realize that hydrothermal synthesis has a sound thermodynamic basis by way of the following phase diagram of water (Figure 3.14). It is usually carried out in an autoclave or Parr reactor (Figure 3.15).

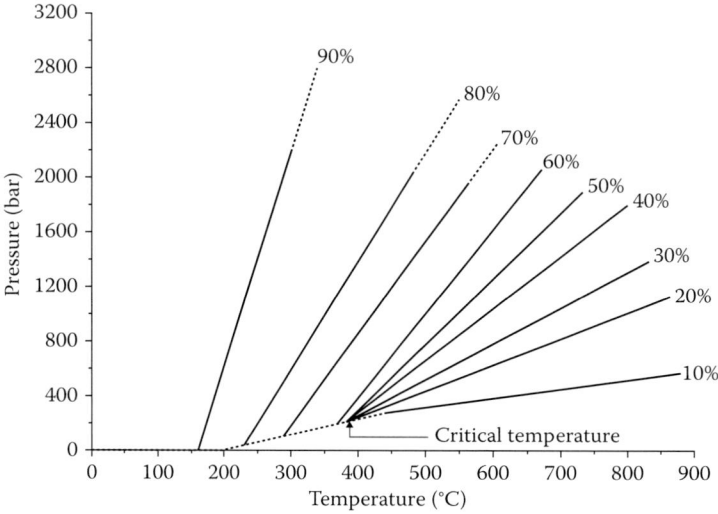

FIGURE 3.14   Pressure versus temperature diagram of water. (From Byrappa, K. and Adshiri, T., *Prog. Cryst. Growth Char. Mater.*, 53, 117, 2007.)

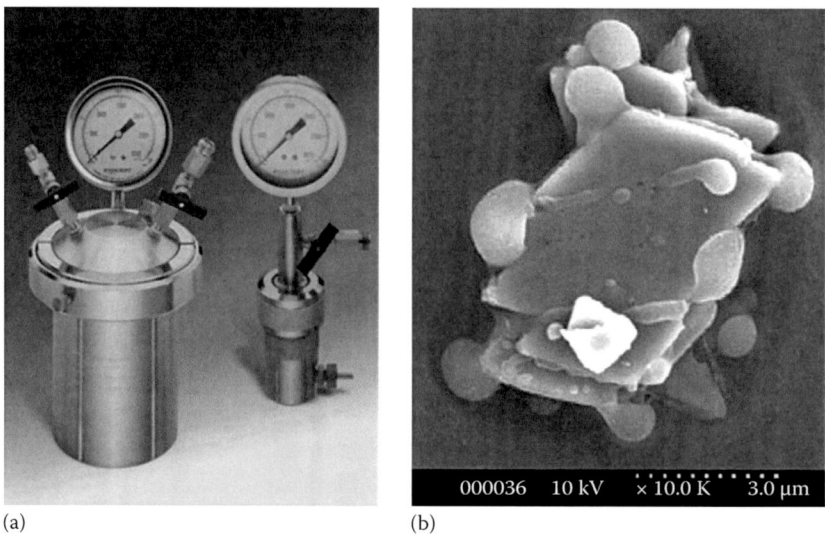

(a)   (b)

FIGURE 3.15   (a) Parr reactor for hydrothermal synthesis. (Courtesy of Parr Instrument Company, St Moline, IL, www.parrinst.com.) (b) A globular carbon coating is formed on $LiFePO_4$ when an excess of ascorbic acid and sugar is added to the reactor. (From Chen, J. et al., *Solid State Ionics*, 178, 1676, 2008.)

We shall illustrate the hydrothermal synthesis by means of three examples: (1) $LiFePO_4$ cathode for the standard Li-ion battery, (2) $Li_4T_{i5}O_{12}$ anode, and (3) sulfur–hollow carbon microsphere composite for the Li–S battery.

1. $LiFePO_4$ may be synthesized in a Parr reactor. The starting materials $FeSO_4 \cdot 7H_2O$ (98% Fisher), H_3PO_4 (85% solution fisher), and LiOH (98% Fisher) are mixed in a molar ratio Li:Fe:P of 3:1:1. A typical concentration of ferrous sulfate is 22 g/L of water. An *in situ* reducing agent is added in the form of sugar/L-ascorbic acid (99% Aldrich), typically 0.8 g/L, to minimize the ferrous-to-ferric oxidation. Also added are multi-walled carbon nanotubes (95% Aldrich). Finally, the autoclave is sealed and heated in the temperature range of 150°C–220°C for 5 h.

Polyethylene glycol (PEG) may be used as a surfactant at 150°C and 175°C maintaining a molar ratio of LiOH:PEG as 1:2.5. The synthesis temperatures for polycrystalline LiFePO$_4$ are also conveniently 150°C and 175°C.

In order to synthesize single crystals 1.8 g/L L-ascorbic acid is to be used. Also addition of 35 mL PEG with constant stirring is necessary. This time the autoclave is sealed and heated at 180°C for 3 h.

LiMnPO$_4$ and LiCoPO$_4$ may also be synthesized by a similar method.

2. Li$_4$Ti$_5$O$_{12}$ or LTO sheets as anodes. A simple low-temperature (60°C) solution—synthesis route to fabricate pure nanostructured Li$_4$Ti$_5$O$_{12}$ or LTO sheets—is suitable for large-scale synthesis mandatory for Li-ion battery applications. When used as an anode material for Li-ion battery, the as-prepared LTO nanosheets present high reversible capacity and good cycling stability even at high current density.

The basic step of this synthesis is the production of TiO$_2$ colloids by the hydrolysis of tetrabutyl titanate. The process is understood by the reaction equation:

$$Ti(C_4H_9O)_4 + 4H_2O \xrightarrow{\text{hydrolysis}} Ti(OH)_4 \downarrow + 4C_4H_9OH \qquad (3.2)$$

In a typical procedure, 4 mmol TiO$_2$ colloids, obtained from the hydrolysis of 1.4 mL tetrabutyl titanate in ethanol/water mixed solution, were mixed with 20 mL 0.2 mol/L LiOH in an Erlenmeyer flask. The Erlenmeyer flask was maintained at 60°C for 7 days and then cooled to room temperature naturally. For comparison, the experiment was also carried out at room temperature (25°C) for 60 days and at 200°C and 240°C for 36 h in autoclave. The resulting white precipitate was recovered by centrifugation, washed with deionized water and ethanol thoroughly, and then dried in an oven at 60°C. Finally, the as-prepared samples were calcinated in a tube furnace at 550°C for 6 h in the air. The four samples were labeled as LTO-60, LTO-25, LTO-200, and LTO-240 corresponding to the synthesis temperatures 60°C, 25°C, 200°C, and 240°C, respectively (Figure 3.16).

A significant aspect of this product is that it goes directly as an anode into the fabrication of a coin cell battery (see Section 3.4.1).

3. A sulfur–hollow carbon (S-HC) cathode material has been synthesized by a hydrothermal process.

In a typical synthesis protocol, monodisperse polystyrene spheres (PS) are dissolved in 6 mL deionized water taken in a glass insert. To this mixture, 0.0080 g sodium dodecyl sulfate (SDS)

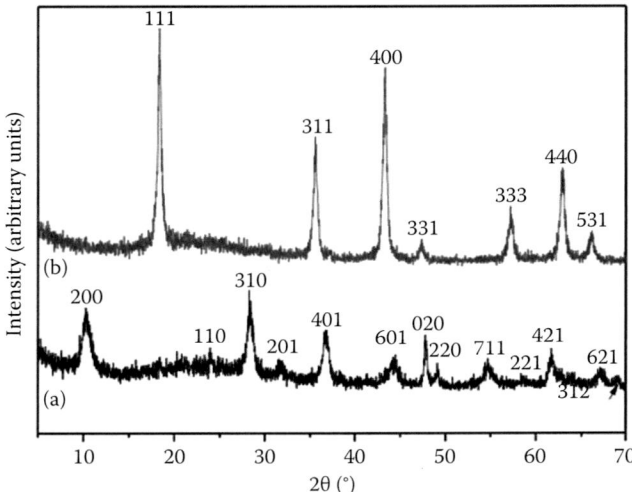

FIGURE 3.16 XRD patterns of Li$_4$Ti$_5$O$_{12}$ calcined at 60°C. (a) Before calcination and (b) after calcination showing monophasic product. (From Wang, J. et al., *Solid State Ionics*, 268, 131, 2014.)

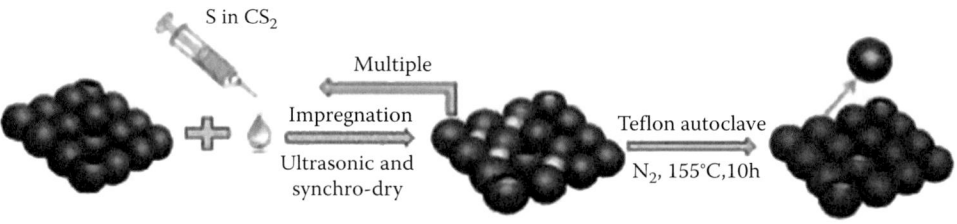

FIGURE 3.17 Hydrothermal synthesis of sulfur–hollow carbon yolk-shell-type cathode. (From Yuan, G. et al., *J. Solid State Electrochem.*, 19, 143, 2015.)

is added and homogenized as a suspension under ultrasonication for 30 min. A further 0.6000 SDS is added under ultrasonication until dissolution of sugar is complete. The glass insert is then transferred into a 50 mL Teflon autoclave filled with sand. The autoclave is then sealed and maintained at 180°C for 20 h and subsequently cooled naturally to ambient temperature. Powder is separated from the suspension with the aid of a centrifuge and then dried at 40°C overnight. Finally, the powder is heated in a furnace at 700°C at the rate of 5°C per minute under argon to remove polystyrene (PS) and obtain hollow carbon microspheres.

The synthesis process is illustrated in Figure 3.17.

To begin with, S was dissolved in carbon.

A direct microscopic evidence of this synthesis is displayed in Figure 3.18.

FIGURE 3.18 Scanning electron micrographs of hollow carbon (a, b) and sulfur-hollow carbon composite (c, d) samples. (From Yuan, G. et al., *J. Solid State Electrochem.*, 19, 143, 2015.)

3.2.6 Microwave Synthesis or Microwave-Assisted Synthesis

Microwave synthesis of materials is a classic example of dielectric heating. In this phenomenon, electric dipoles present in the dielectric materials including solid state ionic materials respond to an applied electric field at microwave frequencies. The reorientation dynamics of the dipoles in the applied alternating electromagnetic fields is mainly responsible for "microwave heating." A convenient measure of this heating effect is $\sin \delta \sim \tan \delta \approx \varepsilon'''/\varepsilon'$ where δ is the phase lag between the current I and the field E. Note that in an ideal dielectric I and E are $90°$ out of phase). (What are ε' and ε''?) This is the energy dissipation factor or loss tangent.

Dipolar species in any medium possesses a characteristic relaxation time τ. Thus, the complex dielectric constant ε^* of the medium (given by the fundamental equation $D = \varepsilon E$) is frequency (ω) *dependent*.

$$\varepsilon^* = \varepsilon' + i\varepsilon'' \qquad (3.3)$$

where
ε' is the real part
ε'' is the imaginary part

In the model of a single relaxation time (τ), ε' and ε'' are given by the Debye equations

$$\varepsilon' = \varepsilon_\infty + \frac{\varepsilon_s - \varepsilon_\infty}{1 + \omega^2 \tau^2} \qquad (3.4)$$

$$\varepsilon'' = \frac{\varepsilon_s - \varepsilon_\infty}{1 + \omega^2 \tau^2} \qquad (3.5)$$

Here
ε_s is the static or zero frequency dielectric constant
ε_∞ is the infinite frequency dielectric constant

Focus on ε''. It varies with ω giving rise to a characteristic peak at $\omega\tau = 1$ or $\omega = 1/\tau$. For water, which is the usual medium for microwave synthesis of materials, the relaxation frequency is ~18 GHz and ε'' is quite large at 2.45 GHz. In a commercial microwave oven employing a dual magnetron oscillator, rapid dissipation of energy and heating of water occurs.

In solid state ionic materials, ions can drift in the applied field giving rise to Joule heating because of ionic currents. In this case $\varepsilon'' = (\sigma_{dipolar} + \sigma_{ionic})/\omega\varepsilon_0$ where $\sigma_{dipolar}$ = conductance due to reorientation current, σ_{ionic} = conductance due to ion drift current, and ε_0 = vacuum dielectric constant.

Microwave power dissipated per unit volume

$$P = \sigma |E|^2 = \omega\varepsilon_0\varepsilon'' |E|^2 = (\omega\varepsilon_0\varepsilon' \tan \delta)|E|^2 \qquad (3.6)$$

In terms of Debye equations

$$P = \left\{ \frac{\varepsilon_0(\varepsilon_s - \varepsilon_\infty)\omega^2\tau}{1 + \omega^2\tau^2} \right\} |E|^2 \qquad (3.7)$$

For negligible heat loss during heating, the rate of heating is given by

$$\frac{\Delta T}{t} = \frac{\sigma|E|^2}{\rho C}$$

(3.8)

where
  ρ is the density
  C is the heat capacity of the material

For a microwave applicator with single-mode resonant cavity, the heating rate

$$\frac{dT}{dt} = \left(\frac{4}{\tan\delta}\right)\left(\frac{1}{\rho C}\right)\left(\frac{\varepsilon''}{\varepsilon'^{1/2}}\right)\left(\frac{1}{V_c}\right)P_0\left(\frac{\xi S}{\rho C}\right)\left[\frac{\text{area}}{\text{volume}}\right]_{\text{sample}}(273+T)^4$$

(3.9)

where
  ρ is the mass density of sample
  V_c is the cavity volume
  P_0 is the microwave power inside the cavity
  S is the Stefan–Boltzmann constant
  ξ is the surface emissivity of the sample

Coupling to microwaves of 2.45 GHz in microwave reactors such as dual magnetron oscillator is possible only if the material has dipole absorption in the region of 2.45 GHz, the reactor frequency. Because

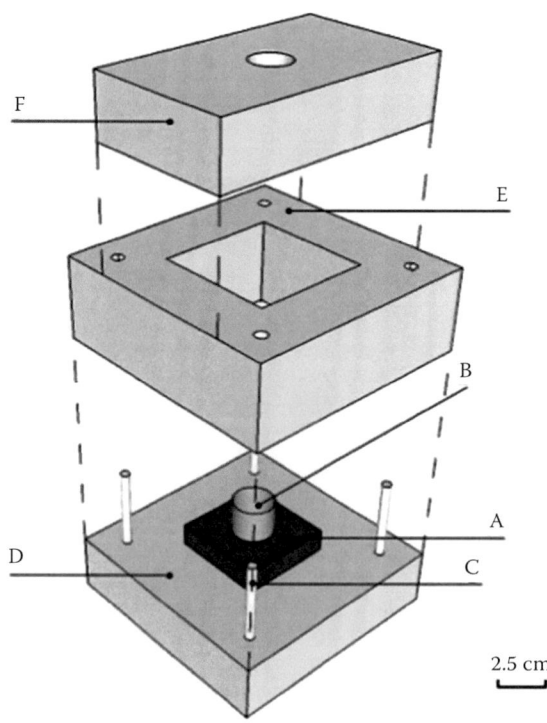

2.5 cm

FIGURE 3.19 Sample chamber to go with a domestic microwave oven (Panasonic Piccolo NNST359WRU, 2.54 GHz, 800 W_{max} power) for microwave synthesis of $LiMn_2O_4$ and other solid state ionic materials. Insulating box shown in its mounting sequence is composed of (A) a SiC plate as the microwave susceptor, (B) a 97%–3% CeO_2 crucible, (C) alumina rods, (D) 80% aluminum oxide–20% zirconium oxide base, (E) walls, and (F) lid. (From Rao, K.J. et al., *Chem. Mater.*, 11, 882, 1999.)

of the dependence of τ on temperature and microwave absorption of the material, there is a need to optimize the power by matching the material relaxation frequency with the microwave reactor frequency. The choice of sample container material is crucial besides frequency and temperature. Typical conditions are 3 min exposure at 1346 K for submicron graphite carbon, 6 min at 1560 K for MnO_2, 5.35 min at 925 K for Ag_2S, 7 min at 1019 K for Cu_2S, and 11 min at 995 K for CuBr. For more details, see Reference 3.23.

Solid state ionic materials synthesized using microwave heating include the following: (1) $LiMn_2O_4$ using a mixture of LiI and MnO_2 irradiated for just 6 min, with I acting as an antioxidation shroud [3.24]; (2) $LiCoO_2$ has been made using metal organic precursors [3.25]. Glasses having NASICON composition have also been made using the rapid heating method [3.26].

An innovative sample chamber to go with the domestic microwave oven (Figure 3.19) suitable for microwave synthesis of solid state ionic materials has been described by Silva et al. [3.27]. The usefulness of this jig has been demonstrated by successful synthesis of high-quality $LiMn_2O_4$.

In the next section, we discuss thin-film synthesis technologies.

3.3 Thin-Film Synthesis Technologies

3.3.1 Physical Vapor Deposition

Especially, vacuum evaporation, sputtering, and pulsed laser deposition, which are all physical vapor deposition (PVD) techniques, are used as a cost-effective method to prepare the inorganic thin film [3.28].

3.3.1.1 Resistive Thermal Evaporation or Thermal Evaporation by Resistive Heating

Thermal evaporation is the most widely used technique for the preparation of thin films of metals, alloys, and also many compounds, as it is very simple and convenient. Here the only requirement is to have a vacuum environment in which sufficient amount of heat is given to the evaporants to attain the vapor pressure necessary for the evaporation. The evaporated material is allowed to condense on a substrate kept at a suitable temperature. When evaporation is made in vacuum, the evaporation temperature will be considerably lowered and the formation of the oxides and incorporation of impurities in the growing layer will be reduced. Evaporation is normally done at a pressure of 10^{-5} Torr. At this pressure, a straight-line path for most of the emitted vapor atoms is ensured for a substrate-to-source distance of nearly 10–50 cm. The characteristics and quality of the deposited film will depend on the substrate temperature, rate of deposition, ambient pressure, etc., and the uniformity of the film depends on the geometry of the evaporation source and its distance from the source.

This is done inside a vacuum chamber where the material, usually in a boat, is heated typically to its melting point and the substrate to be deposited on is positioned facing the source a couple of feet away. A high current flowing through the boat heats it up and causes evaporation. A crystal monitor is mounted close to the substrate, which provides an estimate of how much and how fast the material is being deposited. The distance between the source and the substrate is wide to prevent solid particles reaching the substrate. A schematic of the bell-jar-based equipment is shown in Figure 3.20.

The merits of the vacuum deposition method are as follows:

1. A simple and convenient equipment.
2. The film formation of metal is relatively easy.
3. It is not dependent on the raw material form such as powder, bulk, and wire.
4. The film thickness is controlled by the quantity of the raw material, and the control of the film formation rate is difficult.
5. The adhesion for the substrate of the thin film is comparatively weak.
6. Deposition of high melting point materials such as an oxide (say Li_2O or $LiCoO_2$ or ZrO_2) is difficult.

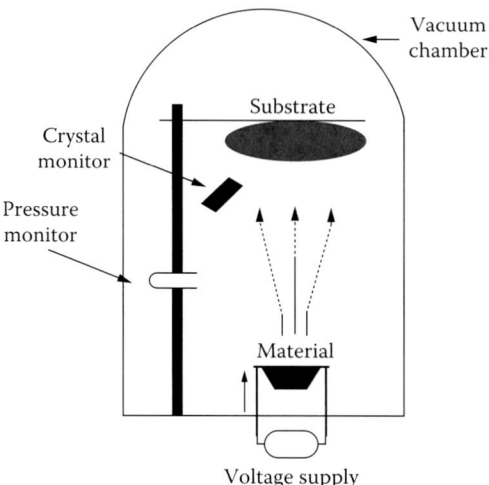

FIGURE 3.20 Vacuum thermal evaporation schematic. The chamber is to be connected to a rotary–diffusion pumping system (see Section 3.3.1.3).

As an example of the ionic conductor fabricated by vacuum deposition method, AgI thin film can be prepared on an insulating substrate. Ag is deposited on the insulating substrates such as SiO_2 glass by the vacuum deposition method. Afterward, AgI thin film is formed by the reaction of Ag with I_2. The thickness of prepared AgI film can be varied by adjusting the quantity of deposited Ag. By x-ray diffraction measurement, the crystal structure of the as-prepared AgI thin films was confirmed to be β-AgI and slightly mixed γ-AgI. Although in this method AgI is not directly deposited, it may yet be classified as a vacuum deposition method.

Vacuum evaporation method is thus a method in which a raw material is heated and melted in the vacuum by a heater, when the evaporated atom or cluster piles up on the substrate. Metals such as silver and copper and also metal halides such as AgI and CuI may be deposited using this method. To melt and evaporate the raw material, not only the resistance heating but also an electron beam and radio frequency sputtering are used.

3.3.1.2 Electron Beam Physical Vapor Deposition

The basic principle of the electron beam physical vapor deposition (EBPVD) technique is that a stream of electrons is accelerated through fields of typically 5–10 kV and focused onto the surface of the material for evaporation. The electrons lose their energy very rapidly upon striking the surface and the material melts at the surface and evaporates. That is, the surface is *directly heated by impinging electrons*. Direct heating allows the evaporation of materials from water-cooled crucibles. These are necessary for evaporating reactive and, in particular, reactive refractory materials to avoid almost completely the reactions with crucible walls. This allows the preparation of high-purity films because crucible materials or their reaction products are practically excluded from evaporation. Electron beam guns can be classified into thermionic and plasma electron categories. In the former type, the electrons are generated thermionically from heated refractory metal filaments, rods, or disks. In the latter type, the electron beams are extracted from plasma confined in a small space. Figure 3.21 shows a schematic of the EBPVD system.

The deposition chamber must be evacuated to a pressure of at least 7.5×10^{-5} Torr to allow passage of electrons from the electron gun to the evaporation material, which can be in the form of an ingot or rod. Electron beams can be generated by thermionic emission, field electron emission, or the anodic arc method. The generated electron beam accelerated to a high kinetic energy is directed toward the evaporation material. Upon striking the evaporation material, the electrons will lose their energy very rapidly. The kinetic energy of the electrons is converted into other forms of energy through interactions with the evaporation material. The thermal energy produced heats up the evaporation material causing it to melt or sublimate. When temperature and vacuum levels are sufficiently high, vapor will result from the melt

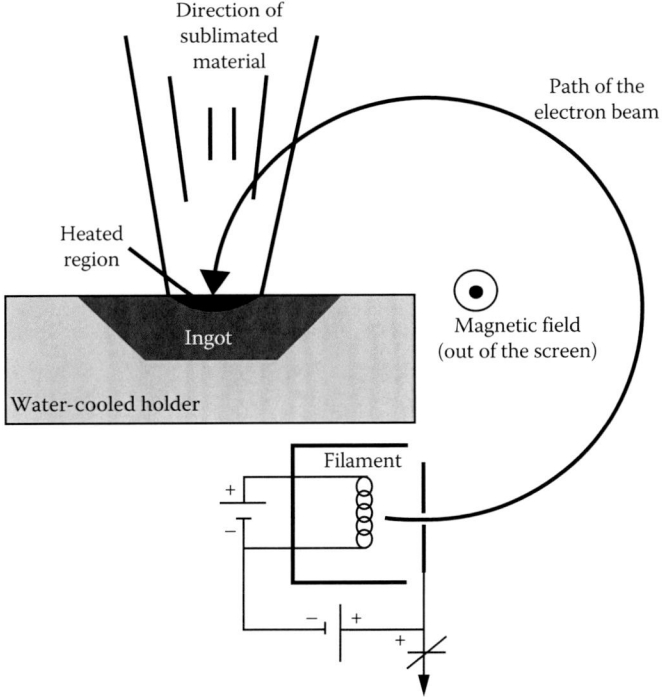

FIGURE 3.21 An e-beam physical vapor deposition (EBPVD) system showing water-cooled sample holder and the magnetron circuit for changing the path of the e-beam. A target anode is bombarded with an e-beam given off by a charged tungsten filament under high vacuum. The e-beam causes atoms from the target to transform into the gaseous phase. These atoms then precipitate into solid form, coating everything in the vacuum chamber (within line of sight) with a thin layer of the anode material. (From Liu, W.Y. et al., *Electrochem. Solid State Lett.*, 7, 136, 2004.)

or solid. The resulting vapor can then be used to coat surfaces. Accelerating voltages can be between 3 and 40 kV. When the accelerating voltage is between 20 and 25 kV and the beam current is a few amperes, 85% of the electron's kinetic energy can be converted into thermal energy. Part of the incident electron energy is lost through the production of x-rays and secondary electron emission.

There are three main EBPVD configurations: electromagnetic alignment, electromagnetic focusing, and the pendant drop configuration. Electromagnetic alignment and electromagnetic focusing use evaporation material in the form of an ingot. The pendant drop configuration uses a rod. Ingots will be enclosed in a copper crucible/hearth while a rod will be mounted at one end in a socket. Both the crucible and the socket must be cooled by water circulation. In the case of ingots, molten liquid can form on its surface, which can be kept constant by vertical displacement of the ingot. The evaporation rate may be on the order of 10^{-2} g/cm^2 s.

In an application to solid state ionics, LiPON or **li**thium **p**hosphorus **oxyn**itride glass films have been fabricated by nitrogen plasma–assisted deposition of e-beam reactive evaporated Li$_3$PO$_4$. A growth rate of ~8.33 nm and an ionic conductivity of 10^{-7}–10^{-8} were achieved [3.29]. Amorphous lithium lanthanum titanate (Li$_x$La$_{(2-x)/3}$TiO$_3$ or LLTO films—as an effective substitute to the moisture sensitive LiPON electrolyte—has been fabricated by Li et al. [3.30]. The continued use of liquid electrolytes in Li-ion batteries gives a major clue that the solid electrolyte can be amorphous as found in this work. The targets for the e-beam evaporation were fabricated by the solid state reaction of Li$_2$CO$_3$, La$_2$O$_3$, and TiO$_2$, mixed in 1:1:4 molar stoichiometry pressed into pellets and calcined at 900°C in air for 8 h.

Molecular beam epitaxy (MBE) method for the preparation of the artificial lattice is also a type of evaporation method. It is carried out in the ultrahigh vacuum—a vacuum regime characterized by pressures lower than about 10^{-7} Pa or 100 nPa (10^{-9} mbar, ~10^{-9} Torr) (see Reference 3.31 for details). LaF$_3$ superionic conducting films have been deposited on Si(111) substrates using MBE [3.32].

3.3.1.3 RF Magnetron Sputtering

When an atom or ion with high kinetic energy collides with a solid, it will lose the kinetic energy by the collision with atoms and molecules that construct the solid. As a result of the momentum transfer from the incident particle, removal and ejection of particles from the bombarded solid surface occur. What is called "sputter" is the phenomenon that the atom, which constructs the solid, is removed from the solid surface by bombardment with high-speed ions.

Sputtering is one of the most versatile techniques used for the deposition of device quality films. Sputtering process produces films with better controlled composition, provides films with greater adhesion and homogeneity, and permits better control of film thickness. The sputtering process involves the creation of gas plasma, usually an inert gas such as argon by applying voltage between a cathode and an anode. The target holder is used as the cathode and the anode is the substrate holder. Source material is subjected to intense bombardment by ions. By momentum transfer, particles are ejected from the surface of the cathode, and they diffuse away from it, depositing a thin film onto a substrate. Sputtering is normally performed at a pressure of 10^{-2}–10^{-3} Torr. There are two modes of powering the sputtering system: DC and RF biasing. In DC sputtering system, a direct voltage is applied between the cathode and the anode. This method is restricted for conducting materials only such as Li and Na. RF sputtering is suitable for both conducting and nonconducting materials; a high-frequency generator (13.56 MHz) is connected between the electrodes of the system. Magnetron sputtering is a process in which the sputtering source uses magnetic field at the sputtering target surface. Magnetron sputtering is particularly useful when high deposition rates and low substrate temperatures are required.

Both reactive and nonreactive forms of DC, RF, and magnetron sputtering have been employed for the deposition of compound semiconductors. In reactive sputtering, the reactive gas is introduced into the sputtering chamber along with argon to deposit thin films. For example, to deposit metal oxide thin films, pure metal target is sputtered in a mixture of argon and oxygen atmosphere. The deposition rates and properties of the films strongly depend on the sputtering conditions such as the partial pressure of the reactive gas, the sputtering pressure, substrate temperature, and substrate-to-target spacing.

The process of RF sputter deposition is made possible due to the large difference in mass, and hence mobility, of electrons and inert gas ions. Because electrons are many times less massive than ions, electrons attain much greater velocities and travel much further than ions during each cycle of the applied RF voltage waveform. Since electrons travel much further, they eventually accumulate on the target, substrate, and chamber walls such that the plasma is the most positive potential in the system. These induced negative voltages or "sheath voltages" cause acceleration of positive ions toward the negatively charged surfaces, which subsequently leads to sputtering events. The volume adjacent to a surface tends to be relatively free of electrons because of the negatively charged surface. This leads to a "dark space" because electrons are not available to excite gas atoms [3.8]. A schematic diagram of the RF sputtering system is shown in Figure 3.22. The target is selectively sputtered by controlling the relative surface areas of the target and the substrate holder. If space-charge-limited current is assumed, the ion current flux J can be estimated by the Child–Langmuir equation,

$$J = \frac{KV^{3/2}}{D^2 m_{ion}} \qquad (3.10)$$

where
  D is the dark space thickness
  V is the sheath voltage
  m_{ion} is the ionic mass
  K is the proportionality constant

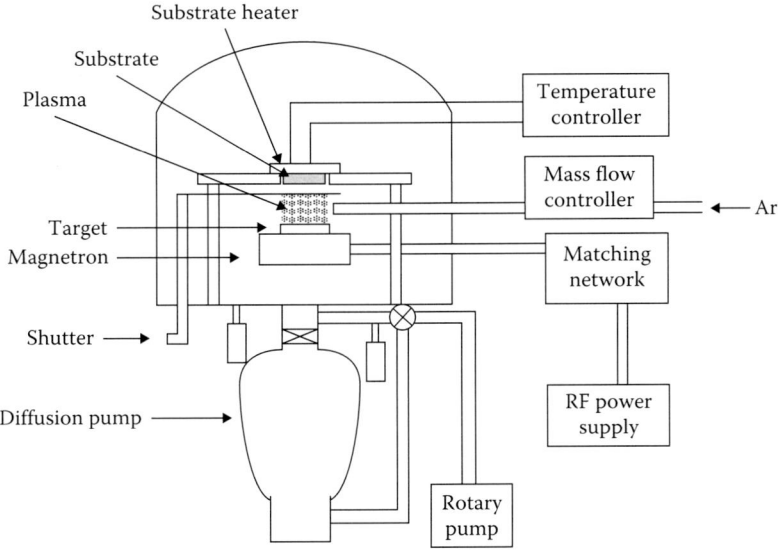

FIGURE 3.22   Schematic of the magnetron for RF sputtering.

Since the positive ion current must be equal at both the electrodes,

$$\frac{A_A V_A}{D_A^2} = \frac{A_B V_B}{D_B^2} \qquad (3.11)$$

where A_A and A_B are the surface areas of electrodes A and B, respectively.

It should be noted that this step differs from the assumption of treating the positive ion current densities equal. If the positive ion current densities were equal, there would be a much greater positive ion current flowing during one half cycle of the applied voltage waveform than the other due to the much greater area of the grounded substrate electrode. Therefore, because this system is assumed to be in the steady state, the total positive ion current per half cycle should be of the relevant quantity.

The glow discharge itself is a region where large quantities of positive and negative charge exist and can be modeled as a wire. Since most of the voltage in the glow discharge is dropped across the dark space, and they have small conductivities, they can be modeled as capacitors such that the capacitances

$$C \propto \frac{A}{D} \qquad (3.12)$$

Furthermore, an AC voltage will divide across two series capacitors such that

$$\frac{V_A}{V_B} = \frac{C_B}{C_A} \qquad (3.13)$$

From Equations 3.10 through 3.12,

$$\frac{V_A}{V_B} = \left(\frac{A_B}{A_A}\right)^2 \qquad (3.14)$$

Thus, smaller area will see larger sheath voltage, whereas larger area will see a smaller sheath voltage by a power of 2. The important inequality $A_B > A_A$ helps selective sputtering of the target. This is done in practice by grounding the substrate holder to the entire chamber resulting in a very large AB. Thus, it is extremely important that the substrate holder and the system are well grounded to ensure that re-sputtering of the growing film does not occur.

Sputtering yield is defined as the quantity of the material sputtered per ion (atoms/ion or grams/ion). To measure the sputtering yield accurately, we need to measure three experimental parameters, namely, the crater volume (cm^3) formed by sputtering, the ion current (A), and the sputtering time (s).

Main features observed for the sputtering techniques are as follows:

1. Sputtering yield is different for different elements. It increases as the reciprocal of the binding energy of the surface atoms.
2. Sputtering yield decreases as the surface damage increases; that is, the sputtering yield of rough surface is lower than that of smooth surface.
3. As the mass of the sputtering species increases, the sputtering yield increases.
4. Light mass ions penetrate deeper into the target than heavier mass ions.
5. As sputtering energy increases, the sputtering yield increases up to 10–100 keV. At higher energies, the sputtering yield again decreases since the ions penetrate into the target.
6. Smaller particles penetrate further into the target. Thus, the energy when the yield starts to decrease is lower for lighter particles.
7. For multicomponent samples, the lightweight particle is usually preferentially sputtered if the binding energies of the components are similar. The sputtering rate of each component increases as the reciprocal of the binding energy and mass of that component.
8. Sputtering of oxide targets result in preferential depletion of oxygen.
9. Sputtering yield of metal oxide is less than the sputtering yield of corresponding metals.
10. For oxide samples, sputtering in an oxygen-rich environment decreases the sputtering yield; the sputtering yield does not vary in other environments (e.g., CO, N_2). *Adsorption without chemical bonding is not enough to reduce the sputtering yield.*

A magnet of 2000 G was used to deflect the ions. A schematic diagram of the magnetron is shown in Figure 3.22. The vacuum system consists of a 6 in. diameter diffusion pump backed by a rotary pump. The RF supply is connected to the magnetron through a capacitive matching network. The flow of argon gas into the vacuum chamber is controlled using a mass flow controller.

RF sputtering is adapted to a fabrication of insulating thin film. The details of the mechanism of sputtering have been discussed by Wehner [3.33] and by Lehmann and Sigmund [3.34].

The sputtering technique has the following principal advantages:

1. The preparation of a thin film of high-melting-point material in a controllable manner is comparatively easy.
2. The thickness of prepared film is comparatively uniform, and the composition is homogeneous.
3. Accurate control of the film thickness (deposition rate).
4. It is possible to prepare a crystalline thin film.
5. Excellent film adhesion for the substrate.

However, the technique has the following demerits:

1. Its small deposition rate.
2. There is a case in which the film is damaged by a re-sputtering.
3. The chemical composition of the target and prepared thin film is often different.

The controlling factors in sputtering technique are as follows:

1. Introduced gas and the pressure
2. The sputtering output
3. The distance between targets and substrate
4. Quality and configuration of the targets
5. Surface state and temperature of the substrates
6. The condition of the post annealing after the film formation

A variety of inorganic oxide ionic conductor thin films may be fabricated by the RF-sputtering method. For example, Li_2S–GeS_2–Ga_2S_3 Li-ion-conducting glassy thin films with a wide composition have been fabricated by Yamashita et al. [3.35]. The conductivity of the thin film was almost identical in bulk glasses of similar composition and increased along with increasing Li_2S content.

3.3.1.4 Pulsed Laser Deposition

Ever since the thin films of high-temperature syperconducing oxide $YBa_2Cu_3O_{7-\delta}$ were made by pulsed laser deposition (PLD), this technique has emerged as one of the premier thin-film deposition technologies and hase ben applied to ferroelectric and superionic materials.

Figure 3.23a shows the schematic of a PLD system. This technique relies—a bit like the sputtering—on material removal through a laser beam. PLD relies on a photon–matter interaction to create an ejected plume of the material in the form of a target. The ejection of the material occurs due to rapid explosion

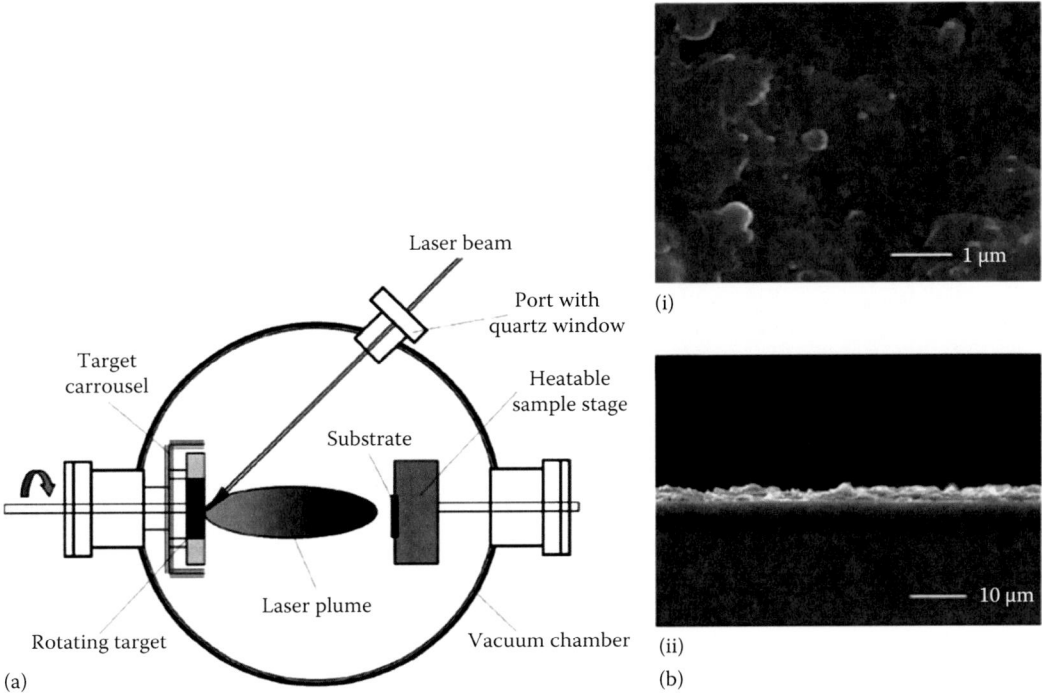

FIGURE 3.23 (a) Schematic of an evacuated pulsed laser deposition system. A laser beam hits a rotating target at 45° angle to the vertical. A laser plume created by a high-power laser pulse-target solid impact travels toward the (heatable) substrate facing the target and a film is produced whose compsotion is roughly that of the target. (b) (i) Scanning electron micrograph of the surface of LiPON film obtained from PLD, (ii) LiPON film cross-sectional image. Laser fluence: 15 J/cm^2, N_2 gas pressure: 200 mTorr. (From Zhao, S. et al., *Thin Solid Films*, 415, 108, 2002.)

of target surface due to superhating. Unlike thermal evaporation, which produces a vapor compsoition that depends on the vapor pressures of the elements constituting the target material, the laser-induced explosion produces a plume of material with stoichiometry similar to the target material. It is generally earier to obtain the required stoichiometry of multielement target materials using PLD than with any other thin-film deposition techniques.

Apart from the laser-material interaction, the mechanism of PLD involves the formation of plasma plume with high energetic species and the transfer of the "ablated" material through the plasma plume onto the heated substrate surface. A four-stage thin-film formation may be enviaged: (1) laser radiation-target interaction, (2) dynamics of the ablated materials on the substrate, (3) deposition of the ablation materials on the substrate, and (4) nucleation and growth of a thin film on the substrate surface. Each of these four stages in the PLD precess is critical to the formation of high-quality, epitaxial, crystalline, stoichiometric, and uniform thin films.

An application of PLD technique to fabricate LiPON thin films is discussed here. A stainless steel vacuum chamber equipped with a rotating holder for the Li_3PO_4 target was used to deposit the LiPON thin films. The Li_3PO_4 target was prepared by compressing Li_3PO_4 powder (99.99%) into a 13 mm diameter pellet and was sintered at 600°C for 2 h. Three hundred and fifty-five nanometer laser beam provided by the third harmonic of a Q-switched Nd:YAG laser (Spectra Physics, GCR-190) was focused and irradiated onto a rotating Li_3PO_4 target at an incidence angle of 45°. The pulsed laser repetition rate and pulse width were 10 Hz and 6 ns, respectively. The laser energy was measured by a power meter with a pyroelectric detector. Three different substrates, Si wafer, Au-coated Si wafer, and Al-coated glass plate, were used in these experiments. The substrate was placed at a distance of 5 cm away from the Li_3PO_4 target. Before deposition, the vacuum chamber was evacuated. N_2 (99.99%) gas with a needle valve. The film deposition was carried out with a continuous flow of the N gas. Figure 3.23b shows the scanning electron micrographs in (A) surface and (B) cross-sectional scans. The crystallinity of the well-formed grains vouches for the successful application of the technique.

Next, we focus on the chemical vapor deposition (CVD) technique.

3.3.2 Chemical Vapor Deposition

In contrast to physical vapor deposition, which essentially creates a vapor phase of a solid precursor by several methods, CVD involves evaporation of a liquid solution (often an aqueous solution) under various conditions. Chemical vapor deposition, chemical bath deposition, and spray pyrolysis are among such techniques of which we discuss the first one, which has the advantage of being combined with sol–gel technique. Let us describe a hybrid CVD–sol–gel (SG) synthesis used to fabricate $LaCoO_3$ thin films. The technique is illustrated in Figure 3.24.

The hybrid CVD–SG synthesis is carried out in three successive steps: (1) deposition of La–O specimens, (2) coverage with a Co–O overlayer, and (3) annealing in air. Both CVD and SG routes are used to prepare La–O and Co–O systems. In the first case (Figure 3.24a), CVD La–O systems deposited at low temperatures are used as substrates. These supports offer, in principle, higher surface area (compared to conventional support with the starting SG solution. A percolation of the SG layer into the CVD layer is thus enabled.

In the second case (Figure 3.24b), the as-prepared SG La–O systems (i.e., $LaO_x(OA)_y$ xerogels where A = R, H) are used as substrates. Their porous structure endowed with non-bridging groups (–OH and –OR) can provide reaction sites for successive chemical modifications on both sides of the surface and subsurface layers. The CVD infiltration power further promotes nucleation into the porous structure. This leads to an intimate intermixing between the two layers already during the deposition process! In both cases, namely, (1) CVD followed by Sg and (2) SG followed by CVD, such intermixing phenomena are expected to favor the subsequent thermally induced La–Co–O reactions, ultimately leading to the formation of $LaCoO_3$. Annealing in air helps further investigations such as the nanocrystallization of $LaCoO_3$ as a function of synthesis and processing conditions.

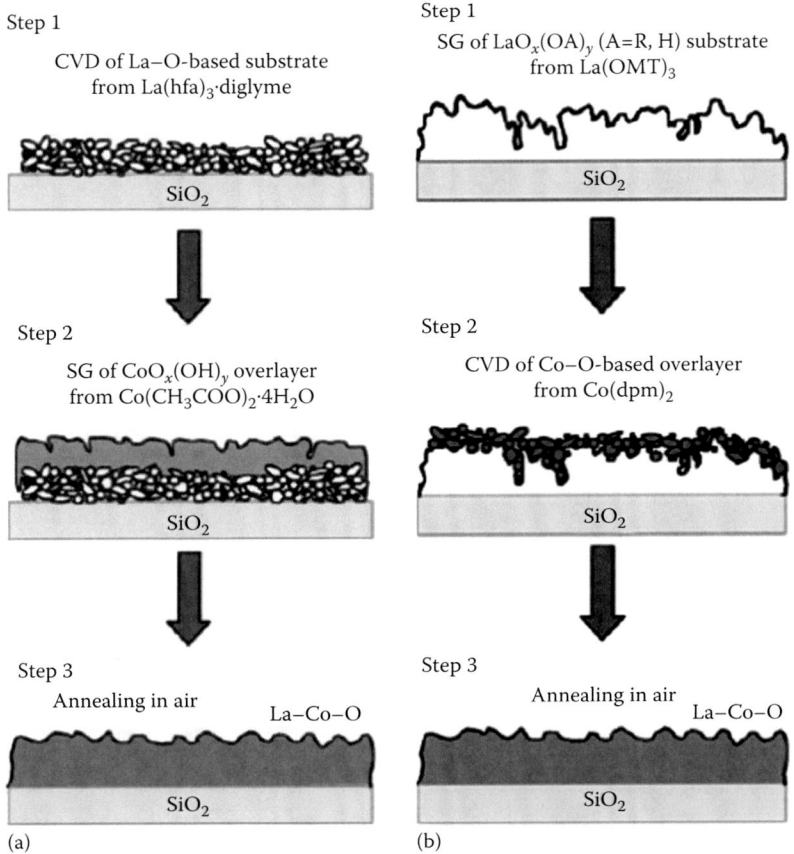

Step 1
CVD of La–O-based substrate
from La(hfa)$_3$·diglyme

SiO$_2$

Step 1
SG of LaO$_x$(OA)$_y$ (A=R, H) substrate
from La(OMT)$_3$

SiO$_2$

Step 2
SG of CoO$_x$(OH)$_y$ overlayer
from Co(CH$_3$COO)$_2$·4H$_2$O

SiO$_2$

Step 2
CVD of Co–O-based overlayer
from Co(dpm)$_2$

SiO$_2$

Step 3
Annealing in air
La–Co–O

SiO$_2$

(a)

Step 3
Annealing in air
La–Co–O

SiO$_2$

(b)

FIGURE 3.24 Illustrating the hybrid CVD–SG method to fabricate LaCoO$_3$ thin films. (a) SG of CoO$_x$ overlayer follows CVD of La–O-based substrate (step 2 follows step 1), (b) CVD of Co–O-based overlayer follows SG of LaO$_x$(OA)$_y$ substrate (step 1 follows step 2). The final annealing in air (step 3) is common to both sequences. (From Armelao, L. et al., *Chem. Mater.*, 17, 427, 2005.)

The actual synthesis of LaCoO$_3$, outlined in Figure 3.24, depends on the technique adopted for the preparation of the La–O substrates. In both cases, the production of monophasic nanoscale LaCoO$_3$ requires optimization of (a) relative amounts of La/Co and (b) annealing conditions. The detailed experimental procedure is as follows:

In the case (a), La–O films are synthesized on silica slides by CVD at 200°C using La(hfa)$_3$-diglyme as a precursor (see Reference 3.37 for details). A five-step cleaning is done before the depositions. The synthesized compound is vaporized at 110°C in each CVD experiment lasting 50 min. Nitrogen is the carrier gas and O$_2$ flow is introduced separately into the reaction chamber after passage through a water reservoir held at 50°C. Optimized pressure/gas flow conditions are total pressure 10 mbar, nitrogen gas flow rate 50 sccm, and oxygen (O$_2$ + H$_2$O) flow rate 100 sccm. Co–O overlayers are produced by SG dip-coating starting from a methanolic solution of Co acetate tetrahydrate through three successive dips at a constant withdrawal speed of 7 cm/min.

In the second case, Figure 3.24b, La oxide xerogels are prepared on cleaned silica slides by dip coating from ethanolic solutions of La(OMT)$_3$ (–OMT = 2-methoxyethoxy) at a withdrawal speed of 50 cm/min. Co-O-based films are fabricated by CVD using Co(dpm)$_2$ as Co source. Precursor is vaporized at 90°C and carried onto the substrate surface ($T = 300$°C) in an O$_2$ flow. The conditions are total pressure 10 mbar, O$_2$ flow rate 150 sccm, and duration 30 min.

In both the cases, the processing conditions are to be chosen after preliminary experiments. The aim is to realize a La/Co atomic ratio close to the stoichiometric one in the final composition.

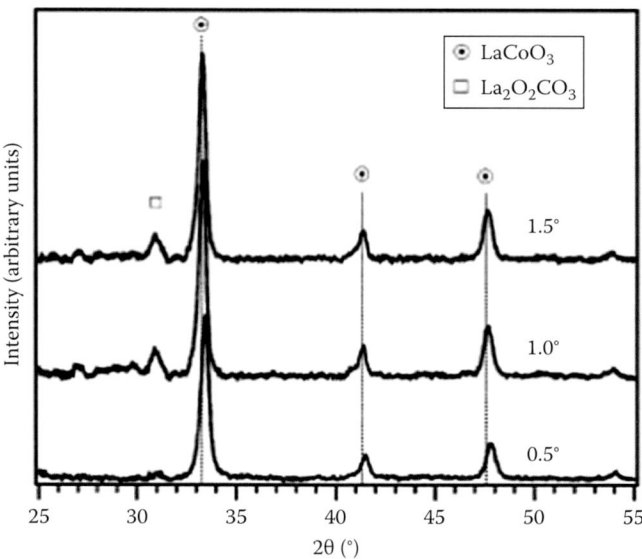

FIGURE 3.25 GIXRD scans on LaCoO₃ collected at angles of incidence 0.5°, 1.0°, and 1.5°. LaCoO₃ was obtained by CVD of Co oxide on LaO$_x$(OA)$_y$ xerogel (CVD followed by SG). Annealing was done in air at 8000^C for 5 h. Peak positions correspond to bulk LaCoO₃. (From Armelao, L. et al., *Chem. Mater.*, 17, 427, 2005.)

The as-prepared samples are to be subsequently annealed in air between 400°C and 900°C for durations from 1 to 8 h. This annealing protocol promotes La–O/Co–O intermixing and reaction processes thus inducing the formation of LaCoO₃.

A glancing angle x-ray diffraction (GIXRD) characterization of the LaCoO₃ (Figure 3.25) establishes the hybrid method.

Last but not the least, we discuss ion-beam-assisted deposition briefly.

3.3.3 Ion-Beam-Assisted Deposition of Thin Films

The technique of ion-beam-assisted deposition (IBAD) involves the bombardment of thin films with energetic particles. It provides thin films and coatings with modified microstructure and properties. The additional energy imparted to the deposited atoms causes atomic displacements in the growing film and enhanced surface atom migration. This can result in improved film properties, which include formation of new phases. Additionally, enhancement of the heteroepitaxy, modification of the residual stress, and better adhesion of thin films are achieved by IBAD.

In the IBAD technique, the ions from an ion source impinge on the substrate simultaneously with the deposited atoms. The majority of ion sources being used for IBAD are of a broad beam design (see later). Other ion sources are the Hall-current source and the electron cyclotron resonance source. Electron beam evaporators, effusion cells, or sputtering targets are used for the physical vapor deposition. A small angle between the ion and vapor sources is the most straightforward geometry. The operating pressure in IBAD systems is typically between 104 and 10 6 mbar. The ion energies have to be chosen according to the application of the films. Typical ion energies range between 50 and 5 keV and ion current densities between 1 and 200 mA/cm². In general, the temperature of the substrate can be chosen between the temperature of liquid nitrogen and some hundred degrees.

Through atomic collisions between incoming ions and target atoms, the atomic mobility and chemical reactivity are athermally enhanced, resulting in lower processing temperatures for epitaxial growth and formation of metastable compounds in the subsurface region. The accurate control over critical

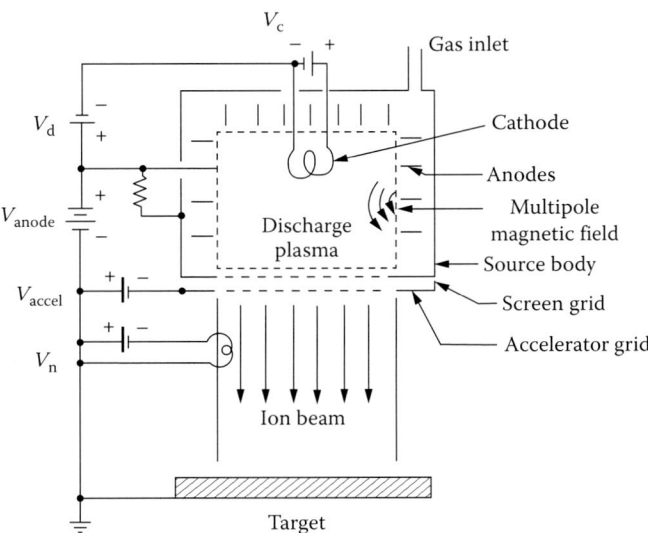

FIGURE 3.26 Broad beam multi-aperture Kaufman-type ion source. (From Mohan, S. and Ghanashyam Krishna, M., *Vacuum*, 46, 645, 1995.)

processing parameters such as ion energy, flux, species, and directionality provides new degrees of freedom to tailor the film properties.

The drawbacks of IBAD are (1) its low growth rate due to the rather low ion fluence, (2) the limitation of the production of low-energy ions with a high current density by the space-charge effect and the limited efficiency of ion sources in the extraction of ions at low voltages, and (3) the generation of defects while bombarding the growing film.

Kaufman broad beam ion source is the most commonly used to do IBAD (Figure 3.26). In this source [3.38] the gas is introduced into the chamber, which has a hot cathode emitting electrons. There is a cylindrical anode surrounding the cathode and gas gets ionized between the two electrodes. A magnetic field is applied transverse to the motion of electrons using permanent magnets to ensure electron confinement and increased ionization efficiency. Two grids, one called the screen grid and the other accelerator grid, are used to extract ions.

After the gas is introduced, the electrons emitted by the cathode strike the gas molecules and impact-ionize them. Some of the ions produced by these collisions reach various surfaces on the chamber and recombine with the electrons on these surfaces. Most of the neutrals return to the discharge chamber while the other ions are formed into small beamlets by passing through the apertures in the screen grid. The ions in these beamlets are attracted by the negative accelerator grid and pass through without striking it due to the alignment of the apertures in the two grids.

The ion beam can be neutralized using a neutralizer cathode that emits electrons. The neutralized beam then has equal densities of ions and electrons and helps in charge compensation at insulating substrates and targets. One of the main advantages of this source is that energy and current density of ions can be independently controlled. For other types of sources such as electron cyclotron resonance (ECR) source, see Reference 3.39.

In an application of IBAD, N-ion-beam-assisted deposition of thermally evaporated Li_3PO_4 has been demonstrated. The growth rate of the so-called LiPON films fabricated by this method can reach up to ~66 nm/min with a good ion conductivity and electrochemical stability [3.40]. A comprehensive review of applications of important physical vapor deposition methods to solid state ionic materials and devices has been given in Reference 3.41.

We now proceed to discuss briefly the fabrication of a Li-ion battery. This subject would be discussed in more detail in Chapter 8.

3.4 Fabrication of the Li-Ion Battery

3.4.1 Coin Battery

Among all consumer batteries, the coin cell (Figure 3.27a) and the flat cell (Figure 3.27b) are most versatile and compact. Let us focus on these two types but the latter one in the form of a thin-film battery. The coin cell design is a straightforward one with the electrodes parallel to each other. The electrodes are separated typically by a polyethylene sheet that contains a liquid electrolyte. A sealable cylindrical metal can hold these materials and a lid is used to seal the can. Electrical short circuiting is prevented by a polymer gasket that seals the lid to the can. The can thus becomes one electrode pole while the lid itself is the other electrode pole. As the casing resembles a coin, this is known as coin cell design.

The wound cell design is the second most common one. Long strips of electrodes and separators are the components of this design. These are wound inside and placed in a long cylindrical can. The sealing procedure in this design is as in the coin cell design. If the wound cell is flat, then it becomes the prismatic cell. In the flat cell, plates of electrodes and separators alternate. Each electrode plate has a tab that is welded to all other tabs of its respective electrodes (Figure 3.27).

The heart of the coin cell is the electrode film. This along with Li metal and a polyethylene separator sheet are punched out as disks. These disks are placed in a coin cell can in the sequence electrode film-separator-Li metal. Each coin cell contains only one of this sequence. Liquid Li-conducting electrolyte is added before the can is sealed. A plastic gasket ring is inserted into the can along with a flat metal disk that sits on top of the electrode stack. A spring is placed on top of the flat disk in order to maintain pressure on the cell stack. Finally, the coin cell lid is crimped into the can using the plastic gasket as a seal and an electric insulator between the lid and the can. The sealed coin cell is now complete with the lid as one pole and the can as the other pole of the electrochemical cell.

3.4.2 Thin-Film Battery Fabrication

A schematic drawing of the layout of a thin-film battery is shown in Figure 3.28a. The substrate is typically $2.54 \times 1.27 \times 0.1$ cm thick polycrystalline alumina plate, although cells could be fabricated on silicon, metal foils, or plastics as well. The latter materials would make the batteries quite flexible. The cathodes are

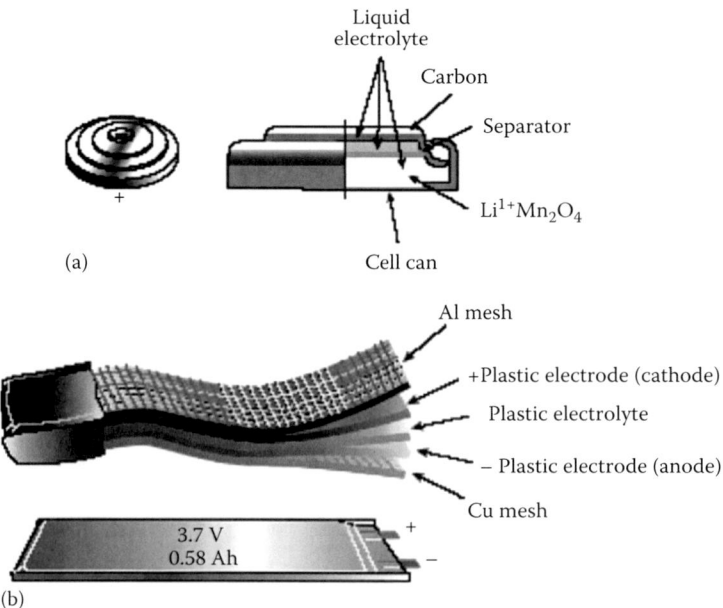

FIGURE 3.27   (a) A coin cell schematic. (b) A flat cell. (From Bates, J.B. et al., *Solid State Ionics*, 135, 33, 2000.)

deposited by RF magnetron sputtering of the parent compounds. These are characterized before deposition of the electrolyte. After deposition of LiPON electrolyte, the Li anode is deposited by thermal evaporation to a thickness that is three to four times overcapacity. The Li-ion cells, however, are cathode heavy.

Cells to be exposed to air for *in situ* XRD measurements or those to be subjected to extended cycling are coated with a protective multilayer of parylene and titanium.

Figure 3.28b depicts the schematic cross section of a Li/amorphous V_2O_5-based thin-film rechargable Li-ion battery. Any of several Li-intercalation compounds that could be deposited in thin-film form could be used in lieu of V_2O_5. Steps for the fabrication of the cell of Figure 3.28b are (a) V current collectors made by DC matnetron sputtering of V in argon, (b) V_2O_5 cathode made by DC matnetron sputtering of V in a mixture of Ar + 14% O_2, (c) Li electrolyte made by RF magnetron sputtering of Li_3PO_4 in N_2, (d) Li anode made by evaporation of Li in 10^{-6} Torr, and (e) protective coating.

Laboratory cells usually have an area of ~1 cm² and a thickness of ~6 μm deposited on 2.54 cm square glass microscope slides. For cells based on vanadia or TiS_2 cathodes, all depositions are at ambient temperature so that any substrate capable of supporting a thin film could be used: Li–V_2O_5 cells have been fabricated on alumina, glass, and 0.1 mm thick polyester.

For sputter depositions, 2 in. magnetrons (torus) may be used. V target could be commercial while Li_3PO_4 may be prepared by pressing commercial Li_3PO_4 powder into a 5 cm diameter × 3-mm-thick disk and sintering it in air at 900°C for 4 h. The substrate may be located 5 cm above the target and the total pressure could be 20 mTorr. The slowest steps in the fabrication process are the cathode and

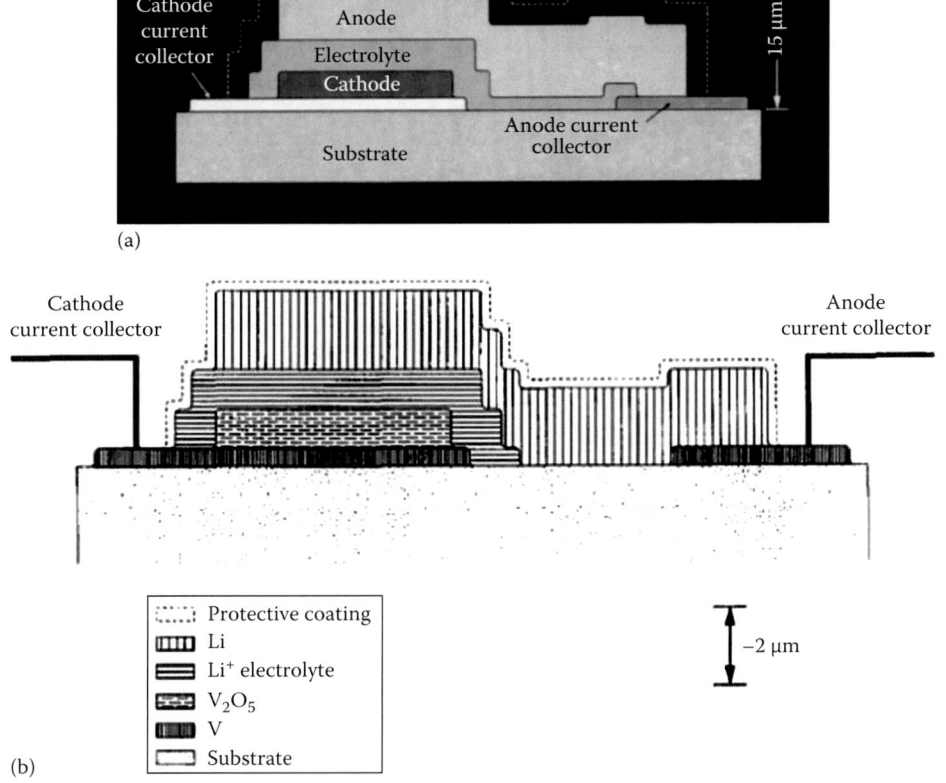

FIGURE 3.28 (a) Schematic layout of a thin-film battery. (From Bates, J.B. et al., *Solid State Ionics*, 135, 33, 2000.) (b) Schematic cross section of a Li/V_2O_5-based thin-film rechargeable Li-ion battery. (From Bates, J.B. et al., *Solid State Ionics*, 70/71, 619, 1994.) *(Continued)*

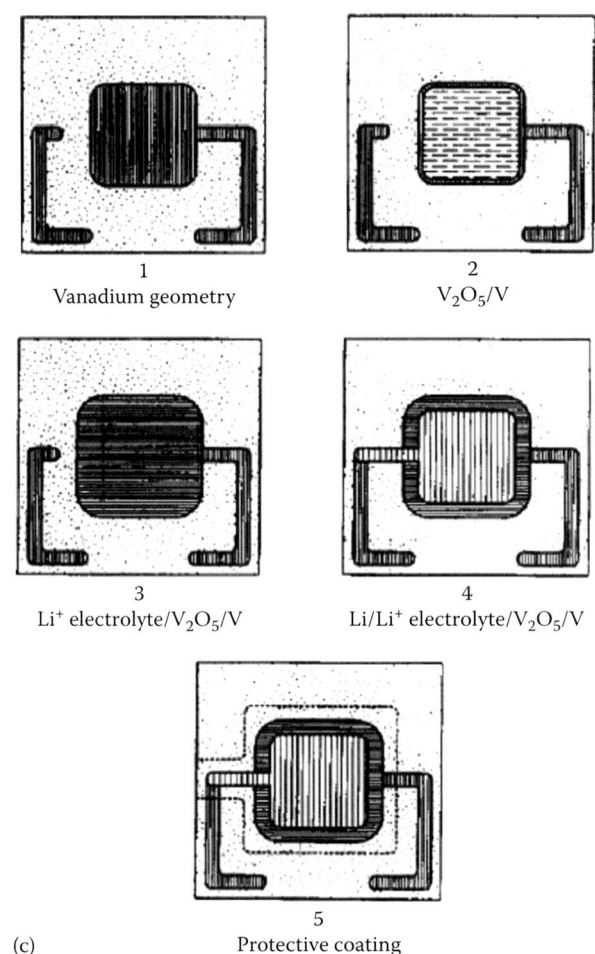

1
Vanadium geometry

2
V_2O_5/V

3
Li^+ electrolyte/V_2O_5/V

4
Li/Li^+ electrolyte/V_2O_5/V

5
Protective coating

(c)

FIGURE 3.28 (*Continued*)   (c) Thin-film Li-ion battery deposition sequence and typical geometry. (From Bates, J.B. et al., *Solid State Ionics*, 70/71, 619, 1994.)

electrolyte deposition. They require deposition rates of ~0.1 µm/h and powers of ~30 W (DC) and ~35 W (RF) applied to the V and Li_3PO_4 targets, respectively, to realize ~1 µm thick films each.

   Li films are deposited at a rate of ~10 µm/h by vacuum evaporation of Li metal contained in a Ti crucible. The substrate has to be just a few mm above a 14 mm tall chimney placed on top of the crucible to minimize Li deposition on the chamber walls and fixtures. Li films are 3–5 µm thick corresponding to about 5–10 times more Li required for full discharge of 1 µm thick vanadia cathode films. All of the operations with Li have to be carried out in a recirculating glove box filled with 99.999 and argon with the O_2 and H_2O content maintained at a few ppm. Before the start of Li depositions, the box has to be purged with Ar to remove any residual N_2. After the deposition of the Li film, the cells have to be transferred in Ar to another deposition system where the protective coating is applied. Figure 3.28c shows the thin-film Li-ion battery deposition sequence along with the typical geometry.

3.5 Summary

In this chapter, methods and technologies to synthesize solid state ionic materials in powder and thin-film forms have been discussed. Among the methods are the energy-saving mechanochemical reaction, sol–gel/sol-combustion, microwave heating, and hydrothermal techniques. Thin-film technologies discussed include thermal evaporation, sputtering, and CVD. This is followed by techniques to fabricate coin-type cells and thin-film batteries. In the next chapter, the focus is naturally on materials characterization.

Problems

3.1 (a) MCR synthesis of Cu_2S. Take commercial powders of copper and sulfur in 2:1 ratio by gram atomic weight so that the total weight of both components is ~5 g. Grind them in an agate mortar (6″ diameter) along with half a gram of sucrose for ~2 h, wearing a mask to avoid inhaling sulfur fumes. Observe color change and homogeneity of the product obtained. Dissolve the sucrose by adding deionized water. Filter, dry filtrate in an oven, and characterize by x-ray diffraction for crystal structure and particle size.

A similar procedure would work for Ag_2S and AgI.

   (b) MCR synthesis of Ag_3SI and Ag_4RbI_5. Also AgCuS. Adopt a similar procedure as before to synthesize the three ternary compounds mentioned previously. You could use AgI made in 1(a) and RbI to make Ag_4RbI_5. Characterize the products as done earlier.

3.2 What is the thermodynamic basis of the hydrothermal synthesis? Apply the method to synthesize $LiFePO_4$. For details of procedure, see Reference 3.44.

3.3 In the inexpensive chemical bath deposition (CBD) technique of fabricating thin films, a suitable solution of the material to be deposited is made in a beaker. With a simple setup to control temperature, a glass slide is covered on the beaker. The glass slide could also be dipped into the solution vertically or at a known angle to the vertical. Use this CBD technique to fabricate Cu_2Se. You would need copper sulfate ($CuSO_4 \cdot 5H_2O$), trisodium citrate, and sodium selenosulfate (Na_2SeSO_3). For detailed procedure, see Reference 3.45.

3.4 Temperature-dependent conductivity measurements of $Ce_{0.8}Gd_{0.2}O_{1.9}$ sintered at 1250°C ($a = 0.54137$ nm, average particle size ~40 nm) yielded the flowing data:

Testing temperature (°C)	500	550	600	650	700	750	800
Electrical conductivity, σ (S/cm)	4.55×10^{-3}	8.37×10^{-3}	1.34×10^{-2}	2.86×10^{-2}	4.09×10^{-2}	5.82×10^{-2}	7.64×10^{-2}

Make an Arrhenius plot ($\sigma T = \sigma_0 \exp(-E/k_B T)$) and deduce the activation energy for oxygen ion transport in $Ce_{0.8}Gd_{0.2}O_{1.9}$. (Ans: 0.87 eV)

REFERENCES

3.1 D. Bharathi Mohan, C. S. Sunandana, *J. Phys. Chem. Solids* 65(2004) 1669.

3.2 E. McCalla et al., *Chem. Mater.* 25(2013) 912.

3.3 P. Senthil Kumar, A. K. Tyagi, C. S. Sunandana, *J. Phys. Chem. Solids* 67(2006) 1809.

3.4 X. Liu et al., *Mater. Res. Bull.* 46(2011) 910.

3.5 C. Julien et al., *J. Mater. Chem.* 11(2001) 1837.

3.6 U. Schubert, N. Husing, *Synthesis of Inorganic Materials*, Wiley VCH, Weinheim, Germany, 2005.

3.7 K. Takahashi et al., *J. Electrochem. Soc.* 159(2012) A342.

3.8 Y. Sundarayya, K. C. Kumara Swamy, C. S. Sunandana, *Mater. Res. Bull.* 42(2007) 2942.

3.9 Y. Shimizu et al.., *J. Mater. Chem.* 7(1997) 1487.

3.10 J. Cheng et al., *J. Chilean Chem. Soc.* 54(2009) 445.

3.11 M. Faraday, *Quart. J. Sci. Lit. Arts* 8(1820) 374.

3.12 K. D. M. Harris, *Nat. Chem.* 5(2013) 12.

3.13 S. Honke et al., *Acc. Chem. Res.* 19(2013) 2376.

3.14 V. V. Boldyrev, *J. Chim. Phys.* 83(1986) 821.

3.15 P. Balaz et al., *Chem. Soc. Rev.* 42(2013) 7571.

3.16 N. V. Kosova et al., *Solid State Ionics* 135(2000) 135.

3.17   L. J. Ning et al., *J. Power Sources* 133(2004) 229.

3.18   G. A. Bowmaker, *Chem. Commun.* 49(2013) 334.

3.19   K. Byrappa, T. Adshiri, *Prog. Cryst. Growth Character Mater.* 53(2007) 117.

3.20   J. Chen et al., *Solid State Ionics* 178(2008) 1676.

3.21   J. Wang et al., *Solid State Ionics* 268(2014) 131.

3.22   G. Yuan, G. Wang, H. Wang, J. Bai, *J. Solid State Electrochem. Lett.* 19(2015) 143.

3.23   K. J. Rao et al., *Chem. Mater.* 11(1999) 882.

3.24   M. H. Bhat et al., *Bull. Mater. Sci.* 23(2000) 461.

3.25   H. Yan et al., *J. Power Sources* 68(1997) 530.

3.26   B. Vaidyanathan, K. J. Rao, *J. Solid State Chem.* 132(1997) 349.

3.27   J. P. Silva et al., *Solid State Ionics* 268(2014) 54.

3.28   K. S. S. Haraha, *Principles of Physical Vapor Deposition of Thin Films*, Elsevier, Amsterdam, The Netherlands, 2006.

3.29   W. Y. Liu et al., *Electrochem. Solid State Lett.* 7(2004) 136.

3.30   C. Li et al., *Thin Solid Films* 515(2006) 1886.

3.31   L. M. Rozanov, M. H. Hablanian, *Vacuum Technique*, Taylor & Francis, New York, 2002.

3.32   K. Koshmak et al., *J. Phys. Chem. C* 118(2014) 10122.

3.33   G. Wehner, *J. Appl. Phys.* 30(1959) 1762.

3.34   C. Lehmann, P. Sigmund, *Phys. Stat. Solidi* 16(1966) 507.

3.35   M. Yamashita et al., *Solid State Ionics* 89(1996) 299.

3.36   S. Zhao et al., *Thin Solid Films* 415(2002) 108.

3.37   L. Armelao et al., *Chem. Mater.* 17(2005) 427.

3.38   S. Mohan, M. Ghanashyam Krishna, *Vacuum* 46(1995) 645.

3.39   B. Rauschenbach, *Vacuum* 69(2003) 3.

3.40   F. Vereda et al., *Electrochem. Solid State Lett.* 5(2002) A239.

3.41   H. Xia et al., *Int. J. Surf. Sci. Eng.* 3(2009) 23.

3.42   J. B. Bates et al., *Solid State Ionics* 135(2000) 33.

3.43   J. B. Bates et al., *Solid State Ionics* 70/71(1994) 619.

3.44   S. Yang et al., *Electrochem. Commun.* 3(9)(2001) 505.

3.45   M. Lakshmi et al., *Thin Solid Films* 386(2001) 127.

4

Materials Characterization

4.1 Motivation

In the last chapter, the procedures and technologies for synthesis of materials and fabrication of devices have been considered. The next natural step, namely, the characterization of solid state ionic materials and devices fabricated from them, is the subject of this chapter.

Characterization of materials in general and solid state ionic materials in particular is threefold:

1. Structural and microstructural
2. Physical
3. Electrochemical (device oriented)

Types 1 and 2 help optimize material composition. Materials characterization naturally follows materials synthesis.

Wave diffraction and spectroscopy are two of the most fundamental and diverse tools for the characterization of materials. Solid state ionic materials mostly come in two forms: "powder" and "thin film." Powder implies polycrystalline aggregates—either on a microscopic scale or on a nanoscopic scale. Thin film implies a 2D atomic/molecular arrangement of the solid either structurally disordered or dynamically disordered or else composite structures. When the disorder is almost complete, one has a noncrystalline or glassy or amorphous solid. Characterization of such materials and forms using, say, X-ray diffraction (XRD) means an application of wide-angle X-ray diffraction (WAXD) and glancing incidence X-ray diffraction (GIXRD), respectively. However, basic thin-film characterization is commonly done using WAXD.

This chapter discusses several important methods of characterization with illustrative examples for each. It must be remembered that although these analytical methods are routinely applied often using standard software, there is scope for innovation as in the case of materials synthesis.

Materials synthesis and characterization, although dealt with here in two successive chapters, are two intimately linked aspects of materials chemistry and physics. Characterization of all materials including solid state ionic materials is the "acid test" for a materials synthesis. This is because however innovative the materials innovation may be, it is the characterization that eventually tells you how useful the test material is for the intended application. The optimally synthesized solid state ionic material must go through a series of well-planned and connected characterization procedures before the synthesis is upscaled for the application—be it battery or sensor, fuel cell, or supercapacitor. Of course, the methods of characterization depend on the type and form of the materials synthesized—particulate (including composite) or 2D (thin film/thick film, hard or soft) or gel/polymer/membrane.

4.1.1 Crystal Structure and Microstructure

Among all the methods that help characterize the structure of solid state ionic matter, scattering experiments are generally and extensively applicable, without any special preparation procedure. These experiments are performed using X-rays (produced from a Roentgen-type tube as well as by the synchrotron), neutrons (produced in a nuclear reactor), and electron beams. The same basic setup is used in all cases as shown in Figure 4.1.

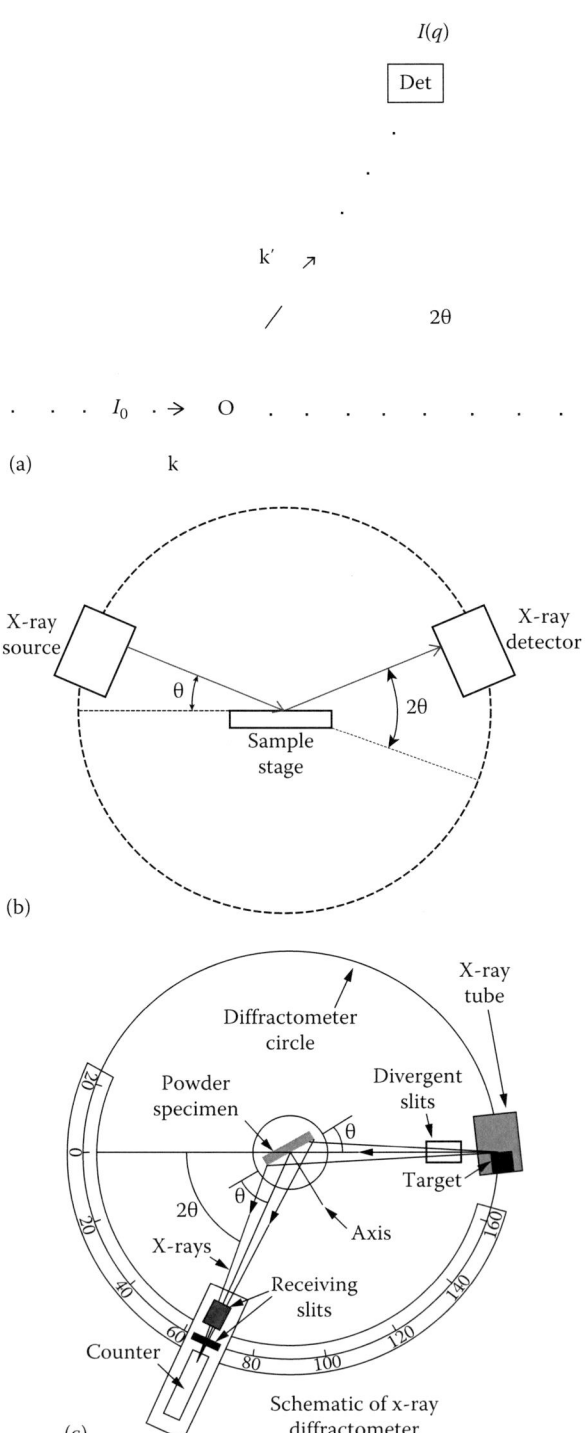

FIGURE 4.1   (a) Basic principle of the scattering experiment valid for both wave (X-ray) and particle (neutron) scattering. R is the distance between the scatterer (O) and the detector. 2θ is the angle between the k' vector and the extended k vector. (b) Principle of X-ray diffractometer. Note the θ–2θ geometry. (c) Schematic of an X-ray diffractometer. (From Cullity, B.D., *Elements of X-Ray Diffraction*, Addison-Wesley, Reading, MA, 1956.)

The primary radiation generated in a radiation source of wavelength λ (which is 0.1548 nm for the K_α radiation from a copper target), frequency ω_0, wave vector $\mathbf{k}$ (magnitude $2\pi/\lambda$), and intensity I_0 hits a sample generating spherical waves. The incident waves are actually scattered by the sample to produce scattered waves. The resulting total scattered intensity I depends generally on the observation direction. It is detected using a suitable detector placed at a distance R, which is large as compared to the sample diameter. The change in angle upon scattering is expressed as 2θ, where θ is the Bragg scattering angle. If the scattered wave vector is $\mathbf{k}'$, then the scattering **vector** is $\mathbf{q} = \mathbf{k}' - \mathbf{k}$. The magnitude of $\mathbf{q}$ is $4\pi/\sin \theta$. $\mathbf{q}$ depends on half of the angle between $\mathbf{k}$ and $\mathbf{k}'$ (4.1). The intensity of the scattered radiation measured by the detector is given by

$$I(q) = \sum_{j,k} f_j f_k \left\langle \exp\left[-iq(r_j - r_k)\right]\right\rangle \tag{4.1}$$

Angular brackets express the fact that the scattering intensity fluctuates during the time experiment is carried out so that what is measured is a time-averaged value. f_j and f_k are the atomic or molecular form factors—the same for the sample with only one type of particle. Therefore, only a sum over all phases of the scattered wave is required:

$$C(q) = \sum_j \exp(-iqr_j) \tag{4.2}$$

so that the scattered intensity $I(q)$ is proportional to "$|C(\mathbf{q})|^2$".

Total scattering intensity is proportional to the number of particles N that cause a scattering. This is represented by the function

$$S(q) = (1/N)\left\langle |C(\mathbf{q})|^2 \right\rangle \tag{4.3}$$

Substituting for $C(q)$ from Equation 4.2

$$S(q) = (1/N)\sum_{j,k=1}^{N}\left\langle \exp\left[-iq(r_j - r_k)\right]\right\rangle \tag{4.4}$$

$S(q)$ is the structure function, which generally characterizes the structure of a solid state ionic material. For the case of crystal diffraction in a 3D solid

$$I(q) \alpha N_1 N_2 N_3 \left|f_c(q)\right|^2 \rho_c \Sigma_{hkl \neq 000} \delta(q - \mathbf{G_{hkl}}) \tag{4.5}$$

where $\mathbf{G_{hkl}}$ is the reciprocal lattice vector defined by

$$\mathbf{G_{hkl}} = h\boldsymbol{a_1} + k\boldsymbol{a_2} + l\boldsymbol{a_3}$$

h, k, and l being integers, and $\boldsymbol{a_1}$, $\boldsymbol{a_2}$, and $\boldsymbol{a_3}$ being the basis vectors of the reciprocal lattice as discussed in Chapter 1.

Using the Ewald construction described earlier, it is possible to determine if a crystal is suitably oriented to reflect the primary beam and also shows the direction of the reflected beam.

How is the atomic arrangement in the unit cell found from the X-ray scattering experiment? This information is contained in f_c of Equation 4.5, which is the cell structure factor. Each reflection with indices *hkl* yields through the corresponding integrated reflection intensity I_{hkl} a value for the cell structure factor. The following applies:

$$I_{hkl} \alpha \left| f_c \left(q = \mathbf{G_{hkl}} \right) \right|$$

$$= \left| \Sigma_j f_j \exp[-i \left(ha_1 + ka_2 + la_3 \right) \left(x_{1j} a_1 + x_{2j} + x_{3j} a_3 \right) \right|^2$$

$$= \left| \Sigma_j f_j \exp\left[-i2\pi \left(hx_{1j} + kx_{2j} + lx_{3j} \right) \right] \right|^2 \qquad (4.6)$$

Measure the intensities for a sufficiently large number of reflections and use Equation 4.6 to arrive at the atomic arrangements.

In a linear diffraction pattern, the detector scans through an arc that intersects each Debye cone (the so-called cones of diffraction emanating from a sample in the Debye–Scherrer method of the past that used a cylindrical camera with the sample mounted along the axis of the cylinder) at a single point. Figure 4.2 illustrates the XRD pattern of $LiCoO_2$ commercial powder and thin films obtained by sputtering.

XRD may also be applied to analyze a device such as an LTO ($Li_4Ti_5O_{12}$) half cell corresponding to a Li-ion battery. Figure 4.3a and b shows the observed and theoretically generated XRD patterns of such a cell.

Figure 4.3c illustrates the meaning and use of the typical XRD peak. For an application to the AgI–CuI nanosystem, see [4.4].

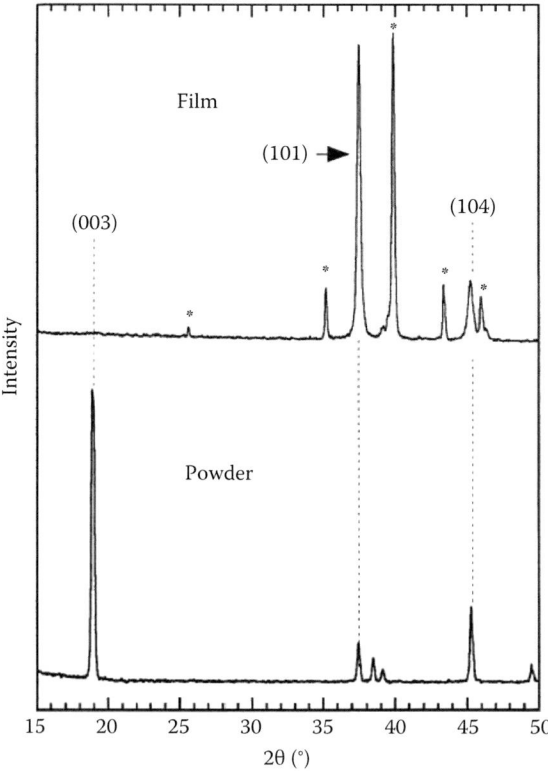

FIGURE 4.2 XRD patterns of 4 μm thick $LiCoO_2$ film made by magnetron sputtering (top) and commercial $LiCoO_2$ powder (bottom). The peaks marked with "*x*" correspond to alumina substrate. Note the most intense (101) peak in the film occurring as a preferred orientation (along with weaker (104) and both parallel to substrate surface) and the most intense (003) peak in random-oriented crystallites in the powder. (From Bates, J.B. et al., *Solid State Ionics*, 135, 33, 2000.)

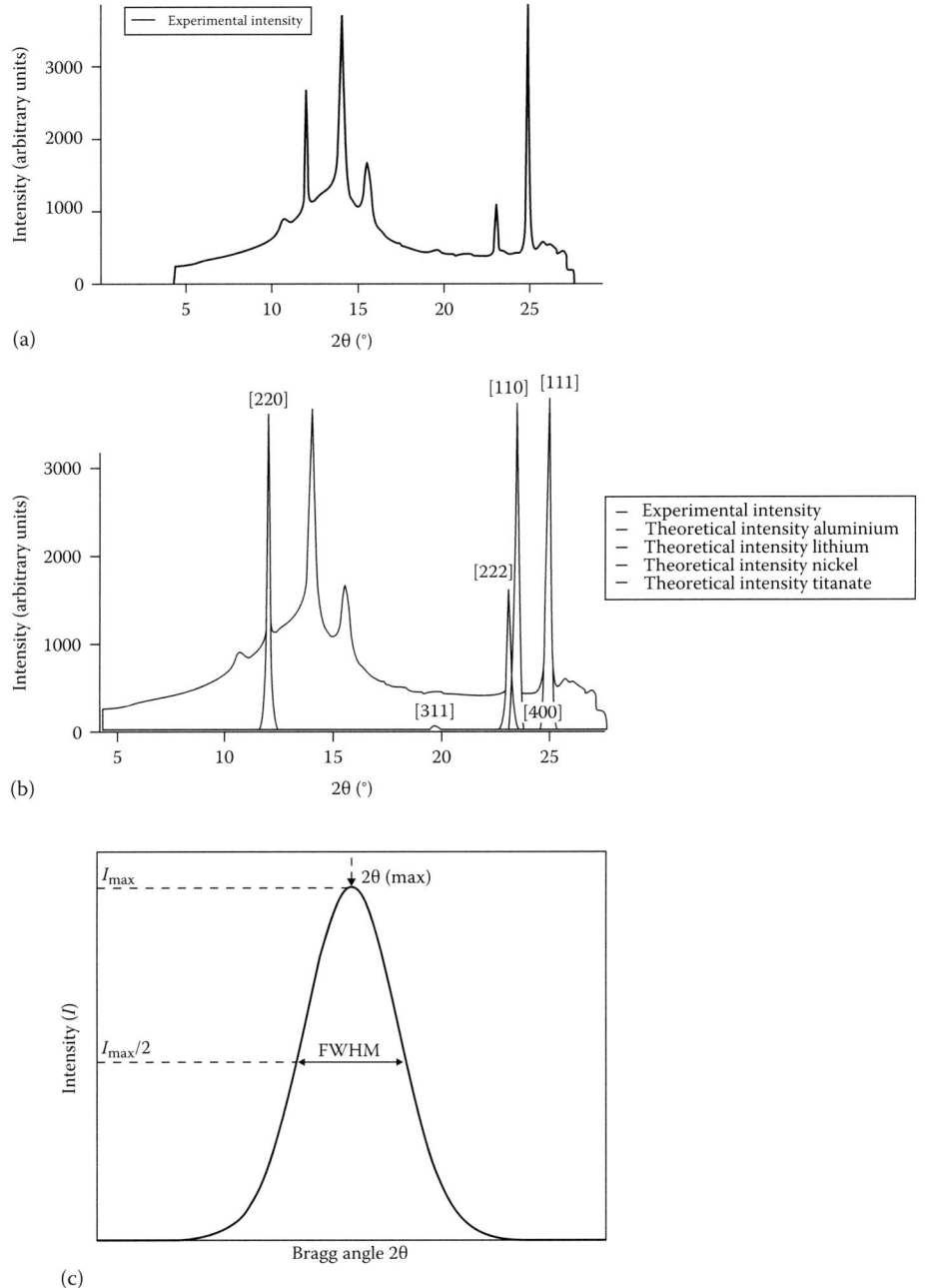

FIGURE 4.3 (a) XRD of LTO generated in situ from an LTO half cell. (b) Continuous plot with several peaks: Experimental. All others are theoretically generated. Peak labeled (220) belongs to titanate; (311), (222), and (400) to Al; (110) to Li; and peak (111) to NI. (From Nordh, T., Thesis. Uppsala University, Sweden, February, 2013.) (c) XRD peak of a powder sample in the Gaussian representation. This representation ideally reflects the random arrangement of crystallites in a powder sample in causing x-ray powder diffraction from a specific atomic plane *hkl* in the crystal, say (111) or (110) in a cubic crystal. The maximum intensity (I_{max}) occurs at $(2\theta)_{maximum}$. Here, θ is the Bragg angle = $\sin^{-1}$ (x-ray wavelength $\lambda/2d$) where $d = a/\sqrt{N}, N = h^2 + k^2 + l^2$. Full width at half maximum (FWHM) corresponds to the *waist* of the XRD peak at $I_{max}/2$. Note that forms other than Gaussian including Lorentzian and Gaussian–Lorentzian could be used to represent the XRD peak. The area under the XRD peak is proportional to the number atoms present in a (*hkl*). The rectangle defined by FWHM and I_{max} as *breadth* and length, respectively, gives the *integral breadth*. These parameters are inputs to the simulation of XRD peaks of a solid state ionic material.

TABLE 4.1

Expressions for *d*-Spacings and Unit Cell Volumes in Seven Crystal Systems

Crystal System	Expression for $1/d_{hk}^2$	Unit Cell Volume (V)
Cubic	$(h^2 + k^2 + l^2)/a^2$	a^3
Tetragonal	$(h^2 + k^2)/a^2 + l^2/c^2$	a^2c
Orthorhombic	$h^2/a^2 + k^2/b^2 + l^2/c^2$	abc
Hexagonal	$(4/3)(h^2 + hk + k^2)/a^2 + l^2/c^2$	$\sqrt{a^2c}/2$
Rhombohedral	$[(h^2 + k^2 + l^2)\sin^2\alpha + 2(hk + kl + hl)(\cos^2\alpha - \cos\alpha)]/[a^2(1 - 3\cos^2\alpha + 2\cos^3\alpha)]$	$a^3[1 - 3\cos^2\alpha + 2\cos^3\alpha]^{1/3}$
Monoclinic	$(1/\sin^2\beta)\{h^2/a^2 + k^2\sin^2\beta/b^2 + l^2/c^2 - 2hl\cos\beta/ac\}$	$abc\sin\beta$
Triclinic	$(1/V^2)(S_{11}h^2 + S_{22}k^2 + S_{33}l^2 + 2S_{12}hk + 2S_{23}kl + 2S_{13}hl)$ where $S_{11} = b^2c^2\sin^2\alpha$, $S_{22} = a^2c^2\sin^2\beta$, $S_{33} = a^2b^2\sin^2\gamma$ $S_{12} = abc^2(\cos\alpha\cos\beta - \cos\gamma)$ $S_{23} = a^2bc(\cos\beta\cos\gamma - \cos\alpha)$ $S_{13} = ab^2c(\cos\gamma\cos\alpha - \cos\beta)$	$abc\{1 - \cos^2\alpha - \cos^2\beta - \cos^2\gamma + 2\cos\alpha\cos\beta\cos\gamma\}^{1/2}$

XRD enables unit cell dimensions to be obtained from an analysis of the diffractograms. As seen in Chapter 2, solid state ionic materials crystallize in all the seven systems from triclinic through monoclinic, orthorhombic, tetragonal, hexagonal, rhombohedral, and cubic, although most of them are to be found in the last three systems.

Table 4.1 gives expressions for d-spacings in the different crystal systems in terms of lattice parameters and (*hkl*) indices. Use of an appropriate expression enables lattice parameters to be derived from experimental data.

Next, we discuss an allied but important technique that focuses on local-level structure in superionic conductors, namely, EXAFS.

4.1.1.1 EXAFS

X-ray absorption fine structure (XAFS) is concerned with the details of how X-rays are absorbed by an atom, at energies close to and above the core-level binding energies of a particular atom. In fact, XAFS is the modulation of the X-ray absorption probability of an atom due to the *chemical* and *physical* state of the atom. Thus, it is particularlty sensitive to the formal oxidation state, coordination chemistry, and the distances, coordination number, and species of atoms in the immediate neighborhood of the selected element.

The X-ray abasorption spectrum is divided into X-ray absorption near edge spectroscopy or XANES and extended X-ray absorption fine-structure spectroscopy or EXAFS. While XANES is strongly sensitive to formal oxidation state and coordination chemistry (such as octahedral or tetrahedral), EXAFS is useful in the determination of distances, coordination numbers, and species of the neighbors of the absorbing atom.

EXAFS is best understood in terms of the wave behavior of the photo-electron created in the absorption process. It is common to convert the X-ray energy to *k*, the wave number of the photo-electron, which has the dimensions of 1/distance and is defined as

$$k = \sqrt{2m(E - E_0)^2} \tag{4.7}$$

where E_0 is the absorption edge energy and *m* is the electron mass. The primary quantity for EXAFS is then $\chi(k)$, the oscillations as a function of photo-electron wave number, and $\chi(k)$ is often referred to simply as "the EXAFS."

EXAFS is fundamentally related to photoelectric effect (Figure 4.4).

In EXAFS, the oscillations well above the absorption edge are of interest. The EXAFS fine-structure function $\chi(E)$ is defined as

$$\chi(E) = \frac{\mu(E) - \mu_0(E)}{\Delta\mu_0(E)} \tag{4.8}$$

where

$\mu(E)$ is the measured absorption coefficient

$\mu_0(E)$ is a smooth background function representing the absorption of the isolated atom

$\Delta\mu_0$ is measured *jump* in the absorption $\mu(E)$ at the threshold energy E_0

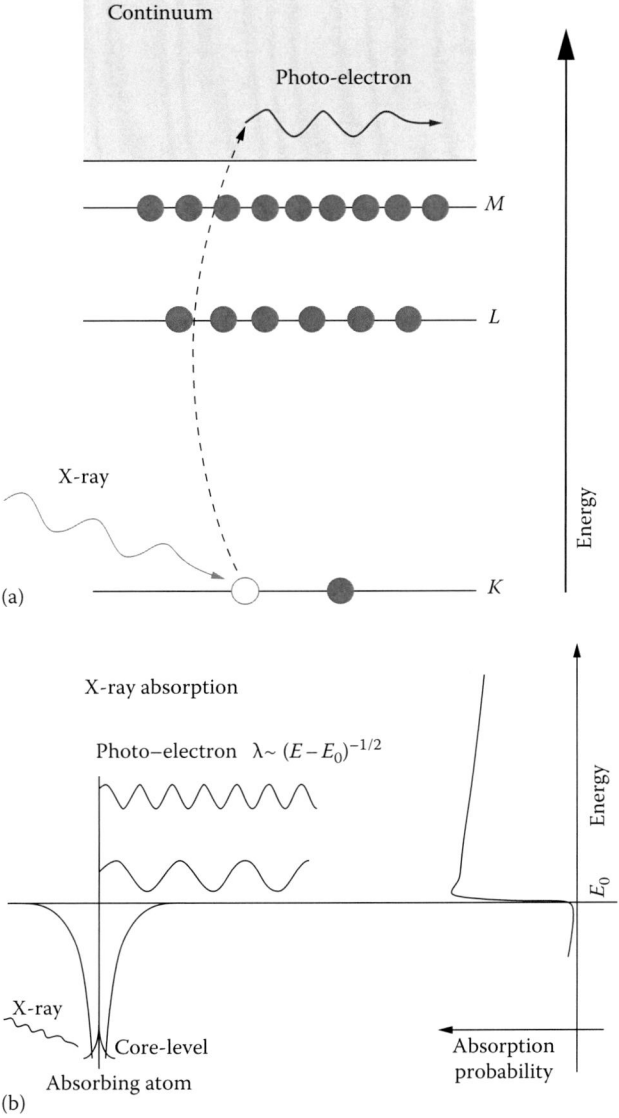

FIGURE 4.4   (a) In photoelectric effect applied to X-rays, an X-ray photon is absorbed and a core-level electron is promoted out of the atom. (b) Basic illustration of the X-ray absorption process and the edge formation. (From Newville, M., *Fundamentals of XAFS*, Tutorial, University of Chicago, Chicago, IL, 2004.)

The different frequencies apparent in the oscillations in $\chi(k)$ (note that E is replaced by wave vector k) correspond to different near-neighbor coordination shells, which can be described and modeled according to the EXAFS equation

$$\chi(k) = \Sigma_j \left\{ N_j f_j(k) \exp\left(-2k^2 \sigma_j^2\right)/kR_j^2 \right\} \sin\left[2kR_j + \delta_j(k) \right] \tag{4.9}$$

Here, $f(k)$ and $\delta(k)$ are scattering properties of the atoms neighboring the excited atom; N is the number of neighboring atoms, R is the distance to the neighboring atom, and σ^2 is the measure of disorder in the neighbor distance. The EXAFS equation enables determination of N, R, and σ^2 once the scattering amplitude $f(k)$ and phase shift $\delta(k)$ are known. Furthermore, as these properties depend on Z of the neighboring atom, EXAFS is sensitive to the atomic species of the neighboring atom.

Let us look at the EXAFS of CuBr as an illustrative example. Figure 4.5a shows the radial distribution function about Br in the zinc blende or γ-CuBr and the superionic α-CuBr. The reason for the drastic difference is seen in Figure 4.5b, where Cu ions are seen to be considerably disordered.

The unusual temperature dependence of EXAFS oscillations and the corresponding radial distribution functions are shown in Figure 4.6a and b, respectively.

The data are analyzed considering Cu and Br ion pairs interacting via a central field potential $V(R)$ where R is an instantaneous distance of the Cu–Br bond. The anharmonic pair potential

$$V(R) = AR\exp(-BR) - C/R \tag{4.10}$$

is adopted. (Anharmonicity is discussed in more detail in Chapter 6.) Here, the first and second terms are the overlap repulsive energy and Coulomb interaction energy, respectively. A, B, and C are constants. Defining u as the displacement of the bond length from the potential minimum position R_0, $R = R_0 + u$, so that

$$V(u) = A\exp(-BR_0)\left[(R_0 + u)\exp(-Bu) - R_0^2 (BR_0 - 1)/(R_0 + u) \right] \tag{4.11}$$

Figure 4.7 shows the temperature dependence of the mean Cu–Br distance deduced from EXAFS measurements successfully explaining the negative thermal expansion observed in CuBr.

Sn-K EXAFS (2.5–12.5 A^{-1}) of (1) $Pb_{0.7}Sn_{0.3}F_2$ at 300 and 77 K and of $PbSn_4F_{10}$ (at 300 K) as well as (2) Pb-L$_3$ EXAFS (2.6–12.0 A^{-1}) of $PbSn_4F_{10}$, β-$PbSnF_4$, and $Pb_{0.7}Sn_{0.3}F_2$ (at 77 K and 300 K) have been measured and analyzed [4.7].

A very interesting correlation exists between ionicity and Debye–Waller factor determined from XRD and EXAFS. In tetrahedral superionic conductors (such as AgI, CuI, and CuBr), one finds that the Debye–Waller factor that is crucial to ion dynamics is controlled by ionicity [4.6]. A critical ionicity was found by Phillips for the occurrence of a structural instability and superionic conductivity that corresponds to the borderline between four- and six-coordinated compounds (see Chapter 7 for details). This number (number line in fact) 0.785 lies just to the right of AgI making it a generic or canonical superionic conductor with a structural phase transition to the superionic phase at 420 K. Unlike AgCl and AgBr, Ag in AgI sits in a tetrahedral cage of I$^-$ ions and vice versa. A good correlation exists between Debye–Waller factor and ionicity, which diverges at 0.785 for tetrahedrally coordinated compounds (Figure 4.8). This divergence is characteristic of structural instability, which when controlled would yield good superionic conductors. Even metastable phases of AgI such as zinc blende or gamma AgI when stabilized by marginal additions of monovalent cations such as Cu could arrest instability and optimize conductivity over an extended range of temperatures.

4.1.1.2 Neutron Diffraction or Elastic Neutron Scattering

Neutron diffraction is the application of neutron scattering to the determination of the atomic and/or magnetic structure of a material. A sample to be examined is placed in a beam of thermal or cold

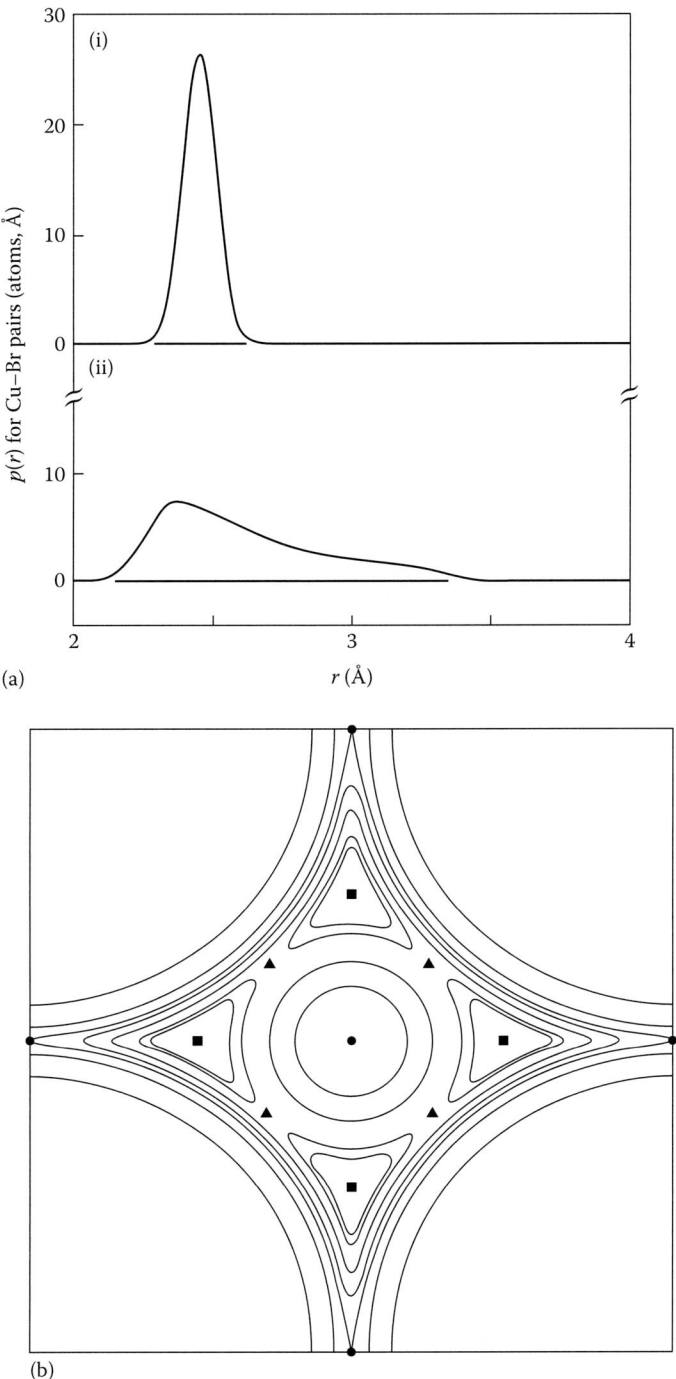

FIGURE 4.5 (a) Radial distribution function of Br ions about each Cu in CuBr in (i) γ-phase at 77 K and (ii) in α-phase at 753 K. (b) Cu ion density in the superionic phase of CuBr at 753 K.

neutrons to obtain a diffraction pattern that provides information of the structure of the material. The technique is similar to XRD but due to their different scattering properties, neutrons and X-rays provide complementary information.

The technique requires a source of neutrons. Neutrons are usually produced in a nuclear reactor or spallation source. At a research reactor, other components are needed, including monochromators and filters to select the desired neutron wavelength. Some parts of the setup may also be movable. At a spallation source, the time-of-flight technique is used to sort the energies of the incident neutrons (higher energy neutrons are faster), so *no monochromator is needed,* but rather *a series of aperture elements synchronized to filter neutron pulses* with the desired wavelength.

The technique is most commonly performed as powder diffraction, which only requires a polycrystalline powder. For single-crystal work, the crystals must be much larger than those used in X-ray crystallography. It is common to use crystals that are about 1 mm^3.

For single-crystal work, the technique requires relatively large crystals, which are usually challenging to grow. The main advantages to the neutron technique are many—sensitivity to light atoms, ability to distinguish isotopes, and absence of radiation damage.

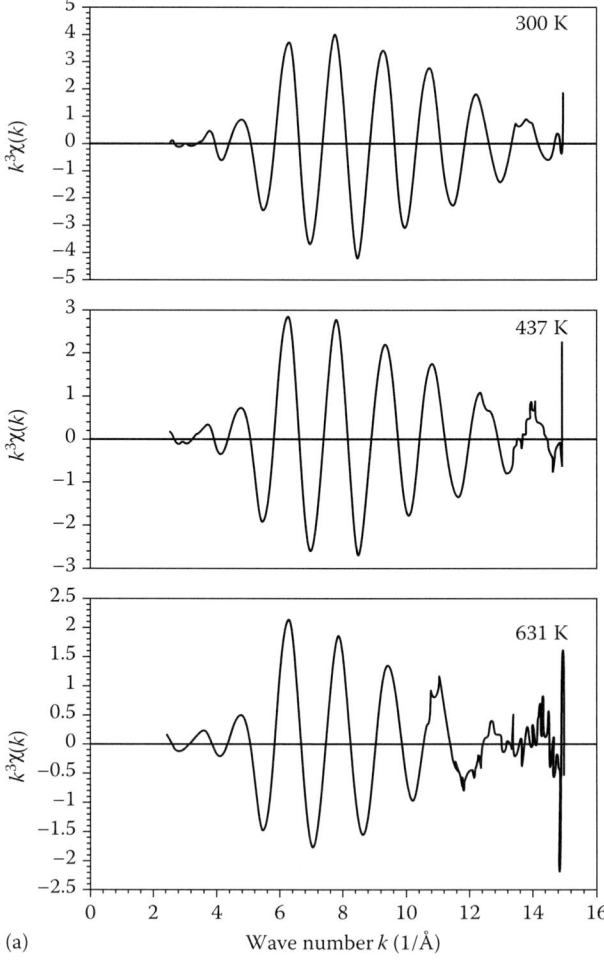

FIGURE 4.6 (a) EXAFS oscillations in CuBr thin films/BN disk at selected temperatures. EXAFS signals $k^3\chi(k)$ above the Cu-K edge of a CuBr thin film sample at 300, 437, and 631 K. Measurements at BL-7C at Photon Factory, KEK Tsukuba, Japan, 2.5 GeV, 250–300 mA. Incident and transmitted intensities measured with ionization chambers filled N_2 and 50 N_2–50 Ar, respectively. *(Continued)*

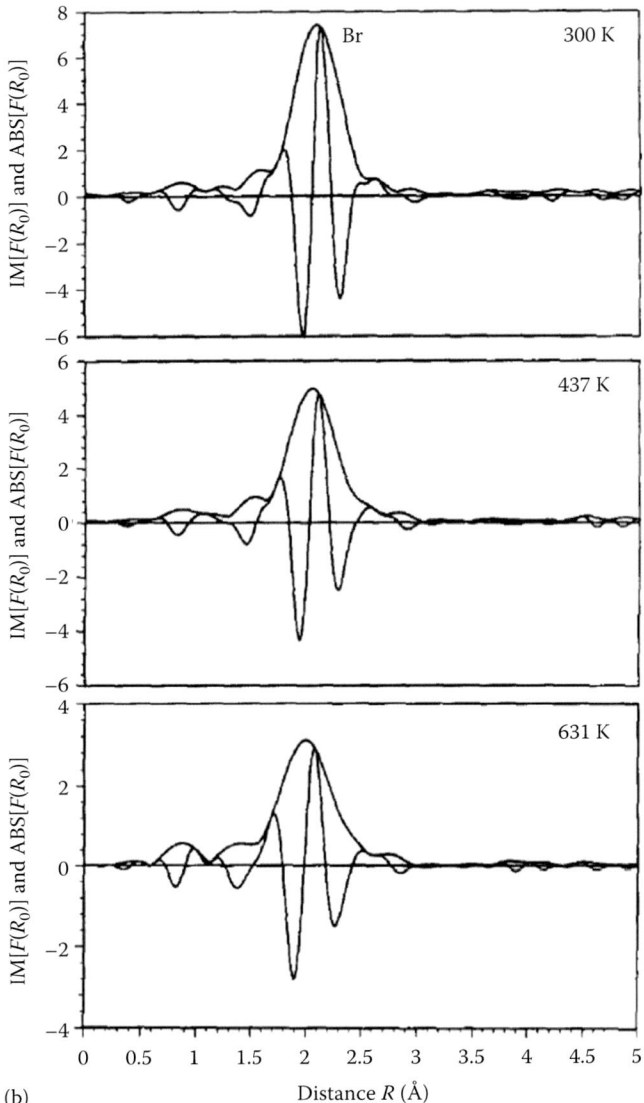

FIGURE 4.6 (Continued) (b) Radial distribution functions. (From Kamishima, O. et al., *Solid State Commun.*, 103, 141, 1997.)

Sources that produce a neutron beam of suitable intensity and speed for diffraction are available at a small number of research reactors and spallation sources globally. The National Facility for Neutron Beam Research (NFNBR) in Mumbai, India, is one such facility [4.9]. A neutron powder diffractometer schematic is shown in Figure 4.9a. Angle-dispersive (fixed-wavelength) instruments typically have a battery of individual detectors arranged in a cylindrical fashion around the sample holder, and can therefore collect scattered intensity simultaneously on a large 2θ range. Time-of-flight instruments normally have a small range of banks at different scattering angles that collect data at varying resolutions.

Let us illustrate the method of refining a structure using the neutron diffraction pattern in α-AgI and the model used to fit it are shown in Figure 4.9b and c.

The function to be minimized is

$$\chi^2 = \Sigma_i w_i \left[y_i(\text{obs}) - (1/c) y_i(\text{calc}) \right]^2 \tag{4.12}$$

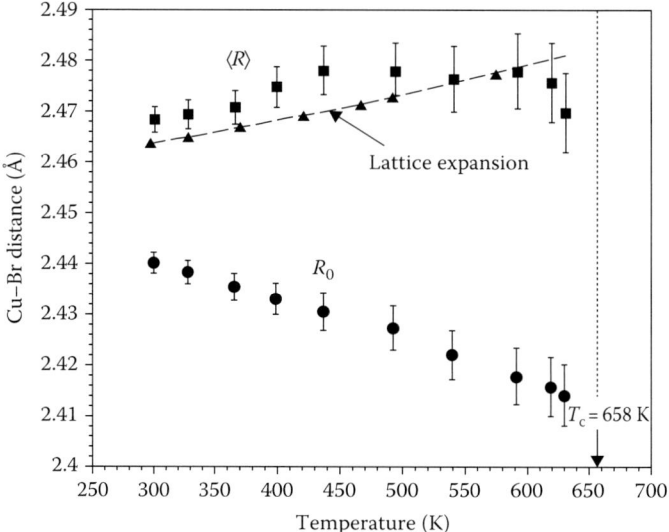

FIGURE 4.7 Mean Cu-Br distance $\langle R \rangle$ {T} and $R_0(T)$ for CuBr deduced from EXAFS measurements. Dashed line from earlier X-ray data. Note lattice contraction and anomaly at the phase transition at 658 K. (From Kamishima, O. et al., *Solid State Commun.*, 103, 141, 1997.)

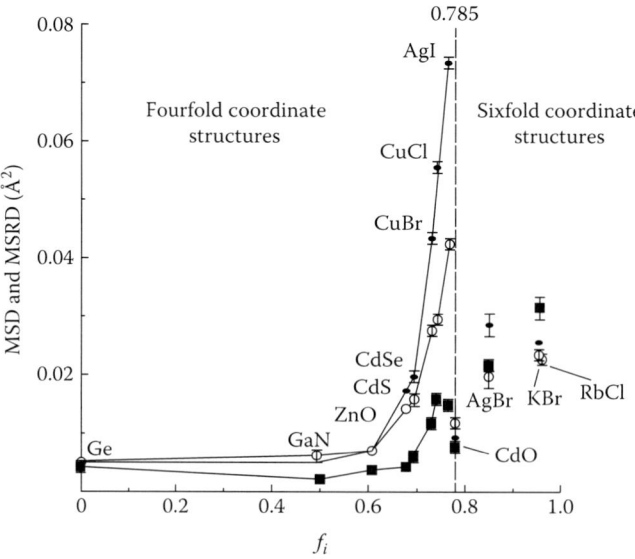

FIGURE 4.8 Ionicity (f_i) dependence of Debye–Waller factors: mean-square relative displacements (MSRD) derived from EXAFS experiments (■) and mean-square displacements for cations (●) and anions (o) derived from diffraction experiments MSRD and MSD correlate well with coordination number and ionicity. MSRDs of four-coordinated materials show a gradual approach to those of six-coordinated ones as the ionicity increases. A diverging curve toward $F_i = 0.785$ is observed for the MSD in the four-coordinated materials indicating lattice instability. (From Yoshiasa, A. et al., *Japanese J Appl Phys.*, 78, 3636, 1997.)

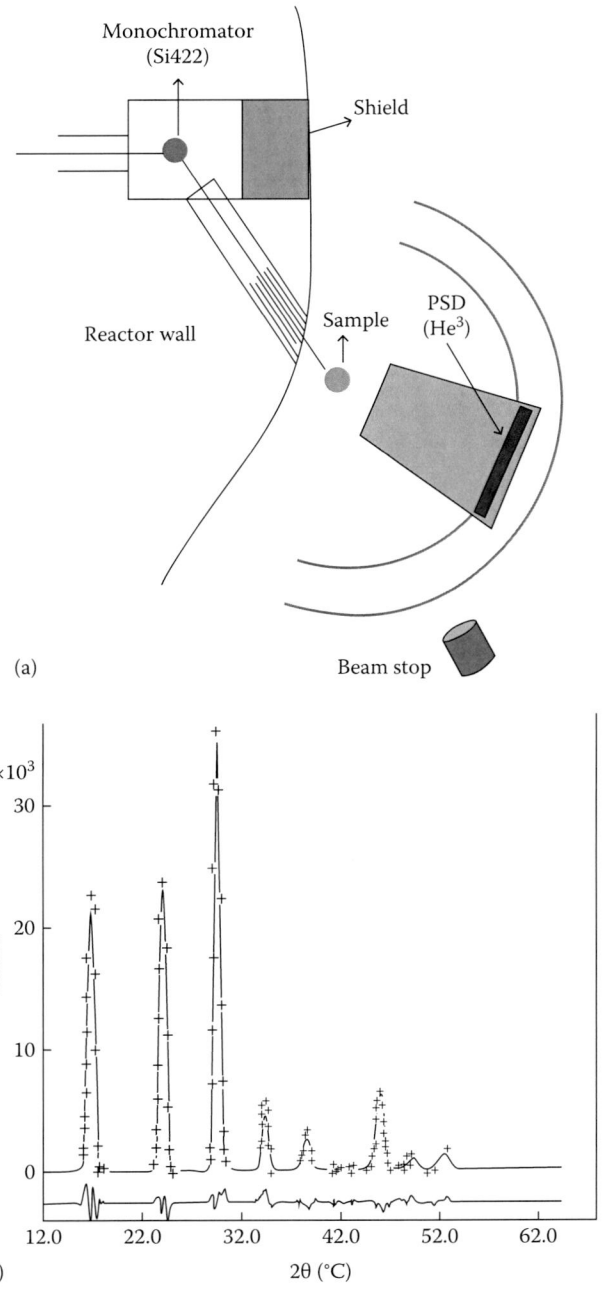

FIGURE 4.9 (a) Schematic of a powder neutron diffractometer. (b) Neutron powder diffraction pattern of alfa AgI at 180°C. Intensity of Bragg reflections diminishes rapidly with increase in scattering angle. The first three peaks correspond to (110), (200), and (211) planes of the bcc lattice. (From Wright, A.F. and Fender, B.E.F., *J. Phys. C Solid State Phys.*, 10, 2261, 1977.)   (*Continued*)

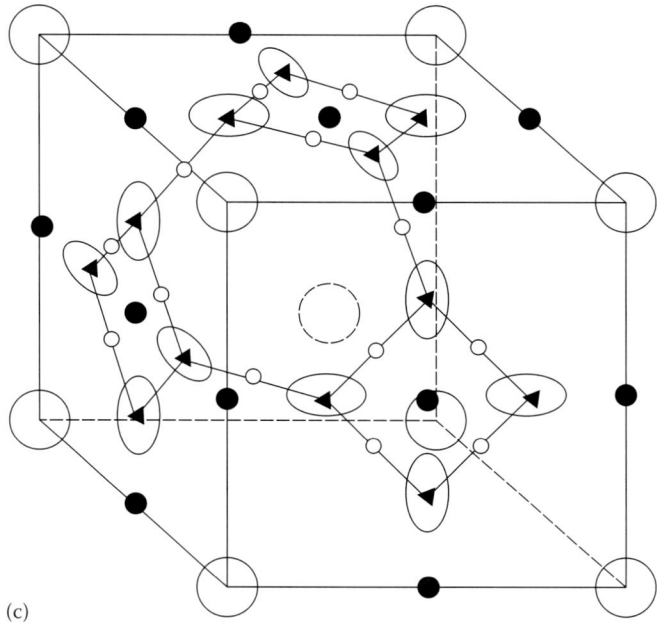

(c)

FIGURE 4.9 (*Continued*)   (c) Bcc structure of alfa AgI (Im3m) based on the Strock model. 42 positions available for occupation by Ag ions are distributed over tetrahedral (●), octahedral (▲), and pyramidal (O) sites. Dashed circle: I, oval: Ag (occupation factor 1/6).

The residual index *R* (or the profile) is given by

$$R = \frac{100\Sigma\left[y_i(\text{obs}) - (1/c)y_i(\text{calc})\right]}{\Sigma y_i(\text{obs})} \qquad (4.13)$$

The results are shown in Table 4.2.

Large values of *B* or temperature factors imply unusually large thermal vibrations.

Table 4.3 gives the powder neutron diffraction data on α-AgI.

$LiFePO_4$ besides being a Li^+ ion conductor is also a magnetic material (antiferromagnetic) at low temperatures whose structure may be determined from neutron diffraction. Figure 4.10 gives the diffraction pattern, the magnetic structure, and the magnetic phase transition.

To give a feel for the detailed structural information obtainable from the neutron powder diffraction experiment, Table 4.4 gives the refined structure parameters for $LiMn_2O_4$ at 230 K. It has an average orthorhombic structure (space group *Fddd*) with lattice parameters $a = 0.8279$ nm, $b = 0.82444$ nm, $c = 0.81981$ nm [4.17].

Phase transitions in $LiMn_2O_4$ are discussed in Chapter 5.

TABLE 4.2

Refinement of Structure of α-AgI at Selected Temperatures Based on Neutron Powder Diffraction Data with Ag Atoms at 12 Tetrahedral Sites at (1/4, 0, 1/2)

Temperature (°C)	Lattice Constant (*A*)	$B_{Ag}(A^2)$	$B_I(A^2)$	$R(I)\%$
180	5.062(1)	11.7(4)	7.5(3)	8.5
255	5.076	11.2(7)	8.3(5)	7.9
350	5.095(5)	15.5(7)	10.9(6)	10.4
450	5.106(2)	14.9(7)	11.4(7)	7.9

TABLE 4.3

Powder Neutron Diffraction Data on bcc AgI at 180°C

hkl	Integrated Intensity I (Counts)	Standard Deviation σ
110	170,568	1189
200	152,696	1083
211	254,296	1172
220	38,424	960
310	23,256	918
222	1,508	903
321	56,644	1089
400	6,708	1066
330	14,676	1053
411		

Let us now discuss quasielastic neutron scattering (QENS), which is a limiting case of inelastic neutron scattering [4.12]. Inelastic scattering refers to large energy is transferred to the system from the incident neutrons while elastic scattering involves negligible energy transfer. Thus, QENS is characterized by energy transfers being small compared to the incident energy of the scattered particles.

QENS spectra reveal two reorientationally distinct populations of BH_4 anions in $LiBH_4$ confined in nanoporous carbon. Such a spectrum at 400 K is shown in Figure 4.11. The quasielastic scattering from this material is best represented by two Lorentzian functions, with line widths differing by nearly an order of magnitude, are associated with the narrower and broader Lorentzian components. These are associated with the reorientational motions of the less mobile, more bulk-like interior, and more mobile interface BH_4 anions. Activation energies for reorientation of 0.166 and 0.11 eV for the relatively less and relatively more mobile BH_4 populations were determined from an Arrhenius plot of the Lorentzian line widths.

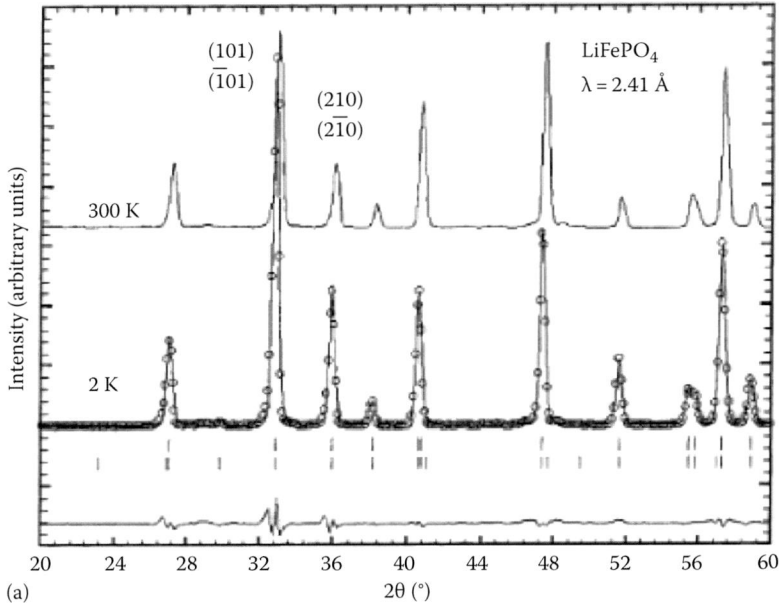

(a)

FIGURE 4.10 (a) Observed (O)-versus-calculated (—) neutron powder diffraction patterns of $LiFePO_4$ (D20, Ï) 2.41 Å at 2 K. The positions of the Bragg reflections are represented by vertical bars (first row, nuclear; second row, magnetic). The difference (obs-calc) pattern is displayed at the bottom of the figure. The pattern recorded at 300 K (paramagnetic phase) is displayed for comparison. *(Continued)*

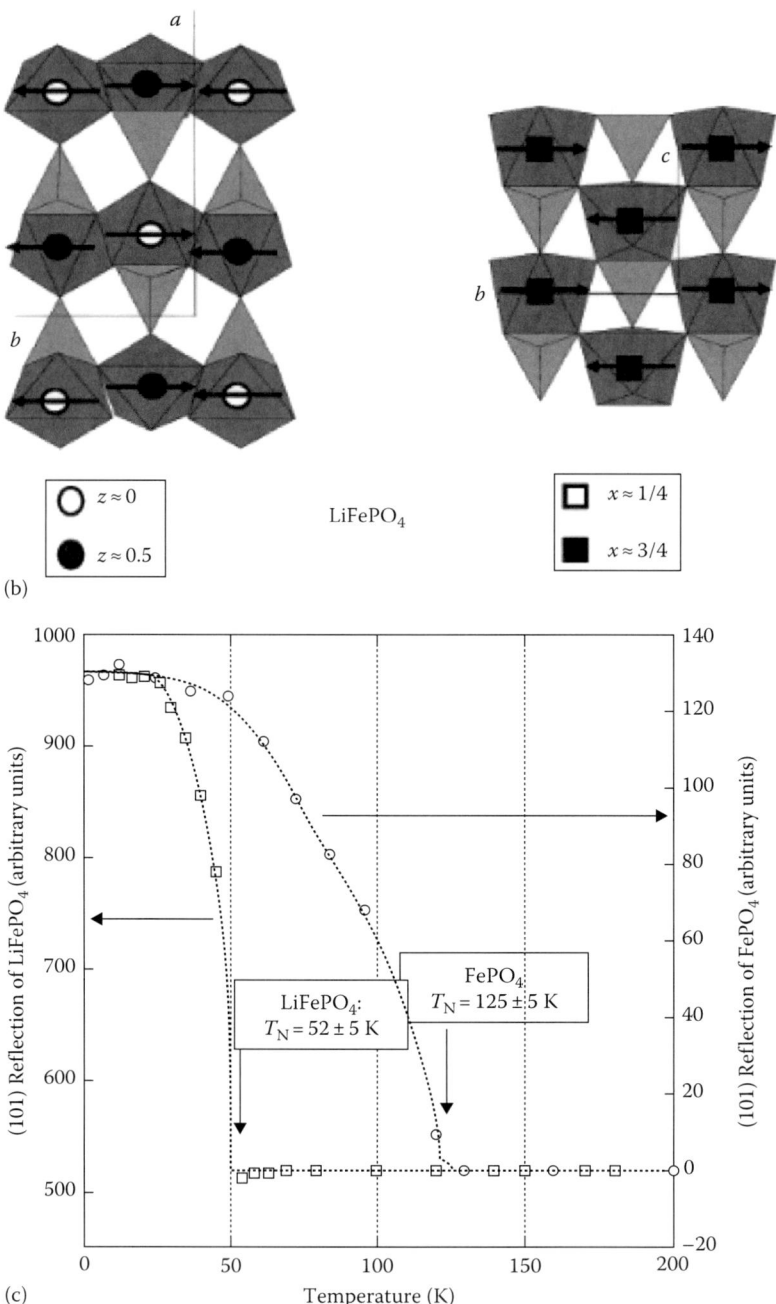

FIGURE 4.10 (*Continued*)   (b) Magnetic structure for LiFePO$_4$. Arrows indicate the magnetic moments on the iron atoms (along the *b*-axis). Refinement of crystal and magnetic structure was done by Rietveld method using FullProf program. (From Rodriguez-Carvajal, J., *Physica B*, 19, 55, 1993; http://www-llb.cea.fr/fullweb/fp2k/fp2k.htm.) (c) Evolution of the integrated intensity of the most intense magnetic reflections for LiFePO$_4$ (101) and for FePO$_4$ (010) as a function of temperature. The Neel temperatures (T_N) for the two compounds are indicated. The dotted lines guide the eye. (Parts (a) and (c) from Rousse, G. et al., *Chem. Mater.*, 15, 2082, 2003.)

TABLE 4.4

Refined Structure Parameters for LiMn2O$_4$ from Neutron Diffraction

Atom	x	y	z	Biso (Å^2)	Wyckoff Site	d(Mn–O) (Å)	Valence Sum	Distortion ($\times 10^{-4}$)
Mn(1)	$\frac{1}{4}$	$\frac{1}{4}$	$\frac{1}{2}$	0.28(2)	16d	2.003(2)	3.20(2)	20.6
Mn(2)	0.0803(3)	0.0855(3)	0.5035(9)	0.28(2)	32h	1.995(4)	3.27(3)	19.4
Mn(3)	0.0839(4)	0.3301(3)	0.2480(16)	0.28(2)	32h	2.021(5)	3.12(5)	36.6
Mn(4)	0.2527(3)	0.1675(3)	0.2491(13)	0.28(2)	32h	1.903(4)	4.02(5)	4.6
Mn(5)	0.1648(3)	0.2447(2)	0.2429(11)	0.28(2)	32h	1.916(4)	3.90(4)	6.1
Li(1)	$\frac{1}{8}$	$\frac{1}{8}$	$\frac{1}{8}$	0.77(6)	8a	1.967(2)	1.033(6)	0
Li(2)	$\frac{3}{8}$	0.2116(10)	$\frac{3}{8}$	0.77(6)	16f	1.984(8)	0.99(2)	0.5
Li(3)	0.2054(11)	$\frac{3}{8}$	$\frac{3}{8}$	0.77(6)	16e	2.016(8)	0.91(2)	0.3
Li(4)	0.2919(8)	0.2953(9)	0.1191(16)	0.77(6)	32h	1.946(9)	1.10(3)	7.5
O(1)	0.1754(2)	0.1682(2)	0.2565(5)	0.522(6)	32h	—	2.02(2)	—
O(2)	0.0786(2)	0.0070(2)	0.4805(8)	0.522(6)	32h	—	1.98(3)	—
O(3)	0.0783(2)	0.3311(2)	0.4795(9)	0.522(6)	32h	—	1.94(3)	—
O(4)	0.2528(2)	0.1732(2)	0.4747(8)	0.522(6)	32h	—	2.06(3)	—
O(5)	0.0055(2)	0.0080(2)	0.2419(6)	0.522(6)	32h	—	1.88(3)	—
0(6)	0.2559(2)	0.0887(2)	0.2375(9)	0.522(6)	32h	—	2.05(3)	—
O(7)	0.1621(2)	0.3225(2)	0.2382(9)	0.522(6)	32h	—	2.04(3)	—
O(3)	0.0908(2)	0.2434(2)	0.2354(8)	0.522(6)	32h	—	2.17(3)	—
O(9)	0.0843(2)	0.1610(2)	0.5150(6)	0.522(6)	32h	—	2.07(3)	—

Notes: Mn(4) and Mn(5) are two Mn sites corresponding to well-defined Mn^{4+} ions with average Mn–O distance 0.191 nm and forbidden for electron hopping. Mn(1), Mn(2), and Mn(3) are not pure Mn^{3+} ions.

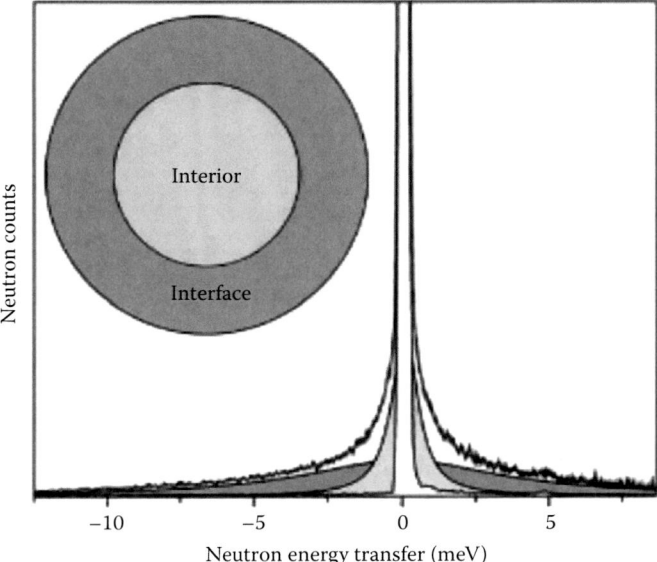

FIGURE 4.11 QENS spectrum of LIBH$_4$ at nanoporous carbon—4 nm at 400 K and 3 A^{-1} momentum transfer. The spectrum is fit to an elastic line with instrumental resolution (the central narrow feature) and two Lorentzian functions (the two shaded tails and their neighboring steeper features). These latter fits reflect BH$_4^-$ reorientational dynamics. The insert is a cross section of a 4 nm pore with two layers of LiBH$_4$. The faster component is dark and the slower light. (From Liu, X. et al., *J. Mater. Chem. A*, 1, 9935, 2013.)

For the former and latter populations, respectively, the reorientation jump rates from the Arrhenius plots varied from 2.6×10^9 s^{-1} and 5.6×10^{10} s^{-1} at 193 K to 3.51×10^{11} s^{-1} and 2.1×10^{12} s^{-1} at 400 K. The reorientation rates and activation energy of the less mobile population are in reasonable agreement with what is observed with QENS for bulk LiBH$_4$ but there is no evidence of the presence of a solid-solid phase transition in this region, in keeping with the confined nature of LiBH$_4$ in the nanoporous carbon matrix.

Next, we briefly discuss a technique based on Rutherford backscattering (RBS), which basically is a high-energy particle scattering phenomenon. RBS is an important near-surface elemental characterization technique for single- and multilayer thin films. In RBS, the sample is bombarded by a beam of high-energy particles (say, 2 MeV He^{++}). While most of them are implanted in the thin-film structure, a small fraction of them undergo a direct collision with near-surface located atoms of the sample. Backscattered particles are measured wrt its energy and direction. The experimental arrangement and a typical RBS spectrum are displayed in Figure 4.12.

4.1.1.3 Electron Microscopy and Diffraction

4.1.1.3.1 Transmission Electron Microscopy (TEM)

Transmission electron microscopy (TEM) involves real space imaging of solid materials at atomic resolution using an electron beam as the probe. The technique is invaluable in obtaining direct information on various length scales.

Electrons have a wavelength (λ) ~0.003 nm or 3 pm, which is much smaller than that of photons ($\lambda >$ 100 nm), and thus provide a much higher resolution. Thus, electron microscopy is far more informative than optical microscopy. Figure 4.13 gives the ray diagram of a TEM, which is usually operated in two modes: (a) bright field imaging mode and (b) selected area diffraction mode.

The cone of diffracted electrons with an aperture of the order of a few tens of milliradians (mrad) can pass through the small pole piece bores of the final lenses provided the back focal plane of the objective lens that contains the first diffraction pattern is focused on the screen [4.15].

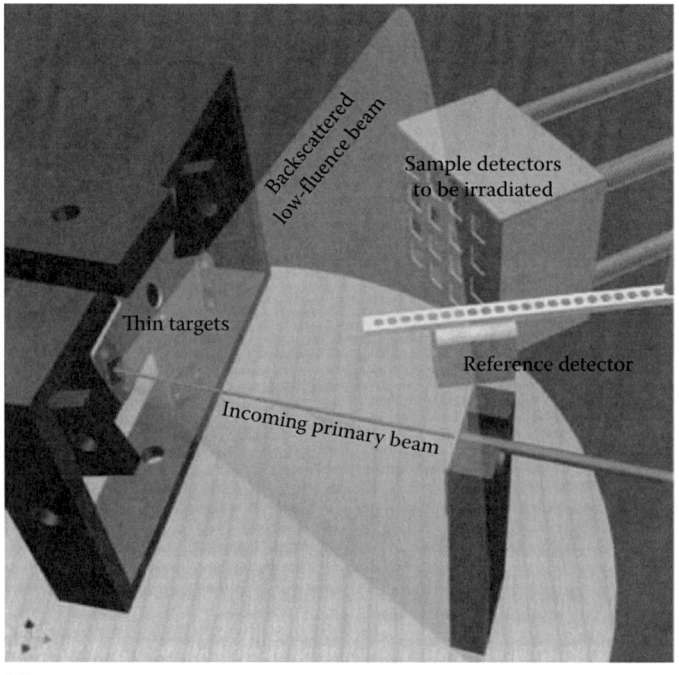

(a)

FIGURE 4.12   (a) Schematic of RBS.   *(Continued)*

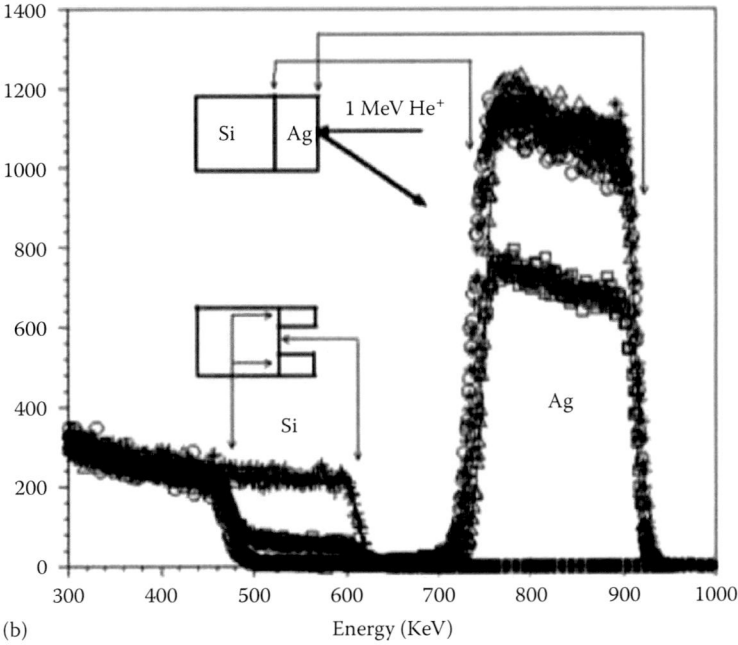

(b)

Energy (KeV)

FIGURE 4.12 (*Continued*) (b) Typical RBS spectra from the Ag/Br-Si (111) samples. As-deposited (Δ), 500°C-(*), 600°C-(o), 700°C-(□), and 800°C-(+) annealed, simulation (–). Note absence of Ag for the 800°C sample demonstrating the sensitivity of RBS. (From Roy, A. et al., *AIP Conf. Proc.*, 1349, 691, 2011.)

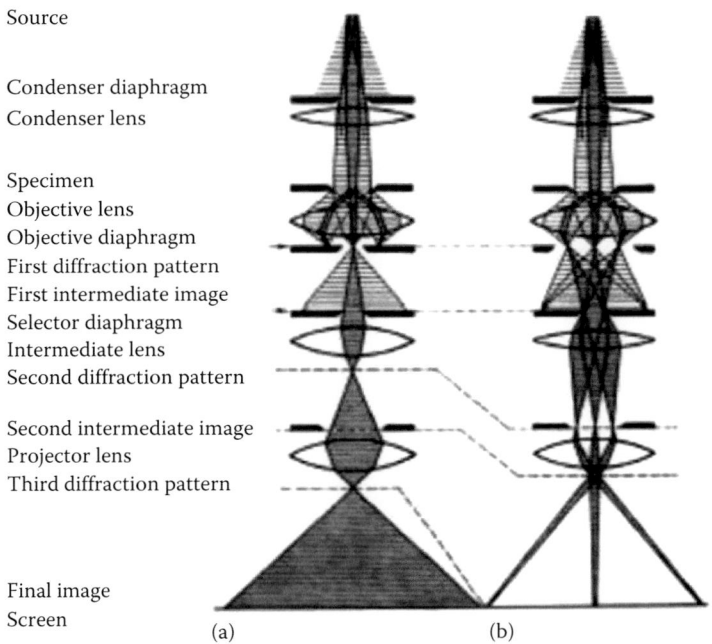

Source

Condenser diaphragm
Condenser lens

Specimen
Objective lens
Objective diaphragm
First diffraction pattern
First intermediate image
Selector diaphragm
Intermediate lens
Second diffraction pattern

Second intermediate image
Projector lens
Third diffraction pattern

Final image
Screen

(a)   (b)

FIGURE 4.13 Ray diagram of a typical transmission electron microscope: (a) bright-field imaging and (b) selected-area diffraction.

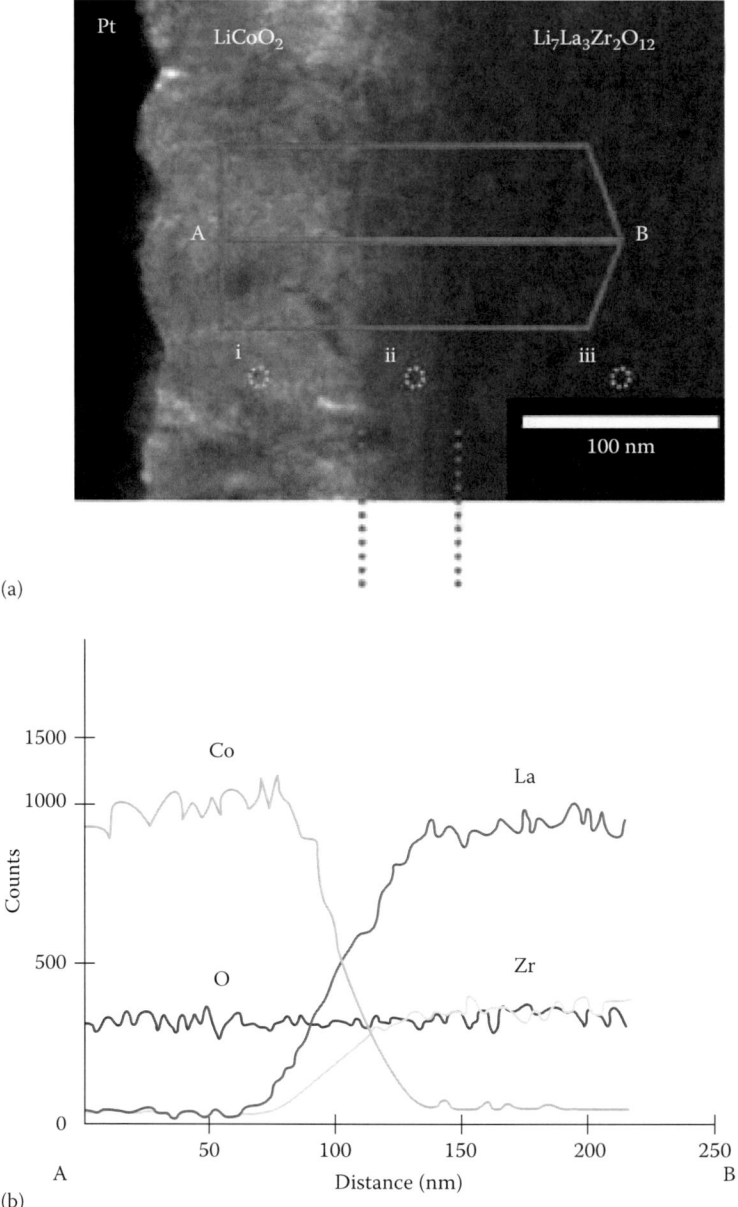

(a)

(b)

FIGURE 4.14   (a) Cross-sectional TEM image of an $Li_7La_3Zr_2O_7/LiCoO_2$ thin-film interface and (b) the energy-dispersive X-ray spectroscopy (EDS) line profile obtained from the region indicated by the red arrow in the direction A–B. The broken red lines indicate the reaction layer at the LLZ/LiCoO$_2$ interface. Points i, ii, and iii indicate locations used for nanobeam electron difraction analysis. (From Kim, K.H. et al., *J. Power Sources*, 196, 764, 2011.)

Most significantly, electron microscopy uses electromagnetic lenses instead of glass lenses employed in optical microscopes.

The major components of a TEM are (1) electron gun, (2) condenser lens system, (3) specimen chamber, (4) objective, (5) intermediate lens, (6) projector system for producing (a) images and (b) diffraction patterns, and (7) viewing screen.

The most important aspect of imaging by TEM in solid state ionics is the electrolyte–cathode interface. Figure 4.14a shows the TEM image of the interface between $Li_7La_3Zr_2O_{12}$ electrolyte and $LiCoO_2$ cathode.

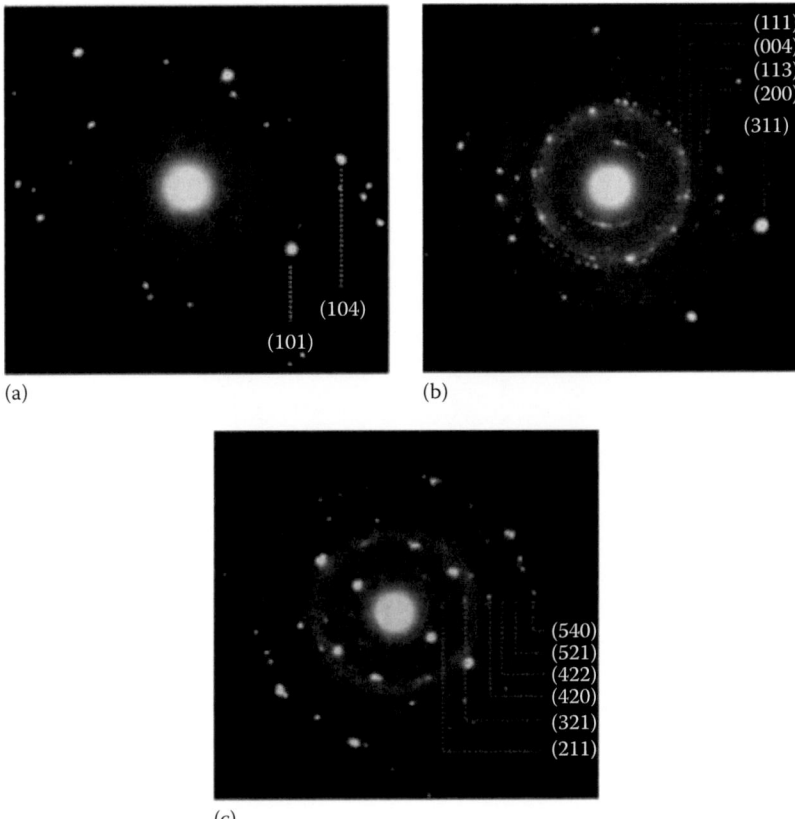

FIGURE 4.15 Selected area nanobeam electron diffraction patterns observed in the (a) $LiCoO_2$ thin film, (b) reaction layer, and (c) LLZ, for regions indicated by broken yellow circles i, ii, and iii, respectively. (From Kim, K.H. et al., *J. Power Sources*, 196, 764, 2011.)

Figure 4.15a shows the selected area electron diffraction pattern observed in a $LLZ/LiCoO_2$ interface.

Figure 4.16b shows an interesting application of electron diffraction to observe two phases in $LiMn_2O_4$ achieved through a phase transition.

4.1.1.3.2 Scanning Electron Microscope (SEM)

A scanning electron microscope (SEM) produces images of a sample by scanning it with a focused beam of electrons [4.18]. The electrons interact with atoms in the sample, producing various signals that can be detected. These contain information about the sample's surface topography and composition. The electron beam is generally scanned in a raster scan pattern, and the beam's position is combined with the detected signal to produce an image. SEM can achieve resolutions better than 1 nm. Specimens can be observed in high vacuum, in low vacuum, in wet conditions (in environmental SEM), and at a wide range of cryogenic or elevated temperatures.

The most common mode of detection is by secondary electrons emitted by atoms excited by the electron beam. On a flat surface, the plume of secondary electrons is mostly contained by the sample, but on a tilted surface, the plume is partially exposed and more electrons are emitted. By scanning the sample and detecting the secondary electrons, an image displaying the topography of the surface is created.

Nonconductive specimens such as fast ion conductors tend to charge when scanned by the electron beam, and especially in secondary electron imaging mode, this causes scanning faults and other image artifacts. They are, therefore, usually coated with an ultrathin coating of electrically conducting material such as gold, deposited on the sample either by low-vacuum sputter coating or by high-vacuum evaporation.

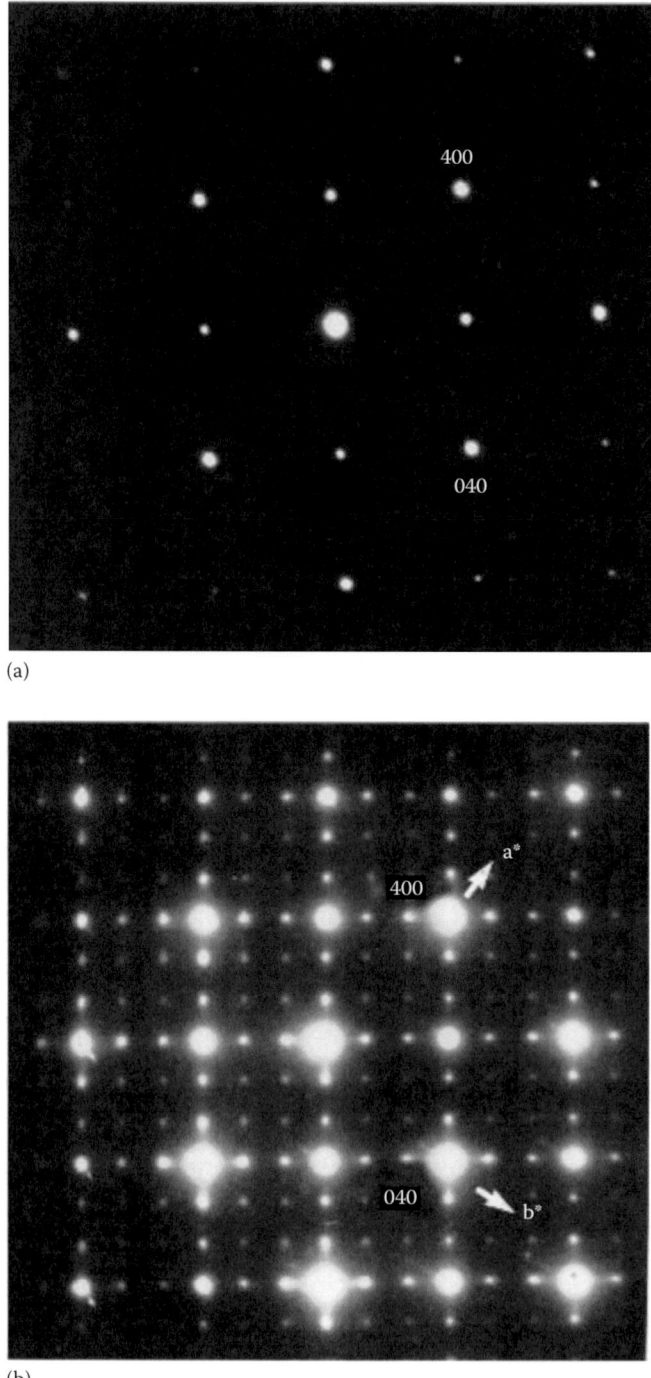

(a)

(b)

FIGURE 4.16 Electron diffraction patterns along [011] of the high-temperature cubic phase (a) and the low-temperature orthorhombic phase (b) of stoichiometric $LiMn_2O_4$ crystal. (From Rodriguez-Carvajal, J. et al., *Phys. Rev. Lett.*, 81, 4660, 1998.)

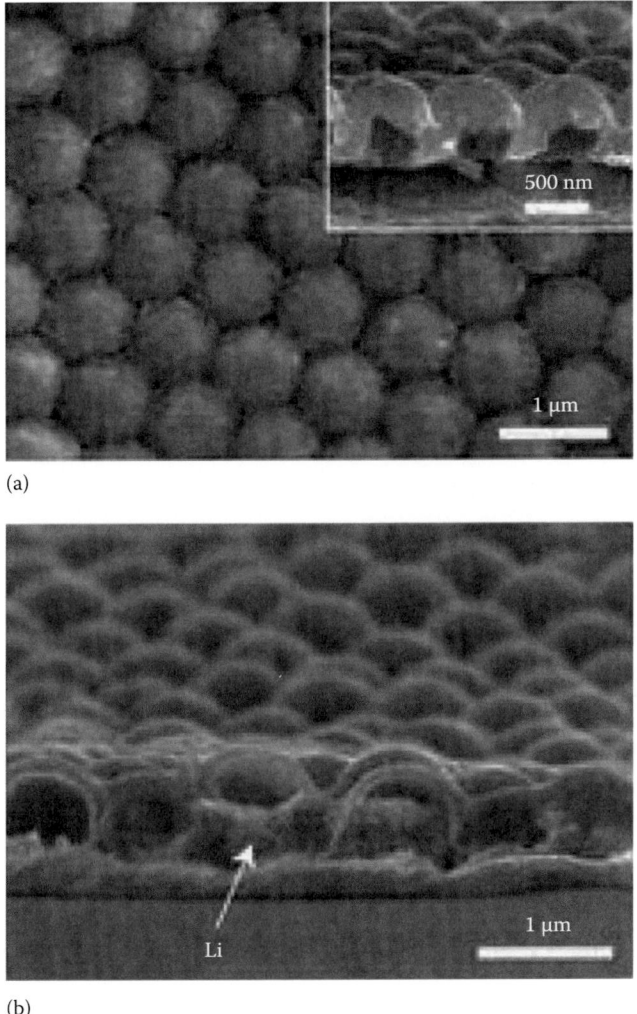

(a)

(b)

FIGURE 4.17 (a) Top-view SEM image of hollow carbon nanospheres after the initial solid electrolyte interface (SEI) formation process. Inset: the hollow carbon nanosphere structure is preserved after SEI coating. (b) Cross-sectional SEM image showing the initial deposition of Li. (From Zheng, G. et al., *Nat. Nanotechnol.*, 9, 618, 2014.)

Figure 4.17 shows the SEM of a hollow carbon nanosphere—solid electrolyte interface (a) before and (b) initial deposition of Li.

Energy-dispersive spectroscopy (EDS) is an analytical technique used for the elemental analysis or chemical characterization of a sample. It relies on an interaction of some source of X-ray excitation and a sample. Its characterization capabilities are due in large part to the fundamental principle that each element has a unique atomic structure allowing unique set of peaks on its X-ray emission spectrum. To stimulate the emission of characteristic X-rays from a specimen, a high-energy beam of charged particles such as electrons or protons, or a beam of X-rays, is focused into the sample being studied. At rest, an atom within the sample contains ground state (or unexcited) electrons in discrete energy levels or electron shells bound to the nucleus. The incident beam may excite an electron in an inner shell, ejecting it from the shell while creating an electron hole where the electron was. An electron from an outer, higher energy shell then fills the hole, and the difference in energy between the higher energy shell and the lower energy shell may be released in the form of an X-ray. The number and energy of the X-rays emitted from a specimen can be measured by an energy-dispersive spectrometer. As the energy of the X-rays is characteristic of the difference in energy between the two shells, and of the atomic structure of the element from which they were emitted, this allows the elemental composition of the specimen to be measured.

SEM of dried ZrO_2 gels prepared from systematically varied polyethelene oxide (PEO) contents are shown in Figure 4.18. The sensitivity of the SEM technique to detect the evolving morphology is dramatically demonstrated. The picture of the monolithic gel achieved by this process is also shown. As a further characterization, pore volume and pore sizes are graphed in Figure 4.19.

Next, we discuss atomic force microscopy.

4.1.1.4 Atomic Force Microscopy

The atomic force microscope (AFM) is an offshoot of scanning tunneling microscope (STM), which images conducting or semiconducting surfaces. The AFM has the advantage of imaging almost any type of nonconducting surface, including polymers, ceramics, composites, gels, and glassy samples, which are the most important material forms for solid state ionics. The original AFM of Binnig, Quate, and Gerber, invented in 1985, has a diamond shard with the sharp tip attached to a strip of gold foil [4.21]. The diamond tip contacts the surface directly. This is also the contact mode of AFM, which has the additional "tapping" or noncontact mode for nonconducting surfaces. The interatomic van der Waals forces provide the interaction mechanism. Detection of the cantilever's vertical movement is done with a second tip—an STM placed above the cantilever (Figure 4.20).

The difference between STM and AFM is that AFM does not require the sample to conduct electricity, whereas STM does. AFM works at room temperature, while STM requires sub-ambient temperature and other stringent conditions.

AFM, apart from looking at "static" surfaces of ionic conductors, offers—more importantly—a direct visual evidence of particles formed during electrochemical processes. An example is given in Figure 4.21.

We now discuss in detail one of the most fundamental characterization tools for solid state ionic materials, namely, diffusion.

4.1.2 Diffusion

4.1.2.1 Basics

Diffusion refers to the process of particle movement driven by a concentration gradient. The process is governed by Fick's first law. This law relates the particle flux j_p to the gradient of concentration c of these particles:

$$j_p = -D^T \nabla c \qquad (4.14)$$

where D^T is the tracer diffusion coefficient or diffusivity. Combine Equation 4.14 with the continuity equation

$$\partial c/\partial t + \nabla j_p = 0 \qquad (4.15)$$

to obtain Fick's second law

$$\partial c/\partial t = \nabla(D^T \nabla c) \qquad (4.16)$$

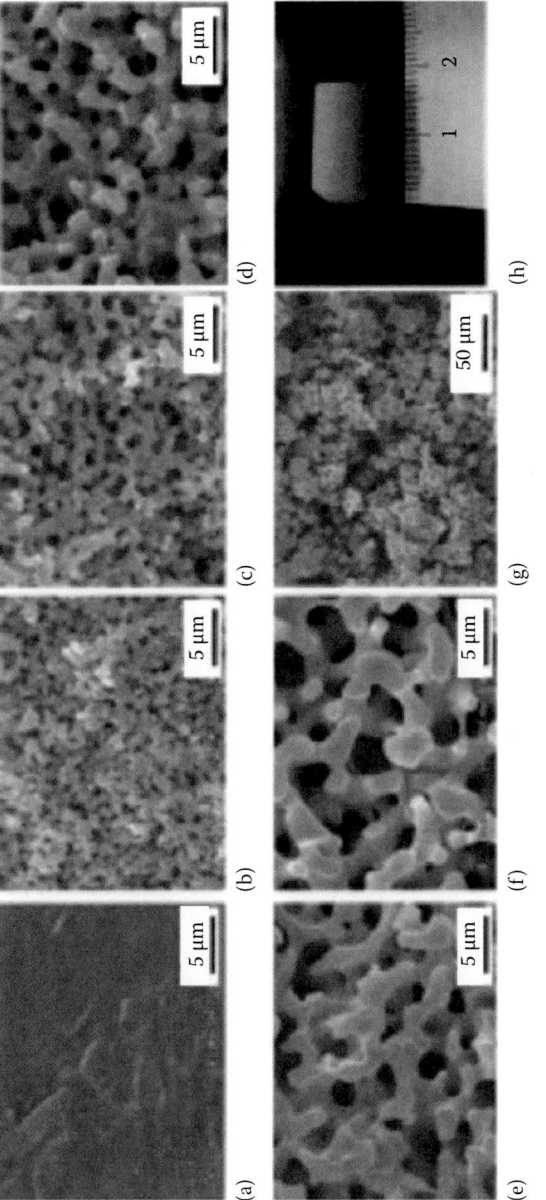

FIGURE 4.18 SEM images of dried ZrO_2 gels prepared by varying PEO content (P): (a) 0.040, (b) 0.090, (c) 0.110, (d) 0.115, (e) 0.120, (f) 0.120, (g) 0.130, and (h) 0.150. (h): Digital picture of monolithic ZrO_2. (From Konishi, J. et al., *Chem. Mater.*, 20, 2165, 2008.)

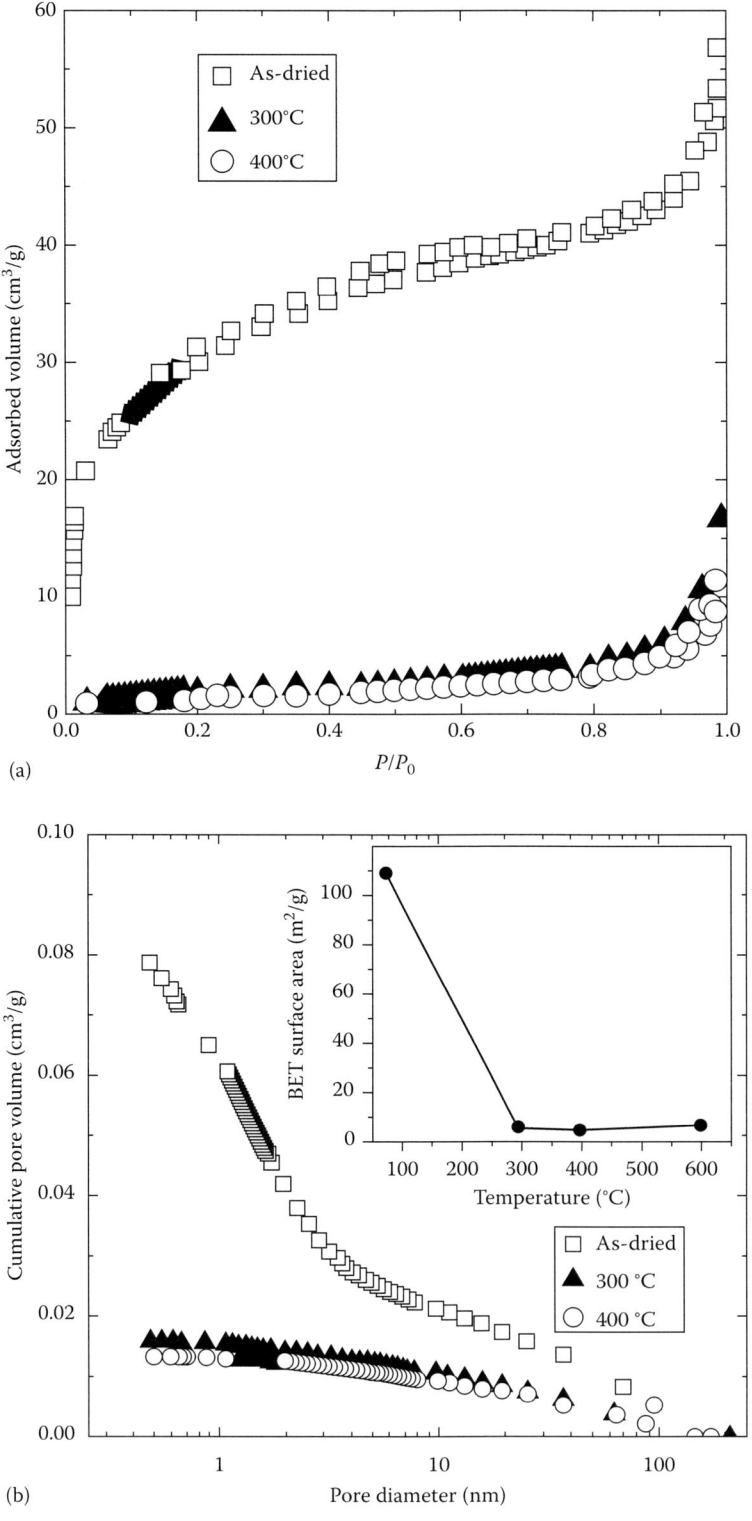

FIGURE 4.19   (a) Pore size of as-dried and heat-treated zirconia gels. (b) Pore volume of as-dried and heat-treated zirconia gels. Inset shows BET surface area as a function of temperature. (From Konishi, J. et al., *Chem .Mater.*, 20, 2165, 2008.)

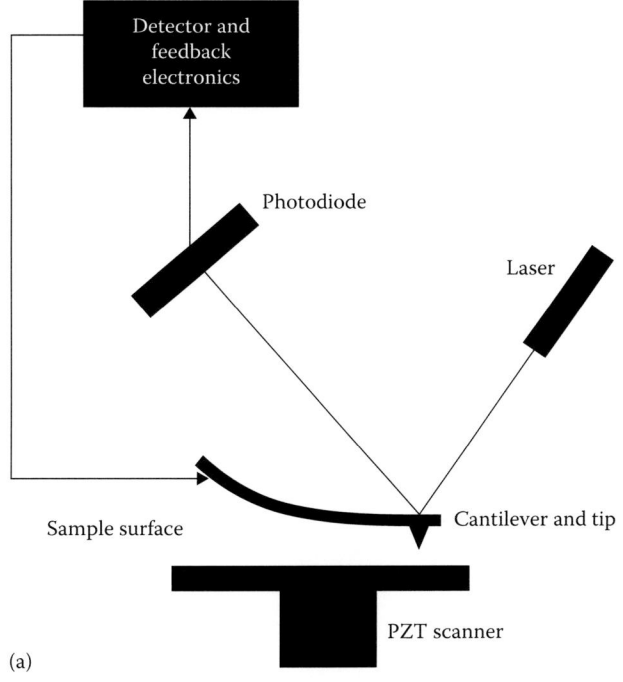

(a)

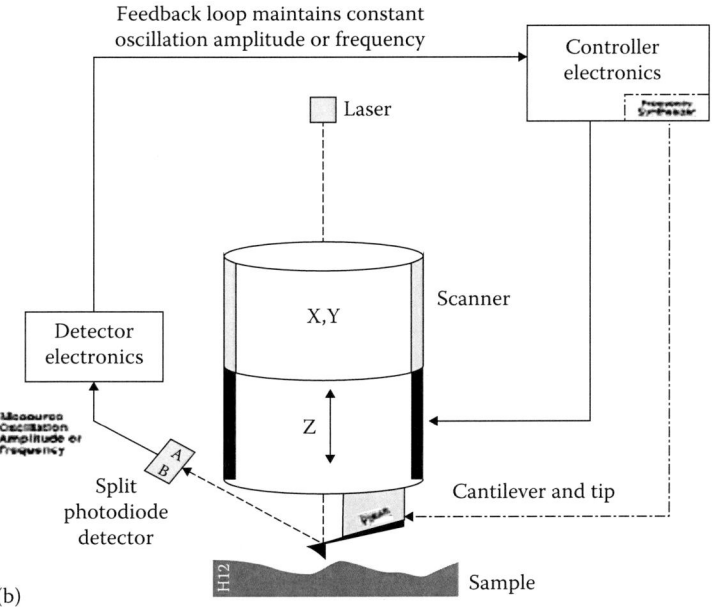

(b)

FIGURE 4.20 (a) Schematic showing principle of AFM. (b) A sketch of the atomic force microscope.

FIGURE 4.21 Li-ion reduction on LICGC (Li-ion conducting glass ceramics with the general composition: $Li_2O-Al_2O_3-SiO_2-P_2O_5-TiO_2-GeO_2$ with a conductivity $\sim 1 \times 10^{-4}$ S/cm at room temperature) using bias sweeps of 0 to -5 V. Li metal particles (white spots) formed after the application of linear sweep from 0 to -5 V (applied to the tip, which is the working electrode) with a sweep rate of 1200 mV/s on a 10×10 grid. (From Kumar, A. et al., *Sci. Rep.*, 3, Article number: 1621.)

This partial differential equation simplifies if D^T is constant and can be solved for particular initial and boundary conditions. Then, one can determine D^T from measurements of concentration profiles $c(r, t)$. The temperature dependence of the diffusion coefficient is described empirically by an Arrhenius relation

$$D^T = D_0^{\,T} \exp\left(-E_A/k_b T\right) \qquad (4.17)$$

where
  E_A is the activation energy for mass transport
  D_0^T is the preexponential factor
  k_b the Boltzmann constant
  T the temperature

Microscopically, D^T can be defined by the Einstein–Smoluchowski relation

$$D^T = \lim_{t \to \infty} \langle r^2(t) \rangle / 2\, dt \qquad (4.18)$$

where
  $\langle r^2(t) \rangle$ is the mean square displacement of the particles after time t
  d is the dimensionality of the movement

An atom moving through a solid will make jumps between different minima in a potential landscape. In crystalline solids, these minima are represented by lattice sites and interstitial sites. Generally, the potential landscape may be time dependent. Consider the situation where the mean jump is short compared to the mean residence time τ in such a minimum. Then, the trajectory of a particle is composed of a sequence of elementary jumps with an average jump length l. In terms of these quantities, a diffusion coefficient D^{uc} for uncorrelated jumps can be defined:

$$D^{uc} = \frac{l^2}{2\,\mathrm{d}t} \tag{4.19}$$

D^T and D^{uc} are related by

$$D^T = f \cdot D^{uc} \tag{4.20}$$

Here, f is the correlation factor. If the movement is pure random hopping, $f = 1$. If the atoms perform correlated motion with enhanced backward hopping probability after completed jumps, $0 < f < 1$. Experimentally determined correlation factor contains information about the diffusion mechanism. In single-crystal samples, point defects are necessary for the movement of atoms or ions. The concentration of such defects in thermal equilibrium is governed by statistical thermodynamics. Starting from a single-crystal introduction of defects and thus disorder leads to increasingly defective systems. These include highly defective single crystals, micro/nanocrystalline materials, and amorphous/glassy systems. Disorder on regular sites to structural or topological disorder arises in such systems. The defects can be zero-dimensional (vacancies or interstitials), one-dimensional (dislocations), or two-dimensional (grain boundaries). The highest degree of disorder arises in amorphous/glassy materials. An important question to investigate is: *Is there a correlation between structural disorder and the appearance of fast diffusion?*

Figure 4.22 displays the nondestructive neutron-activation technique pioneered by Dejus et al. [4.23] may be used to produce the initially activated layer of ^{64}Cu. Among the photon emissions in the nuclear decay process, the 511 keV position annihilation photons are used for the determination of the concentration profile of Cu^+. These photons are emitted in simultaneous pairs at 180° angle to each other enabling coincidence method for detection. The system uses a 3″ × 3″ NaI (Tl) detectors lined up at 180° angle, and two pairs of detectors are used (Figure 4.19). After the amplifier, the signals from the 511 keV photons are screened by timing single-channel analyzer energy discriminators. The coincidence analyzer accepts photon pairs, which fall within the resolving time window set at 23 ns.

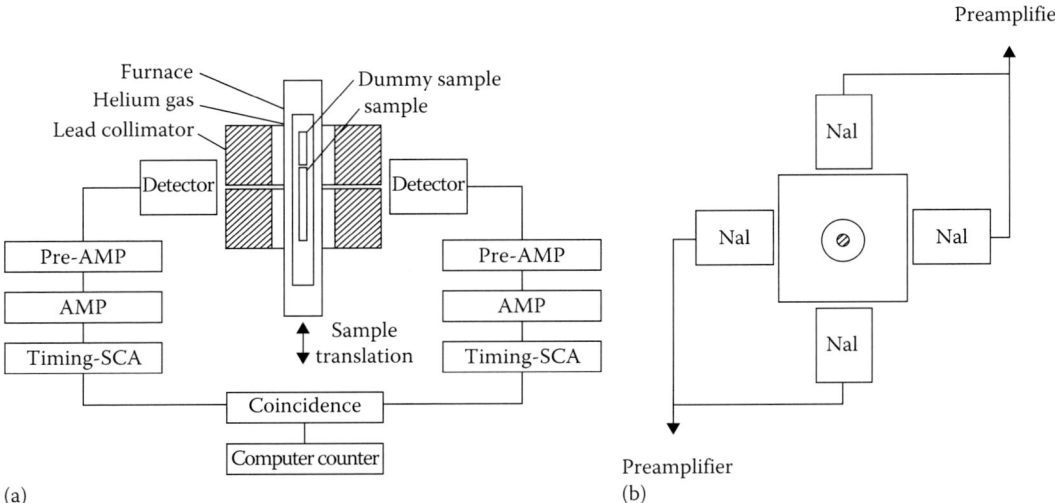

FIGURE 4.22 Coincidence tracer diffusion measurement system. (a) Plan view. (b) Vertical view of the coincidence system. (From Dejus, R. et al, *Solid State Ionics*, 1, 327, 1980.)

The data sets at 379°C and 398°C were collected using the noncoincidence system using two detectors for improved statistics, arranging them to be normal to each other to avoid pulse overlap from pairs of photons. Merck Art 2748 CuI vacuum dried for 2 h at 106°C and slowly heated to 625°C (>melting point (605°C) under 0.3 atm argon) gives yellowish-colored ingot shaped further as a 8 mm diameter, 80 mm long cylinder as the sample taken in a sealed quartz tube for the diffusion experiment. A double-chambered furnace with the inner chamber filled with 1 atm helium gas to maintain good thermal conduction and to prevent the Cu-lined wall of the furnace from oxidizing is used for heating the sample. The outer chamber is evacuated to 1 µTorr. A PCS (personal communications systems) regulator temperature controller along with a chromel–alumel thermocouple completes the setup. Sample temperature is maintainable to better than ±0.6°C while the temperature gradient is 0.038°C/mm within the sample as measured without sample. A stepping motor translates the sample and the furnace relative to the slit in the Pb collimator. During the measurement of diffusion profiles at a constant temperature, the sample rod is scanned continuously. The tracer diffusion measurements of β-CuI cylindrical sample are measured at five temperatures: at 383.5°C, 392.4°C, and 403.5°C by the coincidence system, and at 379°C and 398°C using two detectors at 90° angle (without coincidence).

Figure 4.23a and b give the time-dependent variance of the coincidence system and the log D-versus-$1/T$ plot for β-CuI, respectively. Tables 4.5 and 4.6 give illustrative results on tracer diffusion measurements on β-CuI and D values for all polymorphs of CuI, respectively.

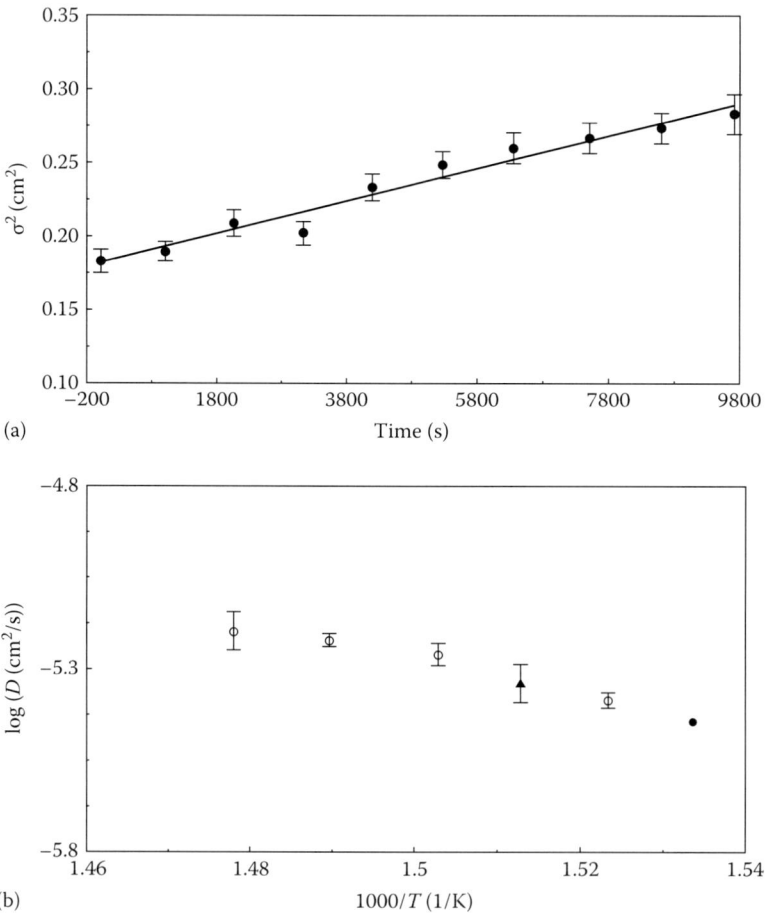

(a)

(b)

FIGURE 4.23 (a) Time-dependent variance of the four-detector coincidence system measured at 392.4°C. (b) Log D-versus-1000/T for β-phase CuI. The hexagonal β-CuI exists as a single phase between 647 and 655 K and shows marked anisotropy in its thermal expansion within this small temperature range. (From Dejus, R. et al, *Solid State Ionics*, 1, 327, 1980.)

TABLE 4.5

Results of Tracer Diffusion Experiments on β-CuI

Temperature (°C)	D (cm²/s)	Error in D (cm²/s)	D_0	E_{act} (cV)
3.79.1 ± 0.2	3.61×10^{-6b}	$\pm 0.03 \times 10^{-6}$	42.4	0.91 ± 0.087
383.5 ± 0.1	4.14×10^{-6a}	$\pm 0.2 \times 10^{-6}$		
388	4.6×10^{-6} [3]	$\pm 0.6 \times 10^{-6}$		
392.4 ± 0.1	5.53×10^{-6ab}	$\pm 0.38 \times 10^{-6}$		
398.3 ± 0.5	6.03×10^{-6b}	$\pm 0.26 \pm 10^{-6}$		
403.5 ± 0.1	6.37×10^{-6ab}	$\pm 0.83 \times 10^{-6}$		

Source: Dejus, R. et al, *Solid State Ionics*, 1, 327, 1980.
[a] Coincidence method.
[b] *Non-coincidence method.*

TABLE 4.6

Diffusion Data on the Three Polymorphs of CuI

Phase	T (°C)	D (cm²/s)		Conductivity σ (Ω cm)⁻¹		E_{act} (eV)	
		Tracer Diff.	e.c.	Tracer Diff.	e.c.	Tracer Diff.	e.c.
γ	315	1.0×10^{-7}	—	5.4×10^{-3b}	$<10^{-2c}$	1.30 [4.3]	1.6c
	369	8.3×10^{-a}	—	4.1×10^{-b}			
β	369	2.8×10^{-6a}	—	0.14^b	0.06^c	0.91	1.1c
	407	7.2×10^{-a}	—	0.35^b			
α	407	1.3×10^{-5a}	—	0.61^b	0.09^c	0.31 [4.3]	0.19c
	600	4.2×10^{-a}	—	1.5^b			

Source: Dejus, R. et al, *Solid State Ionics*, 1, 327, 1980.
[a] Extrapolated value for the phase boundary temperature.
[b] Calculated by Nernst–Einstein relation.
[c] See Chandra [4.110].

The raw data has to be reduced by an exponential function exp (λl) to compensate for the decay of ^{64}Cu during the scan. For further details, see [4.23].

The diffusion profile is described by a Gaussian function for an initially thin layer or an initially Gaussian concentration distribution

$$C(x,t) = \frac{C_0}{\left[2\pi\left(2Dt + {\sigma_0}^2\right)\right]^{1/2}} \exp\left[-\frac{(x-x_0)^2}{2\left(2Dt + \sigma_0^2\right)}\right] \tag{4.21}$$

where
C_0 is the concentration of the initially thin layer
t is the diffusion time
D is the diffusion constant
σ_0^2 is the variance of the profile at $t = 0$

Diffusion during the scan, geometrical resolution of the detector system and its correlation with the diffusion would usually cause the experimental curve to deviate from the ideal Gaussian distribution. But in this case, an overall good fitting is achieved. The observed count rate under a finite geometric resolution function $r(x)$ is expressed by

$$Q(x,t) = \int_{-\infty}^{+\infty} C(u,t)R(u-x)\mathrm{d}u \tag{4.22}$$

TABLE 4.7

Complex Electrical Response Representations

Name	Symbol	Real Part	Imaginary Part
Impedance	Z^*	R_s	$-1/\omega C_s = 1/Y^*$
Admittance	Y^*	G_p	ωC_p
Permittivity	ε^*	C_p	$-G_p/\omega = Y^*/i\omega$
Modulus	M^*	$1/C_s$	$\omega R_s = 1/\varepsilon^* = i\omega Z^*$

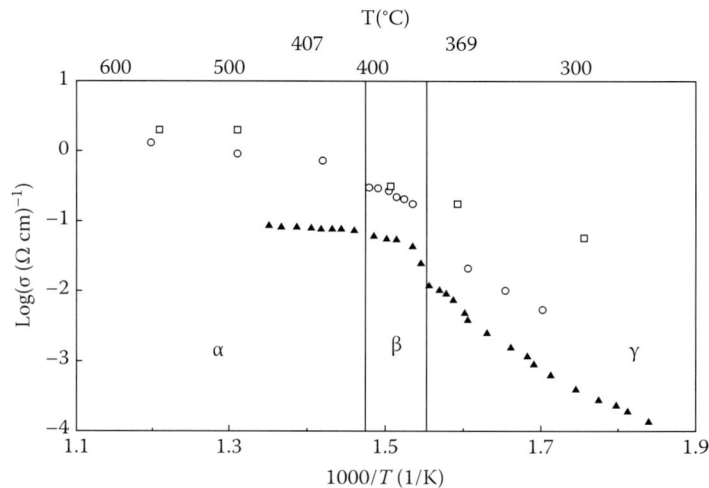

FIGURE 4.24 Log σ versus $1000/T$ for β-CuI from tracer diffusion measurements (O); electrical conductivity measurements by Wagner et al. (Δ); molecular dynamics calculations by Vashishta et al. (□). (From Borg, R.J. and Dienes, G.J., *An Introduction to Solid State Diffusion*, Academic, San Diego, CA, 1988.)

The fitting shown in Figure 4.23 is highly accurate, which is achieved through a large number of measurements made at each temperature. The experimental data are summarized in Table 4.7.

Diffusion generally fits the relation

$$D = D_0 \exp\left(-Ea/k_B T\right) \tag{4.23}$$

And the log D-versus-$1000/T$ plot for β-CuI demonstrates good linearity (Figure 4.17b). The diffusion activation energy E_a is determined to be 0.91 eV from the slope of the straight line. The extracted value of the preexponential factor $D_0 = 42.4 \text{ cm}^2/\text{s}$ is an order of magnitude estimate.

The ionic transport parameters for CuI are given in Table 4.2. The results from different experimental methods including total electrical conductivity as well as from molecular dynamics calculations are plotted in Figure 4.24. Note the close parallel between diffusion data and the molecular dynamics calculations. The conductivity is calculated from tracer diffusion results via the Nernst–Einstein relation $\sigma_D = n(Ze)^2 D/k_B T$ using $n = 1.72 \times 10^{28}/\text{m}^3$ for Cu^+ density, $Z = 1$, and e being electronic charge.

For further discussion, see Johansson et al. [4.24].

We now discuss conductivity briefly, beginning with the transport number experiment.

4.1.2.2 Conductivity

The earliest systematic measurement of ionic conductivity was initiated by Tubandt and Lorenz when they measured the conductivity of the now-famous superionic conductor AgI. As a fundamental demonstration of the exclusive nature of Ag^+ ion conduction in AgI, they devised the transport number experiment briefly described in the following (Figure 4.25a).

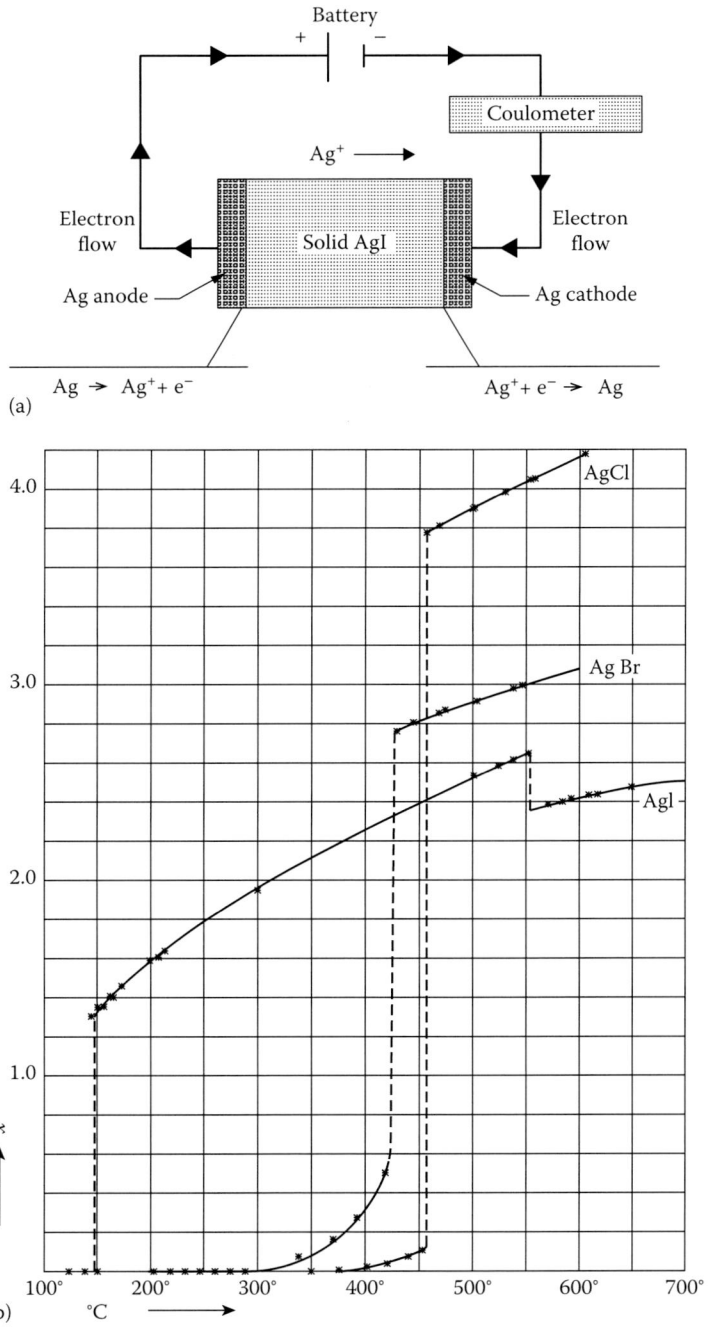

FIGURE 4.25 (a) The Tubandt–Lorenz experiment on AgI. [From Borg, R.J. and Dienes, G.J., *An Introduction to Solid State Diffusion*, Academic, San Diego, CA, 1988.) (b) The original Tubandt–Lorenz plot of ionic conductivity of AgCl, AgBr, and AgI. Note the abrupt *decrease* of conductivity in AgI at the melting point in sharp contrast to the behavior of AgCl and AgBr. The abrupt rise of conductivity at ~150°C in AgI signaled the birth of superionic or fast ion conduction. (Adapted from Funke, K., *Sci. Tech. Adv. Mater.*, 4, 043502, 2013.)   (*Continued*)

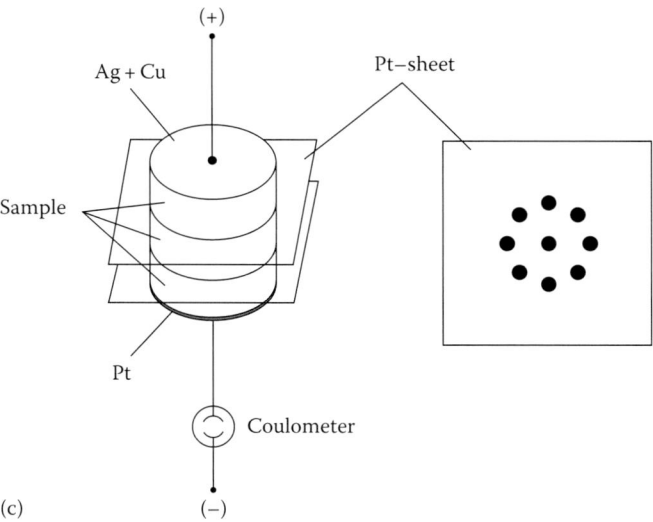

FIGURE 4.25 (*Continued*)   (c) Tubandt's method applied for $Ag_xCu_{1-x}Br$ solid solution to measure transference numbers of charge carriers as the relative contribution to total conductivity. (From Saito, M. and Tamaki, S., *Solid State Ionics*, 60, 237, 1993.)

The method is illustrated for measurements on α-AgI. A pellet of AgI is heated to 150°C. The silver electrodes are accurately weighed before and after the passage of a measurable current. It is ensured by measurement that the amount of Ag gained by the cathode exactly equals that lost by the anode. The total number of ions transferred should be equal to the current registered by the coulometer.

It is instructive to look at the very first ionic conductivity plot for silver halides made by Tubandt and Lorenz (Figure 4.25b).

A more definitive setup for transference measurements is shown in Figure 4.25c.

The measurement cell is constructed from a multi-bored platinum cathode (helps easy peel-off of disks for measurement) sandwiched between three disks of $Ag_xCu_{1-x}Br$ and Ag + Cu anode combination. A constant direct current is passed through the cell held in argon atmosphere. Current density is ~150 μA/m². The mass change of each layer in the cell was determined by weighing it before and after charge flow in a microbalance to a precision of 10 μg. The quantity of electricity was measured by using an electronic coulometer.

Impedance spectroscopy is discussed in Section 4.1.4.1.

For now, we discuss dielectric constant measurement at microwave frequencies.

4.1.2.3 Dielectric Constant Measurement

Usually, the dielectric constants of superionic conductors are not directly measured but are deduced from impedance spectroscopy measurements to be discussed later. However, such measurements are easily done at constant microwave frequencies (~9 GHz). At such high frequencies, the fast ion motions directly respond to the stimulus so that one can not only get a reasonable value for the high frequency dielectric constant but also track phase transitions. We describe microwave dielectric measurements on PbF_2 in the following.

The standing wave method of Roberts and von Hippel [4.29] is used in this measurement. The *unconsolidated* powder sample of thickness d is packed in a rectangular waveguide operated at X-band frequency terminated by a reflecting plane. Microwaves reflected by the specimen set up standing waves in the waveguide. The voltage standing wave ratio or VSWR $= V_{max}/V_{min}$. As indicated in Figure 4.26b, the distance y between the first voltage minimum and the surface of the specimen is determined by the standing wave detector.

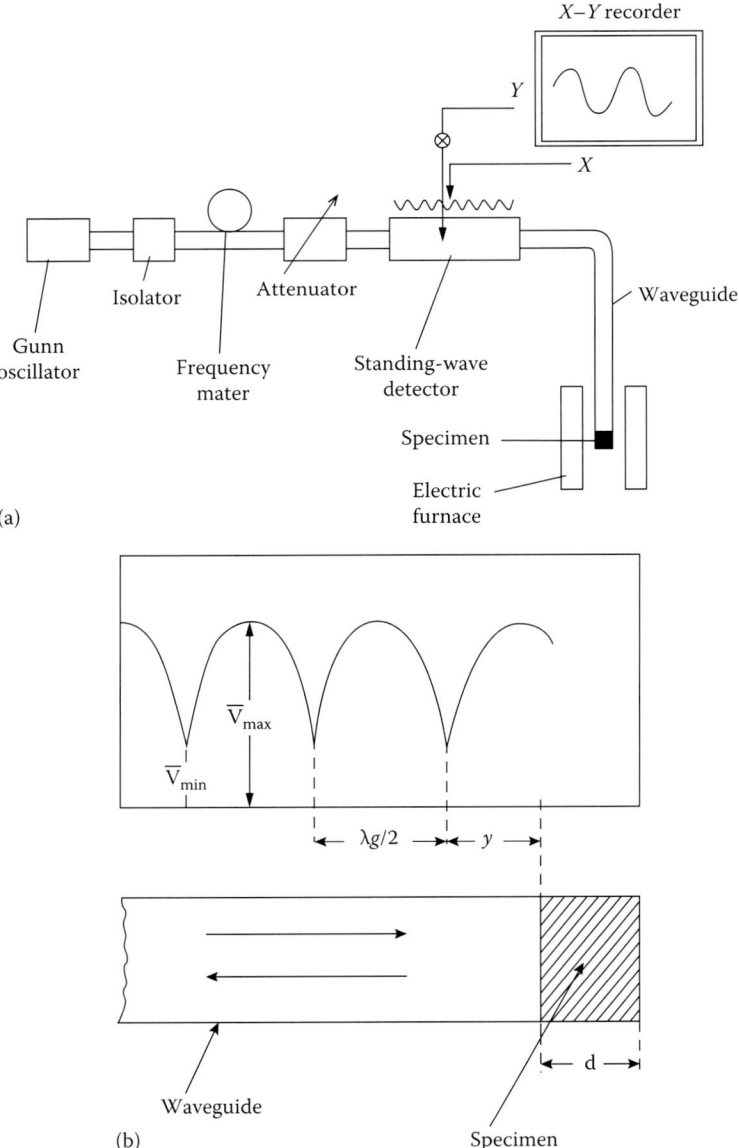

(a)

(b)

FIGURE 4.26 (a) Microwave circuit for dielectric measurements on superionic conductors as a function of tempera-ture ~9 GHz microwaves produced by the Gunn oscillator are guided along the circuit all the way to the sample, which may be heated by pacing in a furnace as shown. Standing wave detector is used to generate the pattern shown on the recorder. (b) Measurement basics. The specimen in the waveguide reflects microwaves and produces standing waves. (From Tateno, J. and Masaki, N., *Solid State Ionics*, 51, 75, 1992.)

From the experimental data, the complex dielectric constant $\varepsilon^* = \varepsilon' - i\varepsilon''$ is calculated using the fol-lowing relations:

$$\varepsilon' = \left(\lambda/\lambda_c\right)^2 - \left(\lambda/2\pi\right)\left(\alpha^2 - \beta^2\right) \tag{4.24}$$

$$\varepsilon'' = 2\alpha\beta\left(\lambda/2\pi\right)^2 \tag{4.25}$$

$$\gamma = \alpha + i\beta \tag{4.26}$$

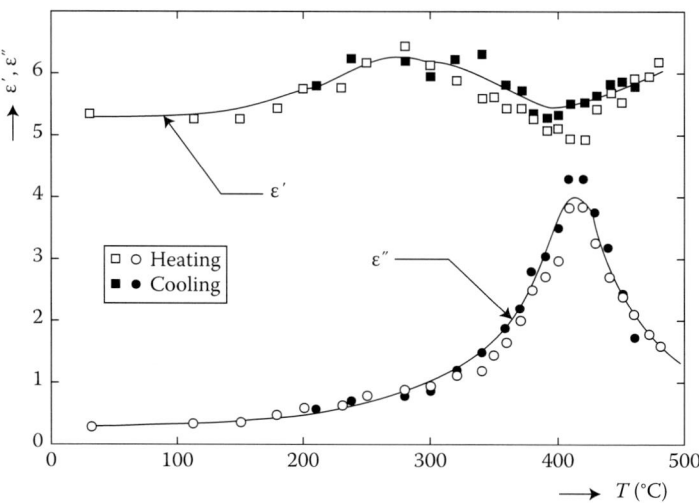

FIGURE 4.27 Dielectric constants (ε', ε'') of PbF_2 with a packing factor 0.41 as a function of temperature. (From Tateno, J. and Masaki, N., *Solid State Ionics*, 51, 75, 1992.)

$$\tanh\left(\gamma d\right)/\gamma d = \left(\lambda_g/2\pi d\right)\left[\tan\left(2\pi y / \lambda_g\right)+i\right]/\left[ir\tan\left(2\pi y/\lambda_g\right)-1\right] \qquad (4.27)$$

where

r is the inverse standing wave ratio or 1/VSWR
λ and λ_g are the wavelength in free space and in air inside the waveguide, respectively
λ_c is the cut-off wavelength

The measurements are accurate to 5% because of the oxide film that invariably forms at high temperature.

Figure 4.27 shows the dielectric constants of PbF_2 in the temperature range 30°C–480°C. The broad peak in ε'' at 420°C is the dielectric signature of the unique superionic phase transition called the Faraday transition [4.26].

Chapter 5 contains a theory of such phase transitions.

Dielectric relaxation has been discussed in Chapter 3 in connection with microwave-assisted synthesis of materials. It is the basis of a problem in this chapter.

Next, we discuss thermal conductivity.

4.1.2.4 Thermal Conductivity

Thermal conductivity is the ability of a material to conduct heat. It arises from Fourier's law for heat conduction. The higher the thermal conductivity of a material, heat is transferred at a higher rate across the material. The phenomenon arises from phonon–phonon collisions in solids due to anharmonicity associated with lattice vibrations. A solid state ionic material such as AgI is thus expected to transfer heat faster than, say, NaI. Changes in ion dynamics in superionic conductors undergoing phase transitions are expected to be reflected in changes in phonon mean free paths above superionic phase transition temperatures. We consider two methods for measurement of thermal conductivity: (1) transient hot wire method and (2) the guarded hot plate method.

1. In the hot wire method, thermal conductivity is measured by monitoring the temperature change ΔT of a heater wire embedded within the silver iodide sample throughout the duration of a current pulse. At any time d, during the action of the pulse, the temperature change approximately fits the expression

$$\Delta T = (q/4\pi k)\ln t + B$$

where q is the instantaneous power dissipated per unit length of the wire. The thermal conductivity k is calculated from the slope of the $\Delta T/q$-versus-ln t data. The previous equation is obtained from the transient solution to the diffusion equation. The latter is solved under the boundary conditions: (a) the heater wire is a perfectly cylindrical heat source, and (b) the sample is large enough so that for $r > r_{wire}$ its temperature is unchanged. The first condition helps measure k.

Goetz and Cowen [4.30] have measured the thermal conductivity of pressed pellets of AgI (99.999%) from 120 to 500 K by the transient hot wire method. The log k (W/m K)-versus-log T (K) plot for AgI is shown in Figure 4.28. The temperature dependence of k changes from $T^{-1.3}$ at the lowest temperatures to $T^{-1.8}$ below the superionic phase transition at 420 K. In the superionic phase of AgI ($T > 420$ K), k exhibits a weak temperature dependence $T^{+0.5}$. This significant result implies shortening of the phonon mean free path at high temperatures due to the high mobility of the Ag$^+$ ions.

2. We now describe the versatile guarded hot plate method (Figure 4.29) suitable for polymer electrolytes and solid state ionic ceramics. Used in the temperature range 80–800 K, it is accurate to 2%. The external plates (say, Al) are held at the constant temperature of a thermal bath. A hot plate supplies a uniform and constant heating power P_0, which is transmitted to the external plates through the sample of thermal conductivity k to be measured. A guard ring around the central zone is heated by a power P_1 such that the guard ring temperature is the *same* as that of the heating plate. Under this circumstance, *all* the power P_0 *crosses* the sample. Note that the guard ring and the hot plate are not in contact. Thermocouples placed on the two sides of the sample measure ΔT_1 and ΔT_2. If L and A are the heating plate area and sample length, respectively, thermal conductivity of the sample k is

$$k(T) = \frac{LP_0}{A(\Delta T_1 + \Delta T_2)} \tag{4.28}$$

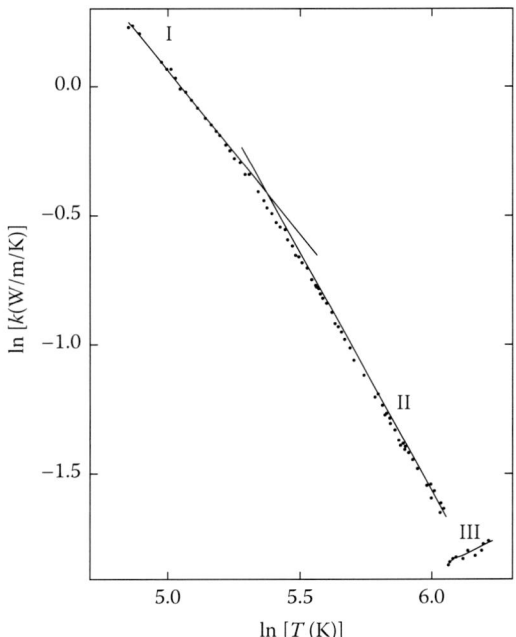

FIGURE 4.28 Log–log plot of thermal conductivity k versus temperature T for AgI. ■: data taken with sample cycled between the anneal temperature and room temperature; ●: data taken after the sample was cooled slowly through the transition. Lines are fits to the data. I, II, and III refer to very low, intermediate, and superionic phases, respectively. Note the change from $T^{-1.8}$ to $T^{+0.5}$ dependence across the II–III regions. (From Goetz, M.C. and Cowen, J.A., *Solid State Commun.*, 41, 293, 1982.)

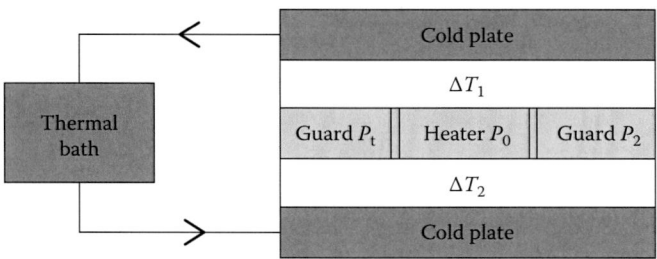

FIGURE 4.29   Schematic of the guarded hot-plate method.

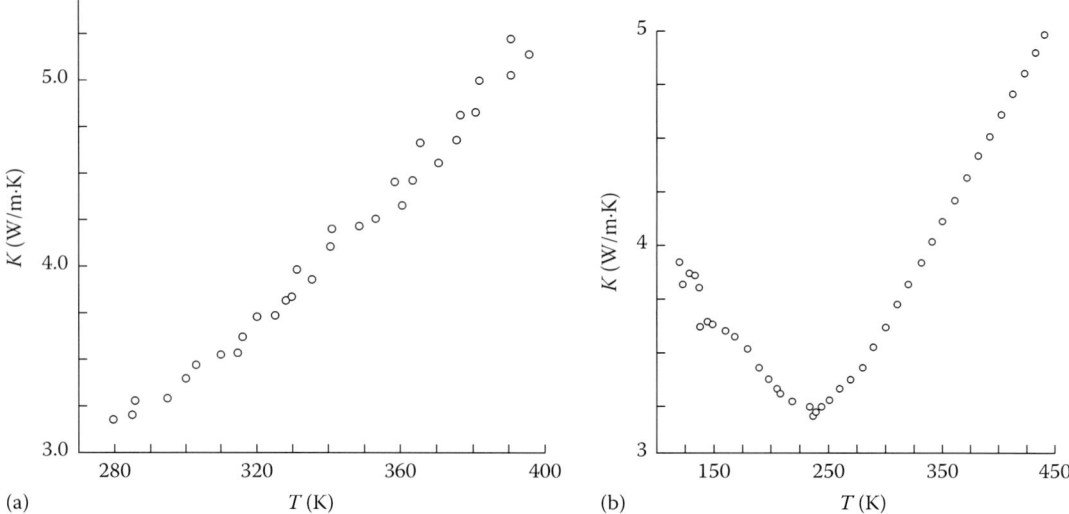

FIGURE 4.30   Temperature dependence of thermal conductivity of (a) LaF_3 and (b) lithium tetraborate. (From Aliev, R. et al., *Phys. Solid State*, 39, 1379, 1997.)

Note that the contact resistance of the sample should be negligible relative to the thermal resistance of the sample.

Examples of thermal conductivity measurements on F^- and Li^+ superionic conductors are shown as follows (Figure 4.30).

Thermal diffusivity is the thermal conductivity divided by density and specific heat capacity at constant pressure [4.32]. It measures the ability of a material to conduct thermal energy relative to its ability to store thermal energy. The laser flash analysis or laser flash method is used to measure thermal diffusivity of a multiplicity of different materials. An energy pulse heats one side of a plane-parallel sample (Figure 4.31). The temperature rise on the backside due to the energy input is detected as a function of time. The higher the thermal diffusivity of the sample, the faster the energy reaches the backside.

In a one-dimensional heat conduction under adiabatic conditions, the thermal diffusivity (a) of a sample is calculated from this temperature rise as follows:

$$a = 0.1388 d^2 / t_{1/2} \tag{4.29}$$

where
   d is the thickness of the sample
   $t_{1/2}$ is the time to the half maximum
   a may be measured by using the Netzsch, LFA427 system [4.34]

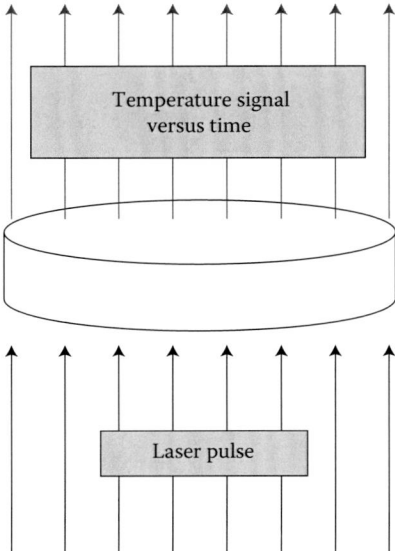

FIGURE 4.31 Parker's flash method of thermal diffusivity measurement. In a vertical setup, a light source (laser or flashlamp) heats the sample from the bottom side and a detector on top detects the time-dependent temperature rise. For measuring the thermal diffusivity, which is strongly temperature dependent, at different temperatures the sample can be placed in a furnace at constant temperature. (From Parker, W.J., et al., *J. Appl. Phys.*, 32, 1679, 1961.)

For a light irradiated ac calorimetric technique for thermal diffusivity measurements on thin films, see [4.35].

We close this section by illustrating a unique simultaneous measurement of thermal diffusivity (λ) and thermal conductivity (κ) of the Na^+ ion conductor Na_3PO_4 by a transient hot disc method (Figure 4.32).

These two properties, thermal conductivity and thermal diffusivity, are important material parameters involved in photoacoustic spectroscopy (PAS) discussed in Section 4.1.4.3.

4.1.2.5 Elastic Property Measurements

Recognizing the important connection between interatomic forces, lattice vibrations, and mechanical properties, we shall briefly consider mechanical characterization of solid state ionic materials. This has relevance to issues such as flexibility of electrolytes in polymer batteries and the use of solid state ionic

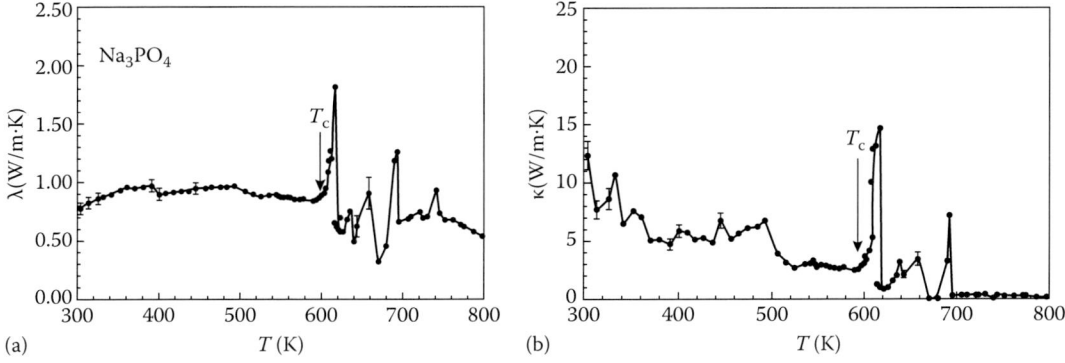

FIGURE 4.32 Temperature dependence of (a) thermal conductivity and (b) thermal diffusivity of sodium orthophosphate simultaneously measured by the transient hot disc method. Arrow indicates the transition temperature. (From Suleiman, B.M. and Lunden, A., *J. Phys. Condens. Matter*, 15, 6911, 2003.)

materials such as CuI as solid lubricants. The Young's moduli of sulfide electrolytes were recently measured to be about 20 GPa, which is an intermediate value between those of typical oxides and organic polymers [4.37]. Such measurements are essential for developing batteries. In a lithium-ion battery, the volumes of the electrode active material particles change during charging and discharging. Contact between the electrodes and solid electrolytes should be maintained during this change in volume. Otherwise, the battery capacity will drop. Solid electrolytes should deform elastically in response to the volume change.

In principle, the elastic constants of a material can be determined by applying a static load and observing the resulting deformation and applying Hooke's law [4.38]. For a cubic crystal, the three elastic moduli, namely, Young's modulus E, the shear modulus G, and the bulk modulus B, can be expressed in terms of the elastic compliance constants c_{11}, c_{12} and c_{44}. E arises from an extension along say x-direction (say e_1) in response to an applied stress T: $E = T/e_1$.

$$E = \frac{(c_{11} - c_{12})(2c_{12} + c_{11})}{c_{11} + c_{12}} \tag{4.30}$$

Decrease the volume of a unit cube by applying equal pressure $T_1 = T_2 = T_3 = T$ to all sides. Fractional change in volume $\Delta V/V$ and thus the bulk modulus B are expressible in terms of P:

$$B = -V\partial P /\partial V = -T/\Delta V$$

$$-\Delta V = 3\Delta V_1 = 3(e_1 + e_2 + e_3)$$

Using solutions of Hooke's law equations, K is obtained

$$B = \frac{c_{11} + 2c_{12}}{3} \tag{4.31}$$

The shear modulus (or rigidity modulus) depends on the plane and the direction of shear. Shear on the {100} cube planes in the ⟨001⟩ directions gives

$$G_{\{100\}} = c_{44}. \tag{4.32}$$

E, B, and G could then be used to derive c_{11}, c_{12}, and c_{44}.

It is to be noted that the energy density of a cubic crystal is $(\frac{1}{2}) B\delta^2$, where δ is three times the uniform dilatation [4.39].

Dynamic methods are used to measure the elastic properties of materials. These essentially involve timing the transit of a pulse of sound between the faces of a sample. Typically, a piezoelectric crystal such as quartz is cemented to one face of the sample to serve as a transducer. The latter will convert radiofrequency (RF) (from 1 to several hundred megahertz) electrical energy into ultrasound energy.

The RF oscillator is modulated so that pulses of sound of the order of 1 μs are produced. These pulses propagate through the specimen, are reflected from the opposite face, and return making a round trip. They are detected by the same crystal transducer connected to a receiver and an oscilloscope. The velocity is obtained by dividing the round-trip distance by the elapsed time. Typically, for an experimental frequency of 15 MHz and a pulse length 1 μs, the wavelength is of the order 3×10^{-2} cm.

Timing techniques such as those used in radar ranging are used to measure the velocity of sound in the crystal. The quartz crystal may be cut to excite longitudinal waves (x-cut) or cut for transverse waves (y-cut). The sample crystal can be orientated in any one of the various ways.

Figure 4.33 gives the block diagram of a typical electronic circuit for carrying out ultrasonic pulse echo measurements of elastic stiffness constants of a solid. More generally, it is used to observe the attenuation or damping of sound in solids. The method is sensitive to point defects and disorder that

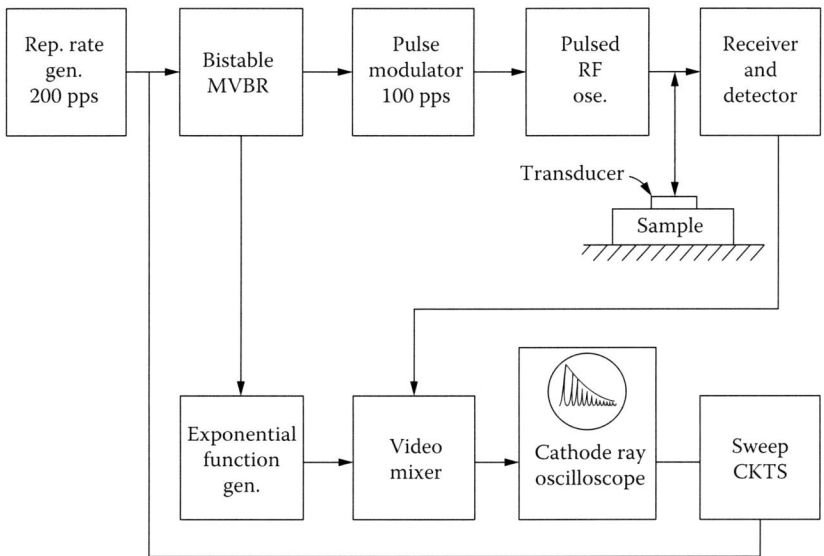

FIGURE 4.33 Block diagram of the electronic equipment for time and attenuation measurements of ultrasound in solids. MVBR: multivibrator, CKTS: circuits. (From Brown, F.C., *The Physics of Solids*, W. A. Benjamin, New York, 1967.)

exist in solid state ionic materials. They are investigated by observing the height of the detected signal as sound propagates in the crystal and is successively reflected at the faces. In a typical case indicated in Figure 4.33, attenuation is determined by matching an exponential waveform on the oscilloscope or the monitor of the computer to the envelope of decaying pulses.

In an important application to solid state ionic materials, Graham and Chang have used an ultrasonic pulse superposition technique to measure the velocities of the three pure-mode waves, which propagate along [110] in a single crystal of rubidium tetra-silver pentaiodide, $RbAg_4I_5$, the room-temperature superionic conductor [4.41]. The temperature was varied from 25°C, through a phase transformation at −65°C, and to a second phase transformation at −151°C. There was a peak in the ultrasonic attenuation at the −65°C transition for the longitudinal and one of the shear waves, and the attenuation and velocity of all of the waves were completely reversible upon cycling the temperature through this transition. The lower-temperature phase was generally several percent softer elastically than the room-temperature phase. Attempts to make measurements through the −151°C transition were not successful. The pressure dependence of the wave velocities was determined at 25°C. The adiabatic second-order elastic constants and their temperature and pressure derivatives at 25°C, which were determined from these measurements, are as follows. The values of C_{ij} (10^{11} dyn/cm²), $C^{-1}_{ij} (\partial C_{ij}/\partial T)$ (10^{-4} K⁻¹), and $\partial C_{ij}/\partial P$ for C_{11} are 1.648 ± 0.002, −3.54 ± 0.12, and +8.73 ± 0.11, respectively; for C_{12}, 0.934 ± 0.002, −3.73 ± 0.13, and +6.29 ± 0.11, respectively; and for C_{44}, 0.4892 ± 0.0005, −4.17 ± 0.05, and +0.884 ± 0.015, respectively. The Debye temperature of 90 K and the thermal expansion coefficient of 0.566×10^{-4} K⁻¹ calculated from these values agree very well with values determined by other means. Figures 4.34 and 4.35 give the temperature variation of c_{11}, c_{12}, and bulk modulus.

Because of the highly asymmetric potential in which Ag^+ ions move in the AgI lattice, the lattice vibrations are highly anharmonic (the interatomic restoring forces are weak), as directly reflected in thermal expansion studies through the phase transition. This aspect has been studied at a microscopic level by monitoring the temperature dependence of the attenuation of longitudinal ultrasonic waves propagated through AgI as done by Page and Prieur [4.42]. Figure 4.36, which dramatically depicts the changes in attenuation behavior near the phase transition temperature, highlights the anharmonic contribution to the elastic constants besides shedding light on the nature of the Ag^+ conductivity relaxation.

We now turn to thermal characterization.

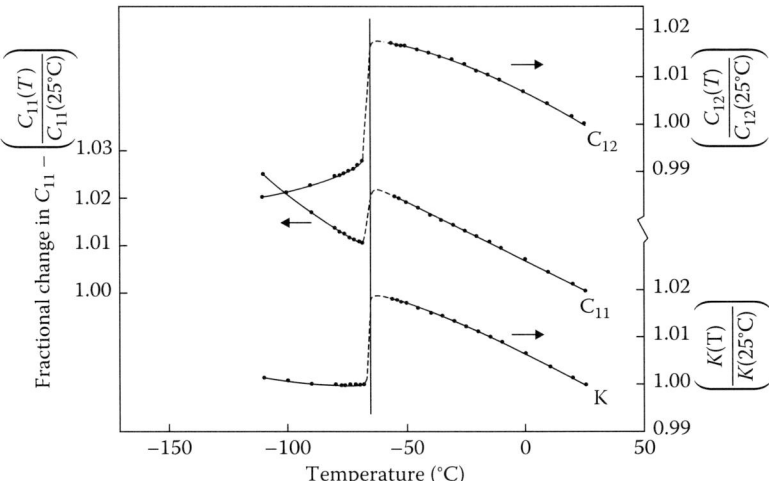

FIGURE 4.34 Temperature dependence of elastic constants c_{11} and c_{12} and bulk modulus K of $RbAg_4I_5$ crystals. (From Graham, L.J. and Chang, R., *J. Appl. Phys.*, 46, 2436, 1975.)

4.1.3 Thermal Characterization

4.1.3.1 Heat Capacity and Thermal Expansion

4.1.3.1.1 Heat Capacity

Heat capacity or specific heat capacity characterizes the material thermally (or thermodynamically) in terms of atomic/molecular vibrations and rotations of the solid and also free electron contributions to the internal energy of the system as a function of temperature. In the case of solid state ionic materials undergoing crystallographic or order–disorder phase transitions, heat capacity measurements provide vital information through anomalies whose analysis helps understand the nature of phase transitions and their connection to ion conduction—an aspect to be discussed in Chapter 5.

The classical solid at thermal equilibrium at a temperature T (>0 K) vibrates as a system of 3D harmonic oscillators, each one taking up a total energy (potential plus kinetic) of $k_B T$. The total vibrational energy of a classical solid is, therefore, $3nN_c k_B T$, where N_c is the number of unit cells making up the crystal and $3n$ is the number of vibrational degrees of freedom. Thus, the specific heat of a Dulong–Petit solid is a temperature-independent quantity $3n\rho_c k_B$, where ρ_c is the "unit cell density" of the material. At room temperature, the value of heat capacity at constant volume C_V defined as $\partial U/\partial T$ at constant V of nearly all solids is ~$3Nk_B$ or 25 J/mol/K.

This classical prediction fails at temperatures below room temperature as low-temperature specific heat measurements have recorded a marked drop approaching zero as T^3 in electronic insulators.

We shall briefly discuss two models for heat capacity from the point of view of characterization: (1) Einstein model and (2) Debye model. The Einstein model is an atomistic solid model while the Debye model is an elastic continuum solid model. Both models are relevant to solid state ionic materials: the first one accounts for the low-energy optical vibrations while the second one exposes the deviations from a typical Debye solid found in the experimental heat capacity of superionic conductors.

Einstein model: Einstein proposed a model for the vibrational specific heat of a solid. In this model all atoms of the solid vibrate with a single frequency. Applying Hooke's law, the normal modes of vibrations of a solid are *independent*. Thus the energy of a lattice mode depends only on its frequency ω and the so-called phonon occupancy n (phonons are discussed in Chapter 6). It is independent of n of other modes so that in thermal equilibrium at temperature T, thermal average occupancy $\langle n \rangle$ is given by the Planck distribution

$$\langle n \rangle = 1/[\exp(\hbar\omega/k_B T - 1] \quad \text{or} \quad \langle n \rangle + 1/2 = \frac{1}{2}\,\text{ctnh}\left(\hbar\omega/2k_B T\right) \tag{4.33}$$

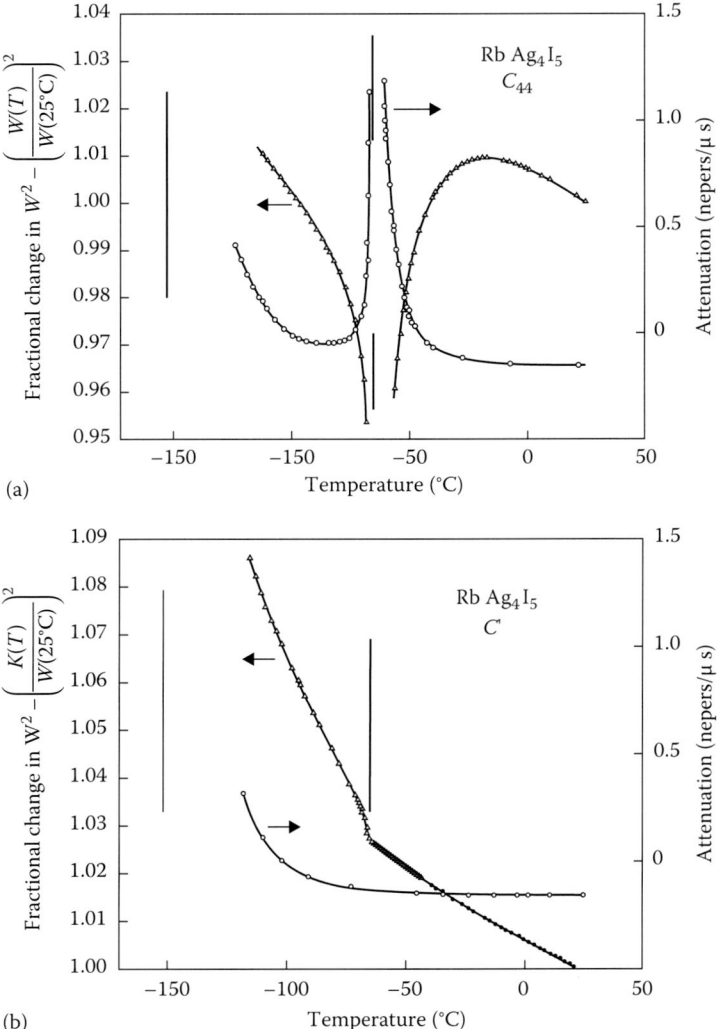

FIGURE 4.35 Shear mode data determining elastic constants (a) c_{44} and (b) C' of $RbAg_4I_5$ single crystals. Phase transitions occur at $-65°C$ and $-171°C$. (From Graham, L.J. and Chang, R., *J. Appl. Phys.*, 46, 2436, 1975.)

The average energy of an oscillator of frequency ω is $\langle n \rangle \hbar \omega$. For N oscillators in one dimension, all having the same resonance frequency, the energy U is

$$U = N\langle n \rangle \hbar \omega = \frac{N\hbar\omega}{\exp(\hbar\omega/k_B T - 1)} \tag{4.34}$$

The heat capacity of the oscillators is

$$C_v = (\partial U/\partial T)_v = \frac{Nk_B(\hbar\omega/\tau)^2 \exp(\hbar\omega/\tau)}{[\exp(\hbar\omega/\tau) - 1]^2} \tag{4.35}$$

where $\tau = k_B T$. Each atom has three degrees of freedom. So in high temperature, the limit of Equation 4.35 is $3Nk_B$, the Dulong–Petit value. At low T, where C_v decreases, $C_v \sim \exp(-\hbar\omega/\tau)$ but experimental C_v

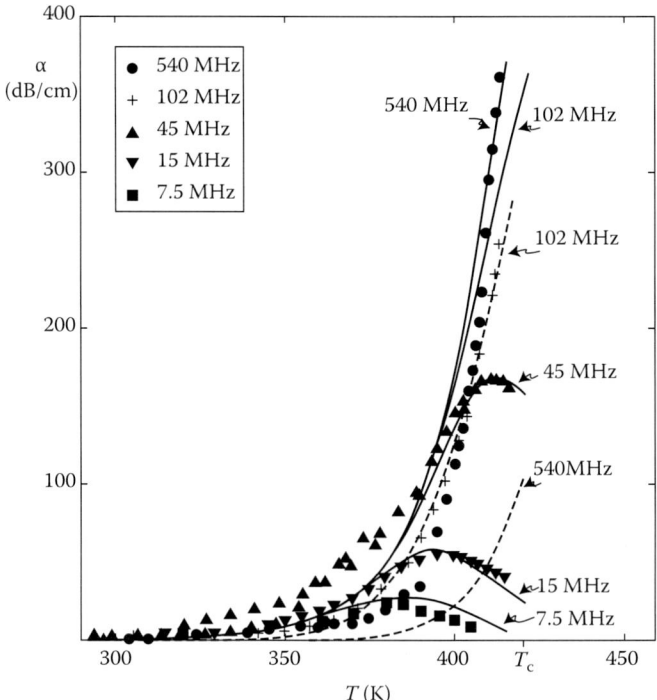

FIGURE 4.36 Temperature dependence of the attenuation of longitudinal ultrasonic waves (α) propagating along the *c*-axis in β-AgI. While points represent experimental data, solid lines represent the equation $\alpha = \kappa^2/v_s[\omega^2\tau_c/(1 + \omega^2\tau_c^2)]$ where $\kappa^2 = \eta^2/\varepsilon\,\varepsilon_0 c_A$, η = piezoelectric tensor element, ε = dielectric constant, ε_0 = vacuum permittivity, c_A = anharmonic contribution to elastic constant, v_s = velocity of sound, τ_c = conductivity relaxation time, and ω = frequency of the ultrasonic wave. Dashed curves are given by $\alpha = \kappa^2/v_s\,[\omega^2\tau_c/\{\omega^2\tau_c + (1 + \omega^2\tau_c\tau)^2\}]$ taking into account of screening and T-dependent Ag$^+$ conductivity relaxation. The atomic nature of Ag and I$_2$ in AgI are highlighted in properties such as Debye temperature, which impact thermal physics of these materials. (From Page, J.H. and Prieur, J.-Y., *Phys. Rev. Lett.*, 42, 1684, 1979.)

goes as ~T^3. The Einstein model is very useful in approximating part of the phonon spectrum, especially the contribution of optical phonons in solid state ionic materials.

Debye model: The experimental variation of $C_v(T)$ can be accounted for by the Debye model. In this model, the solid is made up of harmonic oscillators with *many* frequencies but not a single one as in the Einstein model. Let us now derive the Debye T^3 law for specific heat.

According to quantum mechanics, the energy of a harmonic oscillator changes in quantum steps but not continuously. Thus, a certain minimum amount of energy is already available for the first excitation, and this is only achieved at the appropriate *T*.

The energy eigenvalue of this harmonic oscillator is

$$\varepsilon_{ij} = \hbar\omega_{ij}\left(n_{ij} + 1/2\right) \tag{4.36}$$

or

$$\omega_{ij} = \omega_j\left(k_i\right) \tag{4.37}$$

where n_{ij} = number of vibrational energy quanta occupying the state. Energy quanta of the lattice vibrations are called phonons ("phon" means sound). So, the total vibrational energy

$$U_{\text{vib}} = \Sigma_{ij}\varepsilon_{ij} = \Sigma_{ij}\hbar\omega_{ij}\left(\langle n_{ij}\rangle + \tfrac{1}{2}\right) \tag{4.38}$$

The mean occupation numbers $\langle n_{ij} \rangle$ are calculated using the Boltzmann distribution

$$\langle n_{ij} \rangle = 1/[\exp\,(\hbar\omega_{ij}/k_BT)-1] \tag{4.39}$$

which is recognized as the Bose–Einstein statistics, making phonons Bose particles or bosons.
   So, the average energy is

$$\langle \varepsilon_{ij} \rangle = \hbar\omega_{ij}/[\exp(\hbar\omega_{ij}/k_BT)-1] + \hbar\omega_{ij}/2 \tag{4.40}$$

The total vibrational energy is obtained by a summation over all vibrations of the lattice:

$$U_{\text{vib}} - U_{\text{vib}}(T=0) = \Sigma_{ij}\varepsilon_{ij} = \frac{\Sigma_{ij}\hbar\omega_{ij}}{\exp(\hbar\omega_{ij}/k_BT)-1} \tag{4.41}$$

Debye showed how this sum can be evaluated when the heat capacity at low temperatures is of interest. The important first step is to write the LHS of Equation 4.7 in a continuous (but not discrete) form:

$$U_{\text{vib}} - U_{\text{vib}}(0) = \int_{\omega} D(\omega)d\omega\,\hbar\omega/[\exp(\hbar\omega_{ij}/k_BT)-1] \tag{4.42}$$

Here, $\mathcal{D}(\omega)d\omega$ is the number of lattice vibrations with frequencies in the interval between ω and $\omega + d\omega$. $\mathcal{D}$ is a "spectral density." Equation 4.45 shows that since $\langle n_{ij} \rangle \approx \exp(-\hbar\,\omega_{ij}/k_BT) \approx 0$ at high frequencies, high-frequency vibrations with $-\hbar\omega_{ij} \gg k_BT$ cannot be excited in the crystal. This important result means that only acoustic vibrations, that too in the vicinity of $k=0$ (around the center of the Brillouin zone) can take up energy at low temperatures. Here the dispersion law for sound waves, Equation 4.37, applies:

$$\omega_1(k_i) = c_{\text{sl}}|k_i| \tag{4.43}$$

for longitudinally polarized waves, and

$$\omega_2(k_i) = \omega_3(k_i) = c_{\text{st}}|k_i| \tag{4.44}$$

Debye's important suggestion is to consider only *these* sound waves while calculating specific heats at low temperatures. The number of lattice vibrations for which the wave vector is between k and $k + dk$ is given by

$$D(k)dk = \frac{4\pi k^2 dk V}{(2\pi)^3} \tag{4.45}$$

where $\mathcal{V}$ is the crystal volume. Sum of all three acoustic branches yields for $\mathcal{D}(\omega)$

$$D(\omega)d\omega = \sum_{j=1,2,3} D(k)(dk/d\omega)d\omega = \frac{V}{(2\pi)^2\{(\omega^2/c_{\text{sl}}^2)(1/c_{\text{sl}}) + 2(\omega^2/c_{\text{st}}^2)(1/c_{\text{st}})\}} \tag{4.46}$$

This Debye approximation correctly describes spectral density at low temperatures. But then it also leads to deviations in the form of peaks observed when the dispersion relation of a branch has a horizontal tangent. Note that even in an approximation applied to calculate specific heat, the total number of vibrational degrees of freedom must be maintained at the correct value. Thus as a second step, the distribution of frequencies is truncated at a certain value of ω, requiring that

$$\int_0^{\omega D} D(\omega)d\omega = 3N_c \qquad (4.47)$$

where N_c is the total number of unit cells. This gives us the Debye cut-off frequency ω_D:

$$\frac{V}{(2\pi)^2 \left\{1/3c_{sl}^2 + 2/3c_{st}^2\right\} \omega_D^3} = 3N_c \qquad (4.48)$$

The vibrational energy uptake can now be calculated using

$$\left[U_{vib} - U_{vib}(0)\right]/V = (1/2\pi^2)\left\{1/3c_{sl}^2 + 2/3c_{st}^2\right\}\left\{(k_BT)^4/\hbar^3\right\}\int_0^{xD} x^3 dx/(\exp x - 1) \qquad (4.49)$$

with $x = \hbar\omega/k_BT$, $x_D = \hbar\omega_D/k_BT$. Finally, the Debye T^3 law is

$$C_{v,\text{vibrational}} = \left(\frac{36\pi^4}{15}\right)\rho_c k_B (T/T_D)^3 \qquad (4.50)$$

This law describes very well the T-dependence of specific heat capacity at low temperatures (Figure 4.37) with the Debye characteristic temperature T_D being the material-specific parameter.

Values of T_D are 150 K for AgI, 90 K for RbAg$_4$I$_5$ (from measurements of elastic constants), 165 K for CuI, 586 K for β-alumina, 590 K for ZrO$_2$, and 533 K (computed) for LiFePO$_4$. Note that T_D connects elastic properties to thermodynamic properties—heat capacity, melting temperature, or vibrational entropy of a solid. An important point to note is that the heat capacities of many superionic conductors including AgI and ZrO$_2$ deviate significantly from the Debye model underlining the influence of defect and disorder on the heat capacity of these materials. Heat capacity anomalies accompany superionic phase transitions. Chapter 5 considers this aspect in detail.

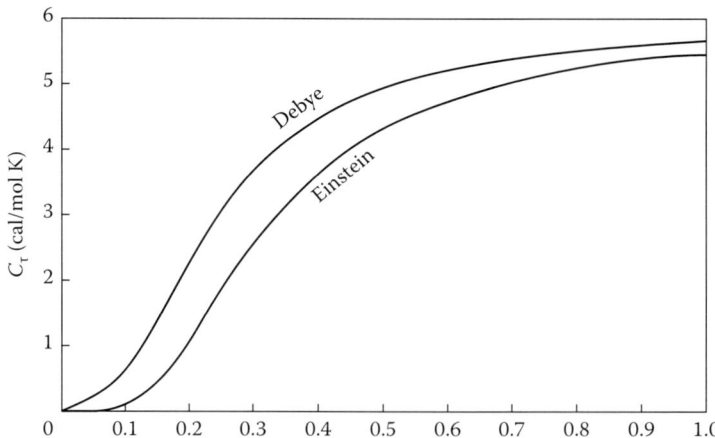

FIGURE 4.37 Heat capacity at constant volume in the Einstein and Debye models plotted against reduced temperature T/T_D.

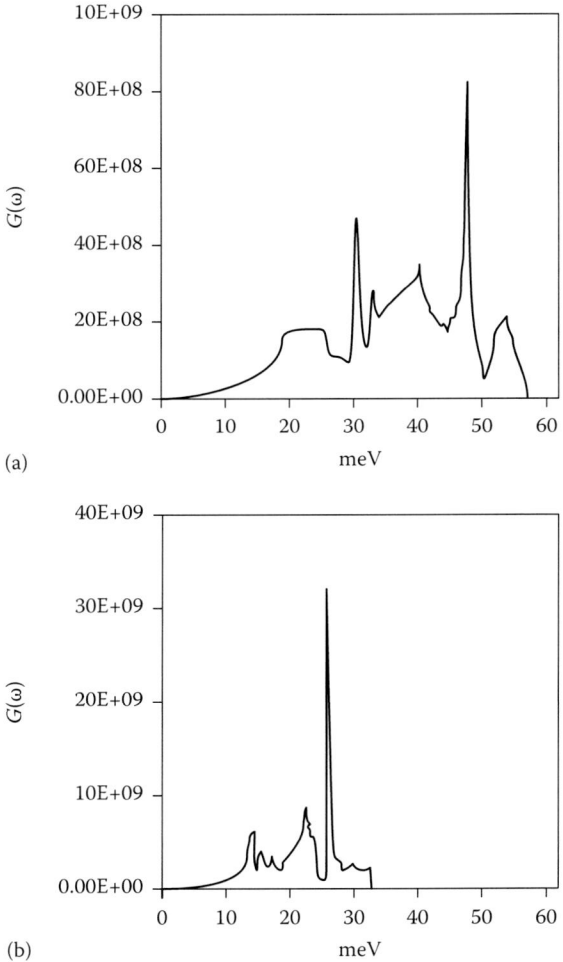

(a)

(b)

FIGURE 4.38 Vibrational density of states for Li_2S and Na_2S obtained from neutron scattering experiments. Deviations from the Debye's parabolic approximation ($\sim\omega^2$) are seen. (a) Li_2S. (b) Na_2S. (From Altorfer, F., Report LNS-152, Paul Scherrer Institute, Villigen, Switzerland, March 1990.)

Figure 4.38 depicts density of states function $\mathcal{D}(\omega)d\omega$ for Li_2S and Na_2S obtained from neutron scattering experiments. Note that Debye approximation for the density of phonon states is a parabola, whereas the experimental plot has many peaks. In the Einstein model, the density of states is a delta function.

Adiabatic calorimetry, ac calorimetry, and also thermal diffusivity measurements are used to measure heat capacity of superionic solid materials. However, even at low temperatures, the heat capacity of AgI to be discussed shows deviations from the T^3 law due to defects and disorder besides anharmonicity. An adiabatic calorimeter designed by Gronvald et al. [4.44] is briefly described in the following (Figures 4.39 and 4.40).

The sample contained is made of a vitreous quartz cylinder, 3 cm outer diameter and 12 cm high, with a central well for resistance, thermometer, and heater. Approximately 50 cm³ sample material goes into the container, which is sealed after filling. It fits into a silver cylinder of 0.5 mm wall thickness with removable bottom. This assembly along with alumina insulated lead wires is surrounded by three shields of silver (top, side, and bottom). The temperature differences between corresponding parts of the calorimeter and the shields are measured by Pt/90% Pt 10% Rh thermocouples, amplified, recorded, and automatically controlled by shield heaters. This is to maintain approximately adiabatic conditions during input and drift periods. The calorimeter and shield systems are surrounded by guard heater bodies of

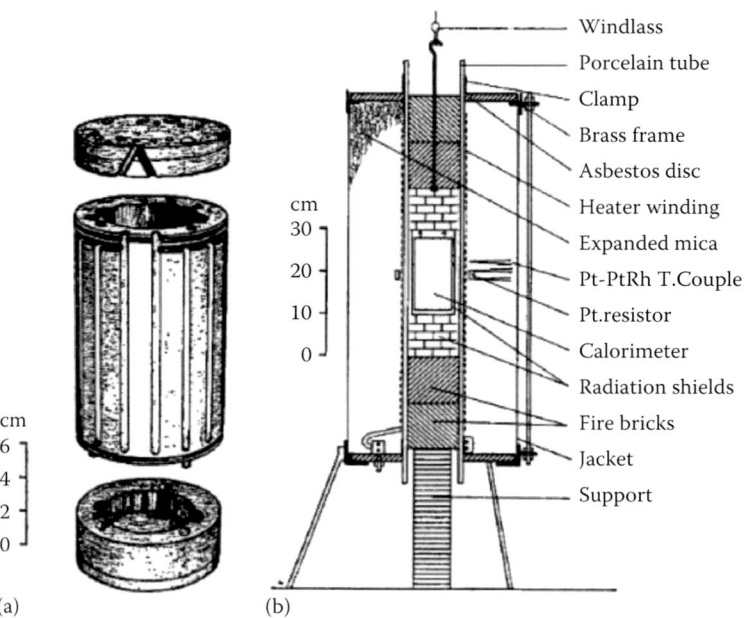

FIGURE 4.39 Adiabatic calorimetry-heat capacity measuring system. (a) Top, side, and bottom guard bodies. (b) Calorimeter thermostat. (From Gronvold, F., *Acta Chem. Scand.*, 21, 1341, 1964.)

silver and placed in a vertical tube furnace. The temperature of the guard bodies is controlled to be 0.4°C below the shields while that of the furnace core 10°C lower.

As an illustrative example, the experimentally determined heat capacity of silver iodide is represented by the sum of three contributions:

$$C_{exp} = C_V + C_{anh} + C_{str} \tag{4.51}$$

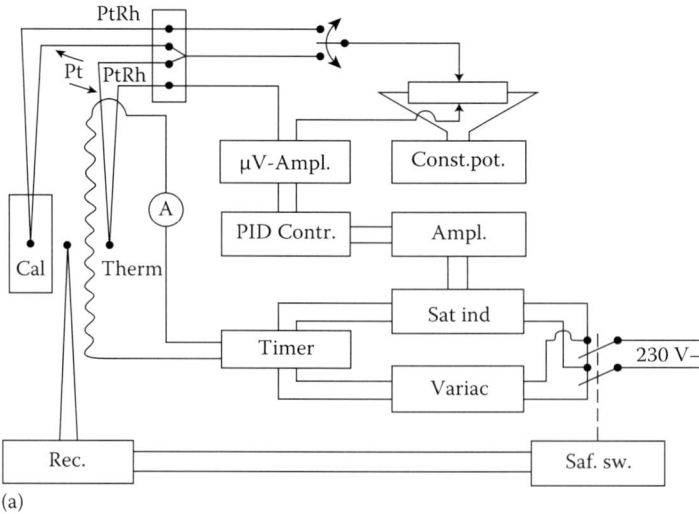

(a)

FIGURE 4.40 (a) Thermostat control circuit. *(Continued)*

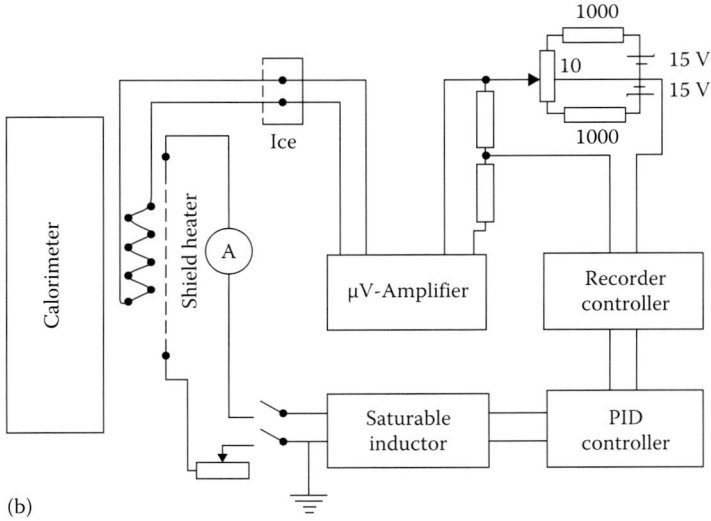

(b)

FIGURE 4.40 (*Continued*)   (b) Shield control circuit. (From Gronvold, F., *Acta Chem. Scand.*, 21, 1341, 1964.)

where
  C_V is the constant volume heat capacity in the harmonic approximation calculated using $T_D = 150$ K
  $C_{anh} = TV\alpha^2/\kappa$ accounting for the anharmonic vibrations
  C_{str} represents the precursor effect contributions to the structural-disordering transition near 420 K

Figure 4.41 is a selection of the heat capacity data on AgI, which highlights features *below* ambient temperature. Figure 4.42a shows the heat capacity *above* ambient temperature focusing on a semi-logarithmic scale the huge anomaly at the wurtzite-bcc structural phase transition. Figure 4.42b shows the "resolved" heat-capacity curves in the pre-transitional, transitional, and post-transitional temperature regimes.

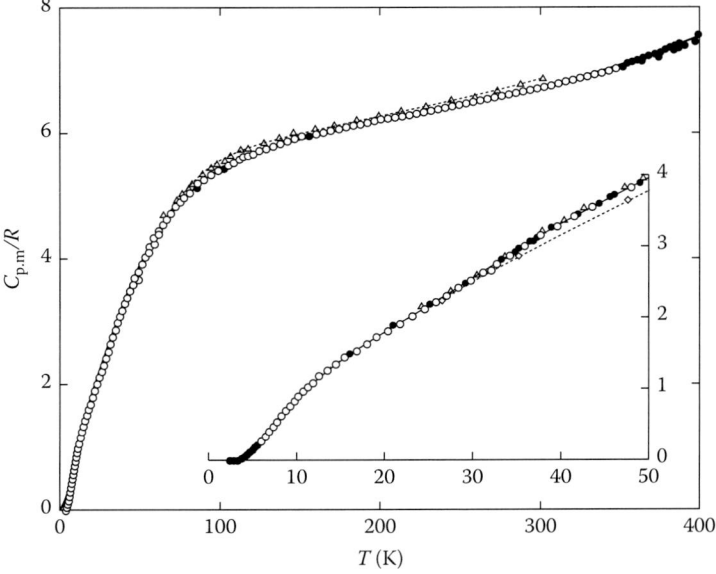

FIGURE 4.41   Heat capacity of AgI below ambient temperature. "o" represent adiabatic calorimetry data of Shaviv et al. [4.45]. Inset shows data between 0 and 50 K. (From Shaviv, R. et al., *J. Chem. Therm.*, 21, 631. 1989.)

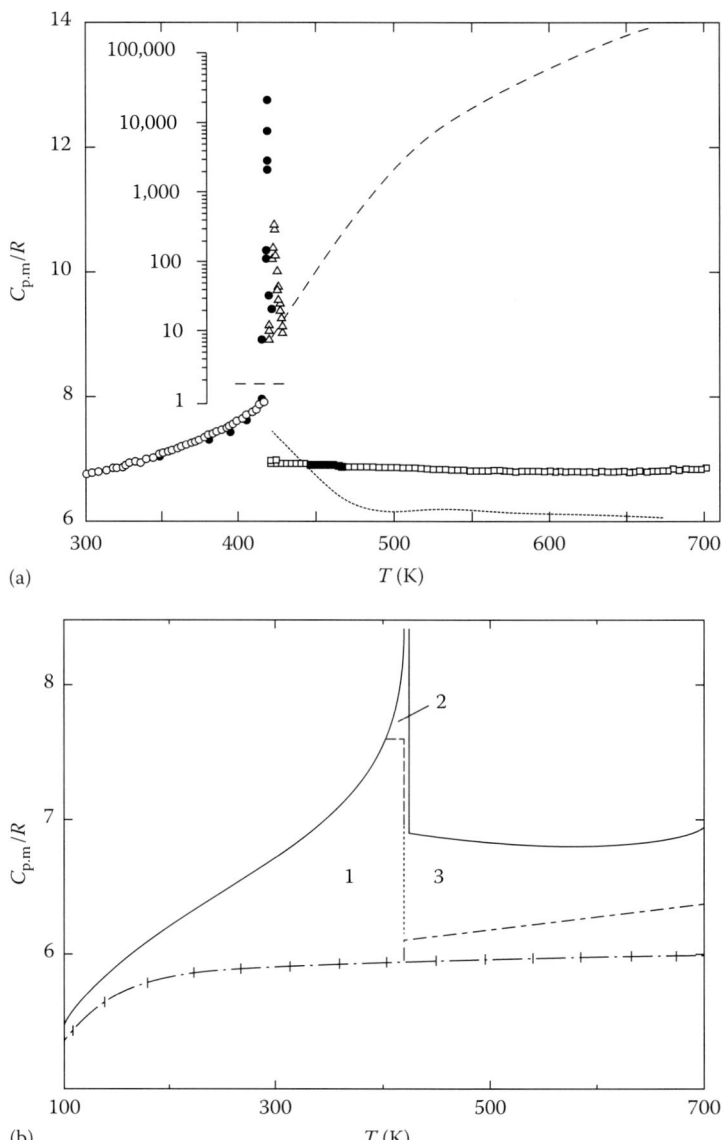

FIGURE 4.42 (a) Heat capacity of AgI above ambient temperature highlighting the anomaly at the wurtzite-bcc structural phase transition through a log-scale plot indicated inside. (b) Resolving the heat capacity data of AgI:+ + +: $C_{v,m}$, _._ _: $C_{p,m}$ (lattice), ____observed $C_{p,m}$; (1) "pre-transitional" area (221R.K); (2) transitional area (759R.K); "post-transitional" area (165R.K). (From Shaviv, R. et al., *J. Chem. Therm.*, 21, 631, 1989.)

Figure 4.43a displays heat capacity of RbAg$_4$I$_5$, which deviates from Debye behavior for $T > \sim 100$ K besides signatures of two-phase transitions. Figure 4.43b shows heat capacity of Ag$_2$S that includes the structural phase transition and a Schottky anomaly.

Figure 4.44 shows heat capacity data on CuBr.

Figure 4.45a displays the unusual low-temperature specific heat behavior of Na-β-alumina. Figure 4.45b shows the excess specific heat (over and above the Debye contribution) for Ag, Na, Rb, K, and Li, which has been fitted to a distribution of two or three Einstein oscillators [4.49].

As a final example, the heat capacity of pure zirconia and stabilized zirconia containing 7.76×10^{-2} mol yttria measured by Tojo et al. [4.50] is presented in Figure 4.46. The yttria-stabilized zirconia showed excess heat capacity at low temperatures, which should be attributed to the so-called low-energy

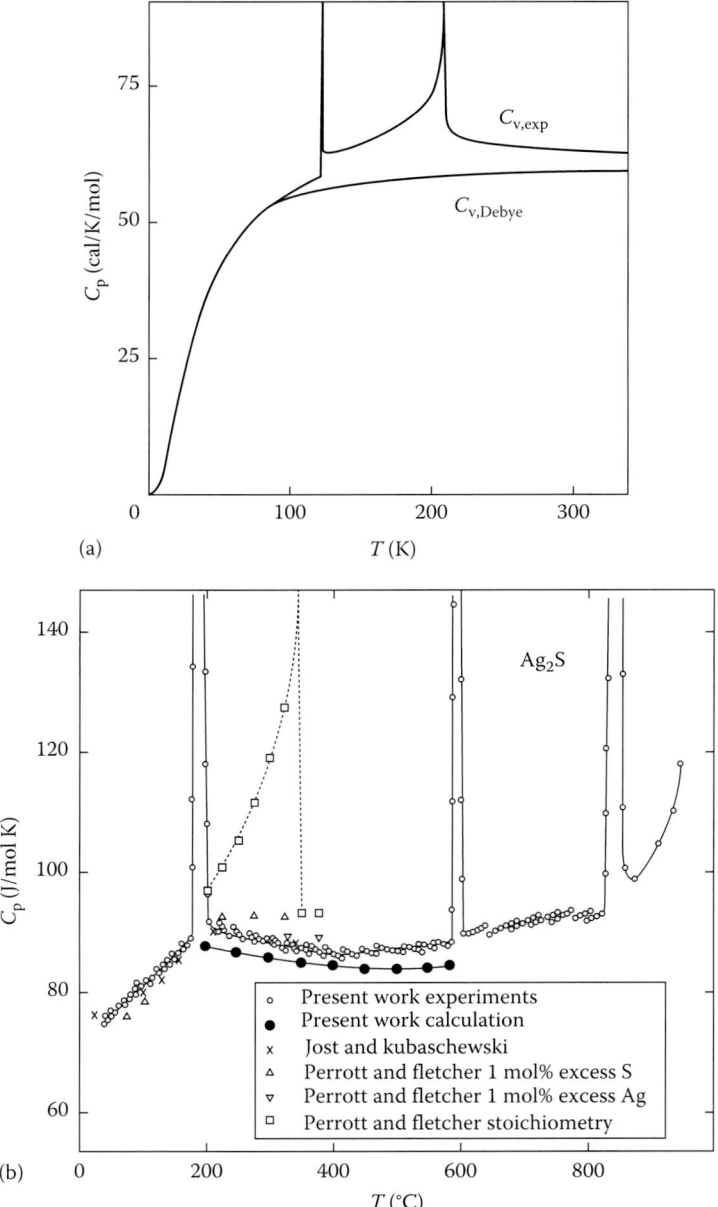

FIGURE 4.43 (a) Constant volume heat capacity of RbAg$_4$I$_5$ ($C_{V\,exp}$ and estimated vibrational contribution to the heat capacity, $C_{V,\,Debye}$. (From Okazaki, H., *Netsu Sokutei*, 21, 212, 1994.) (b) Specific heat of Ag$_2$S from room temperature to melting point (838°C). β–α phase transition occurs at 189°C and another transition occurs at 586°C. O: experimental values, ●: values calculated from the relation.

$$C_p = C_\tau + \frac{9\alpha_l^2 V_m T}{\chi_T} + \Delta C_p$$

where C_v is constant volume heat capacity, second term is due to anharmonic contribution (α_l is the linear thermal expansion, V_m is the molar volume, and χ_T is the isothermal compressibility. Third term represents Schottky-type excess heat capacity caused by cation jumps from equilibrium to activated positions. (From Okazaki, H. and Takano, A., Z. *Naturforsch.*, 40a, 986, 1985.)

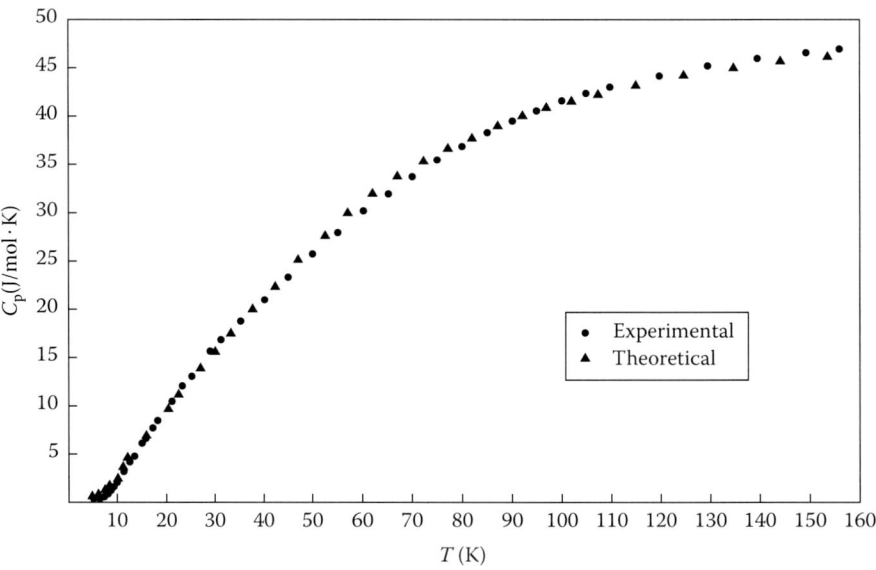

FIGURE 4.44 Theoretical and experimental molar specific heats of CuBr as a function of temperature. (After Vardeny, Z. et al., *Phys. Rev. B*, 18, 4487, 1978.)

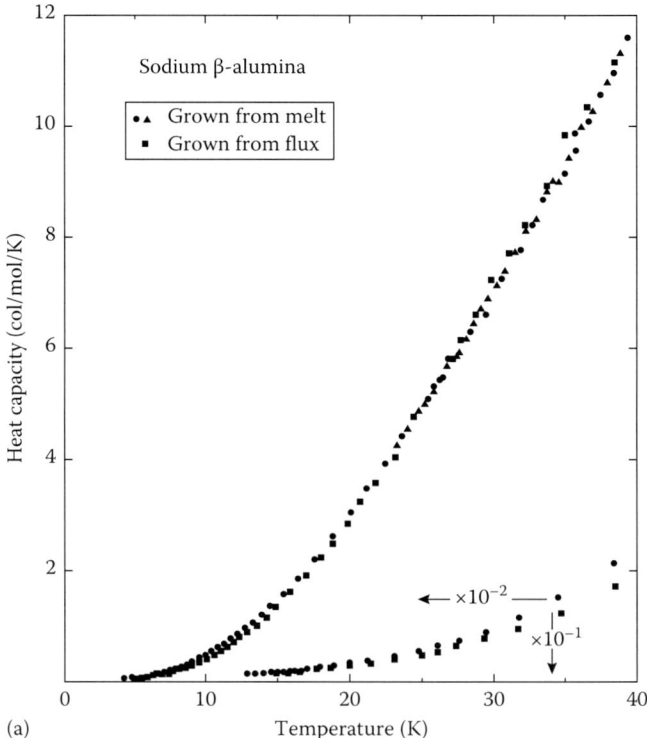

FIGURE 4.45 (a) Low temperature specific heat of Na-β-lumina single crystals grown from melt (●, ▲) and flux. Note the rapid rise in heat capacity.   *(Continued)*

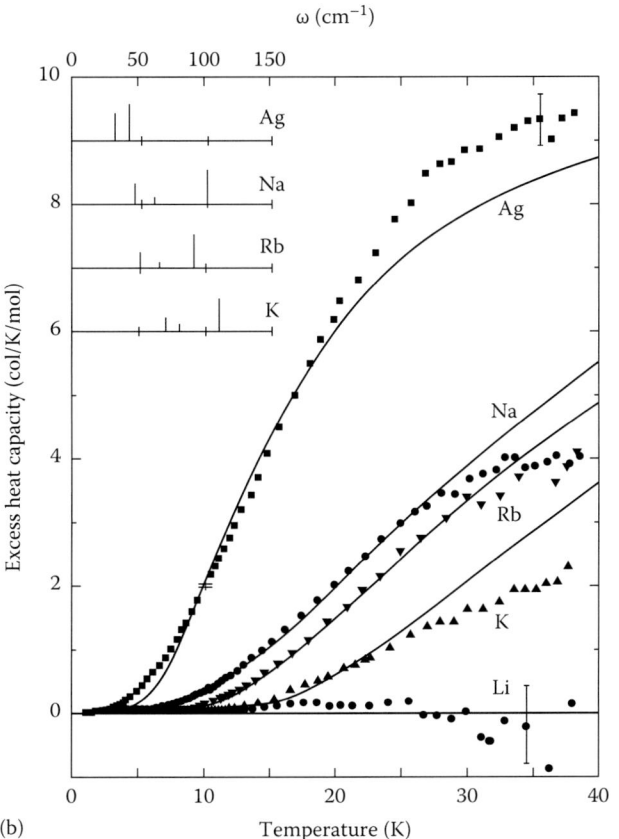

(b)

FIGURE 4.45 (*Continued*) (b) Excess specific heat (over and above Debye model) fitted to a distribution of two or three Einstein oscillators whose frequencies are shown in inset. (From McWhan, D.B. et al., *Phys. Rev. B*, 15, 553, 1977.)

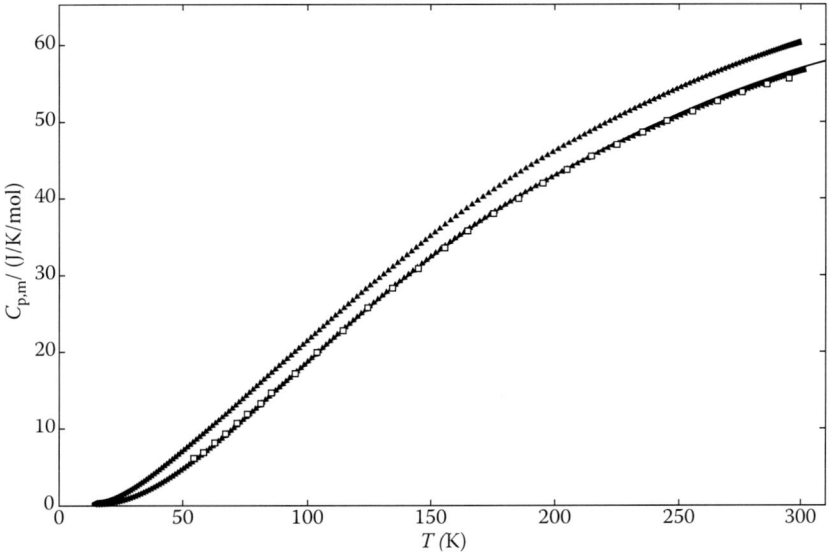

FIGURE 4.46 Molar heat capacity $C_{p,m}$ of zirconia. ●: pure zirconia, ▲: stabilized zirconia containing 7.76×10^{-2} mol yttria. For explanation of other symbols. (From Tojo, T. et al., *J. Chem. Therm.*, 31, 831, 1999.)

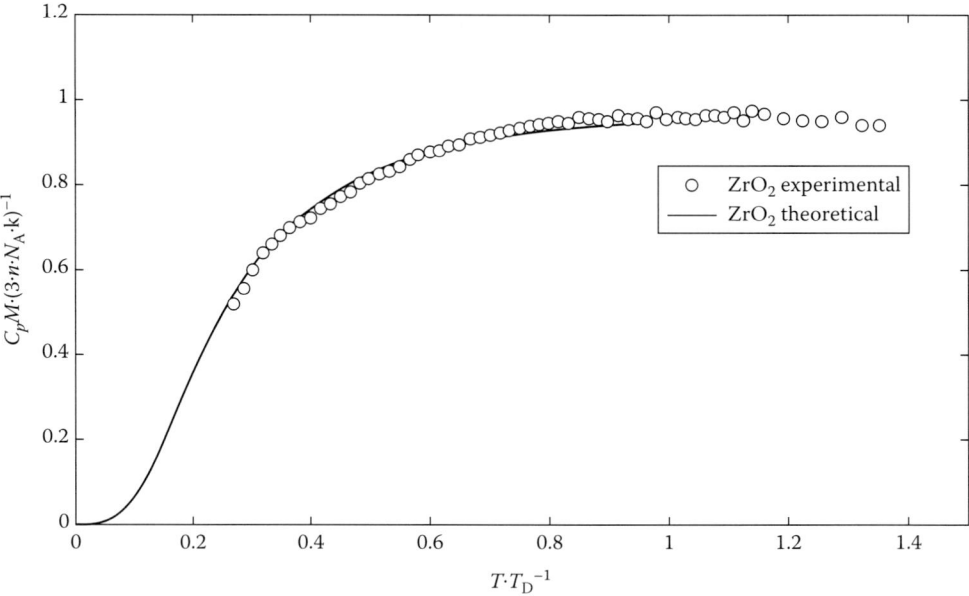

FIGURE 4.47 Experimental (DSC-based) and theoretical molar heat capacity of ZrO_2 ($n = 3$) normalized by its theoretical value versus temperature normalized for $T_D = 590$ K. The output signal of the DSC recorded as raw data set must be evaluated using a separate computer unit with the calculation method. Saphire was used for calibration. Temperature range: 128–823 K (± 0.5 K) Temperature program was 15 min at 128 K, 1 K/min up to 133 K, 15 min at 133 K, 10 K/min up to 823 K; 99.995% N_2 gas at 1 bar as gas flow and liquid nitrogen as coolant for low temperatures. Sample held in a crimped Al pan to ensure good thermal contact with instrument (Sieko 220). C_p values better than 5%. (From Degueldre, C. et al., *Thermochim. Acta*, 403, 267, 2003.)

modes caused by the defects in the crystal. The standard enthalpy and entropy values for pure zirconia are 8.711 kJ/mol and 49.79 J/K·mol, respectively. For yttria-stabilized zirconia, they are 9.595 kJ/mol and 56.47 J/K·mol.

Heat capacity may also be measured using the technique of differential scanning calorimetry [4.51] as shown for the case of zirconia in Figure 4.47.

Next, we deal with thermal expansion.

4.1.3.1.2 Thermal Expansion

As will be seen in Chapter 6, thermal expansion is controlled by anharmonic interactions involving many phonons (typically three or four), which basically arises from a potential expansion involving interatomic displacements raised to third and fourth powers [4.39]. The coefficients of these terms represent the anharmonic interactions. We shall briefly describe two techniques to characterize thermal expansion of solid state ionic materials (Figure 4.48).

1. *High-temperature X-ray diffraction* data on four compositions in AgI–CuI system enable convenient measurement of thermal expansion and determination of partial phase diagram [4.52]. Figure 4.49 displays (a) the sample mounting geometry, (b) diffractograms of $Ag_{0.95}Cu_{0.05}I$ tracing the evolution of the zinc blende–bcc phase transition in AgI, and (c) lattice parameter variation of four AgI–CuI compositions across the phase transition.

    A typical setup for the collection of X-ray powder diffraction data at super-ambient temperatures is a Philips X'pert PRO with an Anton Parr high-temperature attachment. A sealed-tube X-ray generator (40 kV, 30 mA) produces Cu K_α ($\lambda = 0.15418$ nm) radiation, which is used for diffraction experiments in the Bragg–Brentano mode employing a proportional counter. A platinum heater may be used as the heating stage for the sample. A Eurotherm temperature programmer helps collect data to 1 K accuracy. About 100 mg of ground powder is sufficient.

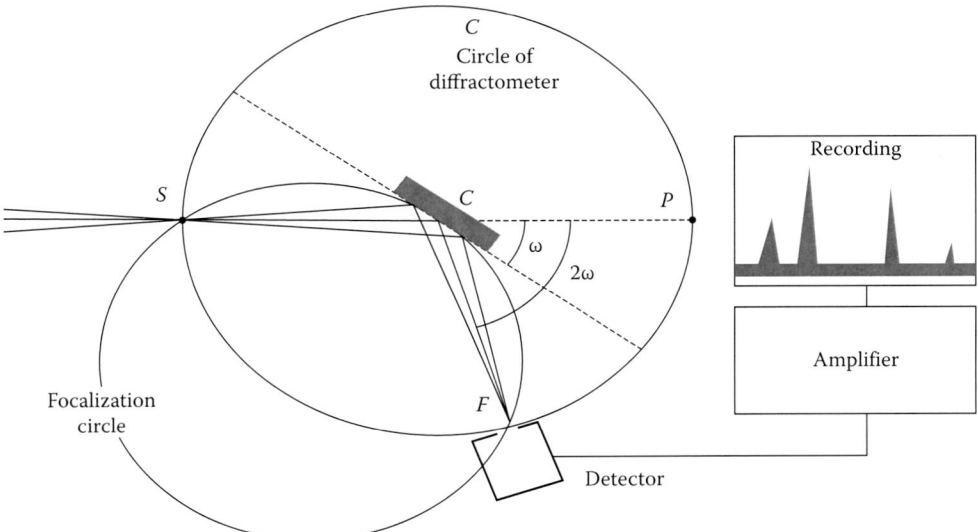

FIGURE 4.48 Illustration of the Bragg–Brentano focus geometry used for thermal expansion by X-ray powder diffractometry. (From Newville, M., *Fundamentals of XAFS*, Tutorial, University of Chicago, Chicago, IL, 2004.)

It is mounted on the Pt heater and the XRD patterns are recorded in a flowing helium atmosphere (to prevent any possible oxidation of the sample) in the Bragg angle range $20° < 2θ < 80°$ at selected temperatures in the range 300–723 K covering all possible phase transitions in AgI and CuI. A convenient scanning rate is 2°/min while the holding time at each temperature is typically 10 min. Silicon is used as the reference material for calibration of observed XRD intensities. The unit cell parameters are extracted from the observed high-temperature XRD patterns using a least-squares refinement program.

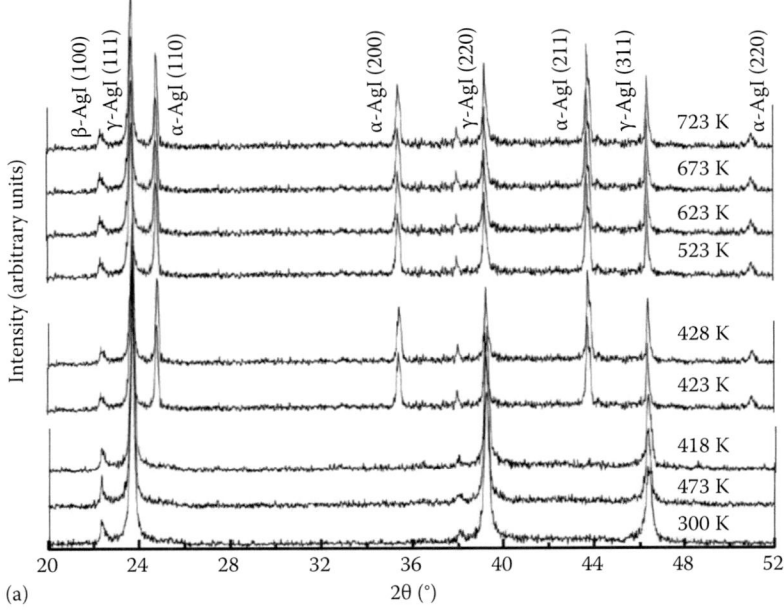

FIGURE 4.49 (a) XRD patterns of $Ag_{0.95}Cu_{0.05}I$ depicting clearly the transformation from the zinc blende phase of AgI to the bcc superionic phase at and above 150°C. *(Continued)*

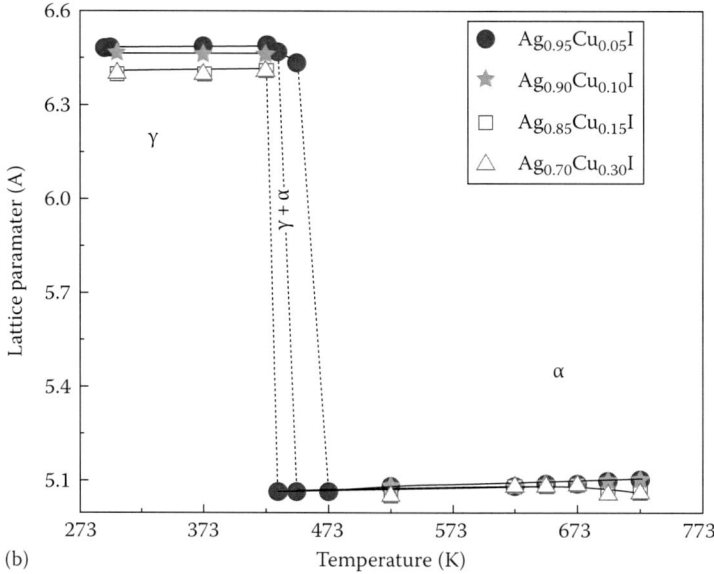

(b)

FIGURE 4.49 (Continued) (b) Evolution of the lattice parameter of $Ag_{0.95}Cu_{0.05}I$, $Ag_{0.90}Cu_{0.10}I$, $Ag_{0.85}Cu_{0.15}I$, and $Ag_{0.70}Cu_{0.30}I$, with increasing temperature. Note the abrupt jump in the lattice parameter at the zinc blende–bcc structural transition. From this data, one can draw the phase diagram for the AgI–CuI binary system. (From Senthil Kumar, P. et al., *J. Phys. Chem. Solids*, 67, 1809, 2006.)

2. *Dilatometry.* The bulk thermal expansion measurements may be made by using high-temperature dilatometer [4.27,4.54]. The change in the length of the sample, typically kept in a quartz cell, may be detected using a linear variable differential transformer (LVDT). Sample temperature may be varied from 25°C to 800°C at the typical rate of 2°C/min during the heating and cooling cycles. The apparatus, calibrated using quartz and OFH copper, has a sensitivity of $0.1 \times 10^{-6}/°C$ for the measurement of a. A PC-based data acquisition system records thermal expansion data at intervals of 120 s averaged over 10 s at each point. Figure 4.50 shows the typical apparatus: (a) the dilatometer and (b) block diagram of the experimental setup, and (c) a typical dilatometric trace of CuI for a heat-cool cycle. AgI possesses negative thermal expansion while CuI exhibits positive thermal expansion. Using dilatometry, a near-zero-thermal expansion material has been established in the AgI–CuI system corresponding to the composition $Ag_{0.25}Cu_{0.75}I$ [4.56].

Next, we consider characterization by Seebeck coefficient and Hall coefficient measurements.

4.1.3.2 Seebeck Coefficient and Hall Coefficient

4.1.3.2.1 Seebeck Coefficient

The Seebeck coefficient (also called thermopower, thermoelectric power, and thermoelectric sensitivity) of a material is a measure of the magnitude of an induced thermoelectric voltage in response to a temperature difference across that material, as induced by the Seebeck effect. Most generally, it is the portion of the electric current driven by temperature gradients.

It is the voltage built up when a small temperature gradient is applied to a material, and when the material has come to a steady state where the current density is zero everywhere. If the temperature difference ΔT between the two ends of a material is small, then the absolute Seebeck coefficient is

$$S = \frac{\Delta V}{\Delta T} \tag{4.52}$$

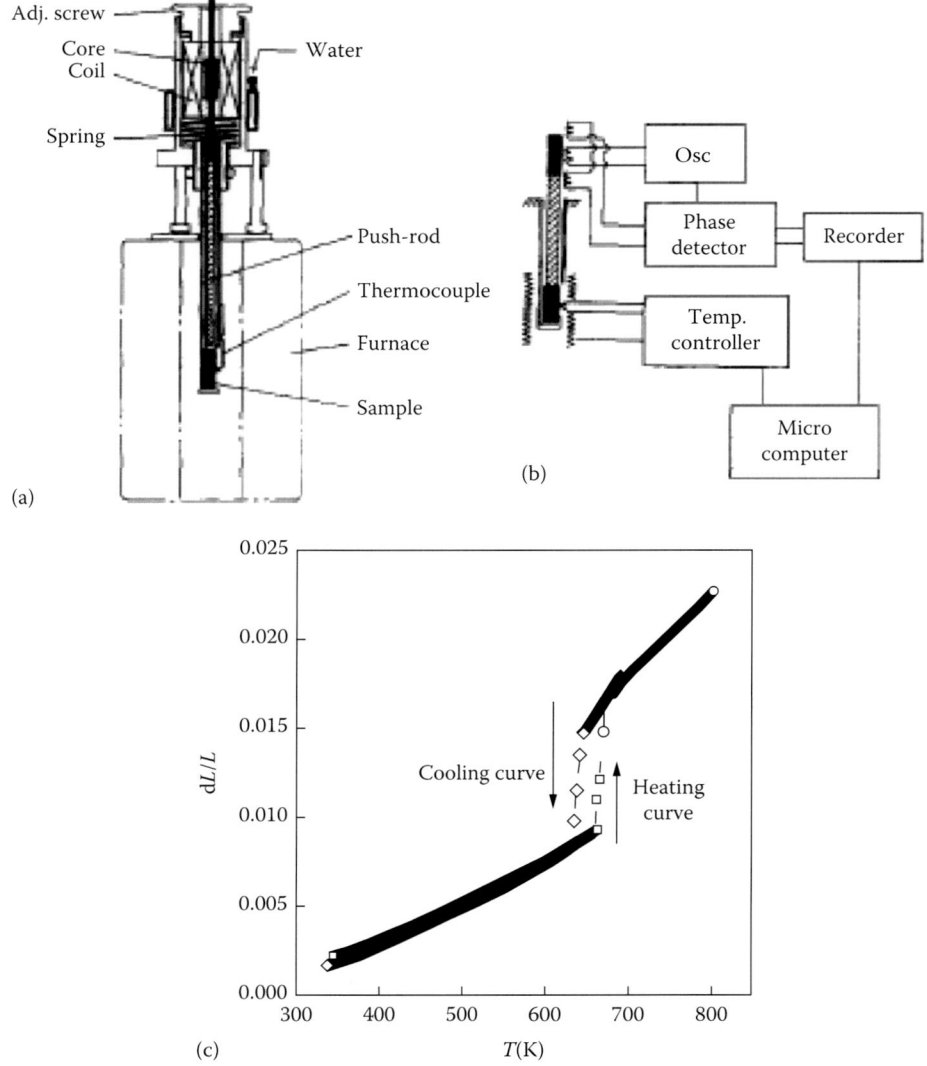

FIGURE 4.50 A differential transformer-type push-rod dilatometer. (a) Dilatometer. (b) Block diagram of measurement setup. For measurements on superionic samples with a well-defined melting point cylindrical samples of 10 mm diameter and 3 mm height are conveniently prepared with suitable vacuum annealing. (After Saito, M. and Tamaki, S., *Solid State Ionics*, 60, 237, 1993.) (c) Typical dilatometric trace for CuI as it undergoes a phase transition from zinc blende to hexagonal phase transition. Load applied by the rod can vary between 0.03 and 0.5 Newton with a contact area ~0.5–1 mm^2. (From Sunandana, C.S. et al., *Indian J. Pure Appl. Phys.*, 37, 325, 1999.)

where ΔV is the thermoelectric voltage seen at the terminals. It can have + or − sign, which is decided by the expression

$$S = \frac{V_{\text{left}} - V_{\text{right}}}{T_{\text{left}} - T_{\text{right}}} \tag{4.53}$$

If S is positive, then the end with higher temperature has lower voltage and vice versa. The absolute Seebeck coefficient is difficult to measure directly, since the voltage output of a thermoelectric circuit, as measured by a voltmeter, only depends on *differences* of Seebeck coefficients. Thus, the measured

Seebeck coefficient is a contribution from the Seebeck coefficient of the material of interest and the material of the measurement electrodes.

$$S_{AB} = S_B - S_A = \frac{\Delta V_B}{\Delta T} - \frac{\Delta V_A}{\Delta T}. \tag{4.54}$$

The temperature, crystal structure of materials, and impurities influence the value of thermoelectric coefficients. The Seebeck effect can be attributed (1) charge-carrier diffusion and (2) phonon drag.

Figure 4.51 shows a commercial setup for measurement of S.

Figure 4.52 shows the temperature dependence of S for superionic conductors for Cu_2Se and $Cu_{1.98}Se$ through the low-temperature (α) to high-temperature (β) phase transition.

Thermopower measurements are important from (1) fundamental and (2) applied viewpoints.

1. S measurements in $Ag_7I_4AsO_4$ superionic conductor have established a straight-line relation $S = 0.20(10^3/T) + 0.06$, where the heat of transport 0.20 eV is very close to the activation energy (0.24 eV) obtained in the free-ion model of Rice and Roth [4.59].

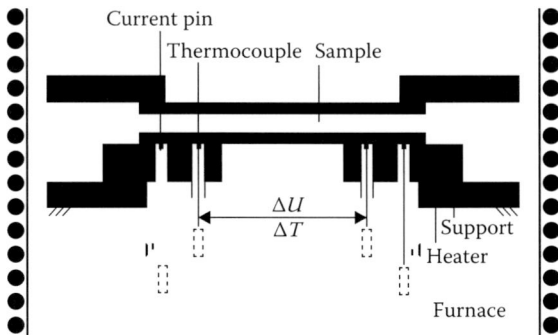

FIGURE 4.51 Measurement setup for Seebeck coefficient–electrical conductivity monitoring system. The system allows different sample geometries and is suitable for the temperature range from room temperature to 800°C. Heating is carried out in an alternating pattern for each selected temperature step. It allows the temperature gradient to be varied in both directions across the sample. (From Netzsch, SBA 458 Nemesis data sheet, SBA_458_Nemesis_E_0814.pdf.)

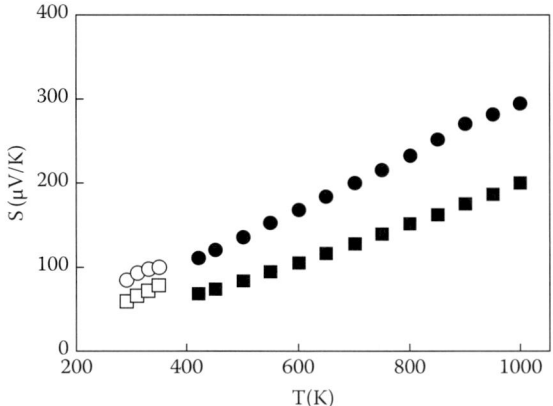

FIGURE 4.52 S versus T for the thermoelectric material $Cu_{2-x}Se$. α-Cu_2Se (O), β-Cu_2Se (●) and α-$Cu_{1.98}Se$ (□), β-$Cu_{1.98}Se$ (■). (From Liu, H. et al., *Nat. Mater.*, 11, 422, 2012.

2. Solid state thermoelectric technology uses electrons or holes as the working fluid for heat pumping and power generation. Thermal-to-electrical energy conversion technology can generate electricity from waste industrial heat. The conversion efficiency is governed by the dimensionless thermoelectric figure of merit $zT = S^2T/\rho(\kappa_L + \kappa_c)$, where S is the thermopower (Seebeck coefficient), ρ is the electrical resistivity, T is the absolute temperature, κ_L is the lattice thermal conductivity, and κ_c is the carrier thermal conductivity. $Cu_{2-x}Se$ shows an unusual liquid-like behavior of Cu ions around a crystalline sublattice of Se, leading to an intrinsically very low lattice thermal conductivity and a high zT, which gives high zT of 1.5 at 1000 K [4.58].

4.1.3.2.2 Hall Effect

Hall effect is well known in metals and semiconductors and is electronic in origin [4.69]. As superionic conductors are ionic rather than electronic materials, how does the ionic Hall effect arise? And how is it measured? The Hall signal in the superionic conductor α-AgI is a result of the strong correlations between the hopping ions. It reflects a fundamental qualitative difference between "normal" (AgBr) and superionic (AgI) conductors.

The Hall voltage U_H is related to the Hall mobility μ_H and the applied electric and magnetic fields E and B:

$$U_H = \mu_H b \cdot E \cdot B \qquad (4.55)$$

where b is the width of the sample.

μ_H is the product of the Hall constant R_H and electrical conductivity σ:

$$\mu_H = R_H \cdot \sigma \qquad (4.56)$$

σ measures drift mobility μ_D as σ/ne. It is found from a careful experiment that

$$\mu_H = \mu_D. \qquad (4.57)$$

Thus, a *correlated* measurement of σ and R_H establishes the ionic Hall effect. However, special restrictions imposed on the movement of ions by the superionic crystal structure prevent the ions from following the Lorentz force *all the time*. Thus, $\mu_H = \mu_D$ for Ag^+ ions in α-AgI and α-RbAg$_4$I$_5$.

We give two experimental setups for ionic Hall-effect measurements. The first one by Liu et al. [4.61] is shown in Figure 4.53.

The second setup for Hall-effect measurement by Stuhrmann et al. [4.62] is shown in Figure 4.54.

The volume available for the sample in this setup is $4 \times 25 \times 35$ mm^3. The Hall voltage is measured by a phase-sensitive amplifier technique not using any active filters. An acceptable signal-to-noise ratio is achieved after rectification and subsequent signal accumulation. The conductivity of the sample is recorded simultaneously with the measurement of the Hall voltage. The experimental setup (Figure 4.55a) is calibrated by measuring the known Hall mobility of copper.

Figure 4.56 shows Arrhenius plot for α-RbAg$_4$I$_5$ and Hall and drift mobilities in an Arrhenius presentation.

4.1.4 Spectroscopic Characterization

4.1.4.1 Impedance Spectroscopy

AC electrical measurements on solid state ionic materials are necessary in order to remove the need for electrodes reversible to the ionic species of interest. This is done by eliminating interfacial effects at sufficiently high frequencies. It is necessary to convert—through a careful analysis—the generally

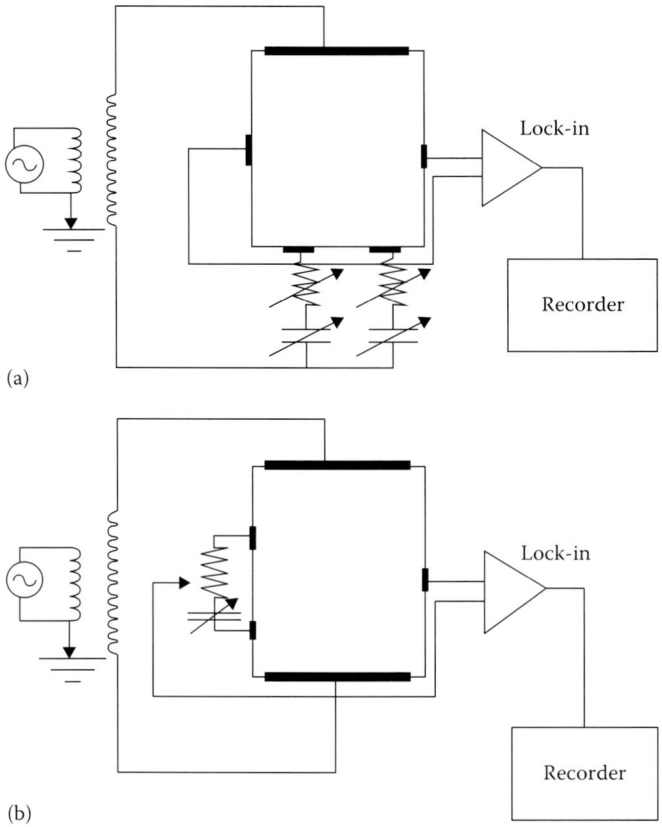

FIGURE 4.53   Five-electrode circuits for Hall-effect measurements on superionic conductors. (a) Split current configuration for thin-film samples. (b) Split Hall probe for bulk samples. (From Liu, Y.L. et al., *Phys. Rev. B*, 41, 10481, 1990.)

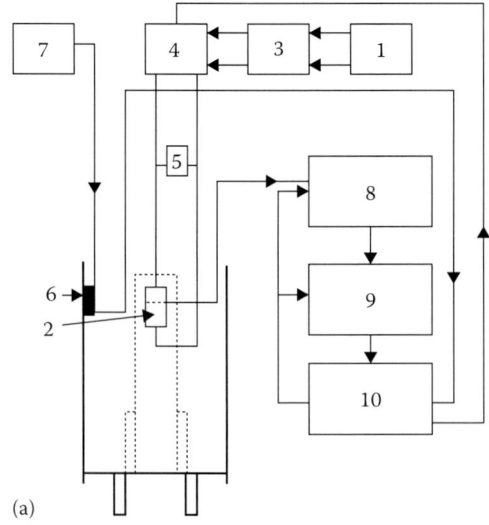

FIGURE 4.54   (a) Setup for measurement of ionic Hall effect in fast ion conductors: (1) 30 kHz generator, (2) sample (sample environment and rotating permanent magnet not shown), (3) ammeter, (4) 180° AC phase switch, (5) voltmeter, (6) rotating magnetic position sensor, (7) DC source, (8) 30 kHz amplifier and demodulator, (9) 1-Hz narrow band amplifier, and (10) personal computer.   *(Continued)*

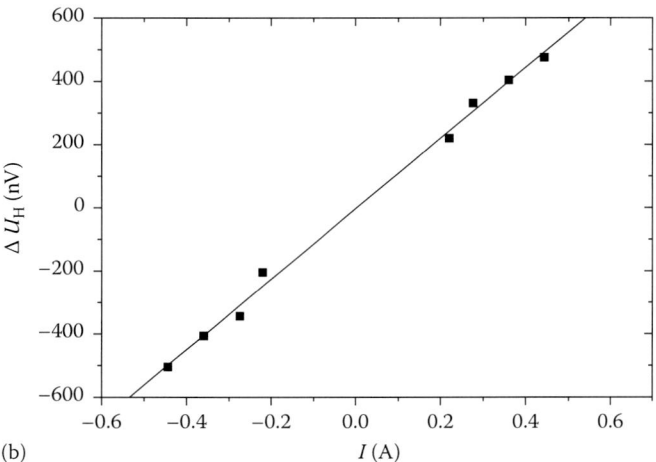

(b)

FIGURE 4.54 (*Continued*)   (b) Linear dependence of Hall voltage on 30 kHz current at 473 K in α-AgI. In this novel experiment involving alternating E and B fields, very different frequencies are used for the periodic electric and magnetic fields, 30 kHz for the former and 1 Hz for the latter. Magnetic field is 1 T, sufficient to generate Hall voltages in the nanovolt range. (From Stuhrmann, C.H. et al., *Solid State Ionics*, 154–155, 109, 2002.)

frequency-dependent AC response experimentally observed into meaningful bulk parameters such as ionic conductivity.

Electrolytic conduction in pure ionic/superionic conductors (such as NaI, LiI, RbAg$_4$I$_5$, and AgI) is due to the motion of charged ions under the influence of applied electric fields, while that of mixed ionic electronic conductors such as Ag$_2$S, ceria, and zirconia is due to the simultaneous motion—in opposite directions of ions and electrons. While the former make good electrolytes, the latter are more suitable as electrodes.

Apply a constant DC potential to an electrolyte sample held between two different sets of electrodes as shown in the following:

| − | Nonblocking electrode | A$^+$B$^-$ | Blocking electrode | + |

Then, the cation starts moving toward the negative electrode. As a result, the right end of the electrolyte suffers a depletion of cations as the supply of cations ceases at the positive electrode. As soon as the circuit is switched on, the instantaneous current and voltage give a measure of the total conductivity (ionic plus electronic). The current decays slowly with time attaining a final stabilized value. This is a measure of electronic conductivity measurements on LMn$_2$O$_4$ [4.63].

These polarization effects are troublesome in a DC measurement unless nonpolarizing electrodes capable of supplying mobile ions at the anode–electrolyte interface are used. Then the electrode configuration becomes where the electrode materials are the same as that of the mobile species. But the problems associated with the choice of an appropriate nonblocking electrode (say, F in PbF$_2$) may be overcome by employing blocking electrodes and measuring as a function of frequency.

| − | Nonblocking electrode | A$^+$B$^-$ | Nonblocking electrode | + |

Impedance measurements are usually done with commercial instruments (see later).

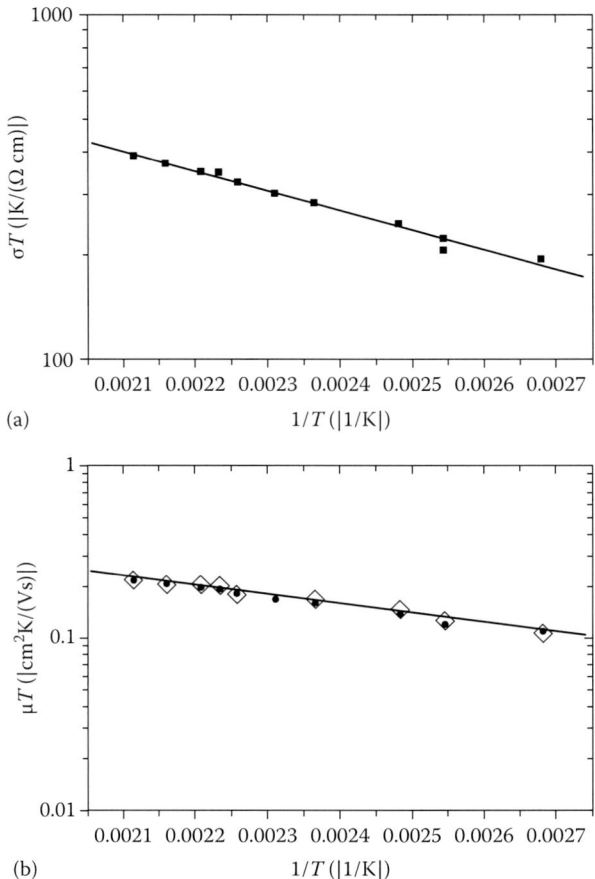

FIGURE 4.55 (a) Log $\sigma_{DC}T$ versus $1/T$ plot for α-RbAg$_4$I$_5$ from 373 to 473 K. (b) Hall and drift mobilities in α-RbAg$_4$I$_5$ in an Arrhenius presentation. Boxes: μ_H, dots: μ_D. Thus the two mobilities are identical. Slope of the line, 0.11 eV, is the activation energy. (From Stuhrmann, C.H. et al., *Solid State Ionics*, 154–155, 109, 2002.)

However, it is possible to build an AC conductivity meter based on the following design concept [4.64].

Thus, arises the frequency-dependent or AC conductivity and impedance spectroscopy. Here, the response is analyzed through complex plane plotting (Figure 4.57 [4.65]). An example is shown in the following and followed by a discussion.

This helps represent the observed frequency-dependent AC response to be analyzed in terms of an *R-C* (and sometimes even *L*) network. Each component of such a network may be linked to a particular atomic process or mechanism within the assembly. These fall into (a) interfacial or (b) bulk microstructural phenomena. The analytical technique used by electrochemists in the study of liquid electrolytes has been adopted for the study of solid state systems. It involves plotting the real and imaginary parts of a complex electrical quantity against one another as a frequency dispersion. Table 4.1 lists the most common types of plot.

Impedance spectroscopy is a valuable method to assess the electrical makeup of both homogeneous and inhomogeneous materials. It is a useful guide to ascertain whether microstructural features of a ceramic as seen by electron microscopy correlate with electrical microstructures. Thus, in some cases grain boundaries may have negligible influence on the electrical properties of a ceramic, while in other cases they may dominate the impedance. Information on the grain boundary geometry, particularly the thickness of the electrically active grain boundaries. A comparison of activation energies helps

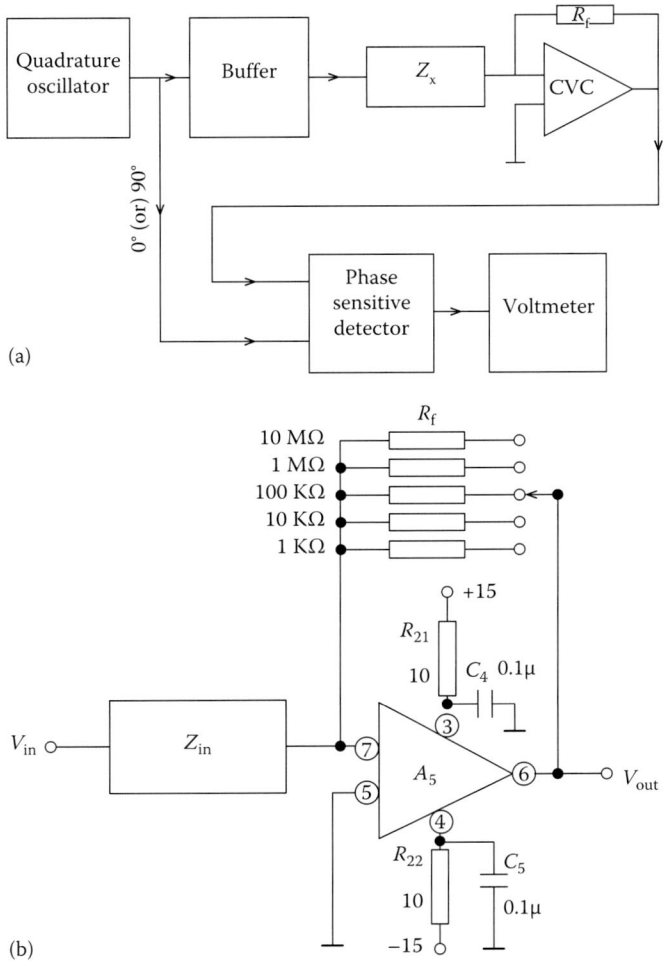

FIGURE 4.56 (a) Block diagram of the phase-sensitive detector–based AC conductivity meter for impedance (Z_x) measurements. (b) CVC (current-to-voltage convertor) circuit. (From Balaya, P. and Sunandana, C.S., *Pramana J. Phys.*, 33, 627, 1989.)

determine whether grains and grain boundaries have similar chemical composition, structure, and conduction mechanisms.

Figure 4.58 presents illustrative examples of impedance spectra from the recent experiments of Beindicho and West on LiFePO$_4$ samples [4.66].

Two things must be borne in mind while analyzing electrical impedance data:

1. Electrical microstructure of practical ceramic samples such as LiFePO$_4$ depends on processing conditions. Data need to be fitted to equivalent circuit models consisting of R, C, and even L as circuit elements. If more than one circuit models fit observed data, it is necessary to zero in one model that truly represents real sample data. This done circuit parameters may be assigned confidently.

2. Departures from ideal R–C–L behavior may exist. It is thus important to analyze these departures, which depend on whether the material is homogeneous or heterogeneous. For homogeneous materials (e.g., single crystals and glasses), the nonideality may arise from the bulk electrical properties. However, for ceramic samples deconvolution of bulk boundary

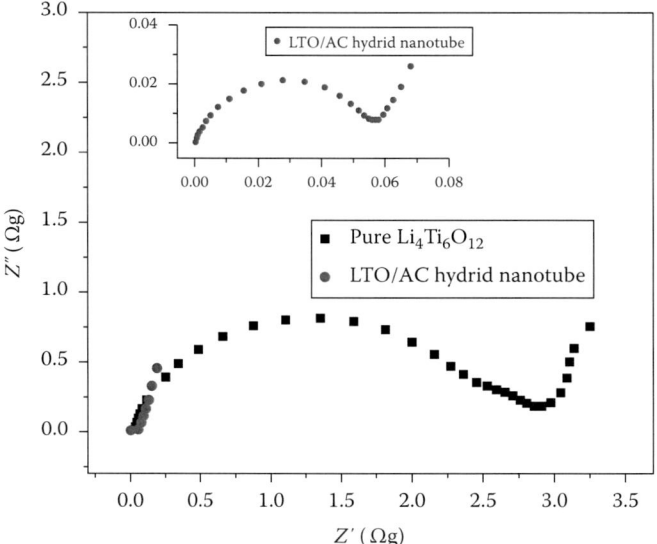

FIGURE 4.57   An impedance plot with real part (Z') plotted on x-axis and imaginary part (Z'') plotted along y-axis. (From Choi, H.S. et al., *J. Mater. Chem.*, 22, 16986, 2012.)

geometry would be necessary. This is because the thickness of the electrically active grain boundaries is involved; by comparing activation energies, one can determine whether grains and grain boundaries have similar chemical composition, structure, and conduction mechanisms.

How are the impedance measurements done? Opposite pellet sample faces are coated with sputtered gold, and they are mounted in a conductivity jig and placed in a tube furnace. Measurements in inert atmospheres are usually done with a Solartron or Hewlett-Packard impedance analyzer in the frequency ranges 1 Hz–1 MHz and 40 Hz–2 MHz, respectively. Isothermal heating steps could be used in these measurements. Corrections for sample geometry, parallel capacitance of the empty jig, and series resistances of leads and electrodes are to be made before data analysis and modeling.

Impedance data are graphically presented using a range of standard formalisms (impedance, admittance, and inverse dielectric) for fitting appropriate equivalent circuits. The ZVIEW software package is used for this purpose. The presentation formalisms interrelate as follows:

$$Z^* = (Y^*) - 1 = j\omega C_0 (\varepsilon^*)^{-1} \tag{4.58}$$

where
   C_0 is vacuum capacitance of conductivity cell
   ε^* the complex dielectric constant

4.1.4.2 EIS Spectrum Analyzer Software

A proper equivalent circuit has to be established before analysis. Two such model circuits are shown in Figure 4.59. For circuit B containing constant phase element (CPE1), the real part of the admittance takes the form

$$Y' = R^{-1} + A\omega^n \tag{4.59}$$

For further details of data interpretation, refer to [4.46].

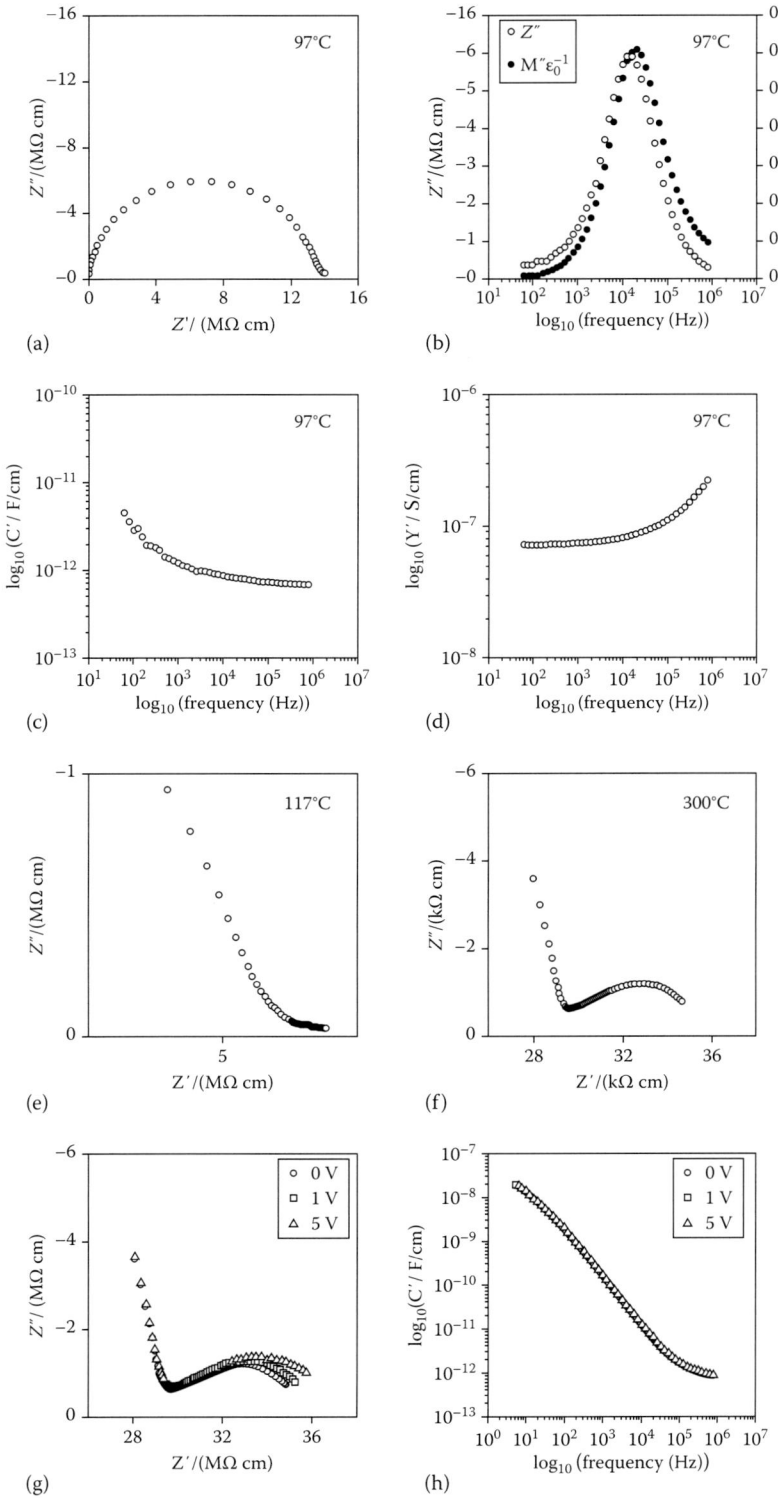

FIGURE 4.58 Typical impedance spectrographs measured at 370 K of LiFePO$_4$ reacted at 1043 K pots of (a) Z^*, (b) Z^*, M^* (c) C, (d) Y (admittance), (e, f) low-frequency impedance measured at 390 K and 673 K, (g, h) Effect of DC bias on low-frequency Z^* and C plots. (From Biendicho, J.J. and West, A.R., *Solid State Ionics*, 226, 41, 2012.)

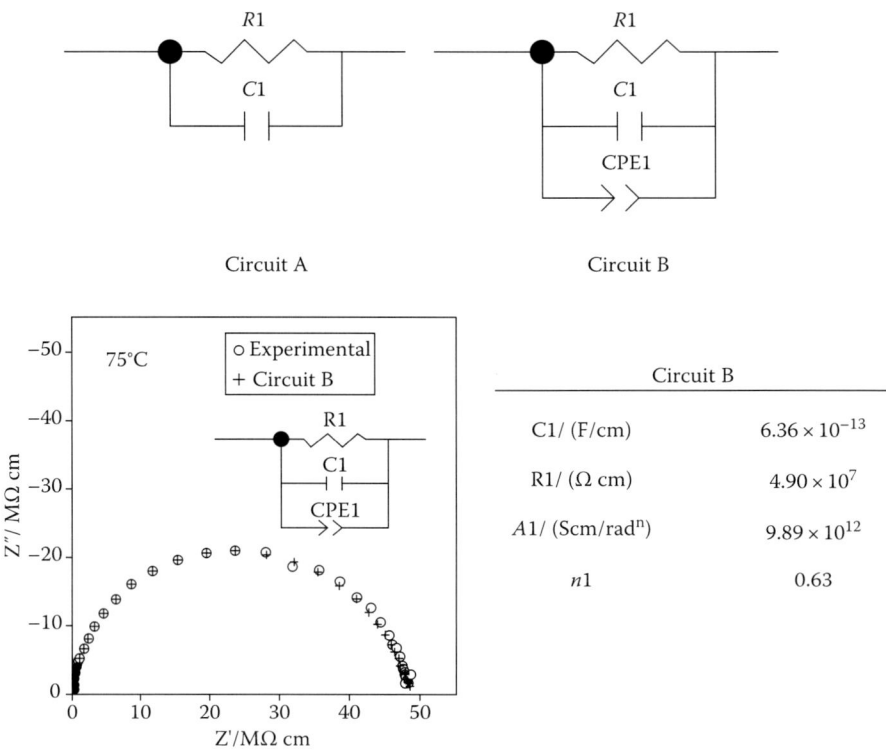

	Circuit B
C1/ (F/cm)	6.36×10^{-13}
R1/ (Ω cm)	4.90×10^{7}
A1/ (Scm/rad^{n})	9.89×10^{12}
n1	0.63

FIGURE 4.59 Two circuits to compare with experimental impedance data. (From Biendicho, J.J. and West, A.R., *Solid State Ionics*, 226, 41, 2012.)

After discussing impedance characterization on pellet samples, we turn to impedance spectroscopy of single crystal samples, which reveal fundamental physics as in the case of Ag-β-alumina (Figure 4.60).

Complex impedance may be measured along the conducting plane of a single crystal such as Ag-β-alumina from 1 Hz to 1 MHz. A frequency response analyzer such as Schlumberger Solartron model 1260.

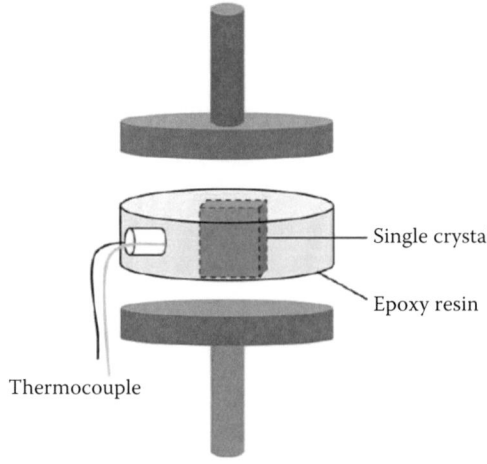

FIGURE 4.60 Measurement system for frequency-dependent conductivity measurements on a single-crystal sample. (From Kamishima, O. et al., *Solid State Ionics*, 262, 495, 2014.)

An oscillation voltage of 0.5 V is suitable for frequencies greater than 100 Hz, but below that frequency a current amplifier may be necessary (Keithley 428). Two-terminal voltages and currents are measured from 100 K to room temperature. Sputtered Ag metal or Ag paste are suitable as reversible electrodes.

Typical conductivity profiles and an Arrhenius plot derived from them are shown in Figure 4.61.

The influence of the electrode–electrolyte interface is reduced upon increasing sample size. When the sample size becomes very large ($\rightarrow$ infinity), real part of σ is proportional to $\omega^{0.11-0.15}$ (the broken line in Figure 4.61a) What happens to this function as $\omega \rightarrow 0$? The DC component of conductivity may be absent! The absence of DC conductivity is posed as a problem at the end of this chapter.

An electrodeless method for measuring ionic conductivity of a superionic conductor is desirable. Such a technique allows us to obtain an intrinsic ionic conductivity free from the ambiguous electrode-specimen interface problem. In principle, the application of alternating magnetic field to the pelletized $RbAg_4I_5$ specimen induces ionic current. The ionic conductivity of the specimen can be evaluated from

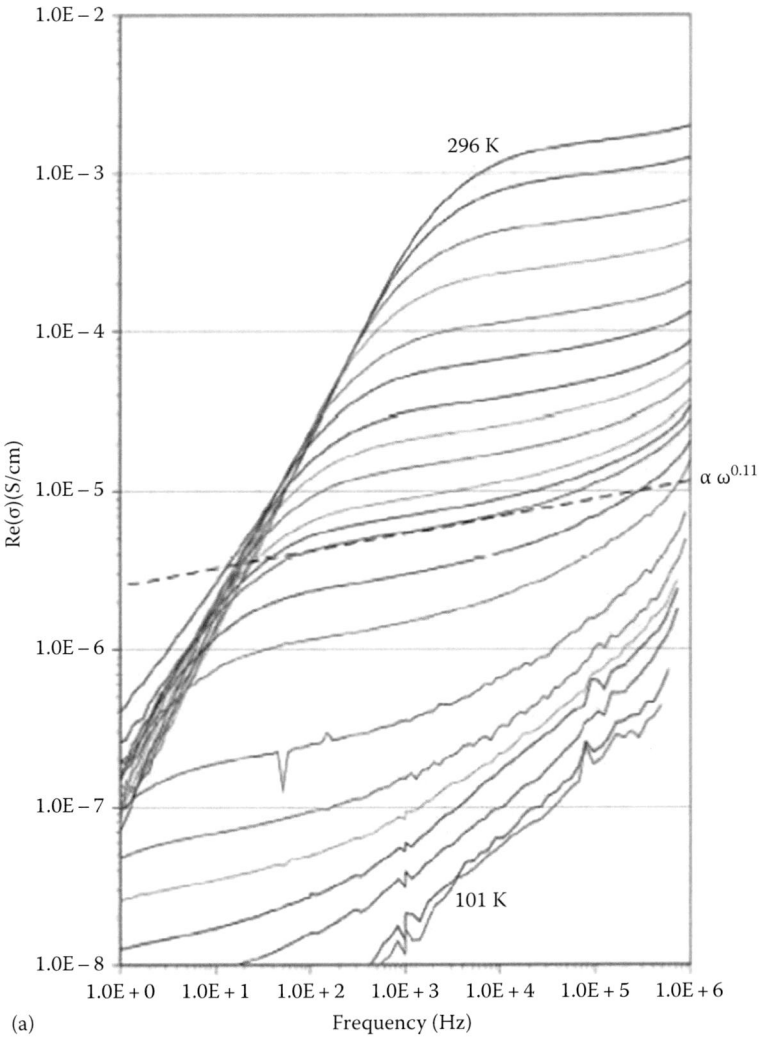

(a)

FIGURE 4.61 (a) Frequency dependence of the real part of complex ionic conductivity (Re(σ, S/cm) versus frequency (Hz) on a log–log scale) for a Ag-β-alumina single-crystal sample measuring $0.925 \times 0.395 \times 0.120$ cm^3 with sputtered Ag electrodes, measured from ~100 K to room temperature. The dashed line indicates a $\omega^{0.11}$ dependence as a guide.

(Continued)

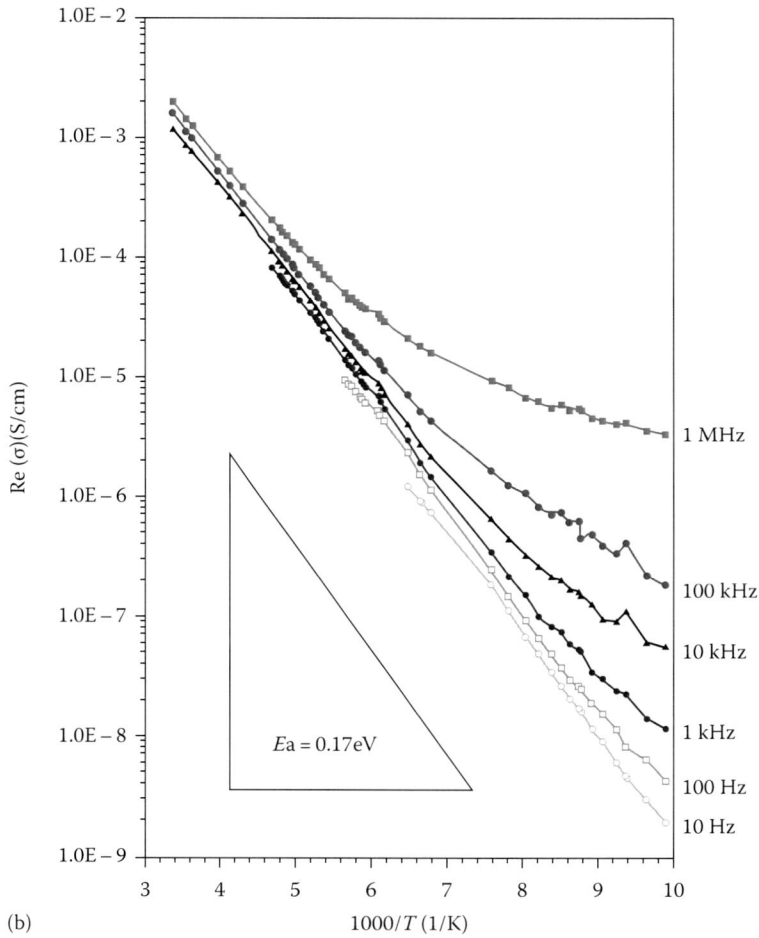

(b)

FIGURE 4.61 (*Continued*)   (b) Arrhenius plot of conductivity (log Re(σ) versus 1000/T (K)) at different frequencies for the sample of (a). The activation energy for Ag$^+$ ion transport is 0.17 eV. (From Kamishima, O. et al., *Solid State Ionics*, 262, 495, 2014.)

the magnetic field generated by the ionic current. An AgI specimen conveniently calibrates the measuring system. An important advantage of this method compared to the time-consuming impedance spectroscopy is that a measurement at a single low frequency gives the intrinsic ionic conductivity. For details, see [4.68].

Finally, the impedance characterization of thin-film structures are of interest both for materials and for device characterization. Figure 4.62 shows the AgI thin-film-based M-I-M structure, the impedance measurements, and Arrhenius plots derived from the impedance plots.

To conclude this section, we give two circuits: (1) for the measuemet of local impedance in an electrode–electrolyte system (Figure 4.63) and (2) schematic of an impedance measurement procedure for (a) sinewave/frequency response analyzer technique and (b) broadband signal/computer technique (Figure 4.64).

Measurements of step and ramp electrochemical time responses on electrochemical systems can lead to parameters useful for tests of practical applicability with good performances rapidly and/or accurately. However, as the choice of these parameters is not imposed by the complete knowledge of the processes involved, that is, a model, their validity has to be proved by a preliminary investigation that compares the practical application to the parameter measurement.

By contrast, techniques involving impedance measurements, that is, AC signal perturbation having a small amplitude, are the most powerful techniques especially when a fundamental investigation is carried out, for example, the search for a model.

Let us move on to optical spectroscopy as a characterization tool.

4.1.4.3 UV–Visible Optical Spectroscopy

Light interacts with matter through absorption, reflection, scattering, and emission. The study of optical properties of solid state ionic materials is an effective means to understand the electronic and atomic structure of mixed ionic–electronic and semiconductor-like ionic solids. Optical properties of these materials, particularly absorption, reflection, and dipersion, arise from electronic excitations. Evaluation of the electronic energy band gap and optical constants over a wide spectral range is an important goal as it leads to elucidation of aspects of electronic band structure.

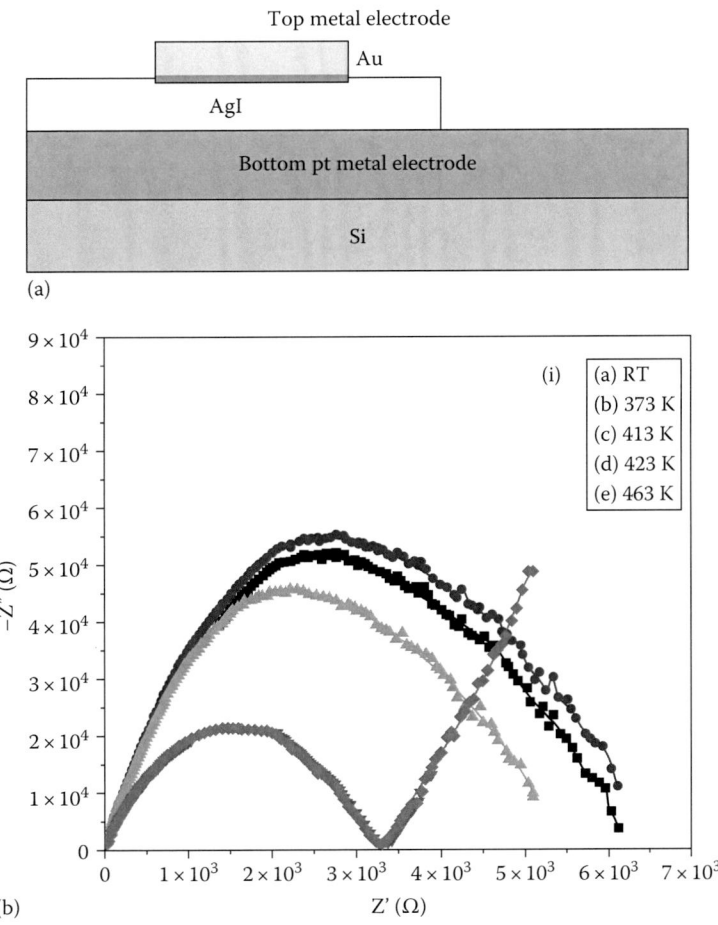

FIGURE 4.62 (a) Cross-sectional view of the M-I-M structure. Ag (and $Ag_{0.9}Cu_{0.1}$, $Ag_{0.8}Cu_{0.2}$) thin films were fabricated by RF sputtering onto Pt/Si substrates and iodized using a figure of eight jig prior to MIM structure fabrication and impedance spectroscopic characterization. Top electrode (300 μm^2 area) was fabricated by the shadow-mask technique. (b) Impedance plots (Z′ vs. Z″) obtained from measurements carried out from 40 Hz to 1 MHz using Agilent 4294A impedance analyzer. The sample AgI is heated from room temperature (RT) to 463 K through the superionic phase transition. Note the clear separation of the plots at 413 and 463 K—below and above the transition temperature (420 K). Similar measurements were done on $Ag_{0.9}Cu_{0.1}I$ and $Ag_{0.8}Cu_{0.2}I$ thin-film structures. *(Continued)*

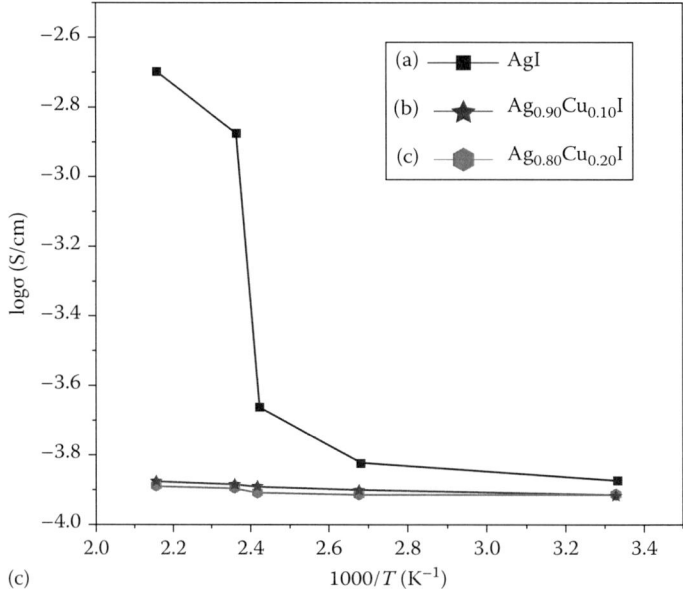

(c)

FIGURE 4.62 (*Continued*)   (c) Arrhenius plots of log σ versus $1000/T$ for AgI, $Ag_{0.9}Cu_{0.1}I$, and $Ag_{0.8}Cu_{0.2}I$ thin-film-based MIM structures showing the sharp phase transition in AgI and the effect of Cu substitution. (From Gnanavel, M. and Sunandana, C.S., *Physics of Iodized Silver & Silver-Copper Thin-Film Nanostructures*, Lambert Academic Publishers, Berlin, Germany, 2013.)

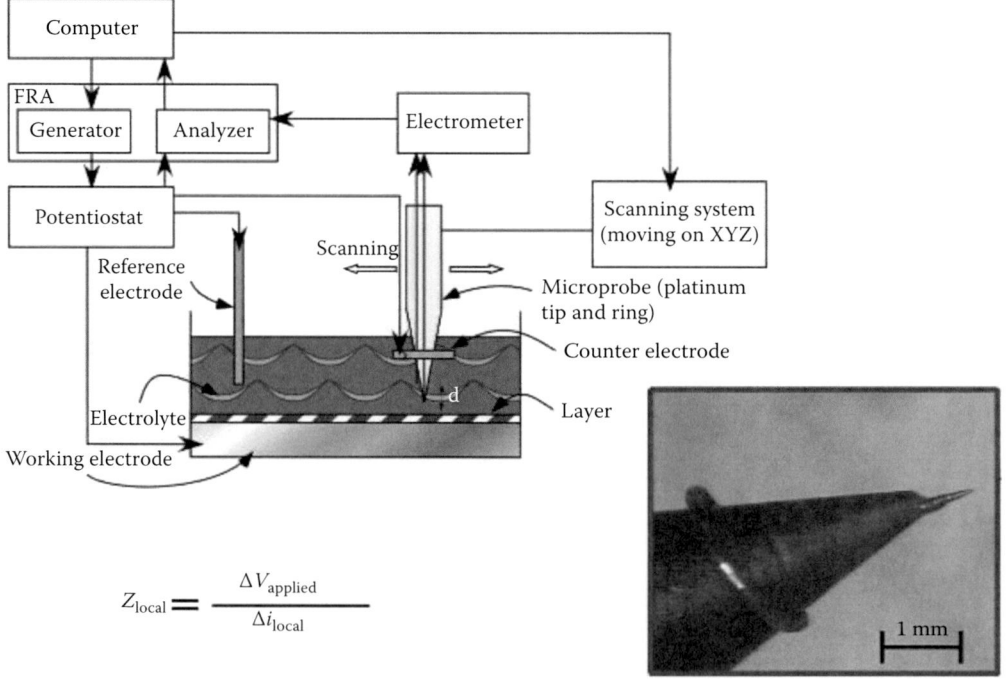

$$Z_{local} = \frac{\Delta V_{applied}}{\Delta i_{local}}$$

FIGURE 4.63   A setup for local impedance measurement in an electrode-electrolyte system. The microprobe is separately shown on the right. (From Orazem, M.E., *Electrochemical Impedance Spectroscopy*, Department of Chemical Engineering, University of Florida, 2008.)

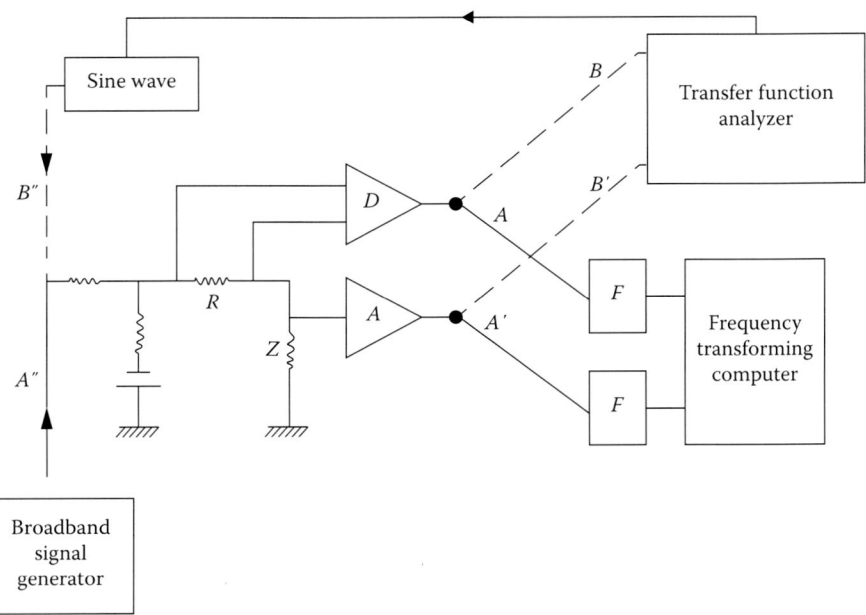

FIGURE 4.64 Schematic of the impedance measurement procedure for sine wave/frequency response analyzer technique and broadband signal/computer technique. The case of galvanostatic (constant current) arrangement is illustrated here. The potentiostatic (constant voltage) case is similar. (From Grabrielli, C., Use and applications of electrochemical impedance techniques, Solartron Technical Report No 24, Issue B, 1997.)

The usual way to determine the optical properties of a solid [4.72] is to shine monochromatic light onto a bulk or thin-film sample and to measure the reflectivity, transmission, or absorption coefficients as a function of photon energy.

Excitation of electrons in atoms or molecules results in light absorption. Such absorption can be quantitatively measured by using Bouguer–Lambert–Beer law: the transmitted light intensity (I) through a nonreflecting medium is given by

$$I = I_0 e^{-\alpha t} \tag{4.60}$$

where
I_0 is the incident light intensity
t is the sample thickness
α is the absorption coefficient of the medium

α is a measure of the energy attenuation or loss as it travels through the medium. If the medium is also partly reflecting,

$$I = \frac{I_0 (1-R)^2 \exp(-\alpha t)}{1 - R^2 \exp(-\alpha t)} \tag{4.61}$$

where R is the reflectivity of the medium.

We briefly introduce the theoretical background for optical characterization [4.73]. Chapters 6 and 7 cover phonons and electron bands.

Generally, the frequency-dependent optical properties of a solid are determined by the relative contribution to the complex dielectric constant tensor ε. These arise from (1) free carriers (ε_{fc}), (2) interband electronic transitions (ε_{ib}), (3) phonon excitations (ε_{ph}), and (4) other processes (ε_{core}) that occur at frequencies higher than those in (1)–(3) and thus regarded as an average. Thus, ε, the complex dielectric tensor, is

$$E = \varepsilon_1 + i\varepsilon_2 = \varepsilon_{fc} + \varepsilon_{ib} + \varepsilon_{ph} + \varepsilon_{core} \tag{4.62}$$

Here, ε_1 and ε_2 are the real and imaginary parts of the ε-tensor. It is normalized such that in a vacuum, ε and ε_{core} reduce to the unit tensor. The ε-tensor helps define the complex index of refraction tensor $n + i\kappa$:

$$(n + i\kappa)^2 = \varepsilon_1 + i\varepsilon_2 \tag{4.63}$$

Solid state ionic materials with hexagonal symmetry such as graphite intercalation compounds and $LiMO_2$ (M = Co, Mn, Ni) require two independent tensor components ε_a and ε_c that arise when the optical electric field is parallel and perpendicular to the layer planes, respectively. In view of the usually very different magnitudes of ε_a and ε_c, these materials may exhibit highly anisotropic optical properties.

$$\varepsilon = \varepsilon_1 + i\varepsilon_2 = \varepsilon_{cariers} + \varepsilon_{interband} + \varepsilon_{lattice} + \varepsilon_{core}, \tag{4.64}$$

$$(n + i\kappa)^2 = \varepsilon_1 + i\varepsilon_2. \tag{4.65}$$

Solid state ionic materials with hexagonal symmetry such as graphite intercalation compounds and $LiMO_2$ (M = Co, Mn, Ni).

Solid state ionic materials with hexagonal symmetry, such as graphite intercalation compounds and $LiMO_2$ (M = Co, Mn, Ni), require two independent dielectric tensor components ε_a and ε_c to be measured. These correspond to the optical electric fields being parallel and perpendicular to the layer planes, respectively. Because the magnitudes of ε_a and ε_c could differ considerably, these materials exhibit highly anisotropic optical properties.

The ultraviolet–visible spectrophotometer uses two light sources, a deuterium (D_2) lamp for ultraviolet region and a halogen lamp for visible region. The light from the source lamp gets reflected from mirror 1 and beam passes through slit 1 and hits a diffraction grating. The grating can be rotated allowing for a specific wavelength to be selected. At any specific orientation of the grating, only monochromatic (single-wavelength) beam successfully passes through slit 2. A filter is used to remove unwanted higher order diffracted beam. The light beam hits a second mirror before it gets split by a half mirror (half of the light is reflected; the other half gets transmitted). One of the beams is allowed to pass through a reference sample (air in the present case), and the other passes through the film coated substrate. The intensities of the light beams are then measured at the end as shown in Figure 4.65. The photometer (not shown) computes the ratio of the sample signal to reference signal (I/I_0) to obtain the transmittance.

An electronic transition consists of the promotion of an electron from an orbital of a molecule in ground state to an unoccupied orbital by absorption of a photon. The molecule is then said to be in an excited state. Experimentally, the efficiency of light absorption at a wavelength λ by an absorbing medium is characterized by the absorbance $A(\lambda)$ or the transmittance $T(\lambda)$, defined as [4.74]

$$A(\lambda) = \log_{10} I_\lambda^0 / I_\lambda = -\log T(\lambda) \tag{4.66}$$

$$T(\lambda) = \frac{I_\lambda}{I_\lambda^0} \tag{4.67}$$

where I_λ^0 and I_λ are the light intensities of the beams and entering and leaving the absorbing medium, respectively.

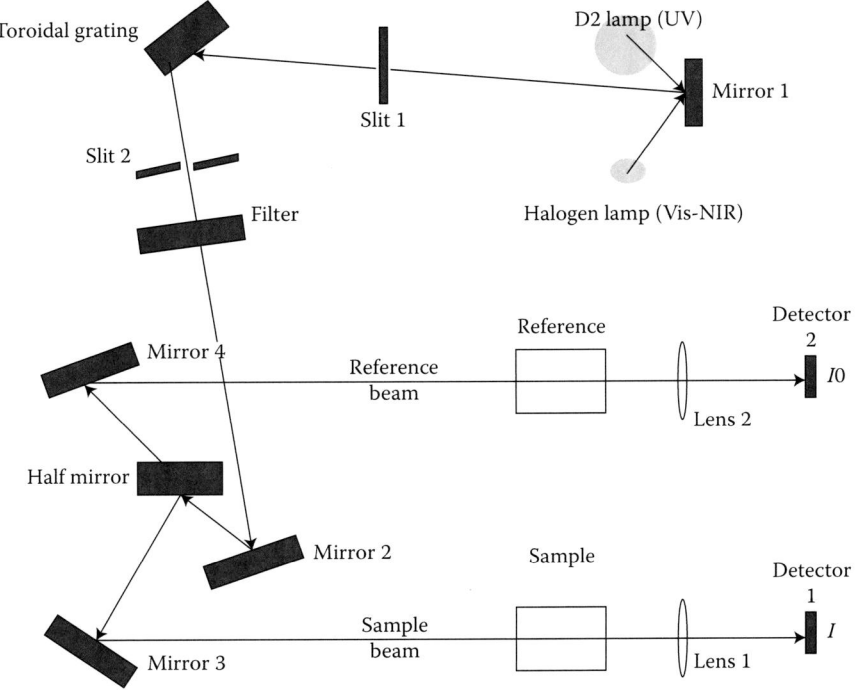

FIGURE 4.65 Schematic diagram of UV–VIS–NIR spectrophotometer.

Usually, the absorbance of a sample follows the Beer–Lambert law:

$$A(\lambda) = \log_{10} \frac{I_\lambda^0}{I_\lambda} = \varepsilon(\lambda) l c \tag{4.68}$$

where
  A is the measured absorbance
  I_0 is the intensity of the incident light at a given wavelength
  I is the transmitted intensity
  l is the path length through the sample
  c is the concentration of the absorbing species

For each species and wavelength, ε is a constant known as the molar absorptivity or extinction coefficient. This constant is a fundamental molecular property in a given solvent, at a particular temperature and pressure, and has units of 1/M·cm or often AU/M·cm.

The absorption spectra give the fundamental absorption. This refers to the band-to-band absorption, that is, the excitations of electrons from the valence band to the conduction band of a semiconductor. A simple analysis of the absorption process yields a fundamental relation for the absorption coefficient, $\alpha \propto (h_\gamma - E_g)^{1/2}$ and $\alpha \propto (h_\gamma - E_g)^{3/2}$ for direct and indirect band gap materials, respectively. Here, E_g is the energy band gap of the material. Conventionally, for inorganic semiconductors, electrons can be excited from the valence band to the conduction band by absorption of incident radiant energy only if this energy is equal to or larger than E_g, the band gap of the system. Thus, absorption of light occurs when a quantum of energy ($h\gamma$) equals E_g:

$$E_g = h\gamma_g = hc/\lambda_{g(nm)} = \left(1240/\lambda_g\right) \text{eV} \tag{4.69}$$

Equation 4.66 helps determine the band gap energy of a solid state ionic material that is optically transparent (to visible and UV). What does one do to measure the optical response of opaque powders and pellets? The next section discusses photoacoustic spectroscopy (PAS), which is applicable to such samples.

4.1.4.4 Photoacoustic Spectroscopy

The term "photoacoustic" generally implies a particular technique or mechanism of detecting and measuring the absorption coefficient of a weekly absorbing or opaque and diffuse materials—such as solid state ionic materials. The technique is particularly useful in cases where conventional photoelectric measurement is not feasible.

The basic principle of PAS is the detection of heat produced in a sample due to *nonradiative* deexcitation processes as a result of the absorption of intensity-modulated light by the sample. The light-induced thermal wave signal is proportional to the light-to-heat conversion efficiency. Therefore, the PA energy conversion is complementary to photovoltaic and photogalvanic processes. This technique was pioneered by Alexander Graham Bell, who called it "spectrophone."

In PAS, the irradiation of the sample contained in a closed gas cell produces temperature and pressure fluctuations in the gas (Figure 4.66) that couple the light to the sample—synchronous with the modulated source signal. A microphone placed in the gas cell picks up these fluctuations and an associated amplifying system produces an electrical signal. The amplitude and phase of this signal monitored over a range of wavelengths contain information about a whole range of optical phenomena including nonlinear absorption, localized defect characteristics, spectroscopy of (de)intercalated systems, semiconductor absorption processes, and phase transitions. The unique advantage of PAS over other conventional spectroscopies is that it does not require *optical detection* of either transmitted or reflected light [4.75].

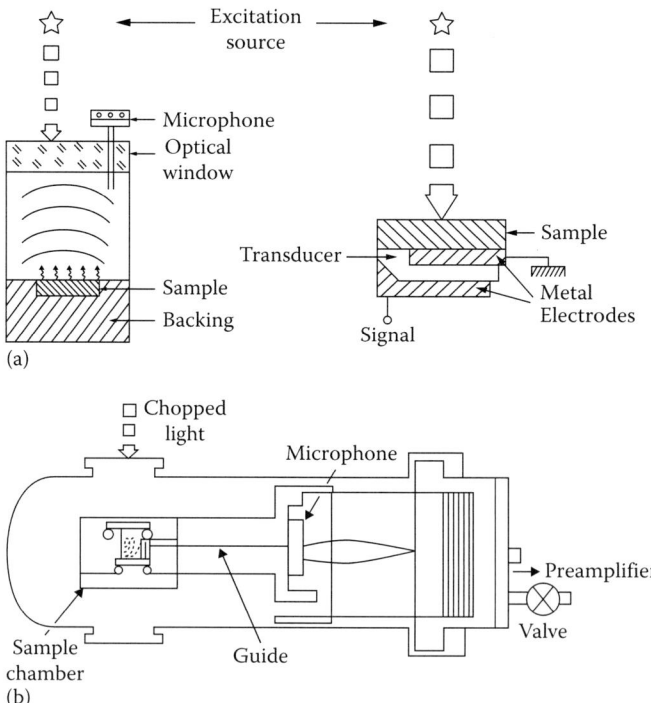

FIGURE 4.66 (a) Two methods of photothermal detection: (1) gas-cell microphone system—pressure (sound) waves detected by microphone before amplification, (2) piezoelectric transducer—direct conversion of pressure waves into photoacoustic signal. (From Sunandana, C.S., *Phys. Status Solidi (a)*, 105, 11, 1988.) (b) Compact and elegant PAS apparatus.

To provide a brief background to PAS, Rosencwaig and Gersho developed an analytical expression for the PA signal [4.76]. They consider (1) the variations of the gas pressure $\Delta P(t)$ as a function of time, as the ultimate consequence of the interrupted illumination of the sample given by

$$I = \left(\frac{I_0}{2}\right)(1 + \cos \omega t) \tag{4.70}$$

which leads to the propagation of a temperature wave from the sample surface, and (2) characteristic parameters of the different media: the solid, the material backing it, and the gas, which couples the solid to the detector:

$$\Delta P(t) = \left(\frac{\gamma P_0 \theta \mu_g}{2^{1/2} l_g T_0}\right) \exp\left[i\left(\omega t - \frac{\pi}{4}\right)\right] \tag{4.71}$$

where γ is the ratio of specific heats of the sample, P_0 the atmospheric pressure, ω the angular frequency of the incident light modulation, θ the temperature distribution, and l_g the length of the gas column. The complex envelope of the sinusoidal pressure variation, which contains the information about the material, is given by

$$Q = \frac{\alpha I_0 \gamma P_0}{2\sqrt{2}k_s l_g a_s T_0 (\alpha^2 - \sigma_s^2)} \frac{(r-1)(b+1)e^{\sigma sl} - (r+1)(b-1)e^{-\sigma sl} + 2(b-r)e^{-\alpha l}}{(g+1)(b+1)e^{\sigma sl} - (g-1)(b-1)e^{-\sigma sl}} \tag{4.72}$$

where
α is the optical absorption coefficient of the solid at a given wavelength
I_0 is the incident light flux (W/cm^2)
k_s, k_g, and k_b are thermal conductivities of sample, gas, and backing material ($J/cm \cdot s \cdot K$)
l_g is the length of gas column (cm)
a_g, a_b, and a_s are thermal diffusion coefficients of gas, backing material, and sample (cm^{-1})
$b = k_b a_b / k_s a_s$ (backing)
$g = k_g a_g / k_s a_s$ (gas)
$\sigma_s = (1 + i)a_s$
l is the sample thickness (cm)
T_0 = ambient temperature (K)

Equation 4.69 is the complex temperature wave propagating across the sample modified by the thermal properties of the backing material, sample, and coupling gas. It gives the amplitude q and the phase φ of the PA signal or the acoustic pressure produced in the cell by the PA effect. In fact, $Q = q \exp(-i\varphi)$. Experiments give this complex signal from which one has to determine q as a function of the wavelength and also of φ if required.

A simple and elegant experimental arrangement for doing PAS is shown in Figure 4.66b. The glass tube enclosed PA cell with residual volume optimized for maximum PA response and fitted with quartz windows permits evacuation and use of any coupling gas. It is adaptable for high/low-temperature studies on polycrystalline and nanocrystalline materials [4.77]. We now consider a few examples of PAS studies on Li-ion and Ag-ion conducting materials. Figure 4.67 shows PAS spectra of $LiFe_5O_8$ and $LiFeO_2$ at 300 K. These spectra have clearly resolved peaks, which are interpreted as due to Fe^{3+} in crystal fields of octahedral and tetrahedral symmetry.

Knowledge of thermal properties of superionic conductors such as $Li_xMn_2O_4$ ($0.8 \le x \le 1.2$) are important from basic and applied viewpoints—structure and thermal runaway issues in electric vehicles, respectively [4.79]. Such knowledge is easily obtained through PAS. Figure 4.68 shows the PA spectra $P(\lambda)$ of the superionic conductors β-Ag_3SI and α-Ag_3SI measured in the range $270 < \lambda < 1800$ nm at room

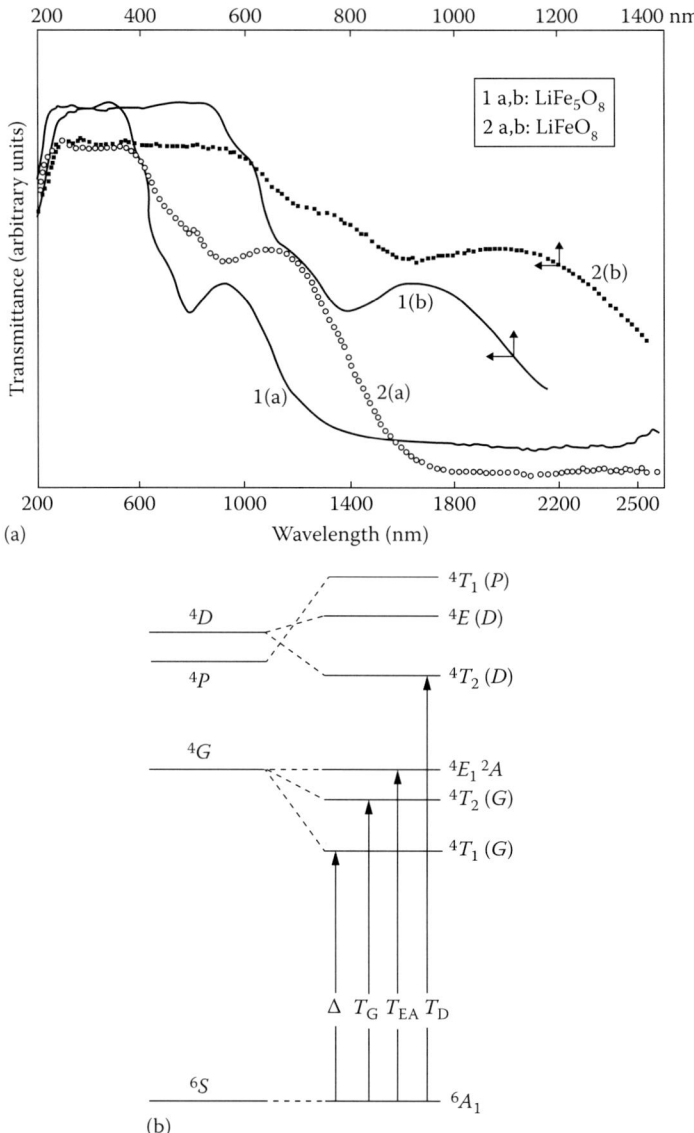

FIGURE 4.67   (a) Photoacoustic spectra of $LiFe_5O_8$ (1a,b) and $LiFeO_2$ (2a,b) at room temperature. Note the clearly resolved peaks in the upper scale spectra (1b,2b). Fe^{3+} is present at both octahedral and tetrahedral sites in $LiFe_5O_8$, which exhibits peaks at 930, 660, 590, 510, 480, 350 and 270 nm. $LiFeO_2$ has peaks at 750, 500, 360, and 300 nm. (b) Energy levels of Fe^{3+} in crystalline electric fields of octahedral and tetrahedral symmetry. In view of the orbital singlet nature of Fe^{3+} ground state, common transitions (as marked) are observed. (From Sunandana, C.S. and Phaninath, D., *Solid State Commun.*, 58, 115, 1986.)

temperature reveal a difference in the band gap observed in the annealed β-phase and in the quenched α-phase [4.80]. A simple band model by Pantelides has shown that the band gap varies approximately with an equilibrium nearest-neighbor distance, d, as d^{-2}. As a result, Ag_3SI ($E_g(\beta) = 1.11$ eV and $E_g(\alpha) = 1.47$ eV) is found to have a small band gap in comparison with γ-AgI ($E_g = 2.85$ eV). Therefore, Ag_3SI resembles a semiconductor and the electrons make a large contribution to the conductivity in comparison with AgI.

The change in the PA spectrum of α-AgI measured by exposing an above-band-gap light as a function of time from 0 to 20 min at 443 K reveals that the PA signal below the band gap of α-AgI increases monotonically up to a factor of 4 with increasing lapse of exposure time. This interesting behavior indicates a photo-induced clustering of Ag^+ ion [4.81].

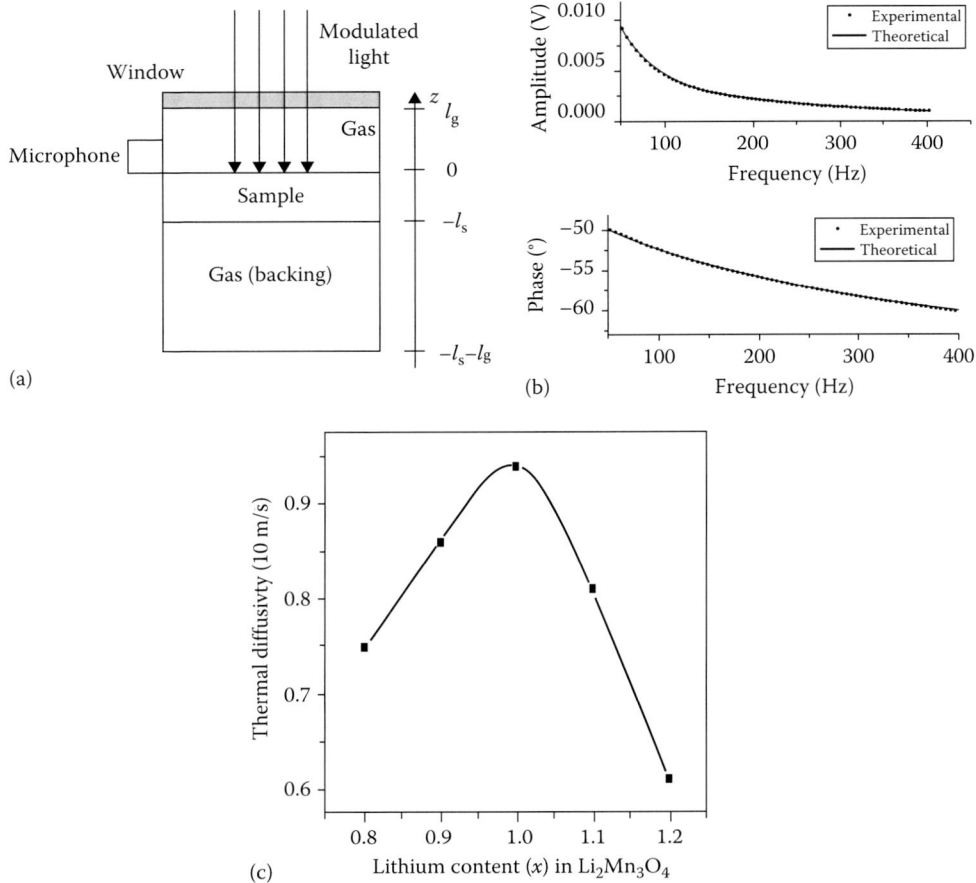

FIGURE 4.68 (a) PA cell (schematic) for thermal diffusivity measurements on thin-film samples. (b) Amplitude and phase variations of PA spectra of $LiMn_2O_4$ as a function of frequency in the range 50–400 Hz. (c) Thermal diffusivity as a function of x in $Li_xMn_2O_4$. (From Raveendranath, K., Li containing metal oxides and metal phosphates: Investigations on their synthesis and various characteristics for rechargeable Li battery applications, PhD thesis, Cochin University of Science and Technology, Cochin, India, July 2008; Raveendranath, K. et al., *Mater. Sci. Eng. B*, 131, 210, 2006.)

Next, we focus on two complementary vibrational spectroscopies, namely, infrared spectroscopy and Raman spectrocopy.

4.1.4.5 Fourier Transform Infrared (FTIR) Spectroscopy

Infrared absorption of a solid sample detects the vibrational characteristics of chemical bonds in it. IR spectroscopy along with the Raman spectroscopy (to be discussed next) provide complementary information about molecular vibrations that are controlled by point group symmetry of the concerned molecular group (e.g., PO_4^{3-} in $LiFePO_4$). The two techniques are different from the viewpoint of the mechanism of interaction of light quanta with the molecule.

Interaction of IR radiation ($200 - 1600$ cm^{-1}) with a vibrating molecule is possible only if the *electric vector* of the radiation field oscillates with the *same* frequency as does the *molecular dipole moment*—a fundamental property of the molecule. A vibration is "IR active' only if the molecular dipole moment μ is *modulated* by the normal vibration, or

$$\left(\frac{\partial \mu}{\partial q}\right)_0 \neq 0 \tag{4.73}$$

where q is the normal coordinate describing the motion of an atom during a normal vibration.

IR absorption spectroscopy is done using a Fourier transform (FT) IR spectrometer. It uses Fourier transform techniques with a Michelson interferometer. To obtain an IR absorption spectrum, one mirror of the interferometer moves to generate interference in the radiation reaching the detector. Since all wavelengths are passing through the interferometer, the interferogram is a complex pattern. The absorption spectrum as a function of wave number (cm^{-1}) is obtained from the Fourier transform of the interferogram, which is a function of mirror movement (cm). This design does not have the reference cell of a dispersive instrument, so a reference spectrum is recorded and stored in memory to subtract from the sample spectrum. A schematic of the instrument is shown in Figure 4.69.

Among the most interesting and informative IR spectra of solid state ionic materials are from the vibrations from phosphate molecular groups in $Li_xTi_2(PO_4)_3$ ($x = 1, 2, 3$) and in $LiFePO_4$. These two cases are briefly discussed (Figure 4.70).

Generally, the vibrational spectra of materials with polyanionic groups such as phosphate and sulfate are classified into internal and external modes. The former have larger force constants than the latter. Internal vibrations consist of intramolecular stretching and bending motions of PO_4^{3-} anions and are described in terms of the four fundamental vibrations of the free ion $\nu_1-\nu_4$. The external modes or lattice phonon modes are composed of Li^+, Ti^{3+}, or Ti^{4+} and PO_4^{3-} translational vibrations and pseudo-rotations of the PO_4^{3-} ions. Li^+ in translator modes are called Li^+ ion "cage" modes. IR spectra (and also Raman spectra) have to be interpreted on this basis ([4.82] and references therein).

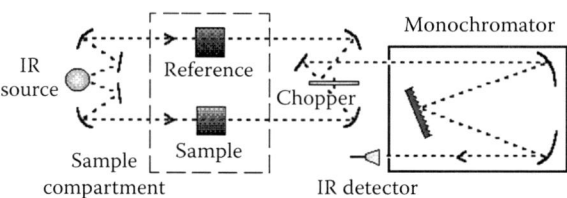

FIGURE 4.69 Schematic of the FTIR spectrometer.

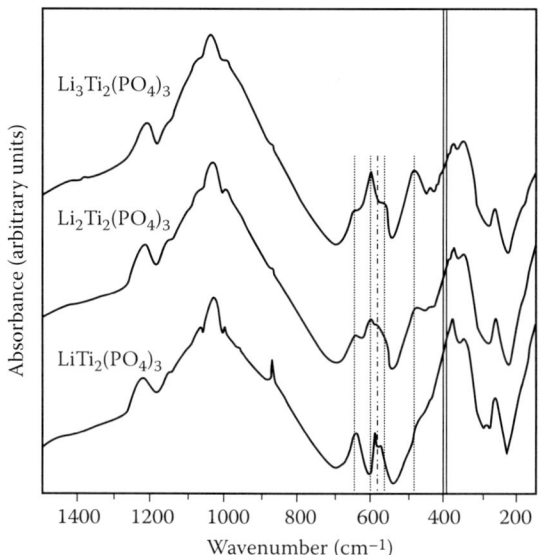

FIGURE 4.70 Infrared absorption spectra (absorbance vs. wave number) of $Li_xTi_2(PO_4)_3$ ($x = 1, 2, 3$). (From Burba, C.M. and Frech, R., *Solid State Ionics*, 177, 1489, 2006.)

The IR spectra of $LiTi_2(PO_4)_3$ are dominated by intense overlapping intramolecular PO_4^{3-} stretching modes (ν_1 and ν_3) in the range 1300–700 cm^{-1}. Note the sharp band at 869 cm^{-1} and the broad features between 700 and 850 cm^{-1}. However, there are no bands near 750 cm^{-1} that identify P–O–P groups that bridge phosphates in the structure. Raman spectroscopy gives intense bands due to P–O–P bending modes (see later).

In $LiTi_2(PO_4)_3$, four bands at 680, 642, 585, and 572 cm^{-1} are assigned to the ν_4 mode. The three intense bands at 464, 381, and 348 cm^{-1} perhaps arise from ν_2 vibrations.

The 1222 cm^{-1} band in $LiTi_2(PO_4)_3$ shifts to 1223 cm^{-1} in $Li_3Ti_2(PO_4)_3$, implying that Li insertion hardly affects the ν_1 and ν_3 (stretching) modes of the phosphate group. Thus, only a small degree of disorder is induced in the structure upon Li insertion. On the other hand, ν_2 and ν_4 (bending) modes are more sensitive to Li insertion as the new band at 600 cm^{-1} in $Li_2Ti_2(PO_4)_3$ shows.

A problem discusses the group theoretical origin of the IR bands in $Li_xTi_2(PO_4)_3$.

Consider next the IR spectrum of $LiFePO_4$ (Figure 4.71). Compared to the spectra of $Li_xTi_2(PO_4)_3$, these are much sharper and the bending (400–700 cm^{-1} range, ν_2, ν_4) and stretching modes (700–700–1300 cm^{-1} range, ν_1, ν_3)are clearly demarcated [4.83], probably reflecting the compact and ordered crystal structure of $LiFePO_4$, as discussed in Chapter 1. The internal modes of phosphate group are split into many components due to correlation effect induced by the coupling with the FeO_6 units in the olivine structure.

At a more fundamental level, IR absorption and reflectivity of single crystals enable investigation of dielectric constant dispersion and lattice dynamics of superionic conductors such as $LiNaSO_4$ [4.84] and Li_3N [4.85], respectively.

Next, we discuss the technique complementary to IR absorption, namely, Raman spectroscopy.

4.1.4.6 Raman Spectroscopy

Raman spectroscopy is based on the phenomenon of Raman scattering [4.86]. It is a two-photon process having net effects of scattering photons but changing their frequency. This changing frequency characteristic of inelastic scattering provides a basis of a form of spectroscopy. The scattering of a photon is caused by a key property of molecule involved, namely, the molecular polarizability, α. α enables an electric field E to induce a dipole moment μ_{in} in an atom or molecule:

$$\mu_{in} = \alpha E \tag{4.74}$$

A light photon can act as an induced electric field, so Equation 4.74 applies when light interacts with atoms and molecules with concurrent changes in the polarization of the molecule—motions like rotations

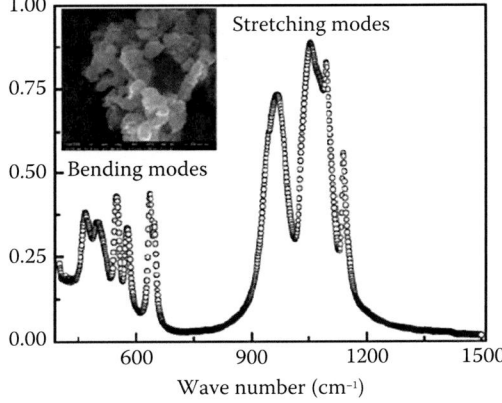

FIGURE 4.71 FTIR spectrum of $LiFePO_4$ (normalized absorbance vs. wave number) recorded at 300 K. It shows the stretching and bending modes of the phosphate ion. Inset shows an FESEM image of $LiFePO_4$. (From Sundarayya, Y. et al., *Phys. Status Solidi B*, 250, 1599, 2013.)

and vibrations. At a molecule's equilibrium geometry, the polarizability has the same value α_0. At some distance Δr away from the previous geometry, the instantaneous polarization α is

$$\alpha = \alpha_0 + \left(\frac{\partial \alpha}{\partial r}\right)\Delta r \tag{4.75}$$

where $(\partial \alpha / \partial r)$ represents the change in the polarizability with change in position. If the molecule is vibrating as a sine wave, Δr can be written in terms of the vibration frequency ν and the time t:

$$\Delta r = r_{max} \cos(2\pi \nu t) \tag{4.76}$$

where r_{max} is the maximum vibrational amplitude. Light of a particular frequency ν_{in} is inducing an electric field E that also has sine wave form:

$$E = E_{max} \cos(2\pi \nu_{in} t) \tag{4.77}$$

where E_{max} is the maximum amplitude of the electric field. Substitute for α, Δr, and E using Equations 4.74, 4.75, and 4.76 into Equation 4.73,

$$\mu_{in} = \alpha_0 E_{max} \cos(2\pi \nu_{in} t) + E_{max} r_{max} \left(\frac{\partial \alpha}{\partial r}\right) \cos(2\pi \nu t) \cos(2\pi \nu_{in} t) \tag{4.78}$$

Using the trigonometric relation $\cos a \cos b = (1/2)[\cos (a + b) + \cos (a - b)]$, Equation 4.78 becomes

$$\mu_{in} = \alpha_0 E_{max} \cos(2\pi \nu_{in} t) + \left[\frac{E_{max} r_{max}}{2}\right]\left(\frac{\partial \alpha}{\partial r}\right)\{\cos(2\pi t(\nu_{in} + \nu)) + \cos(2\pi t(\nu_{in} - \nu))\} \tag{4.79}$$

The first term of Equation 4.79 contains the frequency of the incoming light, and also the outgoing light—the so-called Rayleigh scattering. The second term has two cosines: one has $\nu_{in} + \nu$ relating to an outgoing scattered photon that *increases* in frequency by ν units, which is the frequency of the molecular motion called anti-Stokes frequency. The last term has the variable $\nu_{in} - \nu$ relating to a scattered photon that *decreases* its frequency by the same amount. Thus, the incoming photons will shift their frequencies *up* and *down* by amounts equal to certain motions of the molecule. This is the basis of Raman spectroscopy. Equation 4.77 also gives a gross selection rule for Raman-active modes. Note that the second term has the multiplier $(\partial \alpha / \partial r)$ for the two cosine terms, which is the change in α with position. If $(\partial \alpha / \partial r) = 0$, the entire second term is zero and no Raman scattering is possible. A molecular motion will be Raman active only if the motion occurs with a changing α. To be Raman active, a molecule must have anisotropic polarizability. Symmetric molecules can be Raman active but not infrared active.

Figure 4.72a illustrates Raman scattering.

The original spectrometer used by Raman and the schematic of a modern laser Raman spectrometer are displayed in Figure 4.72b and c followed by a description of the latter.

One may use a Nd:YAG laser with a wavelength of 532 nm and a power of 200 mW as an excitation source for Raman spectroscopy. Two mirrors are used to guide the laser light to a notch filter. Note that Raman light I orders of magnitude weaker than the elastically reflected laser light. So the notch filter helps cut off the laser line. A dielectric coating enables reflection of light of only this particular wavelength namely 532 nm, and is thus used as a mirror for this wavelength. It reflects light toward a microscope objective with 40× magnification. The objective focuses the light onto the sample. This experimental setup performs micro-Raman spectroscopy. If the setup is properly calibrated, the scattered light collected by the objective travels to the notch filter reaching the same positon as the incident

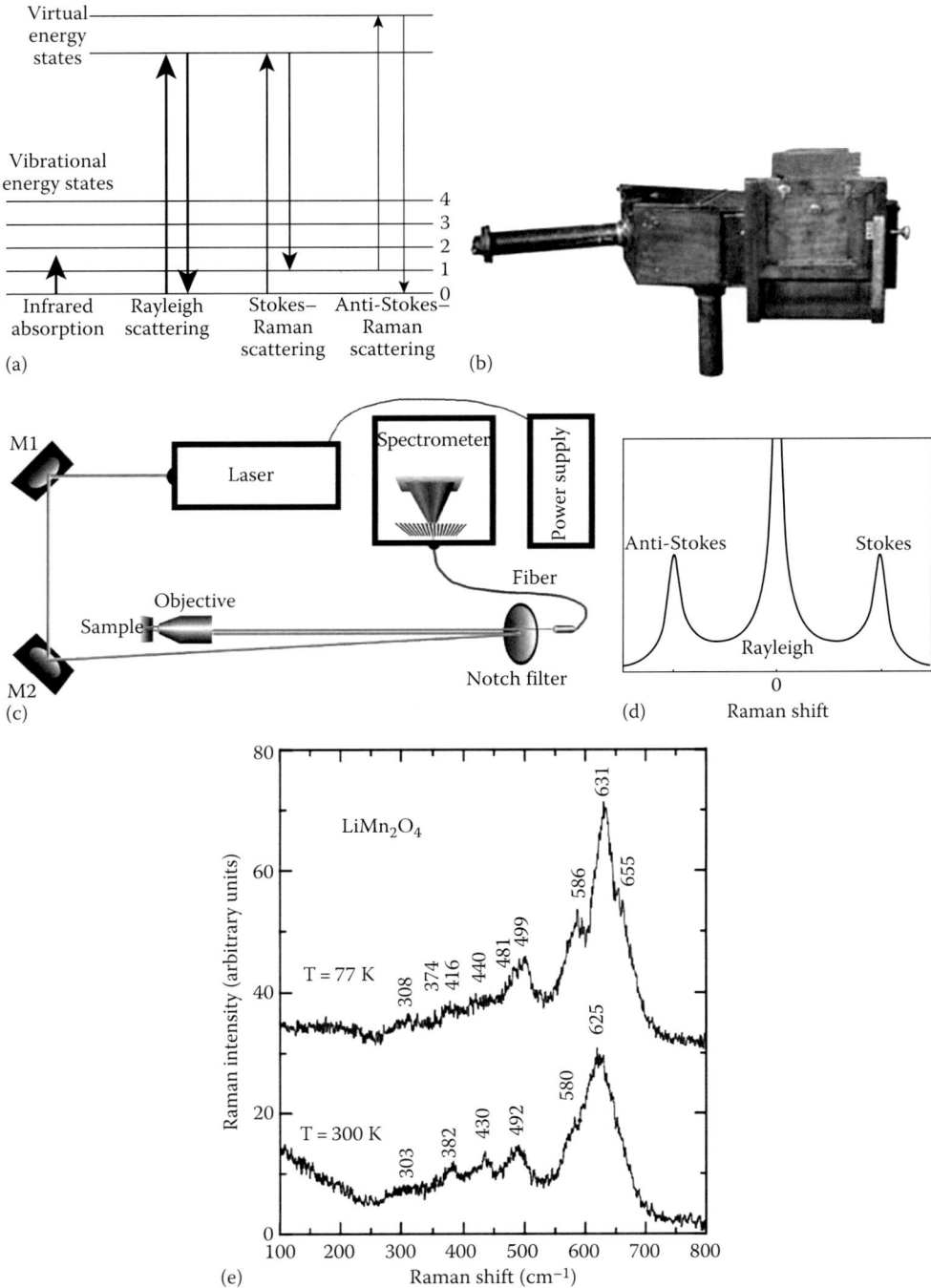

FIGURE 4.72   (a) Raman scattering (anisotropic) showing Stokes, anti-Stokes, and Rayleigh (isotropic) scattering compared with fluorescence. Note that infrared absorption takes place only within the vibrational ground state levels. (b) The original Raman spectrometer. (From www.acs.org/content/acs/en/education/.../landmarks/ramaneffect.html.) (c) Schematic of a typical laser Raman spectroscopy experiment for a solid sample. (d) A typical Raman spectrum. Rayleigh peak is due to elastic scattering of laser light (like Bragg peak of X-ray or neutron scattering of atoms). Stokes and anti-Stokes peaks, which occur in pairs, represent inelastic scattering of light by phonons of crystal. (e) Raman scattering in $LiMn_2O_4$ pellets with mirror-like surface. It is cubic at room temperature ($T = 300$ K) and orthorhombic at low temperature ($T = 77$ K). The strong band at 6205 cm^{-1} is due to the symmetric stretching vibrations of the MnO_6 groups with D_{3d} symmetry. The relative weakness of the peaks is due to the electronic properties of $LiMn_2O_4$. (From Ramana, C.V. et al., *Surf. Interface Anal.*, 37, 412, 2005.)

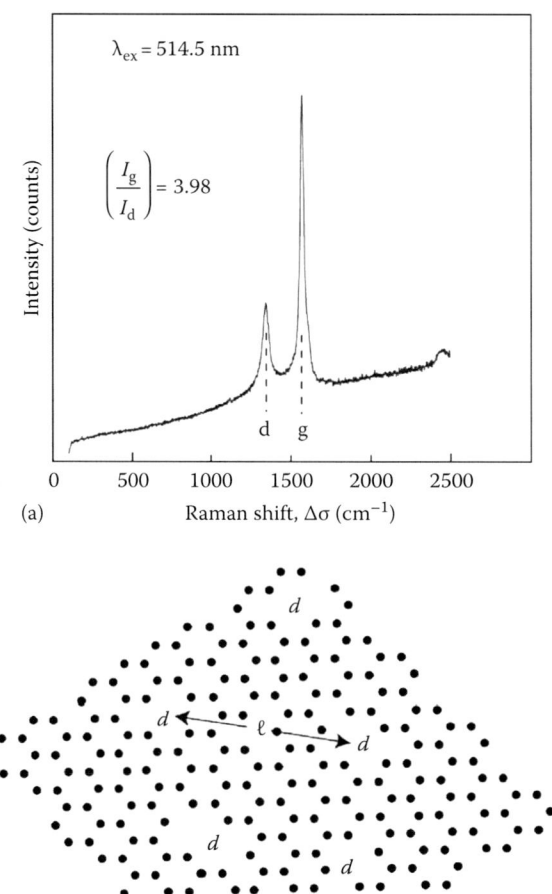

FIGURE 4.73 (a) Characterizing the domain size in nanocrystalline graphite through Raman spectroscopy. There are two peaks—weak one due to disorder (d) and the strong one (g) due to graphitic regions. (b) Characteristic domain size l is given as $l = 44 \ (I_g/I_d) = 17.5$ nm. (From Tuinstra, K., *J. Chem. Phys.*, 53, 1126, 1970.)

beam. The notch filter then cuts off the laser light allowing only the inelastically scattered part to enter the fiber placed behind the notch filter. The fiber guides the light to the micro-Raman spectrometer. The latter consists of a fixed grating and a detector. The grating disperses the Raman light onto the detector. Thus, the spectral distribution of the light is transformed to a spatial distribution recordable by the detector. Finally, a computer program analyzes the data.

Figure 4.72d shows a typical Raman spectrum. Figure 4.72e shows the Raman spectra of MnO_6 groups in $LiMn_2O_4$ [4.88]. Figure 4.73 illustrates an application of Raman spectroscopy to the characterization of domain size in nanocrystalline graphite [4.89]. Figure 4.74, which demonstrates a fuel cell application, proves the sensitivity of Raman scattering to gas diffusion in graphite paper.

Figure 4.75 illustrates the effect of progressive Li insertion into the Ti phosphate framework demonstrating the local disordering around the phosphate ions [4.82].

Before closing this section, consider an important case of Raman scattering applied to study reorientational motion of sulfate ions in $LiAgSO_4$ (Figure 4.76).

In the simple approximation that (1) vibrational (v) and reorientational motions are uncorrelated and (2) dipole–dipole coupling and collision-induce effects are negligible, the reorientational relaxation

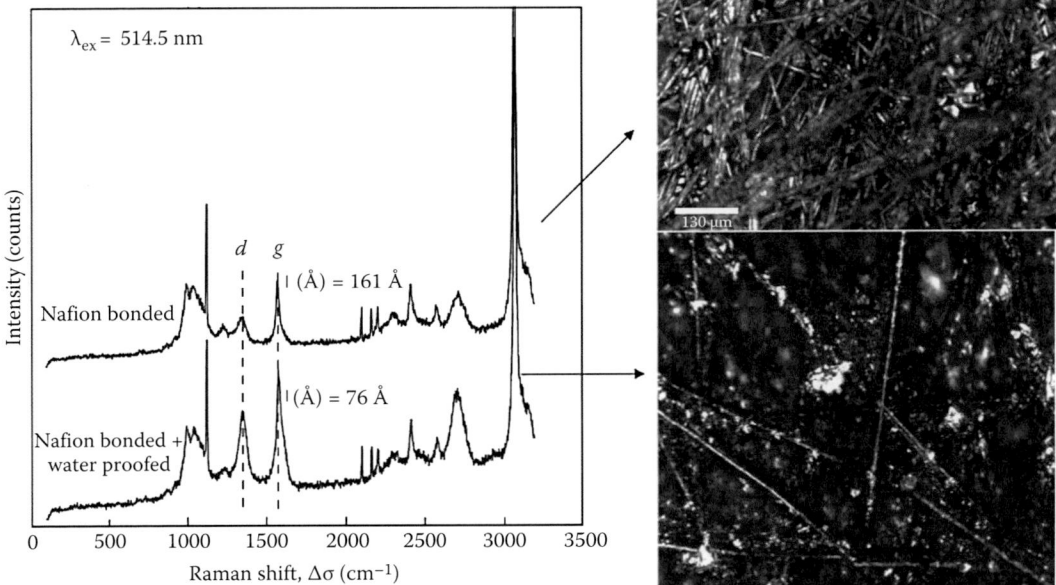

FIGURE 4.74 Raman spectroscopy is sensitive to gas diffusion in graphite paper as directly correlated to fuel cell materials microscopic characterization. (From Schwartz, D.T., Raman spectroscopy: Introductory tutorial, Department of Chemical Engineering, University of Washington, 2005.)

time τ is estimated [4.91] for a totally symmetric Raman vibration from half-widths of the isotropic and anisotropic spectrum:

$$\Gamma_{iso} = \Gamma_v \tag{4.80}$$

$$\Gamma_{aniso} = \Gamma_v + \Gamma_R \tag{4.81}$$

$$\tau = 2\pi c \Gamma_R = 2\pi c \left(\Gamma_{aniso} - \Gamma_{iso} \right) \tag{4.82}$$

where
  c is the speed of light
  Γ is the half-width in cm^{-1}

Width of the isotropic component only due to vibrational relaxation is observed in the polarized spectrum. Anisotropic broadening of the depolarized spectrum arises from both vibrational and orientational relaxation. Thus, orientational width is the difference between observed depolarized spectral width and the intrinsic width. When $\Gamma_{aniso} > \Gamma_{iso}$, the difference comes from reorientational relaxation from which τ may be derived.

4.1.5 Resonance Spectroscopies

We next discuss, one by one, the trinity of resonance spectroscopies operating in three frequency scales, namely, RF (NMR), microwave (EPR), and gamma ray (Mossbauer).

4.1.5.1 Nuclear Magnetic Resonance (NMR) Spectroscopy

Magnetic resonance spectroscopies are sensitive, dynamic, and local characterization probes for solid state ionic materials. Let us focus on nuclear magnetic resonance (NMR) first. In general, nuclei have a non-vanishing total angular momentum and they, thus, possess a magnetic moment. Note that the

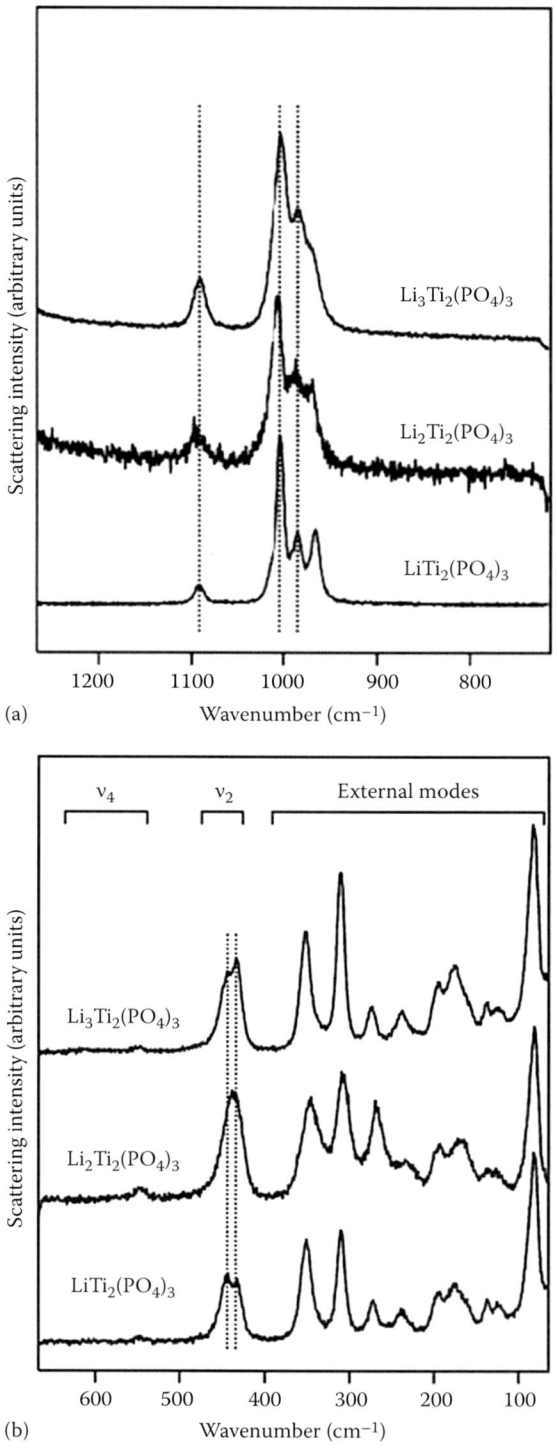

FIGURE 4.75 Raman spectra of three lithium titanium phosphates $Li_xTi_2(PO_4)_3$ ($x = 1$ (space group R_3bar c), 2, 3 (space group R3bar)) (a) 1200–800 cm^{-1} region, (b) 600–100 cm^{-1} region. Note the changes in Raman band intensities and widths as Li is progressively inserted into the Ti phosphate framework. Besides the PO_4^{3-} antisymmetric and symmetric bending vibrational modes, translatory motions of Li^+ ions, or Li "cage" modes arise, introducing a degree of local disordering about the phosphate ions. (From Burba, C.M. and Frech, R., Solid State Ionics, 177, 1489, 2006.)

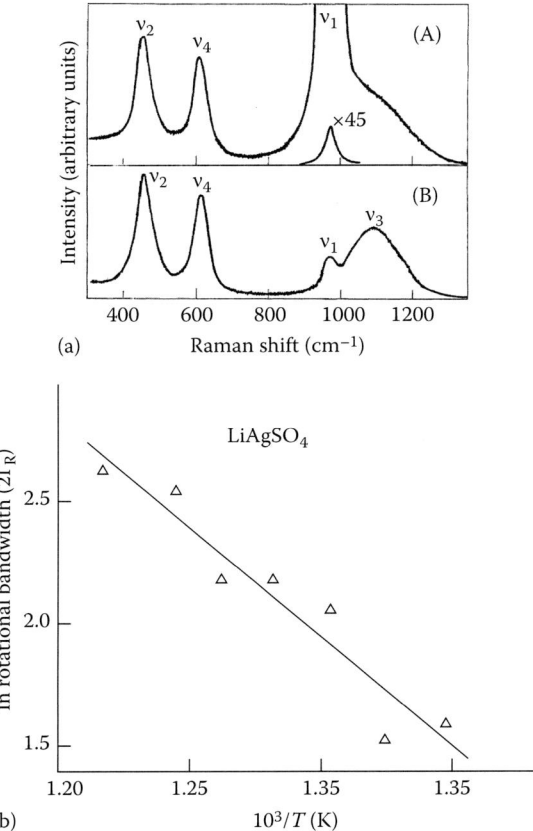

FIGURE 4.76 (a) Raman spectra of superionic bcc LiAgSO$_4$ at 742 K. (A) polarized (isotropic) spectrum, (B) depolarized (anisotropic) spectrum. ν_1 (970 cm^{-1}), ν_2 (453 cm^{-1}), ν_3 (1097 cm^{-1}), ν_4 (629 cm^{-1}) are the vibrational Raman modes of SO$_4^=$ ion. The symmetries of T_d point group are A$_1$, E, T$_2$, and T$_1$, respectively. The symmetric ν_1 mode of SO$_4^=$ in LiAgSO$_4$ whose anisotropic spectral widths have vibrational (Γ_v) and rotational (Γ_R) components. (b) Arrhenius plot of reorientational broadening ($2\Gamma_R$) of the symmetric mode. The slope of the plot gives the activation energy of 0.72 eV quite close to 0.52 eV for cation diffusion. (From Borjesson, L. and Torrel, L.M., *Phys. Rev. B*, 32, 2471, 1985.)

phenomenon of superionic conduction as emphasized in conductivity and diffusion is basically an ion-hopping and particle diffusion phenomenon. The abundant mobile ions (proton, Li$^+$, Na$^+$, Cu$^+$, Ag$^+$, O$^=$, and F$^-$) in solid state ionic materials all possess nuclear spin (I) magnetic moments (μ), and thus NMR of these can probe ion dynamics directly. Let us see how. The DC ionic conductivity given by

$$\sigma_{DC} = \left(\frac{\nu}{6}\right)\rho e^2 l^2 k_B T = \frac{\rho e^2 D}{k_B T} \tag{4.83}$$

where
  ρ is the mobile ion density = mobile ion number N/volume V
  e is the mobile ion charge
  $1/\nu$ is the average time for an ion to hop an average distance l
  D is the bulk diffusion coefficient
  $k_B T$ is the thermal energy at temperature T

The two equations suggest two practical approaches: (1) Know N? Then measure σ. Get product νl^2. (2) Known ion jump mechanism? And thus l? Then get ν. The single conductivity experiment involves two unknowns N and l, which depend on thermodynamics and crystal structure. The unique magnetic resonance advantage is that NMR can give a direct estimate of ν. Unlike conductivity, NMR depends only on how long the nucleus stays at a given lattice site in the superionic crystal. It measures the nuclear relaxation rate as a sample is heated or cooled, thus enabling a direct measurement of diffusion coefficients and activation energies to be directly measured.

The NMR experiment is a sophisticated version of the familiar parallel LCR resonance experiment [4.39]. Instead of the bulk inductance L, we now have the solid sample containing magnetic nuclei, say ^1H ($I = 1/2$, $m_I = \pm\frac{1}{2}$) in $KHSO_4$. NMR occurs when the H sample placed in an RF coil and held in a DC magnetic field (H_0) is excited to its Larmor precession frequency such that the external RF field (H_1) applied normal to H_0 flips the magnetization from $m_I = 1/2$ to $m_I = -1/2$. NMR absorption occurs when the spins return to their ground state by absorbing external RF energy when the condition,

$$\omega_0 = \gamma_n H_0 \tag{4.84}$$

where
  γ_n is the nuclear magnetogyric ratio defined as $g_n e / 2 m_n c$
  g_n is the spectroscopic splitting factor for the nucleus
  m_n, the mass of the nucleus, is satisfied

The nuclear magnetic moment μ arising from the spins of protons and neutrons, which sum to I, is given by

$$\mu = \frac{g_n e}{2 m_p I} \tag{4.85}$$

m_p being the mass of the proton.
  The projection of μ along z is given by

$$\mu_z = \frac{g_n e}{2 m_p} \hbar m_I = g_n \mu_N m_I \tag{4.86}$$

From these relations, one can work out the NMR spectrum expected from a given nucleus.

The NMR spectrometer measures the nuclear magnetic dipole transitions subject to the condition that the change in m_I during a resonance is zero. What one gets is an NMR spectrum with $(2I + 1)$ hyperfine components resolved under special experimental circumstances.

A pulse Fourier transform NMR spectrometer is used to record the NMR spectrum. The basic circuit block diagram to detect NMR is shown in Figure 4.77. The solid state pulse FT NMR spectrometer is shown in Figure 4.78.

In the FT NMR, the free induction decay or FID (akin to a spinning top that loses its ability to spin with time) of nuclear magnetization is subjected to a Fourier transformation that yields the observable NMR signal. FID is analogous to the transient response of a simple LCR circuit to which an AC input voltage is applied momentarily and switched off. Then, the AC output voltage of the LCR circuit decays rapidly to zero with a decreasing amplitude. The *envelope* of the decaying amplitude is usually an exponential decay like that of a damped harmonic oscillator. The FT of this kind of decay leads to the NMR spectrum of the sample in a coil that is empty in an LCR circuit.

To record the FID of a sample, a coil produces a 90° pulse (or a transverse pulse of magnitude H_1) in an imagined frame rotating with an angular velocity $\omega_0 = \gamma H_0$. This pulse acts for a time τ such that $\gamma H_1 \tau$ also serves to pick up the signal caused by the rotating magnetization immediately after the 90° or $\pi/2$ pulse. If coil volume is unity and the coil has a cross-sectional area A and has n turns, then the induced voltage immediately after $\pi/2$ pulse is turned on will be $V_0 = -n/c \eta d\varphi/dt = -1/c \, 4\pi A n \eta dM/dt = -1/c \, 4\pi n \eta A \omega M_0$. Here, $\varphi = BA$ is the flux linking the coil, $B = 4\pi M$, and η is the sample filling factor ($0 < \eta < 1$),

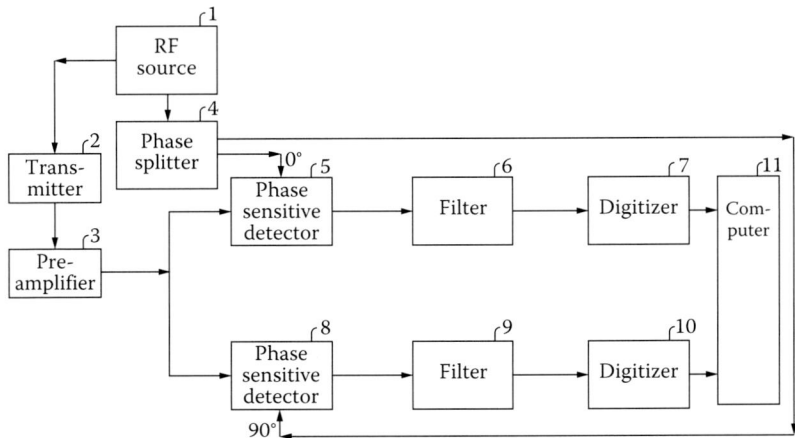

FIGURE 4.77 Device to detect magnetic resonance in the time domain. It consists of two phase sensitive detectors (5, 8), two filters (6, 9), and two digitizers (7, 10), one set for each receiver channel. An RF source (1) feeds a transmitter (2) and a phase splitter (4); the transmitter (2) delivers pulses to a probe preamplifier (3). The probe houses the sample whose magnetic resonance is being investigated. The signal output of (3) is fed to (5) and (8), which get reference inputs from (4), which are in phase quadrature. The output of (5) and (8) is filtered by (6) and (9) and then digitized by (7) and (10) before finally feeding to the computer (11). (From Chandrakumar, N. and Raman, S.V., US Patent 5,973,496.)

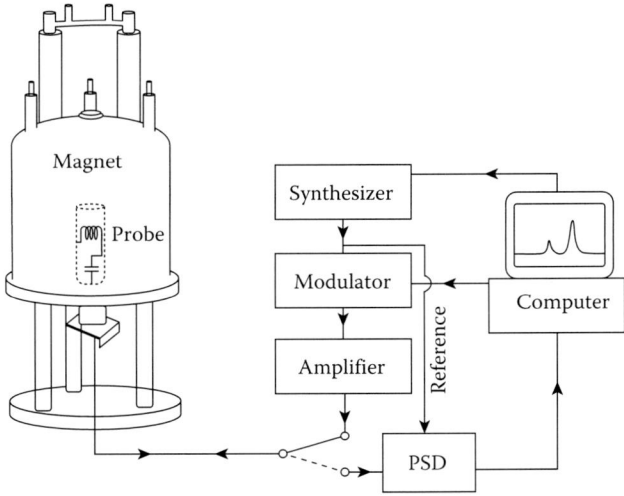

FIGURE 4.78 A typical pulse FT NMR spectrometer that utilizes the phase-sensitive detection technique of Figure 4.77. The sample is located in the coil of an RF resonant part of the probe. Sample and probe sit at the center of a superconducting magnet. Sample nuclei are excited by RF pulses generated in a synthesizer with a modulator and a high-power amplifier. The pulse sequences are controlled by a computer. The relatively weak sample response signal is directed through an RF switch to a phase-sensitive detector (PSD) whose reference signal is derived from the synthesizer. Finally, the computer displays the NMR spectrum of the sample. (From Klinowski, J. (Ed.), *New Techniques in Solid State NMR*, Springer, Berlin, Germany, vol. 246, 2005.)

which accounts for the incomplete flux linkage between the sample and coil. $|V|$ is approximately equal to several millivolts if the coil is part of a resonance circuit of reasonable Q-factor. M_0 is the initial (equilibrium) precessing magnetization fully turned over to the x–y plane of the laboratory Cartesian xyz frame. The transverse magnetization decays exponentially with a time constant T_2 called the spin–spin relaxation time. $T_2 \gg 2\pi/\omega_0$. The signal is contained within a slowly varying envelope. The envelope decays as $V_0 \exp(-t/T_2)$. This is the FID.

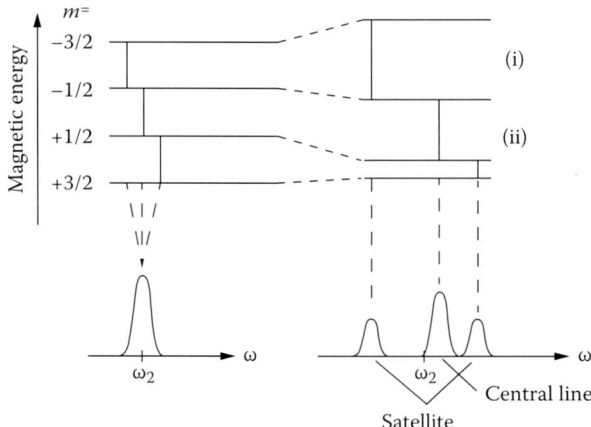

FIGURE 4.79 Energy levels of a nucleus with $I = 3/2$ (^{7}Li, ^{23}Na, ^{63,65}Cu) (i) without and (ii) with quadrupolar interaction. Arrows indicate NMR transitions induced by a radiofrequency field. The resulting NMR spectrum is shown at bottom.

To analyze an experimental NMR spectrum generated by the spectrometer of Figure 4.78 and schematized in Figure 4.79, a Hamiltonian is defined:

$$\mathcal{H}_{NMR} = \mathcal{H}_{Zeeman} + \mathcal{H}_{Quadrupole} \tag{4.87}$$

where

$\mathcal{H}_{Zeeman}$ is the nuclear Zeeman energy term

$\mathcal{H}_{Quadrupolar}$ describes the nuclear electric quadrupolar interaction

By contrast, protons (^{1}H), ^{19}F, ^{31}P, and ^{107}Ag with a spin $I/2$ do not possess quadrupole moment and thus no electric quadrupolar interaction.

An important observation in NMR spectra is the so-called chemical shift, which refers to the reduction in the effective magnetic field at the nucleus of an atom relative to the applied field ($H_{eff} = H_0(1-\sigma)$, where σ is the chemical shift (not to be confused with conductivity)) because of the diamagnetic shielding provided by the enveloping electron cloud. σ for the same atom in different chemical environments will be different.

To illustrate the importance of NMR to superionic conductivity, we consider a few examples.

1. In LiH$_2$PO$_4$ (LDP), a favored candidate for hydrogen fuel cells, the mechanism of its high protonic conductivity remains unclear. Multinuclear (^{1}H, ^{31}P, and ^{7}Li) magic angle spinning NMR is a technique in which the sample is spun at high speeds at an angle relative to the applied magnetic field vector. The spinning removes dipolar broadening of NMR lines to give a well-resolved spectrum. This magic angle is a solution of the equation $3 \cos^2 \theta - 1 = 0$ giving $\theta = 54.7°$, to be arranged between the sample (LDP) and the applied field direction in fields of up to 21.2 Tesla. Well-resolved ^{1}H NMR spectra are observed that are assignable to protons in the short and long O–H···O hydrogen bonds and a peak to physisorbed H$_2$O. The position and intensity for the H$_2$O peak depend on the H$_2$O content. Fast exchange occurs between the adsorbed H$_2$O and the O–H···O protons. ^{31}P and ^{7}Li NMR spectra and spin–lattice relaxation measurements show that the proton hopping/exchange processes involve concerted hindered rotational fluctuations of the phosphate groups. Conductivity data from adsorbed H$_2$O-controlled samples clearly suggest that the mechanism of LDP's protonic conductivity is dominantly the exchange (and hopping) of the adsorbed H$_2$O protons with the short O–H···O hydrogen bonds. This information helps modulate LDP's protonic conductivity by several orders of magnitude via controlling physisorbed water (Figure 4.80).

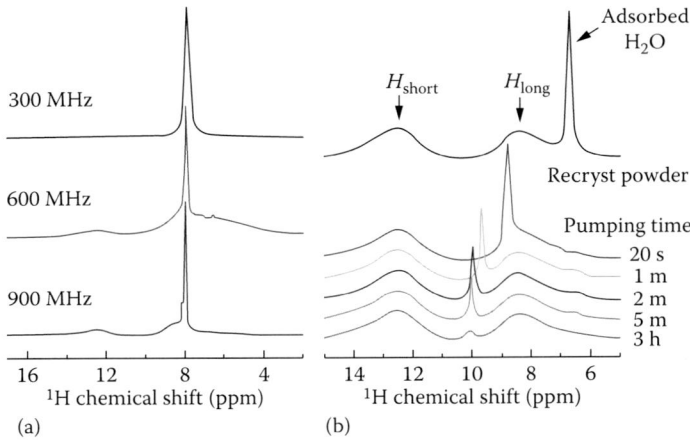

FIGURE 4.80 (a) Proton NMR spectrum of LiH_2PO_4 (LDP) shows a remarkable dependence on adsorbed water and a correlation with proton conductivity. The samples are spun at speeds of 7 kHz (top), 13.5 MHz (middle and bottom) of the recrystallized LDP powder without vacuum pumping. High field enhances resolution. (b) ^1H spectra at 600 MHz demonstrates systematic control of adsorbed water by vacuum pumping time. (From Kweon, J.J. et al., *J. Phys. Chem. C*, 118, 13387, 2014.)

2. An important NMR parameter that is sensitive to structural phase transitions in superionic conductors is the spin–lattice relaxation time. How does it arise? When a nuclear spin system is subjected to a repetitive sequence of strong radiofrequency pulses, a steady state is established. In that state, there is a dynamic balance between the effect of the pulses and spin relaxation. Under certain readily satisfied pulse conditions, the deviation of the intensity of the free induction signal from its thermal equilibrium value is an exponential function of the pulse interval. The time constant associated with this exponential is equal to the spin–lattice relaxation time (T_1).

A commonly used method to determine T_1 is the inversion recovery. Here, a pulse sequence π-t-$\pi/2$ is used. Apply a pulse of the RF field H_1 with a duration t_p, which causes the magnetization to precess at an angle $\gamma_n H_1 t_p$. Choose the pulse length (π-pulse) so as to invert or tilt the magnetization into the x–y plane (x–y plane). This precession will induce a voltage in the coil producing the FID signal. After the initial π-pulse, $M_z(t)$ may be monitored by the *amplitude* of the FID after a $\pi/2$ reading pulse at the evolution time t, which is varied in the experiment. The principle is illustrated in Figure 4.81.

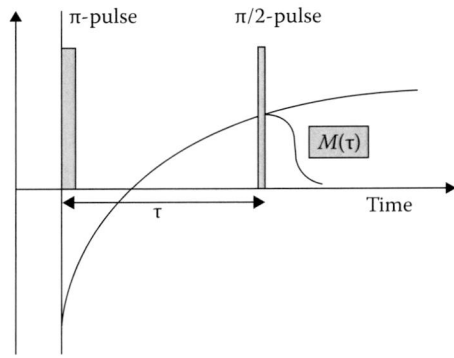

FIGURE 4.81 Inversion recovery profile to determine T_1.

Inversion recovery traces of the proton NMR and the central transition of ^{7}Li NMR follow an exponential function. Fitting the experimental data to the Abragam formula

$$\frac{S(\infty) - S(t)}{2S(\infty)} = \exp(-Wt) \tag{4.88}$$

where

    $S(t)$ is the nuclear magnetization at time t

    W is the transition probability for $\Delta m_1 = \pm 1$ transition

T_1 is given by $1/W$.

This method has been applied to ^{7}Li NMR of LiBH$_4$ [4.95]. Figure 4.82 shows the NMR spectra of LiBH$_4$ from 323 to 533 K. At 388 K, the broad spectrum suddenly narrows down becoming intense. This signals the superionic phase transition from orthorhombic phase to hexagonal phase at 390 K. Motional narrowing arises due to the fast Li$^+$ ion motion at $T > 390$ K. This transition is monitored through $T_1(T)$ shown in Figure 4.83.

Li diffusion in Li$_x$TiS$_2$ is interesting because TiS$_2$ is a layered dichalcogenide supporting 2D motion of Li ions. It is necessary to look at the Li ion motion directly (i.e., without any modeling) over an extended range of frequencies for tracking regimes from ultraslow to fast Li ion diffusion. Spin alignment echo (SAE) NMR probes extremely slow Li ion motions while classical relaxation methods can probe faster motions. We show in Figure 4.84 the remarkable result of Wikening and Heitjans spanning Li ion jump rates obtained from SAE NMR and classical NMR relaxation. For details, see the reference cited in the figure caption.

Next, we discuss electron paramagnetic resonance spectroscopy.

4.1.5.2 Electron Paramagnetic Resonance (EPR) Spectroscopy

Apart from the dynamic processes that NMR successfully probes in solid state ionics, another important aspect of these materials to be probed is the microscopic structure. In this characterization tool, namely, EPR, electronic magnetic moments present or created in a sample are aligned using a stationary magnetic field and then transitions are excited by irradiating with microwaves.

In the NMR technique discussed earlier, the nuclei of mobile ions (Li$^+$, Na$^+$, Ag$^+$)—a major component of the solid state ionic material—all possess magnetic nuclei. Such a situation exists in metals (but not in their iodides!). In Li, Na, and Ag, there are an Avogadro number of conduction electrons, which possess magnetic moments.

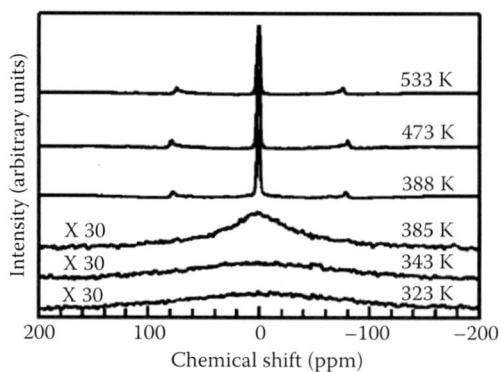

FIGURE 4.82 ^{7}Li ($I = 3/2$) NMR spectra of LiBH$_4$ at selected temperatures above and below the orthorhombic to hexagonal structural phase transition at 390 K. (From Matsuo, M. et al., *Appl. Phys. Lett.*, 91, 224103, 2007.)

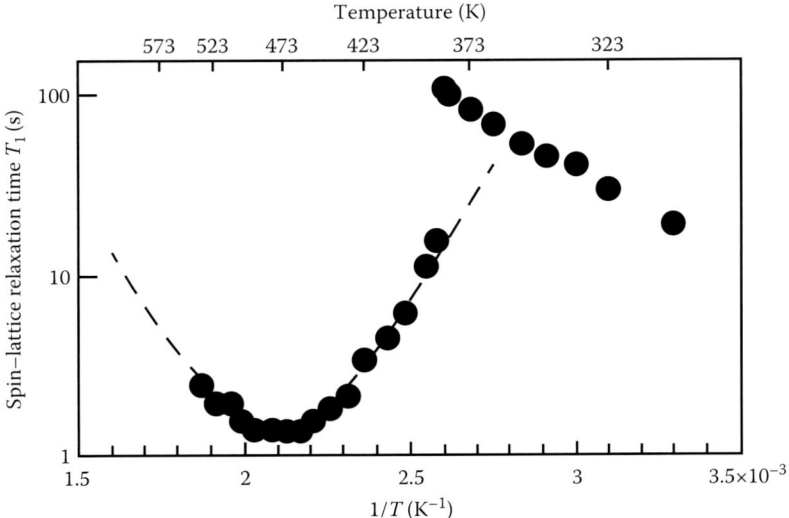

FIGURE 4.83 Temperature dependence of the spin–lattice relaxation time (T_1) of LiBH$_4$. ●: experimental points; - - -: fit to the Bloembergen–Purcell–Pound (BPP) formula for spin–lattice relaxation rate $T_1^{-1} = C[\tau/(1 + \omega^2\tau^2) + 4\tau/(1 + 4\omega^2\tau^2)]$. C is a constant, ω is Larmor frequency, and τ is correlation time for Li$^+$ motion in LiBH$_4$. Assuming $\tau = \tau_0\exp(E_A^{NMR}/k_BT)$ where E_A^{NMR} is activation energy for Li-ion motion and τ_0 the preexponential factor, a straight line is obtained whose slope ($E_A^{NMR} = 0.56$ eV) is in very good agreement with conductivity measurements. (From Matsuo, M. et al., *Appl. Phys. Lett.*, 91, 224103, 2007.)

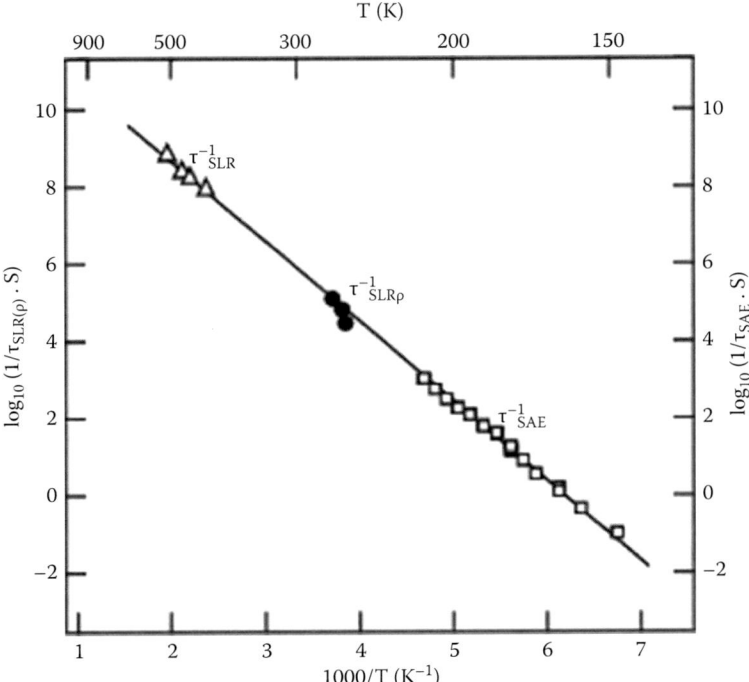

FIGURE 4.84 Li jump rates for Li$_{0.7}$TiS$_2$ plotted against $1000/T$ (K^{-1}). The τ^{-1}_{SAE} rates (□) were measured directly by means of SAE NMR. The rates τ^{-1}_{SLR} (△) and $\tau^{-1}_{SLR\rho}$ were deduced from the maxima positions of the corresponding rate peaks $T_{1diff}^{-1}(1/T)$ and $T_{1\rho diff}^{-1}(1/T)$, respectively. T_{1diff}^{-1} relaxation peaks were recorded at 10, 19.2, 27.9, and 77.7 MHz. $\tau_{SLR\rho}^{-1}$ rates were deduced from NMR spin–lattice relaxation measurements in the rotating frame at $\omega_{1eff}/2\pi = 2.1$, 5.2, and 10 kHz, respectively. The solid line yields an activation energy $E_A = 0.41$ eV and $\tau_0^{-1} = 6.3$ THz. (From Wilkrning, M. and Heitjans, P., *Phys. Rev. B*, 77, 024311, 2008.)

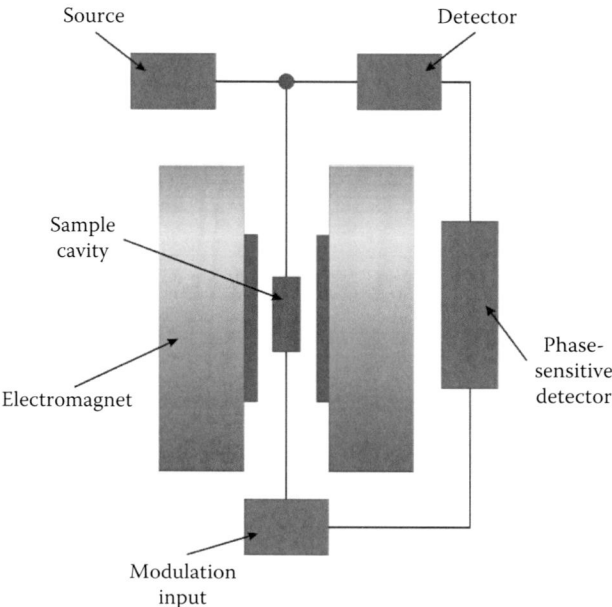

FIGURE 4.85   A typical cw EPR spectrometer operating at X-band or ~9.5 GHz (microwaves of wavelength ~32 mm). The radiation source is a klystron. A stable high-power microwave source with low-noise characteristics and thus giving high sensitivity. The sample is placed in a resonant cavity that admits microwaves through an iris. The cavity is located in the middle of an electromagnet and helps to amplify the weak signals from the sample. Most of the external components, such as the source and detector, constitute the microwave bridge control. Attenuator, field modulator, and amplifier help enhance the performance of the instrument.

Thus, EPR or ESR as a characterization tool in solid state ionics basically relies on the existence of localized paramagnetic impurities created by chemical doping and defects created by annealing or reduction/oxidation, which are not natural constituents unlike in NMR. These are often part of the semi-rigid lattice through which mobile ions move. Thus, EPR gives less information about motion than NMR [4.97]. Let us discuss this technique briefly with the aid of a few examples.

In EPR or ESR too, there is a Larmor precession of electron magnets ($S = 1/2$) as in an H atom (1s electron), which possess a degenerate ground state with $m_S = \pm 1/2$. This is removed by the application of a DC magnetic field (H_0) of about 4 T (for the so-called X-band ESR). Resonance absorption of microwave radiation (~9 GHz frequency) takes place when the sample bathed in a microwave magnetic field (H_1) normal to H_0 satisfies the resonance condition ν (GHz) = 2.80 H_0 (in kG) = 28.0 H_0 (in T).

EPR is done using a continuous wave or CW (the sample receives microwave radiation continuously) spectrometer schematically shown in Figure 4.85.

An EPR spectrum is a plot of the derivative of the microwave power absorbed by the sample versus the DC magnetic field as the latter is slowly scanned through resonance. The resonance conduction is

$$h\nu = g\beta H_0 \tag{4.89}$$

where g is the g-factor or the spectroscopic splitting factor—a numerical measure of the extent to which electron spins in the sample are free. The most important interaction is the electron Zeeman interaction involving the electron spin (S) and the applied field that gives the g-factor or g-tensor. An interaction of the electron magnetic moment and nuclear magnetic moment produces the so-called hyperfine interaction, which determines the microscopic symmetry of the paramagnetic complex. For paramagnetic ions such as Fe^{3+} (or Mn^{++}) with the $3d^5$ outer electron configuration, there is an "electronic splitting" or zero field splitting, which is sensitive to the crystal structure of the superionic conductor (say, zirconia). These interactions may be represented in the form of a spin Hamiltonian convenient for the analysis of experimental EPR spectra:

$$H_{EPR} = \beta SgH + SAI + SDS \qquad (4.90)$$

As EPR spectra are mostly obtained on polycrystalline samples, spectral simulation is usually employed to derive the g-, A-, and zero-field splitting parameters D and E, which occur as components of the D-tensor using SIMFONIA program of Bruker.

We now illustrate the EPR technique through application to (1) zirconia and (2) Mn-doped $Li_4Ti_5O_{12}$.

1. a. Zirconia synthesized by the sol–gel route based on Zr propoxide and subsequently subjected to oxidation, reduction UV irradiation, and oxygen adsorption are investigated by EPR. The results are displayed in Figure 4.86 and Table 4.8.

   Spectrum c is recorded at a reduced magnification.

   b. Fe^{3+} doped into zirconia prepared by calcination of zirconium hydroxide has yielded amorphous, tetragonal, and monoclinic zirconia in which the microscopic environments of Fe^{3+}

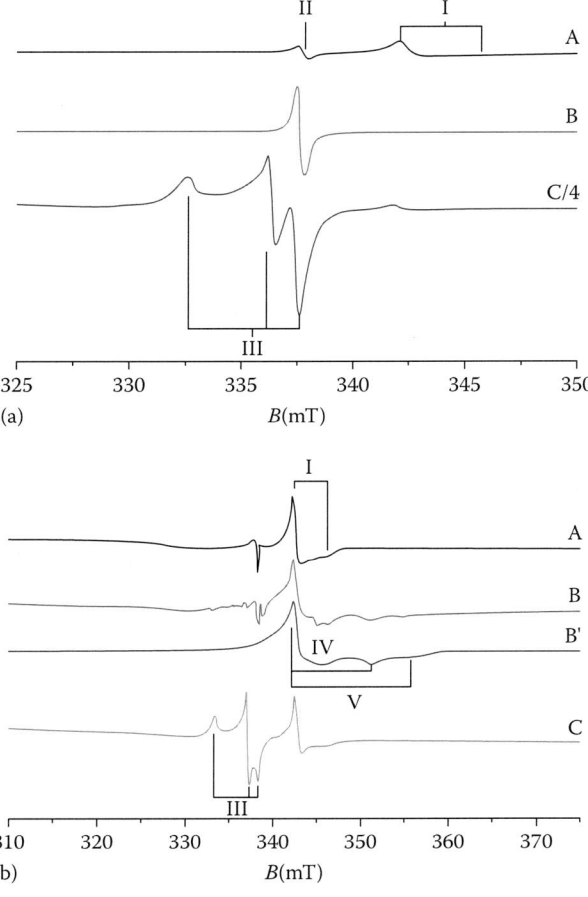

FIGURE 4.86 (a) EPR of zirconia recorded at 77 K temperature. (A) As-prepared sample after activation, (B) as-prepared sample after reduction by vacuum annealing at 873 K, and (C) after contact with oxygen at 400 K. (b) EPR of zirconia recorded at 77 K after (A) oxidative activation, (B) after UV irradiation in H_2 (B′) computer simulation of (B), and (C) after adsorption of O_2 at room temperature. (From Gionco, C. et al., *Chem. Mater.*, 25, 2243, 2013.)

TABLE 4.8

g-Factors and Assignments for Observed EPR Signals in ZrO_2 after Different Treatments

Signal	g_{xx}	g_{yy}	g_{zz}	Assignment
I	1.9768		1.9589	Bulk Zr^{3+}
II	2.0024 (isotropic signal)			Symmetric trapping site
III	2.0025	2.0096	2.033	Surface adsorbed O_2^-
IV	1.9784		1.9288	Surface Zr^{3+}
V	1.9784		1.9060	Surface Zr^{3+}

Source: Gionco, C. et al., *Chem. Mater.*, 25, 2243, 2013.

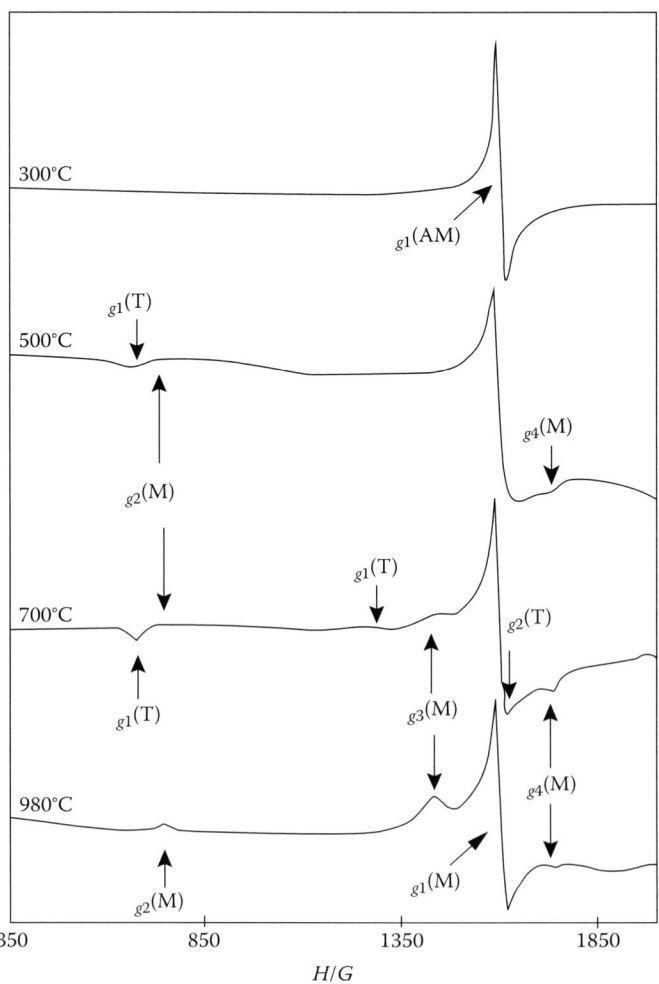

FIGURE 4.87 EPR of Fe^{3+} ions in Fe-doped zironica at 77 K, obtained from calcination of zirconium hydroxide at different temperatures. AM: amorphous, M: monoclinic, T: tetragonal. These spectra demonstrate the sensitivity of zero-field splitting with respect to crystal structure. (From Matta, J. et al., *Phys. Chem. Chem. Phys.*, 1, 4975, 1999.)

TABLE 4.9

Zero-Field Splitting Parameters for Polymorphs of Fe-Doped Zirconia

ZrO_2 Polymorph	$\lvert E/D \rvert$	$\lvert D \rvert$ (cm^{-1})	g_{eff}
Monoclinic (M)	0.31–0.32	~1.7	$g_1(M) = 4.27$, $g_2(M) = 9.2$, $g_3(M) = 4.7$, $g_4(M) = 3.9$
Tetragonal (T)	0.00–0.01	~0.3	$g_1(T) = 10.2$, $g_2(T) = 5.4$, $g_3(T) = 4.15$
Amorphous (AM)	0.333	$\gg$0.3	$g(AM) = 4.27$

Source: Matta, J. et al., *Phys. Chem. Chem. Phys.*, 1, 4975, 1999.

are quite different [4.99]. EPR of Fe^{3+} displayed in Figure 4.87 is found to be very sensitive to the crystallographic phase. A quantification of this sensitivity is through the zero-field splitting parameters E/D and D which clearly distinguish between the amorphous, tetragonal, and monoclinic (Table 4.9).

2. Mn doped $Li_4Ti_5O_{12}$. $Li_4Ti_5O_{12}$ is an important Li battery anode. EPR of an Mn impurity in this spinel lattice helps understand the role of defects in electrochemical phenomena at the anode-electrolyte interface. An important method to increase resolution and sensitivity in EPR is to work at two or three frequencies higher than the usual X-band frequency. Figure 4.88 illustrates this for the case of different concentrations of Mn doped into spinel structured $Li_4Ti_5O_{12}$. The most important results are the following. For $x = 0.01$, Mn^{++} ($S = 5/2$) ions fully substitute Ti^{4+} sites creating oxygen vacancies for charge compensation. Analysis of W-band spectra indicates the presence of Mn^{3+} ($S = 2$) and Mn^{4+} ($S = 3/2$) species in the $Li_4Ti_5O_{12}$ indicating spin–spin interactions.

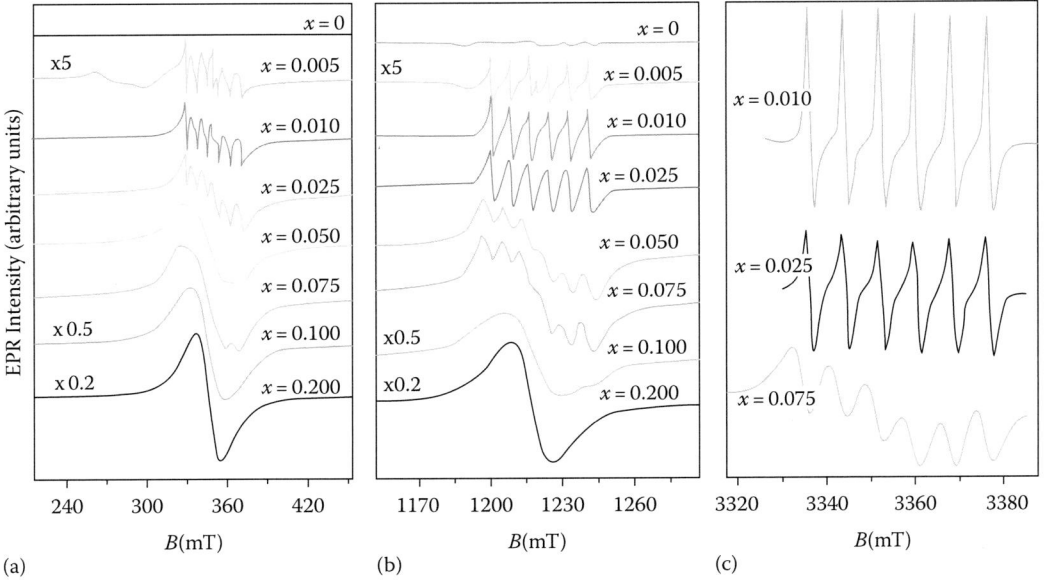

FIGURE 4.88 Room-temperature (a) X-band (9.47 GHz), (b) Q-band (34.20 GHz), and (c) W-band (94.49 GHz) EPR spectra (c) of Li4Ti5O12 samples with different Mn concentrations (x). First-derivative (with respect to the external magnetic field) EPR data are presented. EPR scanning parameters: microwave power, 1 mW; modulation amplitude, 0.2 mT; modulation frequency, 100 kHz; sweep time, 83.8 s; and number of scans 10. (From Kaftalen, H. et al., *J. Mater. Chem. A*, 1, 9973, 2013.)

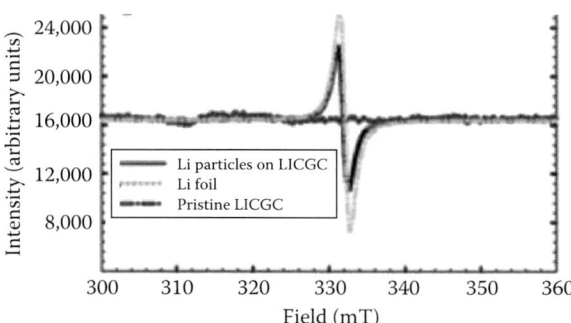

FIGURE 4.89 Electron paramagnetic resonance (EPR) spectra for as-received LICGC (dash-dot line), LICGC (or Li-ion conducting glass ceramic with the general composition $Li_2O–Al_2O_3–SiO_2–P_2O_5–TiO_2–GeO_2$) with Li particles reduced in an Ar-filled glove box (solid line), and Li foil (dotted line). (From Kumar, A. et al., *Sci. Rep.*, 3, Article number:1621.)

To conclude this section, we look at the dynamic application of EPR to the case of reduction of a Li-ion conducting glass ceramic leading to the formation and detection of Li metal particles (Figure 4.89).

We next discuss Mossbauer spectroscopy as perhaps the most sensitive probe for characterization, especially for superionic conductors with Fe (and Sn) as a component.

4.1.5.3 Mossbauer Spectroscopy or Nuclear Gamma Resonance

Mossbauer spectroscopy is a very sensitive characterization technique for solid state ionics based on the Mossbauer effect discovered by Rudolf Mossbauer. It is the recoil-free, resonant absorption and emission of gamma rays in solids.

Like NMR spectroscopy, Mossbauer spectroscopy probes tiny changes in the energy levels of an atomic nucleus in response to its environment. Typically, three types of nuclear interactions may be observed: an isomeric shift, also known as a chemical shift; quadrupole splitting; and magnetic or hyperfine splitting, also known as the Zeeman effect. Due to the high energy and extremely narrow line widths of gamma rays, Mossbauer spectroscopy is a very sensitive technique in terms of energy (and hence frequency) resolution, capable of detecting change in just a few parts per 10^{11}.

Mossbauer essentially showed that gamma photons impinging upon nuclei of atoms in a crystal lattice could excite transitions. In ways that would be forbidden *were the lattice not present*. Free nuclei cannot be excited into their higher energy states by an incoming radiation because the photon absorbed by a nucleus gets both energy and momentum. Conservation laws demand the incoming photon needs to have more energy than required for nuclear excitation. If this excess energy is greater than $\hbar$ divided by the lifetime of the excited state (because of the Heisenberg energy-time uncertainty principle), there would not be any resonant excitation of the nucleus! Mossbauer found that *a gamma ray photon can excite nuclear states if the nuclei sit in a lattice*. The lattice that binds the nuclei takes up the momentum of the photon leaving the energy free to go right to the nuclear excitation. The theory of this process is similar to that of neutron scattering. Here, we need to calculate the probability of exciting the nuclear state by an incoming photon. If q is the wave vector of the incoming photon, then if the nuclear excited state involves an energy $\Delta\varepsilon$ and has a lifetime Γ, the strength of the nuclear gamma resonance or Mossbauer absorption is given by the so-called Breit–Wigner form factor

$$\frac{2\Gamma}{\left(\Delta\varepsilon/\hbar\right)^2 + \Gamma^2} \tag{4.91}$$

multiplied by the analog of the Debye–Waller factor, now called the Lamb–Mössbauer factor

$$f = \exp\left[\frac{-(3/2)q^2\hbar^2}{2M\hbar c k_D}\right] \tag{4.92}$$

where

M is the mass of the atom

k_D is the Debye wave vector

In the Mossbauer spectroscopy experiment, the solid sample is exposed to a beam of gamma radiation, and a detector measures the intensity of the beam transmitted through the sample. The atoms in the source emitting the gamma rays must be of the same isotope as the atoms in the sample absorbing them.

If the emitting and absorbing nuclei were in identical chemical environments, the nuclear transition energies would be exactly equal and resonant absorption would be observed with both materials at rest. The difference in chemical environments, however, causes the nuclear energy levels to shift in a few different ways, as described later. These energy shifts are tiny (≤ 1 μeV). The extremely narrow spectral line widths of gamma rays for some radionuclides make the small energy shifts correspond to large changes in absorbance. To bring the two nuclei back into resonance, the energy of the gamma ray must be slightly changed, and this is done using the Doppler effect.

To apply the method, the source is accelerated through a range of velocities using a linear motor to produce a Doppler effect and scan the gamma ray energy through a given range.

In the resulting spectra, gamma ray intensity is plotted as a function of the source velocity. At velocities corresponding to the resonant energy levels of the sample, a fraction of the gamma rays are absorbed, resulting in a drop in the measured intensity and a corresponding dip in the spectrum. The number, positions, and intensities of the peaks or dips in absorbance/transmission provide information about the chemical environment of the absorbing nuclei and can be used to characterize the sample. There are three experimental parameters of interest: (1) isomer shift, (2) quadrupole splitting, and (3) magnetic hyperfine interaction (important for nuclei such as ^{57}Fe). These parameters help assess the chemical environment around, and magnetic interactions and phased transitions in the solid state ionic material of interest when the Mossbauer spectrum is studied as a function of concentration and temperature [4.101] (Figure 4.90).

Mossbauer effect has been observed in solid state ionic materials including AgI and $LiFePO_4$ through I and Fe resonance, respectively. Figure 4.91a shows the ground and the first excited states of the most studied Fe^{57} nucleus.

Transitions in Mossbauer spectra between appropriate levels are governed by the selection rules $\Delta I = 1$ and $\Delta m_I = 0, \pm 1$. Experimental spectra are recorded as gamma counts versus the velocity of the Mossbauer drive.

Figure 4.92a shows the Mossbauer spectrum of ~30 nm $LiFePO_4$ nanoparticles at 300 K recorded under high gamma count rate. This vouches for the purity of the sample by way of absence of Fe^{3+} ions in the sample. It is instructive to compare this with the spectrum of Figure 4.92b, which serves to distinguish between Fe^{2+} and Fe^{3+}, the latter as an undesirable impurity in $LiFePO_4$.

Figure 4.93a shows the ^{129}I ($I = 7/2$, half life 1.6×107 years) Mossbauer spectrum of AgI nanoparticles at a very low temperature (13 K).

Figure 4.94 shows the ^{119}Sn ($I = 1/2$) Mossbauer spectra of $PbSnF_4$ and $Pb_{0.7}Sn_{0.3}F_4$ at 298 K. Observe the quadrupole splitting in the two cases. Table 4.10 lists the Mossbauer parameters.

In Chapter 7, we describe the electron relaxation study of $LiFePO_4$ through the observation of magnetic phase transition by Mossbauer spectroscopy.

We now discuss briefly characterization of solid state ionic devices.

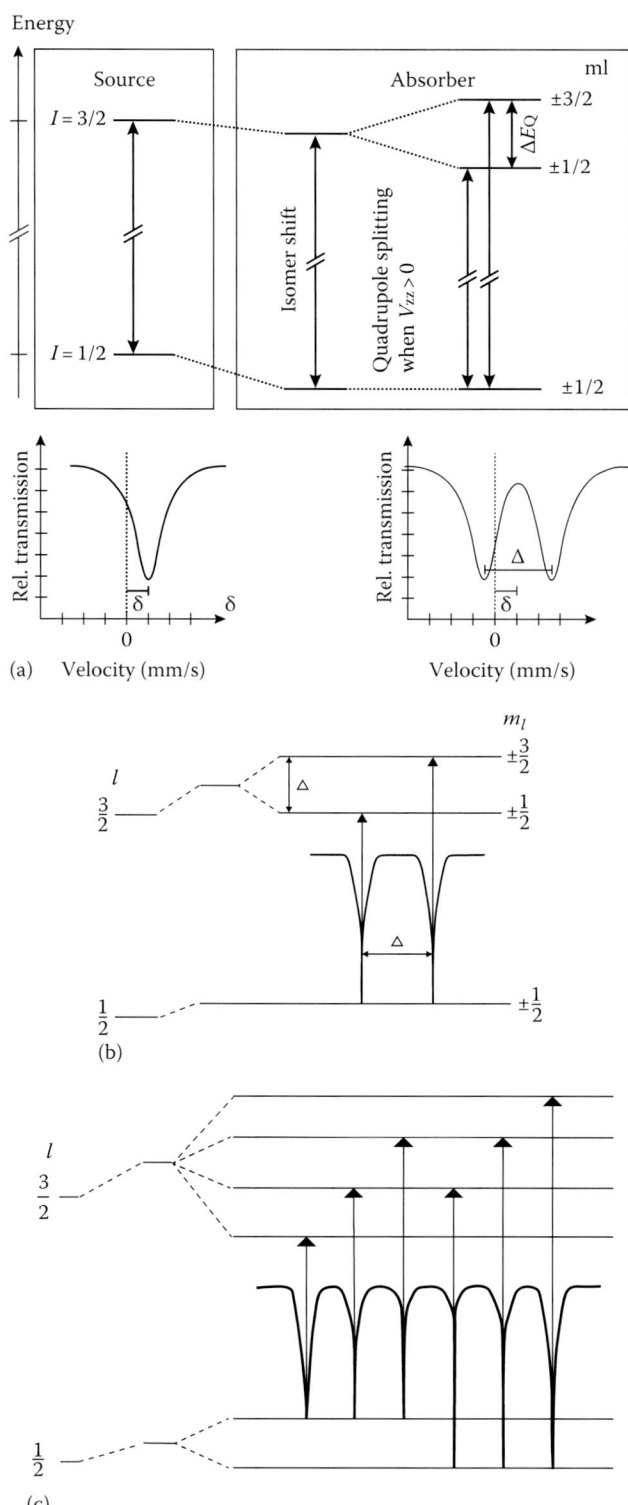

FIGURE 4.90 Illustration of Mossbauer parameters. (a) Isomer shift of an $I = 3/2$ nucleus. (b) Quadrupole splitting of an $I = 3/2$ nucleus (Fe^{57}, Sn^{119}). (c) Magnetic hyperfine splitting of Fe^{57} ground state is a doublet m_1 +1/2 (top), −1/2 (bottom). Excited state is a quartet with m_1 (from top = −3/2, −1/2, +1/2, +3/2).

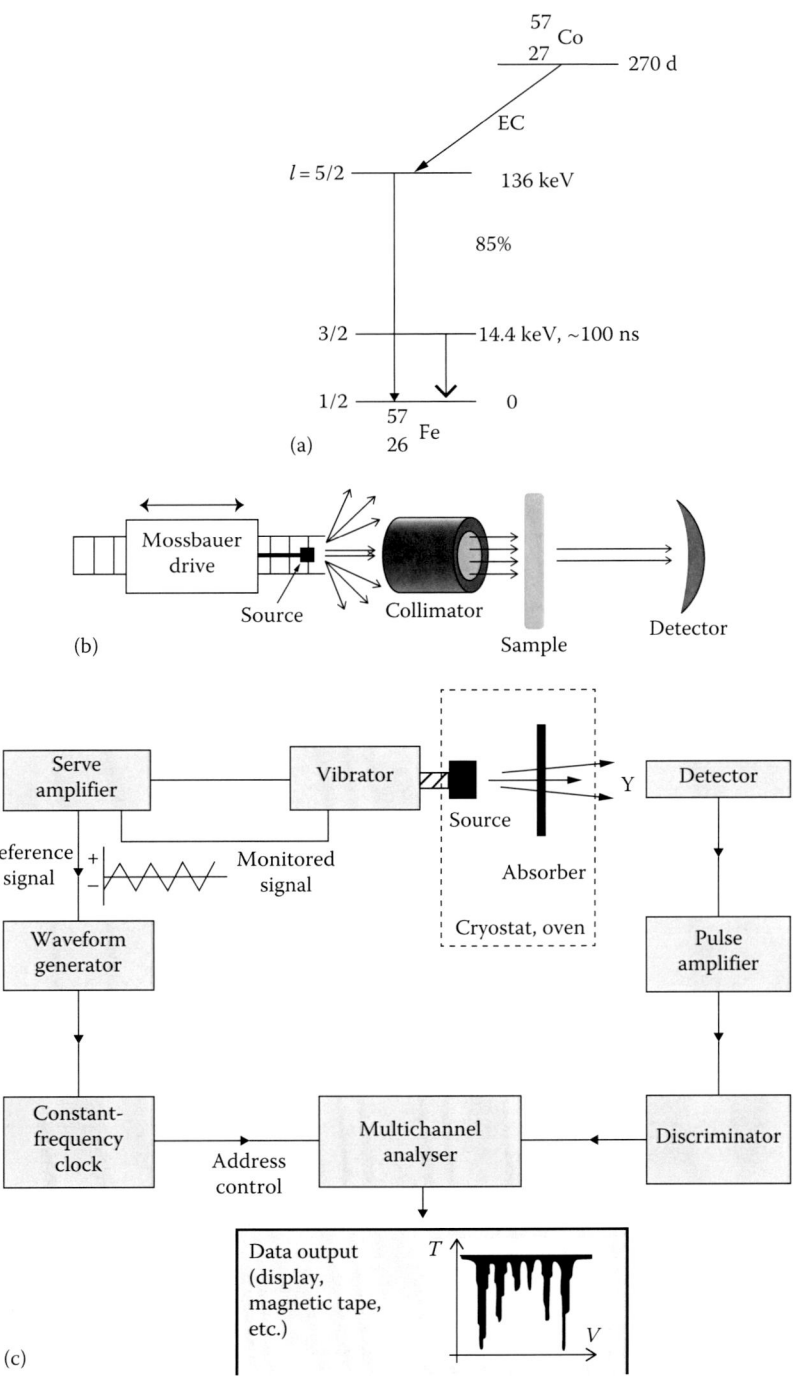

FIGURE 4.91 (a) Fe[57] nuclear decay scheme for Mossbauer spectroscopy. (b) Principle. Mossbauer drive moves the Mossbauer source at velocities in a preset range and the collimator produces a well-defined gamma-ray beam, which is incident on the sample. The Mossbauer effect produces a beam that is received by the detector. (c) Mossbauer spectroscopy instrumentation converts the output of spectrometer into a counts-versus-velocity display, which is analyzed for determining the Mossbauer parameters. (From Santhanam, V., Mossbauer spectroscopy—Principles and applications, SCSVMV University, Kanchipuram, India, 2013.)

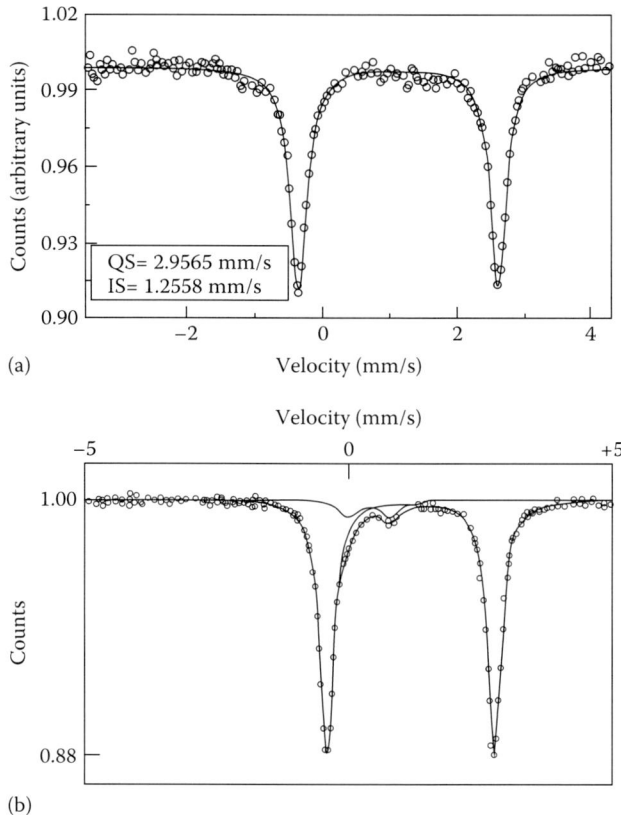

(a)

(b)

FIGURE 4.92 (a) Mossbauer spectrum of LiFePO$_4$ nanoparticles (~30 nm diameter) at 300 K. Circles are experimental points; full curve is a fit to Lorentzian lineshape of the type $I(\nu) = (\Gamma^2/4)[A/\{(\nu - \delta)^2 + \Gamma^2/4\}]$, where Γ = width, δ is the position, and A is the area, and ν is the channel number. The doublet spectrum is due to Fe^{++}, no apparent trace of Fe^{3+}. Parameters isomer shift (IS) and quadrupole splitting (QS) are noted. (From Sundarayya, Y. et al., *Mater. Res. Bull.*, 42, 1942, 2007.) (b) Mossbauer spectrum of LiFePO$_4$ at room temperature. The intense doublet is the same as in (i) but the minor doublet with IS = 0.47 mm/s and QS = 0.74 mm/s are characteristic of Fe^{3+} impurity. This demonstrates the sensitivity of Mossbauer spectroscopy as a diagnostic tool. MOSFIT program of Juraszek, Taillet and Varret (unpublished) was used to fit the experimental data. (From Prince, A.A.M. et al., *Solid State Commun.*, 132, 455, 2004.)

4.2 Devices

The battery and the fuel cell as well as the sensor and the supercapacitor are not only developmental challenges but also characterization opportunities. Dynamic or *in situ* characterization of devices has been initiated. The "half cell" of the lithium-ion battery represents the sample configuration for electrical and allied characterization of solid state ionics.

4.2.1 Electrochemical Characterization of the Li-Ion Battery

Electrochemical testing or characterization involves electrode preparation. The electrode is usually in the form of a foil or sheet. This ensures ease of handling and optimal electrochemical performance. The starting electrode material in the form of powder of <~30 µm particle size is cast onto a thin sheet metal substrate. The latter also serves as a current collector. A polyvinyl difluoride (PVDF) polymer is used as a binder between the particles and the substrate. As practical solid state ionic materials are semiconductors, carbon may be used as an electron conductive agent. Two types of carbon, flaky and spherical particulate, are used. While the flaky carbon is a thin plate forming an electrical network within the electrode, the spherical

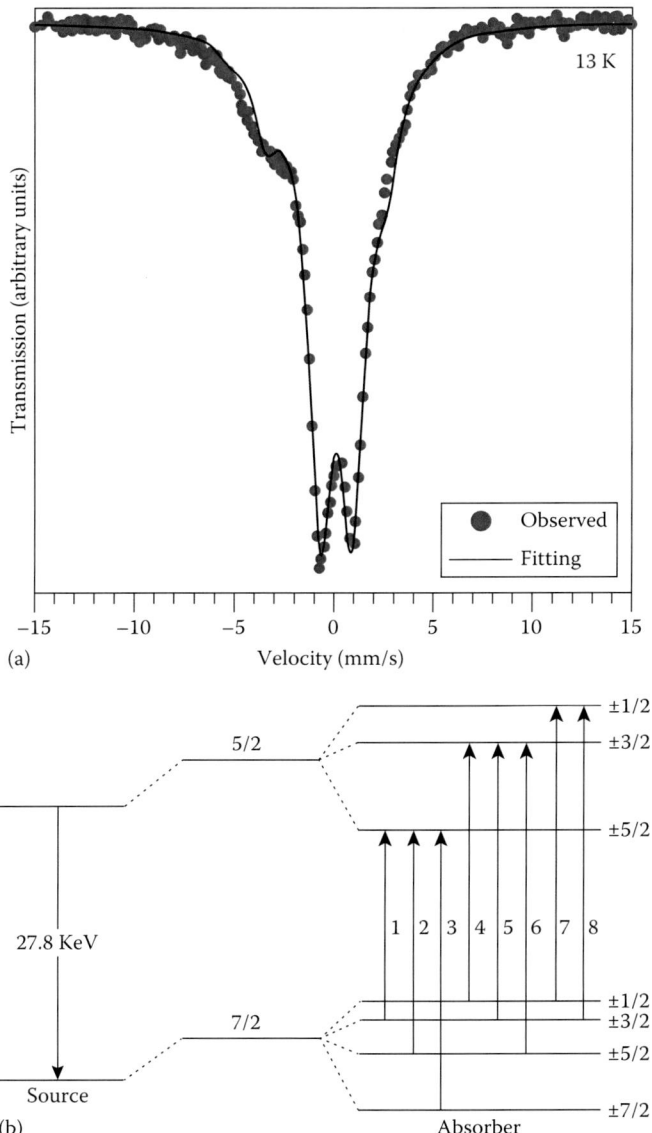

FIGURE 4.93 (a) Mossbauer spectrum of radioactive [129]I in AgI nanoparticles at 13 K. The resolved, quadrupole-split spectra arise due to the large quadrupole moment (−0.55 barn) of the [129]I nucleus. (b) Energy levels of [129]I ground-state nuclear spin is 7/2 and first excited state at 27.8 keV has I = 5/2. Quadrupole moment ratio of excited to ground state is 1.231 and half life 16.8 ns correspond ting to a natural line width of 0.29 mm/s or 6.7 MHz. (From Ohkubo, Y., Project research on science and engineering of unstable nuclei and their uses on condensed matter physics. Research Reactor Institute, Kyoto University, 2012.)

carbon acts as a dispersion agent in order to realize a homogeneous electrode. Carbon would not interfere with the intercalation/deintercalation process because carbon will not intercalate Li ions below 0.1 V. An optimal loading between the amounts of active material and carbon in the electrode makes the electrode active. This loading depends on the particle size and surface area of the active material (Figure 4.95).

The active, nonactive, and binder materials may be mixed using mortar and pestle or mild ball-milling. A slurry results when liquid n-methyl pyrrolidone (NMP) is added to this mixture. NMP acts as a solvent for PVDF binder and evaporates out of the mixture under mild heat. The amount of NMP to be used depends on the amount of carbon and the particle size of the active material. The more viscous the slurry, the thinner the electrode film. How does one cast an electrode film? The doctor-blade technique

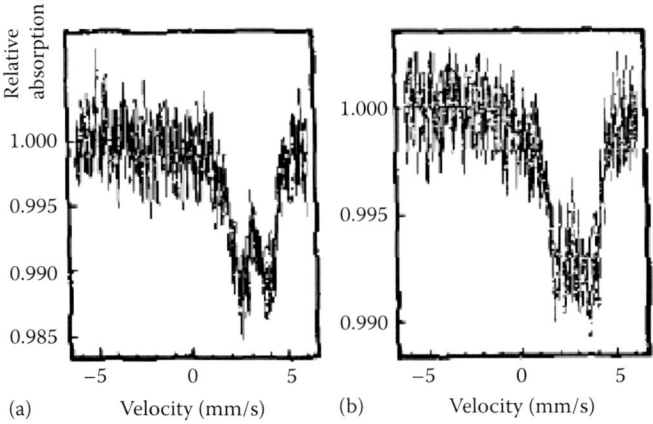

(a)   Velocity (mm/s)   (b)   Velocity (mm/s)

FIGURE 4.94   (a) ^{119}Sn Mossbauer spectra of β-PbSnF$_4$. (b) Pb$_{0.7}$Sn$_{0.3}$F$_2$ at 298 K. (From Denes, G. et al., *J. Solid State Chem.*, 104, 239, 1993.)

TABLE 4.10

Mossbauer Parameters of Pb–Sn–F Compounds

Compound	IS δ^a	QS Δ^b (mm/s)	Line Width (Γ)
Pb$_{0.70}$Sn$_{0.30}$F$_2$	3.07 (2)	1.64 (2)	0.76 (7)
Pb$_{0.60}$Sn$_{0.40}$F$_2$	3.25 (2)	1.69 (3)	0.72 (5)
β-PbSnF$_4$	3.06 (2)	1.66 (2)	0.68 (3)

Source: Denes, G. et al., *J. Solid State Chem.*, 104, 239, 1993.
a   Relative to CaSnO$_3$ as zero shift as reference.
b   5p^x with some likely contribution from 5d orbital. Sn and Pb are disordered in the F$_8$ fluoride cubes.

is helpful. A predetermined thickness of the blade to the metal substrate is to be used. The slurry when poured from a beaker onto the substrate forms a small slurry pool in front of the blade. The blade is then pushed toward and past the slurry. A thin film of slurry is formed on the substrate. The as-cast film is placed in an oven at 120°C to remove the NMP. To ensure better contact between active material and substrate, compression of the foil would be necessary. Overall, a 35% porosity achieves active material to current collector contact and adequate electrolyte-to-active material contact.

The potential stability of Li metal makes it an ideal reference electrode. Additionally, Li metal is a large Li source for the cell and thus serves as a counter electrode too. Thus, all measured potentials for a test material may be measured against the potential of Li metal.

Electrochemical testing is performed using a MACCOR battery tester (a typical model is 4304). This device is programmable and can collect and store performance data. This device contains multi-channels where several cells may be tested at a time. Additionally, the accuracy of the current applied to the cell is specific to a unique channel. Tests can also be carried out by using an electrochemical galvanic/potentiostat device.

This device can provide current (I)-potential (V) plots as well as preforming impedance tests.

Cycling is an important aspect of thin-film battery testing. It may be carried out at constant current between specified voltage limits using two electrometers (say, Keithley) operated under computer control. At the end of each half cycle, the voltage must be held constant until the current decreases to a specific fraction of the charge usually 10%.

Cyclic voltammetry is a basic characterization tool for the Li-ion battery. It is illustrated in Figure 4.96.

The impedance of the cells may be measured at ambient temperature at frequencies from 0.01 to 10 MHz using an impedance bridge. These impedance measurements are carried out at different cell potentials during both charge and discharge cycles. A DC bias equal to the open circuit voltage (OCV) is applied to the cells during the impedance measurements. The AC voltage is 50 mV or less. The OCV

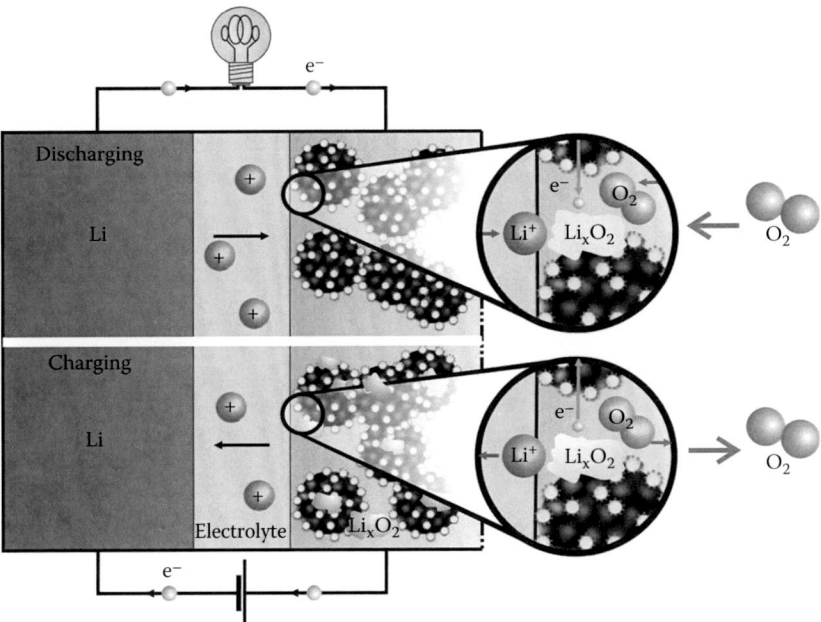

FIGURE 4.95 Schematic illustration of a rechargeable Li-ion–air electrochemical cell. This electrochemical device directly converts fuels to electricity carried by lithium ions at room temperature. It depends on oxygen reduction reaction and its reverse reaction, namely oxygen evolution reaction. (From MIT Electrochemical Energy Lab, Low-temperature electrocatalysis, 2012. http://web.mit.edu/eel/images/EELresearch/LowTFig1b.jpg.)

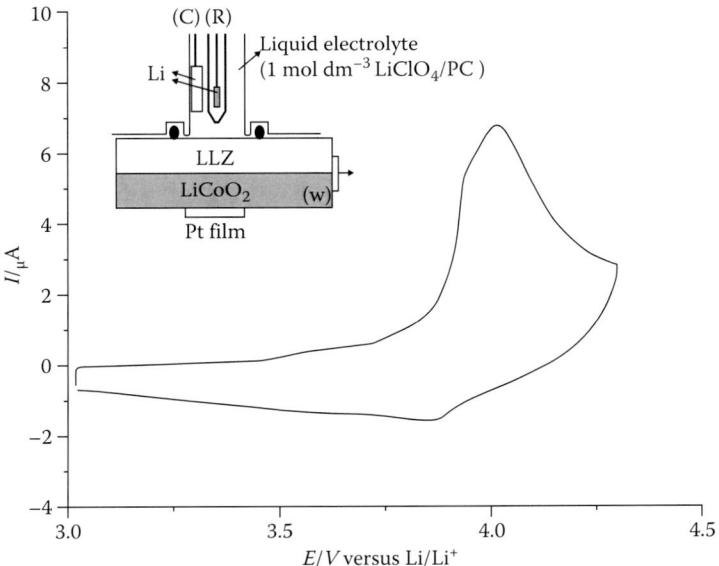

FIGURE 4.96 Cyclic voltammogram characterization of an all-solid state battery: LLZ/LiCoO$_2$ half-cell measured between 3.0 and 4.2 V at room temperature with a scanning rate of 0.1 mV/s. The inset shows the electrochemical cell used to measure these CVs. (From Kim, K.H. et al., *J. Power Sources*, 196, 764, 2011.)

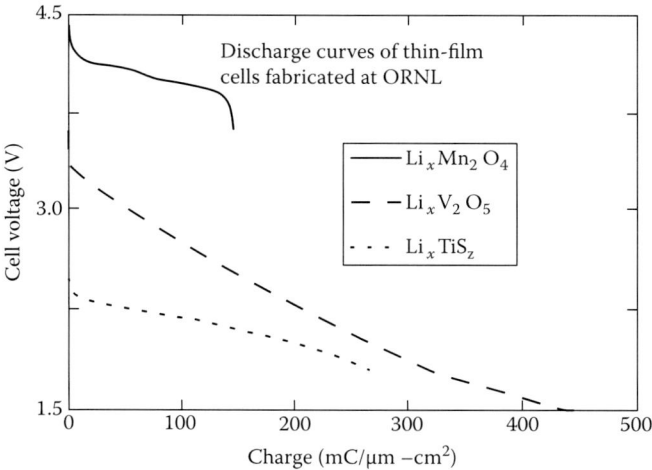

FIGURE 4.97   Discharge curves for Li-TiS$_2$, Li-V$_2$O$_5$, and Li-LiMn$_2$O$_4$ thin-film batteries. (From Bates, J.B., *Solid State Ionics*, 70/71, 619, 1994.)

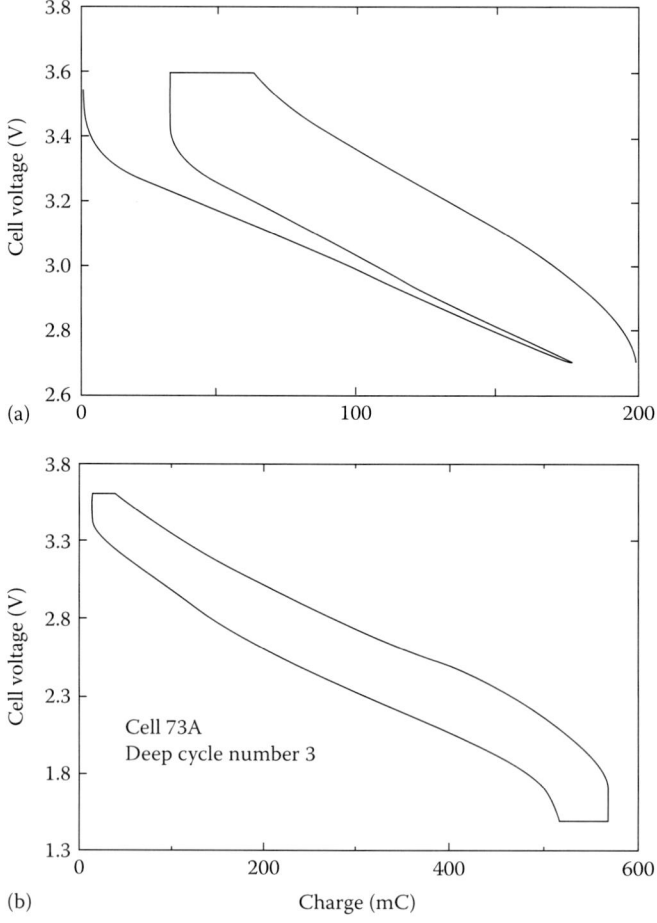

FIGURE 4.98   (a) First discharge to 2.7 V and first few cycles of a Li-V$_2$O$_5$ cell, and (b) deep discharge. (From Bates, J.B., *Solid State Ionics*, 70/71, 619, 1994. Reproduced with permission from Elsevier.)

is measured before and after the impedance measurements agree to within a few mV. Typical discharge curves of thin films using three cathodes are shown in Figure 4.97.

Important cell testing events are (a) first discharge and first few cycles and (b) deep discharge. These are shown in Figure 4.98.

Next, we focus on characterization of supercapacitors.

4.2.2 Electrochemical Characterization of Supercapacitors

Supercapacitors are electrochemical storage devices whose range of energy and power densities can complement the function of batteries and conventional capacitors for storing energy and delivering power. They are systems intermediate between dielectric capacitors and batteries. While batteries are able to store higher energy density than supercapacitors, they deliver less power; as compared to dielectric capacitors, supercapacitors can store higher energy density with less delivered power. Supercapacitors have potential application as electrical storage devices operating in parallel with the battery in an electric vehicle to *transiently* provide *high power*. Three main classes of supercapacitors are metal oxide, electronically conducting polymer, and carbon/carbon supercapacitors.

The most common types of supercapacitor electrodes include activated carbon (AC), and the carbon-based supercapacitor is widely developed because of its low cost, large capacitance, and long cycling life. The electrostatic charges stored at the AC electrode/electrolyte double-layer interface produce a high capacitance because of the highly specific surface area of the AC electrodes.

However, a relatively low electronic conductivity of the AC results in the supercapacitor having a high equivalent series resistance (ESR) and hence lower specific energy (E) and power (P) during charge and discharge. This shortcoming can be resolved by mixing AC with carbon nanotubes (CNTs), which have a higher electronic conductivity [3, 4]. Because the CNTs have a lower specific capacitance (C_{sp}) than the AC, the mixture was subjected to physical/chemical activation in an attempt to increase the surface area and specific capacitance (C_{sp}) of the electrode.

Binderless AC monolith (ACM) electrodes are fabricated from the mixture of self-adhesive carbon grains (SACG) from oil palm empty fruit bunch (EFB) fibers, CNTs, and potassium hydroxide (KOH). Chemical reactions between hydroxides (KOH or NaOH) and carbon during the activation process play an important role in the porosity development of the activated carbon. Green monoliths (GMs) of the aforementioned mixtures were carbonized in a N_2 environment up to various carbonization temperatures to produce carbon monoliths (CMs) prior to the CO_2 activation process used to produce the ACMs. The physical activation is usually carried out by treating the carbonized sample within an oxidizing gas atmosphere (such as CO_2 and water vapor) under a moderately high temperature (800°C–100°C) to improve the internal porosity.

The performance of the supercapacitor cells using ACMs as their electrodes is evaluated by electrochemical impedance spectroscopy (EIS), cyclic voltammetry (CV), and galvanostatic charge–discharge (GCD) methods. An electrochemical instrument interface (Solartron SI 1286 and Solartron 1255HF frequency response analyzer) is used for this purpose.

All of the measurements are carried out at room temperature (25°C). Using the EIS data, the C_{sp} of the electrodes was determined using the equation

$$C_{sp} = \frac{-1}{\pi f_1 Z_1'' m} \tag{4.93}$$

where
f_1 is the lowest frequency
Z_1'' is the imaginary impedance at f_1
m is the mass of the electrode

The EIS data as a function of the frequency may be analyzed using the following three equations:

$$C(\omega) = C'(\omega) - jC''(\omega) \tag{4.94}$$

$$C''(\omega) = \frac{Z'(\omega)}{\omega|Z(\omega)|^2} \tag{4.95}$$

$$C'(\omega) = \frac{-Z''(\omega)}{\omega|Z(\omega)|^2} \tag{4.96}$$

here $Z(\omega)$ is equal to $1/j\omega C(\omega)$, $C'(\omega)$ is the real capacitance, $C''(\omega)$ is the imaginary capacitance, $Z'(\omega)$ is the real impedance, and $Z''(\omega)$ is the imaginary impedance. From the voltammograms, the C_{sp} of the electrodes may be determined using Equation 4.97:

$$C_{sp} = \frac{2i}{Sm} \tag{4.97}$$

where
  i is the electric current
  S is the scan rate
  m is the mass of electrode

From the GCD data (charge–discharge curve) recorded at a selected current density, the C_{sp} of the electrodes may be determined using the following equation:

$$C_{sp} = \frac{2i}{(\Delta V/\Delta t)m} \tag{4.98}$$

where
  i is the discharge current
  ΔV is the voltage
  Δt is the discharge

The values of the specific power (P) and specific energy (E) are deduced from the GCD data using the following equations:

$$P = \frac{Vi}{m} \tag{4.99}$$

$$E = \frac{Vit}{m} \tag{4.100}$$

where
  i is the discharge current
  V is the voltage (excluding the iR drop occurring at the beginning of the discharge)
  t is the time

Typical data on frequency-dependent C_{sp}, C′, and C″ are shown in Figure 4.99.

C–V curves and specific capacitance curves are displayed in Figure 4.100. Galvanostatic charge–discharge curves and the so-called Ragone plot are displayed in Figure 4.101.

Next, we consider characterization of fuel cells.

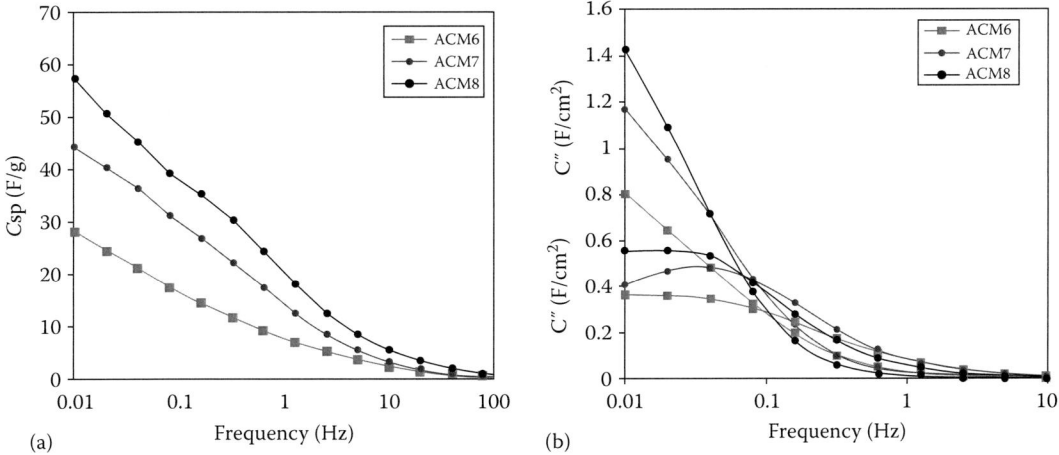

FIGURE 4.99 (a) Specific capacitance versus frequency. (b) Evolution of imaginary and real portions of capacitance for the ACM cells.

4.2.3 Characterization of (a) SOFC and (b) Electrochemical Sensor

4.2.3.1 SOFC Characterization

Like the electrochemical cell, the fuel cell should also be characterized for its performance. For simplicity, we shall focus on a single-chamber SOFC based on an anode-supported thin film electrolyte configuration. Anode powder is a mixture of 60 w/0 NiO and 40 w/0 Gd-doped ceria (CGO). They are made into sintered (1400°C) dry-pressed pellets. Electrolyte CGO thickness is controlled by adding different amounts of CGO during sintering. Cathode powders lanthanum strontium cobaltite ferrite (LSCF-CGO) of four compositions 50-50, 60-40, 70-30, and 80-20) after mixing with a binder (ESL V400, 0.5 g per gram of powder) and solvent (ESL T104) 8 drops per gram, were homogenized and

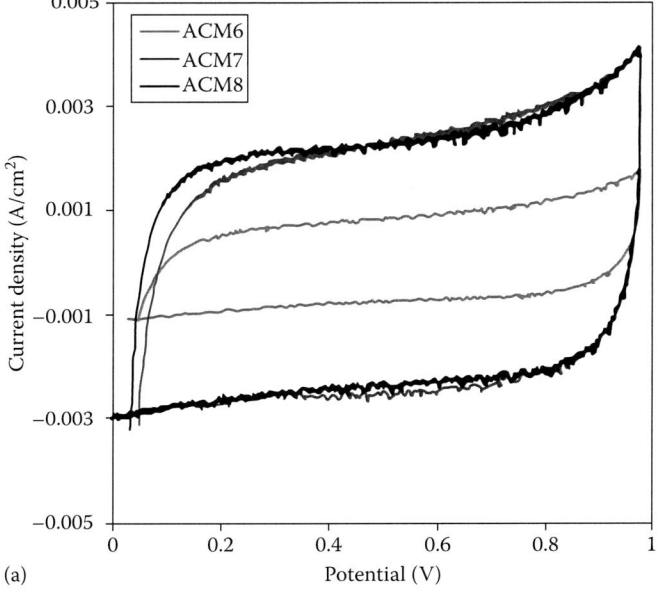

FIGURE 4.100 (a) CV curves at the scan rate 1 mV/s for the ACM cells. *(Continued)*

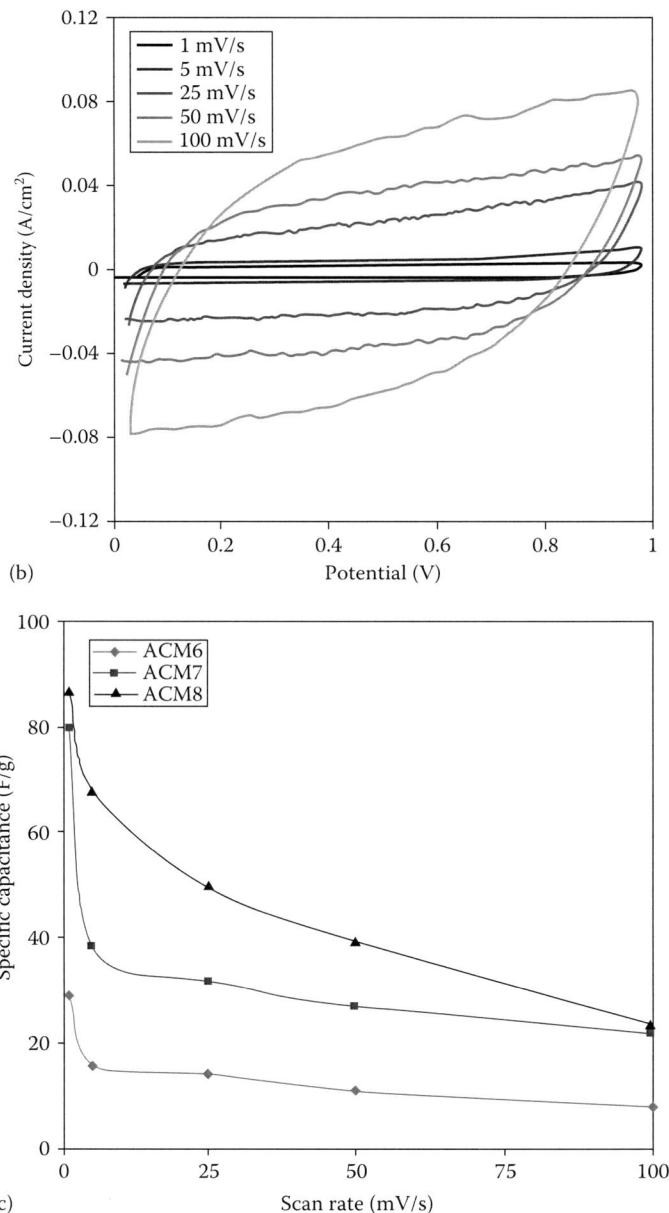

(b)

(c)

FIGURE 4.100 (*Continued*)   (b) CV curves with various scan rates for the ACM8 cell, and (c) specific capacitance versus scan rate for the ACM cell.

deagglomerated in a rolling mill. Resulting ink screen-printed on the electrolyte surface of fresh dual-layer membrane and calcined at 1000°C, 1050°C, and 1100°C for 2 h, respectively. Three cathodes with effective geometric surface areas of 0.36, 0.95, and 2.27 cm² were prepared for evaluation of the influence of electrode area on the cell's power density. A gold mesh screen-printed on the cathode with a commercial gold ink (ESL 8880-H) acted as the current collector. The thicknesses of the fuel cell elements are ~1 mm for anode, 150 μm for electrolyte, and 30 μm for cathode. These cells were tested under methane–air mixtures.

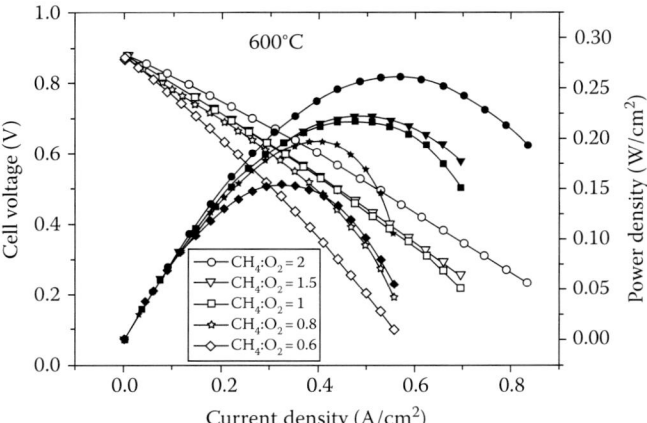

FIGURE 4.101 Current–voltage or I–V curves (open symbols) and current–power or I–P curves (closed symbols) of typical fuel cells testing under different fuel-mixing ratios (R_{mix}). (From Sun, L.P. et al., *Int. J. Hydrogen Energy*, 39, 1014, 2014.)

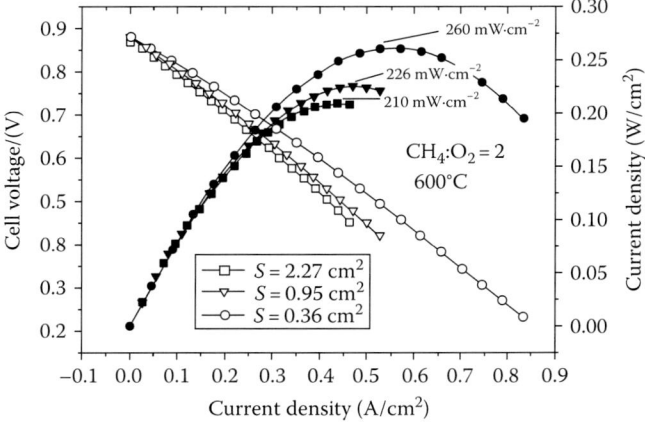

FIGURE 4.102 I–V (open symbols) and I–P curves (closed symbols) for typical fuel cells with different cathode area(s) testing under $R_{mix} = 2$ at 600°C. (From Sun, L.P. et al., *Int. J. Hydrogen Energy*, 39, 1014, 2014.)

To measure the electrochemical properties, gold wire and gold mesh are applied as current collectors on either side of the complete fuel cells. Polarization curves are obtained using a four-probe technique (using Keithley Model 2400 instrument) to eliminate the effect of the wire resistances. Two identical cells are to be prepared and measured separately. Each measurement is usually repeated three times and the average helps evaluate the cell performance.

Typical performance curves are shown in Figure 4.102.

4.2.3.2 Sensor Characterization

Gas sensing—chemical sensing in general and oxygen sensing in particular—is an important application in solid state ionics including oxygen sensing in vacuum systems and more widely in automobiles. We illustrate sensor characterization with respect to a diffusion barrier amperometric sensor (Figure 4.103). Basically, an amperometric sensor is constructed by imposing a diffusion barrier between the test gas

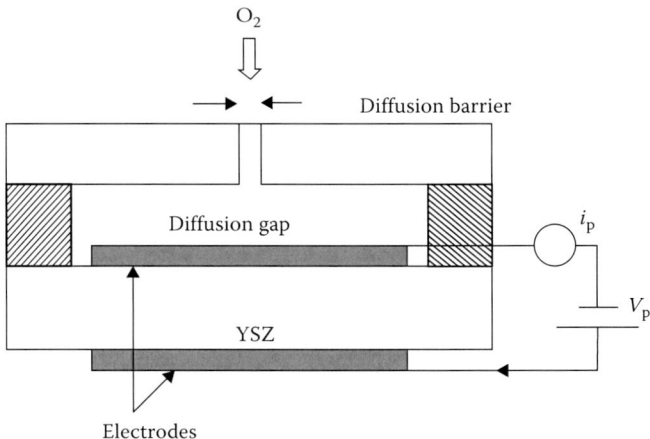

FIGURE 4.103 Current limiting diffusion barrier amperometric sensor based on yttria-stabilized zirconia electrolyte. The diffusion barrier can be either a small aperture or a porous material on top of the cathode. (From Ramamoorthy, R. et al., *J Mater. Sci.*, 38, 4271, 2003.)

flow and the electrode. The electrolyte–electrode cell can be operated in the so-called polarographic mode. Then the current flowing through the cell provides a measure of oxygen. This is the limiting current amperometric sensor.

In the amperometric mode of sensing (in contrast to potentiometric mode), oxygen is pumped from one side of the electrolyte to the other side by the application of an external potential (V_p) to the cell. The resulting ionic current flowing through the electrolyte is a function of the oxygen concentration.

This sensor is simple and does not require reference oxygen column and can respond linearly with oxygen concentration.

The gas-sensing mechanism is controlled by the difusion of oxygen. It can be bulk diffusion or Knudsen diffusion depending on the dimensions of the aperture. The relative magnitude of the Knudsen number Kn = mean free path (l)/hole diameter (d) decides the sensing mechanism. If Kn $\ll$ 1, the rate is determined by gas–gas collisions but not gas–wall collisions. This is bulk diffusion.

For Kn $\gg$ 1, gas–wall collisions are important and Knudsen diffusion results. The current–voltage relationsships and the sensor properties are quite different for the two cases. Limiting current is a function of the geometric parameters of the diffusion barrier. Limiting current varies with oxygen concentration CO_2 as $-\ln(1 - CO_2)$ for bulk diffusion and linearly for Knudswen diffusion.

For a linearly varying limiting current sensor, the diffusion aperture condition is Kn $\gg$ 1 for a given range of total presssure. In this case, the diffusion current of molecular oxygen dNO_2/dt depends on the concentration gradient of oxygen dCO_2/dx, the effective diffusion cross section Q, and the difusion coefficient of oxygen DO_2:

$$\frac{dNO_2}{dt} = -\frac{DO_2 Q}{dx}\, dCO_2 \tag{4.101}$$

The pumping ionic current flowing through the electrolytic cell is given by Faraday's law

$$i_p = \frac{4F\, dNO_2}{dt} \tag{4.102}$$

As the applied pumping voltage V_p is increased, the ionic current also increases until the oxygen pressure near the cathode–electrolyte interface reaches a value near zero. Then, the current saturates at a value

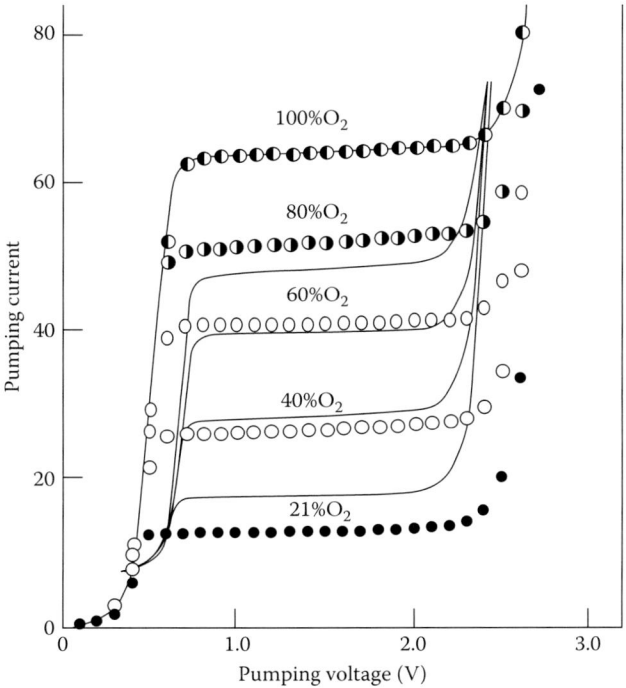

FIGURE 4.104 Limiting current–voltage characteristics of an amperometric sensor for varying pumping voltage and oxygen concentration in O_2–N_2 mixtures held at 450°C at 1 mmHg. (From Ramamoorthy, R. et al., *J Mater. Sci.*, 38, 4271, 2003.)

determined by the molecular current of the oxygen flow along the diffusion barrier. Thus, the limiting current i_p is directly proportional to the oxygen concentration in the test gas and varies linearly. The limiting current–voltage curves are shown in Figure 4.104.

A discussion of potentiometric sensors is provided in Chapter 9.

4.3 Summary

Materials characterization of solid state ionics comprises three aspects: (1) crystal structure and microstructure, (2) physical, which includes ion dynamics, and (3) electrochemical, which is applicable to both materials and devices. In this chapter, materials characterization techniques involving diffraction and microscopy, diffusion and conductivity, thermal properties, and resonance spectroscopies have been covered. Equally important, characterization of electrochemical devices has been considered. This chapter has natural links with the rest of the book and is expected to motivate the reader use this chapter as a basis for understanding and applying the physical and electrochemical methods to solid state ionic materials innovation and device development. As many of the techniques in this chapter have focused on the phase transitions, hopefully, a motivation has been provided for the next chapter.

Problems

4.1 From the data given in the following for β-AgI and γ-AgI, find the lattice constants for the hexagonal and cubic phases of AgI.

β-AgI (hexagonal)			γ-AgI (cubic)		
hkl	$10^5(\sin\theta)^2$	*I*	*hkl*	$10^5(\sin\theta)^2$	*I*
100	3750	1000			
002	4201	571	111	4215	1000
101	4805	652			
			200	5619	3
102	7955	343			
110	11248	792	220	11239	761
103	13211	781			
200	14997	122			
112	15453	465	311	15453	470
201	16048	103			
004	16821	2	222	16858	1
202	19202	76			
104	20570	2			
			400	22477	131
203	24459	261			
210	26245	89			
			331	26692	203

Data taken from Shaviv, R. et al., *J. Chem. Therm.*, 21, 631, 1989.

4.2 Temperature-dependent ionic conductivity of sol-combustion synthesized $Ce_{0.8}Gd_{0.2}O_{1.9}$ pellets sintered at 1250°C ($a = 0.54137$ nm, average particle size ~40 nm) is given in the following table.

Testing temperature (°C)	500	550	600	650	700	750	800
Electrical conductivity, $\sigma(S \cdot cm^{-1})$	4.55×10^{-3}	8.37×10^{-3}	1.34×10^{-2}	2.86×10^{-2}	4.09×10^{-2}	5.82×10^{-2}	7.64×10^{-2}

Make an Arrhenius plot ($\sigma T = \sigma_0 \exp(-E/k_B T)$) and determine activation energy for oxygen ion transport. (Ans: 0.87 eV)

4.3 The complex impedance plane ($Z'-Z''$) plots for $LiMn_{1.95}Ga_{0.05}O_4$ at three temperatures 260, 290, and 320 K are given in the figure below. (From Iguchi et al., *J. Appl. Phys.* 91, 2149, 2002.) Use these data to estimate grain resistance and conductivity. Draw an Arrhenius plot and estimate preexponential factor and activation energy for Li^+ ion conduction.

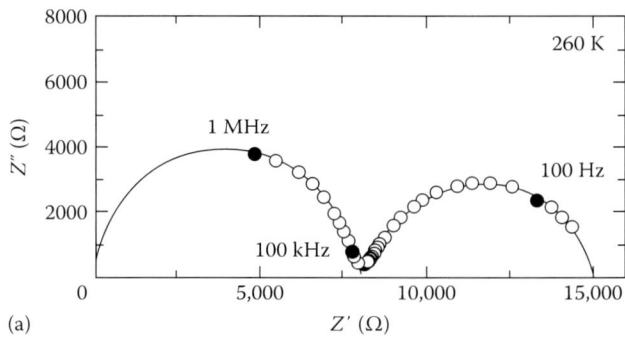

(a)

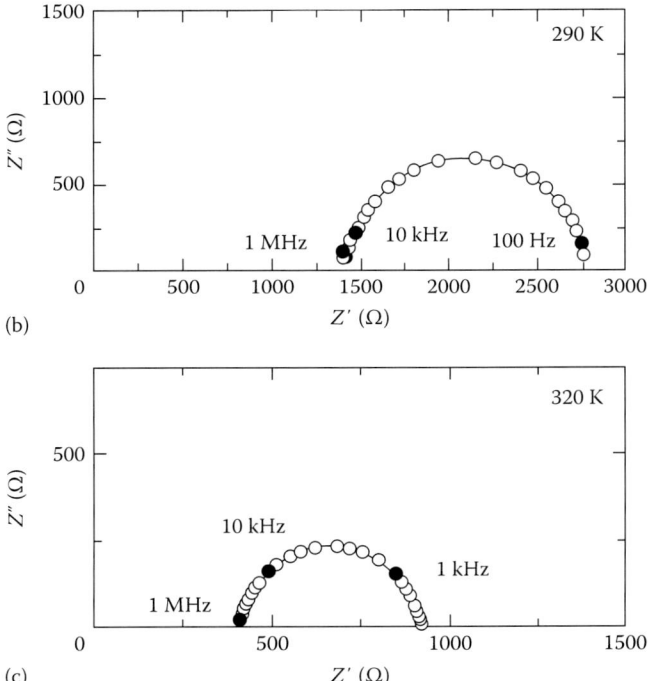

(b)

(c)

4.4 (a) Calculate the Doppler velocity corresponding to the natural line width of the gamma-ray emission from 14.4 keV excited state of Fe-57 nucleus having a half-life of 9.8×10^{-8} s.

(b) Calculate recoil velocity and energy of free Mossbauer nucleus Sn-119 when emitting a gamma ray of frequency $5.76 \times 10{-}18$ Hz. What is the Doppler shift of the gamma-ray frequency to an outside observer?

(c) Mossbauer nucleus Fe-57 makes the transition from the excited state of energy 14.4 keV to the ground state. What is its recoil energy?

4.5 Discuss the group theoretical origin of IR and Raman active vibrational modes of PO_4^{3-} in $Li_xTi_2(PO_4)_2$.

4.6 Fast-Li^+ ion conduction in lithium fulleride (Li_xC_{60}) a polymeric compound. The rare case of an ideal Li^+ conductivity (Debye) relaxation and ^7Li NMR relaxation in the Blombergen–Purcell–Pound model thermal decomposition of Li azide produces this compound, which shows a Li^+ ion conductivity of 0.1 mS/cm at room temperature! The activation energy for DC ion transport is 240 meV. Carefully, look at the crystal structure of this compound, the Arrhenius plot for conductivity, and the NMR relaxation of ^7Li nucleus. Why is the conductivity temperature independent (~0.1 μS/cm) below 130 K? Impedance measurements show that conductivity relaxation is Debye type with an activation energy of 209 meV. Interestingly, the NMR relaxation follows the Bloembergen–Purcell–Pound or the BPP model. NMR-derived activation energy for spin diffusion is 191 meV for a magnetic field of 2 T.

In this model, the spin–lattice relaxation rate R ($=1/T_1$) is given by

$$R = k \left[\frac{\tau}{1 + \omega_0^2 \tau^2} + \frac{4\tau}{1 + 4\omega_0^2 \tau^2} \right]$$

where
k is the constant
$\omega_0/2\pi = 33.10$ MHz
τ is the correlation time

A fit to this model gives 3.0 ps as the correlation time (τ_0) for Li motion in Li_xC_{60}, whereas from AC conductivity this time is 119 ps. Discuss the origin of the observed differences in activation energies obtained from DC conductivity, AC conductivity, and NMR. Also reason out why correlation time from AC conductivity experiment is longer than that from NMR experiment (see Ref. [112]).

4.7 NMR–impedance spectroscopy correlation.

The data in Figure P.7 (a) are analyzed according to the equation

$$A_{echo}(g^2) = A\exp\left[-D\gamma^2\delta^2(\Delta - \delta/3)g^2\right]$$

and are based on a Gaussian fit (curved line). Show that D is $1.8 \times 10^{-12} m^2/s$.

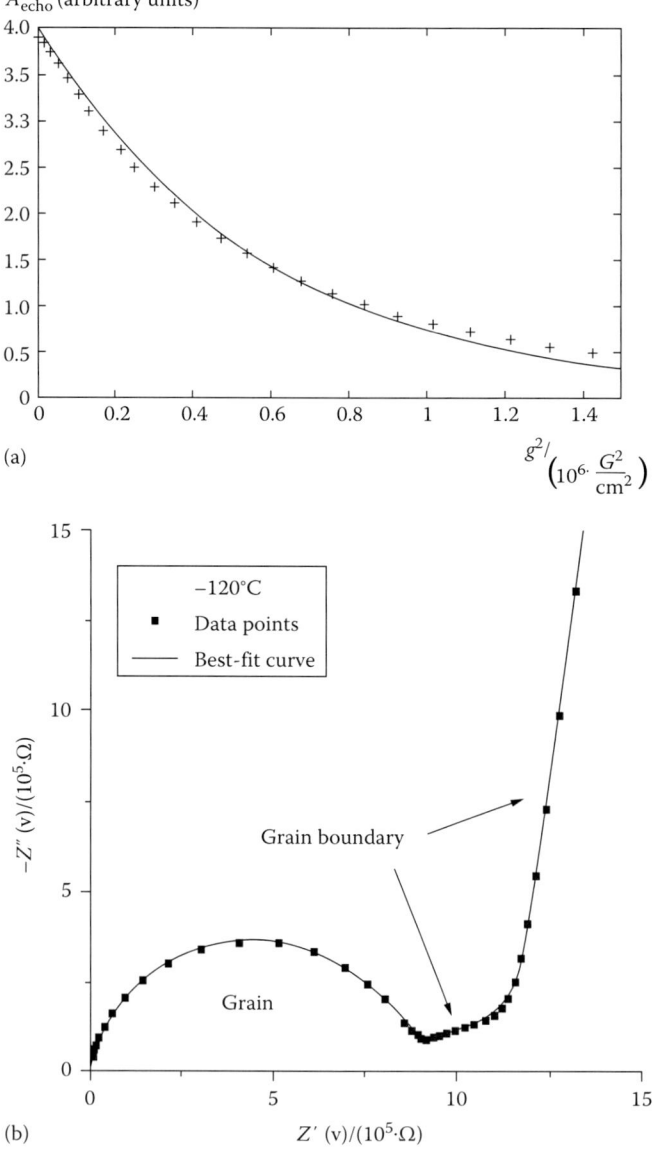

FIGURE P.7   (a) Simulated echo intensity versus pulse field gradient NMR Echo amplitude squared for Li. (b) Nyquist plot of impedance for $Li_{10}SnP_2S_{12}$ at 153 K.   *(Continued)*

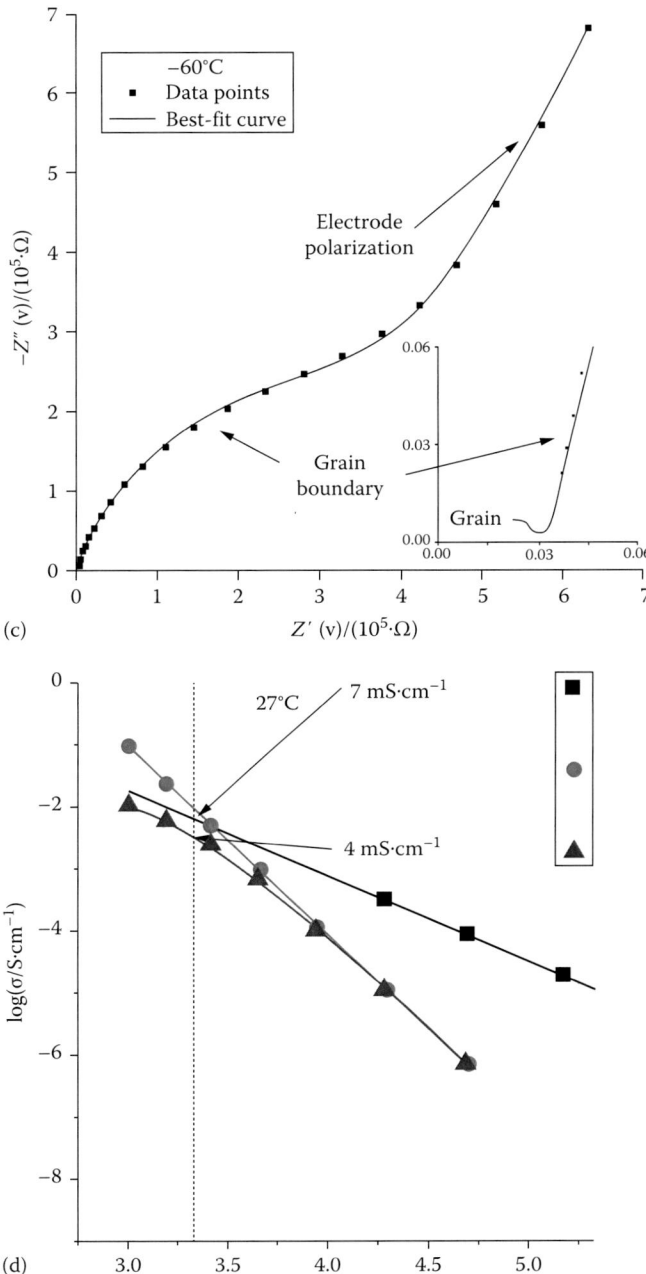

FIGURE P.7 (Continued) (c) Nyquist plots of impedance for $Li_{10}SnP_2S_{12}$ at 213 K. Solid lines correspond to best-fit curves. (d) Truncated Arrhenius plot of the grain conductivity, grain boundary conductivity, and total conductivity of $Li_{10}SnP_2S_{12}$. (From Bron, P. et al., *J. Am. Chem. Soc.* 135, 15694, 2013.)

From the plot in Figure P.7 (d) deduce the activation energies for grain and grain boundary conduction.

4.8 Dielectric relaxation spectroscopy of a polaronic conductor. The following figure (From Bron, P. et al., *J. Am. Chem. Soc.*, 135, 15694, 2013.) shows the frequency dependences of dielectric loss tangent (tan δ defined in Chapter 3) as a parametric function of temperature at 2 K increments in the temperature reanges of 286–326 K for $LiMn_{1.95}Al_{0.05}O_4$. Generally, polaronic conduction involves a Debye-type relaxation (also discussed in Chapter 3). Using the resonance frequency f_{max} in the tan δ-versus-frequency curve at a temperature T from the figure and the relation $f_{\tan \delta}$ $T^{1/2} \sim \exp(-E_{pol}/k_B T)$, where E_{pol} is the activation energy for hopping of non-adiabatic small polarons, compute E_{pol} for $LiMn_{1.95}Al_{0.05}O_4$. (*Hint:* Make a plot of log $f_{\tan \delta}$ $T^{1/2}$ versus $1/T$; for details, see reference of Problem 4.2.)

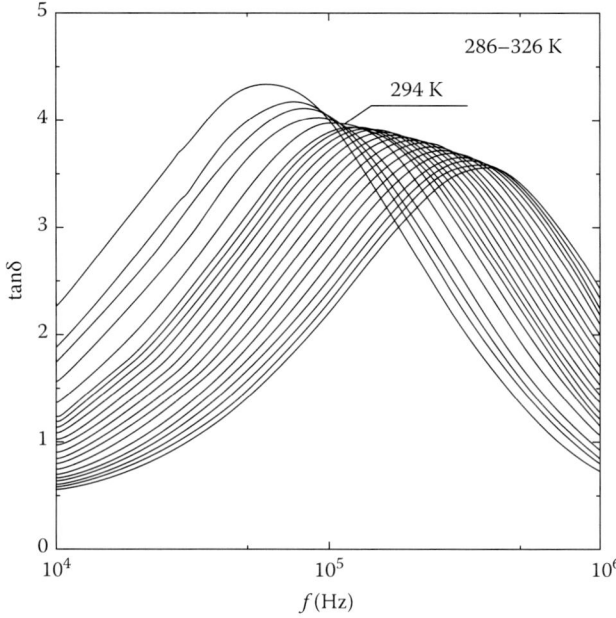

4.9 Sierpinski gasket, scale invariance, and a fractal model for the 2D conductivity of Ag-β alumina. The absence of DC conductivity in Ag-β alumina is a fundamental problem in ion dynamics and ion conduction in a 2D system. According to the scale invariance theory, DC conductance of a 2D system decreases exponentially with inceases in scale size (Anderson localization problem). This could be applied to ionic conduction in Ag-β alumina [4.114]. Look at this 2D triangular object called Sierpinski gasket.

This self-similar fractal has a dimensionality of $d_f = 1.585$. Self-similarity implies scale invariance. The aim is to see the small power-law dependence σ(ω) based on scale invariance.

Consider a random walker(in our cse Ag+ ion) on this fractal structure. The mean-square displacement of the walker $\langle r^2 \rangle$ is given by $\langle r^2 \rangle \sim t^{-2/dw}$ where d_w is the effective dimensonality of the random walk. It may be shown that $d_w = \ln 5/\ln 2 \approx 2.322$ on this gasket. In this case, show that the Fourier–Laplace transform of $\langle r^2 \rangle$ yields $\sigma(\omega)$:

$$\sigma(\omega) \propto \omega^2 \int_0^\infty t^{2/d_w} e^{i\omega t} \; \mathrm{d}t \propto \omega^{1-2/d_w}$$

The small power-law dependence of $\sigma(\omega) \sim \omega^{0.139}$ by assuming $d_w = 2.322$ for Ag-β alumina suggests scale invariance is behind ion dynamics in disordered systems.

4.10 The X-ray and neutron diffraction patterns of the high- and low-temperature modifications of $LiCoO_2$ are similar. So much so when oxygen atoms form an ideal cubic close packing, the two patterns for both modifications are virtually indistinguishable [4.115]. Draw a 3D network of a $2 \times 2 \times 2$ supercell of the fully ordered cubic $LiCoO_2$ structure. Indicate how the initial phase transforms to partially ordered low symmetry phases with space groups Fm3m, Fd3m, and R3bar m.

REFERENCES

4.1. B.D. Cullity, *Elements of X-Ray Diffraction*. Addison-Wesley, Reading, MA, 1956. https://archive.org/details/elementsofxraydi030864mbp.

4.2. J. B. Bates et al., *Solid State Ionics* 135(2000) 33.

4.3. Nordh, T., Li$_4$Ti$_5$O$_{12}$ as an anode material for Li ion batteries in situ XRD and XPS studies. Thesis, Uppsala University, 2013. https://uu.diva-portal.org/smash/get/diva2:609075/FULLTEXT01.pdf.

4.4. D. Bharathi Mohan, C. S. Sunandana, *J. Phys. Chem. Solids* 65(2006) 1669.

4.5. Newville, M., *Fundamentals of XAFS*. Tutorial, University of Chicago, Chicago, IL, 2004. Available at xafs.org.

4.6. O. Kamishima et al., *Solid State Commun.* 103(1997) 141.

4.7. Denes et al., *J Solid State Chem.* 104(1993) 239.

4.8. A Yoshiasa et al., *Japanese J. Appl. Phys.* 78(1997) 3636.

4.9. BARC, National Facility for Neutron Beam Research, Bhabha Atomic Research Centre, Mumbai, nd. http://www.barc.gov.in/publications/tb/nfnbr-book-R.pdf.

4.10. A. F. Wright, B. E. F. Fender, *J. Phys. C Solid State Phys.* 10(1977) 2261.

4.11. J. Rodriguez-Carvajal, *Physica B* 19(1993) 55. http://www-llb.cea.fr/fullweb/fp2k/fp2k.htm.

4.12. R. Hempelmann, *Quasielastic Neutron Scattering and Solid State Diffusion*, Clarendon Press, Oxford, U.K., 2000.

4.13. X. Liu et al., *J. Mater. Chem. A* 1(2013) 9935.

4.14. A. Roy et al., *AIP Conf. Proc.* 1349(2011) 691.

4.15. L. Reimer, H. Kohl, *Transmission Electron Microscopy: Physics of Image Formation*, 8th edn., Springer, New York, 2008.

4.16. K. H. Kim et al., *J. Power Sources* 196(2011) 764.

4.17. J. Rodriguez-Carvajal et al., *Phys. Rev. Lett.* 81(1998) 4660.

4.18. J. Goldstein, *Scanning Electron Microscopy and X-Ray Microanalysis*, Third Edition, Springer, New York, 2003.

4.19. G. Zheng et al., *Nat. Nanotechnol.* 9(2014) 618.

4.20. J. Konishi et al., *Chem. Mater.* 20(2008) 2165.

4.21. Wikipedia.

4.22. A. Kumar et al., *Sci. Rep.* 3, 2013. Article number:1621.

4.23. R. Dejus, K. Skold, B. Graneli, *Solid State Ionics* 1(1980) 327.

4.24. J. X. M. Z. Johansson et al., *Solid State Ionics* 50(1992) 247.

4.25. R. J. Borg, G. J. Dienes, *An Introduction to Solid State Diffusion*, Academic, San Diego, CA, 1988.

4.26. K. Funke, *Sci. Tech. Adv. Mater.* 4(2013) 043502.

4.27. M. Saito, S. Tamaki, *Solid State Ionics* 60(1993) 237.

4.28. J. Tateno, N. Masaki, *Solid State Ionics* 51(1992) 75.

4.29. S. Roberts, A. von Hippel, *J. Appl. Phys.* 17(1946) 610.

4.30. M. C. Goetz, J. A. Cowen, *Solid State Commun.* 41(1982) 293.

4.31. R. Aliev et al., *Phys. Solid State* 39(1997) 1379.

4.32. D. R. Lide (Ed.), *CRC Handbook of Chemistry and Physics*, 90th edn., CRC Press, Boca Raton, FL, 2009, p. 2–65.

4.33. W. J. Parker, R. J. Jenkins, C. P. Butler, G. L. Abbott, *J. Appl. Phys.* 32(1961) 1679.

4.34. H. Liu et al., *Nat. Mater.* 11(2012) 422.

4.35. I. Hatta et al., *Rev. Sci. Instrum.*, 56(1985) 1643.

4.36. B. M. Suleiman, A. Lunden, *J. Phys. Condens. Matter*, 15(2003) 6911.

4.37. A. Sakuda et al., *Sci. Rep.* 3(2013) 2261.

4.38. D. J. Barber, R. Loudon, *An Introduction to Properties of Condensed Matter*, Cambridge University Press, Cambridge, U.K., 1989, Chap. 2.

4.39. C. Kittel, *Introduction to SSP*, 7th edn., Wiley, New York, 1996, Chap. 4.

4.40. F. C. Brown, *The Physics of Solids*, W. A. Benjamin, New York, 1967.

4.41. L. J. Graham, R. Chang, *J. Appl. Phys.* 46(1975) 2436.

4.42. J. H. Page, J.-Y. Prieur, *Phys. Rev. Lett.*, 42(1979) 1684.

4.43. F. Altorfer, Report LNS-152, Paul Scherrer Institute, Villigen, Switzerland, March 1990.

4.44. F. Gronvold, *Acta Chem. Scand.* 21(1967) 1695.

4.45. R. Shaviv et al., *J. Chem. Therm.* 21(1989) 631.

4.46. H. Okazaki, *Netsu Sokutei* 21(1994) 212.

4.47. H. Okazaki, A. Takano, *Z Naturforsch.* 40a(1985) 986.

4.48. Z. Vardeny, G. Gilat, D. Moses, *Phys. Rev. B* 18(1978) 4487.

4.49. D. B. McWhan et al., *Phys. Rev. B* 15(1977) 553.

4.50. T. Tojo et al., *J. Chem. Therm.* 31(1999) 831.

4.51. C. Degueldre et al., *Thermochim. Acta* 403(2003) 267.

4.52. P. Senthil Kumar et al., *J. Phys. Chem. Solids* 67(2006) 1809.

4.53. https://archive.org/details/elementsofxraydi030864mbp.

4.54. N. S. Kini et al., *J. Phys. D Appl. Phys.* 34(2001) 1417.

4.55. C. S. Sunandana, Y. L. Saraswathi, P. Senthil Kumar, *Indian J. Pure Appl. Phys.* 37(1999) 325.

4.56. P. Senthil Kumar, N. S. Kini, A. M. Umarji, C. S. Sunandana, *J Mater. Sci.* 41(2006) 3861.

4.57. Netzsch, SBA 458 Nemesis data sheet, SBA_458_Nemesis_E_0814.pdf.

4.58. H. Liu et al., *Nat. Mater.* 11(2012) 422.

4.59. M. N. Avasthi et al., *J. Phys. C Solid State Phys.* 14(1981) 3521.

4.60. D. Sirdeshmukh et al., *Electrical, Electronic and Magnetic Properties of Solids*, Springer International Publishing, Switzerland, 2014.

4.61. Y. L. Liu et al., *Phys. Rev. B* 41(1990) 10481.

4.62. C. H. Stuhrmann, H. Kreiterling, K. Funke, *Solid State Ionics* 154–155(2002) 109.

4.63. Y. Shimakawa, T. Numata, J. Tabuchi, *J. Solid State Chem.* 131(1997) 138.

4.64. P. Balaya, C. S. Sunandana, *Pramana J. Phys.* 33(1989) 627.

4.65. H. S. Choi, J. H. Im, T. Kim, J. H. Park, C. R. Park, *J. Mater. Chem.* 22(2012) 16986.

4.66. J. J. Biendicho, A. R. West, *Solid State Ionics* 226(2012) 41.

4.67. O. Kamishima et al., *Solid State Ionics* 262(2014) 495.

4.68. T. Ishida et al., *Rev. Sci. Instrum.* 57(1986) 3081.

4.69. M. Gnanavel, C. S. Sunandana, *Physics of Iodized Silver & Silver-Copper Thin-Film Nanostructures*, Lambert Academic Publishers, Berlin, Germany, 2013.

4.70. M. E. Orazem, *Electrochemical Impedance Spectroscopy*, Department of Chemical Engineering University of Florida, 2008. http://www.che.ufl.edu /orazem/pdf-files/Orazem%20EIS%20Spring%20 2008.pdf.

4.71. C. Grabrielli, Use and applications of electrochemical impedance techniques, Solartron Technical Report No 24, Issue B, 1997.

4.72. F. Wooten, *Optical Properties of Solids*, Academic Press, New York, 1972.

4.73. R. B. Barnes, *Phys. Rev.* 38(1931) 328.

4.74. B. Valeur, *Molecular Fluorescence Principles and Applications*, Wiley, New York, 2002.

4.75. C. S. Sunandana, *Phys. Stat. Solidi (a)* 105(1988) 11.

4.76. A. Rosencwaig, A. Gersho, *J. App. Phys.* 47(1976) 64.

4.77. J. Morimoto et al., *Mem. Defense Acad. (Yokosuka, Japan)* 24(1984) 211.

4.78. C. S. Sunandana, D. Phaninath, *Solid State Commun.* 58(1986) 115.

4.79. K. Raveendranath et al., *Mater. Sci. Eng. B* 131(2006) 210.

4.80. M. Kurita et al., *Japanese J. Appl. Phys.* 27(1988) L1920.

4.81. M. Kurita, F. Akao, S. Minomura, *Jpn. J. Appl. Phys.* 4(1992) 1215–1216.

4.82. C. M. Burba, R. Frech, *Solid State Ionics* 177(2006) 1489.

4.83. Y. Sundarayya et al., *Phys. Status Solidi B* 250 (2013) 1599.

4.84. M. Jhang et al., *Solid State Ionics* 188(2006) 37.

4.85. H. R. Chandrasekhar et al., *Phys. Rev. B* 17(1978) 884.

4.86. C. V. Raman, *Nature* 121(1928) 619; *Indian J. Phys.* 2(1927–1928) 388.

4.87. ACS, The Raman effect: International historic chemical landmark. American Chemical Society, Washington, DC, 1998. www.acs.org/content/acs/en/education/.../landmarks/ramaneffect.html.

4.88. C. V. Ramana et al., *Surf. Interface Anal.* 37(2005) 412.

4.89. K. Tuinstra, *J. Chem. Phys.* 53(1970) 1126.

4.90. D.T. Schwartz, Raman spectroscopy: Introductory tutorial, Department of Chemical Engineering, University of Washington, 2005. https://depts.washington.edu/ntuf/facility/docs/NTUF-Raman-Tutorial.pdf.

4.91. L. Borjesson, L. M. Torrel, *Phys. Rev. B* 32(1985) 2471.

4.92. N. Chandrakumar, S. V. Raman, 1992, US Patent 5,973,496.

4.93. J. Klinowski (Ed.), *New Techniques in Solid State NMR*, Springer, Berlin, Germany, vol. 246, 2005.

4.94. J. J. Kweon et al., *J. Phys. Chem. C*, 118(2014) 13387.

4.95. M. Matsuo et al., *Appl. Phys. Lett.* 91(2007) 224103.

4.96. M. Wilkrning, P. Heitjans, *Phys. Rev. B* 77(2008) 024311.

4.97. P. M. Richards, in M. B. Salamon (Ed.), *Physics of Superionic Conductors*, Springer-Verlag, Berlin, Germany, 1979, Chap. 6.

4.98. C. Gionco et al., *Chem. Mater.* 25(2013) 2243.

4.99. J. Matta et al., *Phys. Chem. Chem. Phys.* 1(1999) 4975.

4.100. H. Kaftalen et al., *J. Mater. Chem.* 1(2013) 9973.

4.101. I. S. Lyubutin et al., *Solid State Ionics* 31(1988) 197.

4.102. V. Santhanam, Mossbauer spectroscopy-principles and applications, http://www.slideshare.net/chemsant/mossbauer-spectroscopy-principles-and-applications-26891122?related=2 (accessed October 5, 2013).

4.103. Y. Sundarayya et al., *Mater. Res. Bull.* 42(2007) 1942.

4.104. A. A. M. Prince et al., *Solid State Commun.* 132(2004) 455.

4.105. Ohkubo, Y., Project research on science and engineering of unstable nuclei and their uses on condensed matter physics. Research Reactor Institute, Kyoto University, 2012. www.rri.kyoto-u.ac.jp/PUB/report/PR/ProgRep2012/Project3.pdf.

4.106. Denes et al., *J. Solid State Chem.* 104(2003) 239.

4.107. K. H. Kim et al., *J. Power Sources* 196(2011) 764.

4.108. J. B. Bates, *Solid State Ionics* 70/71(1994) 619.

4.109. T. Usui et al., *Japanese J. Appl. Phys.* 26(1987) L2061.

4.110. S. Chandra, *Superionic Solids*, North Holland, Amsterdam, the Netherlands, 1981.

4.111. Iguchi et al., *J. Appl. Phys.* 91(2002) 2149.

4.112. M. Ricco et al., *Phys. Rev. Lett.* 102(2009) 145901.

4.113. P. Bron et al., *J Am. Chem. Soc.* 135(2013) 15694.

4.114. O. Kamishima et al., *J. Phys. Condens. Matter* 23(2011) 225404.

4.115. K. Kushida, K. Kuriyama, *Solid State Commun.* 123(2002) 349.

4.116. G. Rousse et al., *Chem. Mater.* 15(2003) 2082.

5

Phase Transitions in Solid State Ionic Materials

5.1 Introduction

As mentioned in Chapter 2, phase transition(s) may be recognized as a paradigm of solid state ionics, not only because phase transition is inevitable for the very existence of the superionic conducting materials composition, but more crucially for materials innovation and device development as well. Just consider the daily routine charging and discharging of a Li^+ ion mobile battery to readily appreciate how practical is this paradigm. In this chapter, we will consider the statistical mechanical basis of superionic phase transitions.

The fast ionic or superionic conducting state is essentially a transitioned state with the phase transition from subambient/ambient low temperature (LT) undergoing a structural phase transition bestowing the material with a substantial degree of gas-like ion disorder of Frenkel type—either positive or negative—supported by a rigid framework of the opposite type. Therefore, the theory of phase transitions—particularly the development of Frenkel disorder—is of great interest, both fundamentally and practically.

Phase transitions in matter are as old as thermodynamics [5.1] and as new as nanoscience [5.2]. This statement holds for superionic conductors (SIC) as well ever since Faraday discovered the high Ag^+ conducting phase of silver sulfide and lead sulfide [5.3] and Makiura et al. [5.4] obtained the room temperature superionic in ~10 nm AgI nanoparticles. They connect to the gut-level physics, on the one hand, and the materials synthesis optimization, on the other.

Chandra [5.5] has given a detailed overview of the pre-1981 theories of phase transitions while Salamon [5.6] has provided a critical review of the free energy functional approach to this topic.

Phase transitions in solids can be cooperative and gradual or a sudden phenomenon that involves a global change in crystal structure and physical properties of a system when an external stimulus—usually temperature (T) and/or pressure (P)—is changed continuously [5.7]. With interacting systems, the molecular approximation leads to a phase transition sometimes of the second order and sometimes of the first order [5.8].

Phase transitions in SICs [5.7] are a unique class of phenomena that span the three well-known states: solid, liquid, and gas, and even plasma [5.7a]. They are indeed the primogenitures of the phenomena of (fast) ion diffusion and ionic conductivity. Any attempt to understand the latter necessitates an inquiry into the nature of the phases involved as well as the transitions between them as reflected in the thermal/thermodynamic, electrical/electromagnetic resonance, and relaxation and related responses of specific SICs such as AgI and PbF_2. The decrease in ionic conductivity—usually observed as the superionic transition—approached from above the superionic transition temperature is similar to the structural arrest of a supercooled liquid as it is quenched into an immobile glassy state. The superionic material in its liquid state is analogous to the supercooled liquid state, as demonstrated by the loss of ion mobility as the material is cooled from the melting point where the ionic conductivity has magnitudes approximately of liquid electrolytes [5.7b]. However, AgI is a counter-intuitive exception in that when cooled from its melting point, the conductivity *increases*.

According to Hull [5.9], " ...AgI and beta PbF_2 each of which can be considered a parent of two large families of highly conducting compounds which are related by either structural or chemical means." AgI is a "runaway," non-Landau type [5.10] of extremely "asymmetric" SIC even at 420 K through a huge wurtzite/zincblende to body-centered cubic structural phase transition of first order whose origin continues to be a puzzle and has thus motivated many theoretical studies (Figure 5.la).

The "asymmetry" referred to earlier is related to cation Frenkel defect formation, and the propensity to form an enormous number of Frenkel defects is a unique feature of SICs.

Normal ionic conductors such as NaI and NaF possess thermodynamic point defects (pairs of cation and anion vacancies), while AgI and PbF_2 have structural defects that allow a given mobile ion to occupy through facile hopping motion a large number of vacant sites usually forming part of a "sublattice" (cation sublattice in AgI and anion sublattice in PbF_2). Thus, Frenkel defects in the latter group of compounds really constitute Frenkel disorder in a statistical sense. Theoretical models purporting to deduce physical property anomalies that occur as first-/second-order structural phase transitions must take cognizance of the fact that the concentration of structural defects is far greater than the number of thermodynamic defects, which results in defect–defect interactions. In an intuitive way, phase transitions in superionic conductors could be conceived as "sublattice magic" [5.11].

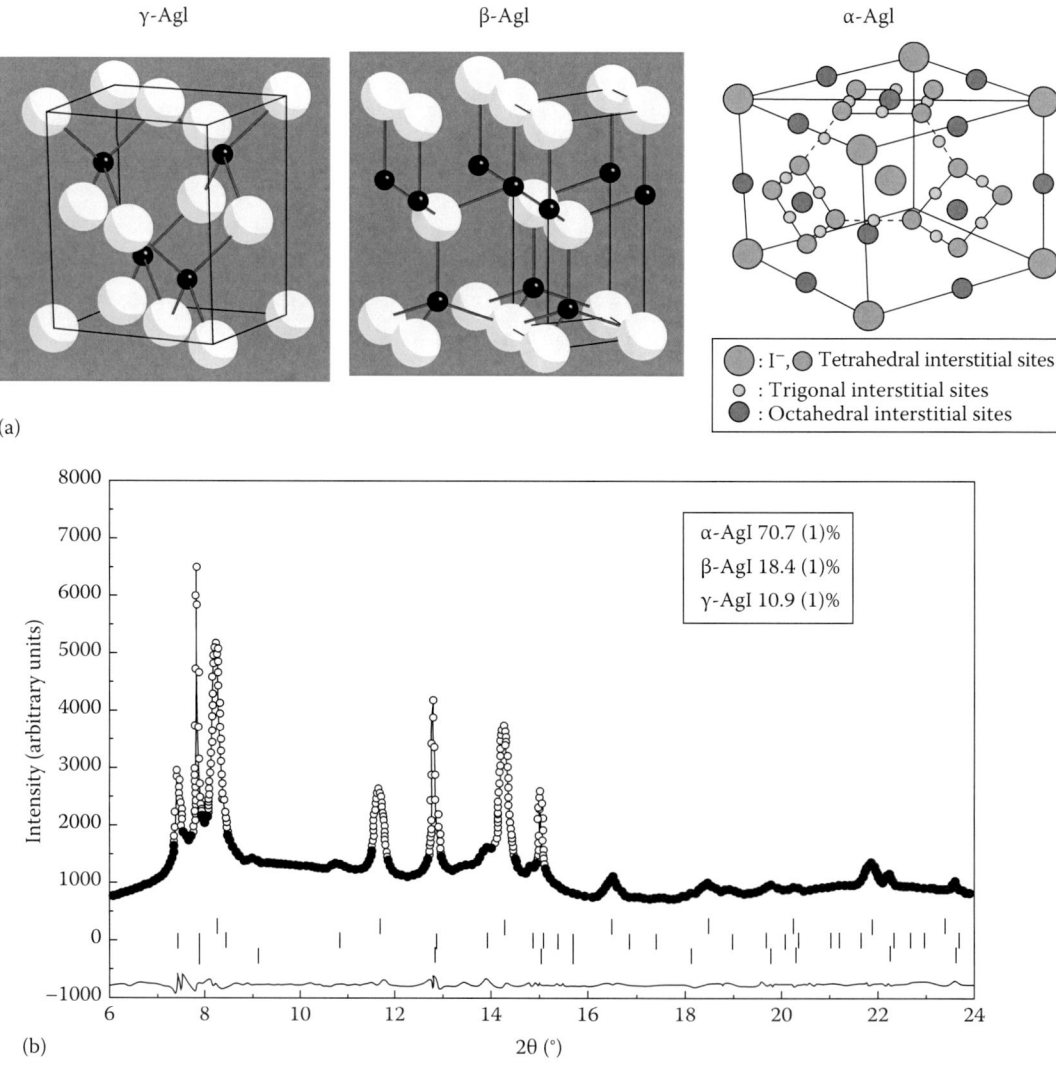

FIGURE 5.1 (a) Unit cells of the three polymorphs of AgI. γ-AgI is the metastable zincblende coexisting with the stable wurtzite (hexagonal) or beta AgI at ambient temperature. β/γ-AgI abruptly transforms to the iodine-based body-centered cubic phase α-AgI at 420 K through a first-order phase transition. (b) Coexistence of BEE (α) a, wurtzite (β), and zincblende (γ) phases of nanoscale silver iodide (11 nm particles) at 311 K. The wurtzite/zincblende to bcc phase transition in bulk AgI occurs at 420 K.   *(Continued)*

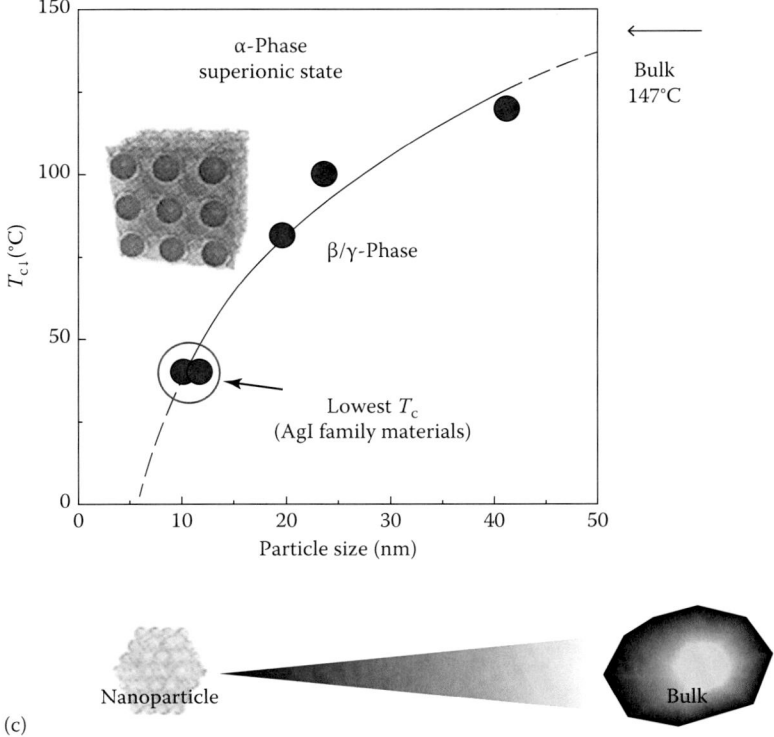

(c)

FIGURE 5.1 (*Continued*)   (c) Particle-size dependence of the wurtzite/zincblende to bcc phase transition temperature (T_c) of AgI on cooling. A temperature of 311 K followed in (a) is the lowest T_c. Note that the phase transition persists on particle downsizing. (After Makiura, R. et al., *Spring-8 Research Frontiers*, 2009, p. 100, JASRI, Hyogo Prefecture, Japan, http://www.spring8.or.jp; Makiura, R. et al., *Nat. Mater.*, 8, 476, 2009.)

In the modern context of nanoscience and nanotechnology, apart from conventional thermodynamic variables T and P, crystallite/particle size particularly in the 1–100 nm range emerges as an important control parameter/variable. The latter has a significant influence over the phase stability and phase transition behavior of materials such as AgI with device applications for nanoionics. An important recent observation is the unusual α–β–γ phase coexistence (Figure 5.1b) and the dramatic reduction in the phase transition temperature upon decreasing the crystallite size (Figure 5.1c).

5.2 Models and Theories

5.2.1 O'Reilly Theory of First-Order Superionic Phase Transitions

An early intuitive theory for superionic conduction through a first-order structural phase transition is that due to O'Reilly [5.12]. Built on the premise that the free energy of a solid electrolyte/SIC is derivable from a configurational entropy term and an energy of promotion of an ion from a lattice site to an interstitial site, it recognizes that the order of the phase transition depends on the degeneracy of the sites for mobile ions. Ionic conduction begins when a greater number of sites become available to the ions to move through the crystal than the number of sites occupied in the LT phase. It is only then that a first-order or second-order transition to the superionic phase can occur.

What is the form of the free energy of a superionic conductor? It must include not only a nearest-neighbor interaction energy term but also a term that explicitly accounts for the degeneracy of sites available to the mobile ion. From an expression for the free energy the mobile ion concentration (that connects to ionic conductivity) and an excess heat capacity are derivable. Comparison with experimental data would then be possible besides a discussion of the question of hysteresis associated with a phase transition approached from opposite directions of T (or P).

Assuming separate lattices for the stationary and mobile ions (that is two sublattices really) and considering (1) vibrational contributions, (2) ordering effects from ions on each (sub)lattice, and (3) the energies of promotion and interaction of the mobile ions, the total partition function for an SIC reads as follows:

$$Z = Z_{vib,S} Z_{vib,M}\, \sigma_S\, G_S \sigma_M G_M \exp(-\beta E) \tag{5.1}$$

The first two factors on RHS are vibrational partition functions for the stationary and mobile ions, respectively, while $\sigma_S\, G_S$ and $\sigma_M\, G_M$ are sums of the distributions of ions over the stationary and mobile sites, E is the disordering energy, and $\beta = 1/k_B T$, where k_B is the Boltzmann constant.

Let us use the intuitive Einstein's model for Z_S which is quite appropriate as the superionic phase exists well above the Debye temperature [5.13]. Assume that E depends only on the concentration of mobile ions—a crucial assumption and so may be pulled out of the summation. The distribution terms are, approximately,

$$\sigma_S G_S = \frac{N_S!}{(N_S - n)!\, n!} \tag{5.2}$$

$$\sigma_M G_M = \frac{N_M!}{(N_M - n)!\, n!} \tag{5.3}$$

The disordering energy in an SIC is made up of (1) energy that promotes ions from stationary (S) to mobile (M) sites and (2) the nearest-neighbor interaction energies resulting from this transfer. So finally

$$E = U_1 n - U_2 n^2/N_S \tag{5.4}$$

Putting E of Equation 5.4 in Equation 5.1

$$Z = Z_O\, (u_S/u_M)^3 n\, N_S!/(N_S - n)!\, n!\, N_M!/(N_M - n)!\, n!\, \exp[-\beta(U_1 n - U_2 n^2/N_S)] \tag{5.5}$$

where Z_O is the vibrational partition function for the perfectly ordered solid.

For a solid, the free energy F is A, so that

$$A = -k_B T \ln Z \tag{5.6}$$

The free energy density

$$F = F_{total}/N_S = U_1 c - U_2 c^2 + 3k_B T\, c \ln(r) + k_B T\, [(m - x) \ln(m - c) + (l - c) \ln(l - c) + 2c \ln c - m \ln(m)] \tag{5.7}$$

where
  $m = N_M/N_S$
  $r = u_M/u_S$
  $c = n/N_S$ is the total number of mobile ions relative to the stationary ions

The vibrational contribution from the perfectly ordered solid Z_O has been subtracted out.

Now what is the equilibrium mobile ion concentration? Set $dF/dc = 0$ to get

$$\ln\left[\frac{(l - c)(m - c)}{c^2}\right] = U_1^* - \left(\frac{2U_1^*}{a}\right)c + 3\ln(r) \tag{5.8}$$

where
  $U_1^* = U_1/k_B T$
  $U_1 = a U_2$

Equation 5.8 may be numerically solved for c which connects to the electrical (ionic) conductivity of the SIC.

What about the excess heat capacity arising from disorder? Note that

$$\text{Entropy } S = -(dF/dT)p \tag{5.9}$$

Heat capacity at constant pressure

$$C_{\text{p}} = T(dS/dT)_p \tag{5.10}$$

so that

$$C_{\text{p}} = -T(d^2F/dT^2)p = -T[28^2F/8c8T) + (8^2F/8c^2)(8c/8T)]8c/8T \tag{5.11}$$

or

$$C_{\text{p}} = k_B U_1^{*2} \frac{\left[1-(2c/a)\right]^2}{\left[\{2m-(m+1)c/C(1-c)(m-c)\}-2U_1^*/a\right]} \tag{5.12}$$

Equations 5.8 and 5.12 enable computations of T dependences of conductivity and heat capacity. They afford a meaningful comparison of theory with experiments. Figures 5.2 and 5.3 display ln c versus $1000/T$ plots for AgI, Ag_2S, Ag_2Se, and $RbAg_4I_5$.

The fair agreement vindicates the essential correctness of the model justifying the reasonable and realistic assumptions thereof. Thus, a canonical statistical approach to superionic conductivity is established. Figures 5.4 and 5.5 display C_{p} versus T plots for the Ag-based binary and ternary SICs.

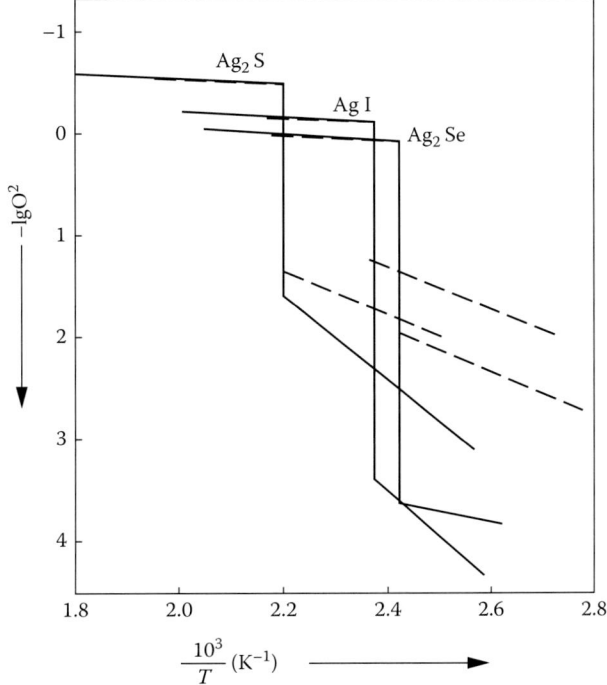

FIGURE 5.2 Testing the O'Reilly model. Experimental (-----------) and calculated (------) ionic conductivity ($-\log c$ versus $10^3/T$) profiles of AgI, Ag_2S, and Ag_2Se which transform to l bcc, S bcc, and Se bcc phases from wurtzite, monoclinic, and orthorhombic structures at 420, 452, and 416 K, respectively. (From O'Reilly, M.B., *Phys. Status Solidi*, 48a, 489, 1978.)

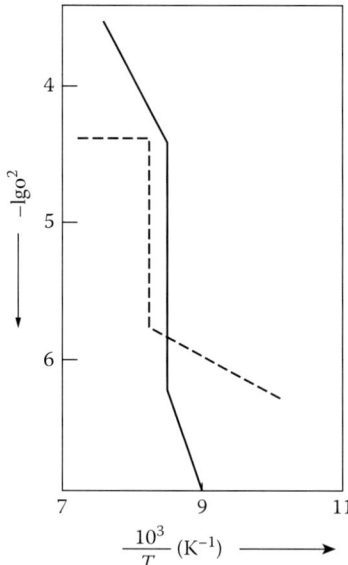

FIGURE 5.3 Testing O'Reilly model for $RbAg_4I_5$, the Ag^+ conductor with the highest of ionic conductivities at room temperature (0.25 S cm). It transforms through a first-order transition from rhombohedral phase to beta Mn structure at 122 K. Experimental (———) and calculated (----) ionic conductivities are shown. T_c of 121.8 K was chosen for calculations compared to 118 K from heat capacity measurements. (From O'Reilly, M.B., *Phys. Status Solidi*, 48a, 489, 1978.)

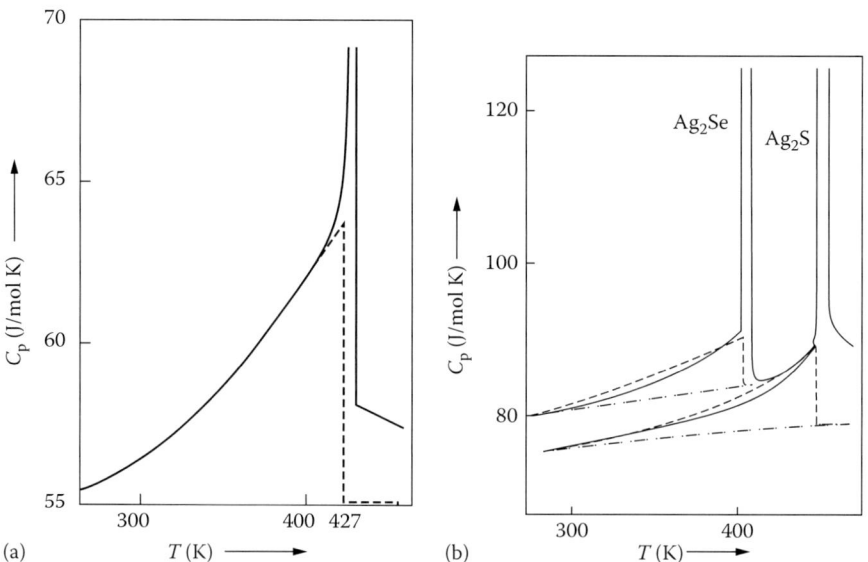

FIGURE 5.4 Experimental (———) and calculated (-·- Debye and (---) Einstein contributions fit to excess heat capacity) heat capacities of (a) AgI, (b) Ag_2S, and Ag_2Se. (From O'Reilly, M.B., *Phys. Status Solidi*, 48a, 489, 1978.)

5.2.2 Mezrin's Approach to the Faraday Transition

Defect formation thermodynamics is intimately linked to the disordering of the mobile ion ensemble in fluorite (CaF_2)-type SICs. This disordering occurs via the Frenkel mechanism involving the formation of anion (F^- ions here) vacancies at tetrahedral sites together with the formation of interstitial anions at octahedral sites of the fluorite lattice (Figure 5.6).

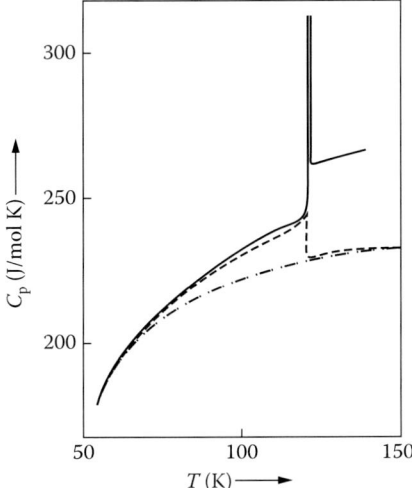

FIGURE 5.5 Heat capacity of $RbAg_4I_5$. ——— is experimental data; (-·-) and (---) are Debye and Einstein model fits to excess heat capacity. (After O'Reilly, M.B., *Phys. Status Solidi*, 48a, 489, 1978.)

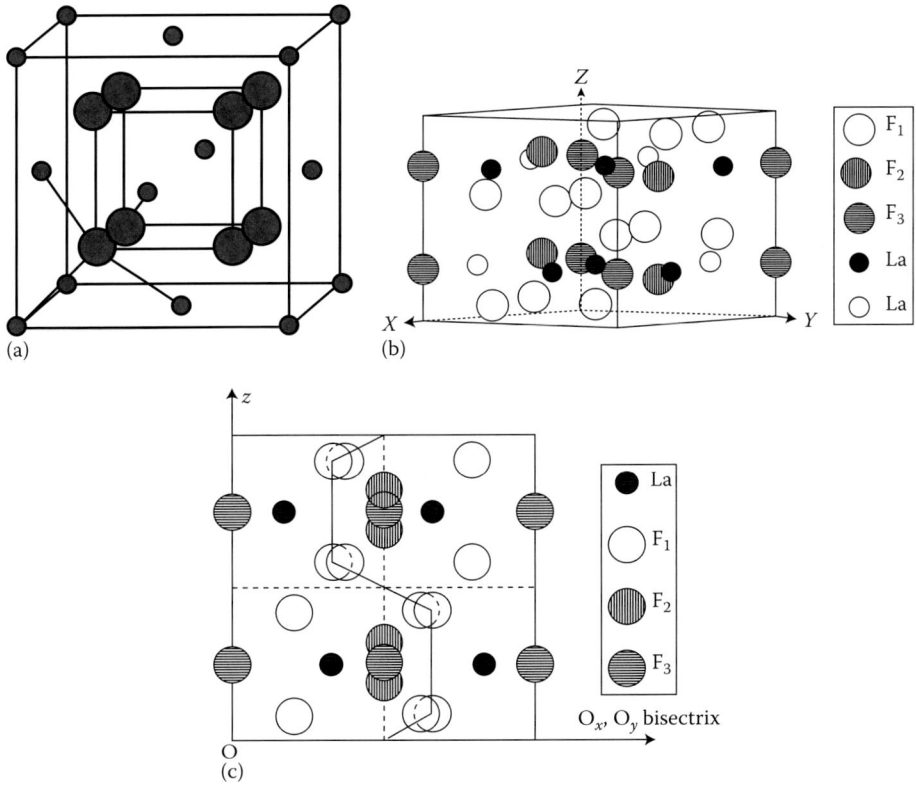

FIGURE 5.6 (a) Schematic of the fluorite lattice. Bigger dots are cations (say, Pb^{++} or Sr^{++}) and smaller dots are halide ions (say, F^- or Cl^-). Cations are tetrahedrally coordinated to anions and Frenkel defects arise from missing halide ions "pushed" to higher energy interstitial positions. (b) A special type of structure—the tysonite structure—adopted by LaF_3. (After Rhandour, A. et al., *Mater. Res. Bull.*, 20, 1309, 1995.) While La^{3+} ions form a rigid cation sublattice, the F^- sublattice may be considered as the superposition of three sublattices referred to as F_1, F_2, and F_3 containing F atoms in the ratios 12:4:2 per unit cell. The motion of F_1 anions in the *ab* plane of the orthorhombic lattice gives superionic conductivity.

The most interesting phenomenon is the persistence of the disordering process right up to the melting point of the SIC and the diffuse nature of the phase transition, also called the Faraday transition, to the superionic phase preceding the conventional melting of the solid.

Mezrin [5.14] has used a Bethe–Kikuchi-type type approximation to describe the all-important short-range order in the arrangement of defects. This approximation allows for the interactions of the mobile ion with its first and third neighbors. The correlations between like-charged defects at second and third neighbor sites are approximately expressed through the correlation between the oppositely charged—the nearest neighbors. Successfully used to deal with the genetically related first- and second-order transitions and diffuse phase transition, this approximation greatly simplifies the problem with only two state variables to deal with.

The defect concentration at short distances may be conveniently described in a mean-field approach while just the Frenkel defect pair formation entropy (leaving out configurational entropy) is handled phenomenologically. As in the O'Reilly model, the defect concentration term contains linear and quadratic terms.

The pairwise approximation focuses on the anion sublattice in the MX_2 fluorite lattice in which vacancies Vx and interstitials XI exist defining the defect short-range order. The asymmetry referred to earlier arises here. What are the defect possibilities? (1) The fraction of vacant site $P(Vx) = x$, (2) the concentration of anions Xx on site 1 $P(Xx) = 1 - x$, (3) the concentration of anions XI in octahedral site 2 $P(XI) = 2x$, and (4) the concentration of vacant octahedral site VI $P(VI) = 1 - 2x$. $x = 0$ define the system at its ground-state energy.

How are the nearest-neighbor defect interactions accounted for? It is done via four interaction energies:

1. E_{12} = Energy of $Vx - X_I$ interaction, the distance being $a\sqrt{(3/2)}$ where a is half the lattice parameter
2. E'_{11} and E''_{11} = energies of $V_x - V_x$ interaction when both vacancies are at one edge of an elementary cube of site I (distance a) and at the opposite face vertices (distance $a\sqrt{2}$)
3. $E_{22} = X_I - X_I$ pair interaction at nearest site 2 (distance $a\sqrt{2}$).

Operationally speaking, pairwise interaction is characterized by the probability of finding, say, V_x and X_I in a random pair of nearest sites 1 and 2. An example of "pairwise correlators" is

$$P(V_xX_1) = P(V_xX_1) \, a\sqrt{3/2} \tag{5.13}$$

The single particle variable $g = P(V_x|X_1) = P(V_xX_1/P(X_1))$ means the a posteriori probability of finding V_x on site 1, the nearest neighbor of site 2, a priori occupied by X_1. The correlators corresponding to the arrangement of V_x, X_x, and X_1, V_1 at the nearest sites 1 and 2 are represented by g and x, respectively.

$$P(VxXI) = 2xg$$

$$P(XxXI) = 2x(1 - g)$$

$$P(VxVI) = x(1 - 2g)$$

$$P(XxVf) = 1 - x - 2x(1 - g) P(Vx|Xf) = g \tag{5.14}$$

$$P(Xf|Vx) = 2g$$

$$P(Vx|Vf) = \frac{x(1 - 2g)}{(1 - 2x)}$$

$$P(Xf|Xx) = \frac{2x(1 - g)}{(1 \quad x)}$$

Anion disorder delineated in terms of correlators (Equation 5.14) contributes to the internal energy of the MX_2 crystal. When calculated per formula unit, it is

$$AU = 2Wx + 2W'x^2 + 8E_{12}P(VxXI) + (6E'_{11} + 12E''_{11})P(XI)x\{[P(Vx|X_1)]^2$$

$$+ P(VI)[P(Vx|V_1)]^2\} + 6E_{22}\{P(Vx)[P(X_1|Vx)]^2 + P(Xx)[P(XI|Xx)]^2 \qquad (5.15)$$

The first term in Equation 5.15 gives point defect formation energy, whereas the second term the defect interaction energies at distances greater than $a\sqrt{2}$. The rest of the terms accounts for the pairwise interactions at short distances.

The configurational entropy of the system taking only the correlations between nearest neighbors (Equation 5.15) is

$$\frac{S_{conf}}{k_B} = 6[P(Vx)\ln P(Vx) + P(Xx)\ln P(Xx)] + 7[P(X_1)\ln P(X_1) + P(V_1)\ln P(V_1)]$$

$$- 8[P(VxX_1)\ln P(VxX_1) + P(VxV_1)\ln P(VxV_1)]$$

$$+ P(XxX_1)\ln P(XxX_1) + P(XxV_1)\ln P(XxV_1) \qquad (5.16)$$

The phenomenological defect formation entropy is

$$\frac{S_{form}}{k_B} = 2sx + 2s'x^2 \qquad (5.17)$$

where
s and s' are dimensionless
$k_B s$ is the formation entropy of a Vx–XI pair as $x \sim 0$

The change in the free energy of the system—the quantity that connects theory to experimentally measured quantities—associated with defect formation and interaction, as a functional of x and g as obtained from Equations 5.16 and 5.17, is

$$AF = Au - TAS = 2x\left\{\begin{bmatrix} W + W'x + 8E_{12}\,g + (3E'_{11} + 6E''_{11})x \\ \left[2g^2 + \dfrac{x(1-2g)^2}{(1-2x)}\right] + 12E_{22}\left[\dfrac{g^2 + x(1-g^2)}{(1-x)}\right] \end{bmatrix}\right\}$$

$$- k_B T\left\{\begin{array}{l} 2sx + 2s'x^2 + 6[x\ln x + (1-x)\ln(1-x)] - (2x)\ln(2x) \\ +7(1-2x)\ln(1-2x) - 8[2x(g\ln g + (1-g)\ln(1-g) + x(1-2g)) \\ +(1-x-2x(1-g))\ln(1-x-2x(1-g)) \end{array}\right\} \qquad (5.18)$$

The problem is to evaluate state variables x and g by minimizing AF by partial differentiation with respect to x and g and equating the first derivative to zero. Thus, a system of nonlinear equations needs to be solved. A typical outcome of the solution is the vacancy concentration (x) and the excess specific heat capacity of the system Ac as functions of the dimensionless quantity T/E. Figure 5.7 shows a family of plots: (1) x versus T/E and (b) Ac versus T/E. Figure 5.7a clearly depicts the diffuse phase transition of the anion subsystem from the ordered to the disordered state as T increases, while Figure 5.7b shows the concomitant peak in the heat capacity which is again a signature of the diffuse transition.

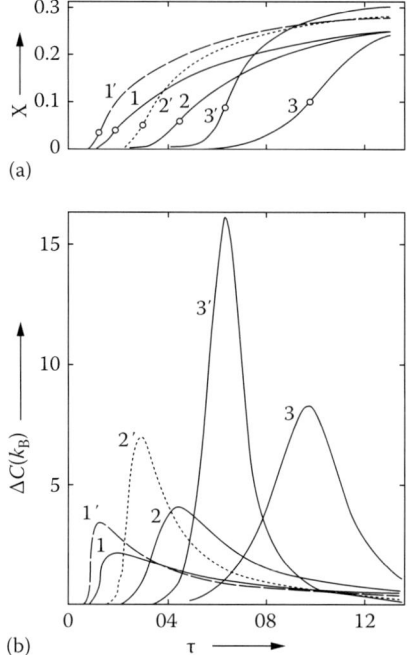

(a)

(b)

FIGURE 5.7 Character and signature of a diffuse transition from normal (ordered) to superionic (Frenkel disordered) state in an MX_2 crystal. (a) x versus τ (T/E) plot showing the gradual movement of the inflection point (o) upon increase of T. (b) The concomitant heat capacity peak (ΔC) versus T/E. (From Mezrin, V.A., *Phys. Stat. Sol.*, 114a, 145, 1998.)

Figure 5.8 depicts the excess heat capacity versus temperature plot for $SrCl_2$ demonstrating the ability of the theory to reproduce the experimental data.

The heat capacity maxima correspond to a Frenkel defect concentration of ~1%–15% that rises to 20%–40% near the melting temperature. The temperature dependence of the defect concentration qualitatively reproduces the temperature dependence of ionic conductivity.

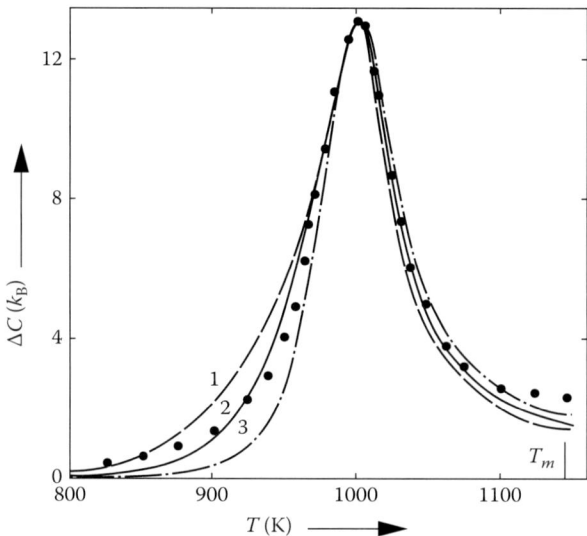

FIGURE 5.8 Excess heat capacity in $SrCl_2$ calculated (curve 2) showing good agreement with experimental data (•). (From Mezrin, V.A., *Phys. Stat. Sol.*, 114a, 145, 1998.)

5.2.3 Coulomb-Driven Transitions in Fluorites

Invoking a spherical cell model, March and Tosi [5.15] visualize the Coulomb interaction as the driving force for the superionic transition in fluorites via interacting Frenkel defects at high concentration. Each defect is assigned its own volume surrounded by a spherical cell of radius R such that $4nR^3/3$ is the volume per defect. Charge neutrality and Gauss theorem require that the electric field is zero across the spherical boundary. With the linear Debye–Hückel equation determining the screened potential within one cell of radius R, the model provides for the overlapping of Debye–Hückel shielding clouds of the individual defect centers essentially occurring when the concentration of Frenkel defects is high.

Thus, this model—unlike the two considered earlier—explicitly accounts for the concentration dependence of Frenkel defects in an ionic system in general.

The basic model is that of charged vacancies and anion interstitials moving in a medium with dielectric constant ε. Vacancy and interstitial defects carry opposite charges. Sit on an interstitial defect at the origin and consider the density of plus and minus charges relative to the reference defect. Assign the interstitial at the origin its own cell of radius R.

The total electrostatic potential $\psi_i(r)$ (i means there is an interstitial at the origin) satisfies, within the cell, the Poisson equation

$$W_i(r) = (-4\pi/\varepsilon)[e_v P_v(r, i) + e_i P_i(r, i)] - (4\pi/\varepsilon)e_i b(r) \tag{5.19}$$

where

e_v and e_i are the equal and opposite vacancy and interstitial defect charges, respectively
$P_v(r, i)$ and $P_i(r, i)$ are the densities of vacancies and interstitials at r conditioned by the interstitial at the origin

The linear Debye–Hückel equation for the potential is

$$(1/r)\mathrm{d}^2/\mathrm{d}r^2\,(rW_i(r)) = K^2[W_i(r) - W_i(R) + A\mu_v/e_v] - 4\pi/\varepsilon\, b(r) \tag{5.20}$$

where

$A\mu_v$ is the shift in the chemical potential due to the finite value of the cell volume $4nR^3/3$
$1/K$ = Debye screening length $=\sqrt{[ek_B T/8nPe^2]}$

Equation 5.20 has to be solved subject to the condition

$$\mathrm{d}W_i(r)/\mathrm{d}r \quad (\text{at } r = R) = 0 \tag{5.21}$$

After R, another length associated with the defects, is the distance of closest approach of the vacancy and the interstitial. For $r < a$ $\psi_i(r)$ satisfies the differential equation

$$(1/r)\mathrm{d}^2/\mathrm{d}r^2(rW_i(r) = -4\pi/\varepsilon\, b(r) \tag{5.22}$$

with the following boundary conditions: solutions of Equations 5.20 and 5.22 for $r > a$ should be continuous with the first derivative across $r = a$. This is the extended Debye–Hückel theory.

Solutions of Equation 5.22 yield the total chemical potential

$$\mu = -e^2 K/2E\,(1 + Ka) + 2e^2 K/ED \tag{5.23}$$

where D is

$$D = (1 + Ka)\,(1 - KR)\exp\,[K(R - a)] - (1 - Ka)\,(1 + KR)\,\exp[-K(R - a)] \tag{5.24}$$

What are the consequences of the enhancement of the concentration c of Frenkel defects due to Coulomb forces?

$$c = c_O \exp(-\mu/k_B T) \tag{5.25}$$

where c_O is the Arrhenius value

$$c_O = \exp[-H_f/2k_BT + S_f/2k_B] \tag{5.26}$$

H_f and S_f are enthalpy and entropy of an isolated Frenkel defect. Enhancement factor

$$F(c) = c/c_O \tag{5.27}$$

is to be determined by using Equation 5.23. From Equation 5.25

$$\ln c = \ln c_O + \mu/k_BT = 0 \tag{5.28}$$

The condition for instability, namely, d(LHS of Equation 5.28)/dc = 0 reduces to

$$\ln F(c_O) = 2(1 + K_c a) + (4e^2 \sim_c/\sim k_BT_c)[c\, d/dc((1 + Ka)/D]T_c \tag{5.29}$$

where
  T_c is the temperature at which instability occurs at a concentration c_c
  $\sim_c$ is the corresponding value of $\sim$

The instability condition implies that beyond T_c Equation 5.27 has no solution for $c(T)$. The second term of Equation 5.29 is due to extended Debye–Hückel theory.
  From Equations 5.25 and 5.27, T_c

$$\ln F(c_c) = -\mu_c/k_BT_c \tag{5.30}$$

c_c is of the form

$$c_c = f\{k_BT_c/(e^2/e_a)\} \tag{5.31}$$

The {…} term being the same for CaF_2, SrF_2, and BaF_2 within experimental error, Equation 5.31 shows that the critical concentration of defects for instability must be the same for these three fluorides but larger for PbF_2 and UO_2 implying larger cc for these latter compounds and possibly for $SrCl_2$ as well.
  Utilizing the extended Debye–Hückel approximation rather than the full solution of Equation 5.31, we get from Equation 5.30 the simple transparent solution

$$k_BT_c = e^2ca/4e_a(1 + ca)^2 \tag{5.32}$$

A sensible use of Equation 5.32 is to choose the crucial parameter a (whose effect is to increase the enhancement factor at T_c beyond 2) to fit the empirical transition temperature.
  What is the excess specific heat at constant pressure (related to the excess Gibbs free energy), $c(g_f + 2\mu)$, due to the interacting Frenkel defect assembly? Use the chemical potential from Equation 5.23 using standard thermodynamics to obtain

$$Ac_p = T[(2S_f - 48\mu/8T)8c/8T - (g_f + 2\mu)8^2c/8T^2 - 2c8^2\mu/8T^2] \tag{5.33}$$

At the instability, as $c \sim c_c$, the slope is vertical at $T = T_c$. But T_c an upper bound to the superionic transition temperature, say, T_s. At T_s the second phase with a practically constant Frenkel defect concentration takes over as $T > T_s$. Taylor-expand $c(T)$ around $T_s < T_c$ where dc/dT has a finite positive slope:

$$C_s - c = A(T_s - T) + 0(2) + \cdots \tag{5.34}$$

with c_s as concentration at T_s. Using Equation 5.34 and neglecting second derivatives, we get

$$Ac_p/T_s = 4k_Bc_s[H_f/2 + \mu_s + (\mu_s/2)/(1 + K_sa)][T_sS_f/2 + (1 + \ln c_s)(\mu_s/2)/(1 + K_sa)]/(k_BT + (\mu_s/2)/(1 + K_sa)^2 \tag{5.35}$$

As $T \to T_c$, the denominator gets progressively small since it vanishes in the instability limit Equation 5.32. Thus, the Coulomb interaction could drive the superionic phase transition leading to a specific heat anomaly at T_c.

5.2.4 Random Variable Theory of Phase Transitions in AgI

Frenkel defect production and ionic conduction in cationic conductors such as AgI, which undergo first-order phase transition from wurtzite/zincblende to a cation-disordered bcc phase, are both random phenomena. Thus, any model dealing with defect interaction and solid electrolyte transition should treat cation fraction as a random variable. The phenomenological model of Burbano et al. [5.16] does that remarkably to come up with an excellent fit to the experimental Arrhenius plot of ionic conductivity. The aim is to simulate the sudden increase in the conductivity at a transition temperature T_t with either first- or second-order phase transition such as those observed in SICs.

The observed conductivity $c_i(T)$ is assumed to be entirely due to the temperature variation of the density of carriers $n_i(T)$ and of their mobility $\mu_i(T)$ so that

$$c_i(T) = n_i(T)Ze\,\mu i(T) \tag{5.36}$$

How is the system defined?

1. The system has well-defined spatial arrangements and interactions among defects; the occupancy of all sites available for ion jumps is equally possible.
2. There is a finite or nonzero probability that a pair of adjacent occupied sites does not exist.

Therefore, such probability may be specified by a single parameter: the fraction p of the defect sites that have been removed or the fraction $q = 1 - p$ of the defects that are present.

The second assumption of the binomial distribution of probabilities suggests that the defect concentration—the result of thermally induced defect–defect interactions and thus temperature dependent—is the order parameter (OP) of a Landau-type free energy expansion. This free energy when minimized would lead to a nontrivial expression for the temperature dependence of defect concentration and eventually to a reasonably accurate temperature dependence of ionic conductivity quantitatively reproducing the anomaly at the transition temperature.

The number of cation sites N_1 and the number of interstitial sites N_2 which enable formation and sustenance of Frenkel defects are truly large so that one can (1) expect vacancy–interstitial interaction energy to be a nonlinear function of defect concentration and (2) sensibly consider the ion transport as a cooperative phenomenon and the system free energy as a function of n_i, $F(n_i)$, serves to handle the phase transition in conductivity and allied effects.

$$F(n_i) = F_O + E(n_i) - n_i\,k_BT \ln X - k_BT \ln\{N_1!/(N_1n_i)!n_i!\ N_2!/(N_2n_i)!n_i!\} \tag{5.37}$$

This equation represents a perfectly ordered system at $T = 0$ K if the activation energy for promoting a (sub)lattice cation (say Ag^+) or anion (say F^-) to the higher energy interstitial site (or Frenkel defect formation energy U_i) is greater than the magnitude of the screening interaction U between the cation at the interstitial site and the vacancy left behind. F_O is the free energy of the completely ordered crystal. $E(n_i)$ is the defect–defect interaction energy expected to be dependent on U and U_i.

$$E(n_i) = n_i(U_i - Un_i/N_2) \tag{5.38}$$

The third term in Equation 5.37 arises from the vibrational entropy

$$1/X = f = (W_i/W)1/3 \tag{5.39}$$

where
    m_i is the localized interstitial phonon frequency
    m is the lattice phonon frequency

The combinatorial entropy involves n_i, the number of disordered ions located at N_2 interstitial positions, and N_1 cation sites. Realizing that $N_1 = N_2 = N$, using Stirling approximation and treating the fraction n/N of cations as the order parameter η (note that this is an important step in the model rooted in statistical mechanics), the free energy density is

$$F(\eta_i) = [F(n_i) - F(O)]/N \tag{5.40}$$

$$F((\eta_i) = U_i\eta_i - U\eta_i^2 + 2k_BT\{n_i \ln \eta_i + (1 - \eta_i)\ln (1 - \eta_i) + (3/2) \eta_i \ln f \tag{5.41}$$

At $T = 0$ K if $U_i > U$ $F(\eta_i)$ is that of a perfectly ordered system. The equilibrium concentration of η_i is found by minimizing $F(\eta_i)$ w.r.t. η_i:

$$dF/d\eta_i = 0 \tag{5.42}$$

Now what is the F that should be minimized?

The model should eventually enable us to see in a general way for all SICs the effects of defect correlations. This would be possible if η_i is considered in its own right as a random variable and not just as an experimentally identifiable number. Associating a binomial probability distribution function

$$P(\eta_i) = p\, b(\eta_i - \eta) + (1 - p)\, b(\eta_i) \tag{5.43}$$

The single parameter that characterizes the probability that a pair of adjacent occupied sites does not occur is the fraction p of defect sites that have been removed or the fraction $q = 1 - p$ of defects that are present. Thus, Equation 5.43 provides a powerful basis for dealing with a system with Frenkel disorder.

A replacement of η by $p\eta$ renders the free energy expression of Equation 5.42 much more analytical, opening up the possibility of generation of theoretical conductivity profiles in the most general way with four adjustable parameters.

The trial free energy of Equation 5.41 is thus rewritten as

$$F((\eta) = pU_i\eta - p2U\eta_i^2 + 2k_BT\{p\eta \ln y\eta + (1 - p\eta) \ln (1 - p\eta) + (3/2) y\eta \ln f\} \tag{5.44}$$

Minimizing $F(\eta)$ w.r.t η

$$\eta = 1/\{p[1 + f^{3/2} \exp((\tau/x)(x/2 - p\eta))\} \tag{5.45}$$

with the four parameters $t = U/k_BT$, $x = U_i/U$ $(x > 1)$ besides f and p defined earlier. These four parameters help analyze the behavior of the system under many circumstances to arrive at the equilibrium behavior of the OP $\langle 1 \rangle$.

In general, a continuous increase in the value of $\langle 1 \rangle$ with T or an increase of $\langle 1 \rangle$ with a discontinuity at a transition temperature T_t (or $1/t_t$), whose value depends on x, is realized.

The transition point is due to the coexistence of two stable configurations $\langle 1 \rangle = \langle 1_1 \rangle$ and $\langle 1_2 \rangle$ $(\langle 1_1 \rangle$ not $= \langle 1_2 \rangle)$. The two 1 values arise as points of inflection when the equation $8^2F/8\ 1^2 = 0$ is solved: Why should these arise? Because the function $F(1)$ has opposite concavities at the two local minima:

$$1 (+/-) = (1+/ - \, `[1 - 4x/t])2p \tag{5.46}$$

Two real roots occur for $t \geq 4x$.

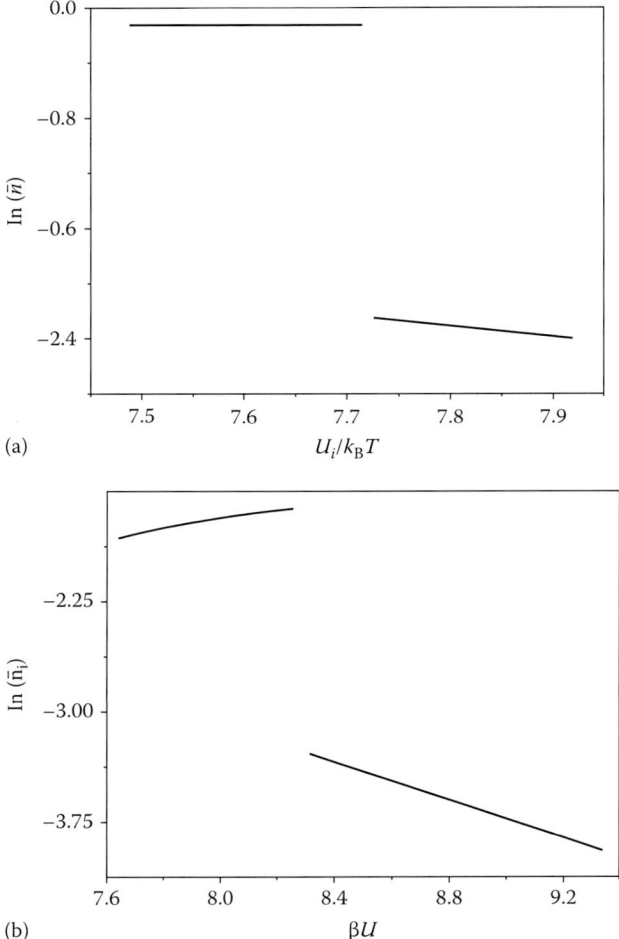

FIGURE 5.9 (a) Random number model for superionic phase transition. Solution of the transcendental equation versus dimensionless quantity $U_i/k_B T$. The model predicts the precipitous drop characteristic of the first-order phase transition. (b) Plot of log $\langle n_i \rangle$ as a function of the dimensionless beta U_i where $\langle n_i \rangle$ is the root of the equation $n_i = [\sqrt{\gamma}$ Gamma $- 3/2$ exp$[0.5 \, \tau(l - 2n_i/x]$, $\tau = \beta \, nU_i$, $x = U_i/U$, $\beta = 1/k_B T$. γ relates to structure deformation due to interstitial sites while Gamma relates to the ratio (perfect lattice frequency/defective lattice frequency). (From Lara, P. et al., *Solid State Ionics*, 175, 451, 2004. With permission.)

An important result of this treatment is an expression for the phase transition temperature—actually an isotherm in the (x, f) plane—which follows when $(\partial F/\partial l)_{l = 1/2p = 0} = 0$ for $T = T_t$:

$$f = \exp[-t_t(x - 1)/3x], \; x > 1 \tag{5.47}$$

Figures 5.9a and b and 5.10 show the results of the random variable model, which reproduce the phase transition. While Figure 5.9 shows numerical simulation depicting the precipitous drop of the OP at the parameter related to transition temperature, Figure 5.10 validates the model—essentially based on the Huberman approach [5.17]—against the experimental conductivity data on AgI–KI.

5.2.5 Other Models

Besides these classical free energy approaches to the phase transitions in SICs with predominantly Frenkel disorder are those pioneered by Welch and Dienes [5.18], Gurevich and Kharkats [5.19], and the more recent cube root model of Hainovsky and Maier [5.20] who show that premelting phenomenon can

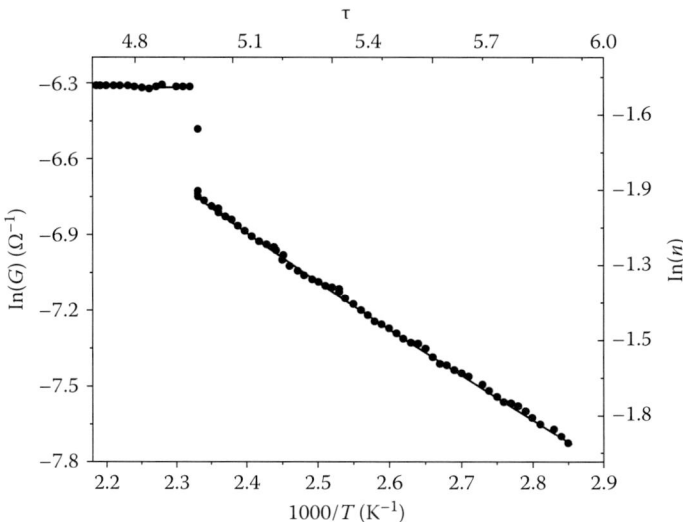

FIGURE 5.10 Validating the random variable model against experimental conductivity plot for AgI–KI near 430 K indicating that the temperature dependence of carrier concentration is the principal contributor to the conductivity anomaly. (From Burbano, J.C. et al., *Solid State Ionics*, 18, 553, 2009.)

be quantitatively described by a cube-root law (the excess ionic conductivity of silver and other halides is described by an excess chemical potential that is proportional to the cube root of defect concentration) not only for the Frenkel disordered silver halides but also for the anti-Frenkel disordered PbF_2. In all cases, the computed defect–defect interaction leads to a phase instability of first order or higher at a temperature very close to the actual transition point. Moreover, the specific heat data can be consistently explained by the same effect as is done in O'Reilly and Mezrin theories. The validity of the cube-root law is discussed in the context of the unexpectedly good prediction of the transition temperatures. The Aniya–Ichihara model [5.12,5.21], however, deemphasizes the functional form of the defect–defect interactions in the determination of phase transition temperatures. Bharathi Mohan and Sunandana have applied this last model to the phase transitions in $Ag_{1-x}Cu_xI$ synthesized by a novel mechanochemical reaction at ambient temperature [5.13,5.14,5.22,5.23].

All these models emphasize disorder and "asymmetry" relative to Frenkel defect components and take only a cursory look at phonons and anharmonicity. Although emphasis on disorder through the statistical mechanics of defect–defect interactions gives a fair understanding of experimental results on ionic conductivity and heat capacity and competently handles the "order" of phase transitions, it leaves the question of mechanism quite open. Thus, the phonons—particularly anharmonic phonons and thus the interatomic potential—deserve a close look. The next chapter would consider this perspective in detail. Apart from anharmonicity, elasticity plays an important role in controlling superionic phase transitions as considered in the following sections.

5.3 Athermal Phase Transitions

The uniqueness of phase transitions in SICs lies in the very origin of a group of compounds exhibiting, say, an apparently strong first-order transition that could be qualitatively different. This feature is spectacularly illustrated in the case of $LiNaSO_4$ (Freheit [5.24]). The bcc to $P31c$ superoinic phase transition at 788 K is termed "athermal" in that it proceeds instantaneously as a function of temperature but not isothermally as in a thermodynamic transition due to thermal activation. At least three processes seem to be at work: (1) lattice thermal vibrations, (2) centre of mass displacements of sulfate groups or "gate percolation," and (3) reorientation of sulfate groups.

The evolution of the rotational disorder of SO_{4-} groups constitutes the OP for phase transition. The transition to the superionic phase itself signals the breakdown of correlated coupling of sulfate groups leading to OP jump. When spontaneous strain couples to OP, the transformation matrix for the bcc to $P31c$ is given by

$$P: \begin{pmatrix} 1 & 0 & 1 \\ -1 & 1 & 1 \\ 0 & -1 & 1 \end{pmatrix}$$

The symmetry change that occurs during the transition is a tripling of the unit cell so that $V = \det(P) = 3$ (Figure 5.11).

What are the strain components allowed by symmetry?

$$e_1 = e_2 = (a/a_O \sqrt{2}) - 1 \tag{5.48}$$

$$e_3 = (c/a_O \sqrt{3}) - 1 \tag{5.49}$$

where a and c are the trigonal lattice parameters while a_O is the cubic phase lattice constant. Through Equations 5.48 and 5.49, the intrinsic thermal expansion (not related to the transition) is excluded by extrapolation.

Linear strain associated with phase transition = symmetry-breaking strain (pure shear) + non-symmetry-breaking strain (pure volume strain)

$$e_t = (1/\sqrt{3})(2e_3 - e_1 - e_2) \tag{5.50}$$

$$e_a = e_1 + e_2 + e_3 \tag{5.51}$$

For a zone boundary transition $e_t \approx (OP)^2$ as in an improper ferroelectric. Spontaneous strain occurs as a secondary response to the driving mechanism.

But what is the "lowest" order coupling?

$e_a \alpha\ Q^2$. Spontaneous strain evolves for $T > T_{tr}$ when thermal expansion is much larger than usual for a high-temperature (HT) phase. To explain the anomalous behavior, one can assume local fluctuations (precursoring) leading to the formation of short-range order in the superionic phase.

If V and V_O are cell volumes in LT and HT phases, then the volume strain (V_s) is

$$e_a = V_s = (V/3 - V_O)/V_O \tag{5.52}$$

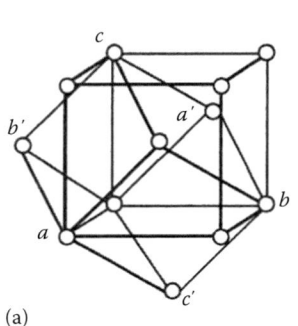 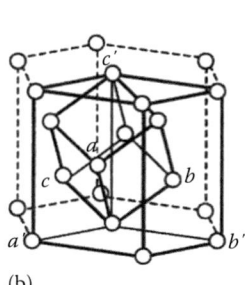 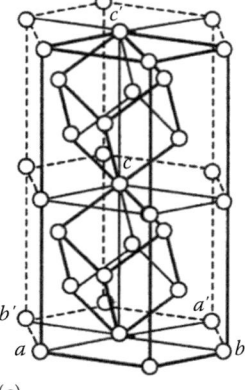

(a) (b) (c)

FIGURE 5.11 (a) Illustrating bcc-to-trigonal (R) structural transformation in LiNaSO$_4$. (b) Trigonal (R)-to-trigonal (P) transformation. (c) Showing the doubled unit cell. The transformation matrix is P (see text). (From Freheit, H., *Solid State Commun.*, 119, 539, 2001.)

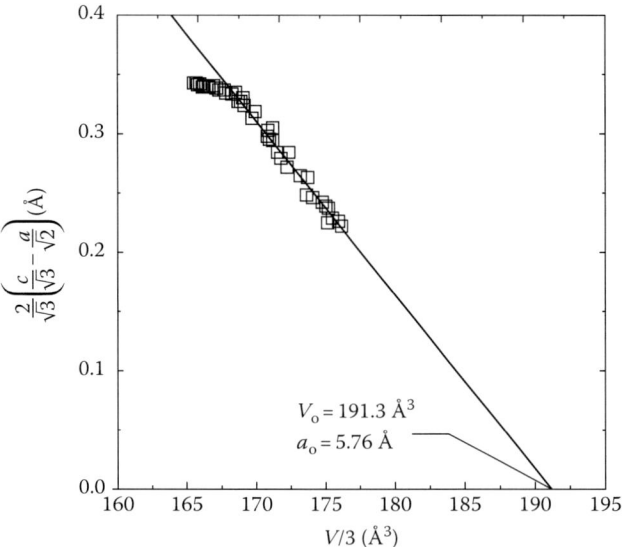

FIGURE 5.12   A plot of the equation $(2/\sqrt{3})(c/\sqrt{3} - a/\sqrt{2}) \sim (V/3 - VO)$ to show variation of shear strain with reduced volume. Extrapolation to zero strain gives volume of cubic phase. This plot demonstrates the role of spontaneous strain in causing the superionic phase transition in LiNaSO$_4$ at 788 K. (From Freheit, H., *Solid State Commun.*, 119, 539, 2001.)

In view of $e_t \propto e_a \propto Q^2$,

$$(2/3^{1/2})(c/3^{1/2} - a/2^{1/2}) \propto (V/3 - V_O) \tag{5.53}$$

Figure 5.12 shows the variation of shear strain with reduced volume (Equation 5.53).

The temperature dependence of thermal and athermal strains, which brings out the ferroelastic or ferroic nature of the phase transition, is shown in Figure 5.13.

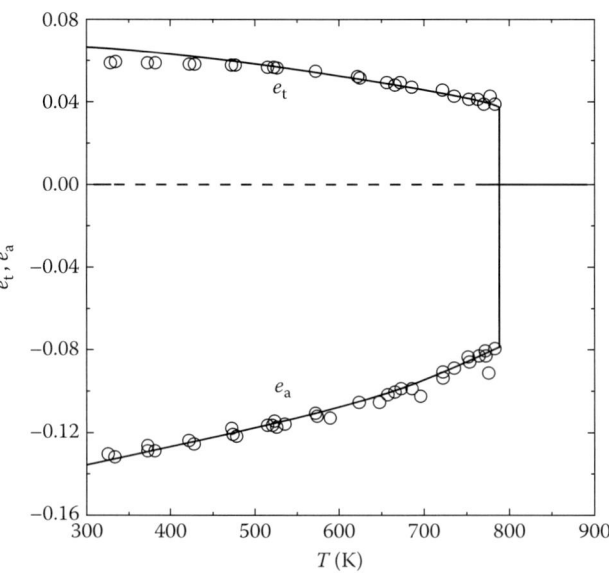

FIGURE 5.13   Temperature dependence of spontaneous strain (e_t, e_a) in LiNaSO$_4$. Note the sudden collapse of the spontaneous strain at 788 K in the manner of a ferroelastic or ferroic phase transition. (From Freheit, H., *Solid State Commun.*, 119, 539, 2001.)

Landau theory gives the difference in the Gibbs free energy between equilibrium phases (LT and HT forms of LiNaSO$_4$) as

$$\Delta G = \Delta G_{\text{Landau}} + \sum_{i,m,n} \lambda_{l,m,n} e_i^m Q^n + \frac{1}{2} \sum_{i,j} C_{ij} e_i e_j \tag{5.54}$$

where the second term is the coupling energy between e and Q and the third term is the elastic energy due to unit cell relaxation. Excess Gibbs free energy for the bcc $\sim$ *P31c* ordering transition in the *absence* of spatial Op fluctuations is

$$\Delta G(Q, e_i) = (1/2)a(T - T_c)Q^2 + (1/4)BQ^2 + (1/6)cQ^6 + \lambda_t e_t Q^2$$

$$+ \lambda_a e_a Q^2 + (1/2)C_t e_t^2 + (1/2)C_a e_a^2 \tag{5.55}$$

where

a, B, and c are Landau coefficients

T_c is the critical temperature

λ_i are the strain–OP coupling coefficients

C_i^O is the symmetry-adapted elastic constants of cubic phase

$$C_t^O = \frac{1}{2} \left(C_{11}^O - C_{12}^O \right) \text{(shear modulus)} \tag{5.56}$$

$$C_a^O = \frac{1}{3} \left(C_{31}^O + 2C_{12}^O \right) \text{(bulk modulus)} \tag{5.57}$$

For elastic equilibrium, $8G/8e_i = 0$ so that

$$e_t = -\left(\frac{\lambda_t}{C_t^O} \right) Q^2 \tag{5.58}$$

$$e_a = -\left(\frac{\lambda_a}{C_a^O} \right) Q^2 \tag{5.59}$$

This leads to the renormalization of fourth-order term in the free energy Equation 5.8

$$G(Q) = (1/2)a(T - T_c)Q^2 + (1/4)B^*Q^4 + (1/6)cQ^6 \tag{5.60}$$

where

$$B^* = B - 2\left(\lambda_t^2 / c_t^O \right) + \left(\lambda_a^2 / C_a^O \right) \tag{5.61}$$

In the LT phase, OP at thermodynamic equilibrium ($8G/8Q = 0$) is given by

$$Q^2 = (2/3)Q_0^2 \{1 + [1 - (3/4)(T - T_c) / (T_{tr} - T_c)]^{1/2}\} \tag{5.62}$$

Inserting experimental values of e_a (or e_t) for Q^2, $T_c = 635$ K, $e_{a,0} = -0.0775$, $e_{t,0} = 0.0382$, the magnitude of the OP jumps at the transition $Q_0 = 0.704$ indicating strong first-order character. The Landau coefficients are $a = 0.107$ kJ/mol K ($a = 2\Delta H_{tr}/T_{tr}Q_0^2$ with delta $H_{tr} = 20.8$ kJ/mol).

A local dynamic probe in the form of infrared (IR) has recently been applied to the phase transitions in LiNaSO$_4$ [5.25]. Interestingly, a Li-7 and Na-23 NMR study has found that Li and Na diffusions both increase on heating (Kanashiro [5.26]).

An interesting observation is that the first-order phase transition in Li$_2$SO$_4$ from the low-temperature insulating monoclinic phase to the high-temperature cubic phase (fcc) quite like the non-Landau type transition in AgI changes remarkably to an athermal phase transition just by replacing one Li by Na. Our early work on Li$_2$SO$_4$ [5.27,5.28] has shown that it is possible to quench the fcc phase from the melt with the aid of a few percent Li$_2$CO$_3$.

The normally unquenchable LiNaSO$_4$ has been quenched using a small percentage of Li$_2$CO$_3$ as an aid to obtain a hard ceramic at ambient temperature with a disordered hexagonal structure ($a = 0.7635$ nm, $c = 0.9861$ nm) and with a log cT of -4.5 at 500 K [5.29] to be compared to -19 obtained from the single-crystal conductivity plot of Mellander et al. [5.30] by extrapolation which strongly suggests substantial quench aided enhancement of conductivity by successfully thwarting the phase transition. The Arrhenius slopes of 0.11 eV ($373 < T < 475$ K) and 0.95 eV ($498 < T < 528$ K) have been deduced as conductivity activation energies.

5.4 Order–Disorder Phase Transitions in Li-Transition Metal Oxides

It is interesting to observe that thermal transitions in LiNaSO$_4$ are driven by competing elastic and thermodynamic forces that favor the high mobility of Li$^+$ (and Na$^+$) with a little help from the fairly high symmetry crystal structure. A natural question to ask is: Is it possible to find structural platforms that facilitate "in-and-out" movement of Li$^+$ ions in the manner of "insertion and deinsertion" into/out of the platform? Are there phase transitions in structures that depend just on the amount of Li they contain? Search for answers brings us to a discussion on transition metal oxides that provide 2D and 3D platforms for Li$^+$ ordering and disordering leading to structural phase transitions of which we discuss two examples.

However, Li intercalation as a unique process of electrochemical energy storage was pioneered by Whittingham who applied it to look at the physics and chemistry of Li$_x$TiS$_2$ [5.31], TiS$_2$ being a member of layered transition metal dichalcogenides [5.32]—predecessors of transition metal oxides in solid state ionics.

5.4.1 LiCoO$_2$ and LiMnO$_2$

Lithium cobaltite (LiCoO$_2$) is an important member of the Li-transition metal oxide family LiMO$_2$ (M = Mn, Fe, Co, Ni) possessing interesting and applicable properties—fast Li$^+$ ion transport, superconductivity, thermoelectricity, and semiconductor-to-metal phase transition. LiCoO$_2$ undergoes several structural phase transitions that are yet to be interpreted in a unified manner. The complexity of this interesting problem arises because the composition (Li and O) and the structure of the phase formed change depending on the conditions and method of synthesis and regions of electrochemical operation (during which the valence state of Co also changes).

There are essentially three modifications of LiCoO$_2$: (1) the LT phase formed at 400°C having a spinel-like structure (space group $Fd3m$) in which the usual spinel positions 8(a) remain vacant while "non-spinel" sites 16(c) are occupied by Li$^+$ ions, (2) a HT modification formed at 800°C and has a layered (alfa NaFeO$_2$ type) structure (space group $R\bar{3}m$) the α-phase, and (3) an unstable (or metastable) phase with NaCl structure, the γ-phase. Note that the space group of the HT phase is not a subgroup of the LT space group like a non-Landau-type phase transition referred to earlier [5.33]. The electrochemical activity suitable for battery application is exhibited by the HT phase while the spinel phase is formed upon intercalation and deintercalation of Li. This peculiar circumstance makes the study of phase transitions in LiCoO$_2$ an experimental and theoretical challenge. From an interesting study addressing this issue, Choi and Manthiram [5.34] revealed that the extraction of Li from the spinel-like LT phase leads to the conventional spinel Li$_{1-x}$Co$_2$O$_4$–delta ($0.44 \leq (1-x) \leq 1.00$.

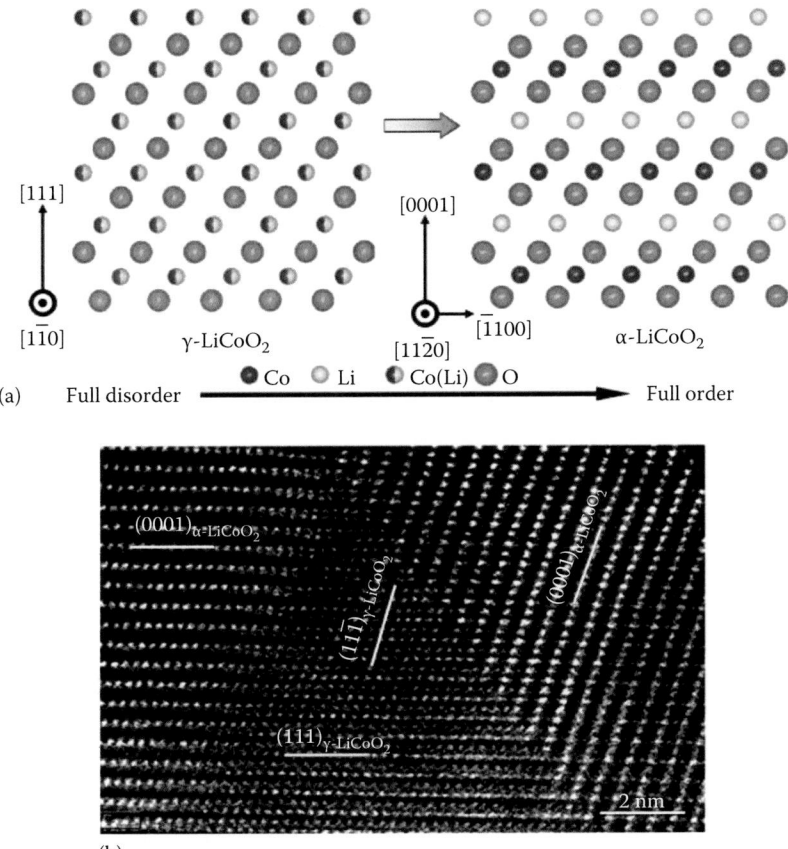

FIGURE 5.14 Relationship between the crystal structures of γ-LiCoO₂ and α-LiCoO₂: (a) a schematic of the model structures; (b) a γ-LiCoO₂ particle wedged between two α-LiCoO₂ particles in a poor-quality thin film deposited on an atomically rough substrate. The basal planes of the two neighboring α-LiCoO₂ particles are aligned with (111) and (1D1D1-) planes of the rocksalt structured γ-LiCoO₂ particle. (From Huang, R. et al., *Mater. Chem. Phys.*, 133, 1101, 2012.)

A recent study by Huang et al. [5.35] has looked at the phase transitions in LiCoO₂ thin films fabricated by pulsed laser deposition onto sapphire substrates. Interestingly, they have characterized the relationship between γ-LiCoO₂ and α-LiCoO₂ (Figure 5.14).

Deposited films with rocksalt structure (gamma phase) transition upon annealing in air at 600°C into a fully ordered α-phase, with the atomically sharp film–substrate interface becoming rougher during annealing and step defects forming, possibly due to a localized reaction at the interface.

What could be the mechanism of the formation of the low-symmetry phase? The higher symmetry LT phase is not the most stable one, while the layered structure (α-phase) is more stable than the spinel phase! The rock salt phase of Huang et al. [5.35] must thus be metastable.

More specifically, what are the thermodynamic conditions for their existence? The types of transformation? And the specific features in the structure of the LiCoO₂ phases? These questions are prompted by the difficulty in the interpretation of x-ray and neutron diffraction patterns.

Thus, when oxygen atoms form an ideal cubic close-packed framework, the diffraction patterns for both LT and HT phases are identical. In both the structures, Li and Co atoms occupy octahedral sites to form LiO₆ and CoO₆ octahedra. In the rhombohedral, HT modification layers of Lui and Co atoms located on opposite sides of oxygen layers alternate with each other. In the ordered phase, Li and Co atoms form Li and Co layers, respectively. In the LT modification, 25% of Li and Co atoms are characterized by a disordered arrangement forming Co-rich (Li₀.₂₅Co₀.₇₅) and Li-rich (Li₀.₇₅Co₀.₂₅) layers.

TABLE 5.1

Low-Symmetry Phases Relevant to Phase Transitions $LiCoO_2$ (Generally $LiMO_2$ [M = Mn, Co, Ni]) Generated from the Group Theoretical Analysis of the Prototype Phase with Space Group *Fm3m*

Number	4D Order Parameter	Derived Group	Volume Ratio	Observed Structure
1	eta eta eta eta	*Fm3m*	8	Unstable/metastable $LiCoO_2$
2	+ − + +	*Fd3m*	8	LT phase $LiCoO_2$
3	0 0 0 +	$R\bar{3}m$	2	HT phase $LiCoO_2$
4	0 + + 0	*Cmmm*	4	
5	eta1 eta1 eta2 eta2	*Immm*	8	
6	1 2 2 2	$R\bar{3}m$	8	HT phase $LiCoO_2$
7	1 −1 2 2	*Ibmm*	8	
8	0 0 1 2	*C2/m*	4	Monoclinic $LiMnO_2$
9	1 2 2 3	*C2/m*	8	
10	1 2 3 4	$P\bar{1}$	8	

Source: Based on Talanov, V.M. et al., *Glass Phys. Chem.*, 33, 596, 2007; See for a comprehensive treatment of phase transitions in the Landau picture Ghozien, M.H.B. and Milk, Y., *J. Phys. C Solid State Phys.*, 16, 4365, 1983.

Talanov et al. [5.36] have used group-theoretical and thermodynamic methods in the Landau phenomenological theory [5.7] of phase transitions to answer the questions raised earlier. They hypothesize that there is a prototype phase generating the entire variety of phase transitions in $LiCoO_2$. This prototype phase is a phase with NaCl structure (*Fm3m*) in which Li and Co atoms are located randomly at the 4(a) sites while O atoms occupy the 4(b) crystallographic sites. This hypothesis is based on two facts:

1. Experiments show that the third modification (apart from LT and HT) formed under special conditions (and also realized in thin films) has a NaCl structure.

2. The structures of the other two $LiCoO_2$ modifications can be constructed as a result of atomic ordering in the NaCl structure.

The main result of this analysis is that the possible critical irreducible representations that generate the entire gamut of phases are 4D representations of the group *Fm3m*. They are listed in Table 5.1. Structures of all the experimentally observed (and yet to be observed) phases of $LiMO_2$, including those of $LiCoO_2$ and also of $LiMnO_2$ (monoclinic phase *C2/m*), are thus predicted in a unified and general manner.

$LiMnO_2$, a good substitute battery cathode material for the more expensive and less safe $LiCoO_2$, is among the many materials possibilities in the Li–Mn–O system with interesting phase transitions. Thus, a recent study of RF magnetron sputtered thin films (on silicon and stainless steel substrates) in the Li–Mn–O ternary has monitored through vacuum annealing the transition from Li_2MnO_3 with Mn^{4+} oxidation state to the stoichiometric o–$LiMnO_2$ (with Mn^{3+}) phase with orthorhombic structure (space group *Pmmm*) through x-ray diffraction and Raman spectra [5.37]. This transition involves a reduction in Li/Mn ratio from 2.0 to ~1.0 with loss of Li. A well-defined XRD pattern and three Raman peaks are due to Mn–O stretching vibrations, O–Mn–O bending mode, and MnO_6 octahedral cage vibrations.

5.4.2 $LiMn_2O_4$

Lithium manganate ($LiMn_2O_4$)—LM—is a cubic spinel ($Fd\bar{3}m$) at room temperature.

Figure 5.15a shows a unique example of a magnetic SIC. Fast Li^+ dynamics, mixed valence of Mn (3+ and 4+), and the chemistry of the Mn–O bond makes the physics of LM exciting. Indeed, Li^+ superionic conductivity and Mn-based magnetism are "connected" by a first-order phase transition at 290 K. Li^+ and Mn ions are accommodated in the cubic close-packed skeleton of $O^=$ ions. Li^+ ions occupy 1/8 of the available tetrahedral sites (interstitial space), while Mn ions are found in one-half of the available octahedral sites. The interstitial space is essentially a 3D network of tetrahedral and octahedral that shares faces facilitating a fairly rapid 3D diffusion of the tiny Li^+ ions through the structure. The octahedral sites in this

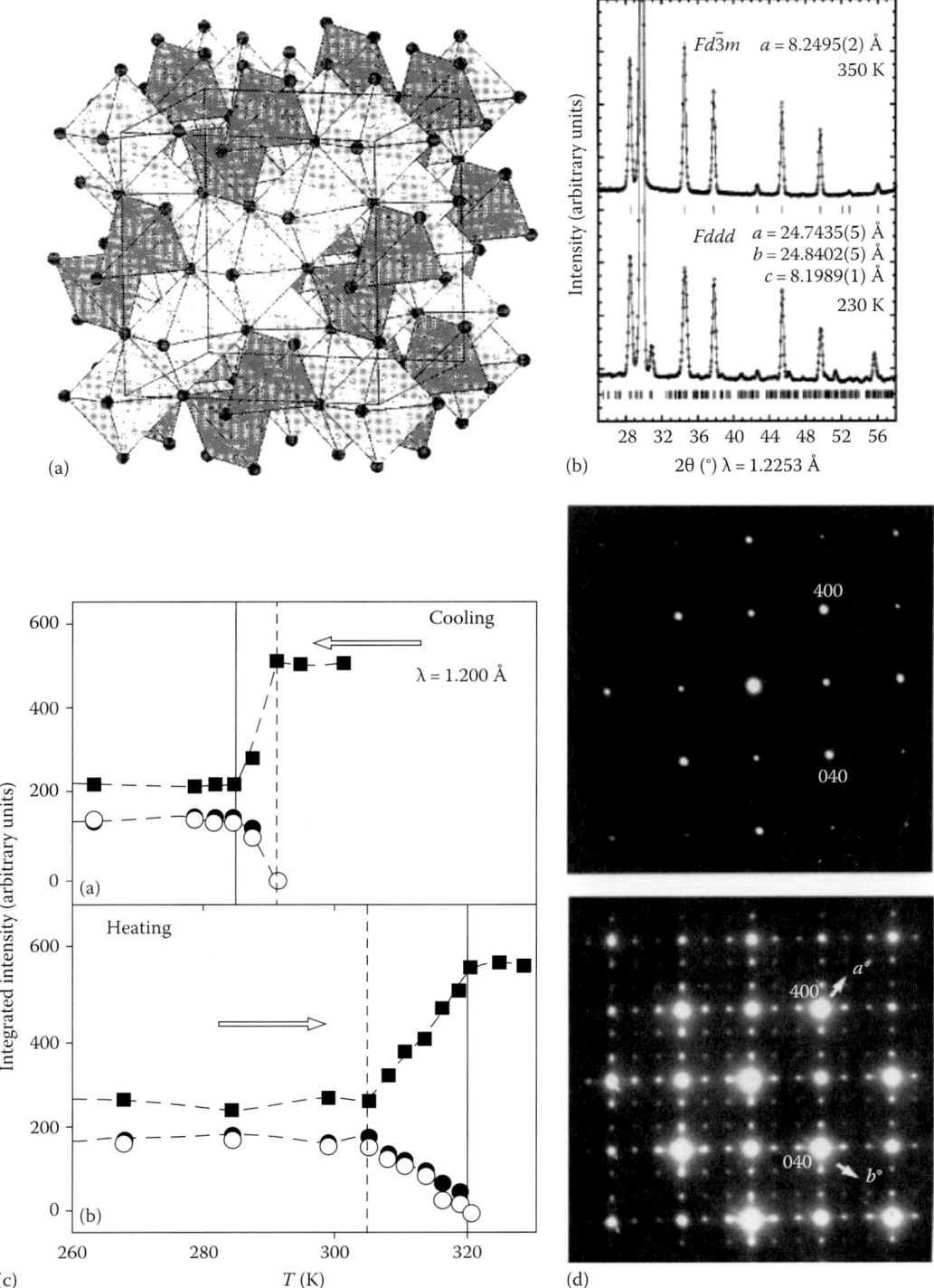

FIGURE 5.15 Diffraction probes of $LiMn_2O_4$. (a) Polyhedral representation of the room temperature cubic structure of $LiMn_2O_4$. (From Wills, A.S. et al., *Chem. Mater.*, 11, 1510, 1999.) (b) Neutron diffraction patterns at 350 and 230 K. Note partial ordering in the low-temperature phase of $LiMn_2O_4$. (c) Intensity versus temperature plot of a triplet of synchrotron x-ray diffraction peaks ($27.5 < 2\theta < 28.5$). Note the hysteresis in the phase transition of $LiMn_2O_4$. (d) Electron diffraction pictures of HT phase (top) and LT phase (bottom) of $LiMn_2O_4$. (Parts (b)–(d) from Rodriguez-Carvajal, J. et al., *Phys. Rev. Lett.*, 81, 4660, 1998.)

unusual structure form an anomalous lattice that essentially leads to perfect short-range order. However, finite entropy is still maintained for the system. This circumstance makes nearest-neighbor forces and even long-range Coulomb forces ineffective in creating long-range order. Focusing on the network of vertex-sharing tetrahedra only and putting spins on the vertices of tetrahedra, the following question arises: how do these spins align themselves so that they are simultaneously antiparallel to each other? The spins experience a confusion and compromise their position leading to a "canted" arrangement of spins in which the vector sum of the magnetic moments over a tetrahedron is zero. What is the magnetic ground state of the system? As the tetrahedra share only one vertex, a unique spin structure cannot be defined!

Is LM a metal or a semiconductor? The average valence of Mn in LM is 3.5, the average of 3 and 4. Mn^{3+} is a Jahn–Teller active ion which means that the ground state (orbital configuration) is degenerate and is unstable against distortions that remove this degeneracy. In the orbital picture, $Mn^{3+}(3d^4)$ ion has a configuration $t_{2g}(3)e_g(1)$. Now the Mn ions are bridged by 0 ions in a 90° arrangement in the spinel structure. This bridging arrangement makes the e_g orbitals of neighboring Mn ions to be orthogonal. More importantly, the e_g electrons are localized. Therefore, at room temperature LM is a semiconductor (conductivity ~100 µS/cm)and not a metal—a direct consequence of Mn^{3+} being a Jahn–Teller ion capable of triggering a phase transition through a structural distortion!

A dynamical Jahn–Teller distortion around Mn^{3+}–$O^=$ octahedral is held responsible for the lowering of the lattice symmetry. Conditions are thereby created for the formation of "small polarons" (an electron coupled to the nearby polarization) around the Mn^{3+} ions, making LM a polaronic semiconductor rather than a degenerate semiconductor or a metal. While the charge carriers are known to be electrons, the band gap and effective masses of carriers as well as the dielectric behavior would be interesting to know.

Stoichiometric LM transforms at 290 K from cubic to orthorhombic or pseudo-tetragonal phase as probed by neutron and x-ray and electron diffraction (Figure 5.15b–d) as determined by Rodriguez-Carvajal et al. [5.38]. Clues to the origin of the phase transition are provided by calorimetry, conductivity, and magnetic susceptibility measurements displayed in Figure 5.16a through c, respectively, taken from the work of Wills et al. [5.39].

Hysteresis observed in differential scanning calorimetric (DSC) scans establish the thermodynamic origin of phase transition as first order with latent heats 5.7 and 6.0 J/g in the heating and cooling runs, respectively, pointing to entropy values of 5.7 J/300.8 g K and 6.0 J/291.2 g K and a hysteresis of 300.8−291.2 = 19.6 K.

The anomaly in conductivity accompanied by hysteresis observed in ionic conductivity plots corroborates the DSC results. The magnetic susceptibility and its inverse have a discontinuous change around

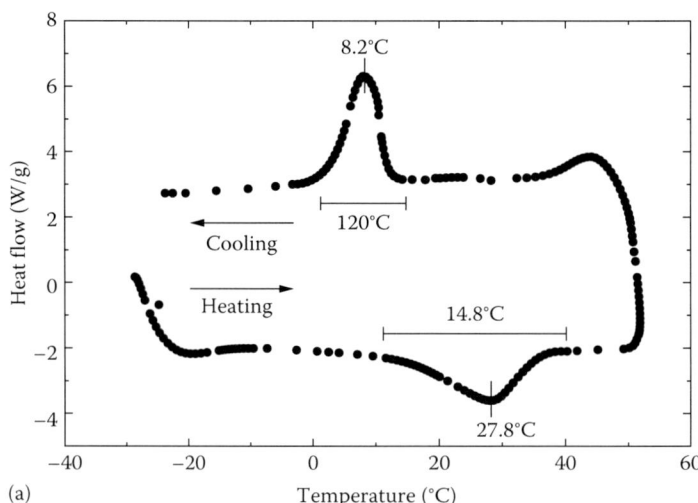

(a)

FIGURE 5.16 Thermal, electrical, and magnetic response of $LiMn_2O_4$. (a) Differential scanning calorimetric (DSC) scans of $LiMn_2O_4$ for one heat–cool cycle. Note hysteresis in the phase transition. Latent heats are 5.7 J/g (heating) and 6.0 J/g (cooling).   *(Continued)*

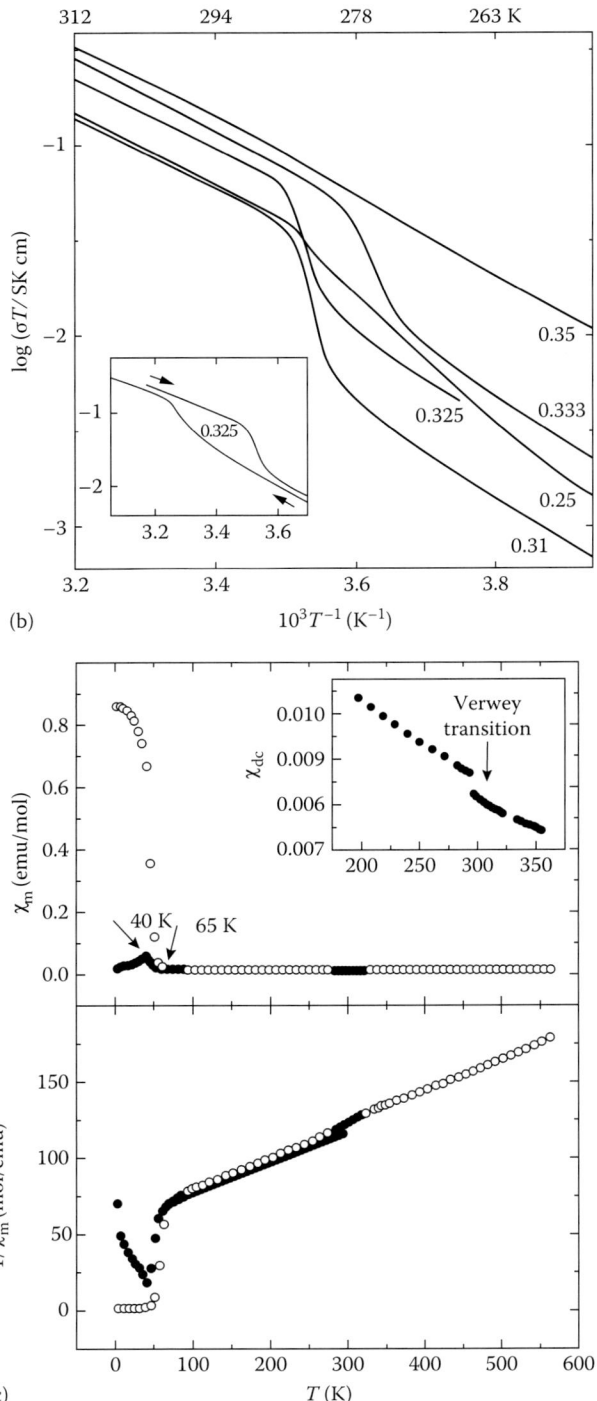

FIGURE 5.16 (*Continued*) Thermal, electrical, and magnetic response of $LiMn_2O_4$. (b) Li^+ ion conductivity of $Li_xMn_2O_4$ in the temperature range including phase transition ($0.25 \leq x \leq 0.35$. $x = 0.33$) corresponds to stoichiometric composition. Note hysteresis for a nearby x (=0.325). (c) Magnetic susceptibility and its inverse of $LiMn_2O_4$ between 2 and 500 K. (•), zero field-cooled measurement; (o), field-cooled measurement. The 400–500 K data can be fitted to the Curie–Weiss law chi $= C/(T -$ theta Weiss) to yield values of the Curie constant C and Weiss temperature (theta Weiss) 4.86(15) emu K mol^{-1} and −300(20) K, respectively. The value of the Curie constant is consistent with Mn^{3+} being in the high-spin configuration and the average Mn valency is 3.50(8). (From Wills, A.S. et al., *Chem. Mater.*, 11, 1510, 1999.)

290 K providing a vital clue to the origin of the discontinuous change in the thermodynamic and ionic/electronic and magnetic properties of LM. The origin of this unusual phenomenon is sought and found in Verwey's work where a transition similar to that depicted in the inset of Figure 5.16c that occurs in familiar magnetite Fe_3O_4 has been explained [5.40]. This Verwey transition refers to the Fe^{2+}–Fe^{3+} ordering in magnetite within the octahedral sites of the spinel network (the ones occupied by Mn in the LM lattice). This occurs when the Coulomb interaction overcomes the kinetic energy of the carriers below the phase transition temperature—a situation referred to as "charge ordering" which also occurs in Mn-based perovskites (La, Ca) MnO_3 or colossal magneto resistance materials [5.41]. The tell-tale haloes seen in the electron diffraction pattern of LM at low temperatures (Figure 5.15d) is visual proof of the existence of charge ordering or "Wigner crystallization." The charge carriers involved in this condensation process are small polarons or Jahn–Teller polarons.

For temperatures less than 100 K, a larger cell tetragonal structure develops with the distributions of Mn^{3+} and Mn^{4+} becoming apparently nonrandom. Long-range magnetic order persists, however, making LM a geometrically frustrated antiferromagnet like its delithiated cousin λ-MnO_2 [5.42]. In a geometrically frustrated system, charge segregation and further (than nearest) neighbor interactions tend to control the magnetic behavior. Geometric frustration results whenever the geometry around a magnetic ion does not allow the ground state to be nondegenerate, that is, only large spin degeneracy occurs for the ground state of the system and the magnetic subsystem is unable to select a unique ground state. Therefore, the system must necessarily evolve through three regimes: a HT paramagnetic regime, a low(est) temperature long-range ordered antiferromagnetic regime, and a short-range ordered critical regime.

5.4.2.1 Pressure-Induced Phase Transitions

Experiments on the dependence of electrical and allied properties of SICs help extract useful information on (1) the mechanism of disordering of ionic crystals leading to a transition to state of superionic conduction and (2) the kinetics of charge transport [5.43]. Such a transition is usually a first-order phase transition, and therefore, the pressure dependence of the transition temperature T_{tr} is described by the Clausius–Clapeyron equation

$$dT_{tr}/dP = (V_\beta - V_\alpha)/(S_\beta - S_\alpha) \tag{5.63}$$

where $V_{\alpha,\beta}$ and $S_{\alpha,\beta}$ are the specific volumes and entropies of the α- and β-phases between which the transition occurs. Different phases generally possess different values of isothermal compressibilities K_α and K_β so that Equation 5.63 may be written as

$$\frac{dT_{tr}}{dP} = \left[V_\beta^0 - V_\alpha^0 - (V_\beta K_\beta - V_\alpha K_\alpha)P \right](S_\beta - S_\alpha)^{-1} \tag{5.64}$$

where $V_{\alpha,\beta}^0$ are the values of the specific volumes at the pressure $P = P_0$ taken as the initial pressure. Gurevich and Kharkhats [5.43] while discussing this aspect have pointed out that normal ion conductors such as AgCl are distinguished from superionic conductors such as α-AgI from their high and low activation volumes, respectively.

Just as temperature, pressure too affects the phase transitions in SICs which is seen through pressure-induced changes in the transition temperatures. We now discuss the case of pressure dependence of phase transitions in $RbAg_4I_5$.

External pressure is believed to help sort out diffusion mechanisms for ionic conduction. Figure 5.17 shows the pressure dependence of first-order (T_{c1}) and second-order (T_{c2}) phase transition temperatures: ~130 and ~200 K, respectively, at ambient pressure in MAg_4I_5 (M = Rb, K, NH_4^+). This is based on conductance (G) measurements over a temperature range 127–150 K at selected constant pressures up to 7 kbar [5.44]. The most interesting observation is that both T_{c1} and T_{c2} show large quadratic pressure dependences: $(p - p_{0i})^2 = A_i(T_{ci} - T_{c0i})$ where p_{0i}, A_i, and T_{c0i} are pressure-independent constants.

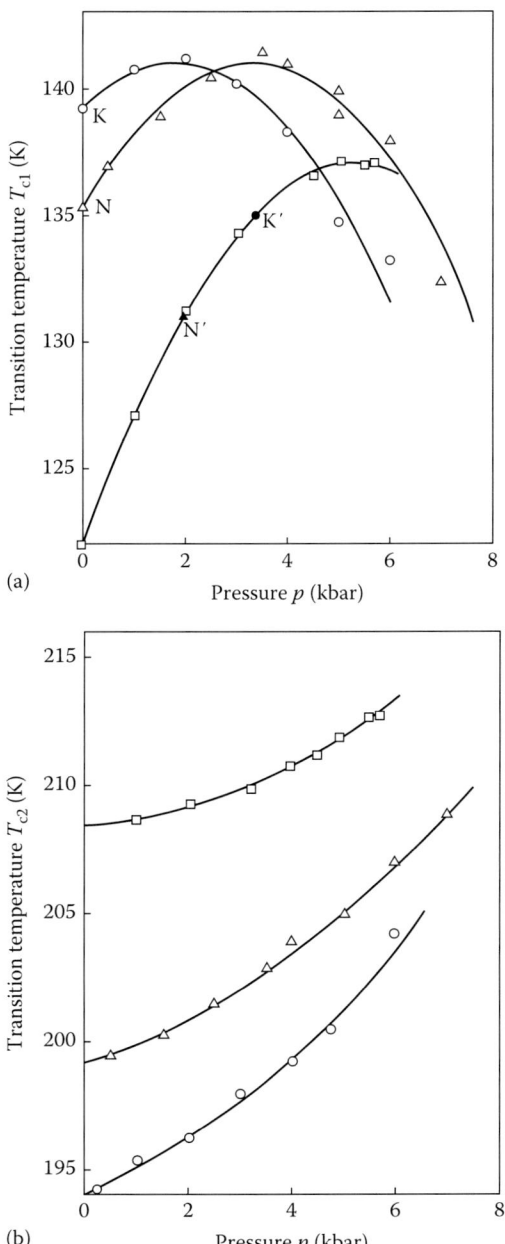

FIGURE 5.17 Pressure (p) dependence of phase transitions in pressed powders of MAg_4I_5 ($M = Rb, K, NH_4$). (a) First-order transition temperature $T_{c1}(p)$ and (b) second-order transition temperature $T_{c2}(p)$. $RbAg_4I_5$: O, K, Ag_4I_5: □; $NH_4Ag_5I_5$: Δ. (From Fujimoto, S. et al., *J. Phys. D. Appl. Phys.*, 13, L95, 1980.)

The large pressure dependence of T_{c1} is explained through the Clausius–Clapeyron equation. The molar volume (V_c) of the conductive phase ($T > T_{c1}$) is larger than that (V_n) of the nonconductive phase ($T < T_{c1}$) ($\Delta V = V_c - V_n > 0$). However, the conducting phase is softer than the nonconducting phase. With increasing pressure, the volume of the conductive phase decreases eventually becoming less than that of the nonconducting phase ($\Delta V < 0$). Applying the relation $dT_{c1}/dp = \Delta V/\Delta S$, where ΔS is the molar entropy change, the sign of dT_{c1}/dp is known to change. From Figure 5.17b, the derivative dT_{c2}/dp at atmospheric pressure increases on replacing M from Rb to K. dT_{c2}/dp increases with pressure indicating the decisive role of lattice constant (1.113 nm for KAg_4I_5 through 1.119 nm for $NH_4Ag_4I_5$ to 1.224 for $RbAg_4I_5$) in the second-order phase transition.

Recent studies on pressure-induced phase transitions are the following:

1. High pressure (12–17 GPa) stabilizes the superionic AgI–V phase with diffusion coefficients $\sim 3.4 \times 10^{-4}$–8.6×10^{-4} cm^2/s at room temperature [5.45].

2. Under pressure Li_3N shows phase transitions between the structure types α-$Li_3N \rightarrow Na_3As \rightarrow Li_3Bi$ [5.46].

3. CaF_2 transforms from the fluorite phase ($Fm\bar{3}m$) to the cotunnite ($PbCl_2$) phase with space group $Pnma$, $Z = 4$ at a pressure of 8–10 GPa. This is a reconstructive phase transition. The transition path has been constructed via molecular dynamics simulations. An artificially prepared transformation route connects fluorite to cotunnite. Subsequent trajectory rectification evolves to a distinct picture of the favored mechanism. The mechanism is characterized by nucleation and growth of the new phase. The overall transformation mechanism is identified as a symmetry lowering step from the cubic to the orthorhombic atomic configuration. The reorganization of one-half of the octahedral voids causes the transformation. At the interface between the cubic and the orthorhombic structure, a pressure-induced local melting of the fluoride sublattice is observed. This produces defects that allow for the reorganization of the calcium sublattice eventually leading to the recrystallization of the fluoride ions fixating one of the stable structures. Variation of the thermodynamic parameters reveals that the mechanism is conserved over the experimentally relevant range. In accord with the experiment, an increasing tendency toward incomplete transformation on lowering the temperature is indicated in computation [5.47].

A pedagogically intuitive group–subgroup relation is derived (Figure 5.18):

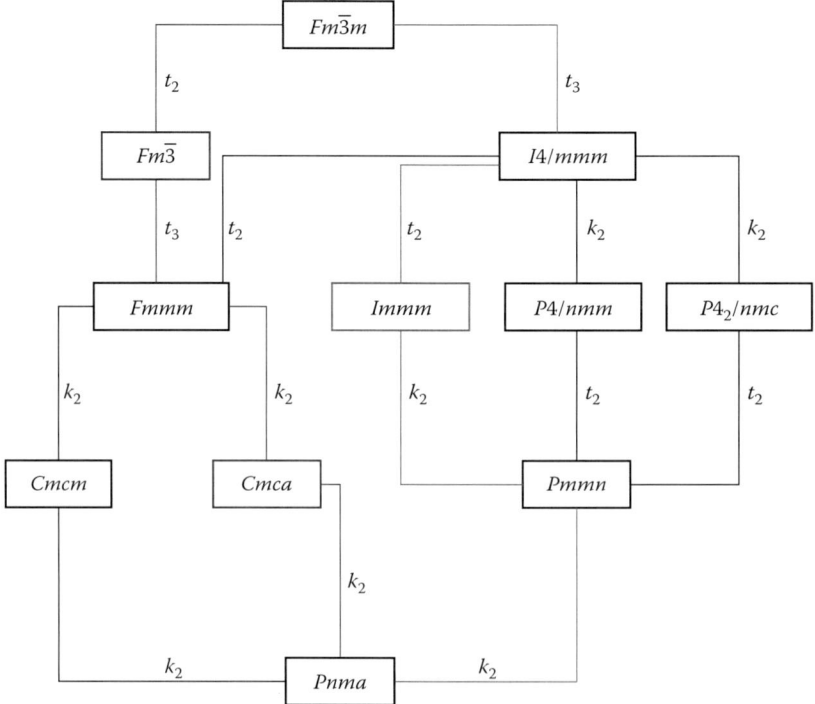

FIGURE 5.18 Group–subgroup relation between the space group of fluorite $Fm\bar{3}m$ and the space group of cotunnite $Pnma$ is depicted. The path used for the geometric modeling is highlighted in light gray and the preferred mechanism in dark gray. The intermediate space groups $Immm$ and $Cmca$ of the first branch and the final branch are highlighted. (From Boulfelfel, S.E. et al., *Phys. Rev.*, B74, 094106, 2006.)

Before closing this section and chapter, a unique and rare pressure-induced solid state synthesis of NASICON-type structures $Na_3Al_2(PO_4)_3$ with high bulk conductivity carried out from 850 to 1273 K at pressures up to 8 GPa is briefly discussed [5.48]. $Na_3Al_2(PO_4)_3$ crystallizes in a NASICON-type structure at 923 K and pressures above 0.5 GPa. It then becomes an ionic conductor with a bulk conductivity of ~50 mS/cm at 600 K. In fact the form recovered at ambient conditions is the monoclinic deformation of the rhombohedral NASICON unit cell. Two non-quenchable phase transitions leading to the true *R-3c* NASICON cell are realized in high-pressure impedance spectroscopy experiments. For details [5.48] may be consulted.

5.5 Summary

Phase transition is an inevitable phenomenon especially for solid state ionics remembering that a perfect ionic solid at absolute zero temperature is a source of infinite electrochemical energy or ionic potential energy. Phase transition is a means of unleashing this ionic power or electrochemical potential through Frenkel defects and their thermal evolution until the ionic solid becomes a liquid.

Phase transitions—especially first order and Faraday as they occur in AgI and PbF2, respectively— enable insights and applicable strategies to be obtained of the superionic state in the general free energy functional picture of Landau through an OP linking theory to the experiment. Anomalies in electrical conductivity and heat capacity are successfully understood and accounted for satisfactorily through the statistical mechanical approach as illustrated O'Reilly and Mezrin theories, and more fundamentally by a probability attached to the OP now treated as a random variable as done by Huberman [5.17] and Burbano et al. [5.16].

However, the Landau formalism can only serve as a point of departure in solid state ionics where the effects of phonons and bonding electrons—to be considered in the next two chapters—are inextricably linked to the conduction phenomena.

Reduction in the physical size of the solid state ionic materials (typically AgI and PbF2) has a profound influence on the phase transitions that need to be understood in a generalized manner—an aspect considered in a later chapter.

Athermal phase transitions in $LiNaSO_4$ and related materials bring in elasticity as a competing force dictating the superionic phase transition connecting solid state ionics to multiferroics.

Order–disorder phase transitions that occur in Li-based superionics $LiCoO_2$ and $LiMn_2O_4$ (a magnetic superionic) are fundamentally interesting as well as a materials challenge to Li^+ ion battery development. A battery cathode must be robust with respect to dynamic electrochemical environments existing in the device at work. Thus, the suppression of phase transition(s) while still retaining beneficial electrical properties of the oxide materials becomes the development agenda and motivates the search for newer materials.

Problems

5.1 (a) AgI transforms from wurtzite to bcc phase at 145°C accompanied by a volume change of −2.2 cm^3 and an enthalpy change of 6.145 kJ/mol.

(b) Li_2SO_4 transforms from monoclinic structure to fcc structure at 590°C undergoing a volume change of +3.81 cc and an enthalpy change of 28.842 kJ/mol.

Calculate the entropies of the two first-order phase transitions. Comment on the relative magnitudes.

5.2 A two sublattice model for phase transition in solid electrolytes based on Feynman's variational principle is interesting. Sublattice melting, as it occurs in AgI, may be understood as a phase transition of (I) the mobile ion (Ag^+) system from solid to liquid phase and (II) no phase transition of the cage system of I^- ions. Develop a model for this phase transition by considering two models for the two systems (I) and (II). Model I: there is only one particle per cell of the mobile ion sublattice, which has no equivalent cells; particle is unable to go out of its own cell (cell is closed for both ground and excited states). Model II: there is at most one particle on the ground states of a cell; the particle is able to go out of the cell and occupy any available excited states (open cell for excited states but closed cell for ground states). Write the Hamiltonian H for the system. The exact free energy of the system Fexact is $-k_BT \ln T_r[\exp(-H/k_BT)]$, where T_r is the trace of the Hamiltonian. Derive the expressions for the entropy of the systems I and II by applying the Feynman's variational principle: $F_{exact} \leq F_{approximate} = F_0 + \langle H - H_0 \rangle_0$ where H_0 is a trial Hamiltonian with a free energy F_0 and $\langle -- \rangle_0$ represents the thermal average of the canonical ensemble for H_0. The problem can be solved numerically in terms of two OPs x and y whose temperature dependence reproduces the features of the first-order phase transitions in AgI-like superionic conductors. (see Reference 5.49).

5.3 The diagram here graphically connects the two phases of zirconia: monoclinic (*m*) and tetragonal (*t*). (From Chevalier, J. et al., *J. Am. Ceram. Soc.*, 92, 1901, 2009.)

Can the $m \rightarrow t$ transition be driven by the application of pressure?

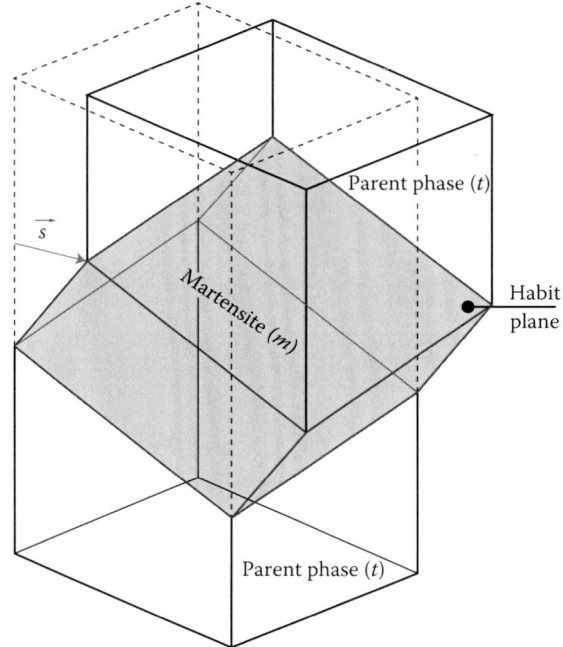

5.4 *Premelting and sublattice melting.* Superionic conductivity is essentially an ionic defect- and disorder-controlled phenomenon. Cation and anion Frenkel defects (vacancy–interstitial combination) and Fenkel disorder with considerable defect–defect interactions are usually involved. "Premelting" refers to a sudden buildup of Frenkel defect concentrations just before the bulk solid melts to become liquid. "Sublattice melting" means a process whereby one of the two (or more) of cation or anion substructures in the overall structure suddenly (or gradually in steps) becomes unsustainable and the crystal undergoes a first-/second-order structural phase transition. The first phenomenon occurs in AgBr with Ag^+ Frenkel defects and in CaF_2 with F^- Frenkel defects. The second phenomenon occurs in AgI with Ag^+ Frenkel defects and in PbF_2 with F^- Frenkel defects. Is there a correlation between defect concentration and metal–nonmetal bonding and observable physical properties such as ionic conductivity, magnetic susceptibility, and heat capacity in these materials? Discuss this with the aid of the three references given later: References 5.11, 5.20, and 5.51.

REFERENCES

5.1   D. Kondepudi, I. Prigogine, *Modern Thermodynamics*, Wiley, Chichester, U.K., 2004.

5.2   J.-C. Niepce, L. Pizzagalli, *Structure and Phase Transitions in Nanocrystals*, in *Nanomaterials and Nanochemistry*, Brechignac, P. Houdy, M. Lahmani (Eds.), Springer, Berlin, Germany, 2007.

5.3   M. Faraday, *Experimental Researches in Electricity*, Taylor & Francis, London, U.K., 1839.

5.4   R. Makiura, H. Kitagawa, K. Kato, *Spring-8 Research Frontiers*, JASRI, Hyogo Prefecture, Japan, 2009, p. 100, http://www.spring8.or.jp; R. Makiura, T. Yonemura, T. Yamada, M. Yamauchi, R. Ikeda, H. Kitagawa, K. Kato, M. Toakta, *Nat. Mater.* 8(2009) 476.

5.5   S. Chandra, *Superionic Solids*, North Holland, Amsterdam, the Netherlands, 1981.

5.6   M. B. Salamon, Ed., *Physics of Superionic Conductors*, Springer, Berlin, Germany, 1979, Chap. 7.

5.7   M. El-Batanouny, F. Wooten, *Symmetry and Condensed Mater Physics: A Computational Approach*, Cambridge, U.K.: Cambridge University Press, 2008, Chap. 17; F. Duan, J. Guojun, *Introduction to Condensed Matter Physics*, World Scientific Publishing Co., Singapore, 2005, Vol. 1, Chap. 15.

5.7a C. S. Sunandana, *AIP Conf. Proc.* 286(1994) 261.

5.7b A. Gray-Weale, P. A. Madden, *J. Phys. Chem. B* 108(2004) 6634.

5.8   S. Strassler, C. Kittel, *Phys. Rev.* 139(1965) A758.

5.9   S. Hull, *Rep. Prog. Phys.* 67(2004) 1233.

5.10 C. S. Sunandana, P. Senthil Kumar, *Bull. Mater. Sci.* 27(2004)1; R. S. Elliott, J. A. Shaw, N. Triantafyllidis, *J. Mech. Phys. Solids* 50(2002) 2463.

5.11 C. S. Sunandana, *Indian J. Pure Appl. Phys.* 51(2013) 296.

5.12 M. B. O'Reilly, *Phys. Status Solidi* 48a(1978) 489.

5.13 J. B. Boyce, B. A. Huberman, *Phys. Rep.* 51(1970) 189.

5.14 V. A. Mezrin, *Phys. Status Solidi* 114a(1998) 145; R. Kikuchi, *Phys. Rev.* 81(1951) 988.

5.15 N. H. March, M. P. Tosi, *J. Phys. Chem. Solids* 42(1981) 809.

5.16 J. C. Burbano, R. A. Vargas, D. Pena Lara, C. A. Lozano, H. Correa, *Solid State Ionics* 180(2009) 1553.

5.17 B. A. Huberman, *Phys. Rev. Lett.* 32(1974) 1000.

5.18 D. O. Welch, G. J. Dienes, *J. Phys. Chem. Solids* 38(1977) 809.

5.19 Ya. Ya. Gurevich, Yu. I. Kharkats, *Sov. Phys. JETP* 45(1977) 968; *J. Phys. Chem. Solids* 39(1978) 751.

5.20 N. Hainovsky, J. Maier, *Phys. Rev. B* 51(1995) 15789.

5.21 M. Aniya, S. Ichihara, *J. Phys. Chem. Solids* 66(2006) 288.

5.22 D. Bharathi Mohan, C. S. Sunandana, *J. Phys. Chem. Solids* 65(2004) 1669.

5.23 D. Bharathi Mohan, C. S. Sunandana, *AIP Conf. Proc.* 1349(2011) 147.

5.24 H. C. Freheit, *Solid State Commun.* 119(2001) 539.

5.25 M. Zhang, A. Putnis, E. J. H. Salje, *Solid State Ionics* 177(2006) 37.

5.26 T. Kanashiro et al., *J. Phys. Soc. Jpn.* 63(1994) 3488.

5.27 P. Balaya, C. S. Sunandana, *Solid State Commun.* 70(1989) 581.

5.28 P. Balaya, C. S. Sunandana, *J. Phys. Chem. Solids* 55(1994) 39.

5.29 S. Rama Rao, C. S. Sunandana, *Solid State Commun.* 98(1996) 927.

5.30 B. E. Mellander, B. Graneli, J. Roos, *Solid State Ionics* 40/41(1991) 162.

5.31 M. S. Whittingham, in *Physics and Chemistry of Materials with Layered Structures*, Vol. 6, F. Levy (Ed.), D Ridel Publishing Co., Dordrecht, Germany, 1979.

5.32 G. V. Subba Rao, C. S. Sunandana. Layered chalcogenides and intercalated materials, in *Preparation and Characterization of Materials*, J. M. Honig, C. N. R. Rao (Eds.), Academic Press, New York, 1981, pp. 261–282.

5.33 M. Yoshio, H. Noguchi, A review of positive electrode materials for lithium-ion batteries, in *Li-Ion Batteries Science & Technology*, M. Yoshio, R. J. Brodd, A. Kozaum (Eds.), Springer, New York, 2009, Chap. 2.

5.34 S. Choi, A. Manthiram, *J. Electrochem. Soc.* 149(2002) A1157.

5.35 R. Huang et al., *Mater. Chem. Phys.* 133(2012) 1101.

5.36 V. M. Talanov, V. B. Shirokov, V. I. Torgashev, G. A. Berger, V. A. Burtsev, *Glass Phys. Chem.* 33(2007) 596; See for a comprehensive treatment of phase transitions in the Landau picture M. H. B. Ghozien, Y. Milk, *J. Phys. C: Solid State Phys.* 16(1983) 4365.

5.37 J. Fisher et al., *Thin Solid Films* 528(2013) 217.

5.38   J. Rodriguez-Carvajal et al., *Phys. Rev. Lett.* 81(1998) 4660.

5.39   A. S. Wills et al., *Chem. Mater.* 11(1999) 1510.

5.40   F. Walz, *J. Phys. Cond. Matter* 14(2002) R285.

5.41   A. J. Millis, B. I. Shraiman, R. Mueller, *Phys. Rev. Lett.* 77(1996) 175.

5.42   J. E. Greedan, N. P. Raju, A. S. Wills, C. Morin, S. M. Shaw, J. N. Reimers, *Chem. Mater.* 10(1998) 3058.

5.43   Yu. Ya. Gurevich, Yu. I. Kharkhats, *Sov. Phys. Uspekhi.* 25(1982) 257.

5.44   S. Fujimoto, N. Yasuda, S. Kameyama, *J. Phys. D: Appl. Phys.* 13(1980) L95.

5.45   Y. H. Han et al., *J. Chem. Phys.* 140 (2014) Article number 044708.

5.46   H. J. Beister et al., *Angew. Chem.* 100(1988) 1116; *Angew. Chem. Int. Ed. Engl.* 27(1988) 1101.

5.47   S. E. Boulfelfel et al., *Phys. Rev. B* 74(2006) 094106.

5.48   F. Brunet et al., *Solid State Ionics* 159(2003) 35.

5.49   T. Ishii, Y. Kondo, T. Kawabe, *Solid State Ionics* 5(1981) 109.

5.50   J. Chevalier et al., *J. Am. Ceram. Soc.* 92(2009) 1901.

5.51   W. Hayes, *Contemp. Phys.* 27(1986) 519.

5.52   P. Lara et al., *Solid State Ionics* 175(2004) 451.

6

Phonons

6.1 Motivation

"Atoms" of a crystalline solid at a temperature greater than absolute zero (0 K) execute vibrations about their mean positions. The bulk solid is represented by its unit cell that contains a small number of atoms. The solids classically speaking are elastic objects that obey Hooke's law and possess elastic moduli (Young's shear and bulk moduli) when they are subject to deformations. When deformations have wavelengths of the order of interatomic distances, however small the amplitude of deformations, classical (continuum) elasticity fails qualitatively! A quantitative understanding of these microscopic deformations can only be obtained by considering the vibrations of crystalline lattices. In the quantum mechanical description of the solid state, the elastic vibrational energy of the crystal is quantized. A quantum of such vibrational energy is called a phonon. Why should one study phonons of solid state ionic materials? The study of phonons deals with enumerating and naming the ways in which the lattice vibrates, the so-called modes of vibration. Then, why should one worry about how to calculate their frequencies and find how phonons interact with mechanical, electromagnetic, and other forces?

Simply put, phonons are traveling waves in crystals. So phonon spectrum represents the frequency versus wave number relationships of the lattice which could be considered in one-, two-, and three dimensions.

Where do phonons occur? Phonons occur everywhere since they manifest the presence of temperature. In crystalline materials, such a phononic motion can be quite well classified. Phonons play an important role in quite a number of microscopic and bulk phenomena such as inelastic coherent and incoherent neutron scattering, coherent inelastic x-ray scattering, inelastic nuclear absorption, infrared absorption, and Raman scattering. A large part of the temperature dependence of the thermodynamic functions comes from phonons. Hence, phonons influence phase diagrams, chemical reactions, atomic diffusion, thermal expansions, Debye–Waller factors temperature dependence of mechanical properties like elastic constants. A deep understanding of the phonon behavior, especially that of new materials, is a necessary condition for future technological developments.

Phonons can be quite reliable when obtained directly from the ab initio calculations using standard ab initio codes and the phonon code bases on the direct method. One can obtain sufficiently accurate phonon frequencies for insulator, semiconductors, metals, molecular crystals, including systems containing the 3d elements and crystals showing magnetic properties, enabling an understanding of the phonon modifications in these systems, and hence the changes in the temperature dependences of many related properties.

Phonons of solid state ionic materials are important because such solids contain large concentrations of point defects and are inherently disordered. Thermodynamic phase stability of such materials (with respect to temperature, pressure, and defect concentration) is crucially determined by the nature of phonons observed in these materials. Apart from defects and disorder, the presence of anharmonicity—that is, the potential energy curve is asymmetric about the equilibrium position so that the solid state ionic lattice is usually an anharmonic lattice—makes the phonons in these materials quite different from those in harmonic lattices.

6.2 Interionic Potential and Force: Basis for Lattice Dynamics

A simple intuitive model of interatomic forces is shown in Figure 6.1a.

The balance of attractive and repulsive forces that lead to the electrostatic stability of an ionic crystal essentially arises from the existence of an interionic or more generally interatomic potential.

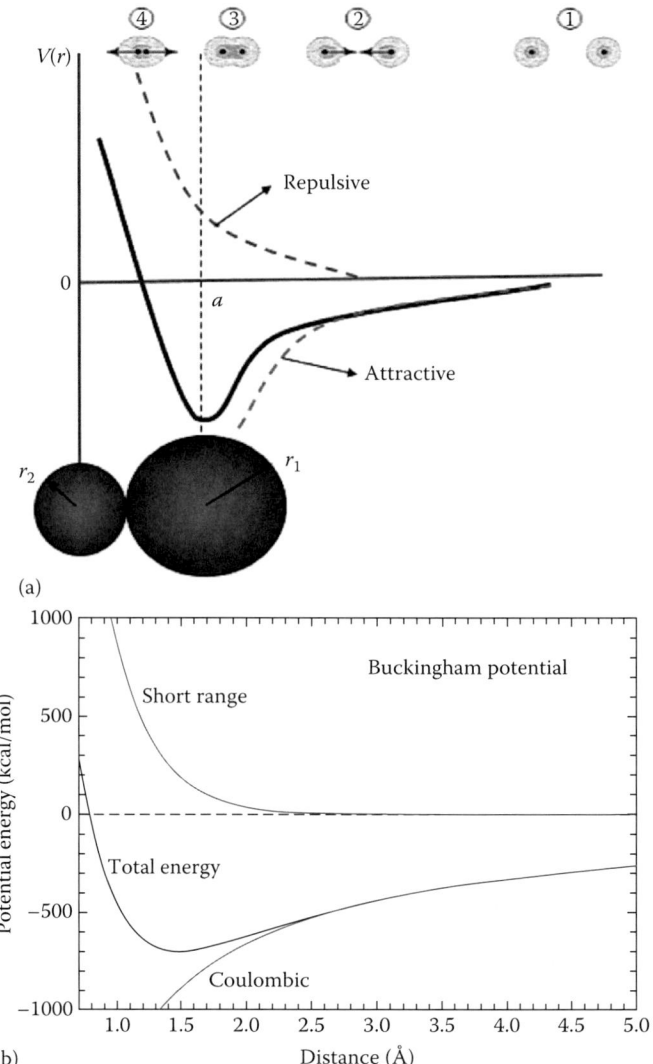

FIGURE 6.1 (a) Attractive and repulsive forces between spherical particles of radii r_1 and r_2 ($r_1 > r_2$). (1) At large interparticle separations, there is no force. (2) Particles approach causing attractive interaction. (3) At equilibrium separation, the attractive and repulsive forces balance. (4) At very close separation, repulsive forces dominate. (b) Buckingham potential. This potential used extensively in molecular dynamics simulations represents Pauli repulsion and van der Waals interaction $U_{12}(r)$ between two not directly bonded atoms as a function of the interatomic distance r. $U_{12}(r)$ is a unique interatomic potential formulated as $Ae^{-Br} - C/r^6$ where A,B,C are constants. The first term accounts for repulsion while the second attraction because their first derivatives, with respect to r, are negative and positive respectively. Repulsion essentially due to interpenetrating closed electron shells is modeled as an exponential. (From Cygan, R.T., *Rev. Mineral. Geochem.*, 42, 1–35, 2001.)

An understanding of this basic interaction is important not only for the measurement and computation of solid state ionic phenomenological properties but also to optimize solid state ionic device performance.

A potential suitable for modeling interionic interactions in superionic conductors is the Buckingham potential sketched in Figure 6.1b.

To focus on the importance of the interatomic potential to the most important application, namely solid state battery components, Table 6.1 gives the specific nature of the involved basic interactions.

TABLE 6.1

Interaction Potentials and Bond Parameters in Cell Components

Cell Component	Bond Type	Interaction Potential	Bond Length (Å)	Bond Strength (kJ/mol)
Anode (graphite, in-plane)	Covalent	—	1.46	374
Anode (graphite, inter-plane)	van der Waals	Lennard–Jones potential $U(r) = 4\varepsilon\left\{\left(\dfrac{\beta}{r}\right)^{12} - \left(\dfrac{\beta}{r}\right)^{6}\right\} \cdots (1)$	3.35	5.9
Cathode (spinel, $LiMn_2O_4$)	Ionic	Coulombic interaction $U(r) = \dfrac{q_1 q_2}{4\pi\varepsilon r r} \cdots (2)$	1960(Li-0)	426.48 (Li-0)
Liquid electrolyte	Coulombic	Coulombic interaction $U(r) = \dfrac{q_1 q_2}{4\pi\varepsilon r r} \cdots (3)$	—	—

Source: Park, M. et al., *J. Power Sources*, 2010.

The type of potential most commonly used for ionic materials consists of a Coulomb term to describe the electrostatic interactions between point ions and a Buckingham potential to describe the short-ranged (repulsive) interactions. The potential energy between ions i and j separated by a distance r_{ij} with ionic charges q_i and q_j is then given by

$$U_{ij}(r) = \frac{q_i q_j}{r} + A_{ij}\exp\left(-\frac{r}{\rho_{ij}}\right) - \frac{C_{ij}}{r^6} \tag{6.1}$$

The parameters in this equation can be determined by fitting to experimental properties or energy surfaces determined by first principles calculations, or a combination thereof.

The interionic potential in an ionic solid AB arises from the precise nature of the A–B bond. If the solid is made up of ions A^+ and B^-, then the interionic potential consists of three pairs of interactions namely A–A, A–B, and A–B. The lattice dynamics is controlled by this potential (*in toto*), and ion diffusion, elastic, thermal, electronic, and electrochemical properties are in turn dictated by lattice dynamics.

Microscopic understanding of all the properties of solid state ionics including phonons and thermal properties (and electronic structure too!) stems from the potential energy of the mobile ions as a function of temperature and position in the crystal. This is enabled by looking at the pair potential. For any two ions i and j, the pair potential is

$$V_{ij}(r) = A_{ij}\exp\left[\frac{(9r_i + r_j - r)}{\rho}\right] + \left(\frac{q_i q_j}{r}\right)e^2 - \left(\frac{1}{2}\right)(\alpha_i q_j + \alpha_j q_i)\left(\frac{e^2}{r^4}\right) \tag{6.2}$$

Here, the first term represents the overlap between closed shell ions with $r_i = i$th ion radius and r is the distance between ion i and ion j. The second term is the Coulomb potential with q_i being the fraction of charge on the ith ion. The third term gives the polarization self energy of the ith ion with polarizability α_i. This many-parameter many-body potential is the starting point for all calculations on ionic and polar covalent crystals.

Two questions set the pace for further discussion: (1) How does one treat the motion of the ions within a crystal structure? (2) How would the motion of a bound ion affect its immediate neighbors and those beyond?

Now consider a simple cubic lattice (or a primitive cubic structure with lattice constant a) along the (100) direction (Figure 6.2) [6.1]. Focus on the displacement of the virtually infinite number of lattice planes (at a $T < 0$ K).

This is just one of the ways in which the displacements generate an elastic wave—the longitudinal way. But there is also the transverse way in which such waves can be generated and propagated. Transverse excitations appear as shown in Figure 6.3.

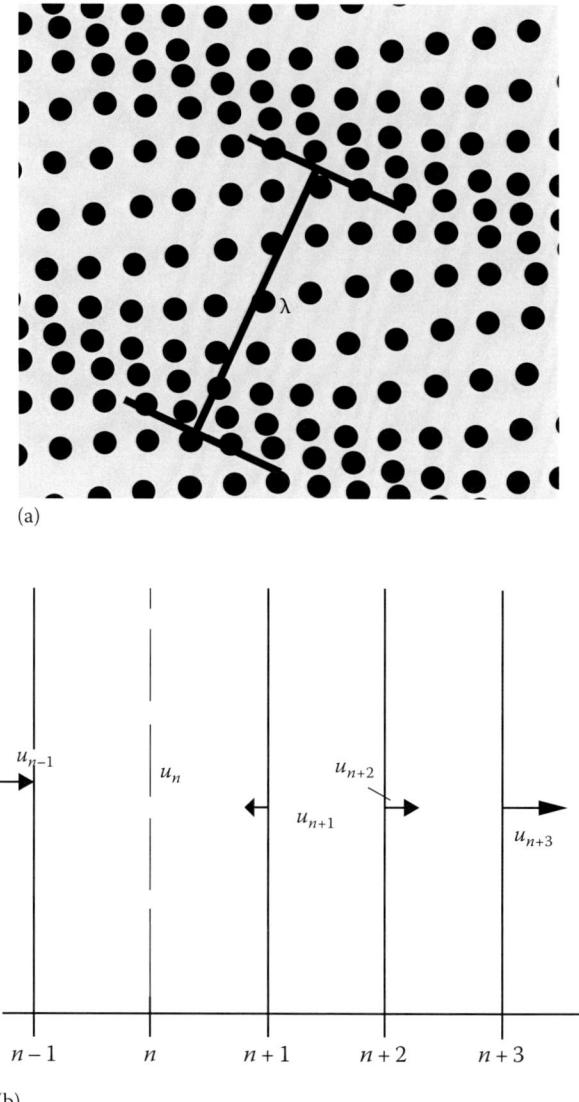

(a)

(b)

FIGURE 6.2   (a) Lattice wave of wavelength λ. (b) Planar lattice model. Atomic planes of a simple cubic lattice displaced along (100) direction. These displacements generate longitudinal elastic waves. Particle motion is in the same direction as the wave motion. Note there is also a particle-wave duality.

Note that considering simple cubic lattice is not a trivial exercise. Indeed, it helps visualize sublattice dynamics in bcc, fcc, spinel, and perovskite structures adopted by solid state ionic materials. Even the layered structure mimics the situation of transverse elastic waves.

Now the way to calculate the dynamics of atoms in crystal lattices is to assume that elastic forces are linear with respect to atomic displacements. In other words, Hooke's law is strictly obeyed by the elastic crystal. Thus,

$$\mathbf{F} = -k\mathbf{x} \qquad (6.3)$$

where
 $\mathbf{F}, \mathbf{x}$ are vectors
 k is the force constant

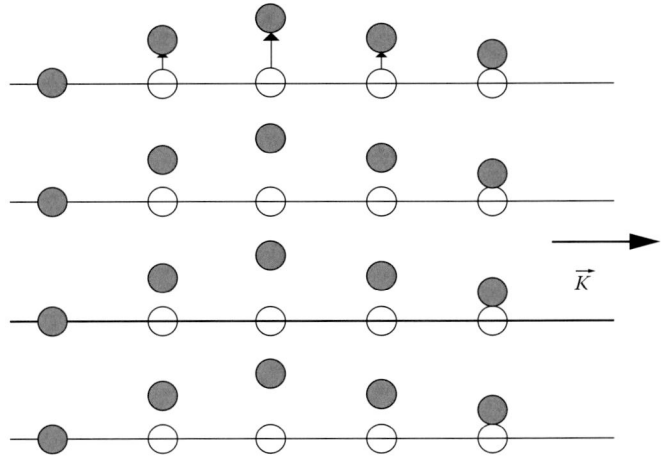

FIGURE 6.3 Transverse elastic waves in a simple cubic lattice. The wave propagates along *x*-axis but the atomic displacements are in a plane normal to it.

The minus sign indicates that the force acts in a direction opposing the displacement. The idea of restoring force is contained in Equation 6.1. Thus, the force due to displacements u_n, u_m (Figure 6.1) is

$$F_{nm} = -c_{nm}(u_n - u_m) \tag{6.4}$$

Here, C is the force constant; C_{nm} is the force constant matrix connecting two vectors force and displacement. Consider only the nearest neighbors (although in solid state ionics, next nearest neighbors are quite important and must be considered in formulating the interatomic force).

The total force acting on an atom within a lattice plane (Figure 6.4) is

$$F_n = c(u_{n+1} - u_n) + c(u_{n-1} - u_n) \tag{6.5}$$

Apply Newton's second law of motion to this problem. The equation describing the motion of the atoms of mass M of the lattice plane is

$$M\frac{\mathrm{d}^2 u_n}{\mathrm{d}t^2} = c[u_{n+1} - u_{n-1} + 2u_n] \tag{6.6}$$

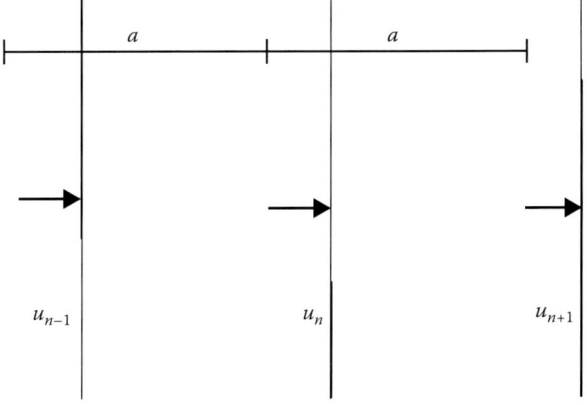

FIGURE 6.4 Displacements of three successive lattice planes.

Let all displacements have an identical frequency dependence ~$\exp(-i\omega t)$. That would mean a simple equation of motion

$$-M\omega^2 u_n = c[u_{n+1} + u_{n-1} - 2u_n] \qquad (6.7)$$

Under this assumption, we get normal mode solutions to the problem of lattice dynamics.
   Assume a wave-like solution with (a) wave vector k and (b) an additional phase factor. This factor is

$$U_n = u \exp(-inka) \qquad (6.8)$$

A fixed relationship exists between any two neighboring planes. Substitution of Equation 6.6 into Equation 6.3 leads to

$$-\omega^2 M u \exp(inka) = cu[(\exp i(n+1)ka + \exp i(n-1)ka) - 2\exp(inka)] \qquad (6.9)$$

Simplifying and recognizing the cosine expression on the RHS, we can write

$$\omega^2 M = 2c[1 - \cos(ka)] \qquad (6.10)$$

or

$$\omega^2 = \left(\frac{4c}{M}\right)\sin^2\left(\frac{ka}{2}\right) \qquad (6.11)$$

This is the relation between the elastic wave vector k and its frequency ω. It is known as the dispersion relation, which is fundamental to wave physics. Quantized elastic waves give rise to phonons. So, this leads to the phonon dispersion relations to be discussed later in this chapter.
   A plot of ω versus k exhibits maxima at $(4c/M)^{1/2}$, which could be normalized to unity for convenience. Figure 6.5 shows this plot for $-\pi/a \le k \le \pi/a$
   The rocksalt-type LiI or NaI structure or the zincblende-type CuBr structure has two atoms per primitive cell unlike the monatomic Li. Thus, we could expect new features from a consideration of the

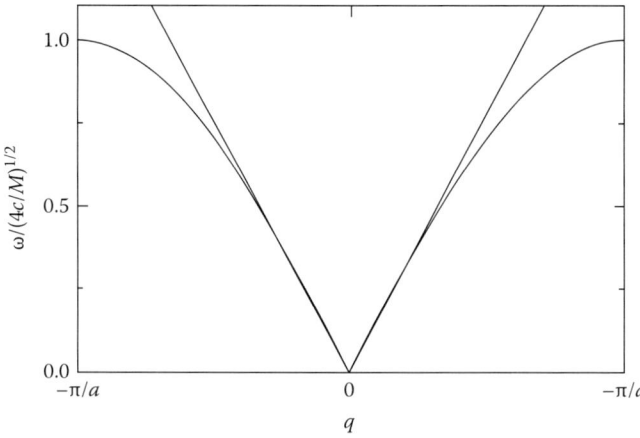

FIGURE 6.5 Dispersion relation (ω vs. q) of a monatomic lattice. There is only one phonon branch—the acoustic branch. The slope of the straight line gives the velocity of sound propagation.

diatomic lattice. A useful concept is to regard the ionic crystal to be composed of ionic molecules. Now the ω versus K relation develops two branches, for each polarization mode in a given direction of wave propagation. These two branches are called acoustic and optical branches so that we could now speak of acoustic phonons and optical phonons. These are longitudinal acoustic (LA), transverse acoustic (TA), longitudinal optical (LO), and transverse optical (TO) phonons (Figure 6.6). In general, if there are p atoms in the primitive cell, then there will be $3p$ branches to the phonon dispersion relation: *three* acoustical branches and $3p - 3$ optical branches. The mathematical justification for this assertion is as follows: We write the equation of motion for each atom in the cell, which results in p equations. Since these are vector equations, they are equivalent to $3p$ scalar equations, which have $3p$ roots. Three of these roots always vanish at $q = 0$, which results in three acoustic branches. The remaining $(3p - 3)$ roots, therefore, belong to the optical branches. The nature of these special features is illustrated in a series of figures that follow [6.2a,b].

Consider a cubic lattice where ions of mass M ($m < M$) are on one set of planes while ions of mass m are on planes in between those of the first set. Consider waves that propagate in asymmetry direction. In such a case, a single plane contains a single type of ion, for example, [111] directions in the NaI or hypothetical AgI cubic crystal and [100] in a bcc structure. Assume that the plane interacts only with its nearest neighbor planes. Furthermore the force constants (or "spring" constants) are identical between *all* pairs of nearest neighbor planes. The Newton's equations of motion are

$$M \frac{d^2 u_n}{dt^2} = C(v_n + v_{n-1} - 2u_n) \tag{6.12}$$

$$M \frac{d^2 v_n}{dt^2} = C(u_{n+1} + u_n - 2v_n) \tag{6.13}$$

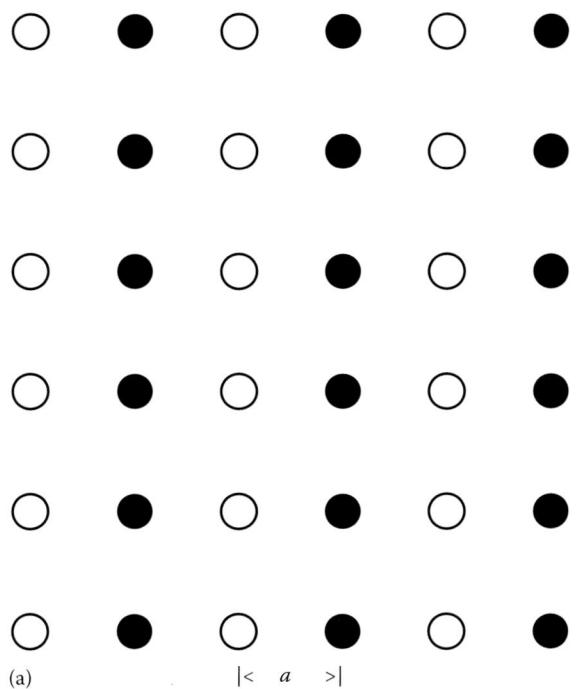

(a) |< a >|

FIGURE 6.6 Lattice and its dynamics. (a) A linear diatomic chain made up of an ion such as Na⁺ or Ag⁺ (●) with mass m and I⁻ (○) with mass M ($M > m$) when repeated in two dimensions generates a plane lattice. The lattice parameter is a. The displacements of successive ○ are u_{n-1}, u_n, and u_{n+1} while those of ● are v_{n-1}, v_n, and v_{n+1}. *(Continued)*

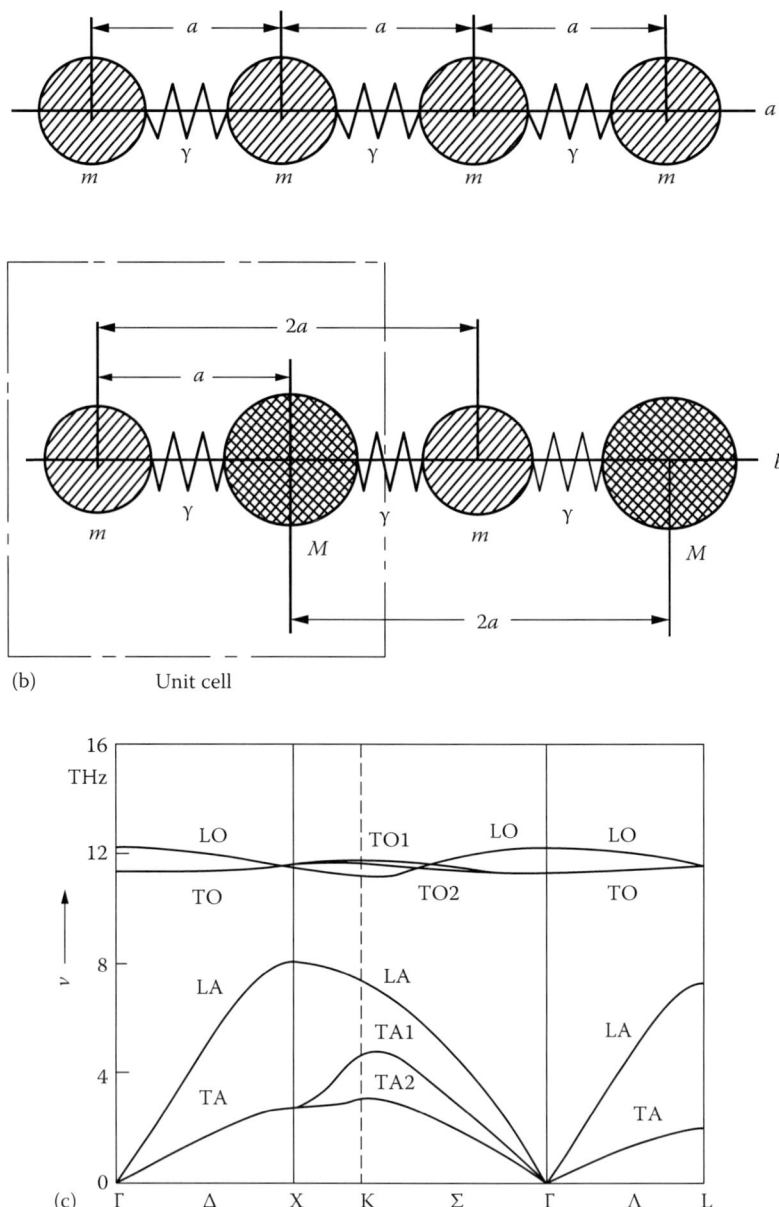

FIGURE 6.6 (*Continued*) Lattice and its dynamics. (b) Mass-spring model; $M > m$. (c) Typical phonon dispersion for a diatomic solid showing LA, TA, LO, and TO phonon branches. The symbols on the x-axis refer to special points on the Brillouin zone.   (*Continued*)

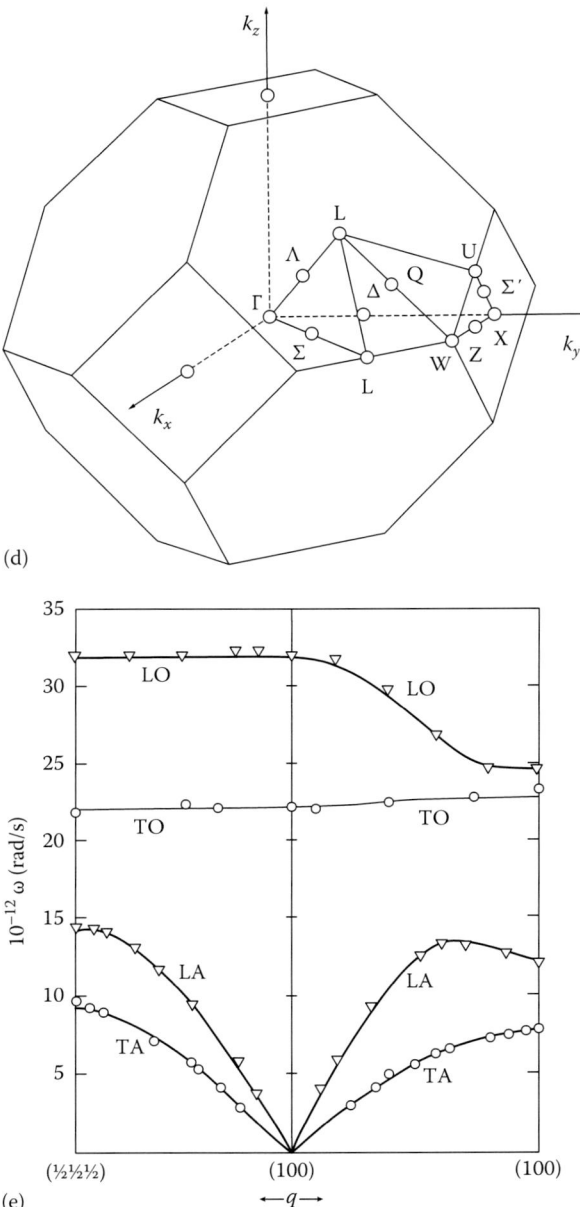

(d)

(e)

FIGURE 6.6 (Continued)   Lattice and its dynamics. (d) Brillouin zone of the fcc crystal. (e) Phonon dispersion in NaI. The acoustic branches LA and TA and optical branches LO and TO are clearly seen for this diatomic lattice ($m_I \gg m_{Na}$). Experimental points are fitted to a theoretical shell model allowing for the deformability of the nominally spherical ions and for some covalency in Na-I bonding. Solid state ionic materials such as AgI and CuBr too have partial covalency. (After Dick, B.G. and Overhauser, A.W., *Phys. Rev.*, 112, 90, 1958; Hayes, W. and Stoneham, A.M., *Defects and Defect Processes in Nonmetallic Solids*, Wiley, New York, 1985, Chap. 2.)

Equations 6.12 and 6.13 are a set of mutually coupled differential equations. Are there traveling-wave or plane-wave solutions with different amplitudes u and v on alternate planes? Try

$$u_n = u \exp(i[nKa - \omega t]) \tag{6.14}$$

$$v_n = v \exp(i[nKa - \omega t]) \tag{6.15}$$

Substitute Equation 6.11 in Equation 6.10 to obtain

$$-\omega^2 M u = Cv(1 + e^{-iKa}) - 2Cu \tag{6.16}$$

$$-\omega^2 m u = Cu(e^{iKa} + 1) - 2Cv \tag{6.17}$$

Equations 6.16 and 6.17 have a nontrivial solution only if the determinant of the coefficients of u and v vanishes:

$$\begin{vmatrix} 2C - M\omega^2 & -C(1 + e^{-iKa}) \\ -C(1 + e^{iKa}) & 2C - m\omega^2 \end{vmatrix} \tag{6.18}$$

The quartic equation in ω reads

$$Mm\omega^4 - 2C(M + m)\omega^2 + 2C^2(1 - \cos Ka) = 0 \tag{6.19}$$

To make contact with the Brillouin zone, let us look at the approximate solutions for ω^2 (Exercise: Solve Equation 6.16 exactly). To do that, consider the limiting cases $Ka \gg 1$ and $Ka = \pm\pi$ at the Brillouin zone boundary. For small Ka, $\cos Ka \cong 1 - 1/2\,K^2a^2 + \cdots$. The two roots of Equation 6.16 are

$$\omega^2 \cong 2C\left(\frac{1}{M} + \frac{1}{m}\right)(\text{optical branch}) \tag{6.20}$$

$$\omega^2 \cong \left[\frac{\frac{1}{2}C}{(M + m)}\right]k^2 a^2 (\text{acoustical branch}) \tag{6.21}$$

Brillouin zone ranges from $-\pi/a$ to π/a for K or $\pi/a \leq K \leq \pi/a$, where a is the lattice parameter. The roots of Equation 6.19 at $K_{max} = \pm\pi/a$ are

$$\omega_+^2 \cong \frac{2C}{M}; \quad \omega_-^2 \cong \frac{2C}{m} \tag{6.22}$$

Figure 6.5 shows the dispersion curve for $M > m$ which schematically corresponds to LiI, NaI, CuI, and AgI among others.

What is special about the optical branch at $K = 0$? From Equations 6.16 and 6.20

$$\frac{u}{v} = \frac{-m}{M} \tag{6.23}$$

The unit cell also moves within itself! Atoms move but center of mass remains fixed as the ions carry opposite charges (Na^+ or Ag^+ and I^-).

The electric field of an electromagnetic wave can excite this type of motion.

Note that frequencies *between* the values given by Equation 6.22 are not excited. Thus a frequency gap exists (akin to an electronic energy gap to be discussed in Chapter 7). This situation is mathematically expressed by saying that solutions for frequencies within the gap occur only for complex K values so that the wave would be damped spatially. Another type of waves arise (Equation 6.21) in which center of mass (of M and m) moves along with the masses. This motion generates acoustic branches, with the slope of the linear portion of dispersion curve giving the velocity of sound in the given crystal.

What is the effect of lattice periodicity on the modes of the diatomic chain? To understand that compare the dispersion relations of monatomic and diatomic chains one above the other. Observe that for the monatomic chain the period in $k = \pi/a$, while that for the diatomic chain it is $k = \pi/2a$, where a is the repeat distance. Therefore, for the diatomic lattice there is a "doubling" of the period so that the dispersion relation can be restricted from $0 \leq k \leq \pi/2a$.

Acoustic modes of the diatomic chain correspond to *sound waves* in the long-wave limit. Hence the name "acoustic." In this limit, $\omega \to 0$ as $k \to 0$.

In the same limit, optical modes interact strongly with the electromagnetic radiation in polar covalent crystals such as Ag and Cu halides. The name "optical" arises because of this interaction. Such modes when present give rise to strong optical absorption when photons are annihilated and phonons are created. An important characteristic of optical modes is $\omega \to$ finite value as $k \to 0$. These modes essentially arise from folding back the dispersion curve as the lattice periodicity is doubled (or halved in k-space).

At the boundary of the Brillouin zone, all modes are standing waves so that $d\omega/dk = 0$, a necessary consequence of lattice periodicity. In a diatomic chain, the frequency gap between the acoustic and optical branches depends on the mass difference so that in the limit of identical masses ($M = m$) the gap tends to zero.

What is the response of a diatomic crystal to infrared waves? This could be answered if we include a local electric field in Equations 6.12. The crystal absorbs or reflects infrared light. It may be shown that the crystal exhibits a resonance at $\omega = \omega_T = 2C/$reduced mass. The electric field polarizes the crystal. We can then calculate the ionic polarization and a frequency-dependent dielectric constant. Optical reflectivity of solid state ionic materials enables the determination of the dielectric function.

Now the question arises: what are the models for solid state ionic crystals?

Solid state ionic crystals adopt zincblende or wurtzite or fluorite structures as in CuBr, β-AgI, and PbF_2 (or ZrO_2), respectively. Thus, one has to consider two sublattices that are occupied by different atoms. We shall now outline the dynamics of such systems. How do elastic waves propagate in such systems?

6.3 Phonon Dispersion in Typical Solid State Ionic Materials

Phonons in a crystal arise as a special manifestation of the harmonic oscillator, so fundamental to quantum physics [6.3]. Indeed, the quantum state of the crystal is specified by giving the number of lattice vibrational quanta, called phonons. There are $3N_0$ oscillators for a crystal of N_0 atoms labeled by $3N_0$ values of wave vectors k and wavelengths λ with frequencies omega (k, λ). The crystal then is a sum of oscillator pieces represented by a Hamiltonian. The phonon concept proves useful if we consider deviations from this picture: taking into account *nonleading* terms (third- and occasionally fourth-order terms) in the Taylor expansion of the crystal potential or the interaction between the crystal and some external probe. This leads to phonon–phonon interaction. Interaction of electromagnetic field with the crystal is the basis of the optical modes of vibration studied by infrared and Raman spectroscopy.

This indeed is the theme of this chapter.

In Chapter 4, we have seen that diffraction of x-rays, neutrons, and electrons tells us about the equilibrium positions of the atoms in a solid. This phenomenon is governed by Bragg's law. What about atomic motions about these positions? At a temperature greater than 0 K, dynamic fluctuations of the atoms about their equilibrium positions arise. These are the phonons. How are the phonons determined? They are determined by the interatomic force constants which are in turn dependent upon the nature of the atoms of the solid. They are different in NaI, LiI, CuI, and AgI; PbF_2 and ZrO_2, $LiCoO_2$; and $LiMn_2O_4$ and $LiFePO_4$. How are phonons measured experimentally? They are measured by the inelastic (which means energy and momentum non-conserving) light or neutrons. Neutron scattering can be used to measure the phonon frequencies of crystalline solids throughout the Brillouin zone. Optical techniques restrict such measurement of just the center of the Brillouin zone. Let us focus briefly on the neutron scattering basics.

6.3.1 Inelastic Scattering of Neutrons

Thermalized neutrons have wavelengths in the range ~0.1 < λ < 0.5 nm and energies from 3 to 100 meV. Note that 1 meV ≡ 8.07 cm^{-1} ≡ 11.6 K ≡ 0.26 THz. Thus, 300 K corresponds to 26 meV. These numbers very well correspond to the range of interatomic spacings and energies of collective excitations of most solids including solid state ionics. Equally importantly, the activation energies for fast ion transport in solid state ionics fall in the 10–100 meV range.

Conservation laws give momentum and energy of neutrons given up to the solids in the form of an excitation of wave vector Q and of energy $\hbar\omega$. Momentum conservation is

$$Q = k_i - k_f \tag{6.24}$$

where k_i and k_f are the initial and final wave vectors. Energy conservation is

$$\hbar\omega = \frac{\hbar^2}{2m_N}\left(k_i^2 - k_f^2\right) \tag{6.25}$$

where
$|k| = 2\pi/\lambda$
m_N is the mass of neutron

$$Q = q + G \tag{6.26}$$

where
G is the reciprocal lattice vector
Q is measured within one Brillouin zone

Equations 6.11 and 6.12 emphasize the advantages of neutron over light and x-rays. For light, lambda is ~500 nm, so $Q = 0.01$ nm^{-1}. Only excitations near the center of the Brillouin zone can be probed. Then what about x-rays? Photon energy being ~keV for x-rays, the energy resolution $\Delta E/E$ necessary to probe lattice dynamics is ~10^{-5} which cannot be achieved by conventional x-ray sources.

In neutron scattering, relatively short wavelength and low energy enable the study of the entire Brillouin zone. A resolution of only 10^{-1} required is easily achieved in neutron spectroscopy. The crucial parameter is the cross section for inelastic scattering of neutrons

$$\frac{d^2\sigma}{d\Omega d\omega} = \left(\frac{k_f}{k_i}\right)\omega F(q,\omega)^2 \tag{6.27}$$

where

$$F(q,\omega) = (2\pi)3\left(\frac{N}{V}\right)\Sigma p[g_p(Q)]^2[n(\omega)+1]\delta(Q+q-G)\delta(\hbar\omega+\hbar\omega_p) \tag{6.28}$$

for creation of phonons
N is the number of unit cells in the crystal
V is the volume of the unit cell
p is the mode of wave vector q and branch j
$n(\omega)$ is the Bose–Einstein population factor $=[\exp(\hbar\omega/kT) - 1]^{-1}$

The dynamical structure factor

$$g_p(Q) = \frac{\Sigma i \, b_i (Q \cdot \xi_p)}{2Nm_i \omega_p} \exp i(Q \cdot R) \exp(-WQ^2) \tag{6.29}$$

where
  the sum being taken over i atoms in the unit cell
  b_i is the scattering length of atom i with mass M and position R_i within the cell
  ξp is the atomic displacement of atom i moving with p (mode of polarization)
  W is the Debye–Waller factor describing the attenuation of x-ray scattering or coherent neutron scattering caused by thermal motion (also called the B factor or the temperature factor)

The experiment measures the intensity of the phonons for different *directions* of Q and in different Brillouin zones to determine (1) the displacement of the atoms and (2) the symmetry of the vibrational modes.

Triple axis spectrometer (Figure 6.7) is the main instrument used in the measurement of phonon dispersion. Neutrons of all energies from a high-flux–beam reactor leave the reactor through a beam tube. They are monochromatized by the first crystal (the first axis). Then they are diffracted by the sample (the second axis) when they lose or gain energy. The energy of the scattered neutrons is analyzed at the third axis. Energy scans are performed by varying the final energy (E_f) with the incident energy (E_i) fixed or vice versa. Scans may be done with fixed momentum transfer (i.e., constant Q) and variable energy transfer to the solid.

How does one analyze phonon spectra? This analysis starts with the Born–von Karman (B–vK) theory of lattice vibrations. The equation of motion in the harmonic approximation (retaining terms up to quadratic in the crystal potential energy expansion) is

$$M\frac{d^2 u(l)}{dt^2} = -\Sigma \varphi(l, l') u(l') \tag{6.30}$$

where
  $l = (m, n)$ refers to the nth atom in the mth cell
  $u(m, n)$ is the displacement vector of the atoms with equilibrium positions
  $r(m, n) = r(m) + r(n)$

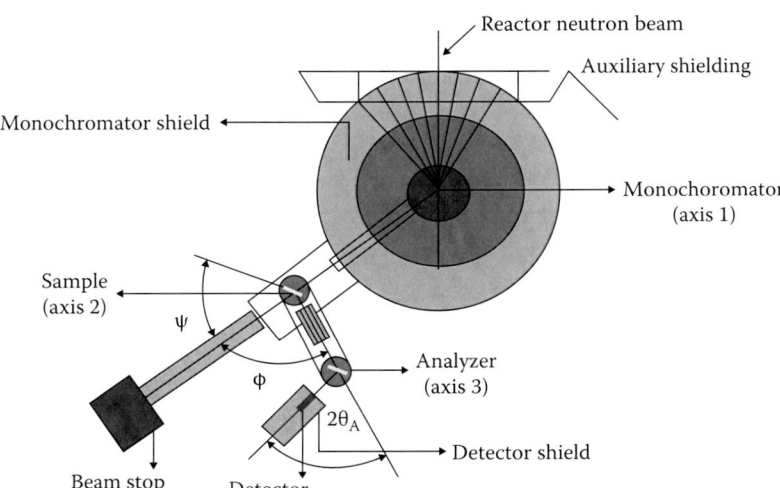

FIGURE 6.7 Schematic of the triple axis (monochromator–sample–analyzer) neutron spectrometer used for phonon dispersion studies in solids. Important parameters follows: ΔE range = 3–200 meV, q range 1–10 Å^{-1}, incident wavelength range 0.6–2.3 Å. This facility at BARC, Mumbai, India, uses a Cu(111) mono chromator and a BF$_3$ counter as detector. Neutron flux at the sample is $10^6 n$/cm^2/s. Pyrolytic graphite PG(0002) is the analyzer. The scattering angle can be varied from 10° to 100°. (From BARC, National Facility for Neutron Beam Research, Bhabha Atomic Research Centre, Mumbai, nd, http://www.barc.gov.in/publications/tb/nfnbr-book-R.pdf.)

The 3×3 force constant matrices $\varphi(l, l')$ must obey the following:

1. The laws of conservation of energy, momentum, and angular momentum, and
2. The restriction of space group symmetry.

How does one solve Equation 6.28? Assume a plane wave solution

$$u_l = \xi_l \exp i[q \cdot r(l) \quad \omega(q_l)t] \tag{6.31}$$

Next obtain a secular equation of the determinantal form

$$\det | D(q) - Ma^2 I | = 0 \tag{6.32}$$

where
  $D(q)$ is the dynamical matrix defined by

$$D(q) = \Sigma_{l,l'}(\varphi(l,l'))\exp i\{q \cdot [r(l) - r(l')]\} \tag{6.33}$$

  M is a mass tensor
  I is a unit matrix

The eigenvalues of Equation 6.31 are the normal modes of the system. ξ_i are the eigenvectors associated with each mode of vibration.

From the measured dispersion curves to be discussed later in this section, the components of $\phi(l, l')$ are determined *by adjusting* their values to get the *best fits* to the measured curve. In the B–vK analysis, $\phi(l, l')$ is the short-range interaction. However, in ionic solids including solid state ionics, a long-range Coulomb interaction is added to $\phi(l, l')$. This is the rigid ion model. Force constants are determined from dispersion curve.

How does one make the contact with the macroscopic thermal and elastic properties of solids such as the heat capacity and Debye temperature and bulk modulus? For that one has to find the phonon density of states (PDOS).

Note that solid state ionic materials with polar covalent bonds have highly polarizable cations and anions. These possess an electronic polarizability. To account for this, we need to extend the rigid ion model and arrive at the shell model. What is this shell model?

In the shell model introduced by Dick and Overhauser [6.2a,6.2b], each ion in a crystal such as NaI or AgI is not a point charge but deforms in such a way as to adopt an electric dipole moment. The adiabatic approximation is relaxed. This changes the forces between the ions and allows a long-range dipole–dipole interaction. The polarization on each ion arises because of (1) the electric fields associated with displacement and (2) the deformation associated with neighboring ions. Figure 6.8 gives an intuitive idea of the (valence) shell model. The essence of this basically harmonic model is that the outer electrons of each ion constitutes a massless shell of charge $Y|e|$, while the nucleus plus inner electrons provide a massive core of charge $X|e|$ so that the total ionic charge is $(X + Y)|e|$. The core and shell are thought to be coupled by harmonic springs of spring constant K.

The shell model represents the interatomic forces better than the rigid ion model. Later we give examples of shell model–derived phonon dispersion curves for solid state ionic materials.

How do neutrons enable phonon dispersion to be deduced experimentally?

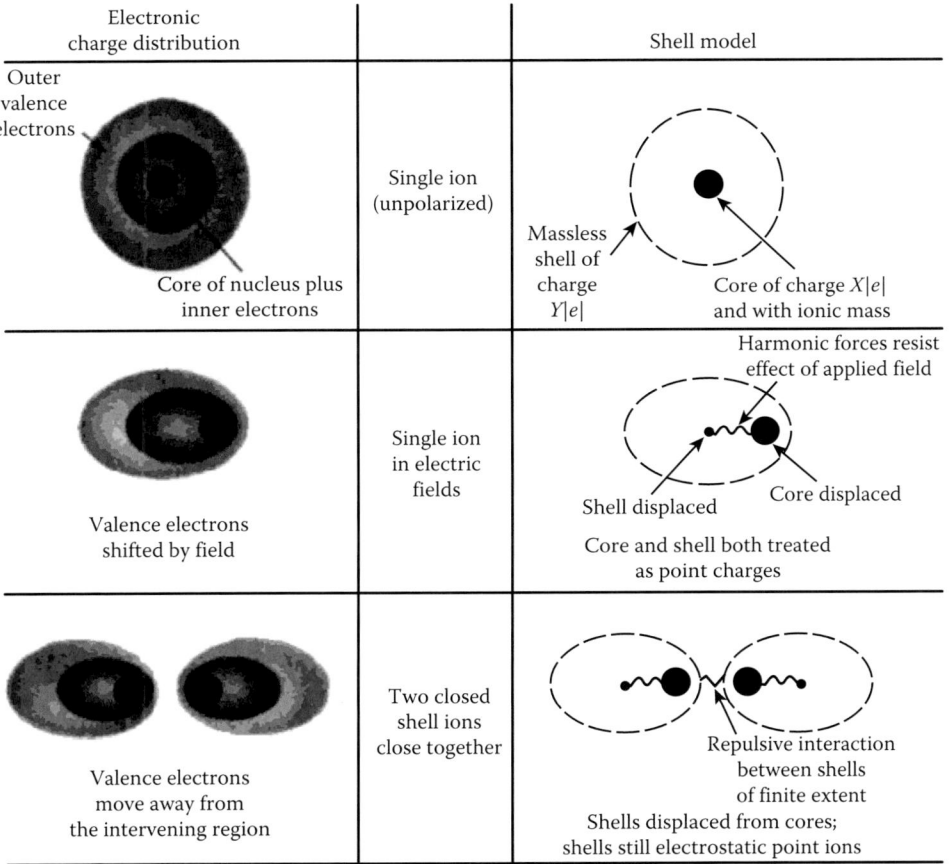

FIGURE 6.8 Schematic of the shell model. Note the difference between electron densities and the features of the shell model. The atom is represented by a core consisting of a nucleus and inner electrons and an outer electron shell. The shell and the core are coupled and can distort giving rise to a small dipole moment. (From Hayes, W. and Stoneham, A.M., *Defects and Defect Processes in Nonmetallic Solids*, Wiley, New York, 1985, Chap. 2.)

The two parts of the scattering of neutrons by phonons are coherent and incoherent. Coherent scattering gives phonon dispersion curves. The incoherent part enables measurement of the PDOS. Consider the basic equation

$$\frac{d^2\sigma}{d\Omega dE_1}\bigg|_{\text{incoh}} = \left(\overline{b^2} - \overline{b}^2\right)\Sigma\left(\frac{k_1}{k_0}\right)$$

$$\cong \sigma_{\text{inc}}\left(\frac{k_1}{k_0}\right)\exp(-2W)\times\int\exp[\langle Q\cdot u_0(0)Q\cdot u_0(t)\rangle]\exp(-i\omega t)dt \qquad (6.34)$$

where $4\pi(\overline{b^2} - \overline{b}^2)$ is the incoherent cross section. Expand exponent in a power series where the *n*th term corresponds to the *n*-phonon process.

6.3.2 Basic Lattice Dynamics

Lattice vibrations of three-dimensional solids go beyond the 1D diatomic lattice discussed earlier. A solid may be modeled as a continuum only if the probe wavelength is very long so that $k = 2pi/\lambda \to 0$. If $\lambda <$

interatomic spacing, then the continuum model fails. Also, the elastic solid now considered as a 3D network of masses and springs is a quantum object whose vibrations are quantized. The lattice vibrational energy quantum is a phonon. Phonon being a wave possesses a wave vector (k) and a frequency omega. Lattice dynamics means solving the problem of lattice vibrations assuming a suitable model and obtaining theoretically the relation between omega versus k, which is a called a phonon dispersion curve. Experiments done using neutrons and light help map out lattice vibrations at high symmetry points on the first Brillouin zone.

We shall briefly discuss the lattice dynamics with Li metal as an illustrative example. Then, we consider the experimental phonon dispersion of important solid state ionic materials.

The phonon dispersion relation between frequency v (or ω) and the wave vector **q** for a Bravais lattice is given by the eigenvalues of the dynamical matrix $D\alpha\beta(\mathbf{q})$

$$\Sigma_\beta D_{\alpha\beta}(\boldsymbol{q})e_{j\beta} = v_j^2 e_{j\alpha}, \quad j = 1,\dots, 3 \tag{6.35}$$

For a simple metal, D may be taken as the sum of three terms: D^C, the Coulomb interaction between ions of charge $+Ze$ immersed in a uniform compensating negative charge; D^E, the ionic interaction via the conduction electrons; and D^R, a repulsive interaction due to the core wavefunction overlap (not significant for Li).

D^E can be related to a scalar function $G(q)$, which is equal to the Fourier transform of the effective interionic potential due to the conduction electrons divided by $-4\pi Ze^2/\Omega q^2$. Application of second-order perturbation theory gives the dynamical matrix as

$$D_{\alpha\beta}^E(\vec{q}) = -v_j^2 \sum_3 \left(\frac{(q_\alpha + h_\alpha)(q_\beta + h_\beta)}{\left|\vec{q} + \vec{h}\right|^2} G\left(\left|\vec{q} + \vec{h}\right|\right) - \frac{h_\alpha h_\beta}{h^3} G(h) \right), \tag{6.36}$$

where

h is the reciprocal lattice vector
v_p is the ion-plasma frequency:

$$v_p^2 = \frac{Z^1 e^2}{\pi M \Omega} \tag{6.37}$$

Z, M, and Ω being the number of conduction electrons, mass, and volume per ion, respectively. For a local pseudopotential, $G(q)$ is related to the bare electron–ion psuedopotential $w(q)$ and the dielectric function $\varepsilon(q)$, which describes the conduction electron screening.

$$G(q) - \left| \frac{\pi 4q}{-4\tau Zc^2/\Omega q^2} \right|^2 \left(\frac{c(q) - 1}{c(q)} \right) \tag{6.38}$$

An important step in phonon dispersion of Li is the calculation of $\varepsilon(q)$ for which there are many routes. These are shown in Figure 6.9 which depicts the calculated phonon dispersion in Li. Among these, the calculations based on the dielectric function devised by Singwi agree the most with experimental results. What is this dielctric function? A mechanical model answers this question for metals and insulators.

Typical wavelength of light passing through a solid is greater than or equal to 10 nm, much larger than unit cell dimensions or electron mean free paths. In such a case, the applied electric fields depend only on frequency but not on wave vector. So the response functions also depend on frequency alone. Then one can solve the problem of damped harmonic oscillators subject to an applied electric force $-eE$ to obtain a complex function $\varepsilon(\omega)$ called the dielctric funtion. This function consists of two parts: the real part $\varepsilon_1(\omega)$ and imaginary part $\varepsilon_2(\omega)$. The two parts have to be obtained by Kramers–Kronig analysis before they can be applied to study the optical behavior of solid state ionic materials. The dielectric function owes its origin to the presence of conduction electrons in metals Li, Na, Cu, and Ag and valence electrons in AgI, CuBr, and CuI. Dielectric function is intimately connected with the concept of screening. Screening idea originally comes from the behavior of electrolytic solutions such as in a lead–acid battery! Place a charged

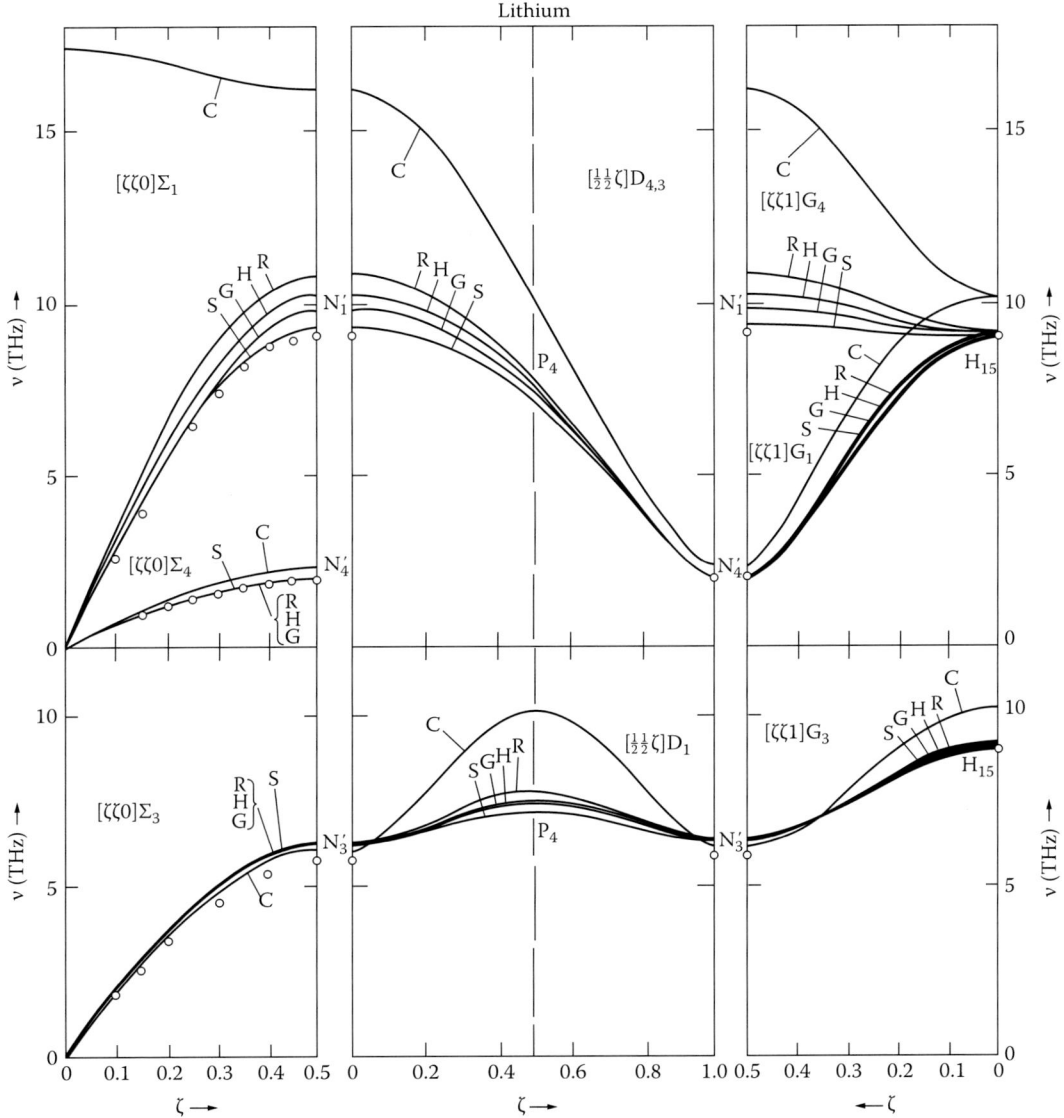

FIGURE 6.9 Phonon dispersion in lithium calculated and compared with experiment. Calculations are based on dielectric function approach. Curves are separately plotted for three different directions in Brillouin zone as indicated. C, Coulomb interactions only; R, dielectric function obtained from random phase approximation; H, Hubbard's modified dielectric function; G, Geldart–Vosko dielectric function; S, self-consistent dielectric function; "o," experimental points. Calculations based on S agree the most with experiment. (From Price, D.L. et al., *Phys. Rev. B*, 2, 2983, 1970.)

ion into such a solution. Initially the potential due to the added ion extends its influence far out, dying off slowly as $1/r$. But mobile ions nearby rapidly react to the "intruder." The motion they make in response results in an almost complete cancellation of its electric field, except within a characteristic distance called the screening length. This phenomenon occurs quite generally for any assembly of charged particles. It is as fundamental as the Thomas–Fermi theory of the atom with nuclear charge Z and Z electrons.

LiI and NaI are two alkali iodides whose lattice vibrations have been interpreted in terms of shell model. Their phonon dispersion curves are shown in Figures 6.10 and 6.11.

Shell model successfully accounts for phonon spectra and density of states of antifluorite Li^+ conductor Li_2O as shown in Figure 6.12.

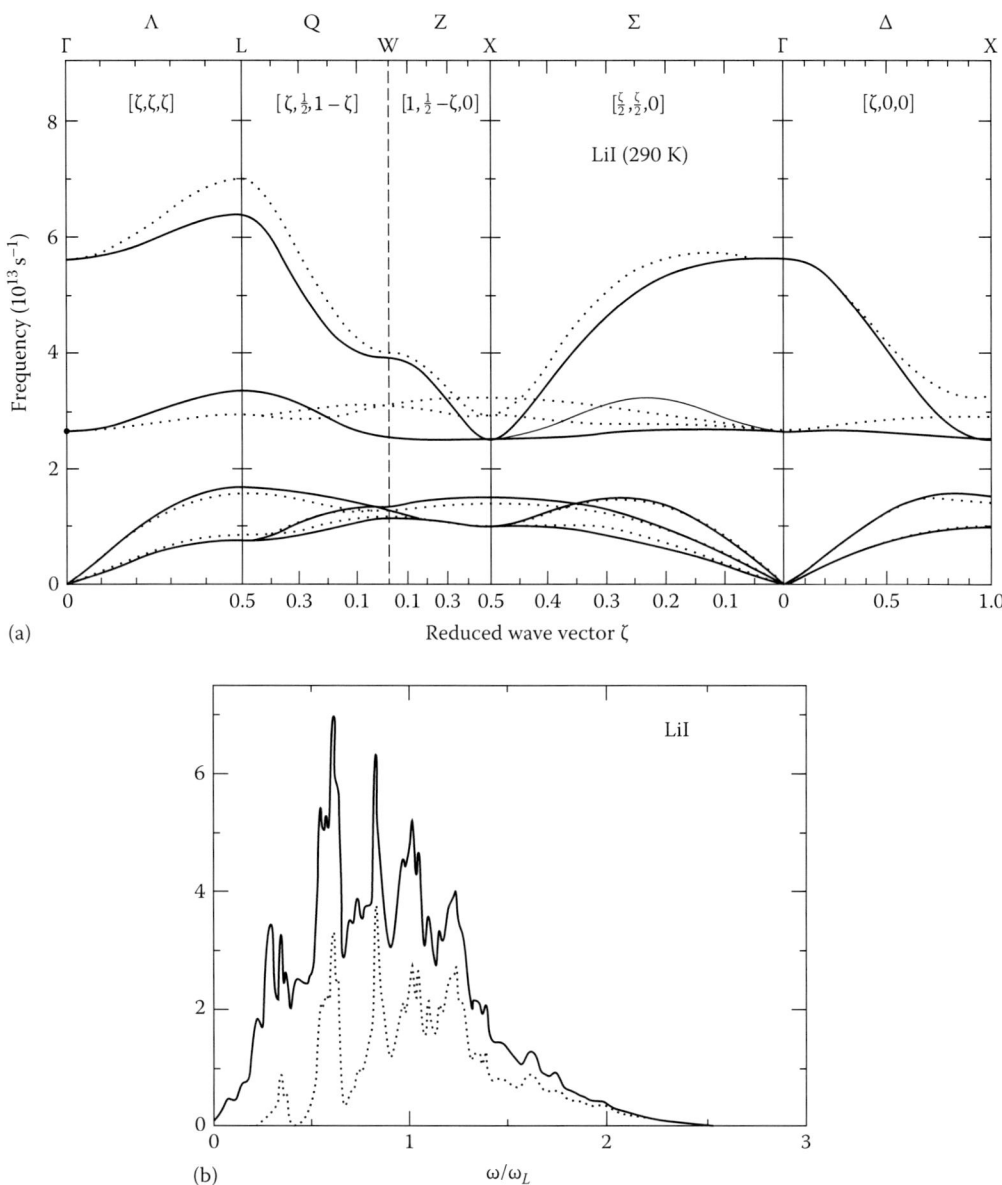

FIGURE 6.10 (a) Calculated phonon dispersion curves for LiI for breathing shell model (::) and basic deformation dipole (DD(A)) and next nearest neighbor intreactions between anions (NNN(A)) (a kind of three body force) (.....) models. (b) Calculated frequency dependence for LiI at 5 K(.....) and 100 K(_____) using DD(A) and NNN(A) 3B input data of Γ(0, 1, ω). (From Rastogi, A. et al., *Phys. Rev. B*, 9, 1938, 1974.)

In calculating the dispersion curves, the interatomic potential used is made up of long- and short-range terms:

$$V(r) = \frac{e^2}{4\pi \in_0} \frac{Z(k)Z(k')}{r^2} + a\exp\left[\frac{-br}{R(k) + R(k')}\right] \qquad (6.39)$$

where a and b are empirical parameters. Oxygen atoms are considered in the shell model.

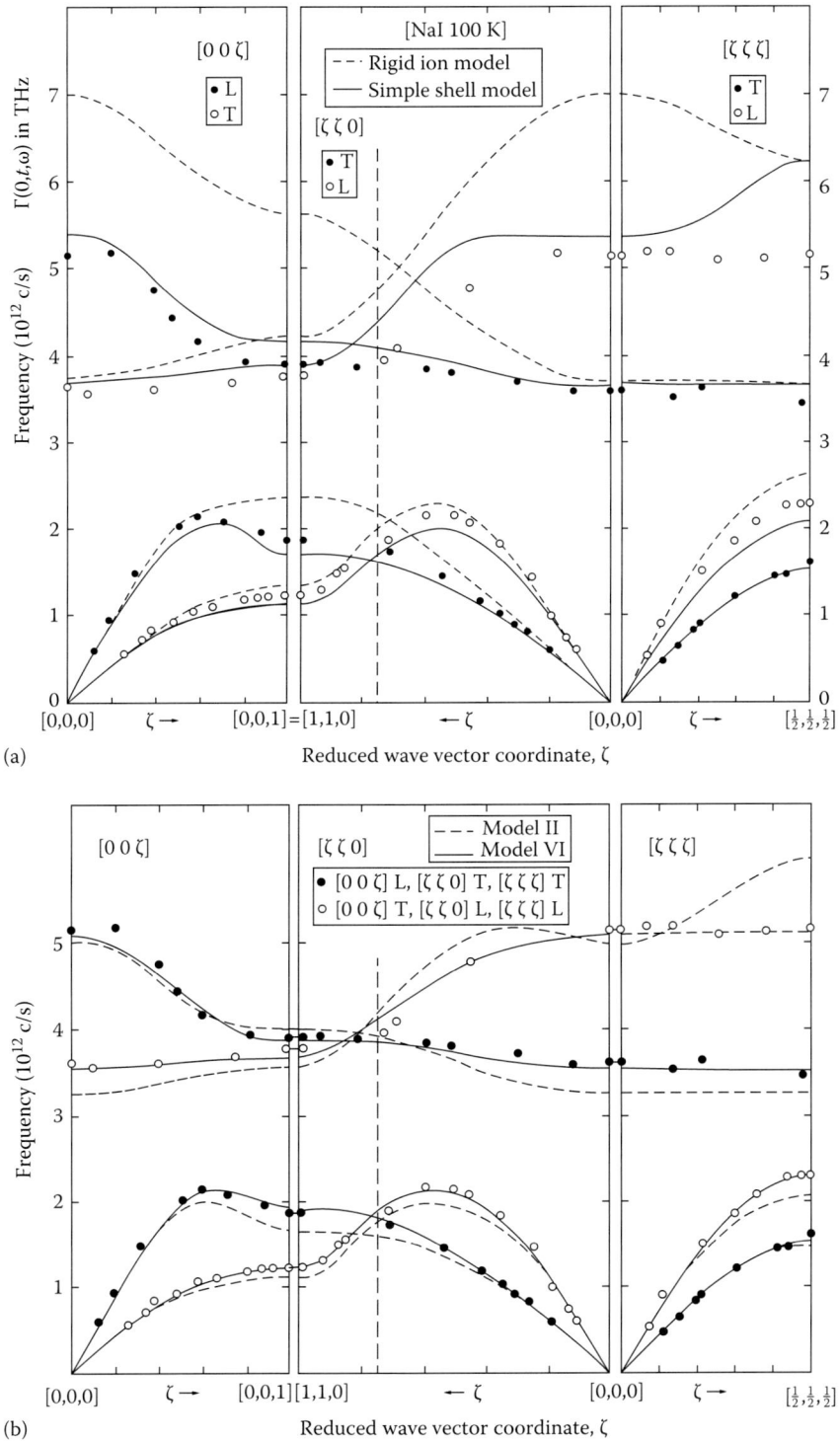

FIGURE 6.11 (a) Dispersion curves for lattice vibrations of NaI measured at 100 K. T, Transverse; L, Longitudinal. (After Woods, A.D.B. et al., *Phys. Rev.*, 131, 1025, 1963.) (b) Phonon dispersion of NaI fitted to improved shell models. Model I refers to inclusion of second neighbor forces and non-central first neighbor forces. Model VI refers to the one in which all parametres are varied giving the best fit. (After Cowley, R.A. et al., *Phys. Rev.*, 131, 1030, 1963.)

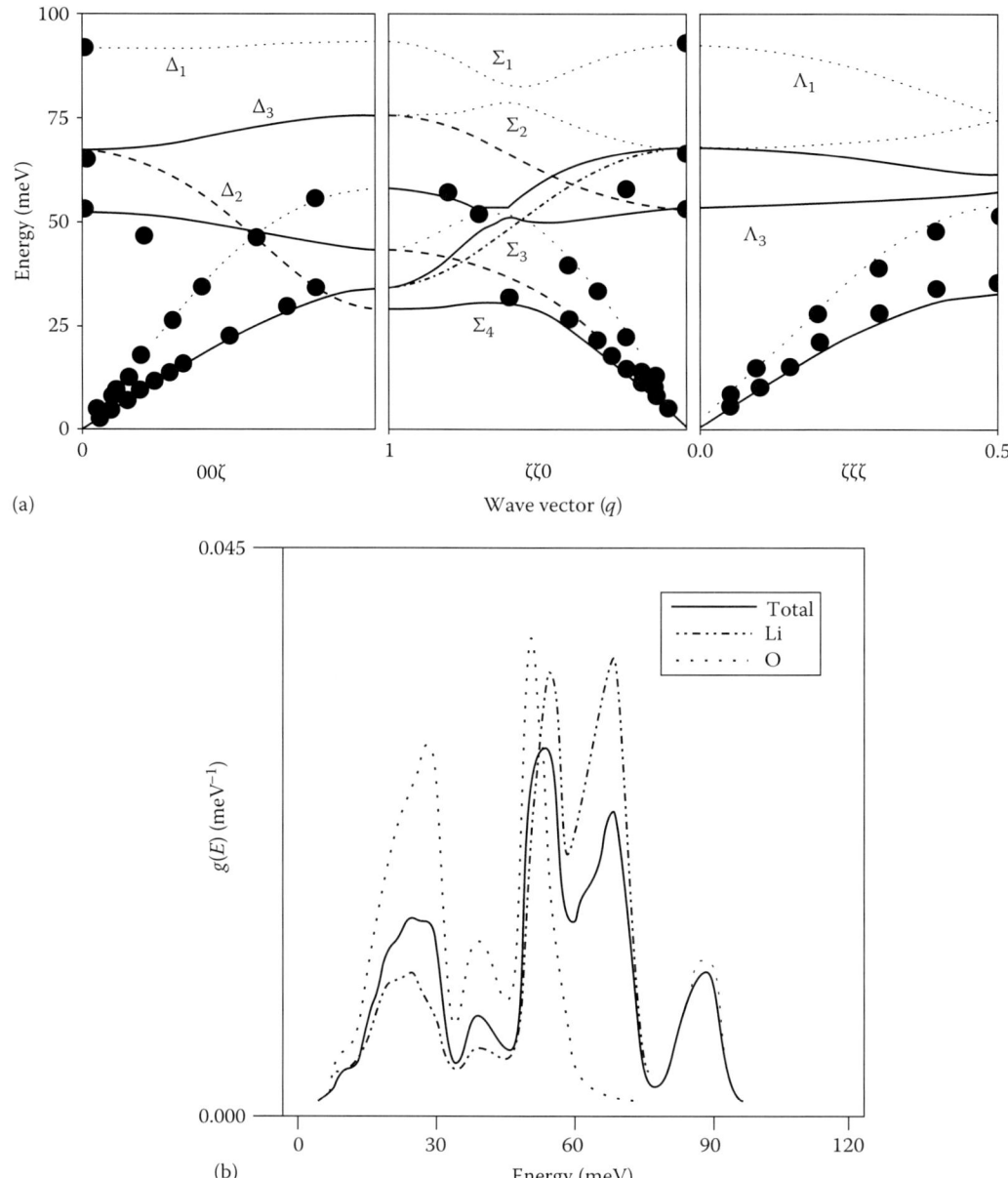

FIGURE 6.12 (a) Phonon dispersion relations in lithium oxide. Points are experimental values from neutron scattering experiments. Full and dashed lines represent calculations (From Goel, P. et al., *Pramana J. Phys.*, 63, 409, 2004), (b) Phonon density of states in lithium oxide. Both Li and O contribute over the entire Brillouin zone.

Now let us deal with the lattice dynamics of silver and cuprous halides that form the canonical superionic conductors.

6.3.3 Lattice Dynamics of Ag and Cu Halides in the Quadrupolar Deformability Model [6.9]

Lattice dynamics of Ag and Cu halides is quite unlike that of alkali halides. This is because the latter contain the unshielded 4d- and 5d-electron shells, respectively. The presence of the d shell makes Ag^+ and

Cu⁺ ions give rise to core polarization effects and the so-called quadrupolar deformability (QD). In Ag, virtual d–s excitations are supposed to cause this QD. This has a marked effect on the lattice vibrations which must be taken into account while considering the lattice dynamics of AgCl, AgBr, AgI, CuBr, and CuI. Note that in the last three compounds, tetrahedral coordination prevails. Let us look at the influence of QD on phonon dispersion. We consider four short-range force constants, four shell parameters, and the *quadrupolar shell parameters*. Note that because of QD, we have to consider the ion interactions of first neighbor and second neighbor ions. The shell parameters come from the electronic and ionic polariz-abilities caused by valence electrons. QD parameters are determined by considering the d–s excitation of Ag and Cu ions. The QD model goes beyond the shell model but is pedagogic and intuitive.

Consider the *l*th unit cell where an ion λ undergoes a displacement u_l^λ. The lattice dynamics in such a situation is presented as

$$M_\lambda \omega^2 u_l^\lambda = \Sigma_{l'} \Sigma_{\lambda'} \Sigma_\beta \varphi_{\alpha\beta}^{\lambda\lambda'} \qquad (6.40)$$

where
$M\lambda$ is the mass of ion λ
$\varphi_{\alpha\beta}^{\lambda\lambda'}$ is the inter ionic force constant
α and β denote the Cartesian components

Let us assume the adiabatic approximation for the ion–electron system. Then, the force constant, which is the second derivative of the total energy with respect to the position vector, is given by

$$\varphi_{\alpha\beta}^{\lambda\lambda'}(ll') = \left[\frac{\partial^2}{\partial R_{l\alpha}^\lambda \partial_{l'\beta}^{\lambda'}} \right] [V_{ii}(\{R\}) + E_e(\{R\})] \qquad (6.41)$$

where
$E_e(\{R\})$ is the total energy of electrons
R_l^λ is the position vector of λ ion in the *l*th lattice

Adopting the adiabatic approximation for this coupled electron–ion vibration problem, the shell model for Equation 6.40 is formulated as

$$M\omega^2 \mathbf{u} = (\phi^R + \mathbf{Z}\phi^C\mathbf{Z})\mathbf{u} - \Sigma_\Gamma \mathbf{T}_\Gamma \mathbf{S}_\Gamma^{-1} \mathbf{T}_\Gamma^+ u \qquad (6.42)$$

where
M is the mass matrix
ϕ^R is the short-range force constant matrix of the rigid ion model for lattice vibrations
$\mathbf{Z}$ is the effective ionic charge matrix
ϕ^C is the Coulomb force constant matrix
$\mathbf{T}\Gamma$ is the electron–ion coupling matrix
$\mathbf{S}\Gamma$ is the electron–electron coupling matrix

For a shell model based on quadrupolar deformability of, say, Ag⁺ ion, the electronic matrix terms are taken as

$$\mathbf{T}_\Gamma = \phi^R + \mathbf{Z}\phi^C\mathbf{Y},$$

$$\mathbf{S}_\Gamma = \mathbf{k} + \phi^R + \mathbf{Y}\phi^C\mathbf{Y} \qquad (6.43)$$

where
$\mathbf{k}$ is the ion core–shell coupling matrix
$\mathbf{Y}$ is the shell-charge matrix

It is important to realize that electron–phonon coupling arises in a fundamental way in this problem. Thus, the lattice dynamics and phonon dispersion of solid state ionics are connected to the electronic band structures of these materials. As the mathematical details are too technical, we shall present a semi-qualitative account of this many-body calculation. $E_e(\{R\})$ of Equation 6.41 is the sum $\Sigma_k f_{nk} E_{nk}$, where f_{nk} and E_{nk} are the distribution function and the energy of the Bloch state ψ_{nk} of the band electron, whose wave vector is k on the band n. Significantly, the QD is derived from the second-order perturbation term $\delta E_{nk}^{(2)}$ caused by the ionic displacements. QD thus derived helps set up the dynamical matrix:

$$\omega_q^2 e_\alpha^\lambda = \Sigma_{\lambda'} \Sigma_\beta D_{\alpha\beta}^{\lambda\lambda'}(\mathbf{q}) e_\beta^{\lambda'}(\mathbf{q}) \tag{6.44}$$

where
  ω_q is the phonon frequency
  $e\lambda(\mathbf{q})$ is the unit vector of the modes with wave vector $\mathbf{q}$

Note that $D_{\alpha\beta}^{\lambda\lambda'}(\mathbf{q})$ contains the electron–phonon coupling matrix element.

What are the electronic band structures that Ag and Cu halides are composed of? Basically, they are made up of halide anion-p and cation d-valence bands and empty cation-s conduction band. The anion displacements in these compounds cause virtual d–s excitation of Ag or Cu ions. Tight-binding approximation is applicable to this problem. Thus the wavefunctions of the d band and s band are represented as

$$\Psi_{d\mu k}(\mathbf{r}) = \left(\frac{1}{\sqrt{N_0}}\right) \Sigma_l \exp\left(ik \cdot \mathbf{R}_l^A\right) \varphi_{d\mu}\left(\mathbf{r} - \mathbf{R}_l^A\right),$$

$$\Psi_{sk}(\mathbf{r}) = \left(\frac{1}{\sqrt{N_0}}\right) \Sigma_l \exp\left(ik \cdot \mathbf{R}_l^A\right) \varphi_s\left(\mathbf{r} - \mathbf{R}_l^A\right), \tag{6.45}$$

where
  $\mathbf{R}^A$ ($\mathbf{R}^B$) are the position vectors of cation (anion)
  ϕ_s is the s atomic orbital
  $\phi_{d\mu}$ refers to the five d atomic orbitals
  ϕ_{x2-y2}, ϕ_{3z2-r2}, ϕ_{xy}, ϕ_{xz}, and ϕ_{yz}

Now how does the QD dynamical matrix arise? Assuming a flat band model in the virtual d–s excitation between Ag-d band and Ag-s band, QD dynamical matrix between anions is realized. Finally, the QD dynamical matrix is (multiplying by M_B the mass of B)

$$D_{\alpha\beta}^{BB}(\mathbf{q}) = \sum_{\mu=1}^{5} \frac{2 f_{d\mu}(1 - f_s)}{(E_{d\mu} - E_s) V_{sd\mu}^{B\alpha}(\mathbf{q}) V_{d\mu s}^{B\beta}(\mathbf{q})} \tag{6.46}$$

$$V_{sd\mu}^{B\alpha}(\mathbf{q}) = \sum_{\delta=1}^{z} \mathrm{del}_\alpha E_{sd\mu}^{B}(d_\delta) \exp iq \cdot d_\delta \tag{6.47}$$

$$E_{sd\mu}^{B}(d_\delta) = \int \varphi_s^*(r) V_B(r + d_\delta) \varphi_{d\mu}(r) dr \tag{6.48}$$

where
  $V_{sd\mu}^{B\alpha}(\mathbf{q})$ is the s–d coupling matrix element
  z is the coordination number
  $d\delta$ is the position vectors of the δth nearest neighbor anions around the cation

To calculate $D_{\alpha\beta}^{BB}(\mathbf{q})$, one needs to find the matrix elements $E_{sd\mu}^{B}(d_\delta)$ that describe the chemical bonding between anions and cations.

Equations 6.46 through 6.48 provide the essential apparatus to find the phonon dispersion of Ag and Cu halides. We will illustrate it for the case of γ-AgI.

As discussed in Chapter 2, the low-temperature or γ-phases of AgI and CuCl, CuBr, and CuI all crystallize in the zincblende structure. Let us consider the ionic deformability of this structure, which belongs to the tetrahedral symmetry group T_d. The modes of vibration are shown in Figure 6.13. The filled circle at the center of the cube is the Ag(or Cu) cation while the open circles at the corners of the tetrahedron are I⁻ (or halogen) ions. The doubly degenerate $E(\nu_2)$ and the triply degenerate $T_2(\nu_4)$ are the quadrupole modes. The crystal field matrix element V_Γ^α arises due to the normal coordinates, and the corresponding dynamical matrix elements $D_{\alpha\beta}^{BB}(\Gamma)$ are of interest. The structure factors g_j are

$$g_1(\mathbf{k}) = e^{ik \cdot d1} + e^{ik \cdot d2} - e^{ik \cdot d3} \quad e^{ik \cdot d4}$$
$$g_1(\mathbf{k}) = e^{ik \cdot d1} - e^{ik \cdot d2} + e^{ik \cdot d3} - e^{ik \cdot d4} \qquad (6.49)$$
$$g_1(\mathbf{k}) = e^{ik \cdot d1} - e^{ik \cdot d2} - e^{ik \cdot d3} - e^{ik \cdot d4}$$

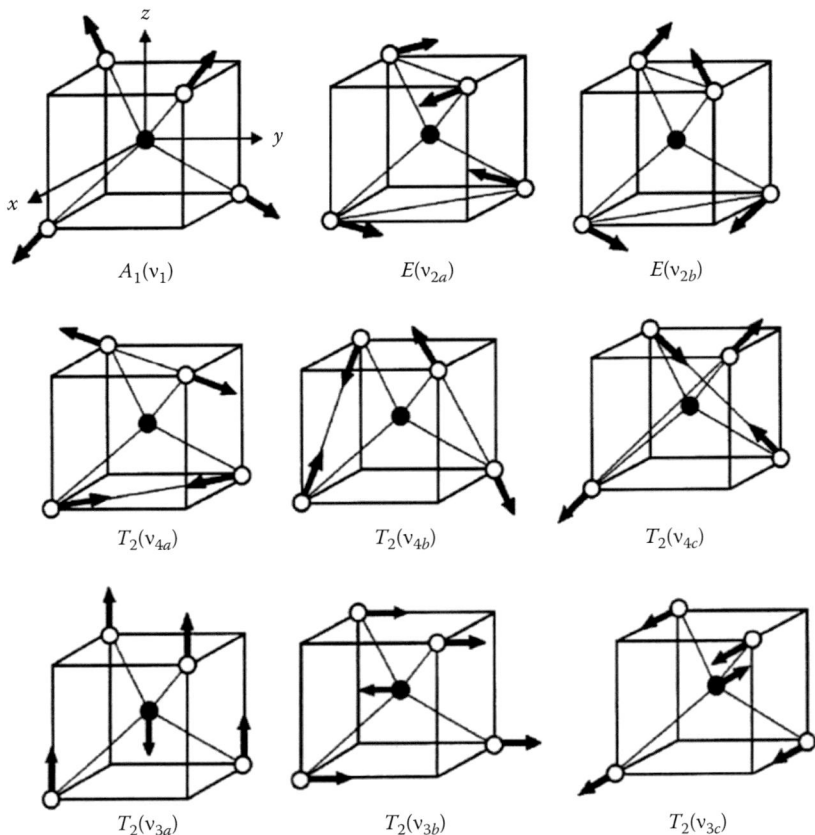

FIGURE 6.13 Different vibrational modes of nearest-neighbor ions with point group symmetry T_d. Solid circle, Ag ion; open circles, I ion; $A_1(\nu_1)$, breathing mode; $T_2(\nu_3)a$, dipole mode; $E(\nu_2)$ and $T_2(\nu_4)$, quadrupole modes. (From Tomoyose, T. et al., *Solid State Ionics*, 167, 83, 2004.)

In the tetrahedral structure, the *radial* differential elements of E_{sd1} and E_{sd2} matrices vanish. Only the *angular* differential elements remain. This implies that E_{sd1} and E_{sd2} matrices give rise to bond-bending modes $E(2)$. Similarly the QD mode of $T_2(\nu_4)$ arises. The matrix $D^{BB}(q)$ is

$$D^{BB}(q) = -S_E D_E(q) - S_T D_T(\mathbf{q}) \tag{6.50}$$

where
$$S_E = 4V_{sd\alpha}^2/[3a^2(E_s - E_E)]$$
$$S_T = (V_{sd\sigma}/3 + V_{sd\sigma}/3a)^2/(E_s - E_T)$$

Finally, the QD matrices D_E and D_T are given by

$$D_E(q) = 2|g_1|^2 - g_1 g_2^* - g_1 g_3^*$$
$$- g_1^* g_2 2|g_2|^2 - g_2 g_3^*$$
$$- g_1^* g_3 - g_2^* g_3 2|g_3|^2 \tag{6.51}$$

$$D_T(q) = |g_1|^2 + |g_2|^2 g_1^* g_2 g_1^* g_3$$
$$g_1 g_2^* |g_2|^2 + |g_3|^2 g_2^* g_3$$
$$g_1 g_3^* g_2 g_3^* |g_2|^2 + |g_3|^2 \tag{6.52}$$

Phonon dispersion of γ-AgI is obtained by estimating the parameters of the short-range force constant ϕ^R and using the optimized parameters: first-neighbor cation–cation force constant parameters and second-neighbor anion–anion force constants

Figures 6.14 and 6.15 show dispersion curves of AgCl and AgBr. The agreement between theory and experiment shows the importance of the effect of core 4d electrons and QD in providing a realistic view of phonons in these two compounds.

AgI has a unique place in the phonon physics of superionic conductors because of the marginal covalence of the Ag-I bond and the resultant tetrahedrality and polymorphism leading to four phases namely wurtzite, zincblende, bcc, and rocksalt phases among others. The wurtzite or beta phase is also a topological insulator. The crystal structure of this phase emphasizing relevant interatomic force constants is shown in Figure 6.16.

Figure 6.17a shows the phonon dispersion in β-AgI calculated from rigid ion and valence shell modes. Note the low-lying (~2.0 meV) excitation showing almost no dispersion. This feature observed in many

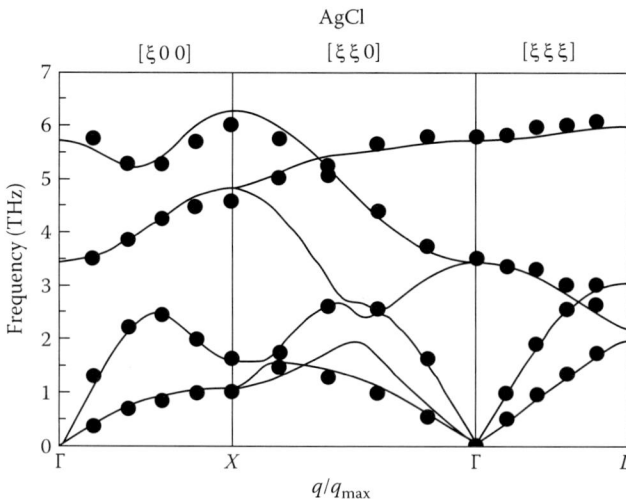

FIGURE 6.14 Phonon dispersion relation for AgCl. _____, theoretical values based on quadrupole deformable shell model; ●, experimental values. (From Tomoyose, T. et al., *Solid State Ionics*, 167, 83, 2004.)

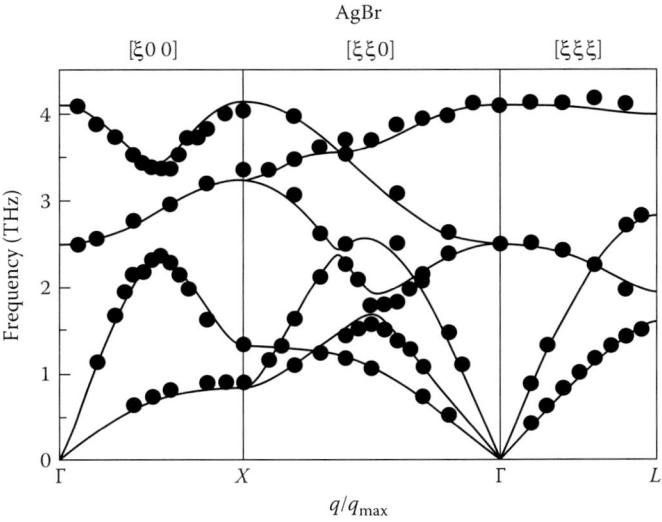

FIGURE 6.15 Phonon dispersion relation for AgBr. _____, theoretical values based on quadrupole deformable shell model; ●, experimental values. (From Tomoyose, T. et al., *Solid State Ionics*, 167, 83, 2004.)

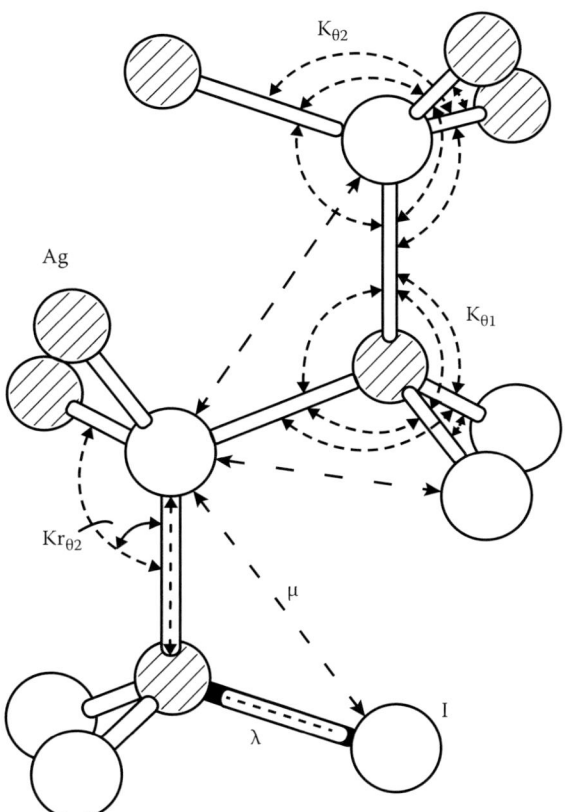

FIGURE 6.16 Crystal structure of wurtzite or β-phase of AgI. Space group $P6_3mc$. Each Ag ion is tetrahedrally surrounded by four I ions and vice versa. Two formula units in the primitive cell give 3×4 or 12 phonon branches. Note the internal force constants which involve non-central forces arising from covalency. The structure zone is 24 times larger than the Brillouin zone which allows a good separation of the phonon branches. (From Buhrer, W. et al., *Phys. Rev. B*, 17, 3362, 1978.)

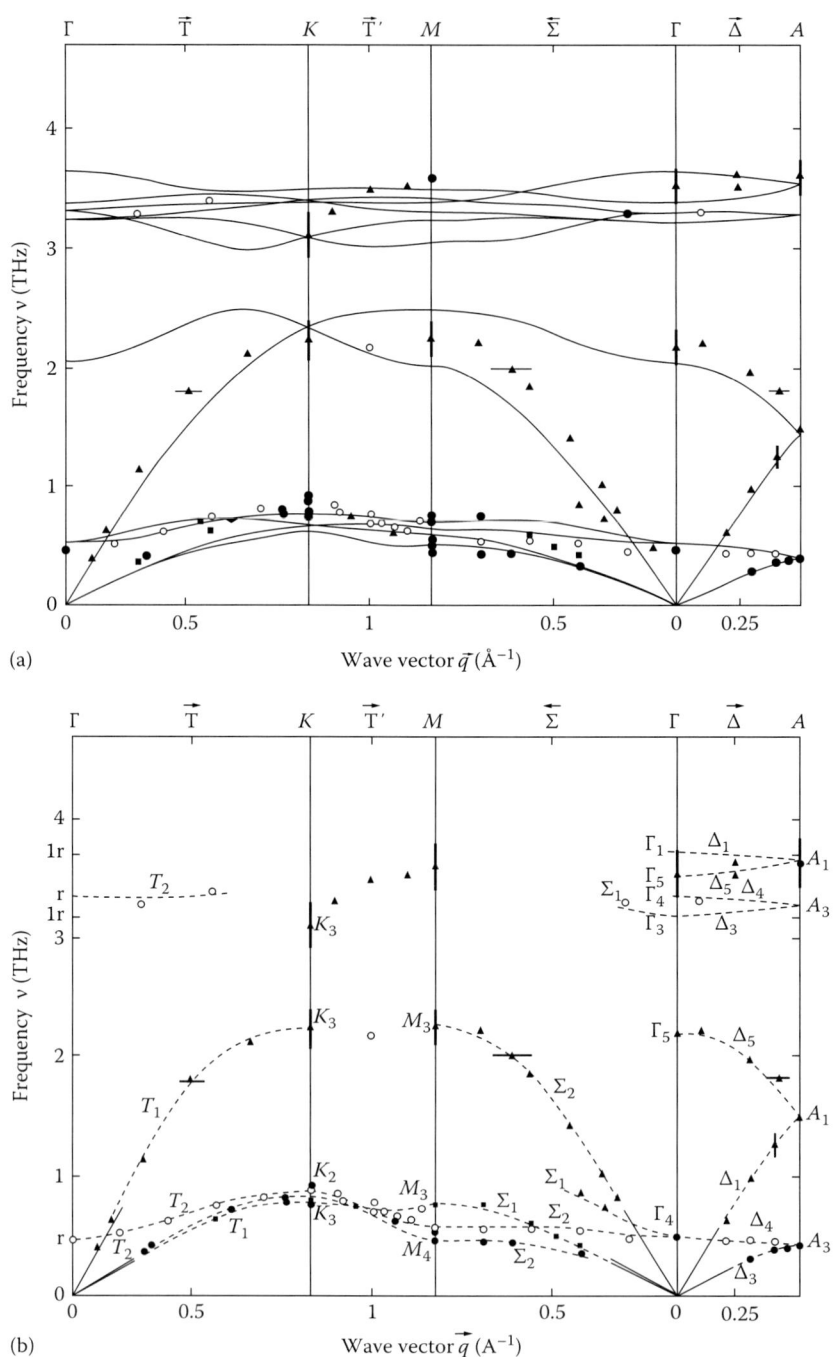

(a)

(b)

FIGURE 6.17   (a) Calculated phonon dispersion in β-AgI based on rigid ion model (dotted lines) and valence shell model (solid lines). (From Buhrer, W. et al., *Phys. Rev. B*, 17, 3362, 1978.) (b) Phonon dispersion curves of single crystal β-AgI measured at 160 K from coherent inelastic neutron scattering using a triple axis spectrometer. Polarization vectors are as follows: Filled triangles, LA; TA perpendicular to *c*; ■, TA parallel to *c*; empty triangle, LO; ◯, TO perpendicular to *c*; ◻, TO parallel to *c*; ◊, unknown; Ir-R, infrared Raman. Dashed lines guide the eye. Solid lines are velocities of sound. (After Buhrer, W. et al., *Phys. Rev. B*, 17, 3362, 1978.)   (*Continued*)

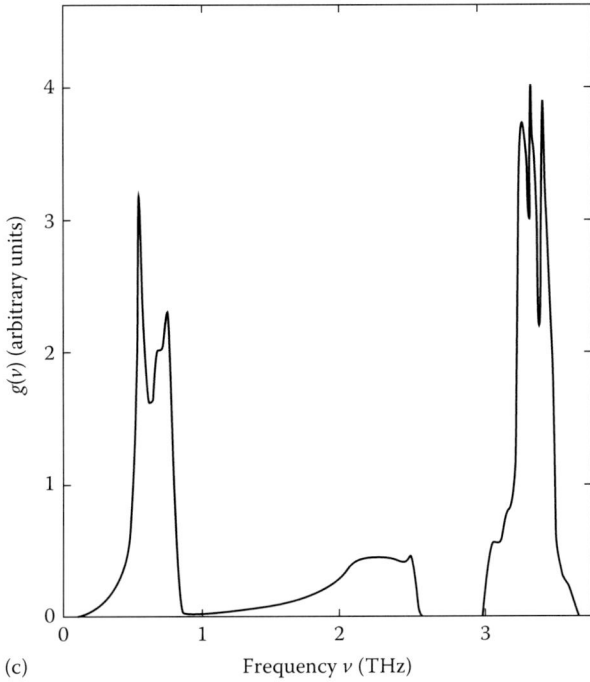

(c)   Frequency ν (THz)

FIGURE 6.17 (*Continued*)   (c) Phonon density of states $g(\nu)$ for β-AgI plotted as a function of frequency ν calculated with the valence shell model. (From Buhrer, W. et al., *Phys. Rev. B*, 17, 3362, 1978.)

solid state ionic materials at temperatures far below their superionic transition temperatures characterizes them from ordinary ionic solids. Figure 6.17b displays the experimental dispersion curves from coherent inelastic neutron scattering measurements, while Figure 6.17c shows the unique density of states function exhibiting two maxima, one each in the low-frequency and high-frequency regions.

The case of phonons in metastable γ-AgI is interesting. Single crystals of the zincblende phase cannot be grown precluding independent measurement of phonon spectra. Therefore γ-CuBr has been used as a model to "reduce" experimental values of γ-AgI as shown in Figure 6.18.

The rocksalt phase of AgI is not stable at normal pressures (Figure 6.19). Therefore its calculated phonon dispersion (Figure 6.20) is sensibly compared with experimental data on rocksalt AgBr at 0.1 MPa. This should guide future experiments.

Now we discuss phonons in cuprous halides. Figure 6.21 shows the dispersion curves on CuCl to be compared with calculations using quadrupole deformable shell model. The satisfactory agreement between theory and experiment justifies the use of QD as an essential basis for generating phonons in this compound.

Figures 6.22 and 6.23 show phonon dipersion curves for γ-CuBr calculated from rigid ion model and QD model. They give an opportunity to consider careful comparisons with experiment. The temperature dependence of phonon density of states and of heat capacity of AgCuS shown in Figure 6.26 helps an intuitive understanding of the temperature dependence of phonon damping.

Figure 6.24a shows the experimental phonon dispersion in γ-CuI along with a rigid ion model calculation which fits experimental data fairly well. In Figure 6.24b, the PDOS calculated in the same model is presented. As a simple calculation of elastic constants of this zincblende cubic crystal, the initial slopes of the three acoustic branches give estimates of c_{44}, c_{12}, and c_{11} as 0.158, 0.305, and 0.405, respectively, in units of 10^{12} dynes/cm^2.

Results of phonon dispersion measurements on Ag$_3$SI single crystals using a triple axis neutron spectrometer are shown in Figure 6.25 for the three phases of Ag$_3$SI. A unique feature of this compound is the observation of a low energy (~2 meV) vibrational peak that provides a clue to fast ion conduction (Figure 6.25a). This peak has been observed in all the three phases, α, β, and γ, and analyzed as a local

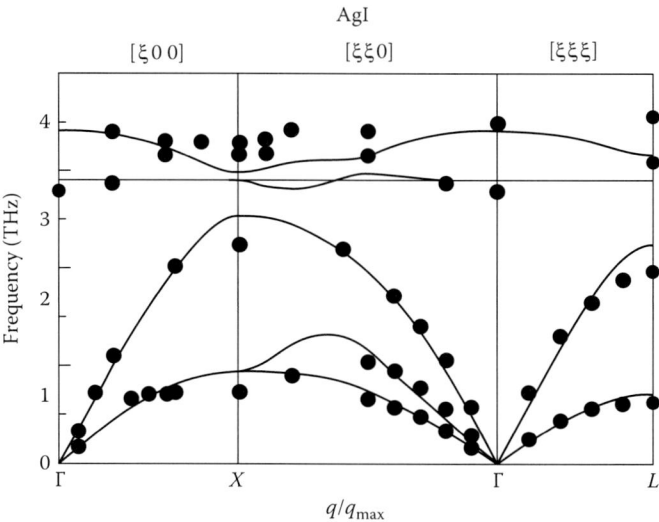

FIGURE 6.18 Phonon dispersion relation for γ-AgI. _____: theoretical values based on quadrupole deformable shell model. ●: experimental values reduced from γ-CuBr. (From Tomoyose, T. et al., *Solid State Ionics*, 167, 83, 2004.) The bcc superionic conductor α-AgI at 300°C has a distinct phonon spectrum displaying a flat transverse acoustic branch (Figure 6.19). Only the low energy transfers less than ~3 meV of TA and LA phonons are seen. The damping of phonons due to anharmonicities in α-AgI means that only a part of the phonon dispersion can be seen.

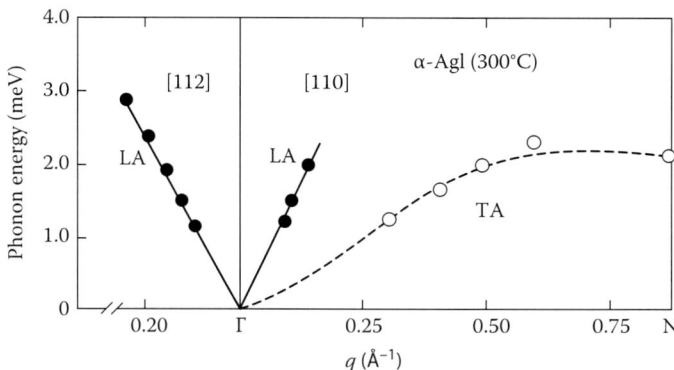

FIGURE 6.19 Phonon dispersion of α-AgI. (From Bruesch, P. et al., *Phys. Rev. B*, 22, 970, 1980.)

vibrational plus a TA mode highlighting the translational diffusion of Ag ions. In high-temperature α-phase, A and I show disordered arrangement similar to binary alloys, besides the disordering of Ag. This helps retain α-phase at low temperature when the crystal is cooled rapidly from a temperature above β–α transition point. The dispersion relations of Figure 6.25b highlight two features: the damping of phonons in α-Ag$_3$SI even at 90 K! and observation of a dispersionless mode ~5 meV besides the mode at 2.0–2.5 meV common to other Ag conductors.

Apart from the halides of Ag and Cu, chalcogenides of Ag and Cu are of interest, and so are mixed cation chalcogenides such as AgCuS. The room-temperature β-phase of a material that has a mineralogical analogue stromeyerite is built from a distorted hexagonal packing of S atoms resulting in an orthorhombic structure. It transforms to the hexagonal superionic phase at 366.5 K. At $T \sim 250$ K, it transforms to a closely related orthorhombic γ-phase. Lattice vibrations of α and β phases are of interest [6.17]; because ion diffusion in AgCuS is a process connected with the low-energy optical (LEO) mode observed in the

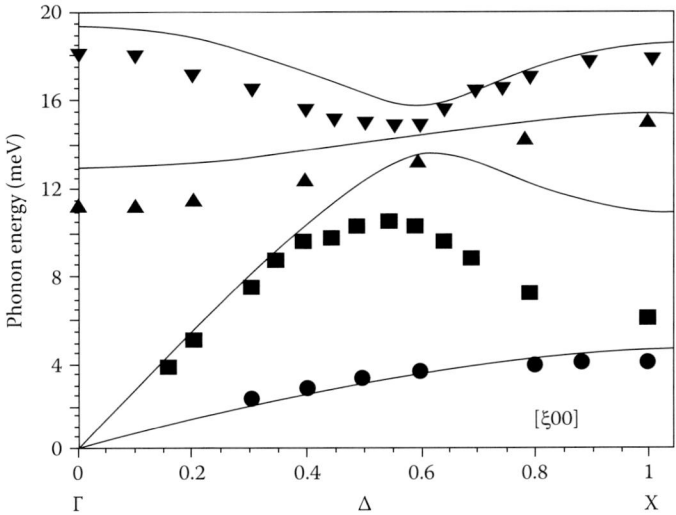

FIGURE 6.20 Calculated phonon dispersion in rocksalt phase of AgI at 6.0 GPa (full curves). For comparison, experimental data on rocksalt AgBr at 0.1 MPa is used (black symbols). Curves are derived from x-ray absorption fine structure (EXAFS) experiments. (From Yosihasa, A. et al., *J. Phys. Conf. Ser.*, 121, 102211, 2008.)

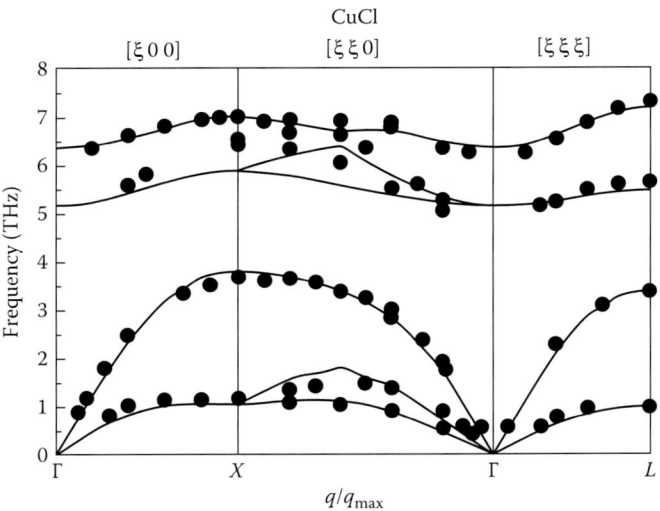

FIGURE 6.21 Phonon dispersion relation for γ-CuCl. _____: theoretical values based on quadrupole deformable shell model. ●: experimental values. (From Tomoyose, T. et al., *Solid State Ionics*, 167, 83, 2004.)

lattice vibrational spectra. PDOS tracks the LEO mode and gives an indication of any mode softening taking place. Figures 6.26 and 6.27 display the PDOS data.

Important thermal properties related to PDOS use lattice heat capacity C_v and Debye temperature.

Debye temperature can be obtained from $G(\varepsilon)$ function under the assumption that $G(\varepsilon)$ shows Debye-like behavior in low-energy limit. In this model, PDOS extends up to a Debye cut-off energy ε_D as

$$G_D(\varepsilon) = \begin{cases} \alpha\varepsilon^2, & \varepsilon \leq \varepsilon_D \\ 0, & \varepsilon \leq \varepsilon_D \end{cases} \tag{6.53}$$

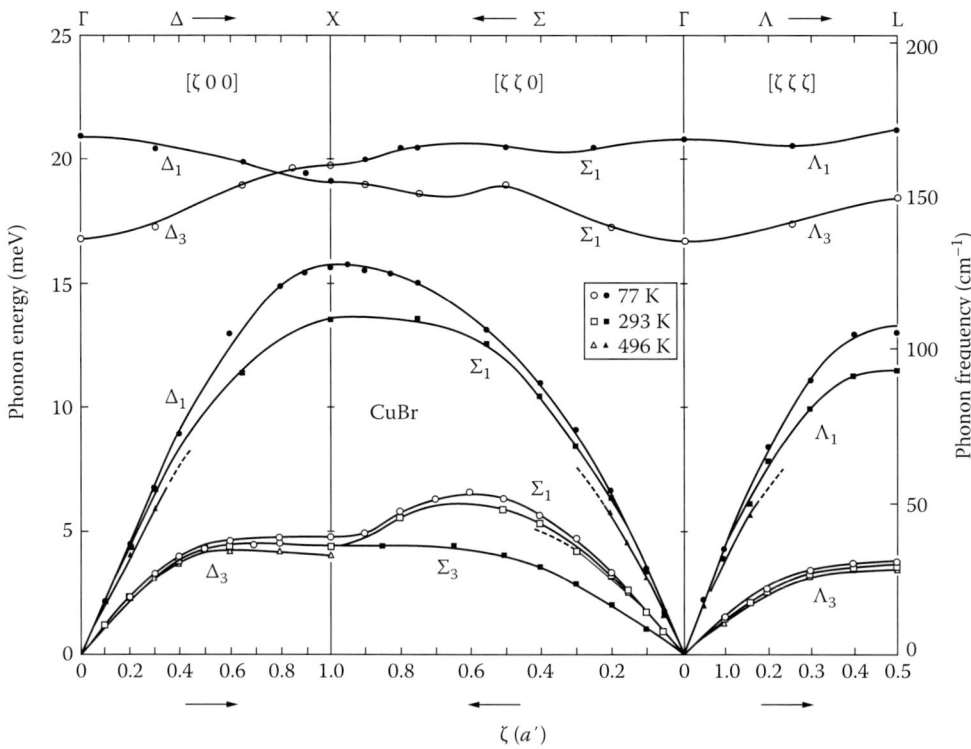

FIGURE 6.22 Phonon dispersion of γ-CuBr at three different temperatures in the rigid ion model. (From Hoshino, S. et al., *J. Phys. Soc. Jpn.*, 41, 965, 1976.)

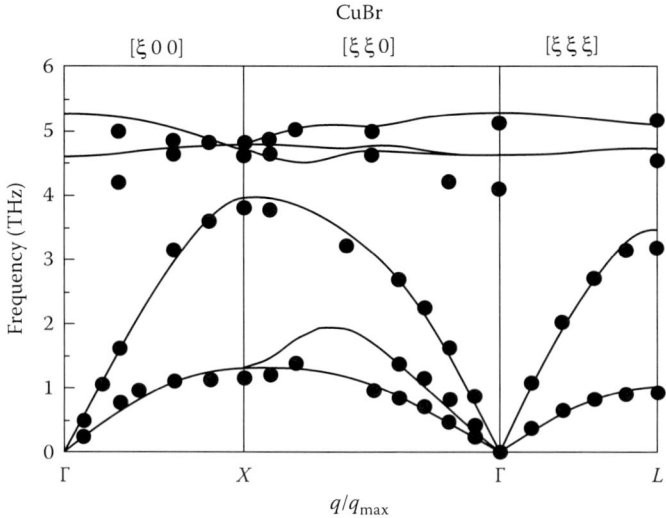

FIGURE 6.23 Phonon dispersion relation for γ-CuBr. _____, theoretical values based on quadrupole deformable shell model; ●, experimental values. (From Tomoyose, T. et al., *Solid State Ionics*, 167, 83, 2004.)

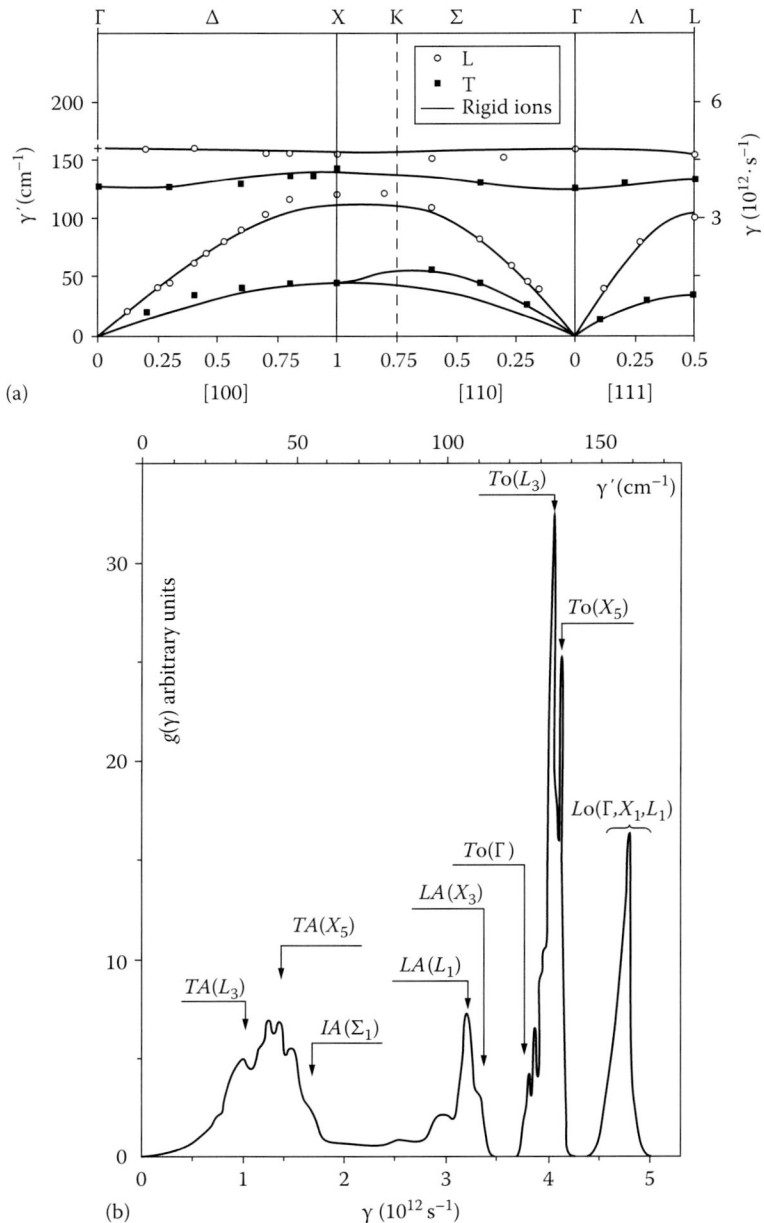

FIGURE 6.24 (a) Phonon dispersion in CuI at room temperature. Experimental points for longitudinal (o) and transverse (●) branches compare favorably with the calculations based on rigid ion model using six parameters for short-range potential. (From Hennion, B. et al., *Phys. Rev. Lett.*, 28, 964, 1972.) (b) Phonon density of states in CuI according to rigid ion model. Note the clear separation of the acoustic and optical branches. (After Hennion, B. et al., *Phys. Rev. Lett.*, 28, 964, 1972.)

The first and second moments of $G(\varepsilon)$ correspond to mean- and means square energies. The latter quantity gives the mean force constant $(M/\hbar^2 \int_0^\infty G(\varepsilon)\varepsilon^2 d\varepsilon$. If we neglect the temperature dependence of the density of states (quasi-harmonic approximation), we can obtain the constant volume lattice specific heat per atom as

$$C_V = 3k_B \int_0^\infty G(\varepsilon)\left[\frac{\beta\varepsilon}{(e^{\beta\varepsilon}-1)}\right]^2 e^{\beta\varepsilon} \qquad (6.54)$$

Now. we will discuss about the phonons in Li battery electrode materials. Batteries were introduced in Chapter 1 and would again be discussed in detail in Chapter 8.

6.3.4 Phonon Dispersion in Li Battery Electrode Materials

Battery cathode materials systems such as $LiCoO_2/Li$ half cell operate in dynamic environments so that basic properties such as lattice vibrations influence the very design of such materials. Therefore, it is

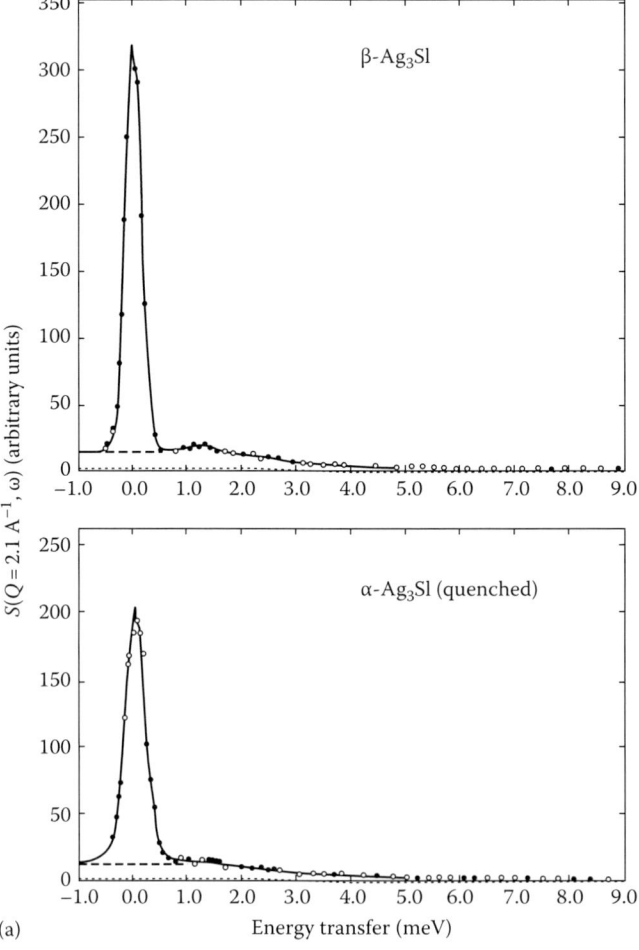

FIGURE 6.25   (a) Dynamic neutron scattering function $S(Q, \omega)$ with $Q \approx 2.1$ A^{-1} of β- and α-Ag$_3$SI at 295 K from neutron time-of-flight spectra. The low-energy peak characteristic of superionic conductors is seen. (After Shibata, K. and Hoshino, S., *J. Phys. Soc. Jpn.*, 54, 3671, 1985.)   *(Continued)*

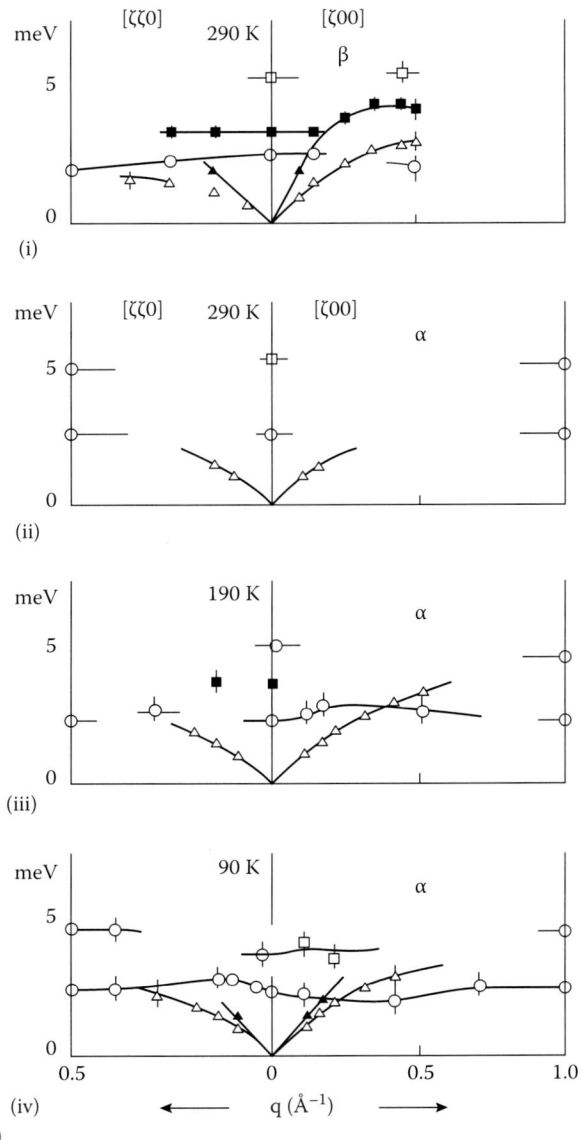

FIGURE 6.25 (*Continued*) (b). Phonon dispersion in Ag$_3$SI using triple axis spectrometer. (i) β-phase at 293 K. (ii–iv) Quenched α-phase at 290, 190, and 90 K, respectively. (After Hoshino, S. et al., *J. Phys. Soc. Jpn.*, 57, 4199, 1988.)

necessary to include vibrational energy while evaluating their temperature-dependent thermodynamic properties. These have an influence on their performance, namely discharging that involves gradual removal of Li from LiCoO$_2$. A first-principles quasi-harmonic approach specifically taking the lattice vibrational contribution to the Helmholtz free energy of the system $F(V, T)$ can solve the problem. This approach developed by Shang et al. [6.18] starts by defining $F(T, V)$ as

$$F(T,V) = E(V) + E_{elec}(V,T) + E_{vib}(V,T) \qquad (6.55)$$

where

$E(V)$ is the static energy at 0 K (without taking zero point energy)
$E_{elec}(V, T)$ is the thermal electronic contribution at V, T significant only at low temperatures
$E_{vib}(V, T)$ is the lattice vibrational contribution

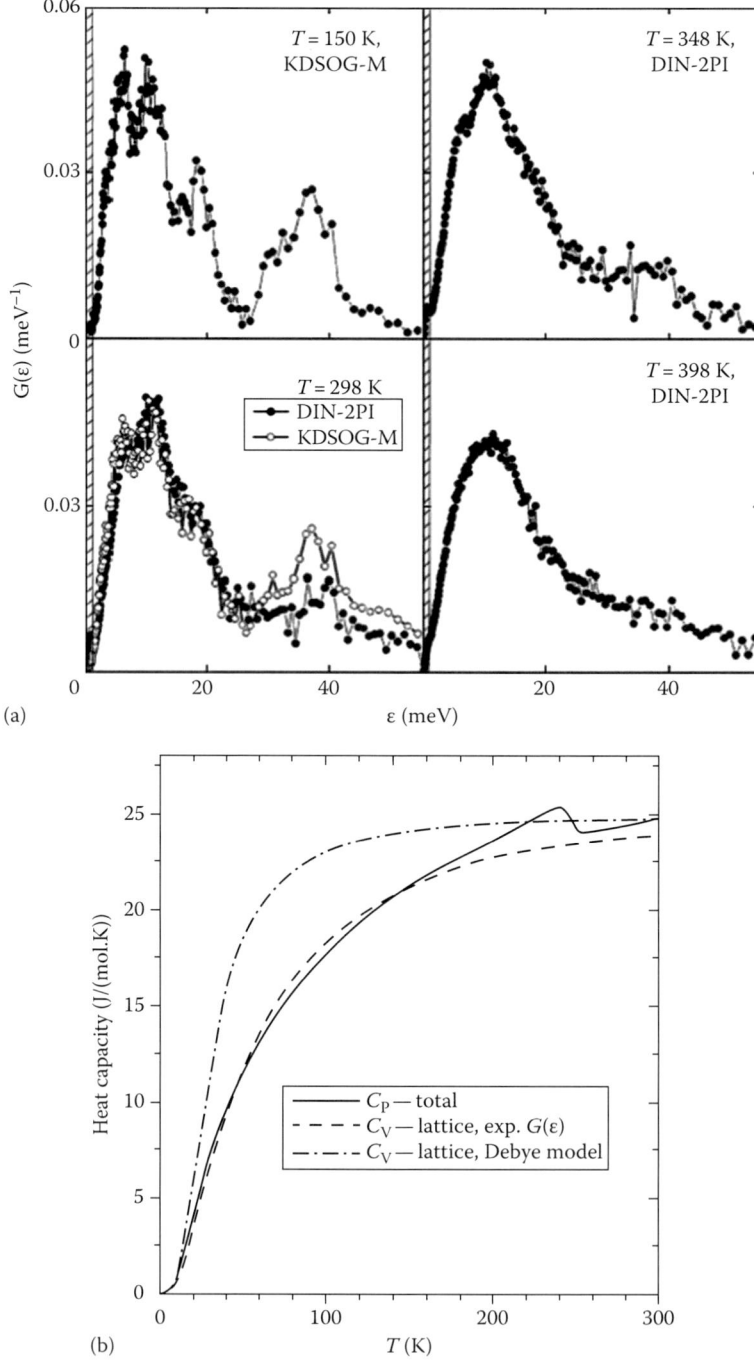

FIGURE 6.26 (a) PDOS spectra ($G(\varepsilon)$ vs. ε) of AgCuS measured by inelastic neutron scattering in the wave vector range 1–7 7 A^{-1} (much wider than the first Brillouin zone of the reciprocal lattice) at selected temperatures covering α (398 K), β (348, 298 K), and γ (150 K) phases measured at two neutron facilities. Significant softening of the well-resolved peaks at 3, 6, 10, 16, and 18 meV at 150 K occurs at 398 K when the low-energy mode becomes overdamped. This makes the PDOS function at low energy non-Debye in nature with $G(\varepsilon) \sim \varepsilon$. (b) Heat capacity of AgCuS. Full curve: experimental $C_p(T)$ from calorimetry experiments showing an anomaly at ~250 K (γ–β phase transition); - - -: $C_v(T)$ derived from G(ε) in quasi-harmonic approximation and - - - $C_v(T)$ calculated in the Debye model with $\theta_D = 130$ K. Note that the Debye model overestimates the low-temperature heat capacity.

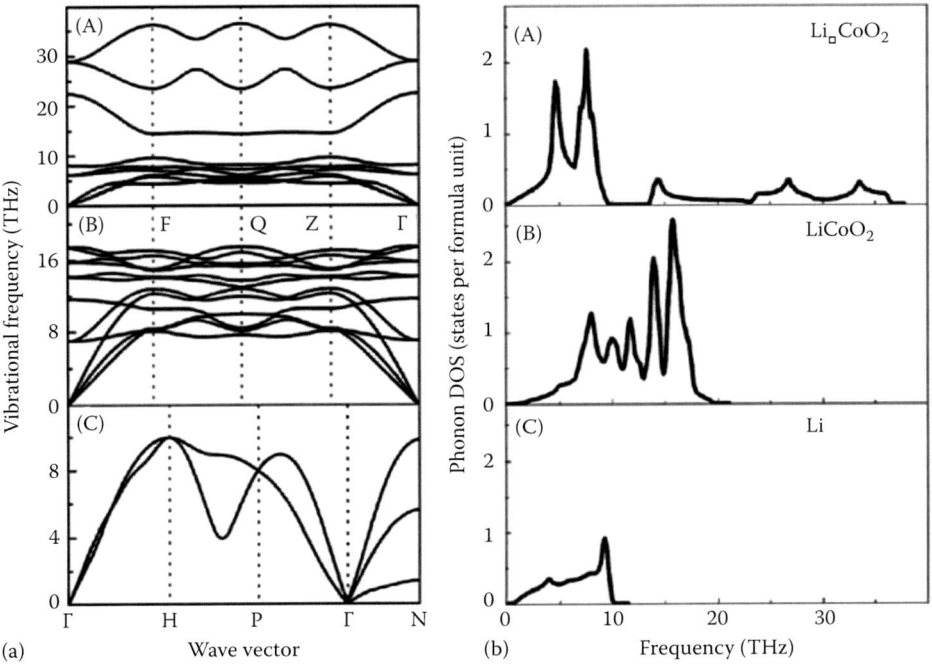

FIGURE 6.27 (a) Phonon dispersion curves for (A) Li$_\square$CoO$_2$, (B) LiCoO$_2$, and (C) Li metal in the first Brillouin zone. Dotted lines refer to high symmetry special wave vector points. (b) Phonon density of states of (A) Li$_\square$CoO$_2$, (B) LiCoO$_2$, and (C) Li metal. (After Gong, X. et al., *Int. J. Electrochem. Sci.*, 8, 10549, 2013.)

$E_{vib}(V, T)$ can be calculated from phonon free energy according to Bose–Einstein statistics:

$$F_{vib} = \frac{1}{2}\sum_i^{3N} \varepsilon_i + k_BT\sum_i^{3N} \ln(1 - e^{-\beta\varepsilon_i}) \tag{6.56}$$

where
$\varepsilon_i = h\nu_i$ is the phonon energy at different vibrational modes
ν_i is the vibrational frequency
$\beta = 1/k_BT$

The ground state energy and the forces for a given cathode material (here LiCoO$_2$) may be calculated by the Vienna *ab initio* simulation package (VASP) [6.19]. The ground state of the electronic structure is described with density functional theory (DFT) (to be discussed further in Chapter 7) and the generalized gradient approximation plus the Hubbard U. The on-site Coulomb energy term U for the Co-3d state is 4.91 eV suitable for the layered LiCoO$_2$. The phonon dispersion curves are calculated through the force-constant method while the atomic forces are calculated with VASP.

The body-centered cubic metal Li is modeled with the primitive cell and a $9 \times 9 \times 9$ Monkhorst-Pack scheme k-points mesh used in the vibrational frequency calculation [6.20]. Primitive cell models are also useful for LiCoO$_2$ and its delithiated version Li$_\square$CoO$_2$ together with $5 \times 5 \times 5$ Monkhorst-Pack scheme k-point meshes for the integration in the irreducible Brillouin zone. Energy cut-off for the plane waves is chosen to be 600 eV.

Figure 6.28a shows the phonon dispersion curves for Li$_\square$CoO$_2$, LiCoO$_2$, and Li metal in a unified way. Let us discuss it one by one. Li metal has a body-centered cubic lattice and the primitive cell contains only one Li atom. Therefore, there are only three acoustic vibration modes, which are completely degenerate at the Γ, H, and P points in the Brillouin zone. Also, the dispersion curves along lines between those

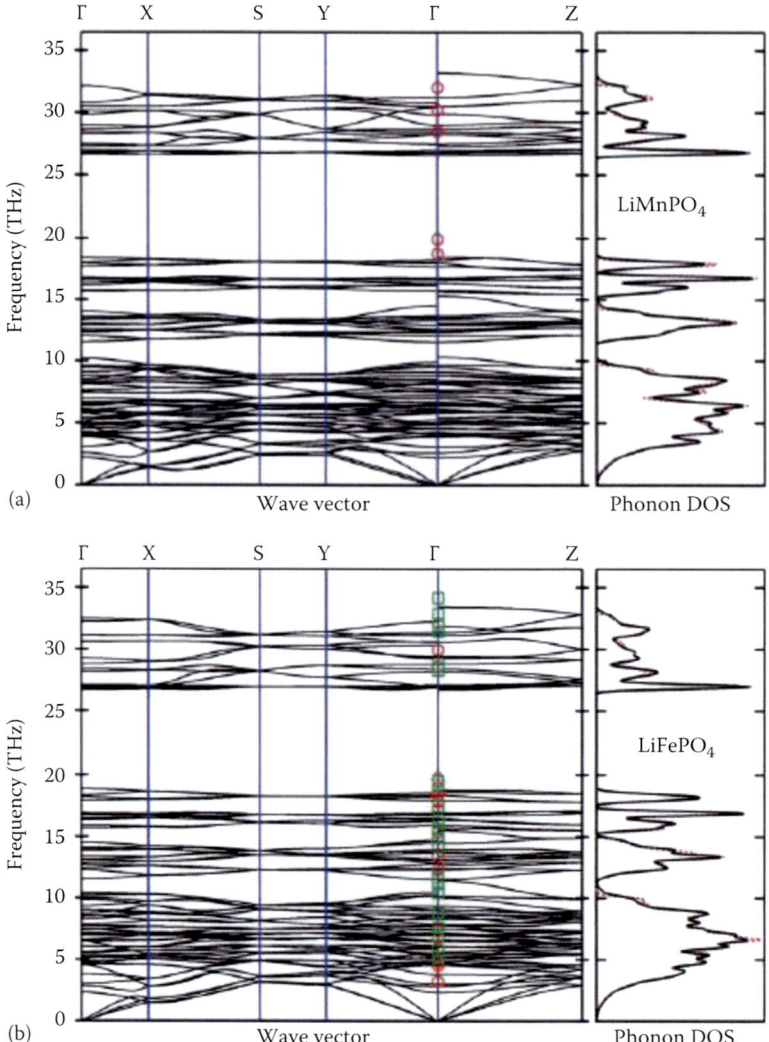

FIGURE 6.28 Predicted phonon dispersion and density of states for (a) $LiMnPO_4$ and (b) $LiFePO_4$ in their antiferromagnetic state from first-principles study. Red circles refer to experimental Raman data and green circles to experimental infrared data at the Γ-point. (After Shang, S.L. et al., *J. Mater. Chem.*, 22, 1142, 2012.)

high symmetry points are also degenerated leading to two independent dispersion curves. The highest vibrational frequency is about 9.98 THz, which is observed at the *H*-point.

The primitive cell of $LiCoO_2$ contains four atoms giving *nine* optical branches and *three* acoustic branches in the phonon dispersion curves, as it is shown in Figure 6.28b. The three acoustic vibration modes are completely degenerate at the Γ point ($q \rightarrow 0$, $\omega \rightarrow 0$), which represents the vibrations of all atoms as a whole. The $LiCoO_2$ lattice has the point symmetry of $D_{3d}(-3m)$. According to this symmetry, the nine optical vibration modes are partly degenerate and only five frequencies are shown at the Γ point. For the delithiated compound $Li_\square CoO_2$, the point symmetry is retained. Vibrational modes have a similar degeneracy. Interestingly, no imaginary frequency appears when all Li atoms are removed from the lattice, so the bulk structure is stable upon lithium removal. Two vibration modes look separated from the rest, and its vibration frequency is close to 30 THz at the Γ point. These two modes may represent the vibration of Co atoms within the ab plane (CoO_2 layer plane). After lithium is extracted from the lattice, the oxidation state of Co changes from Co^{3+} to Co^{4+} and the Co–O bond lengths become shorter within the CoO_6 octahedral structure. Consequently, the Co–O interaction becomes stronger, which enhances the vibration frequency substantially. On the other hand, the vibration frequency along the *c*-axis direction shifts to the low-frequency region because the lithium layer is completely removed from the lattice and the Coulomb interaction between the Li layer and the CoO_2 layer has vanished.

A better comparison of the vibration frequency can come from the PDOS and is given in Figure 6.28b. Phonon DOS data enable the calculation of the vibrational contribution of the thermodynamic quantities, which directly connects to intercalation potential of electrode materials for lithium ion batteries, and decreases appreciably with increasing temperature. The intercalation potential originating from the Gibbs free energy changes the battery system during the lithium intercalation process. For more details, see Reference 6.18 and references therein.

$LiMnPO_4$ and $LiFePO_4$ with olivine structure discussed in Chapter 1 are cathode materials for Li ion batteries. In addition to the favorable ionic conductivities they possess, they are also antiferromagnetic. The phonon spectra of these two phosphates based on first-principles calculations in the so-called mixed-space approach are shown in Figure 6.29 along with PDOS. Features at the center of the Brillouin zone are to be compared with the Raman and infrared spectral data. For details of the calculation procedure and detailed interpretation, see Shang et al. [6.21].

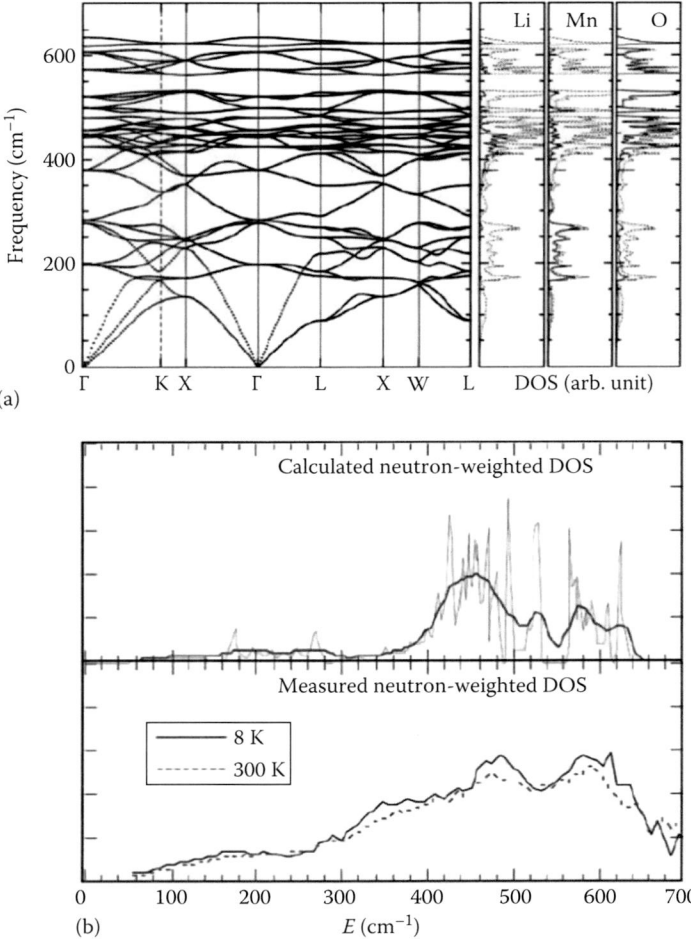

FIGURE 6.29 (a) Dispersion curves of $LiMn_2O_4$ spinel along high-symmetry directions in the Brillouin zone. Also shown are partial and total density (dotted line) of states of phonons in $LiMn_2O_4$. (b) Top: Neutron-weighted phonon density of states: calculated (thin line) and resolution broadened (thick line); bottom: measured neutron-weighted density of states from inelastic neutron scattering experiment. *(Continued)*

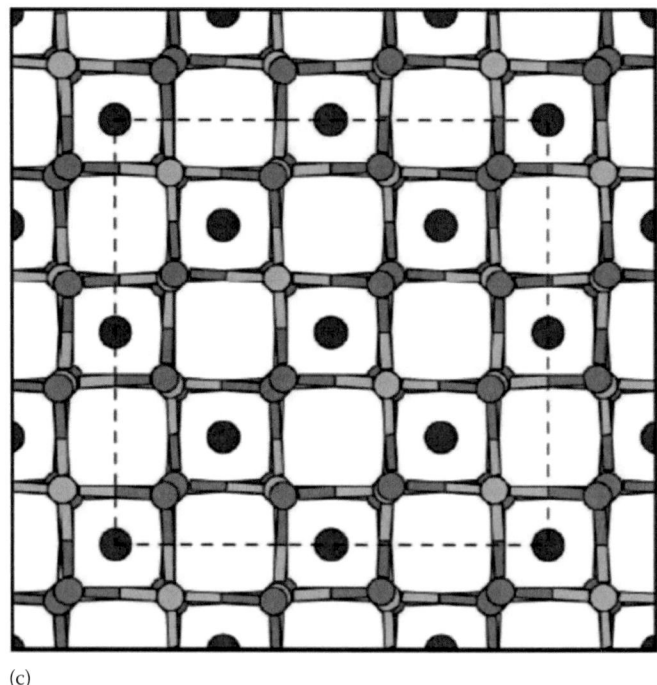

(c)

FIGURE 6.29 (*Continued*)   (c) The projected relaxed structure of $LiMn_2O_4$ used for calculations of phonon dispersion. Conventional structure is shown as a dashed square. (After Fang, C.M. et al., *J. Mater. Chem.*,17, 4908, 2007.)

Phonon dispersion curves and PDOS of the spinel $LiMn_2O_4$ are given in Figure 6.29. Above the acoustic phonon branches, the heavy Mn ions and the light Li ions separately contribute to making optical phonons of low (~180 to 280 cm^{-1}) and high (>400 cm^{-1}), respectively. The O ions participate in lattice vibrations over a wider range of frequencies (180–644 cm^{-1}). Measured PDOS of anhydrous powder of $LiMn_2O_4$ are in accord with the first-principles calculations [6.22].

Next we discuss the important issue of anharmonicity in the context of physical properties of superionic conductors taking Li_2S as an example.

6.4 Anharmonicity and Its Consequences

What is anharmonicity? In classical mechanics, anharmonicity is the deviation of a system from being a simple harmonic oscillator. We can speak of the anharmonic oscillator as a realistic model for fast ion conductors. In a typical simple harmonic oscillator (SHO), the displacement of the particle is a linear function of the force driving the motion of the particle. Such an oscillator has a single frequency of oscillation whatever the amount of energy pumped into it. The fundamental frequency of vibrations of a harmonic oscillator is thus independent of the vibrational amplitude. A practical analogy is the "philharmonic orchestra."

Therefore, an oscillator not oscillating as an SHO is anharmonic. In such a case, the oscillating system can be approximated to a harmonic oscillator and the anharmonicity can be calculated using perturbation theory. An important effect of anharmonicity is the appearance of oscillations with frequencies 2ω, 3ω, … where ω is the fundamental frequency of the oscillator.

If the frequency ω deviates from the frequency ω_0 of the harmonic oscillator, then, to a first approximation, frequency shift $\Delta\omega = \omega - \omega_0 \sim$ square of the oscillation amplitude $A \sim A^2$.

The phenomenon of thermal expansion of solids including solid state ionics is related in a basic way to the anharmonicity of the lattice interaction potential.

From thermodynamics, the pressure P is related to the Helmholtz-free energy (F) of a thermodynamic system through

$$P = -\left(\frac{\partial F}{\partial V}\right)_T \qquad (6.57)$$

F itself is defined as

$$F = U - TS \qquad (6.58)$$

When applied to a solid, Equation 6.57 emphasizes that the equilibrium state for a solid (e.g., at normal atmospheric pressure) is reached when the external pressure exactly balances the volume derivative of the free energy at a finite temperature T.

F of an electronic insulator has two contributions:

1. Equilibrium energy U_0 of the crystal lattice, that is, energy of the "springs" of their equilibrium positions.

   $P_0 = \partial U_0/\partial V$ is T-independent.

   Therefore,

$$\frac{\partial P_0}{\partial T} = \frac{\partial^2 U_0}{\partial V\, \partial T} = 0 \qquad (6.59)$$

2. Energy and entropy terms for the phonon system. The volume derivative for this part of H can be considered as the pressure of the phonon gas P_{ph}.

Now the linear thermal expansion coefficient

$$\alpha = \left(\frac{1}{3V}\right)\left(\frac{\partial V}{\partial T}\right)_P = -\frac{(1/3V)(\partial P/\partial T)_V}{(\partial V/\partial T)_T} \qquad (6.60)$$

Equation 6.105 is obtained by recognizing that T, P, and V are linked by the thermodynamic equation of state

$$F(T,P,V) = 0 \qquad (6.61)$$

Take the total derivative with respect to each variable. Solve the determinantal equation.

Note that the bulk modulus of a solid (the inverse of isothermal compressibility) is

$$B = -V\left(\frac{\partial P}{\partial V}\right)_T \qquad (6.62)$$

Using this in Equations 6.60 and 6.57 thermal expansion, we get

$$\alpha = \left(\frac{1}{3B}\right)\left(\frac{\partial P}{\partial T}\right)_V = \left(\frac{1}{3B}\right)\left(\frac{\partial P_{ph}}{\partial T}\right)_V \qquad (6.63)$$

What is P_{ph}. It is made up of two terms: (1) volume derivative of the T-independent zero-point energy of lattice vibrations and (2) the T-dependent term through phonon population.

$$n_s(k) = [\exp \beta \, h \, \text{cut} \, \omega_s(k) - 1]^{-1} \tag{6.64}$$

The first term is ignored as being very small. The second term is nonzero if at least some of the phonon frequencies are V-dependent.

If the lattice is harmonic, phonon frequencies are V-independent, so thermal expansion is zero at all Ts. Therefore only anharmonicity can produce a nonzero thermal expansion of a solid. It is given by

$$\alpha = \frac{\gamma C_V}{3B} \tag{6.65}$$

where γ is a dimensionless, model-dependent parameter that depends on phonon frequency and volume.

For typical materials, α is positive. Springs become stiffer as the volume is reduced and thus ~1. It can be $\gg 1$. Sometimes, they might be negative too as in the case of AgI. CuI with positive α and AgI with negative α form an interesting pair whose solid solutions include a composition with zero thermal expansion [6.23].

Anharmonicity (both cubic and quartic) directly affects integrated intensities of Bragg reflections as seen in the temperature-dependent neutron diffraction of single crystals of zincblende CuBr which has been analyzed in terms of an anisotropic Debye–Waller factor [6.24]. Powder neutron diffraction studies over an extended range of temperatures on zincblende CuCl and CuBr have revealed that the thermal motion of Cu is strongly elongated along the tetrahedral diagonals, due to the effect of quartic anharmonicity [6.25].

We now illustrate the tremendous effect of anharmonicity on the physical properties of a Li^+ fast ion conductor Li_2S.

Bertheville et al [6.26] have measured—in a systematic and comprehensive Raman spectroscopy experiment—the temperarure dependence of the one allowed T_{2g} Raman phonon mode in single crystals of Li_2S over the temperature range 7–1183 K (Figure 6.30). Aided by x-ray diffraction experiments and Raman pressure measurements, the effects of the *quasi-harmonic* part associated with thermal expansion on the temperature dependence of the phonon frequency are studied. The results show that these effects could not totally explain the experimental temperature dependence of the phonon frequency ω_j and linewidth Γ_j above 400 K. **Fourth-order anharmonicity must be considered.** Cubic anharmonicity allows for the decay of one LO phonon into *two* LA acoustic phonons in a *three-phonon interaction* process and leads to a linear temperature dependence of linewidth. However, quartic anharmonicity may lead to the creation of *three* LA acoustic phonons in a *four phonon process* resulting in the quadratic temperature dependence of linewidth. For more details see [6.26].

Anharmonicity allows for the decay of one LO phonon into *two* LA acoustic phonons in a *three-phonons interaction* process and leads to a linear temperature dependence of linewidth. However quartic anharmonicity may lead to the creation of *three* LA acoustic phonons in a *four-phonons process* resulting in the quadratic temperature dependence of linewidth. For more details, see Reference 6.26.

6.5 Optical Phonons in AgI-Type Compounds

Optical phonon does not migrate readily to the surface of a crystal because their non-dispersive nature near the Brillouin zone center gives them a very low group velocity. Instead, their energy is transferred to acoustic modes.

An important question in solid state ionics is the connection between phonons and the nature of the superionic conducting phenomenon. Are there any special phonons in solid state ionic compounds that could be responsible for the establishment of the fast ion conducting state at not too high temperatures with respect to ambient temperature? The answer to this question takes us to the role of the lowest frequency or energy (LEO) optical phonons and its role in mobile defect formation and ionic (Ag^+, Cu^+) conductivity, say, in zincblende and wurtzite forms of AgI and zincblende CuBr. In this

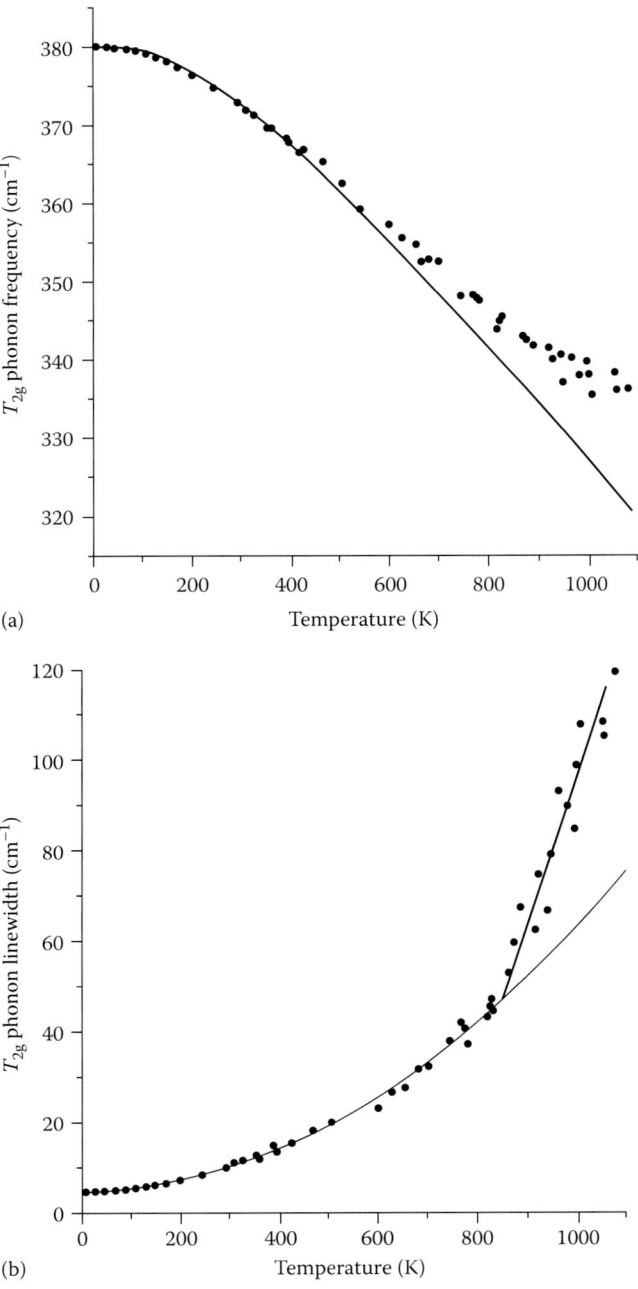

(a)

(b)

FIGURE 6.30 (a) Temperature dependence of the T_{2g} phonon frequency in Li_2S single crystal. Full circles, experimental points; full line, intrinsic volume contribution due to crystal thermal expansion. (b) Raman linewidth Γ_j of the T_{2g} phonon mode in Li_2S as a function of temperature. Full circles, experimental data; solid line: fit of the full width at half maximum to $\Gamma_j(T) = \Gamma_0 + a(n_a + 1/2) + b[(n_b + 1/2)^2 + 1/12]$ where Γ_0 is the residual linewidth due to, say, crystal defects and $a/2$ and $b/3$ are third- and fourth-order coupling constants producing phonon line broadening at absolute zero due to cubic and quartic anharmonicities, respectively. n_a and n_b are the phonon occupation numbers given by Bose–Einstein statistics. Anharmonic constants obtained are $a = 0.572$ cm^{-1} and $b = 1.862$ cm^{-1}. Residual linewidth due to disorder in crystal $\Gamma_0 = 4.21$ cm^{-1}. Theory and experiment agree the temperature range 0–850 K. Inclusion of quartic anharmonicity predicts a quadratic temperature dependence of linewidth as observed. This figure strongly demonstrates that the frequency and width (besides intensity) of the Raman phonon line strongly depend on crystal temperature as a consequence of the presence of anharmonic parts in the crystal potential energy.

discussion, the concept of Brillouin zone established in Chapter 1 and the crystal structure ideas of Chapter 2 are utilized.

As discussed earlier in connection with the diatomic lattice, application of an electric field (E) to NaI or AgI or CuBr crystal causes the displacement of positive and negative charges, thereby polarizing the crystal. The ionic polarization is given by

$$P_{ionic} = Ne(u - v) = \frac{[Ne^2/\mu]}{\left(\omega_T^2 - \omega^2\right)E} \tag{6.66}$$

where $\omega_T^2 = 2C/\mu$, μ being the reduced mass. Resonance occurs at $\omega = \omega_T$. The E here is the local electric field rather than the macroscopic average field given by Maxwell's electromagnetic theory. Equation 6.102 may be generalized to obtain a frequency-dependent dielectric function

$$\varepsilon(\omega) = \varepsilon(\infty) + \left[\frac{\omega_T^2}{(\omega_T^2 - \omega^2)}\right][\varepsilon(0) - \varepsilon(\infty)] \tag{6.67}$$

What is the significance of Equation 6.66? Firstly it gives the frequency-dependence of dielectric constant in terms of infinite and zero frequency εs. It sets frequency limits between which electromagnetic waves do not propagate through the crystal but are reflected from its boundaries. This property gives a powerful experimental tool namely optical reflectivity for solid state ionic materials. For our immediate discussion, Equation 6.67 forms the basis of the optical phonons observed in solid state ionic materials.

Electromagnetic waves will not pass a forbidden frequency band. This is a profound consequence of Equation 6.67 or the dielectric function. It has a powerful analogy in the electronic energy band theory of solids where the concept of band gap arises as we shall see in Chapter 7.

The forbidden frequency region is defined by the inequality

$$\omega_T^2 < \omega^2 < \omega_L^2 \text{ identically equal to } \omega_T^2 \varepsilon(0)/\varepsilon(\infty) \tag{6.68}$$

ω_L is defined by the term to the left of identical equality sign.

Now Equation 6.67 may be written as

$$\varepsilon(\omega) = \varepsilon(\infty)\left\{\frac{\left(\omega_L^2 - \omega^2\right)}{(\omega_T^2 - \omega^2)}\right\} \tag{6.69}$$

Note that at frequencies between ω_T and ω_L, the $\varepsilon(\omega)$ function is negative!

But the frequency and wave vector for electromagnetic waves in a dielectric are related as follows:

$$c^2 K^2 = \varepsilon(\omega)\omega^2 \tag{6.70}$$

If ω is real and $\varepsilon(\omega)$ is negative, then K must be imaginary! The wave then has the form

$$\exp(iKx) \to \exp{-\text{mod }Kx} \tag{6.71}$$

This is a damped, non-propagating wave! It can only be propagated through a thin slab which is 1/mod K unit thick. Wave propagation is forbidden in this sense. This implies reflection of the electromagnetic waves from the boundaries of the slab which is the basis of reflectivity measurements in solid state ionics.

How are the optical modes derived? Let us illustrate this by taking the zincblende structure (adopted by γ-AgI, γ-CuBr, and γ-CuI) as an example. The space group is T_d. The question to ask is: What are the transverse and longitudinal modes at the center of the Brillouin zone corresponding to the irreducible representation Γ_{15}? These modes are threefold degenerate (i.e., they possess same energy). The long-range

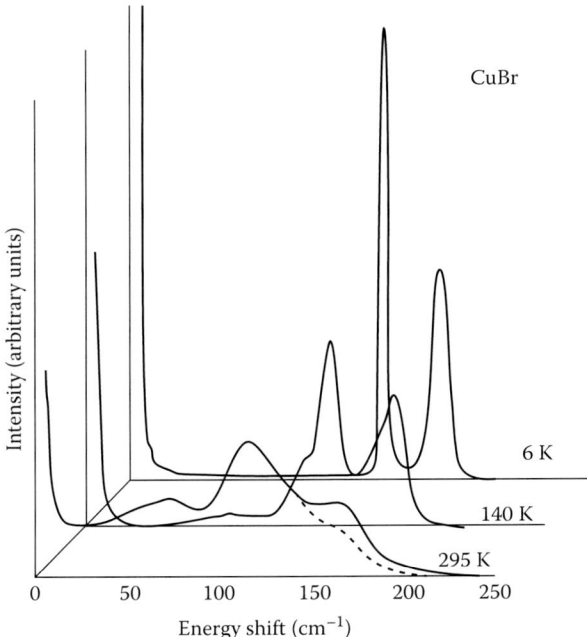

FIGURE 6.31 Raman stokes spectra of CuBr at 295, 140, and 6 K. The geometry $x'(z, x')z$ implies $x'(110)$ as the direction of propagation of the incident Ar^+ laser beam while $z(001)$ is the polarization direction, the scattered beam being observed along the z direction. This geometry permits both LO and TO modes to be seen. The dashed curve corresponds to the $x'(z, y')z$ geometry which allows propagation of TO modes only thus indicating changes observed at 295 K. (From Potts, J.E. et al., *Solid State Commun.*, 13, 389, 1973. Reproduced with permission from Elsevier.)

Coulomb interaction of the LO mode lifts the degeneracy such that the frequency of the non-degenerate LO mode is raised above that of the doubly degenerate LO mode.

The ratio of these frequencies is related to the low- and high-frequency dielectric constants through the Lyddane–Sachs–Teller relation

$$\frac{\varepsilon_0}{\varepsilon_\infty} = \left(\frac{\omega_{LO}}{\omega_{TO}}\right)^2 \tag{6.72}$$

If Equation 6.71 is satisfied, then the mode assignments are correct. For the case of Raman spectra of CuBr, the LHS and RHS of this equation correspond to 1.34 ($\varepsilon_0 = 6.5$, $\varepsilon\infty = 4.84$) and 1.30, respectively [6.27]. So the assignments $\omega_{LO} = 151$ cm^{-1} and $\omega_{TO} = 133$ cm^{-1} are justified (Figure 6.31). We shall briefly consider the transverse optical phonons of Cu halides measured by Raman and neutron scattering experiments and calculated based on valence shell models that include third order or cubic anharmonicity.

Figure 6.32 shows the TO phonons at the Γ point (center) of the Brillouin zone in CuCl, CuBr, and CuI. The calculated anharmonic lineshape is given by

$$I(\lambda, \omega) = \frac{\Gamma(\lambda, \omega)}{[\omega(\lambda) + \Delta(\lambda, \omega) - \omega]^2 + \Gamma^2(\lambda, \omega)}[n(\omega) + 1] \tag{6.73}$$

Raman scattering refers to light scattering from optical phonons, just as Brillouin scattering happens from acoustic phonons. However, neutron scattering probes both acoustic and optical phonons. While Raman scattering measures a frequency shift of ≥ 1 cm^{-1} (1 meV = 8 cm^{-1}), through the use of a grating spectrometer, Brillouin scattering measures much smaller shifts ranging from 0.001 (30 MHz) to 1 cm^{-1} (30 GHz) by means of a Fabry–Perot interferometer.

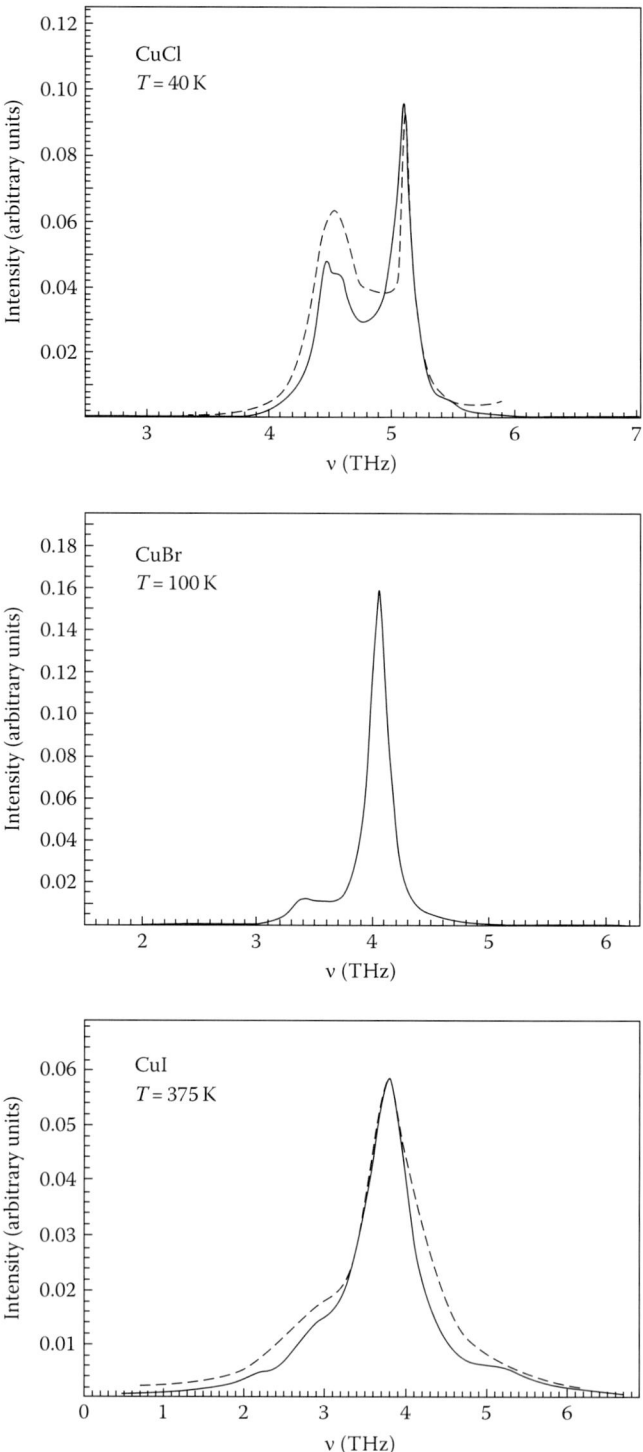

FIGURE 6.32 Brillouin zone center TO Raman phonons in CuCl, CuBr, and CuI. Dashed lines: experimental data; full lines: fit to lineshape function I(λ,ω) using the valence overlap shell model including third order or cubic anharmonicity. Note that for CuBr only the theoretical curve is given. The anomalous lineshapes are attributed to Fermi resonance. (From Kanellis, G. et al., *Phys. Rev. B*, 33, 8733, 1986.)

The following parameters about elementary excitations in solid state ionic materials may be determined by Raman measurements:

1. Energy ($\hbar\omega$) from $\omega_i - \omega_s$, the difference between the frequencies of incident and scattered light waves;
2. Momentum ($\hbar_k$) from the difference in the wave vector $k_i - k_s$;
3. Lifetime (τ) of the excitation which is 1/spectral linewidth Γ;
4. Symmetry of an elementary excitation from the polarization selection rules of the scattering process, that is, the scattering intensity is finite only for certain combinations of the polarization of the incident light (e_i) and the scattered light (e_s).

The Raman scattering intensity is basically determined by the coupling constant of the elementary excitation to the incident photons. In a typical Raman scattering experiment, the incident laser beam interacts very weakly with the sample; only one in 10^{14} photons is scattered and detected.

Raman scattering experiments have been carried out on three principal classes of solid state ionic materials: (1) copper and silver metal halides and some of their ternary compounds with RbI and Ag_2S, (2) the β-aluminas especially Na- and Ag-beta alumina, and (3) the fluorites.

Let us take the example of Ag-beta alumina to establish the important connection between fast ion conduction and phonons.

In a superionic conductor, a mobile ion vibrates at the equilibrium site during finite time. Thereafter, it jumps to the neighboring metastable site. Generally, the mean time of flight is reasonably assumed to be negligible compared to the mean residence time. So the ion motion is "jump diffusive." The relaxation time of a low-energy phonon of Ag-beta alumina is such that the attempt frequency for the jump diffusion process is strongly coupled to such a phonon.

The chemical formula for Ag beta alumina which has the same structure as Na-beta alumina discussed in Chapter 2 is $Ag_2O \cdot 11Al_2O_3$. Its space group is $P6_3/mmc$ with one formula unit per primitive hexagonal cell. Thus there are 58 atoms in the stoichiometric unit cell giving rise to 174 independent lattice vibrational modes. These modes can be characterized as Raman active and infrared active translational modes by a factor group analysis of the crystal. The important modes arise as the irreducible representation Γ of the phonon modes of Ag beta alumina at $k = 0$ A^{-1}, which is the center of the Brillouin zone. The decomposition yields:

$$\Gamma = \mathbf{\mathit{10A_{1g}}} + 3A_{2g} + 12B_{1g} + 3B_{2g} + \mathbf{\mathit{13E_{1g}}} + \mathbf{\mathit{15E_{2g}}} + 3A_{1u} + 13A_{2u} + 3B_{1u} + 11B_{2u} + 16E_{1u} + 14E_{2u} \quad (6.74)$$

Here 38 (10 + 13 + 15) modes written in bold italics are Raman active. Twenty seven are infrared active ($12A_{2u} + 15E_{1u}$) besides ($A_{2u} + E_{1u}$) which are translational modes. The experimental Raman spectra have to be assigned to these modes predicted by group theoretical analysis originally formulated by Bhagavantam and Venkatarayudu [6.29] who wanted to interpret Raman effect in relation to crystal structure. The analysis was applied to sodium beta alumina by Frech and Bates [6.30]. See Appendix 6.B for a group theoretical analysis of vibrational spectra for the T_d symmetry group.

Out of the 38 Raman active modes, the low-lying phonon band at 23 cm^{-1} identified as possessing E_{2g} symmetry. The eigenvector of this phonon corresponds to Ag ions vibrating along the conduction plane with the Al and O ions almost fixed at their equilibrium positions. Most significantly, vibrational analysis confirmed that the out-of-phase vibrational motion of Ag ions contributes to this phonon mode by more than 99% making it possibly the strongest Raman mode [6.31] (Figure 6.33).

Wakamura has traced the fundamental connection between phonons and superionic conduction. We shall briefly discuss two aspects: (1) phonon amplitude and low-energy optical phonon connection to superionic conduction or phonon-assisted ionic conduction [6.32] and (2) optical dielectric constant—short-range and long-range force connection [6.33].

A point ion model helps to estimate the vibrational amplitudes of mobile and cage ions below the superionic transition temperature (T_c) of a AgI-type superionic conductor. A point ion has mass m, charge Z^*e, spring force f averaged over constituent ions, and a restoring force proportional to the speed of the

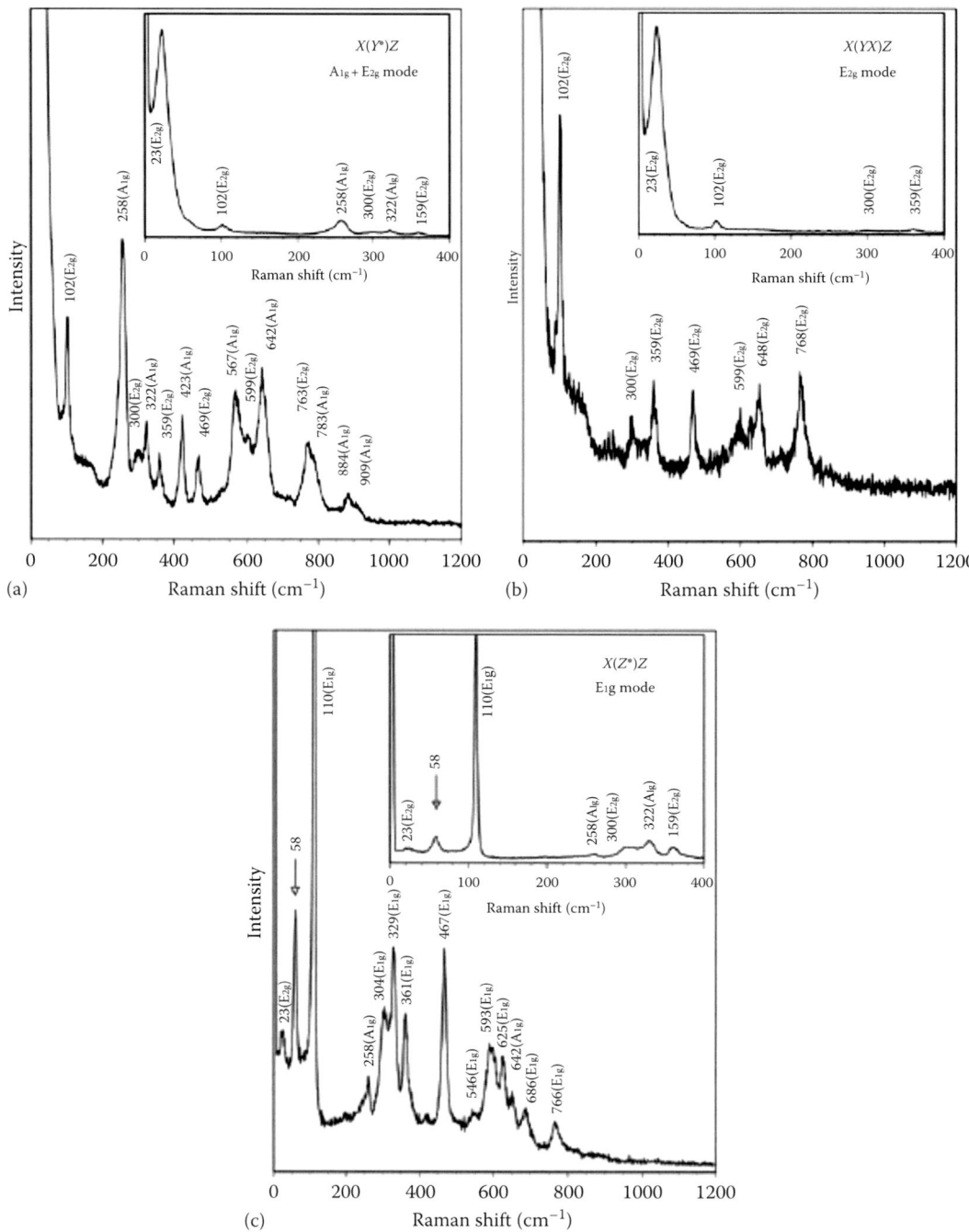

FIGURE 6.33 Polarized Raman spectra of phonons in Ag-β-alumina single crystals. Polarizations are set to $X(Y^*)Z$, $X(YX)Z$, and $X(Z^*)Z$ in a right-angle scattering geometry to observe $A_{1g} + E_{2g}$, E_{2g}, and E_{1g} modes, respectively. "*" means a non-polarized condition of the scattered beam. The excitation source was a 514.5 nm line of an Ar$^+$ laser with a power level of 160 mW focused in the sample to 130 μm diameter at the beam waist. Inset is an enlarged view of low-energy (<400 cm^{-1}) spectra. (After Kamishima, O. et al., *JPCM.*, 23, 225404, 2011.)

vibrating ion. The proportionality constant or the damping constant has different values for the cation and anion because of their different polarizabilities.

The equation of motion for the A ion in a linear chain is

$$m_A u_{A''} - \gamma_A m_A u_A + f_A u_A = Z * eE \tag{6.75}$$

where

m_A, u_A, γ_A, and f_A are atomic mass, displacement, damping constant, and spring force of the A ion, respectively

$Z*e$ is the ionic charge, approximately equal in magnitude for cation and anion under charge neutrality conditions

E is the electric field around the A ion

Inserting the time dependences for u_A and E, $u_A = X_A e^{-i\omega t}$, $E = E_V e^{-i\omega t}$, Equation 6.75 becomes

$$-m_A \omega^2 u_A X_A + i\gamma_A \omega\, m_A X_A + f_A X_A = Z * eE_V \tag{6.76}$$

At resonance, let $\omega = \omega_{TO} = (f_A/m_A)^{1/2}$. Then Equation 6.76 becomes $iX_A = Z*eE_V/m_A\omega_{TO}\gamma_A$. u_{A0}, the absolute value of X_A, is

$$u_{A0} = \left(\frac{Z*e}{m_A \omega_{TO} \gamma_A} \right) E_0 \tag{6.77}$$

A nonzero value of $Z*e$ implies an optical phonon giving a finite value for the vibration amplitude.

From Equation 6.77,

$$u_A - u_{X0} = Z * eE_0 R_G \omega_{TO} \tag{6.78}$$

where $R_G = (\gamma_X m_X - \gamma_A m_A)/\gamma_A \gamma_X m_A m_X$. From now on, we need to consider only R_G. Assume that at constant temperature, the damping coefficients γ_A and γ_X are proportional to the ionic radius:

$$\gamma_A = br_A \quad \text{and} \quad \gamma_X = br_X \tag{6.79}$$

where b is a constant of proportionality. The damping constant of the phonon mode shows a rapid **decrease** above T_c when the mode is constructed from only the cage ion vibration (e.g., I^- vibration in AgI), but **increases rapidly** for the mode involving mobile ion vibration (Ag^+ in AgI). Thus the phonon band showing **rapid broadening** connects to the mobile ions even as it shows **rapid narrowing** while relating to the immobile ions of the cage. From Equation 6.79, R_G of Equation 6.78 becomes

$$R_G = \frac{r_X m_X - r_A m_A}{r_A r_X m_A m_X b} \tag{6.80}$$

Figure 6.34 illustrates the trends in differential phonon amplitudes between mobile and cage ions R_G/ω_{TO} for superionic and non-superionic conductors. Figure 6.34a is for all compounds while Figure 6.34b is for silver and copper halides and silver and copper halides.

6.5.1 Phonon-Assisted Superionic Conduction

The correlation between the phonon frequency of the low-energy optical or LEO phonon ω_{LEO} (which is actually acoustic like) and the activation energy for mobile ion conduction E_{ac} establishes the important paradigm of phonon-assisted superionic conduction:

$$E_{ac} = C\omega_{LEO}^2 \qquad (6.81)$$

Figure 6.35 shows the plot of Equation 6.81 for a large number of ionic and superionic compounds.

This elegant model predicts two phenomena characteristic of superionic conductors: (1) bowing of pressure-dependent ionic conductivity and (2) absence of superionic conduction in gold chalcogenides.

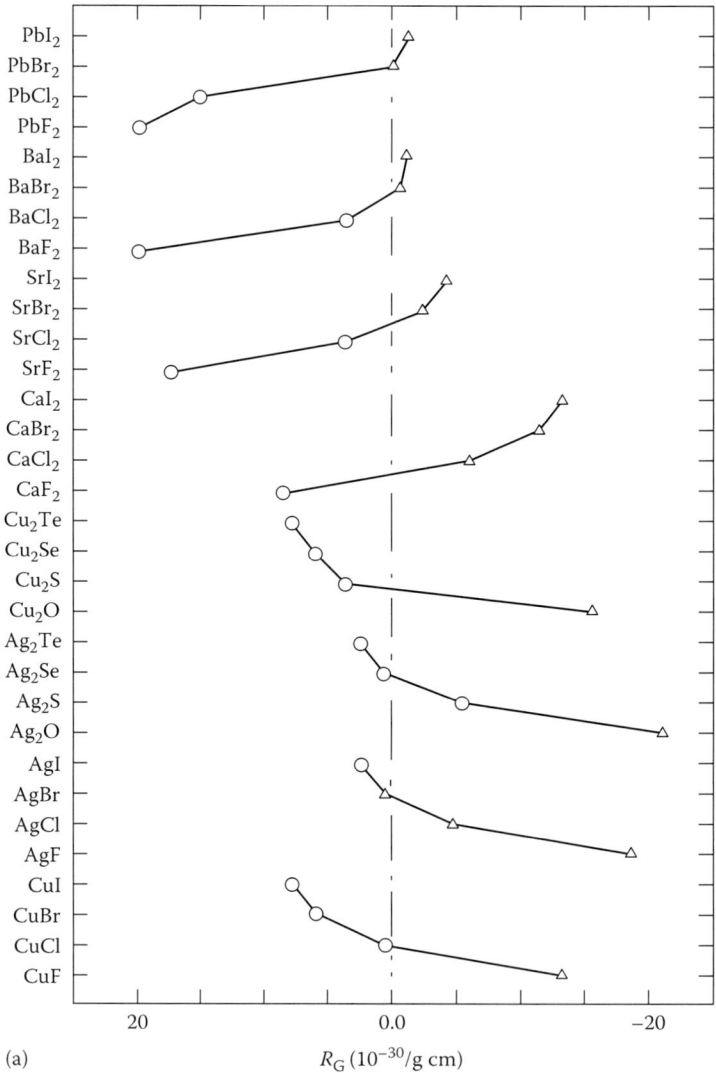

(a)

R_G (10^{-30}/g cm)

FIGURE 6.34 (a) The difference in vibration amplitude between mobile and cage ions in superionic (○) and non-superionic (△) compounds. The vertical dash–dot–dash line separates superionics from non-superionics.   (*Continued*)

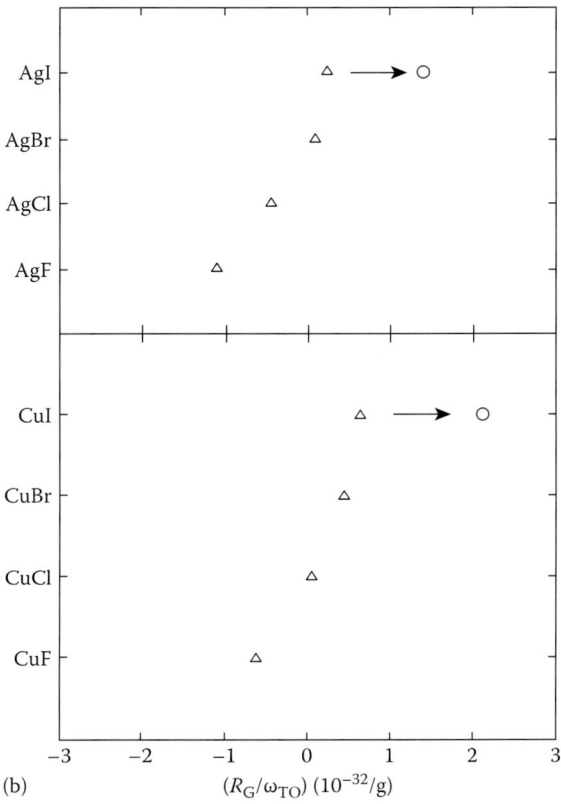

FIGURE 6.34 (Continued) (b) Trends in R_G/ω_{TO} for silver (top) and copper halides. For AgF, AgI, and CuI, the values 106, 200, and 128 cm^{-1} are indicated by △. For AgI and CuI, the values 17 and 38 cm^{-1}, characteristic of wurtzite-type structure, are shown by ○. (From Wakamura, K., *Phys. Rev. B,* 56, 11593, 1997.)

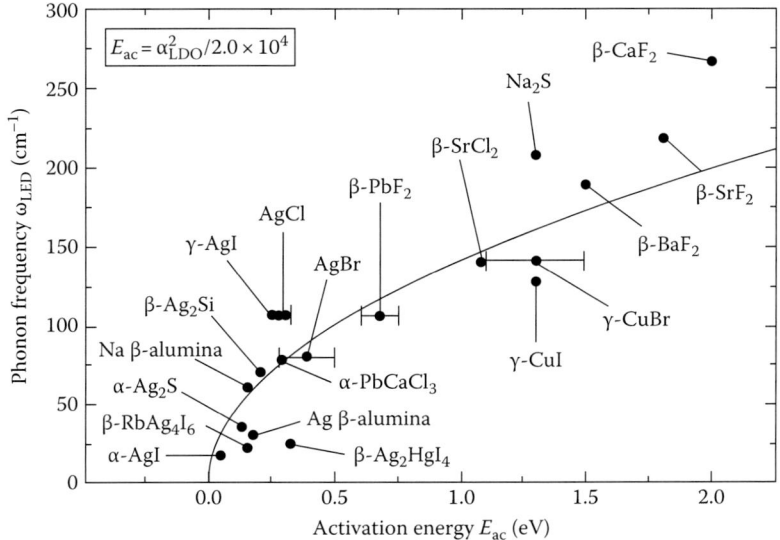

FIGURE 6.35 Plot of activation energy E_{ac} versus the lowest energy of optical phonon ω_{LEO}. Solid line represents equation $E_{ac} = \omega_{LEO}^2/20{,}000$. (From Wakamura, K., *Phys. Rev. B,* 56, 11593, 1997.)

An intuitive picture of the connection between the high-frequency (optical) dielectric constant ($\varepsilon\infty$) and the interatomic forces (both short range and long range) in cubic and near-cubic ionic (and semiconducting) compounds is obtainable using the rigid ion model. The TO and LO modes with frequencies ω_{TO} and ω_{LO} enable the calculation of the interionic force:

$$\omega_{TO}^2 = \left(\frac{f}{\mu}\right) - \frac{2\pi(Z*e)^2}{3\mu V} \tag{6.82}$$

$$\omega_{LO}^2 = \left(\frac{f}{\mu}\right) + \frac{4\pi(Z*e)^2}{3\mu V} \tag{6.83}$$

Here f, μ, $Z*e$, and V are the short-range force constant, reduced mass, effective ionic charge, and the unit cell volume, respectively. Using measurable values, f/μ and long-range force can be estimated from Equation 6.82:

$$\frac{f}{\mu} = \frac{2\,\omega_{TO}^2 + \omega_{LO}^2}{3}$$

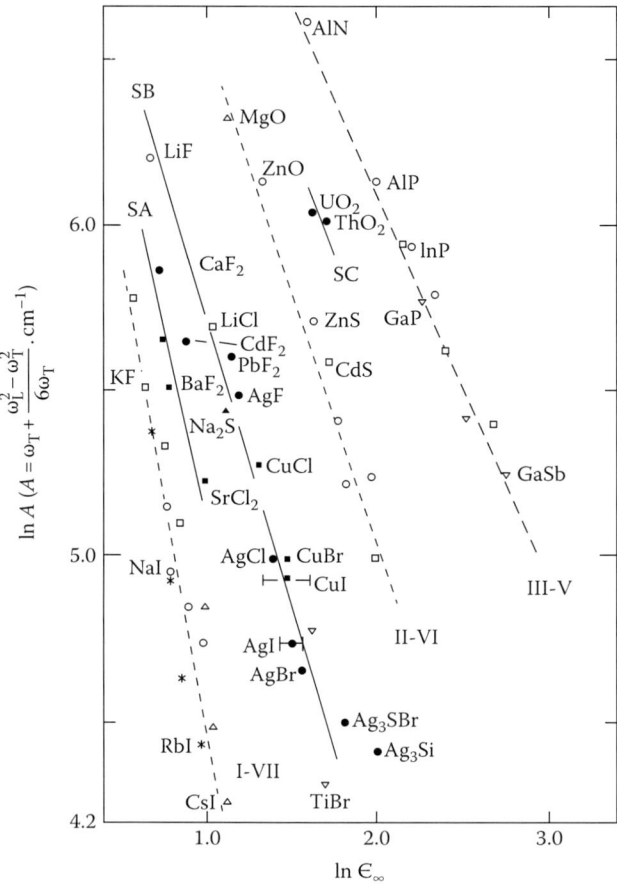

FIGURE 6.36 A linear relationship of ln A (short-range potential) vs. ln $\varepsilon\infty$ for superionic conductors and normal compounds. ———, through experimental points for fast ion conductors; ------, through experimental points of normal compounds. Isoelectronic compounds are shown by the same symbols. For ternary compounds only the intense phonon band is used. (From Wakamura, K., *Solid State Commun.*, 86, 503, 1993.)

$$\frac{\pi(Z*e)^2}{3\mu V} = \frac{\omega_{TO}^2 - \omega_{LO}^2}{6} \tag{6.84}$$

The short-range potential A $(=(f/\mu)^{1/2})$ is evaluated approximately from Equation 6.83:

$$\left(\frac{f}{\mu}\right)^{1/2} \approx \omega_{TO} + \frac{\omega_{TO}^2 - \omega_{LO}^2}{6\omega_{TO}} = A \tag{6.85}$$

This last equation enables a plot to be made between $\ln A$ and $\ln \varepsilon_\infty$ for several families of ionic and semiconducting compounds. Figure 6.36 shows such several plots each of which is linear.

Another important relationship exists between the superionic conducting transition temperature (T_c) and $\varepsilon\infty$: $\varepsilon_\infty^2 T_c = \exp(9.0)$. This approach could be tested for all the superionic conductors to obtain insights into the phenomenon.

Next we will consider the connection between phonons and the universal unsolved problem of absolute rate theory known as the Kramers problem that has special significance in superionic conductivity.

6.6 The Kramers Problem

Chemical reactions and diffusion processes in solids are conventionally described by absolute rate theory (ART). In this theory, a classical particle in a potential well hops at the rate ν given by

$$\nu_{ART} = \nu_0 \exp(-\beta U) \tag{6.86}$$

where
ν_0 is the attempt frequency associated with intra-well excitations
$\beta = 1/k_B T$
U is the height of the potential barrier

Within this framework, fast ion conductors such as CuI and PbF_2 possess a ν_0 determined from infrared measurements which is of the order of 1–10 THz, a typical phonon frequency. This is how phonons connect to ion conduction mechanism. What about Naβ–Al_2O_3 and other defect-structure ionic conductors? Surprisingly ν_0 extracted from (Equation 6.65) is consistently lower by several orders of magnitudes than typical optical phonon frequencies! Thus the fast ion conduction problem connects to the Kramers problem. What is this Kramers' problem?

Kramers pointed out in a landmark paper in 1940 [6.34] that Equation 6.86 is not valid in the extreme cases of (1) very low or (2) very high frictional forces acting on the moving particle which is the mobile ion Na$^+$ in, say, Naβ–Al_2O_3. For ART to be applicable, a stringent requirement is the existence of an equilibrium distribution for the states of the particle. Therefore the validity of ART is limited to a range of viscosities where the random fluctuations that push the particles over the energy barrier are compensated by the dissipation of energy that restores the distribution to its thermal equilibrium value. What is the situation in systems with very small dissipation? Kramers' showed that the actual rate is proportional to the frictional forces acting on the particle. In this case, the prefactor of ν can be considerably smaller than typical vibrational frequencies. Therefore, if the dissipative processes are properly taken into account, we can explain the low values of prefactors observed in defect structure ionic conductors.

Let us now outline the treatment of this problem by Huberman and Boyce [6.35] model, which explains the low prefactors and yields microscopic information on the viscous forces at play. The model used is that of random walk with absorbing walls, which enables ν to be obtained in terms of elementary forces acting on the mobile particle.

Consider a gas of ionic charge carriers (classical particles) in thermal equilibrium with the crystalline cage through which they move. The flux, J, of ions escaping from a deep potential, $\beta U \gg 1$, is determined by ART [6.36]. How? Assume that the equilibrium distribution for the particles extends to *all* energies, including those *above* the top of the potential well. Apply kinetic theory to the problem.

$$J_{ART} = \int_0^\infty \frac{v\,dp}{hf_0(\varepsilon)} \tag{6.87}$$

where
  v is the velocity of the particle
  p its momentum
  ε its energy
  $f_0(\varepsilon)$ its thermal distribution

$$f_0(\varepsilon) = \exp(\beta(\varepsilon - \mu)) \tag{6.88}$$

where μ is the chemical potential of the particles.
  Substituting in Equation 6.87 and integrating,

$$J_{ART} = \left(\frac{1}{\beta}\right)\exp(-\beta\mu) \tag{6.89}$$

What is the number of particles n inside a deep potential well of depth U with a vibrational frequency ν_0 near the *bottom of the well*?

$$n = \left(\frac{1}{\beta h\nu_0}\right)\exp\{-\beta(\mu + U)\} \tag{6.90}$$

Note that the rate at which particles leave the well, ν, is given by $J = n\nu$. Therefore, Equation 6.86 is obtained.

In the "random walk with absorbing walls" model, it is natural to ask what the probability is that the particles "stick to the wall" or get trapped in the well. To work that out, let us assume that (1) the equilibrium states of particles in the potential well extend only up to the top of the well and (2) these trapped particles are in thermal equilibrium with a gas phase above the well. From the perspective of the solid state ionics, consider an equilibrium between ions in the lattice and ions in flight with thermal velocities from site to site. Let a particle of the gas phase with energy $\varepsilon \approx k_B T$ above the top of the well collide with the potential well. If the collision is inelastic and it loses an energy $> k_B T$, it will get trapped. Otherwise it will be reflected back to the gas phase. Let $P(\varepsilon)$ be the trapping probability of the particle with energy $\varepsilon \approx k_B T$. As particles of the gas phase fall into the wells, their flux must be balanced by the outgoing flux of particles. The outgoing flux J_{out}, in thermal equilibrium, is given by

$$J_{out} = J_{incident}\langle P \rangle = J_{ART}\langle P \rangle \tag{6.91}$$

where $J_{ART}\langle P \rangle$ is the fraction of the flux colliding with the well which gets trapped and falls into the well. The sticking coefficient $\langle P \rangle$ is given by

$$\langle P \rangle = \int_0^\infty \beta\,P(\varepsilon)\exp(-\beta\varepsilon)d\varepsilon \tag{6.92}$$

From Equation 6.86, the rate at which particles leave the well is

$$v = \langle P \rangle v_0 \exp(-\beta U) \tag{6.93}$$

It follows that in the limit $\langle P \rangle \rightarrow 1$ (when every particle that collides with the well gets trapped), the ionic charge carriers are in equilibrium with the crystalline cage *at all times*, and ART applies in this situation. Generally, however, the actual prefactor gets reduced from v_0 by the factor $\langle P \rangle$. Let us apply the general Equation 6.72 to the physics of ionic hopping. Here ions collide with the lattice. Assume that such collisions are Markovian in nature. Let $W(\varepsilon, \varepsilon')$ be the probability that a particle with energy ε' will end up with energy $\varepsilon < \varepsilon'$ in one round trip between the turning points of the potential. Then $P(\varepsilon)$ is given by the integral equation

$$P(\varepsilon) = \int_{-\infty}^{0} W(\varepsilon, \varepsilon') P(\varepsilon') d\varepsilon' \tag{6.94}$$

The ion-crystalline cage system here is a random system that changes states according to a transition rule that only depends on its current state. Therefore, in the course of one oscillation of the mobile particle, the atoms of the crystalline cage retain no memory of previous collisions. If the crystalline cage vibrational frequencies are much larger than those of the mobile ions near the top of the well, the ion hopping process would indeed be Markovian. The object is to solve Equations 6.92 and 6.94 in the limit of small dissipation. Suppose the average energy per collision is δ, then

$$\delta = \int_{-\infty}^{0} (\varepsilon - \varepsilon') W(\varepsilon, \varepsilon') d\varepsilon\, d\varepsilon' \tag{6.95}$$

From Equations 6.93 and 6.95, we get

$$v = \delta \beta v_{ART} \tag{6.96}$$

What do Equations 6.86 and 6.96 mean? A measurement of v and a knowledge of the absolute rate theory prefactor v_0 are sufficient to determine δ. This elemental energy relates to a frictional force acting on the particle, $F = mv\eta$, where m is the mass of the particle, v is its velocity inside the well, and η is the friction coefficient:

$$\delta = \int \eta m v^2 dt \tag{6.97}$$

Thus, the Kramers problem is formulated in terms of a mobile ion collision problem whose results are applicable to experiments on solid state ionic materials.

6.7 Soft Modes

What is a soft mode? Perhaps the best answer to this question has been given by Fleury [6.37]:

> In its simplest embodiment,the soft mode is a particular normal mode of the crystal lattice which exhibits an anomalous reduction in its characteristic frequency as the transition is approached, and whose eigenvector (array os atomic displacement) represents the distortion that the new structure impose upon the old. The soft mode thus carriers both dynamic and static information

on the developing transition. In addition, intersections between the soft mode and other (vibrational, electronic, magnetic, etc) degree of freedom in the material provide a microscope basis for changes in those physical properties that are indirectly associated with the phase transition. These interactions, as well as the singularities directly associated with the transition, might be utilized for new kinds of devices, as well as for new types of materials (e.g., metastable phases) engineered for specifically enhanced physical properties.

Soft mode idea is invoked to provide a picture of the mechanism of displacive phase transitions in ferroelctric materials such as $SrTiO_3$ in terms of the softening of a TO phonon. Figure 6.37 shows schematially the development of the soft mode in the temperature-(frequency)^2 plane. Below the temperature T_c, the frequency is imaginary making the phase unstable against a phase transition. The frequency at 0 K temperature is the harmonic value [6.38]. Lattice dynamical studies of Cu-based superionic conductors in the quasi-harmonic approximation [6.39] suggest several peculiarities in their physical properties. The soft mode behavior is a characteristic feature which may be a general feature signaling the superionic transition. In CuBr, for instance, the transverse acoustic modes exhibit softening.

Let us consider soft phonon modes in the superionic conductor $Cu_{1.85}Se$. In fact, $Cu_{2-x}Se$ is a mixed ionic–electronic conductor with a superionic transition at 414 K for the stoichiometric composition. At room temperature, the superionic phase exists in the concentration range from $x = 0.15$ to 0.25. The characteristic structural features of copper selenide are (1) the ordering of Cu atoms in the low-temperature phase and (2) a random distribution of Cu over interstitial sites in high-temperature superionic phase. Figure 6.38 the phonon dispersion of this room-temperature superionic conductor obtained from neutron inelastic scattering measurements on single crystal samples. A notable feature is the softening of the acoustic phonon branch in the [111] direction. This softening (frequency tending to zero as q tends to zero) is somewhat like the Kohn anomaly. This is by definition an *anomaly* in the dispersion relation of a phonon branch in a metal. For a specific wavevector, the frequency—and thus the energy—of the associated phonon is considerably lowered, and there is a discontinuity in its derivative. In the case of $Cu_{1.8}Se$, it is actually an instability of acoustic modes directly related to the ordering of Cu atoms observed in copper selenide. Thus a soft mode drives the ordering process.

We close this section and the chapter by looking at the most interesting case of soft-mode driven phase transitions in zirconia. Pure zirconia adopts a fluorite cubic structure (*C*) close to its melting point 2700 K. On cooling it undergoes two phase transitions: from cubic to tetragonal (*T*) to monoclinic (*M*) phases, the last one being stable at room temperature. The *C–T* transition does not involve much volume change and is thus second order, while *T–M* transition is supposedly first order involving dramatic changes in the coordination number of Zr atoms from 8 to 7, accompanied by a 4.5% volume increase in a spontaneous shear (~9°) deformation. The *C–T–M* structural evolution has been examined by a detailed group theoretical analysis according to which *both* transitions may be induced by the condensation of soft modes [6.41]. These soft modes belong to the special boundary points in the Brillouin zone (BZ). For the *C–T*

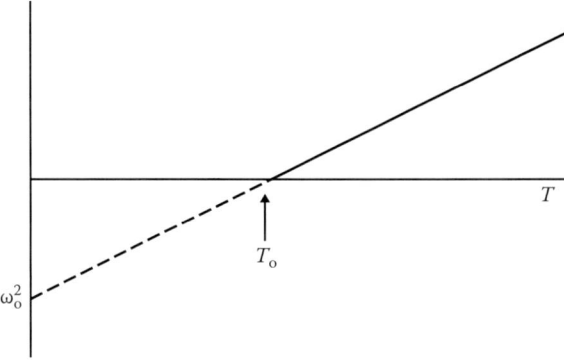

FIGURE 6.37 The soft mode visualized as causing a lattice instability leading to a phase transition. (From Dove M.T., *Introduction to Lattice Dynamics*, Chapter 8, Cambridge University Press, Cambridge, U.K., 1993.)

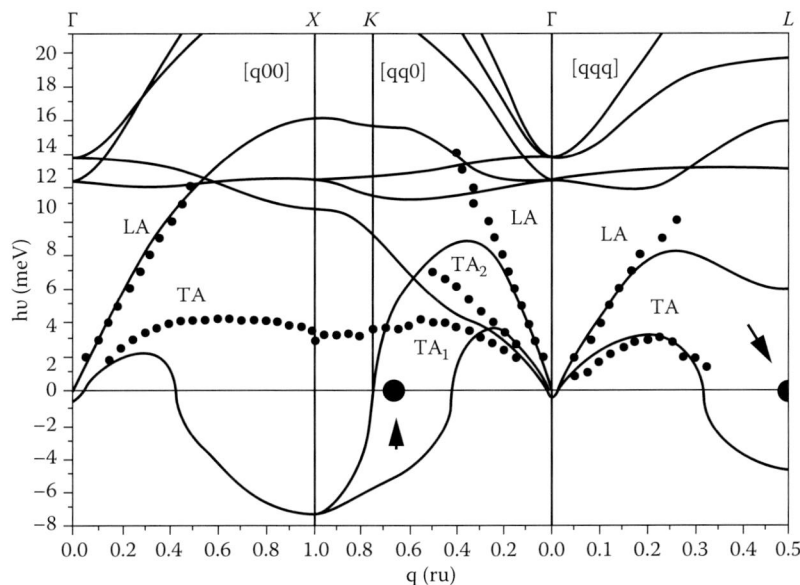

FIGURE 6.38 Phonon dispersion in single crystal of room-temperature superionic phase (α-phase with cubic structure) $Cu_{1.8}$Se showing experimental (●) and density functional theory calculated (_____) curves. The experimental points were obtained from neutron inelastic scattering experiments. Arrows show positions of superstructure reflections. Initial slope – a measure of sound velocity agrees well with DFT calculations. Significantly, in [111] direction at a q of ~0.1 reciprocal lattice unit (rlu), the phonon branch softens somewhat like the Kohn anomaly. (From Danilkin, S.A. et al., *J. Phys. Soc. Jpn.*,79(Suppl A), 25, 2010.)

inversion such a point is the *X*-point in *C*–BZ. However, thorough experimental work on Raman spectra covering the *C–T–M* transformation regime can only help test these expectations.

The direct observation of a soft mode in the experimental Raman spectra of pressurized zirconia is of interest (Figure 6.39). The soft mode at 260 cm^{-1} is the strongest feature in the spectrum at zero pressure, with its intensity *twice* as large as the combined intensities of the two high-frequency modes. This is so both in the experimental spectrum (Figure 6.39a) and in the calculated spectrum (Figure 6.39b). The intensity of the soft mode *decreases* monotonically under *compression*, falling to zero upon transition to the cubic modification. At the same time the intensity of the high-frequency features *increases* due to the monotonic *increase of the intensity of the B_{1g} line* combined with the *jump* of the E_g mode intensity at the transition point. The pressure dependence of the *combined intensity* of these two modes (Figure 6.39b) is qualitatively similar to the experimental observation.

We now focus on a Raman spectral study of tetragonal (*t*) zirconia single crystal grains stabilized in epitaxial thin films grown on monocrystalline alumina. This study correlates the Raman active modes of *t*-zirconia and the theoretical phonon dispersion of cubic zirconia. Figure 6.40 shows this correlation. A_{1g} mode emerges as a soft mode driving the tetragonal–cubic phase transformation. The other modes are also matched to specific points at the *X*-point of the Brillouin zone of cubic zirconia.

For a recent critical review on soft modes, see Cowley [6.45].

As discussed earlier, a soft mode refers to a particular mode of vibration of specific point symmetry whose (frequency)$^2 \rightarrow 0$ near a transition point. A very practical approach to gain insight into the mechanism of phonon softening in the context of structural phase transitions has been suggested by Schober [6.46]. It is useful to focus on the determinant of that block of the dynamical matrix (rather than the dynamical matrix itself) which will be dominated by the soft-going modes at temperatures sufficiently close to the transition temperature. If the softening occurs around points of high symmetry, then the determinant of the block corresponding to the irreducible representations of the soft-going mode only need be considered but not the determinant of the complete dynamical matrix, as illustrated using the example of quartz. As the regularity of the tetrahedral of β-quartz is emphasized as the origin of its

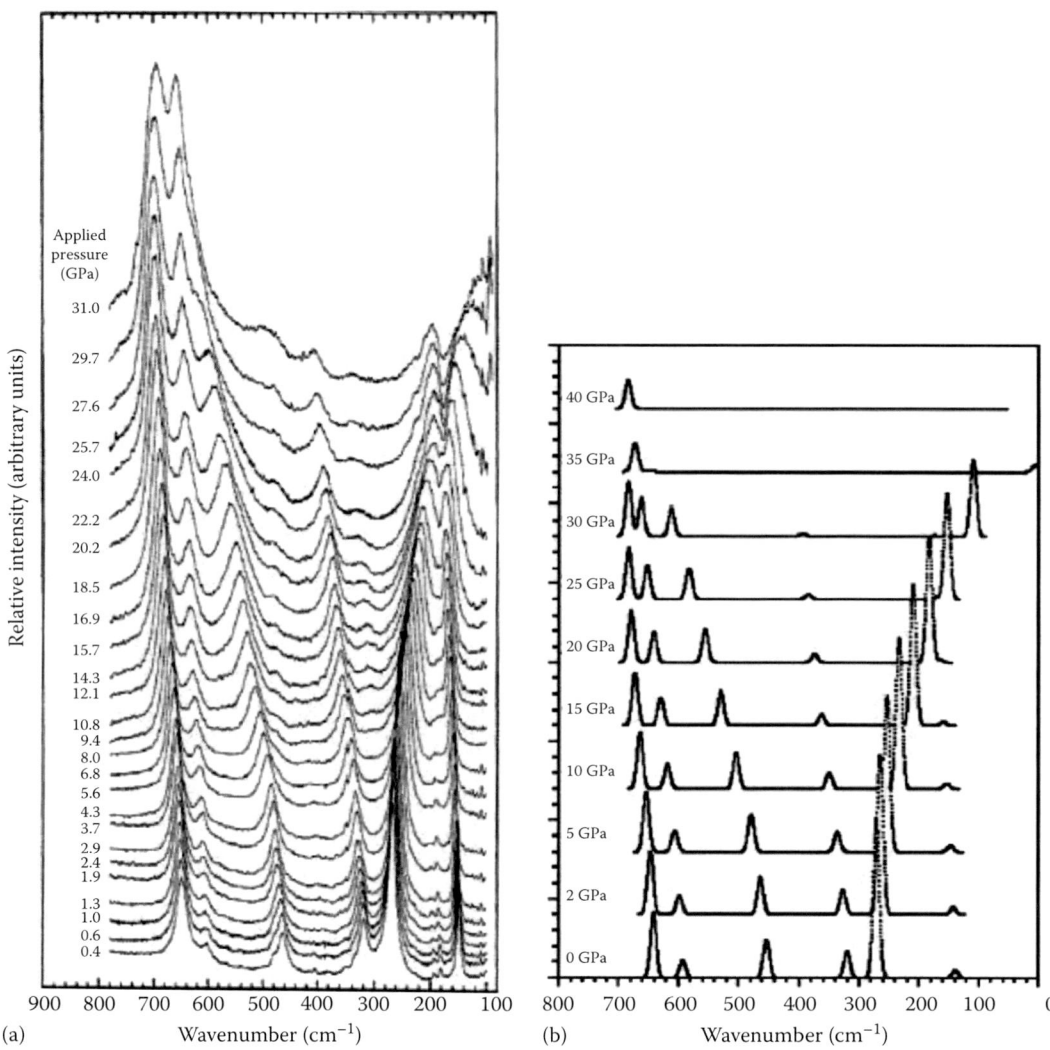

FIGURE 6.39 Direct observation of a soft mode (A_{1g}) at 260 cm^{-1} in pressurized tetragonal zirconia. Evolution of Raman spectra of ZrO_2 as a function of pressure. (a)Experimental data. Note the global loss of intensity above 31 GPa besides frequency tending to zero. Softening is fitted to a relation $v = a(P_c - P)^b$ with $P_c = 38$ GPa (From Bouvier, P. and Luccazeau, G., *J. Phys. Chem. Solids*, 61, 569, 2000.) (b) Predicted spectra based on *ab initio* density functional theory calculations in the local density approximation. (From Milman, V. et al., *J. Phys. Condens. Matter*, 21, 485404, 2009.)

inherent stability, this approach could be adopted for those superionic conductors that have tetrahedral building blocks such as AgI.

6.8 Summary

Phonons in solid state ionic materials have been discussed in this chapter as they connect to the phenomenon of fast ion conduction. Starting from simple models for monatomic and diatomic lattices, we have discussed phonon dispersion in a variety of ion-conducting compounds including LiI, Ag, and Cu halides with the aid of shell model and QD models. Phonons in Li battery electrode materials such as $LiCoO_2$ and $LIFePO_4$ are best discussed in a quasi-harmonic thermodynamic approach taking vibrational energy explicitly and applying the formalism of DFT. Anharmonic vibrations in Li_2S has been discussed as a prominent example

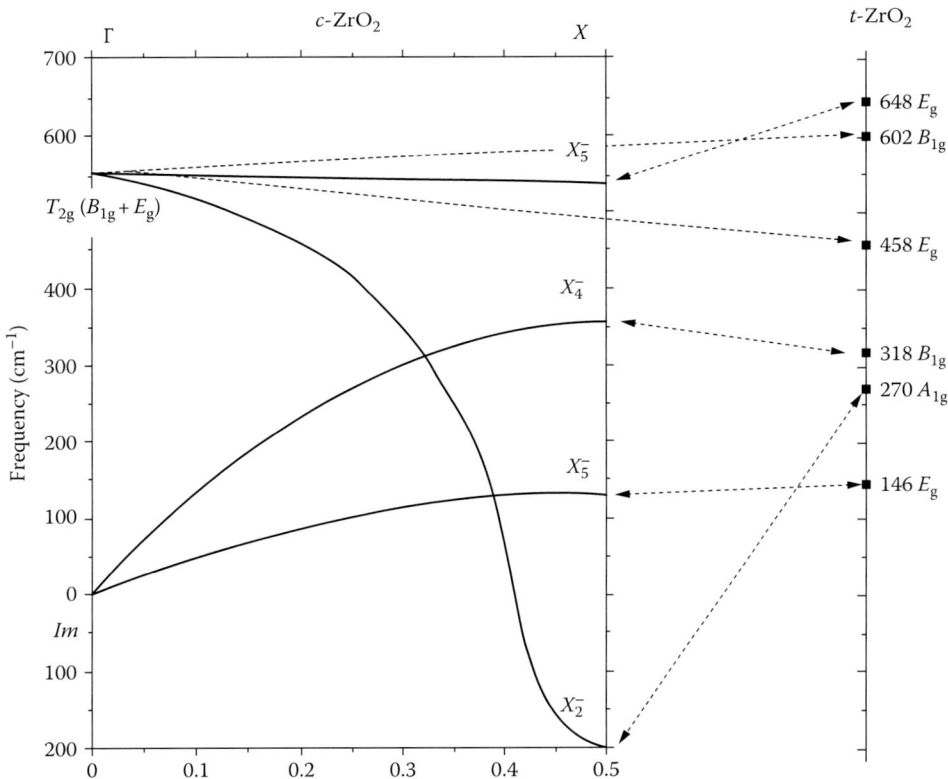

FIGURE 6.40 Visualizing a correlation between the experimental Raman-active modes of single crystal grains of tetragonal (*t*)-ZrO$_2$ realized in epitaxial films and the theoretical dispersion branches of cubic ZrO$_2$. Only the branches related to the Raman spectra are shown. Dotted arrows represent the symmetry rule which interrelates the Raman-active modes of *t*-ZrO$_2$ and the relevant vibrations of *c*-ZrO$_2$. The interrelations between the Raman spectrum of *t*-ZrO$_2$ and the relevant phonons of *c*-ZrO$_2$ are considered, implying the A1g vibration as a soft mode driving the tetragonal–cubic transformation. (From Merle, T. et al., *Phys. Rev. B*, 65, 144302, 2002.)

where quartic terms in the interatomic potential play a significant role in deciding the temperature dependence of physical properties including Raman spectra. Optical phonons in AgI-type compounds and silver beta alumina and soft modes in zirconia provide valuable insights into the nature of interatomic forces and their role in triggering phase transitions in these superionic conductors. The next chapter carries forward this discussion in the domain of electrons as a subsystem in solid state ionic materials.

6A.1 Appendix

In this appendix, the group theoretical analysis of vibrational spectra with special reference to the case of the olivine structure LiMPO$_4$ (M = Ni, Co, Fe) is given.

The space group for this structure is *Pnmb*. Seven atoms per formula units and four formulas in the Brillouin unit cell result in a total of 84 degrees of freedom. These are the normal modes vibration of the crystal. These 84 zone center normal modes are distributed on the irreducible representation of the D$_{2h}$ point group as

$$\Gamma_{vib} = 11A_g + 7B_{1g} + 11B_{2g} + 10A_u + 14B_{1u} + 10B_{2u} + 14B_{3u}.$$

The terms with subscript *g* represent 36 symmetrical Raman-active optical modes. The remaining representations belong to the three acoustic vibrations and the 45 antisymmetric modes.

Among the antisymmetric modes, B_{1u}, B_{2u}, and B_{3u} correspond to the infrared active vibrations. These modes are classified as internal and external. Internal modes arise from PO_4^{3-} tetrahedral while external modes arise from pseudo-rotations and translations of the units. The translations include motions of the center of mass of PO_4^{3-}, Li^+, and Co^{2+}. This separation just aids the discussion because vibrations may be coupled. The internal modes representation are

$$\Gamma_{int} = 6Ag + 3B_{1g} + 6B_{2g} + 3B_{3g} + 3A_u + 6B_{1u} + 3B_{2u} + 6B_{3u}$$

$$\Gamma_{trans} = 4A_g + 2B_{1g} + 4B_{2g} + 2B_{3g} + 5A_u + 6B_{1u} + 4B_{2u} + 6B_{3u}$$

$$\Gamma_{rot} = A_g + 2B_{1g} + B_{2g} + 2B_{3g} + 2A_u + B_{1u} + 2B_{2u} + B_{3u}$$

What are the basic principles of solving the molecular vibration problem? First of all fundamental quantities and relations are defined. From these, the secular equation is obtained in a form convenient for numerical calculation. An infinitesimal amplitude of vibration is considered to allow for potential energy to be simplified. Higher order terms in the coordinate are neglected. Let S_t be one of the $3N - 6$ internal coordinates, where N is the number of atoms in the molecule and ξ_i is one of the $3N$ Cartesian displacements. Then the relation between S_t and ξ_i is

$$S_t = \sum_{i=1}^{3N} B_{ti}\xi_i \quad t = 1,2,\ldots, 3N - 6 \tag{6A.1}$$

The coefficients of atomic displacements B_{ti} are related to a set of quantities $G_{tt'}$ as

$$G_{tt'} = \sum_{i=1}^{3N} \frac{1}{m_i} B_{ti}B_{t'i} \quad t,t' = 1,2,\ldots, 3N - 6 \tag{6A.2}$$

where m_i is the mass of the atom to which the subscript refers.

The kinetic energy of vibration in terms of the internal coordinates is

$$2T = \sum_{tt'}(G^{-1})_{tt'}\dot{S}_t\dot{S}_{t'} \tag{6A.3}$$

Here the dot symbols represent the derivatives. $(G^{-1})_{tt'}$ are related to $G_{tt'}$ through the expression

$$\sum_{t'}G_{tt'}(G^{-1})_{t't''} = \delta_{tt''} \tag{6A.4}$$

where the symbol on the right is Kronecker delta function.

The potential energy in terms of the internal coordinates is

$$2V = \sum_{tt'}F_{tt'}S_tS_{t'} \tag{6A.5}$$

where $F_{tt'}$s are the force constants. Then the secular equation for the vibrational problem is

$$\left|F - G^{-1}I\lambda\right| = 0 \tag{6A.6}$$

$$\text{with } \lambda = 4\pi^2v^2 \tag{6A.7}$$

Equivalently,

$$\left|FG - \sigma^2E\right| = 0 \tag{6A.8}$$

where

$\sigma = 2\pi v$

E is the unit matrix

Numerical solutions of Equation 6A.8 represent the vibrations present in the system considered.

Problems

6.1 Work out the dynamics of a one-dimensional lattice of Ag^+ and I^- ions considering only nearest-neighbor interionic forces. What is the effect on the equations of motion if we consider next-nearest neighbors to interact? What is the effect of partial covalency?

6.2 As introduced in Chapter 1, the (first) Brillouin zone of a crystal forms the basis for the discussion and computation of the phonon- and electron-related properties of a given solid state ionic material. Look at the template later, which is made up of interconnected hexagons and squares. When cut and pasted on a cardboard and folded along the dashed lines joined together, you get the Brillouin zone of the fcc lattice—essentially relevant for important superionic conductors adopting zincblende and rocksalt structures. The letters K, L, U, W, X, and Z shown refer to the high-symmetry points/lines on the zone at/along which phonon spectra show special features that connect to the physical properties.

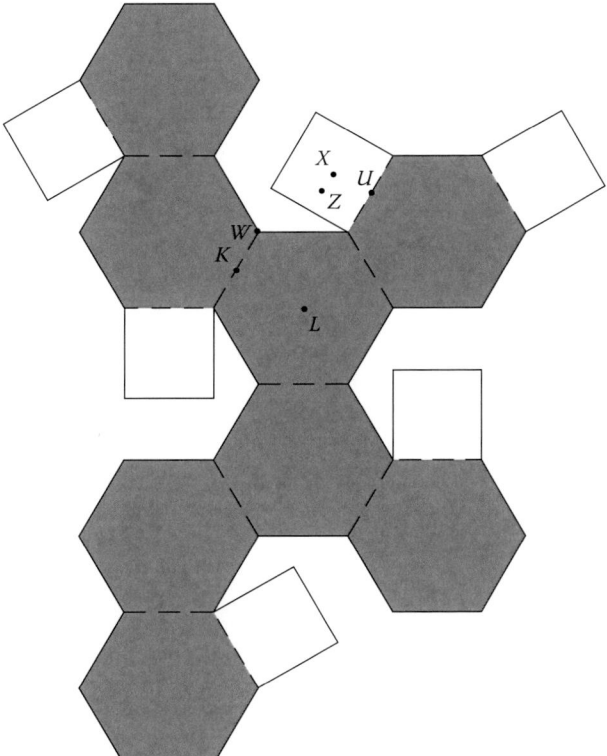

6.3 Superionic conductors adopt (a) bcc, (b) fcc, and (c) hcp structures. Given the primitive vectors or "generators" of the three lattices here, find their Brillouin zones and sketch them.

(a) *bcc*: $\mathbf{a}=(a/2)(\mathbf{i}+\mathbf{j}+\mathbf{k})$, $\mathbf{b}=(a/2)(-\mathbf{i}+\mathbf{j}+\mathbf{k})$, $\mathbf{c}=(a/2)(-\mathbf{i}-\mathbf{j}+\mathbf{k})$

(b) *fcc*: $\mathbf{a}=(a/2)(\mathbf{i}+\mathbf{j})$, $\mathbf{b}=(a/2)(\mathbf{i}+\mathbf{k})$, $\mathbf{c}=(a/2)(\mathbf{j}+\mathbf{k})$

(c) *hcp*: $\mathbf{a}=(a/2)\mathbf{i}+(3^{1/2}a/2)\mathbf{j}$, $\mathbf{b}=(-a/2)\mathbf{i}+(3^{1/2}a/2)\mathbf{j}$, $\mathbf{c}=c\mathbf{k}$

6.4 Consider the effect of anharmonicity in the potential energy ($U(x)$) of a 1D classical oscillator. Anharmonicity affects the mean separation of a pair of atoms at a temperature T. At a displacement x from their equilibrium separation at absolute zero temperature, $U(x) = cx^2 - gx^3 - fx^4$ with c, g, and f all positive. The term of order 3 in x or cubic term represents the asymmetry of mutual repulsion of atoms while the term of order 4 in x or the quartic term represents the softening of the vibrations at large amplitudes. For small oscillations, $U(x)$ is an adequate representation of an interatomic potential. Show that the average displacement (assuming anharmonic energy $< k_B T$) is $\langle x \rangle = (3g/4c^2)k_B T$.

(*Hint*: Use the Boltzmann distribution function. This function weights the possible values of x according to their thermodynamic probability).

Using the same $U(x)$, show that the approximate heat capacity of the 1D classical anharmonic oscillator is $C \approx k_B[1 + (3f/2c^2 + 15g^2/8c^3)k_BT]$. (Useful Reference 6.1)

6.5 Low-energy-optical (LEO) phonon exists only in superionic conductors and thus characterizes a compound like AgI from normal ionic conductors such as AgF, AgCl, and AgBr. In the latter three, they are not observed. Also identified as the low energy or LE phonon, these have frequencies as follows: 17 cm^{-1}(for wurtzite and zincblende AgI), 38 cm^{-1} for CuBr, 34 cm^{-1} for CuI, 16 cm^{-1} for α-Ag_2S, $27cm^{-1}$ for β-Cu_2Se, 22 cm^{-1} for Ag–Al_2O_3, and 40–48 for Na–Al_2O_3. Contrast this with 146 cm^{-1} for CaF_2 and 93 cm^{-1} for SrF_2. (1) Negative pressure dependence and (2) an extremely small temperature dependence characterize LE or LEO mode. The E_2 mode of AgI is a unique example. It is actually a Brillouin Zone Edge Acoustic phonon dominated by shear force constant which accounts for property no. (1) above. Negative thermal expansion of AgI arises from this property.

Make a plot of ω_{LEO} versus $1/(\text{mobile ion atomic mass})^{1/2}$ for superionic conductors of the Ag, Cu, and alkaline earth halides. Discuss the basic origin of this relationship.

6.6 Point group character tables for three most important crystal classes adopted by superionic conductors are given later.

These are important for understanding and interpreting not only phonon spectra but also in assigning experimental infrared and Raman spectral features.

1. D_{3d} (Rhombohedral)

$D_{3d} = D_3 \otimes i(\bar{3}m)$			E	$2C_3$	$3C'_2$	i	$2iC_3$	$3iC'_2$
$x^2 + y^2, z^2$		A_{1g}	1	1	1	1	1	1
	R_z	A_{2g}	1	1	1	1	1	1
$(xz, yz), (x^2 - y^2, xy)$	(R_x, R_y)	E_g	2	−1	0	2	−1	0
		A_{1u}	1	1	1	−1	−1	−1
	z	A_{2u}	1	1	−1	−1	−1	1
	(x, y)	E_u	2	−1	0	−2	1	0

2. C_{6v} (Hexagonal)

C_{6v} (6mm)			E	C_2	$2C_3$	$2C_6$	$3\sigma_d$	$3\sigma_v$
$x^2 + y^2, z^2$	z	A_1	1	1	1	1	1	1
	R_z	A_2	1	1	1	1	−1	−1
		B_1	1	−1	1	−1	−1	1
		B_2	1	−1	1	−1	1	−1
(xz, yz)	(x, y) (R_x, R_y)	E_1	2	−2	−1	1	0	0
$(x^2 - y^2, xy)$		E_2	2	2	−1	−1	0	0

$C_{6h} = C_6 \otimes i$ (6/m) (hexagonal); $S_6 = C_3 \otimes i \bar{3}$ (rhombohedral)

3. T_d (Cubic)

T_d ($\bar{4}$3m)		E	$8C_3$	$3C_2$	$6\sigma_d$	$6S_4$
$x^2 + y^2 + z^2$	A_1	1	1	1	1	1
	A_2	1	1	1	−1	−1
$(x^2 - y^2, 3z^2 - r^2)$	E	2	−1	2	0	0
(R_x, R_y, R_z) $yz, zx, xy)$	T_1	3	0	−1	−1	1
(x, y, z)	T_2	3	0	−1	1	−1

(a) Name three fast ion conductors that adopt each of the three classes given before.

(b) What is the crystal class belonging to cubic zirconia that gives the soft mode discussed in Section 6.6. Write down its character table and irreducible representation.

(c) Condensation of X_2^- normal vibrational mode in the cubic zirconia lattice is supposed to be the origin of the transformation to the tetragonal phase. Discuss this point. (see Reference 6.47).

6.7 The two optical mode frequencies ω_{TO} and ω_{LO} are related to the effective charge e^* of a superionic conductor through the equation

$$\sum_j \left(\omega_{jLO}^2 - \omega_{jTO}^2 \right) = \left\{ \frac{1}{\varepsilon_v V} \sum_k \frac{e_k^{*2}}{m_k} \right\}$$

In this equation, j runs over all infrared active optical modes, while k runs over all atoms in the unit cell volume V. ε_v is dielectric constant in vacuum and m_k the atomic mass. This equation has been applied to analyze the phase transition behavior of Ag_3SI and AgI through $e^*(T)$. A steplike discontinuity is observed for the γ-β superionic phase transition in Ag_3SI at 157 K. The step width increases in the order I < Ag < S. Discus the origin of this observation. Also the $e^*(T)$ for AgI exhibits a change of slope from negative to positive at the wurtzite–bcc superionic phase transition at 420 K. Z too shows an abrupt step at the phase transition temperature. Discuss this interesting observation (see References 6.48 and 6.49).

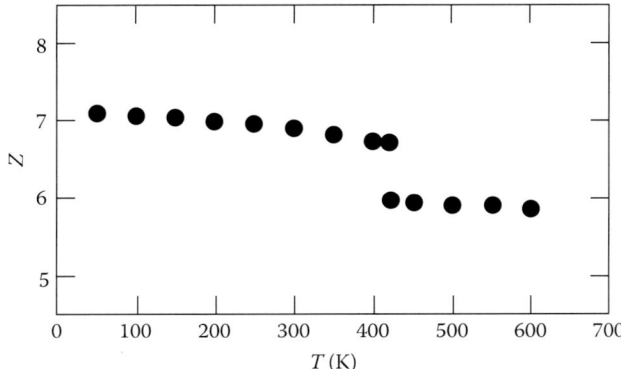

6.8 Rahman–Vashishta potentia (RVP) and ion motion in α-AgI. Rahman and Vashista have constructed effective pair potentials; simply describe, with reasonable accuracy, the structural and dynamical properties of α-AgI. For AgI, it is given by

$$V_{ij} = \frac{A_{ij}(\sigma_i + \sigma_j)^n}{\gamma^n} + \frac{Z_i Z_j e^2}{r}$$

$$-\frac{1}{2}\left(\alpha_i Z_i^2 + \alpha_j Z_j^2\right)\frac{e^2}{r^4} - \frac{W_{ij}}{r^6},$$

where
    i, j denote ion type
    A_{ij} denotes the repulsive strength
    σ_i, σ_j the particle radii
    α_i, α_j the electronic polarizabilities

Knowing σ_s, α_s, and W_s five parameters, A_{ij}'s, n, and $|Z_i| = |Z_j|$ are to be determined. Suppose $A_{ij} = A$, then the situation becomes simple. Repulsive term means each ionic "contact" contributes

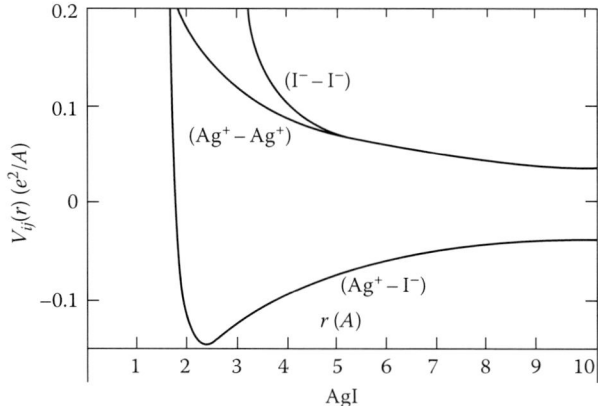

FIGURE P.7   Plot of interionic potential $V_{ij}(r)$ in units of e^2/A versus $r(A)$ for AgI.

energy A (coefficient of r^{-n} is "scaled" according to sum of particle radii). Experimental data on crystal structure at low temperature, cohesive energy, and compressibility help get these three parameters. But AgI is not purely ionic. For AgI, compressibility is $\sim 1 \times 10^{-11}$ cm^3/erg, Phonon dispersion measurement fixes $|Z| = 0.6$. Pauling's concept of ionic radii helps to fix σ_s (bond lengths): $\sigma_{Ag} + \sigma_I$ = Ag–I distance and $2\sigma_I$ = I–I distance. Size of I– $\gg$ size of Ag$^+$. So Ag–I and I–I interactions together fix the crystal structure. $\sigma_I = 2.2$ A, $\sigma_{Ag+} = 0.63$ A. σ_{Ag} being less than the ionic radius of Ag$^+$ implies the presence of covalent interaction. Impose a pairwise additive effective interaction scheme, which makes molecular dynamics relatively easy. Choose $n = 7$. Impose a condition that crystal energy minimum occurs at a lattice distance of 5.06A with $\alpha_{Ag} = W_{AgAg}$, $W_{AgI} = 0$, $\alpha_I = 6.52$, $W_{II} = 6.93$, $H_{AgI} = 17.893$, $H_{AgAg} = 0.062$, and $H_{II} = 394.934$. Here A is the unit of length and $e^2/A = 14.39$ eV is the unit of energy. Thus we get the RVP. This is shown in Figure P.7. Carefully examine this plot. Also look at the fundamental, molecular dynamics calculations on α-AgI from which Rahman and Vashishta have deduced ion motion [6.50].

REFERENCES

6.1   C. Kittel, *Introduction to Solid State Physics*, Wiley, New York, 1991.

6.2a   B. G. Dick and A. W. Overhauser, *Phys. Rev.* 112(1958) 90.

6.2b   W. Hayes, A. M. Stoneham, *Defects and Defect Processes in Nonmetallic Solids*, Wiley, New York, 1985, Chap. 2.

6.3   R. Shankar, *Quantum Mechanics*, 2nd edn., Springer, New York, 1980, Chap. 7, p. 198.

6.4   D. L. Price et al., *Phys. Rev. B* 2(1970) 2983.

6.5   A. Rastogi et al., *Phys. Rev. B* 9(1974) 1938.

6.6   A. D. B. Woods et al., *Phys. Rev.* 131(1963) 1025.

6.7   R. A. Cowley et al., *Phys. Rev.* 131(1963) 1030.

6.8   P. Goel et al., *Pramana J. Phys.* 63(2004) 409.

6.9   T. Tomoyose et al., *Solid State Ionics* 167(2004) 83.

6.10   W. Buhrer et al., *Phys. Rev. B* 17(1978) 3362.

6.11   P. Bruesch et al., *Phys. Rev. B* 22(1980) 970.

6.12   A. Yosihasa et al., *J. Phys. Conf. Ser.* 121(2008) 102211.

6.13   S. Hoshino et al., *J. Phys. Soc. Jpn.* 41(1976) 965.

6.14   B. Hennion et al., *Phys. Rev. Lett.* 28(1972) 964.

6.15   K. Shibata, S. Hoshino, *J. Phys. Soc. Jpn.* 54(1985) 3671.

6.16   S. Hoshino et al., *J. Phys. Soc. Jpn.* 57(1988) 4199.

6.17   A. N. Skormorokhov et al., *J. Phys. Condens. Matter* 19(2007) 186228.

6.18   S. L. Shang et al., *Phys. Rev. B* 83(2011) 144204.

6.19 G. Kresse, J. Furthmuller, *Phys. Rev. B* 54(1996) 11169.

6.20 X. Gong et al. *Int. J. Electrochem. Sci.* 8(2013) 10549.

6.21 S. L. Shang et al., *J. Mater. Chem.* 22(2012) 1142.

6.22 C. M. Fang et al., *J. Mater. Chem.* 17(2007) 4908.

6.23 P. Senthil Kumar et al., *J. Mater. Sci.* 41(2006) 3861.

6.24 K. Harada et al., *J. Phys. Soc. Jpn.* 41(1976) 1707.

6.25 F. Altorfer et al., *J. Phys. Condens. Matter* 6(1994) 9949.

6.26 B. Bertheville et al., *J. Phys. Condens. Matter* 10(1998) 2155.

6.27 J. E. Potts et al., *Solid State Commun.* 13(1973) 389.

6.28 G. Kanellis et al., *Phys. Rev. B* 33(1986) 8733.

6.29 S. Bhagavantam, T. Venkatarayudu, *Proc. Indian Acad. Sci.* 9A(1939) 224.

6.30 R. Frech, J. B. Bates, *Spectrochim. Acta* 35A(1979) 685.

6.31 O. Kamishima et al., *J. Phys. Condens. Matter* 23(2011) 225404.

6.32 K. Wakamura, *Phys. Rev. B* 56(1997) 11593.

6.33 K. Wakamura, *Solid State Commun.* 86(1993) 503.

6.34 H. A. Kramers, *Physica* 7(1940) 284; V. I. Mel'nikov, *Phys. Rep.* 209(1991) 1.

6.35 B. A. Huberman, J. B. Boyce, *Solid State Commun.* 25(1978) 843.

6.36 S. Glasstone et al., *The Theory of Rate Processes*, McGraw Hill, New York, 1968.

6.37 P. A Fleury, *Annu. Rev. Mater. Sci.* 6(1976) 157.

6.38 M. T. Dove, *Introduction to Lattice Dynamics*, Cambridge University Press, Cambridge U.K., 1993, Chap. 8.

6.39 S. Ghosh et al., *Phys. Status. Solidi* 123(1984) 445.

6.40 S. A. Danilkin et al., *J. Phys. Soc. Jpn.* 79(Suppl A)(2010) 25–28.

6.41 K. Negita, H. Takao, *J. Phys. Chem. Solids* 50(1989) 325.

6.42 P. Bouvier, G. Luccazeau, *J. Phys. Chem. Solids* 61(2000) 569.

6.43 V. Milman et al., *J. Phys. Condens. Matter* 21(2009) 485404.

6.44 T. Merle et al., *Phys. Rev. B* 65(2002) 144302.

6.45 R. A. Cowley, *Int. Ferroelec.* 133(2013) 109.

6.46 H. Schober, *Phys. B* 219–220(1996) 599.

6.47 K. Negita, *Acta Metall.* 37(1989) 313.

6.48 M. Aniya, K. Wakamura, *Phys. B* 219–220(1996) 463.

6.49 M. Aniya, *Rec. Res. Dev. Phys. Chem. Solids* 1(2002) 99.

6.50 P. Vashishta, A. Rahman, *Phys. Rev. Lett.* 60(1978) 1337.

6.51 M. Park et al., *J. Power Sources* 2010, 195(2010) 7904.

7

Electronic Structure

7.1 Perspectives

In this chapter, we shall consider the problem of electrons in solid state ionics. The electron problem is different from the phonon problem discussed in the last chapter. In the phonon problem, there are no counterparts of ion motion (elastic filed) outside the first Brillouin zone (BZ). In the electron problem that is akin to the x-ray diffraction problem discussed in Chapter 4, the electromagnetic field is omnipresent in the crystal and not just in the ions.

A non-superionic conductor such as NaCl or NaI is an electronic insulator. A superionic conductor is often a poor compound semiconductor. Examples include AgI, CuBr, and $LiCoO_2$. This could serve as a fundamental distinction and an intuitive introduction to the electronic structure of solid state ionic materials.

Just as we call GaAs a III–V semiconductor and CdS a II–VI semiconductor, we could term AgI or CuBr a I–VII semiconductor. Thus a simple model for such a type is shown in Figure 7.1.

Consider the following selection of materials: Lithium. Lithium iodide-alumina composite. Graphite. C_6Li. LiPON. $Li_4Ti_5O_{12}$. AgI. $RbAg_4I_5$. CuI. YSZ, YSZ, PbF_2, $LiCoO_2$, $LiMn_2O_4$, and $LiFePO_4$. All of these, including a metal (Li) and a semimetal (graphite), are used in the fabrication of batteries, fuel cells, sensors, and supercapacitors. They possess ionic and mixed ionic–electronic conductivities and adopt high-symmetry crystal structures. How does one know that Li is a metal—a highly reactive one; graphite is a semi-metal; AgI is a polymorphic material with at least four structures zincblende, wurtzite, bcc, and fcc; CuI is a large-gap p-type semiconductor of temperature up to 200°C and a Cu^+ ion conductor at higher temperatures? What role does electronic conductivity play in oxygen ion conductors such as YSZ? Beyond the preliminary discussion in Chapter 1, answers to these and more questions come from electronic band structure theory and related experiments, the latter discussed in Chapter 4. The dielectric theory of the chemical bond is firmly based on the Penn's dielectric screening model that implies that some form of band model is relevant for the solid state ionic materials based on Cu and Ag halides possessing zincblende structure.

1. Metal–nonmetal (or hetero-polar) bond covalency.
2. Ionic polarizability

These are the two concepts that guide us and lead us systematically to the electronic structure of solid state ionic materials of various families discussed in Chapter 2 through their role in promoting the facile movement defects and creating optimal ion-conduction pathways in the solid state ionics crystal structure. The ion movement in solid state redefines the concept of bonding. This is because dynamic bonds are made and broken during ion transport. It is easy to appreciate that it is indeed so in covalency-stabilized open structure adopted by silver iodide and copper halides.

Bond fluctuations couple to phonons (discussed in Chapter 6) and play their crucial role in phase transitions to the fast-ion conducting phase. In the fast ion conducting phase, these fluctuations sustain fast diffusion of ions through jump diffusion and other mechanisms.

The nature of the superionic bond is determined by the fundamental concepts 1 and 2 mentioned earlier and helps derive additional criteria for transport of ions through properties such as electronegativity,

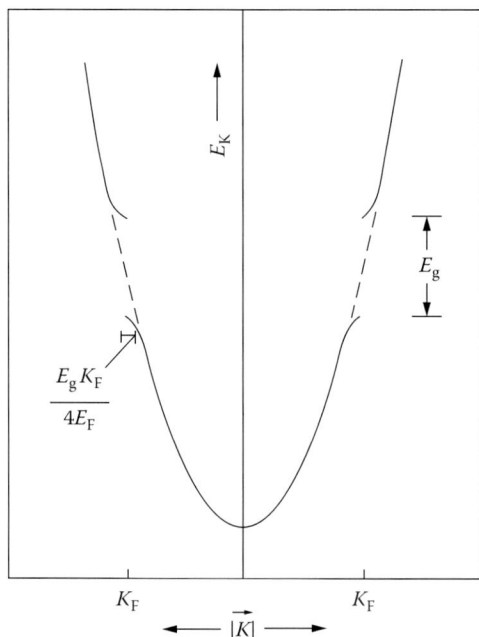

FIGURE 7.1 A simple band model for an isotropic semiconductor calculated in the nearly free electron approximation. For such bands a single reciprocal lattice vector is chosen. This approach is relevant for the discussion that follows later in this chapter on the nature of the superionic bond. (After Rashenov, V.K. et al., *Phys. Stat. Sol. (b)*, 54, 355, 1972.)

hardness, and fractional ion transfer that are firmly rooted in density functional theory (DFT) to be discussed later in this chapter. The fluctuating character of the "superionic bond" leads to a new approach to the problem of fast ion conduction involving the well-known concepts of solid state physics such as pseudopotentials.

7.2 Nature of the Superionic Bond

Why is chemical bonding in solid state ionic materials so important? It is because these materials, somewhat like organic compounds, contain unsaturated chemical bonds in whose breaking and remaking lies the basic origin of the phenomenon of fast ionic conduction.

This could as well be titled *Chemical Bonding and Electronic Energy Bands in Superionic Conductors.*

"The cases of a molecule, a solid, a crystal are not really different" said Schrodinger in his book *What Is Life?* [7.2]. We begin this chapter with a question What is the connection—no matter how remote—between the nature of the superionic bond and the parameters that determine the electrical (ionic and electronic) conductivity of solids that possess such bonds? The motivation to find such a connection is the reasonable belief that in the transition from the atom to the molecule to the cluster to the structured solid, the last state of matter does retain some features of the first and second states. This is basically an empirical approach to the problem of superionic or fast ionic conductor. The empirical models that have worked are the following:

  1. The critical ionicity approach of Phillips [7.3]. In this model, all ion conductors possessing high symmetry structures have been considered as $A^N B^{8-N}$ solids where N is the number of electrons involved in the outermost subshell of the elements A and B forming the binary solid AB. 8 is characteristic of the "octet" that needs to be completed for stability. For III–V compounds such

as GaAs, $N = 3$, while for II–VI compounds such as CdS, $N = 2$. For halides of noble metals Cu, Ag, Au, $N = 1$, making these I–VII compounds with just one metal electron being shared by four nonmetallic ions in a weakly covalent tetrahedral coordination. The concept of "critical ionicity" separates the ionic conductors into two groups separating normal ion conductors of NaCl group from fast ion conductors of the noble metal halides. An important result of this model is the prediction that a critical iconicity of 0.785 arises from a metal–nonmetal bond with mixed ionic–covalent character. At this iconicity, the transition occurs between covalently bonded four-coordinated solid and the ionically bonded six-coordinated solid. This mixed bond character leads to the relative stability of four- and six-coordinated ionic compounds. It is significant that AgI possesses an iconicity of 0.77 closest to the critical iconicity. Ag prefers to coordinate with four I ions in AgI while Ag coordinates with six Br ions in AgBr, which has an iconicity of 0.85.

In the Phillips model, critical iconicity f_c is given by

$$f_c = \frac{\left(\text{covalent energy}\right)^2}{\left(\text{covalent energy}\right)^2 + \left(\text{ionic energy}\right)^2} \tag{7.1}$$

It is significant that the difference in Gibbs free energy for the four- and six-coordinated compounds is nearly zero for AgI. This accounts for a large number of ternary compounds formed by AgI due to the flexibility of the Ag–I bond.

2. Wakamura model [7.4] connects the high conductivity of a fast ion conductor with the dielectric anomaly.
3. The Catlow–Stoneham model [7.5] looks at the physical origin of iconicity. This model essentially integrates the two aforementioned models.

These three empirical models have to be contrasted with two theoretical approaches namely

1. The pseudopotential approach [7.6] inspired by the theory of metals; this emphasizes, as in AgI, the metallic nature of the Ag sublattice and
2. The density functional theory approach [7.7]. Here, the density profiles of mobile ions are calculated from first principles.

We shall consider these two approaches in detail later in this chapter.

How does one connect concepts in crystal structure and chemical bonding in the context of ionic conductivity?

Consider the simple expression for the ionic conductivity of a solid

$$\sigma_i = n_i \left(Ze\right)_{\text{eff}} \mu_i \tag{7.2}$$

The first factor on the right, n_i, is the mobile ion concentration that depends on the structure and thermodynamics; the second factor, the effective charge $(Ze)_{\text{eff}}$, depends on the nature of the chemical bond; the third factor, μ_i, is the ion mobility that depends on statistical and structural factors [7.8,7.9]. Now what are the key concepts involved in the bonding of a fast ion conductor? With respect to an ionic molecule MX made from a metal M and a nonmetal Z, we can consider two concepts: 1. electronegativity first introduced by Linus Pauling [7.10] and 2. softness/hardness of the molecule M. Note that M could be K or Ag, and also M could be I.

Another important concept to be discussed later is the complex nature of the electronic energy band gap of $A^N B^{8-N}$ semiconductors.

Introduced by Pauling in an empirical manner, electronegativity is the capacity of a metal atom to attract electrons toward itself. It is the property of an atom in a molecule and thus not directly measureable.

The difference in electronegativity between atoms A and B is given by:

$$\chi_A - \chi_B = (eV)^{-1/2}[E_d(AB)] - \left\{\left(E_d(AA) + E_d(BB)\right)/2\right\}^{1/2} \tag{7.3}$$

where the E_ds are dissociation energies of A–B, A–A, and B–B bonds expressed in eV with factor $(eV)^{-1/2}$ making the result dimensionless.

Mulliken provided the connection between electronegativity and basic atomic properties namely ionization energy (E_i) and electron affinity (E_{ea}):

$$\chi_M = \frac{E_i + E_{ea}}{2} \tag{7.4}$$

or for E_i and E_{ea} in eV,

$$\chi_M = 0.187\left(E_i + E_{ea}\right) + 0.17 \tag{7.5}$$

An important connection between χ_M and chemical potential of thermodynamics μ is that $\chi_M = -\mu$ so that

$$\mu = -\frac{\left(E_i + E_{ea}\right)}{2} \tag{7.6}$$

Equation 7.6 helps us to make the connection between electronegativity in the context of DFT. What is DFT? Simply put, it considers the total electron density ρ of a system (molecule or crystal) as a functional just as ordinary quantum mechanics considers wave function. A functional is nothing but a function of a function; here energy of the system is a function of ρ and thus $E(\rho)$ is the energy density functional or simply density functional. In DFT, electronegativity is defined as the negative of the chemical potential μ. μ is obtained from the equation

$$\delta E(\rho)/\delta\rho = \mu \tag{7.7}$$

Table 7.1 is a partial periodic table that gives average values of electronegativities of elements that form a large number of solid state ionic materials. This is constructed from the numbers obtained from DFT calculations [7.11] (Figure 7.2).

Let us consider the (super)ionic molecule as a chemical system [7.12]. Define its electrochemical potential as

$$\mu = \left(\partial E / \partial N\right)_V = -\chi \tag{7.8}$$

Here, E is the total energy of the system consisting of N electrons and V is the total potential due to nuclei and any external potential.

The "absolute hardness" of the earlier system is given by

$$\eta = \left(1/2\right)\left(\partial\mu / \partial N\right)_V = \left(1/2\right)\left(\partial^2 E / \partial N^2\right)_V \tag{7.9}$$

η is a local property while χ is a global property for an equilibrium system such as the one considered here.

What happens when M and X combine in a wet chemical reaction or a dry solid state reaction or even a dry mechanical reaction such as rubbing a silver button on an iodine crystal to form AgI? There occurs an electron flow from M with lower χ to X with higher χ. This electron transfer occurs until their μ values

TABLE 7.1

Partial Periodic Table of Elements Relevant to Solid State Ionics Showing Average Values of Electronegativity (χ), and Hardness (η), in eV Constructed from DFT Data

H					
Li	**N (3−)**	**O**	**F**		
χ = 3.14	7.38	8.03	11.3		
η = 2.53	6.54	5.69	6.72		
IR = 0.060 nm	0.171	0.140	0.136		
Na	**Mg(++)**	**P(3−)**	**S**	**Cl**	
3.07		5.22	6.07	8.18	
2.38		4.53	3.81	4.41	
0.095	0.065	0.212	0.184	0.181	
K	**Ca**	**Ti(4+)**	**Cu(+)**	**Se**	**Br**
2.57	3.20		4.71	5.70	7.45
1.98	2.80		4.22	3.52	3.98
0.133	0.099	0.068	0.096	0.198	0.195
Rb	**Sr**	**Zr**	**Ag**	**Te**	**I**
3.35	2.62	3.36	4.28	5.27	6.71
0.94	2.93	2.85	3.12	3.12	3.48
0.148	0.113	0.080	0.126	0.221	0.216
Cs	**Ba**	**Hf**	**Au(+)**		
2.23					
1.72					
	0.135	0.078 (Ahrens)	0.126		

Note: Ionic radii (IR) given are due to Pauling unless stated otherwise.

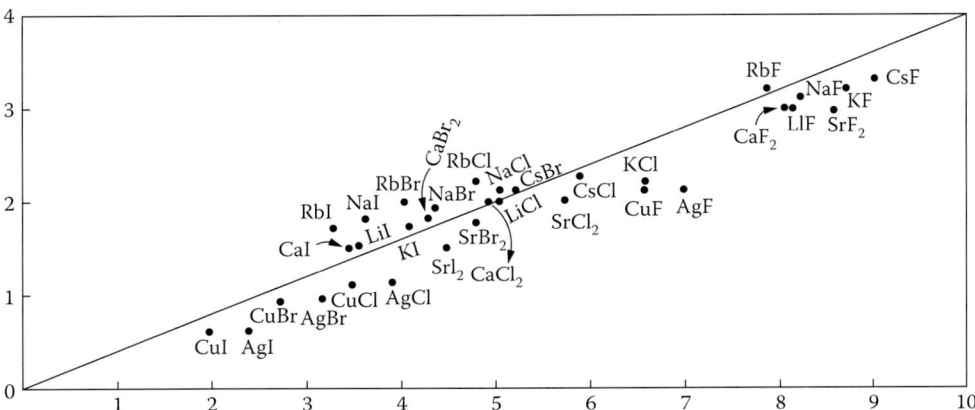

FIGURE 7.2 A plot of Pauling electronegativity difference $\Delta\chi_{Pauling}$ versus $\Delta\chi_{DFT}$ for alkali-, noble metal-, and alkaline earth metal halides. DFT data for χ of elements. (From Sunandana, C.S., *J. Phys. Chem. Solids*, 58, 1359, 1997.)

are equalized. A useful first approximation, that works for materials such as AgI where the electron transfer from Ag to I is not complete, is the concept of fractional electron transfer. This is defined by

$$\Delta N = \frac{\chi_X - \chi_M}{2(\eta_X + \eta_M)} \tag{7.10}$$

Note that $\chi_X - \chi_M$ is the "force" that drives electron transfer. What about $\eta_X + \eta_M$? It acts as a resistance to such transfer. Most importantly, covalent bonding as well as ionic bonding depends on ΔN. Equation 7.10

is thus a useful measure of the chemical bonding in the superionic or solid state ionic molecule. Table 7.2 displays values of χ, η, and ΔN for monovalent halides of those elements for which χ values derived from DFT are available. Before we discuss Table 7.2, let us take a look at the interesting plot of $\Delta\chi$ due to Pauling versus $\Delta\chi$ derived from DFT calculations (Figure 7.1). A careful look at the plot reveals the following:

1. Solid state ionic (or superionic) halides all tend to have a rather low χ. Therefore, the electron transfer from metal to halogen is essentially incomplete.

2. The "conversion factor" between the Pauling and the DFT-based electronegativities is the slope of the "diagonal" of the plot passing through the origin; the value of this conversion factor is 0.4. Thus, $\Delta\chi_{DFT} = 2.5\ \Delta\chi_{Pauling}$. This rationalizes the present approach to the discussion of the superionic bond.

Now focus on Table 7.2. There are three quantities namely χ, η, and ΔN—the last of which has a bearing on the iconicity of the M–X bond. The question is: is it possible to "split" ΔN, considering it as the

TABLE 7.2

Electronegativity Difference ($\Delta\chi$), Chemical Hardness Sum ($\eta A + \eta B$) and Fractional Electron Transfer (ΔN) for Ion Conducting Halides Based on Table 7.1 and Equation 7.10

Halide	$\Delta\chi$	$\eta_X + \eta_M$	ΔN	χ_{MX}[a]	η_{MX}[b]
LiF	8.17	9.25	0.44	5.96	4.03
LiCl	5.04	6.94	0.36	5.07	3.72
LiBr	4.31	6.51	0.33	4.84	3.62
LiI	3.57	6.01	0.30	4.59	3.51
NaF	8.24	9.10	0.45	5.89	3.96
NaCl	5.11	6.79	0.38	5.01	3.66
NaBr	4.38	6.36	0.34	4.78	3.57
NaI	3.64	5.86	0.31	4.54	3.45
KF	8.74	8.70	0.50	5.39	3.43
KCl	6.61	6.39	0.52	4.58	3.21
KBr	4.88	5.96	0.41	4.38	3.13
KI	4.14	5.46	0.38	4.15	3.05
RbF	7.93	7.66	0.52	6.18	4.27
RbCl	4.80	5.35	0.45	5.28	3.92
RbBr	4.07	4.89	0.42	5.02	3.81
RbI	3.3	4.22	0.38	4.76	3.69
CsF	9.08	8.44	0.54	5.02	3.05
CsCl	5.95	6.13	0.48	4.27	2.87
CsBr	5.22	5.70	0.46	4.07	2.81
CsI	4.48	5.20	0.43	3.87	2.74
CuF	6.60	10.94	0.30	7.30	5.45
CuCl	3.47	8.63	0.20	6.21	4.9
CuBr	2.74	8.20	0.17	5.92	4.73
CuI	2.00	7.70	0.13	5.62	4.51
AgF	7.03	9.84	0.36	6.96	5.09
AgCl	3.90	7.53	0.26	5.92	4.61
AgBr	3.17	7.10	0.22	5.65	4.11
AgI	2.43	6.60	0.18	5.36	4.28
NH$_4$I			0.10		
C$_5$H$_5$NHI			0.044		

Source: Sunandana, C.S., *J. Phys. Chem. Solids*, 58, 1359, 1997.
Notes: Also given are molecular electronegativity (χ_{MX}) and molecular hardness (η_{MX}). All quantities except ΔN are in eV.
[a] Square root of $X_M X_X$.
[b] Calculated from Equation 7.13 of Chattaraj [7.58]. $\eta_{AB} = 1.64X_{AB}/(X_A + X_B)$. Data for K-, Li-, Na-, and Rb-halides are also from the aforementioned reference. Rest of the data is from [7.12].

total effect of electron sharing and electron transfer? In view of the mixed ionic–covalent nature of these bonds, it seems reasonable to split ΔN into a covalent and an ionic part:

$$\Delta N = \Delta N_{\text{covalent}} + \Delta N_{\text{ionic}} \tag{7.11}$$

Such a clear-cut division derives support from microscopic experimental probes such as electron paramagnetic resonance hyperfine coupling constants [7.13]. The magnitudes of isotropic and anisotropic hyperfine coupling constants in favorable cases could be related to ΔN_{ionic} and $\Delta N_{\text{covalent}}$, respectively. A careful survey of Table 7.2 reveals that for the group of halides examined, $0.54 < \Delta N < 0.04$. ΔN covers an entire gamut of compounds ranging from very ionic CsF to nearly covalent C_5H_5NHI—the precursor of pyridinium silver iodide—a room-temperature superionic conductor [7.14,7.15]. The trend among alkali halides is that LiI has the lowest ΔN of 0.30; among silver and cuprous halides, CuI has the lowest ΔN of 0.13 which makes it a semiconductor! Indeed, as noted from Chapter 2, these three halides, LiI, AgI, and CuBr, with the first one in composite form are among the solid state ionic materials with largest ionic conductivities.

How do we connect ΔN with electrical conductivity? Note that conductivity involves hopping of ions over an electrostatic potential barrier. This barrier is decided by the ease with which the mobile ion (Li^+ or Ag^+ or Cu^+) gets detached from its immediate bonding partner in a labyrinth of bonds—be it in a glass/polymer/composite or a polycrystal or a monocrystal. Imagining the solid to be made up of a virtually infinite number of MX bonds, a broken bond could mean that the breakaway mobile ion has still got enough charge to remake another bond. Thus, ΔN has a rough inverse correlation to the average effective charge held by the mobile ion in an ionic/superionic conductor. Realize that ΔN could be a function of temperature owing to anharmonic vibrations whose amplitudes increase with increasing temperature and the associated thermal expansion effects. Thus, the metal–nonmetal bond could weaken, thus affecting electron transfer. In the solid state, this would affect features of electronic band structure (to be discussed later in this chapter) and the electronic contribution to conductivity [7.16]. What is the importance of chemical hardness η? It is related to polarizability and the energy gap. The energy gap is in turn related to the dielectric constant and finally to conductivity. In fact Heyne has given a band gap criterion for a superionic conductor: $E_g > (T(K)/300)$ eV which means at 300 K, a material with a band gap of >1 eV is likely to be a superionic conductor.

Examination of Table 7.2 shows that $2.74(CsF) < \eta < 5.65(CuF)$. The bromides and iodides of copper and silver have a narrow range of hardness values: 4.11 (AgBr), 4.28 (AgI), 4.54 (CuI), and 4.73 (CuBr) just as they have a narrow range of dielectric constants [7.4].

A fundamental principle based on hardness that has implications for materials synthesis as much as for chemical bonding is the hard–soft acid–base or HSAB principle. "Hard acids prefer to coordinate with hard bases and soft acids to soft bases" [7.17]. Among the cations that form ionic conductors, Li^+ and Na^+ are hard Lewis acids with rather large electronegativities and hardness values. Thus, they can only form σ bonds with their nonmetallic partners. However, Cu^+ and Ag^+ are soft cations with relatively small hardness values but with not too large electronegativities. Therefore, π bonding could occur when these cations form compounds.

Using this HSAB principle, let us discuss bonding trends in normal ionic conductor KI and fast ion conductor AgI. Why do hard acid cations prefer to bond to hard anions? The hard K^+ does not have d or unfilled p shells ($1s^2 2s^2 2p^6 3s^2 3p^6$) while the soft Ag^+ has a filled d shell ($1s^2 2s^2 2p^6 3s^2 3p^6 3d^{10} 4s^2 4p^6 4d^{10}$). Thus, the KI "molecule" made of point charges is less stable than the AgI molecule made of rather extended charges. In AgI, a repulsion is possible between two filled π orbitals formed by d–p hybridization. But this repulsion is reduced by a mutual polarization. That is, the filled d or p orbitals on each atom are hybridized with empty p or d orbitals on the same atom, resulting in an energy lowering. Consequently, it is easier to promote Ag^+ from $3d^{10}$ to $3d^9 4p$ than to promote K^+ from $3p^6$ to $3p^5 3d$. Similarly, it is easier to promote I^- than $F^=$. The crucial point is that these promotion energies correlate with hardness values and with inverse polarization. Thus, emerges the role of d electrons in silver and copper in controlling the bonding and dielectric behavior of their compounds. The d–p hybridization complicates band structure calculations as we shall see later in this chapter.

Finally, the HSAB principle offers a recipe for making a good solid state ionic material: let a soft acid combine with a soft base.

To wind up this section, consider qualitatively the connection between bonding and ionic conduction in two well-known solid state ionic materials AgI and Li_2SO_4. In AgI, the tetrahedral coordination of Ag by I and of I by Ag is a direct consequence of the hybridization of the 5p orbitals of I and the 4d orbitals of Ag. The result is a pseudo-ionic or iono-covalent Ag–I bond associated with the AgI molecule. The softness of this acid–base pair leads to shallow potential energy minima, the ease of formation of AgI and also the polymorphism exhibited by this compound. The significant conductivity of ~0.1 mS/cm of wurtzite (hexagonal) phase AgI (β-AgI) at room temperature is the result of the easy movement of Ag^+ ion from one tetrahedral site to another. With increase in temperature, the Ag–I bond progressively weakens (or the "dipole" associated with a Ag–I cluster gradually unbinds, from a dielectric/ferroic viewpoint). At the structural phase transition from the wurtzite to bcc phase (α-phase), the bond breaks almost completely (the dipoles unbind) at 147°C. Most significantly, the Ag-coordination number of Ag suddenly changes from 0 below 147°C to as many as 14 at 147°C when Ag^+ breaks away from the lattice in a process called "sublattice melting"! We could also call it covalency collapse from the chemical bonding viewpoint. Suddenly Ag^+ ions that have broken free of the lattice assume a pseudo-metallic bonding leading to the formation of an ion plasma that persists even in the molten state! The mobility and concentration of Ag^+ ions even in the wurtzite phase (where the mobile ion concentration is not excessively large) and much less in the bcc phase do not conform to the usual Arrhenius type of thermally activated ion transport. Thus the role of Ag–I bonding assumes significance and motivates us to understand it in depth.

The case of the Li^+ conductor Li_2SO_4 is that it transforms from monoclinic β-phase to the fcc or α-phase at 575°C. Let us discuss this case from a chemical bonding viewpoint. As a "molecule," Li_2SO_4 is more like H_2SO_4 rather than Na_2SO_4. However, the bonding between Li and O of sulfate is more like that in Li_2O which has an antifluorite structure. Why? Because replacement of $O^=$ by $SO_4^=$ accompanied by distortion would give us the precursor of the Li_2SO_4 structure. The hard acid Li^+ would be prone to form an ionic bond as it does in LiF. But the covalency of the tetrahedral sulfate unit and the unusually small size of Li^+ would render the overall binding in lithium sulfate quite covalent. The unusual thermal stability of the β-phase until 575°C and its poor conductivity (<1 femto S/cm) bear this out. Thus as in the case of AgI, the structural phase transition from the β- to the α-phase in Li_2SO_4 is also a sudden covalent-to-ionic bonding transition where the motion of the sulfate ions would cancel out the effects of their covalent interactions with Li^+ ions. Without such a dynamic equilibrium between sulfate and Li^+ ion subsystems, the cubic structure would not exist! Thus, the high conductivity of α-Li_2SO_4 and the phase transition both manifest from the abrupt bonding changes in the lithium sulfate "molecule." This discussion sets the pace for quantum mechanical considerations of the superionic bond in terms of the electronic structure of these and other solid state ionic materials.

7.3 Bond Fluctuations and Phase Transitions

Chemical bonds in solid state ionics, such as the Li–I, Ag–I, Cu–Br, Li–O in $LiCoO_2$ and Zr–O in zirconia or Pb–F in PbF_2 or La–F in LaF_3 or Ti–O in compounds of Li–Ti–O system, are specialized to provide a defect-ridden and disorder-prone structure favorable for fast ion diffusion. After getting a qualitative picture of the "superionic bond," let us ask what the nature of the bond is in the superionic state attained through a phase transition.

The MX_4 polar covalently bonded tetrahedron (M=Cu, Ag, X=Cl, Br, I, S, Se, Te) serves quite well as the building block of a large group of solid state ionic materials. With open crystal structures and 3D connectivity, these are weak M–X bonds with just one valence electron per M atoms shared by 4X atoms. This weakness leads to structural instability and polymorphic phase transitions to the Frenkel-defect ridden/disorder prone superionic phases. This "static" view of the "superionic bond" gets extended when we consider the temperature dependence of the bond characteristics such as the complex nature of mobile ion charge and its local coordination. This view is directly relevant to the (super)ionic conduction process. The structural instability refers to the instability of the four-coordinated ZB/W structure with respect to the phase transition to the rocksalt/bcc phase with the increase of pressure/temperature, respectively, in the case of AgI. The "dynamic bond" refers to the fluctuating bond in a very broad sense, covering the phenomenology of fast ion conduction in (nano)crystals, glasses, and melts. The "passage" of the mobile (Cu^+) ion from tetrahedral $\rightarrow$ octahedral $\rightarrow$ tetrahedral pathway as in CuI gives the concept of double-minimum potential and opens the door to anharmonicity and all its consequences for the solid state ionics phenomenology. In this section, we shall discuss the empirical but powerful bond fluctuation model of Aniya [7.18].

Open structures with weakly tetrahedral bonds as the Ag–I bond in AgI with low coordination number (4) for mobile ion are the hallmarks of such superionic solids. The question is how do bond fluctuations arise? In AgI at room temperature, the zincblende/wurtzite structure is stabilized by many configurations of the Ag–I bond. The lattice parameter of beta AgI practically does not change from ~4 to 420 K when an abrupt phase transition occurs—a very extended range of temperatures indeed! The many configurations of the Ag–I bond responsible for this "dynamic stability" of the beta phase of silver iodide could be termed "bond fluctuations." The "amplitude" of such fluctuations become so large suddenly at 420 K that the zincblende/wurtzite structure cannot be supported at all! Thus, a phase transition occurs at such a pressure/temperature temperature when the bond fluctuations become critical so that the low pressure/temperature crystal structure cannot be sustained. The origin of these "bond fluctuations" is essentially local and electronic originating from the 3d and 4d electrons of Cu and Ag halides, respectively. A combination of d-, p-, and s-wave functions that produce an average tetrahedral geometry for AgI locally constitutes the "fluctuating bond." A direct experimental demonstration of these bond fluctuations comes from the temperature dependence of the effective number of valence electrons in AgI (Figure 7.3).

Aniya's model is based on three criteria a material should satisfy to become a fast ion conductor:

1. An open structure with 3D connectivity with a low coordination number for a mobile ion with a small radius (note that such a criterion holds for a glass as well!);

2. A structural phase transition occurs as relatively low pressure; phase transition occurs at 3 kbar for AgI and RbI but at >100 kbars for LiI; and

3. For a material AB composed of atoms A and B, with valences Z_A and Z_B, an empirical parameter defined as $Z^* = (Z_A Z/\varepsilon_0)^{1/2}$ is small. Here, ε_0 is the static dielectric constant; CuI has a Z^* of 0.37, CuBr 0.39 and Ag_2S 0.26, Cu_2Se 0.27, AgI 0.38, while LiI has a Z^* of 0.24. NaCl has a Z^* of 0.41.Ternary fast ion conductors $RbAg_4i_5$ and Ag_3SI have Z^* values 0.38 and 0.17, respectively, while PbF_2 has 0.17.

These criteria have much in common with Phillips iconicity concepts.

Consider a solid state ionic material in which ions are vibrating around their equilibrium lattice sites. As the ion core moves, the associated electrons try to follow this movement causing a deformation of the electronic cloud distribution to take place. Equally significantly, the deformed electron cloud creates a new field of forces to move the ions. In this reasonable picture of ion dynamics, the characteristic feature of fast ion conductors is that the electronic cloud deformation is easily induced. It only costs a small disturbance in the electronic subsystem (of course coupled to the ionic subsystem) to produce a drastic change in the *local* structure of the ionic subsystem and *vice versa* in a reversible manner. This is nothing but a local fluctuation of the chemical bond occurring within a superionic solid. Figure 7.4 is a schematic illustration of this model. Note that the concepts of a mutually coupled ionic and electronic subsystems and ideas of shell model are ingrained in the bond fluctuation model.

Phenomenologically, the bond fluctuations may be described as follows. Let A and B (Figure 7.4b) represent stable and metastable bonding configurations. (A metastable configuration is a long-lived configuration unlike an unstable one.) At site B, bond fluctuations have occurred. At a given thermodynamic

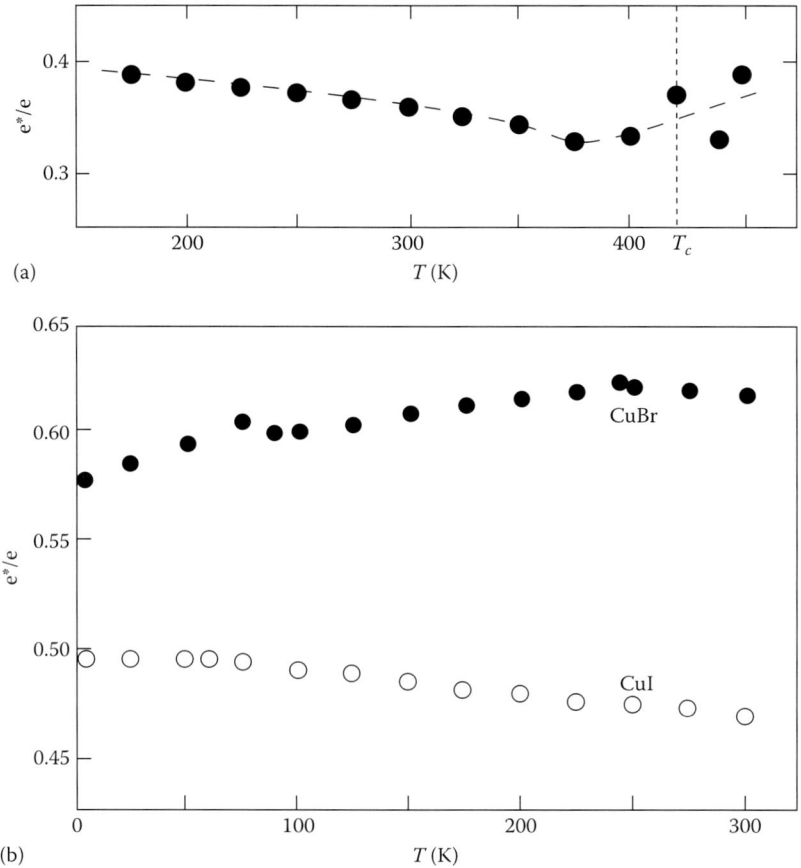

FIGURE 7.3   Temperature dependence of effective charge in (a) AgI, (b) CuBr and CuI.

state, the bond fluctuations are described as A ↔ B. Assuming [A] + [B] =1, the rate of bond fluctuations is given by

$$d[B]/dt = k(i - [B]) \exp(-\Delta G^*/RT) - k[B] \exp\left[-\left(\Delta G^* - \Delta G^0\right)/RT\right]$$

$$= k \exp(-\Delta G^*/RT)\left\{1 - [B]\left(1 + \exp\left(-\Delta G^0/RT\right)\right)\right\} \qquad (7.12)$$

where
  k is the proportionality coefficient
  R is the gas constant
  T is the absolute temperature
  ΔG^* and ΔG^0 are the Gibbs free energies as indicated in Figure 7.4b

The equilibrium condition is d[B]/dt =0, leading to

$$[B]_\infty = \left[1 + \exp\left(-\Delta G^0/RT\right)\right]^{-1} \qquad (7.13)$$

which gives the average concentration of sites in the configuration B in a given thermodynamic state. At low temperatures, Equation 7.13 predicts a nearly zero [B]$_\infty$, implying that bond fluctuations cease at such temperatures.

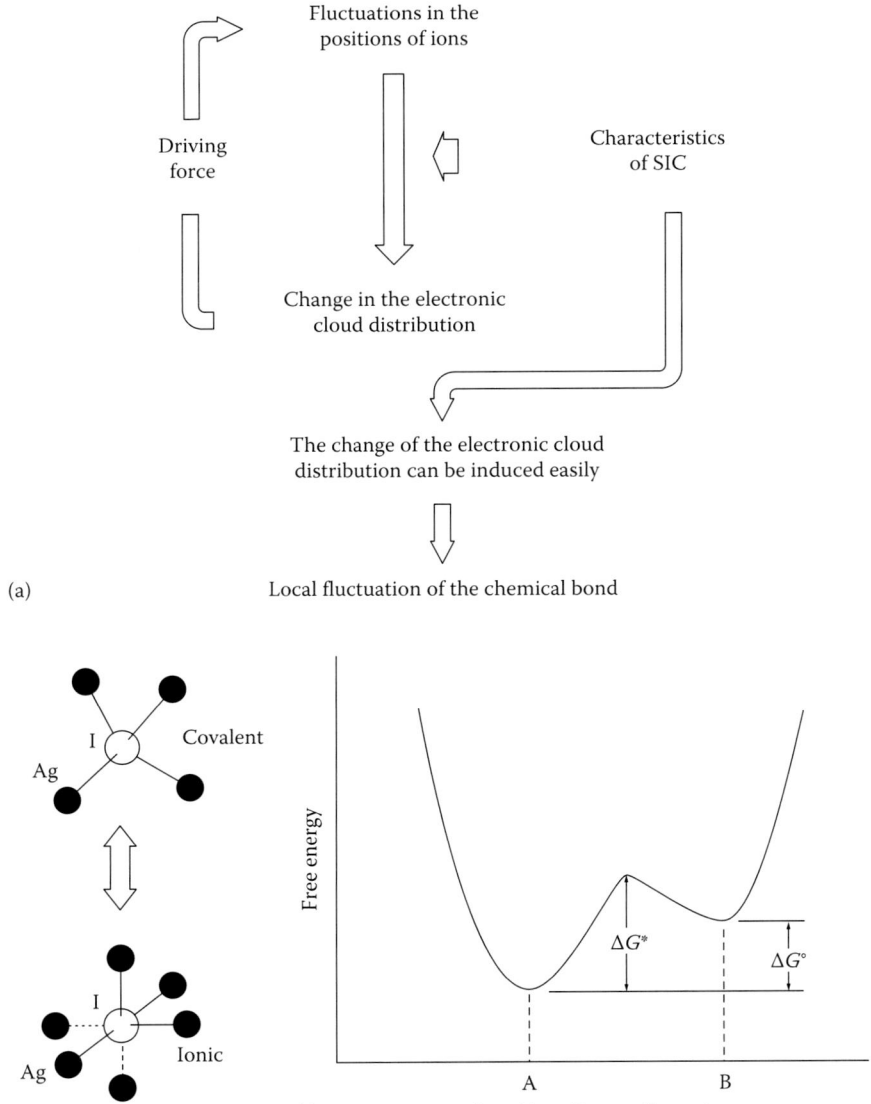

(a)

(b)

(c)

FIGURE 7.4 (a) An intuitive visualization of the basic stimuli and responses leading to fluctuation of the polar covalent bond. (b) AgI_4 (covalent, four-coordinated) to AgI_6 (ionic, six-coordianted) reversible bond switching in AgI. Illustrating local bond fluctuation. (c) Free-energy model using double minimum potential representing the local bonding configurations A (stable) and B (metastable) and the Gibbs free energy difference ΔG_0 between them.

How long do bond fluctuations last? To answer that put Equation 7.13 into 7.12 and write $d[B]/dt = -([B]-[B]_\infty)/\tau$ and obtain τ. By applying this model to AgI, one can estimate that the lifetime of bond fluctuations to be 1–2 periods of optical phonons thus linking phonons with superionicity. How are chemical bond fluctuations seen in the structure? They are perceived as the local coordination number at any instant of time. In Ag and Cu halides, the non-superionic state at low temperature is characterized by tetrahedrally coordinated mobile ions Ag^+ and Cu^+. In the superionic state, the mobile ion instantaneously takes a higher coordination (say octahedral) as it moves between sites provides by the cage ion sublattice Br in CuBr and I in AgI. The ion transport takes the path tetrahedral $\rightarrow$ octahedral $\rightarrow$ tetrahedral site occupancy. This in bond fluctuation language means covalent $\rightarrow$ ionic $\rightarrow$ covalent bond fluctuations. This discussion hopefully motivates a more rigorous investigation extended to ternary halides and chalcohalides of Cu and Ag. It is pertinent to mention that bond fluctuations occur in polymers as a trigger for phase transitions so that polymer electrolytes are another class of materials to look for bond fluctuations.

7.4 Critical Ionicity Marginal Covalency and Electronic Structure

7.4.1 Critical Ionicity and Pseudopotentials

It is interesting that the simple NFE model (Figure 7.1 and Problem) applied to semiconductors and ionic crystals has yielded a demarcation between the two classes of materials at a critical ionicity of 0.72. Table 7.3 shows the data. This connects to the similar concept of Phillips iconicity discussed in the following.

The concept of critical iconicity separates tetrahedral semiconductors with four-coordinated cations (e.g., AgI) from electronic insulators/ionic conductors with six-coordinated cations (NaI). Phillips [7.3] approached the problem empirically and produced a map in which a critical ionicity f_c of 0.785 separates the two classes of materials. It is the fact that ionicities of silver and copper halides lie close to this value which connects the domains of chemical bonding, crystal structure, and electronic band structure of fast ion conductors. Interestingly, the concept of complex energy gap also introduced empirically has fundamental significance to an understanding of solid state ionic phenomena. The missing link of crystal structure in Phillips model was supplied by Van Vechten [7.19]. An important property of the diatomic crystals of the type $A^N B^{8-N}$ is the real, static electronic dielectric constant $\varepsilon(0)$. Among the iodides, LiI has measured $\varepsilon(0)$ of 3.8, RbI 2.7, AgI 4.9, and CuI 6.5, while LiBr has a value of 3.2 and AgBr and CuBr 4.4. In tetrahedrally coordinated crystals—AgI, CuBr, and CuI—the average energy gap E_g based on $\varepsilon(0)$ represents the energy difference between bonding and antibonding hybridized orbitals. E_g is decomposable into contributions due to the symmetric and antisymmetric parts of the potential within the unit cell.

The ability of Ag and Cu to form compounds such as halides and chalcogenides with a low coordination (usually 4d and 3d bands) stems from the occurrence of 4d and 4d bands in their electronic structure. Unlike the semiconductors GaAs and CdS, the covalency in the former compounds is weak. Four-coordination and marginal covalency ensure stereochemical flexibility of AgI, CuBr, CuI as well as chalcogenides of Ag and Cu. Ease of bond breaking and making and low activation energy for Frenkel defect migration and thus fast cation transport complete the picture. The unhindered motion of cations is apparently linked to their facility in polarizing the anions especially I⁻. The outer d-electrons provide ineffective shielding of the nuclear charge so they readily attract electrons. This ability allows them to form

TABLE 7.3

Dielectric Constant, Bandgap and Iconicity for Semiconducting and Ionic Materials

Crystal	Type[a]	Lattice const. (arb. units)	ε_o	N_{eff}	E_g (eV)	C (eV)	f_i
ZnTe	Z	11.510	7.3	4.9	6.16	4.88	0.628
ZnS	Z, W	10.222	5.2	4.3	8.45	6.77	0.641
ZnSe	Z, W	10.710	5.9	4.7	7.62	6.16	0.653
CdTe	Z	12.246	7.2	5.2	5.83	4.85	0.693
CdS	Z, W	11.047	5.2	4.5	7.69	6.47	0.708
CdSe	W	11.489	5.2	4.9	7.08	5.99	0.716
AgBr	R	10.912	5.0	4.7	8.20	7.00	0.728
AgCl	R	10.482	4.2	4.3	9.32	8.02	0.741
LiI	R	11.338	3.8	4.1	8.65	7.72	0.797
LiBr	R	10.396	3.2	4.1	11.1	10.0	0.810
LiCl	R	9.693	2.7	4.0	13.9	12.6	0.823
NaI	R	12.232	3.0	4.1	9.13	8.53	0.874

Source: Alarashi, R.A. et al., *Phys. Status Solidi b*, 54, K5, 1972.

Notes: The separation of these two classes tetrahedral and octahedral occurs at an ionicity (*fi*) of 0.72. Critical ionicity of Phillips is 0.785.

[a] Z, Zincblende; W, Wurtzite; R, Rocksalt.

bonds with covalent character at each intermediate position through which they move, thereby reducing the energy of this transition state.

Kobayashi et al. [7.20] have discussed the problem from a fundamental viewpoint and provided a microscopic understanding of critical iconicity. Let us outline this approach. The dielectric properties of A^NB^{8-N} are responsible for chemical bonding, which is the result of a competition between the covalent and ionic bonding. In this dielectric picture, the basic physical concept involved is the polarizability of the ion and the electron at a local level. Also, note that these two components are involved in the complex energy gap (E_g). E_g is generally defined as

$$E_g = E_h + iC \tag{7.14}$$

where
  E_h is the homopolar energy (Ag–Ag bond in AgI) band gap
  C is the heteropolar energy band gap

To begin with, consider that the zincblende structure adopted by AgI, CuBr, and CuI and the rocksalt structure adopted by LiI, NaI, and AgBr are both constructed from fcc sublattices of A and B atoms. Therefore, the electronic structures of both classes of compounds may be described by using the same BZ (discussed in Chapter 1) for the fcc structure. We shall use the pseudopotential model to discuss the energy gap of these A^NB^{8-N} compounds which possess eight valence electrons per two atoms. All valence electrons can occupy the inner energy states in the *fourth* BZ whose number of states is four times that of the first BZ. The fourth BZ, unfortunately, is not isotropic and complicated. What is the alternative then? There is this Jones zone, devised by Heine and Jones, whose number of states is just four times that of the first BZ. Not just that. It is also more isotropic than the fourth BZ. Figure 7.5 shows the Jones zone of an fcc structure. It is a polyhedron constructed from twelve equivalent Bragg planes dividing the reciprocal lattice vector $G = (2\pi/a_0)(220)$, where a_0 is the lattice constant. Now what is the average energy gap at the X points of $k = (2\pi/a_0)(110)$ and $k = (2\pi/a_0)(-1-10)$ on the Jones zone surfaces? X point $k = (2\pi/a_0)(110)$ corresponds to the center on the Bragg cross section dividing equivalently the reciprocal lattice vectors $G = (2\pi/a_0)(220)$. Check that another X point $k = (2\pi/a_0)(-1-10)$ does the same to $-G = (2\pi/a_0)(-2-20)$. In the nearly free electron approximation applicable to A^NB^{8-N} semiconductors, the average energy gap at the X-point is given by $E_g = 2|V(G)| = 2|V(220)|$, where $V(G)$ is the Fourier component of the pseudopotential $V(r)$. What is a pseudopotential? The concept of a pseudopotential arises from the energy band

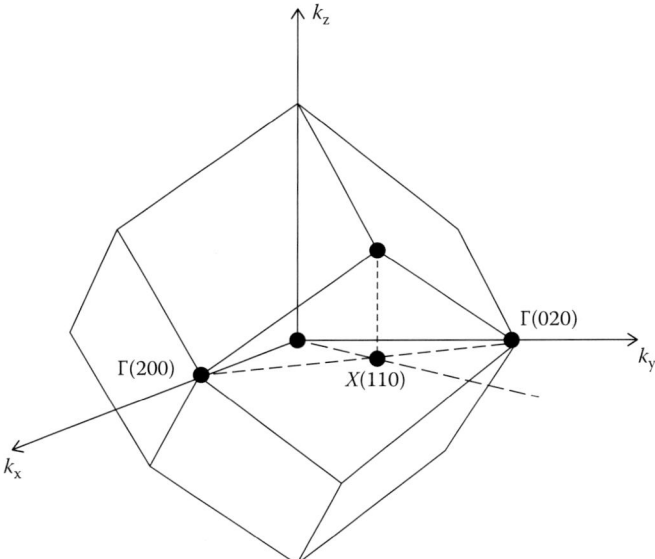

FIGURE 7.5 Jones zone representation for fcc cubic crystals.

theory of metals [7.21]. Put simply, it is a theoretical means of replacing the actual potential energy in the core region of a metal (In Ag the core is made up of all the electrons up to $5s^1$.) by an effective potential energy or the pseudopotential. The advantage of using the pseudopotential is that it gives the same wavefunctions outside the core as does the actual potential. The amazing fact about the pseudopotential is that it is actually zero in the core region!

But in semiconductors, the component $|V(111)|$ is considerably larger than $|V(220)|$ because it arises from a crystal plane with the maximum number of atoms; thus we cannot ignore $V(111)$ in the calculation of the energy gap. Therefore, we need to set up and solve the Schrodinger equation for the valence electron to obtain the energy gap for $A^N B^{8-N}$ semiconductors that include superionic conductors. The equation to be solved is

$$\left[-\hbar^2/2m_e\right) + V(\boldsymbol{r})\right]\psi_k(\boldsymbol{r}) = E(k)\psi_k(\boldsymbol{r})\tag{7.15}$$

To the first order in perturbation, the solution is

$$\psi_k(\boldsymbol{r}) = \varphi_k(r) + \Sigma_{k'\neq k}\left[\langle k'|V|\varphi_k\rangle\right]/\left(E_0(k) - E_0(k')\right)|k'\rangle,\tag{7.16}$$

where m_e is the mass of electrons, $\varphi_k(r) = a_k|k\rangle + b_k|-k\rangle$ mixes two plane waves $|k\rangle$ and $|-k\rangle$.

From the Schrodinger equation, we obtain the following secular equations (What is a secular equation?):

$$\left[E_0(k) - E(k) + A(k)\right]a_k + V_{\text{eff}}(220)b_k = 0,$$

$$V_{\text{eff}}(220)a_k + \left[E_0(-k) - E(k) + A(k)\right]b_k = 0,\tag{7.17}$$

$$V_{\text{eff}}(220) = |V(220)| + \Sigma_k, \left[\langle k|V|k'\rangle \langle k'|V|-k\rangle\right]/\left[E_0(k) - E_0(k')\right]$$

From Equation 7.16, we get the energy gap as $E_g = 2V_{\text{eff}}(220)$. $A(k)$ does not contribute to the energy gap E_g (Why?).

$V(111)$ is the largest potential term so that we can restrict to the terms $\boldsymbol{k}' = (2\pi/a_0)(001)$ and $\boldsymbol{k}' = (2\pi/a_0)(00\text{-}1)$ in the sum of Equation 7.16. Thus, the energy gap is presented as

$$E_g = 2V_{\text{eff}}(220) = 2|V(220)| + 2V^2(111)/\Delta T\tag{7.18}$$

where $\Delta T = (2\pi\hbar/a_0)^2/2m_e$.

The generalized complex energy gap of $A^N B^{8-N}$ compounds is derived from the Fourier component $V(G)$ of the pseudopotential of these compounds. Note that the pseudopotential is defined in the real space while the Fourier component is in the reciprocal lattice vector space.

$$V(G) = 1/\Omega \int \Sigma_l \Sigma_{\lambda=A,B} V_\lambda (\mathrm{r} - \mathrm{R}_{l\lambda})\exp(iG\cdot\mathrm{r})d\mathrm{r}$$

$$= \Sigma_{\lambda=A,B} V_\lambda(G)\exp(-iG\cdot\mathrm{R}_\lambda)\tag{7.19}$$

where

Ω is the volume of the solid

$V\lambda(\mathbf{G})$ is the pseudoatom form factor

Ag atom in AgI crystal or Cu atom in CuBr crystal is viewed as a pseudoatom in a semiconductor crystal. An electron wave in a crystal is scattered by pseudoatoms in a manner analogous to the way X-rays, for example, are scattered by real atoms. Thus, pseudoatom form factor effectively represents what structure factor does in X-ray scattering. A detailed discussion is available in [7.3].

Select the origin of the coordinate system at the midpoint between cation (atom A) and anion (atom B): $R_A = -\tau$ and $R_B = \tau$. This is because the shared valence electrons forming the A–B bond tend to "pile up" midway between A and B to form the covalent bond. In fact, the band structure and the gap are essentially due to these electrons. Now $V(\mathbf{G})$ is rewritten as

$$V(\mathbf{G}) = V_s(\mathbf{G})\cos(\mathbf{G}\cdot\mathbf{r}) + iV_a(\mathbf{G})\sin(\mathbf{G}\cdot\mathbf{r}) \tag{7.20}$$

Note that $V_s(\mathbf{G})$ and $V_a(\mathbf{G})$ are the symmetric (real) and antisymmetric (imaginary) parts of the complex pseudopotential that translate to the real and imaginary parts of the complex energy gap. In fact, Kittel (ISSP) [7.21] gives a graphic and motivating account of this connection. $\tau = (a_0/8)(111)$ for the zinc-blende structure where a_0 is the lattice parameter. Sine factor vanishes for $\mathbf{G} = (2\pi/a_0)(220)$. Therefore, $V(\mathbf{G}) = -V_s(\mathbf{G})$ and $V(\mathbf{G})$ is real. However, for $G = (2\pi/a_0)(111)$, $V(\mathbf{G})$ is a complex number because $|\cos(\mathbf{G}\cdot\tau)| = |\sin(\mathbf{G}\cdot\tau)|$. Therefore, while considering E_g ($=E_h + iC$) in the complex number representation, we find that instead of Equation 7.17 the complex energy gap at the X-point on the Jones' zone is

$$E_h = 2|V_s(220)| + 2V_s^2(111)\Delta T \tag{7.21}$$

and

$$C = 4V_s(111)V_a(111)/\Delta T \tag{7.22}$$

The real energy gap is thus made up to two contributions: $E_g^2 = E_h^2 + C^2$ where the homopolar energy gap is related to $V_s(G)$ and the heteropolar (A–B) energy gap C is proportional to $V_a(G)$. Note that $V_a(G)$ vanishes for monatomic semiconductors such as Si for which $V_A(\mathbf{G}) = V_B(\mathbf{G})$. **Therefore, C can be used to define the ionicity** $f_i = C^2 + E_g^2$ in the Phillips model. Si is completely covalent with $f_i = 0$. But completely ionic compounds ($f_i = 1$) are not found in the $A^N B^{8-N}$ family because these compounds are segregated at critical iconicity $F_i = 0.785$. A fundamental thermochemical phase diagram due to Phillips (Figure 7.6) [7.22] demarcates covalent and ionic materials at the critical iconicity line.

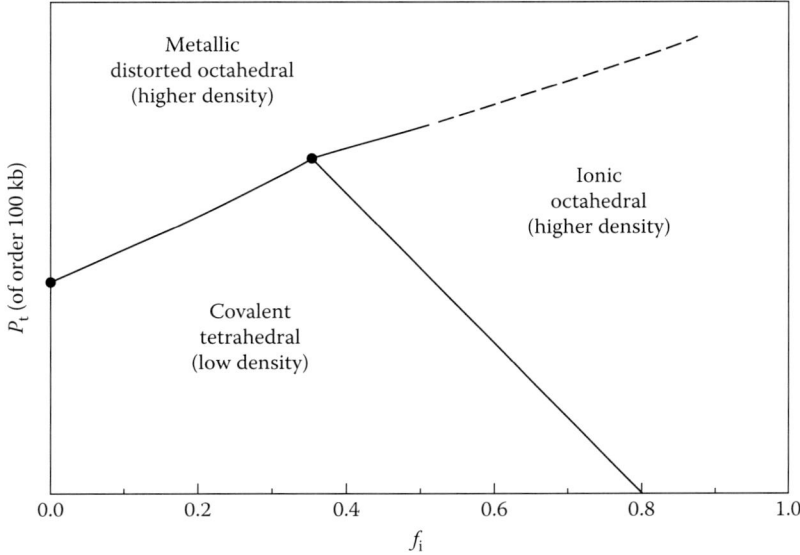

FIGURE 7.6 Schematic thermochemical phase diagram (transition pressure P_t versus ionicity (f_i) clearly demarcating covalent tetrahedral, low density, and ionic octahedral high density $A^N B^{8-N}$ binary semiconductors. Ionicity of tetrahedral AgI and allied superionic conductors are close to the covalent–ionic phase boundary. Note the importance of density which changes from low to high when AgI transforms from wurtzite to bcc structure at 147°C. (From Phillips, J.C., *Bonds and Bands in Semiconductors*, Academic Press, New York, 1973.)

As noted earlier, $V_s(111)$ is large enough and negative for zincblende compounds. Therefore, $V_s(111)$ is mainly responsible for E_h. Use the definition of f_i and combine Equation 7.20 and 7.21 to obtain

$$f_i = 4V_a^2(111)(1-\delta)^2 \Big/ \Big[V_s^2(111) + 4V_a^2(111)(1-\delta)^2 \Big] \tag{7.23}$$

where $\delta = 2$. $\big|V_s(220)\big| = E_h$ is the correction term. What happens when the potentials satisfy the condition $\big|V_a(111)\big| = V_s(111)$? Equation 7.22 gives the critical ionicity f_c:

$$f_c = 4(1-\delta)^2 \Big/ \Big[1 + 4(1-\delta)^2 \Big] \le 0.8 \tag{7.24}$$

Exercise: Derive Equation 7.23. Also, show that covalent zincblende compounds become unstable at $f_i = f_c$ and transform to ionic rocksalt compounds.

For $\delta = 0$, $f_c = 0.8$, a value slightly larger than the Philllips value $f_i = 0.785$. Taking account of correction term δ of several covalent compounds, the average critical iconicity $<f_c> = 0.78$ is very close to F_i.

A question naturally arises at this stage: How do the valence electrons of $A^N B^{8-N}$ compounds respond to ionic displacements? Answer to this question takes us to the concept of dynamic effective charge that includes, just as in the energy gap discussed earlier, both ionic and electronic contributions. Thus, effective charge is complex. Thus, chemical bonding is involved here too. We shall briefly outline the approach of Kobayashi et al. in the framework of linear response theory. The electrons in solids respond to an external perturbation with the wave vectors $\mathbf{q}$ so that the electron wave vector $\mathbf{K} = \mathbf{q} + \mathbf{G}$, $\mathbf{G}$ being the reciprocal lattice vector. The external fields induce an electron density which must be considered in a self-consistent manner. The important physical consequence is the polarizability of the solid. The microscopic effective charge is derived from the electric polarization field caused by the ionic displacements. Linear response theory helps calculate the induced electron charge density.

The important results of the theory are simple expressions for complex effective charges on atoms A and B based on the pseudopotential:

$$Z(A) = Z_A - 2N_v V_A(G)\exp(-iG \cdot R_A)/V(G) \tag{7.25}$$

$$Z(B) = Z_B - 2N_v V_B(G)\exp(-iG \cdot R_B)/V(G) \tag{7.26}$$

These two equations tell us that valence electrons $N_v = 8$ are proportionally redistributed depending on the strength of the potentials $V_A(G)$ and $V_B(G)$ with $G = (2\pi/a_0)(111)$.

$Z(A)$ and $Z(B)$ are theoretical quantities but what is actually observed is the transverse effective charge is

$$Z_T = \text{Re}[Z(A) - Z(B)] \tag{7.27}$$

where Re denotes real part. From Equations 7.25 and 7.26, Z_T comes out as

$$Z_T = \Delta Z/2 - 2N_v \text{Re}\Big[V_A(G)\exp(-iG \cdot R_A) - V_B(G)\exp(-iG \cdot R_B)/V(G) \Big] /$$

$$\Big[V_A(G)\exp(-iG \cdot R_A) + V_B(G)\exp(-iG \cdot R_B)/V(G) \Big] \tag{7.28}$$

where $\Delta Z = Z_A - Z_B$. As was done for the complex energy gap,

$$Z_T = \Delta Z / 2 - N_v V_s(G) V_a(G) / \left\{ \left[V_s(G)\cos(G \cdot \tau) \right]^2 + V_A(G)\sin(G \cdot \tau)]^2 \right\} \tag{7.29}$$

where $\tau = (a_0/8)(111)$ for zincblende structure and $\tau = (a_0/4)(111)$ for rocksalt structure and $G = (2\pi/a_0)(111)$. Derivation of Z_T^{ZB} and Z_T^{RS} is left as an exercise.

Finally, Z_T connects to the ionicity f_i through the equations

$$Z_T^{ZB}(f_i) = Z_T^{I}(f_i) = \Delta Z/2 + 4N_v \left[f_i(1-f_i) \right]^{1/2} / (4 - 3f_i) \tag{7.30}$$

$$Z_T^{ZB}(f_i) = \Delta Z/2 + 2N_v \left[(1 - f_i) / f_i \right]^{1/2} \tag{7.31}$$

where the second terms give the valence electron contribution to transverse charge. Check what would you get if you equate $Z_T^{ZB}(f_i)$ and $Z_T^{ZB}(f_i)$.

Experimentally, the transverse effective charge Z_T is proportional to the difference of the optical modes $\omega_{LO}^2 - \omega_{TO}^2$, which is related to the long-range Coulomb interaction. Thus, zincblende compounds in the vicinity of critical ionicity (F_i) increased by long-range Coulomb interaction. AgI and Cu halides are distributed near F_i.

An important final point is to consider the effect of core electrons on Z_T. This comes in the form of d-electron contribution promoted to the valence band introducing an effective number of valence electrons $N_{eff} = N_v + N_d$ where N_d is due to the d-band excitation. Thus, the general expression for Z is

$$Z(\lambda) = Z_\lambda - 2N_{eff} V_\lambda(G)\exp(-iG \cdot R_\lambda) / V(G) \tag{7.32}$$

fluctuating in time. Note the change in the electronic configurations of Ag and I in the covalent and ionic bond types (Figures 7.7 and 7.8).

A microscopic mechanism for fast ion conduction in solids must invariably involve chemical bond dynamics especially in Cu and Ag halides and chalcogenides and their ternary and quaternary compounds, which form a large family. As Aniya has proposed, reversible change of bonding from covalent-to-ionic and back in these polar covalent solids occurring locally and fluctuating in time is perhaps the

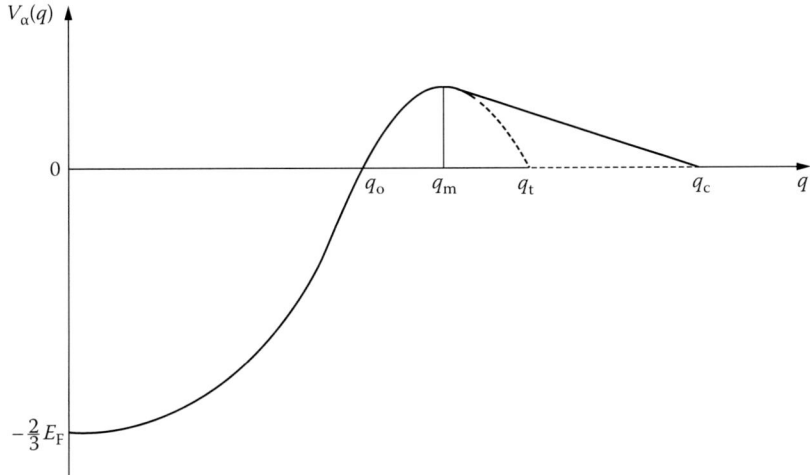

FIGURE 7.7 Schematic of a typical pseudopotential $V(q)$ plotted against q. At q_0, V crosses zero and reaches a maximum at q_m. q_c is the tail off (to zero) value, whereas q_t is a form cutoff while fitting band structure data. (From Tsang, Y.W. and Cohen, M.L., *Phys. Rev. B*, 9, 3541, 1974.)

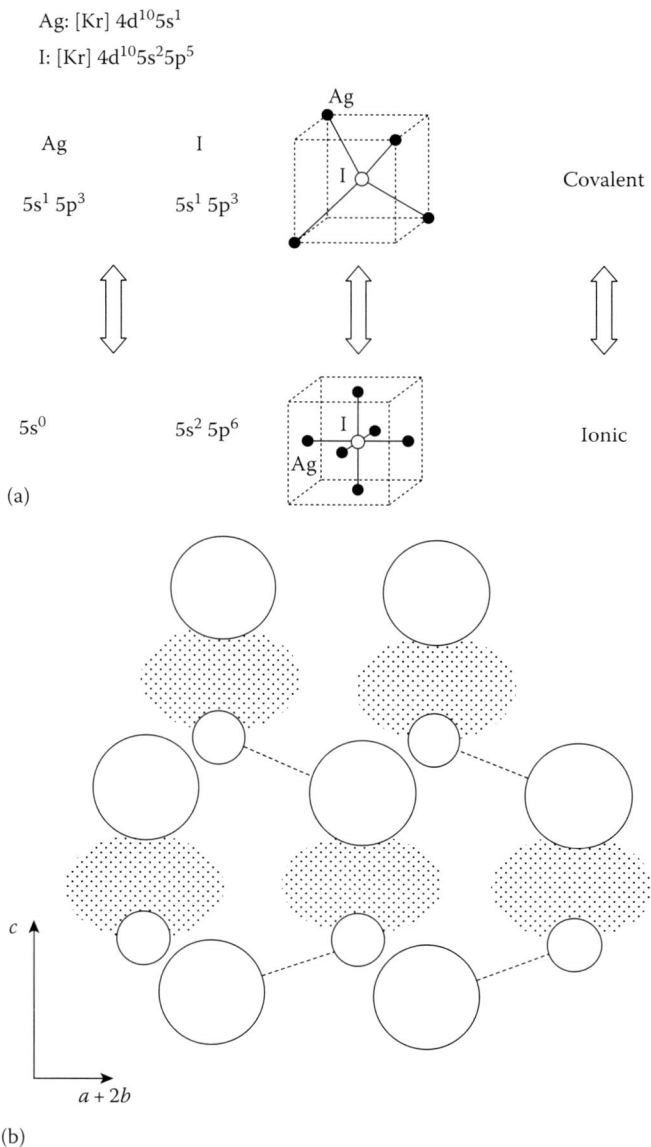

FIGURE 7.8 (a) Schematic showing local change of the Ag–I bond. (b) Visualization of the Ag–I bond in β-AgI. Covalent and ionic charge distributions in β-AgI. O: I⁻ ions, o: Ag⁺ ions. Clusters of dots between Ag and I along *c*-axis: molecular-like covalent bond; broken line: ionic bond. Note charge accumulation along *c*-axis. (From Kobayashi, M., in *Physics of Solid State Ionics*, Sakuma, T. and Takahashi, H. (Eds.), 2006, pp. 1–15.)

key mechanism for superionicity. A strong reason for this is the presence of d-electrons in the mobile ions and the polarizability of the anions that are brought into play when the ions move. Note that Cu and Ag halides are strategically placed in the periodic table between traditional ionic solids (I–VII alkali halides) and covalent semiconductors (III–V and II–VI). This implies that an overlap of wavefunctions of Cu/Ag and the halogens (Figure 7.9) determines the nature of the chemical bond, being sensitive to the interatomic separation. This mechanism intuitively and naturally accounts for important features of superionic conductors including the correlated motion of mobile ions Cu^+ and Ag^+.

The pseudopotential method used in solid state physics for the calculation of electronic and other properties of metals and semiconductors can be adopted to obtain the electron density distribution in real space

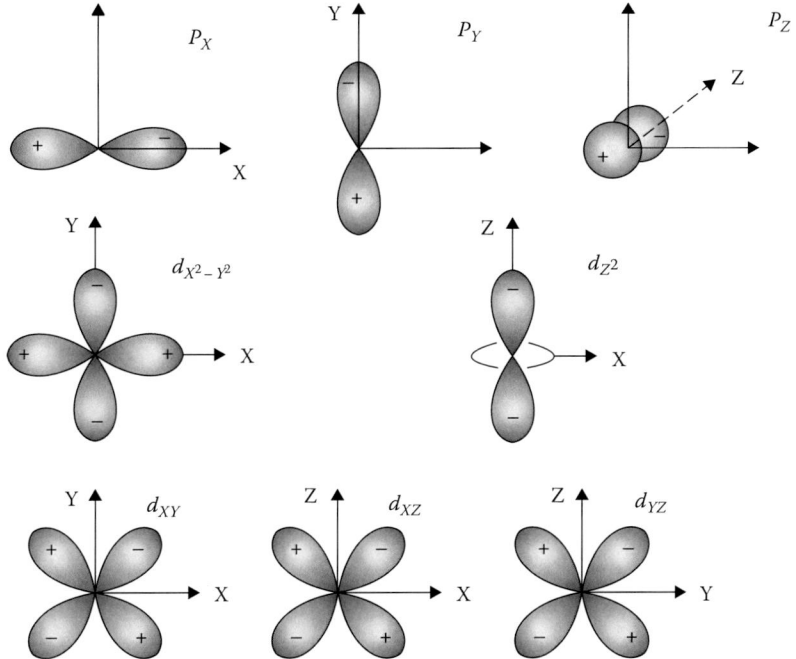

FIGURE 7.9 p-Orbitals of halogen (X) and d-orbitals of noble metals (M) whose overlap is directly responsible for the formation of MX_4 tetrahedra which are the building blocks for deriving crystal structure, phonons, and electronic band structures of MX.

of binary compounds AgI, CuBr, and CuI. This provides a direct and useful information about the chemical bond. This method is based on the idea of switching *back and forth* between the *pseudopotential form factors* in the *reciprocal space V(G)* and the *crystal pseudopotential in the real space*, $V_p(r)$. Note that $V_p(r)$ is defined in terms of the atomic pseudopotentials, $v\alpha$ as $\Sigma_j \Sigma_\alpha v \, (r-R_{j\alpha})$, where $R_{j\alpha} = R_j + \tau_\alpha$ are the sum of lattice and basis vectors, respectively. Thus, one can derive the following general equation for obtaining the crystal pseudopotential:

$$V_p(r) = (1/N_a) \sum_\alpha \sum_G S_\alpha(G) V_\alpha(G) \exp(iG \cdot r) \tag{7.33}$$

where

N_a is the number of basis atoms
$S_\alpha(G)$ is the structure factor
$V_\alpha(G)$ is the pseudopotential form factor
G being the reciprocal lattice vector

Diamond, zincblende, and rocksalt structures all have a two-atom basis per lattice site. If τ_1 and τ_2 are the positions of these basis atoms, then Equation 7.31 becomes

$$V_p(r) = \sum_G \times \left[V^S(G) \cos(G \cdot \tau) + iV^A(G) \sin(G \cdot \tau) \right] \times \exp(iG \cdot r) \tag{7.34}$$

where $V_S(G)$ and $V_A(G)$ are the symmetric and antisymmetric form factors defined as

$$V^S(G) = \left(\frac{1}{2}\right) \left[V_1(G) + V_2(G) \right] \tag{7.35}$$

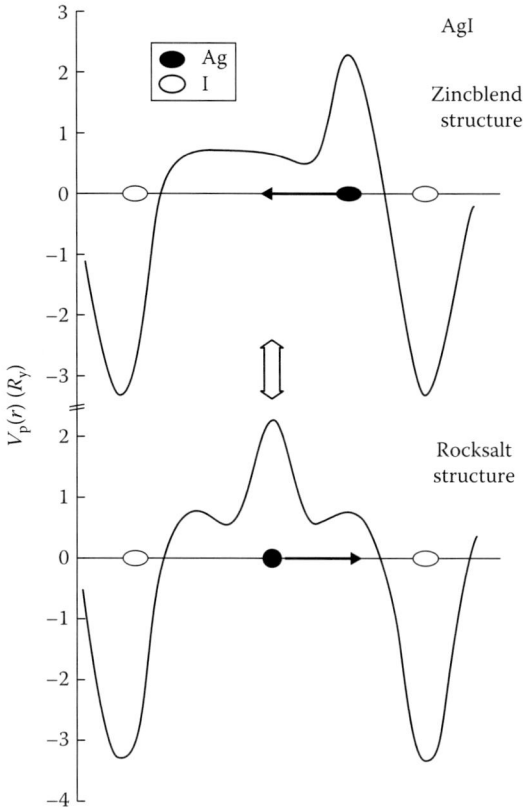

FIGURE 7.10 Pseudopotentials in zincblende and rocksalt AgI. Migration of Ag ions (●) is accompanied by the valence electron distribution. Refer back to Figure 7.4a.

and

$$V^A(\boldsymbol{G}) = \left(\frac{1}{2}\right)\left[V_1(\boldsymbol{G}) - V_2(\boldsymbol{G})\right] \tag{7.36}$$

The real part of Equation 7.32 contains information about the electron density distribution in real space. To focus on the zincblende and rocksalt phases of AgI, Figure 7.10 displays their $V_p(r)$. This figure helps demonstrate a link between ion transport and electronic structure of superionic conductors. The microscopic mechanism that initiates the migration of a mobile ion could be traced back to *local electronic cloud deformation*. The latter is driven by the lattice vibrations. Thus, we see an intimate connection between structure, phonons, and the d-electrons in AgI-type superionic conductors.

Next we will look at the band structure, density of states, and projected electron localization functions (ELF) of CuI (Figure 7.11).

A change in the coordination number (of Cu by I) from four (zincblende) to five (orthorhombic) occurs due to the application of pressure. This change comes along with a volume contraction of 33%. The nearest neighbor distance R_{Cu-I} decreases from 2.63. A for zincblende phase to 2.53 A for the orthorhombic phase. Predictably, the nature of bonding also changes in the zincblende-to– tetragonal –to- orthorhombic phase change. The evidence comes from the projected ELF shown in Figure 7.12. Note that the maximum value of ELF is ~0.77 in orthorhombic CuI. Also $0 < ELF < 1$ represents the complete delocalization and complete localization of electron density.

The real and imaginary parts of the dielectric function of the three polymorphs namely zincblende, tetragonal, and orthorhombic CuI are given in Figure 7.13.

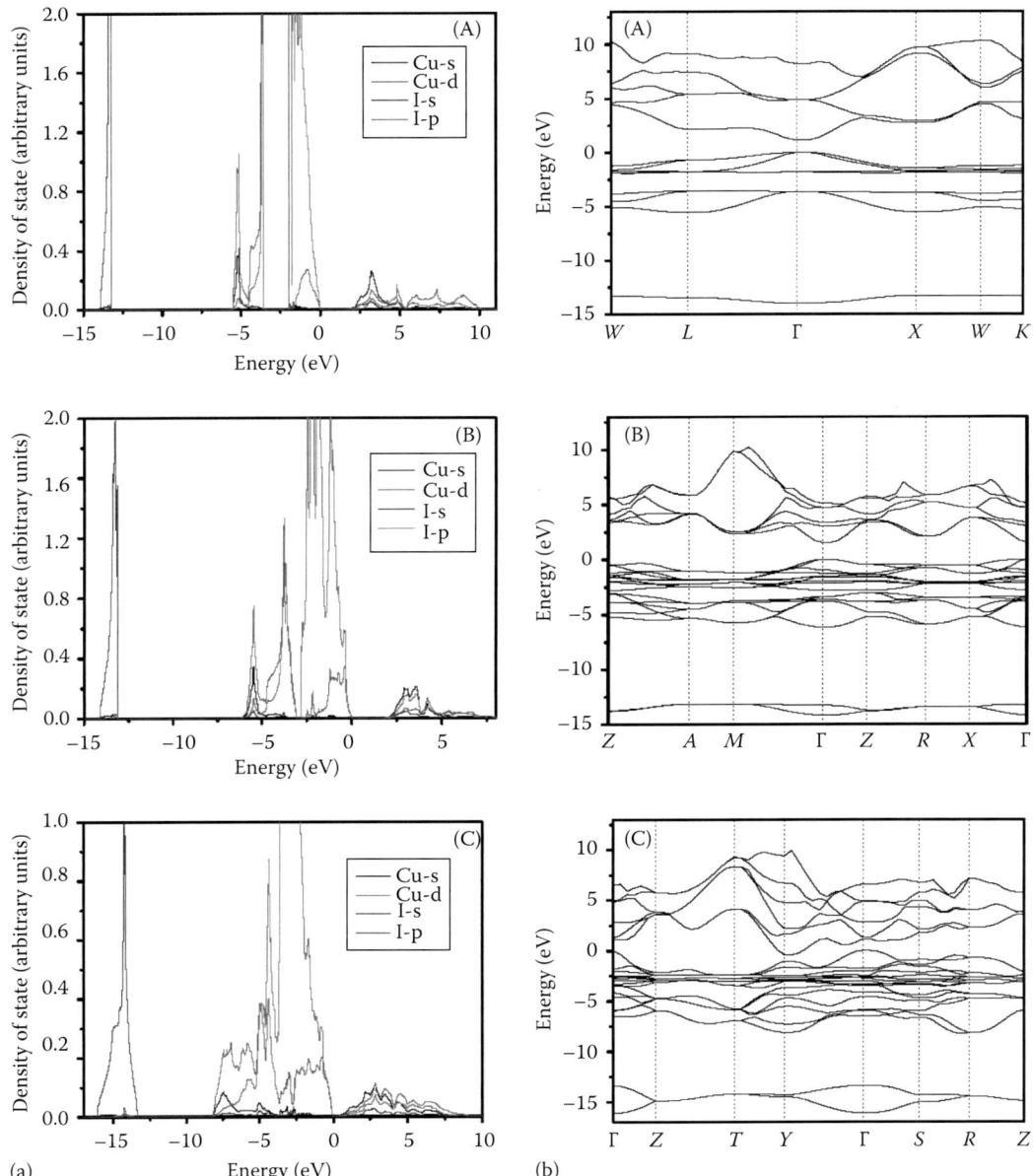

FIGURE 7.11 (a) Electronic density of states calculated for (A) zincblende, (B) tetragonal, and (C) orthorhombic phases of CuI. (b) Calculated band structures of (A) zincblende, (B) tetragonal, and (C) orthorhombic CuI. (From Jhu, J. et al., *J. Phys. Condens. Matter*, 24, 2012, 475503.)

7.4.2 Marginal Covalency: Tight-Binding Approximation

Marginal covalency of the metal halogen bond in Ag and Cu halides provides a natural entry into their electronic structure.

1D diatomic lattice consisting of alternating cation and anion (discussed in Chapter 6) introduces us to single gap ideas. This involves atoms (hetero) of same/different rows. Van Vechten has clarified the significance of band gap from a rigorous quantum mechanical viewpoint. He has underlined the importance of crystal structure for arriving at the band structure of a zincblende solid with metal and nonmetallic atoms. In solid state ionics, refer to AgI, CuBr, and CuI, he has indicated how to go from the diatomic

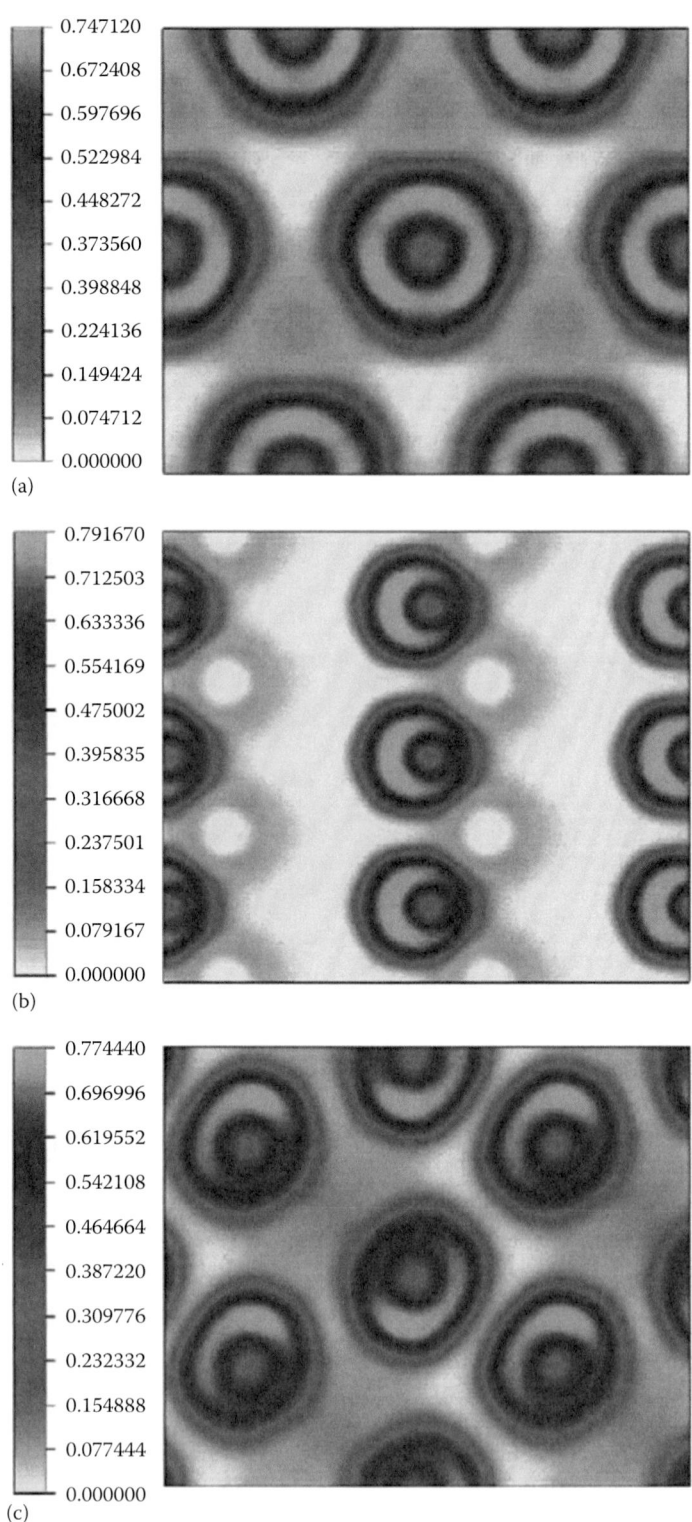

FIGURE 7.12 Projected electron localization functions of CuI: (a) zincblende, (b) tetragonal, and (c) orthorhombic CuI in the local density approximation.

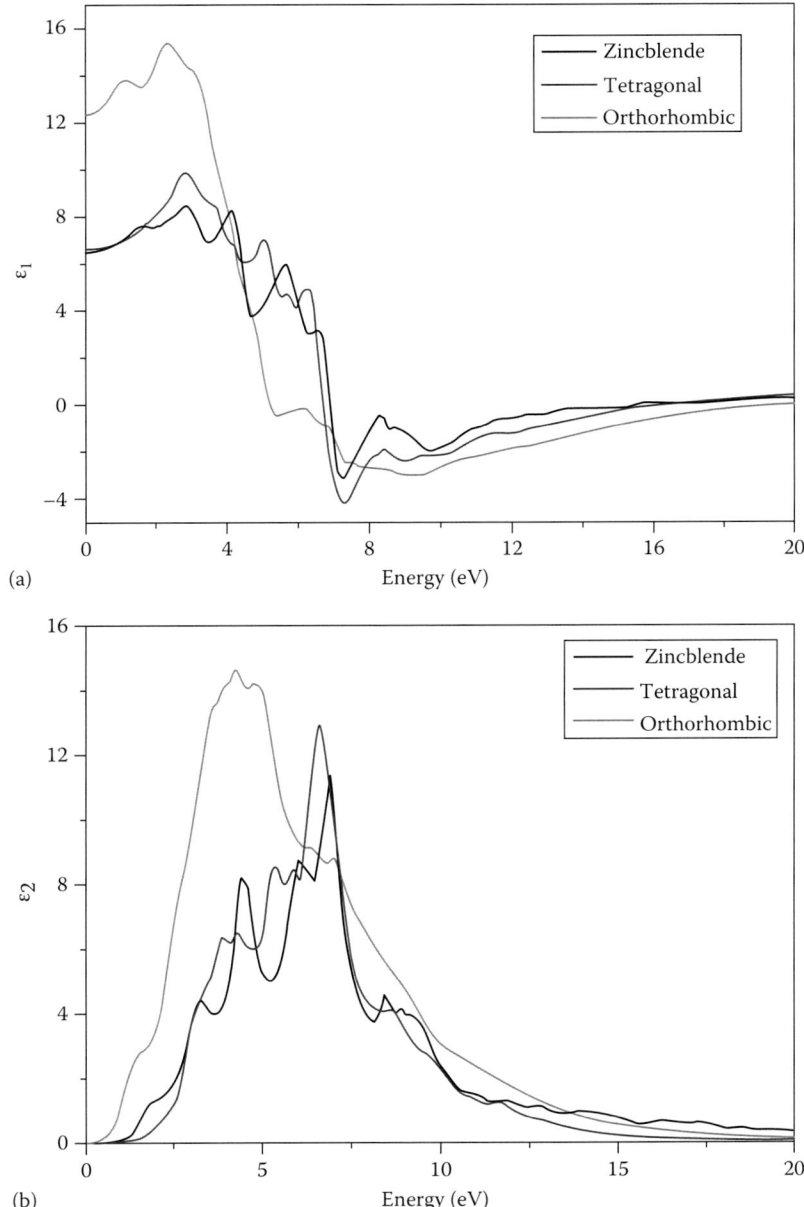

FIGURE 7.13 (a) Real part (ε_1) and (b) imaginary part (ε_2) of the complex dielectric function ε^* for zincblende, tetragonal, and orthorhombic polymorphs of CuI. (From Jhu, J. et al., *J. Phys. Condens. Matter*, 24, 2012, 475503.)

heteroatom lattice to the 1D electronic band structure and introduced the idea of dielectric function going on to discuss the dominant effects of d-electronic states on the static dielectric constant.

The total energy of stable solids including superionic solids is made up of electrostatic, vibrational, and electronic band-structure energies. Just as an atom or molecule has an electronic configuration representing its ground and higher-energy states, a solid has an electronic band configuration consisting of a "ground" energy band or the valence band and many higher energy bands referred to as "conduction" bands. The valence band is fully occupied by electrons at absolute zero temperature while the conduction bands are essentially empty. Electronic excitation is possible from valence band to higher bands through optical and other radiation. The difference in energy between the valence band and the

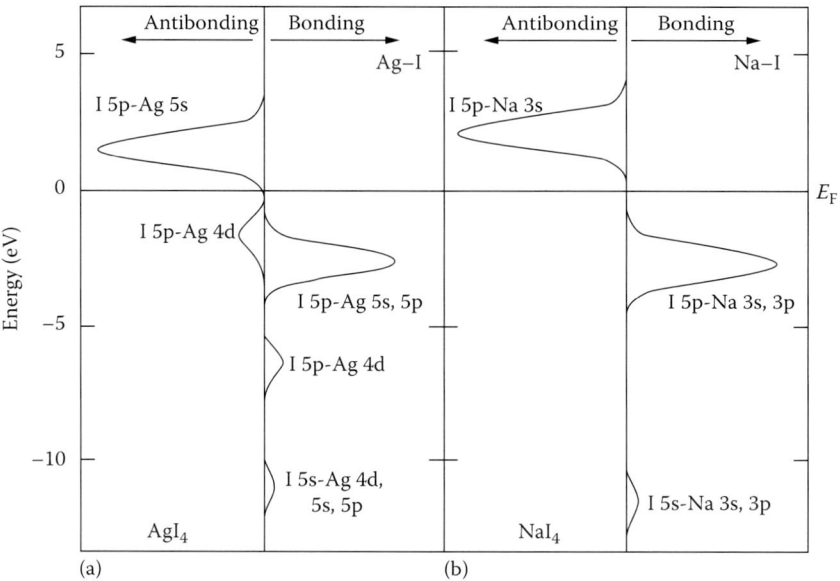

FIGURE 7.14 The nature the chemical bond differentiates a superionic conductor AgI from a non-superionic conductor NaI. In this figure, energy distribution of bond overlap populations is plotted. (From Kowada, Y., *J. Non-Cryst. Solids*, 232–234, 497, 1998.)

immediate conduction band just above it is the electronic energy gap or the band gap. The band gap is the most important characteristic of a solid state ionic material of the $A^N B^{8-N}$ type. Band structures of such solids are unique because of the involvement of 3d and 4d bands as in Ag- and Cu-halides and chalcogenides, or in reduced dimensionality (2D) solids as Ag-beta alumina and $LiCoO_2$ and the presence of 1D channels as in $LiFePO_4$.

The role of d-bands as seen in the discussion on complex energy gap is to introduce marginal covalency to the ionic bond. This same d-band connection introduces us to the electronic band structure of solid state ionic materials. Consider AgF for instance. The Ag^+ 4d valence band is found to lie above the F^- 2p valence band and is hence the highest occupied band! [7.26].

A comparison of bond overlap population (BOP) distributions for AgI4 and NaI4 clusters gives another insight into the role of d-bands in determining the band structure. *There is no anti-bonding contribution in the upper valence band of NaI4*. The nearly filled 4d band of Ag and the 5p band of I in the AgI_4 cluster *considerably overlap in the anti-bonding manner* to bring about repulsive interactions (Figure 7.14).

There exist *both* components of the anti-bonding and bonding in the BOP for AgI4, while there exists *only the bonding component* in the BOP for NaI_4. The crucial difference of the BOP of AgI_4 and NaI4 is in the existence of the *4d band* of Ag^+ for AgI_4. There is thus a crucial connection between BOP and Phillips ionicity.

The calculation of the band structure of a crystal involves only the values $V(G)$ of the Fourier coefficients of the potential at the reciprocal lattice vectors. Only a few coefficients suffice for the determination of the band structure to good accuracy.

Let us now consider the fundamental tight binding model or approximation as a prelude to the discussion on band structure.

Developed by Bloch, the tight-binding approximation (TBA) works to derive a major part of the total electron energy from the periodic potential. All atoms are sufficiently widely separated that there is very little overlap of the wave functions of the neighboring atoms. Thus the mutual interaction between neighboring electrons is weak. So the wave functions and energies of the crystal would be close to those of the individual atoms (Figure 7.15). As we shall see, this model is quite relevant to I–VII compounds such as AgI and its polymorphs. Following Dreselhaus [7.27], we shall briefly consider the tight binding approximation (TBA), which is particularly appropriate for the three superionic halides CuBr, CuI, and AgI.

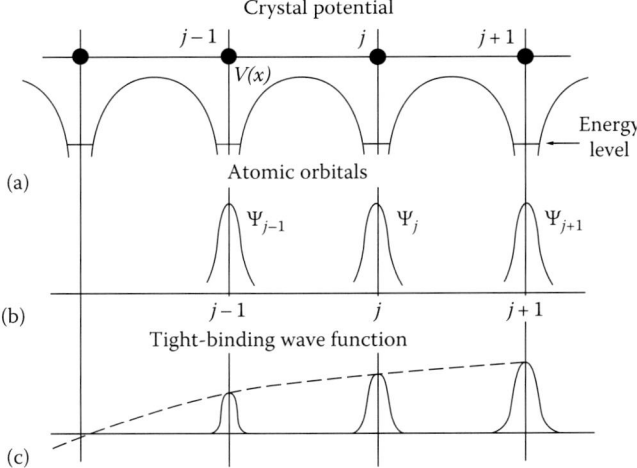

FIGURE 7.15 In the tight-binding approximation, the crystal potential (a) very closely resembles the atomic orbitals (b). Therefore, the tight-binding wave function (c) is almost the atomic orbital modulated by the periodic potential.

In TBA, the following points are assumed:

1. Energy eigenvalues and eigenfunctions are known for an electron in an isolated atom.
2. As atoms are brought together to form a nonmetallic solid, they remain sufficiently apart so that each electron can be assigned to a particular atomic site.
3. The periodic potential is approximated by a superposition of atomic potentials.
4. Perturbation theory can then treat the difference between the periodic potential and the atomic potential around which the electron is localized.

Let us now focus on the major features:

Let $\phi\,(\mathbf{r} - \mathbf{R_n})$ be the atomic wave function for an atom at a lattice position $\mathbf{R_n}$ measured with respect to the origin (Figure 7.16a). The Schrodinger equation for an isolated atom is

$$\left[-\left(\hbar^2/2m \right)\nabla^2 + U\left(r - R_n \right) - E^{(0)} \right]\phi\left(\mathbf{r} - R_n \right)\psi\left(r \right) = 0 \tag{7.37}$$

Here, $U(r - R_n)$ is the atomic potential and $E^{(0)}$ is the atomic eigenvalue (Figure 7.16a). As atoms are brought together to form the crystal, periodic potential $V(r)$ develops. Let $\psi(r)$ and $E(k)$ be the wave function and energy eigenvalue for the electron in the crystal:

$$\left[-\left(\hbar^2/2m \right)\nabla^2 + V\left(r \right) - E \right]\psi\left(r \right) = 0 \tag{7.38}$$

In TBA, we write $\psi(r)$ as a sum of atomic potentials

$$V\left(r \right) = \Sigma_n U\left(\mathbf{r} - \mathbf{R_n} \right) \tag{7.39}$$

If we ignore interatomic interactions, then each state has a degeneracy of N, the total number of atoms in the crystal. But the interatomic interactions lift this degeneracy. Now the energy eigenstates $E(\mathbf{k})$ in TBA for a non-degenerate s-state is simply

$$E(\mathbf{k}) = \left\langle k \left| H \right| k \right\rangle / \left\langle k | k \right\rangle \tag{7.40}$$

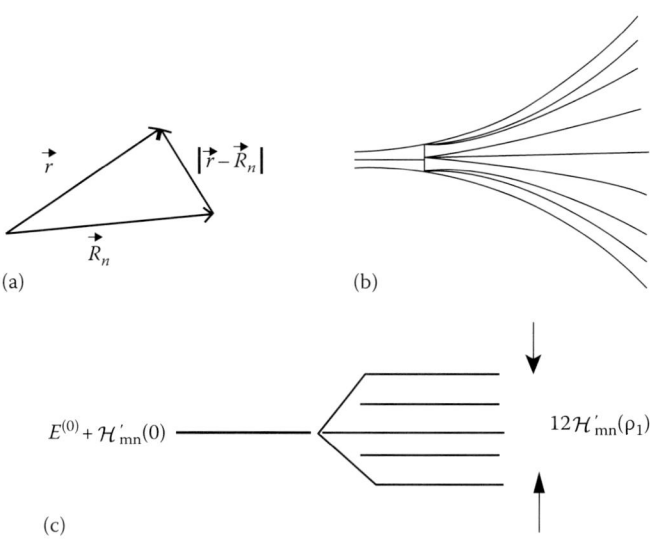

FIGURE 7.16 TBA schematics. (a) Vector diagram, (b) definition of ρ_{nm}, the distance between two atoms at R_m and R_n, and (c) relation between atomic levels and broadened levels in TBA.

Denominator is the normalization factor that normalizes the TBA wave function $\psi_k(r)$. The TBA Hamiltonian is written as

$$H = -(\hbar^2/2m)\nabla^2 + V(r) - \left\{ -(\hbar^2/2m)\nabla^2 + V(r) - U(\mathbf{r} - \mathbf{R_n}) + U(\mathbf{r} - \mathbf{R_n}) \right\} \qquad (7.41)$$

$$H = H_0 + H' \qquad (7.42)$$

where H_0 is the atomic Hamiltonian at site n

$$H_0 = -(\hbar^2/2m)\nabla^2 + U\left(r - R_n \right) \qquad (7.43)$$

More importantly, H' is the *difference* between the *actual* periodic potential at the lattice site and the *atomic* potential at the lattice site n. This is the *perturbation* in the TBA calculation. Construct the wave functions for the unperturbed system as a linear combination of atomic functions $\varphi_j(\mathbf{r} - \mathbf{R_n})$ labeled by quantum number j.

$$\psi_j\left(r \right) = \Sigma_{n=1}^{N} C_{1n}\phi_j\left(\mathbf{r} - \mathbf{R_n} \right) \qquad (7.44)$$

$\psi_j(r)$ is an eigenstate of the Hamiltonian satisfying the lattice periodic potential. Assume that $\psi_1(r)$ can be identified with a single atomic state φ_j (to be relaxed while dealing with degenerate levels). $\Psi_j(r)$ must satisfy the Bloch's theorem:

$$\psi_j\left(r + R_m \right) = e^{ik \cdot Rm}\psi_j\left(r \right) \qquad (7.45)$$

where R_m is an arbitrary lattice vector. Now the coefficients C_{1n} assume a special form. Substitute for $x\psi_j(r)$ from Equations 7.44 in 7.45 to get

$$\psi_j\left(r + R_m \right) = \Sigma_n C_{j,n+m}\phi\left(r - R_n \right) \qquad (7.46)$$

For the RHS of Equation 7.45, we have

$$e^{ik \cdot Rm} \psi_j(r) = \Sigma_n C_{j,n+m} e^{ik \cdot Rm} \phi(r - R_n) \tag{7.47}$$

Non-arbitrary coefficients $C_{1,n}$ relate $\psi_1(r)$ to $\phi_1(r - R_n)$ through

$$C_{j,n} = \xi_j e^{ik \cdot Rn} \tag{7.48}$$

Here, ξ_j is independent of n. Therefore,

$$\psi_{j,\mathbf{k}}(r) = \xi_j \Sigma_n \exp^{i\mathbf{k} \cdot \mathbf{Rn}} \phi_j(r - R_n) \tag{7.49}$$

where j labels the particular atomic state of degeneracy N and $\mathbf{k}$ is the quantum number for the translation operator labeling the Bloch state ψ_j, $\mathbf{k}(r)$.

Let us calculate TBA of s-bands (similar to p-bands) and drop j. Find the matrix elements of the Hamiltonian. Write

$$\left\langle \vec{k}' | \mathcal{H} | \vec{k} \right\rangle = |\xi|^2 \sum_{n,m} e^{-(\vec{k} \times \vec{R}_n - \vec{k}' \times \vec{R}_m)} \int_\Omega \phi^*(\vec{r} - \vec{R}_m) \mathcal{H} \phi(\vec{r} - \vec{R}_n) d^3 r \tag{7.50}$$

where the integration is over the crystal volume Ω. H has the periodicity of the lattice. So the only significant distance is $\mathbf{R}_n - \mathbf{R}_m = \rho_{nm}$. Now the integral of Equation 7.50 is

$$\left\langle \vec{k}' | \mathcal{H} | \vec{k} \right\rangle = |\xi|^2 \sum_{\rightarrow} e^{i(\vec{k} - \vec{k}') \cdot \vec{R}_m} \sum_{\rightarrow} e^{i\vec{k}\vec{\rho}_{nm}} \mathcal{H}_{mn}(\vec{\rho}_{nm}) \tag{7.51}$$

The matrix element $\mathcal{H}_{mn}(\rho_{nm})$ is

$$\mathcal{H}_{mn}(\vec{\rho}_{nm}) = \int_\Omega \phi^*(\vec{r} - \vec{R}_m) \mathcal{H} \phi(\vec{r} - \vec{R}_m - \vec{\rho}_{nm}) d^3 r = \int_\Omega \phi^*(\vec{r}') \mathcal{H} \phi(\vec{r}' - \vec{\rho}_{nm}) d^3 r' \tag{7.52}$$

where the integral depends only on ρ_{nm}. The first term in Equation 7.51 is

$$\sum_{\vec{R}_m} e^{i(\vec{k} - \vec{k}') \cdot \vec{R}_m} = \delta_{\vec{k}',\vec{k}+\vec{G}} N \tag{7.53}$$

where $\mathbf{G}$ is the reciprocal lattice vector. Restricting wavevector $\mathbf{k}$ to lie within the first Brillouin zone, one can count the electronic states according to periodic boundary conditions on a crystal of dimension d on a side:

$$k_i d = 2\pi m_i \tag{7.54}$$

where m_i is an integer in the range $1 \le m_i < N_i$, where $N_i \approx N^{1/3}$. Note that the maximum values assumed by m_i and k_i are N_i and $2\pi/a$ at the BZ boundary since $N_i/d = 1/a$. Thus, k and k' must both lie within the first BZ and cannot differ by any other reciprocal lattice vector other than $G = 0$.

The form of matrix elements for $\mathcal{H}$ (and $\mathcal{H}_0, \mathcal{H}'$ too) is

$$\left\langle \vec{k}' | \mathcal{H} | \vec{k} \right\rangle = |\xi|^2 N \delta_{\vec{k},\vec{k}'} \sum_{\vec{\rho}_{nm}} e^{i\vec{k} \cdot \vec{\rho}_{nm}} \mathcal{H}_{mn}(\vec{\rho}_{nm}) \tag{7.55}$$

leading to the result

$$E(\vec{k}) = \frac{\langle \vec{k}|\mathcal{H}|\vec{k}\rangle}{\langle \vec{k}|\vec{k}\rangle} = \frac{\sum_{\vec{\rho}_{nm}} e^{i\mathbf{k}\cdot\vec{\rho}_{nm}} \mathcal{H}_{mn}(\vec{\rho}_{nm})}{\sum_{\vec{\rho}_{nm}} e^{i\mathbf{k}\cdot\vec{\rho}_{nm}} \mathcal{S}_{mn}(\vec{\rho}_{nm})} \tag{7.56}$$

In Equation 7.55,

$$\langle \vec{k}'|\vec{k}\rangle = |\xi|^2 \delta_{\vec{k},\vec{k}'} N \sum_{\vec{\rho}_{nm}} e^{i\mathbf{k}\cdot\vec{\rho}_{nm}} \mathcal{S}_{mn}(\vec{\rho}_{nm}) \tag{7.57}$$

where the matrix element $\mathcal{S}_{mn}(\rho_{nm})$ measures the *overlap of atomic functions* on different sites. The overlap integral which is *central to TBA* will be close to 1 when $\rho_{nm} = 0$ and falls off rapidly as ρ_{nm} increases. Equation 7.42 now becomes

$$\mathcal{H}_{mn} = E^{(0)} \mathcal{S}_{mn}(\vec{\rho}_{nm}) + \mathcal{H}'_{mn}(\vec{\rho}_{nm}) \tag{7.58}$$

Leading to the general result of TBA:

$$E(\vec{k}) = E^{(0)} + \frac{\sum_{\vec{\rho}_{nm}} e^{i\mathbf{k}\cdot\vec{\rho}_{nm}} \mathcal{H}'_{mn}(\vec{\rho}_{nm})}{\sum_{\vec{\rho}_{nm}} e^{i\mathbf{k}\cdot\vec{\rho}_{nm}} \mathcal{S}_{mn}(\vec{\rho}_{nm})}. \tag{7.59}$$

The second term of Equation 7.59 is supposed to be small, which is justified if the overlap of the atomic wave functions is small. According to the distance between site m and site n, the sum over ρ_{nm} is classified as (1) zero distance, (2) nearest-neighbor distance, and (3) next-nearest neighbor distance.

$$\sum_{\vec{\rho}_{nm}} e^{i\mathbf{k}\cdot\vec{\rho}_{nm}} \mathcal{H}'_{mn}(\vec{\rho}_{nm}) = \mathcal{H}'_{nn}(0) + \sum_{\vec{\rho}_1} e^{i\mathbf{k}\cdot\vec{\rho}_{nm}} \mathcal{H}'_{mn}(\vec{\rho}_{nm}) + \cdots \tag{7.60}$$

Here, the first term leads to a constant additive energy independent of **k**. The sum over the nearest neighbors gives a *k*-dependent perturbation. This is of vital importance in calculation of the energy band structure. $\mathcal{H}'_{00}(0)$ and the nearest-neighbor terms are of comparable magnitude. While treating the denominator of the perturbation sum of Equation 7.58, we must sum

$$\sum_{\vec{\rho}_{nm}} e^{i\mathbf{k}\cdot\vec{\rho}_{nm}} \mathcal{S}_{mn}(\vec{\rho}_{nm}) = \mathcal{S}_{nn}(0) + \sum_{\vec{\rho}_1} e^{i\mathbf{k}\cdot\vec{\rho}_{nm}} \mathcal{S}_{mn}(\vec{\rho}_{nm}) + \cdots \tag{7.61}$$

Here, the leading term $\mathcal{S}_{nm}(0) \approx 1$ and the overlap integral $\mathcal{S}_{mn}(\rho_{nm})$ over nearest-neighbor sites is small and can be neglected to the lowest order compared to unity. The nearest neighbor term of Equation 7.61 is of comparable magnitude to the next-nearest neighbor terms arising from $\mathcal{H}_{mn}(\rho_{nm})$ in Equation 7.60.

Let us apply the TBA method to explicitly evaluate $E(k)$ for the bcc lattice, which is relevant to superionic conductors. Assume that overlap of atomic potentials on neighboring sites is sufficiently weak so that only nearest-neighbor terms are important in the sum $\mathcal{H}'_{mn}$ and only the leading term in the sum of $\mathcal{S}_{mn}$.

The eight ρ_1 vectors for the nearest-neighbor distances in the bcc structure are $(\pm a/2, \pm a/2, \pm a/2)$. Thus, there are eight exponential terms, which combine pairs such as

$$\left[\exp(ik_x a/2)\exp(ik_Y a/2)\exp(ik_Z a/2)\right] + \left[\exp(-ik_x a/2)\exp(ik_Y a/2)\exp(ik_Z a/2)\right] \tag{7.62}$$

yielding

$$2\cos\left(k_x a/2\right)\exp\left(ik_y a/2\right)\exp\left(ik_z a/2\right) \tag{7.63}$$

Thus, we obtain the following dispersion relation for the bcc structure:

$$E\left(k\right) = \text{constant} + 8\mathcal{H}'_{mn}\left(\rho_1\right)\cos\left(k_x a/2\right)\cos\left(k_y a/2\right)\cos\left(k_z a/2\right) + \cdots \tag{7.64}$$

where $\mathcal{H}'_{mn}\left(\rho_1\right)$ is the matrix element of the perturbation Hamiltonian taken over nearest-neighbor atomic orbitals.

The case of simple cubic structure is particularly simple but illustrates the features of dispersion relation very well. In this case, there are $6\rho_1$ vectors (try it) and the $E(k)$ is

$$E(\vec{k}) = E^{(0)} + \mathcal{H}'_{nn}(0) + 2\mathcal{H}'_{mn}(\vec{\rho}_1)\left[\cos k_x a + \cos k_y a + \cos k_z a\right] + \cdots \tag{7.65}$$

Check that Equation 7.65 satisfies three properties caharacteristic of energy eigenvalues in such periodic structures:

1. Periodicity in k space under translation by a reciprocal lattice vector $k \rightarrow k + G$.
2. $E(k)$ is an even function of k.
3. $\partial E/\partial k = 0$ at the BZ boundary.

In Equation 7.65 for $E(k)$, the maximum value in the brackets is ± 3. Thus, for a simple cubic lattice in the TBA, we obtain a bandwidth of $12\mathcal{H}'_{mn}\left(\rho_1\right)$ from the nearest-neighbor interactions (see Figure 7.16c).

$E(k)$ is different for simple cubic, bcc, and fcc lattices because of different locations of nearest neighbors in the three lattices. Thus, TBA takes explicit account of the crystal structure.

7.4.3 Electronic Structure in the Tight-Binding Approximation

Smith [7.28] has applied the TBA suitably modified to take account of the 4d electrons of Ag to derive the energy band structures of AgCl and the four polymorphs of AgI: (1) fcc or rocksalt phase stable only at high pressures, (2) zincblende or γ AgI, (3) bcc or α-phase, and 4. wurtzite or β-phase. These are shown along with the density of states in Figures 7.17 through 7.20.

Figure 7.21 compares the band structures and density of states of AgX (X = F, Cl, Br).

Akopyan [7.30] has considered the optical spectroscopy of superionic conductors. The BZ symmetry of both zincblende and wurtzite structures of AgI allows the p–d mixing throughout the zone. This is in contrast to the rocksalt structure of AgCl and AgBr where mixing cannot occur at $k = 0$. The p–d hybridization in AgI is due to the Γ_{15} components of the d-states, while the $\Gamma_{12}-$ components remain isolated. The degree of the p–d hybridization estimated from the spin–orbit splitting of E_0 excitons is 32% for Ag-4d states and 68% for I-5p states. The changes in the valence band in AgI associated with the phase transition to the bcc (α-) phase lead to a broadening of the I-50 band by ~0.3 eV with an insignificant effect on Ag-4d states. The temperature dependence of the p-band width is shown in Figure 7.25a. The changes in the photoemission spectra are ascribed to the breakdown of symmetry selection rules of the p–d hybridization upon Ag$^+$ ion sublattice melting. In the α-phase, p–d admixture is allowed for the entire d-band.

Figure 7.22a is a spectacular display of the p–d hybridization collapse at the superionic phase transition. Figure 7.22b shows the reciprocal susceptibility determined by X-band EPR spectroscopy of AgI powders produced by mechanochemical reaction showing an abrupt sharp rise at the phase transition echoing the changes in the photoemissions spectra.

Collapse at 147°C [7.30] (a) Temperature dependence of reciprocal susceptibility $(1/\chi)$ obtained from EPR of mechanochemical reaction synthesized AgI powder depicting the sharp transition at ~420 K [7.31]. The band structure and density of states of β Ag$_2$S are shown in Figure 7.23. Here too the dominating effect of Ag(4d) and S(3p) hybridization may be seen.

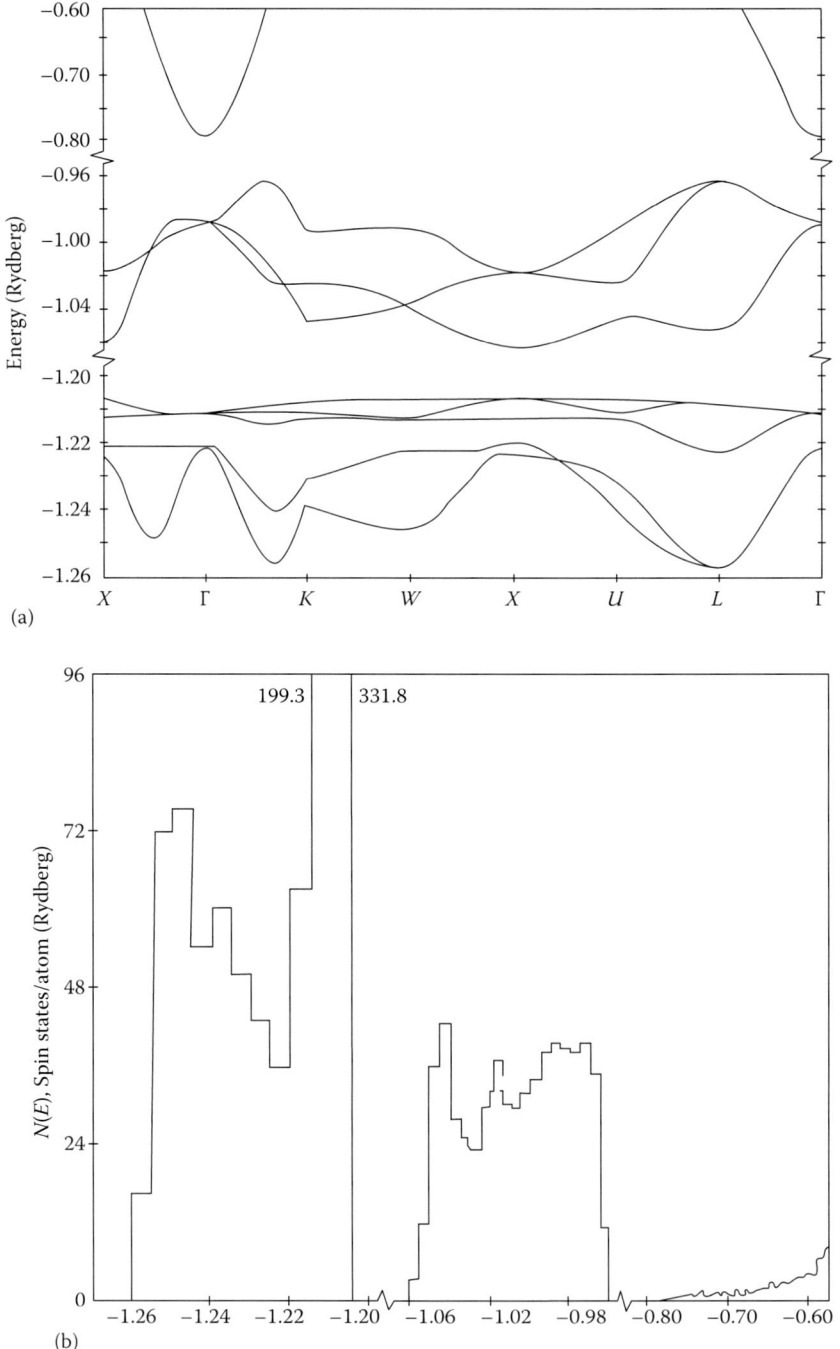

FIGURE 7.17 (a) Tight binding approximation (TBA) band structure of fcc AgI. (b) Density of states of fcc AgI corresponding to 2500 random wave vectors in the 1/48th section of the Brillouin zone.

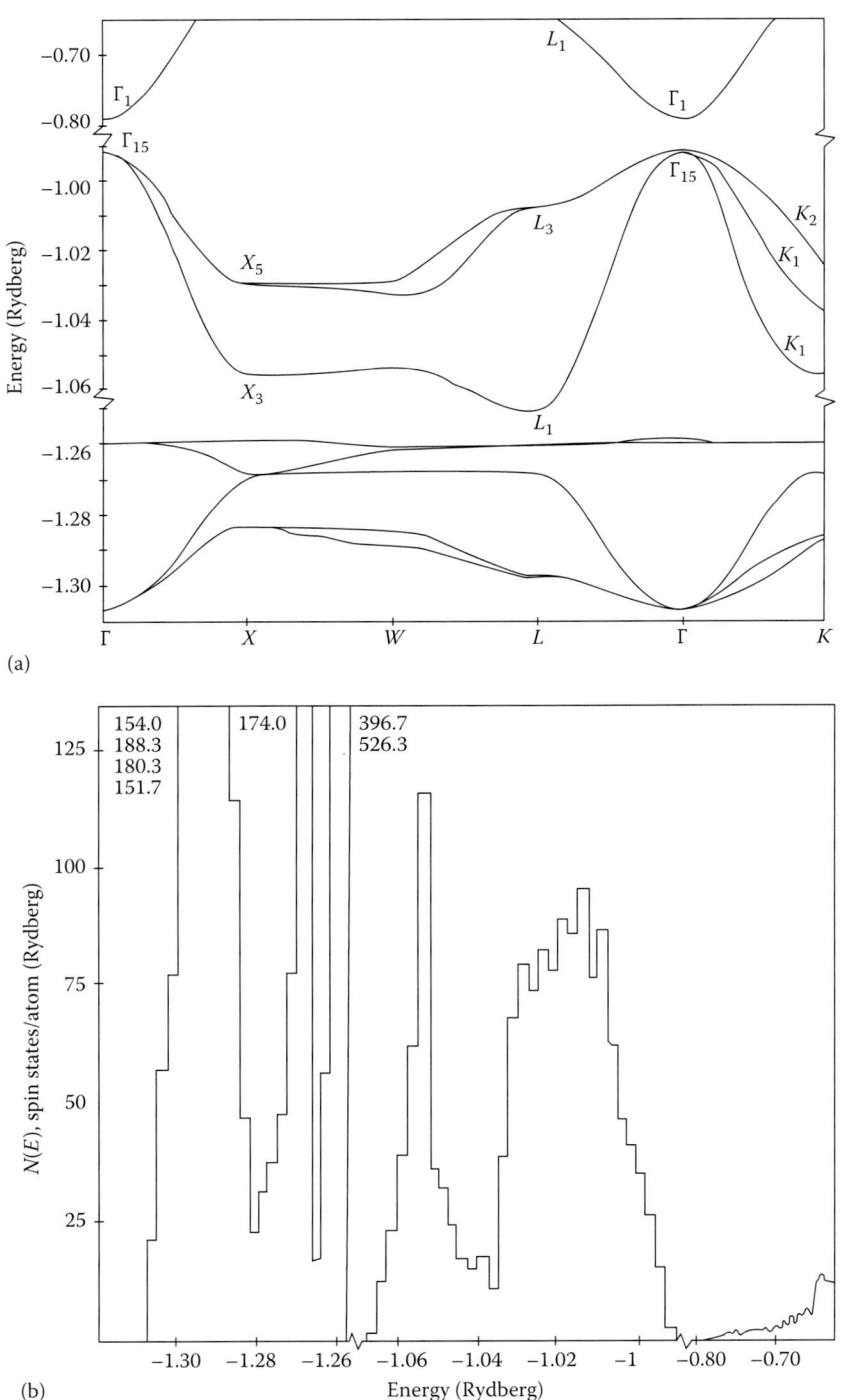

(a)

(b)

FIGURE 7.18 (a)TBA band structure of zincblende (γ) AgI and (b) density of states of γ AgI corresponding to 1500 random wave vectors in the 1/48th section of the Brillouin zone.

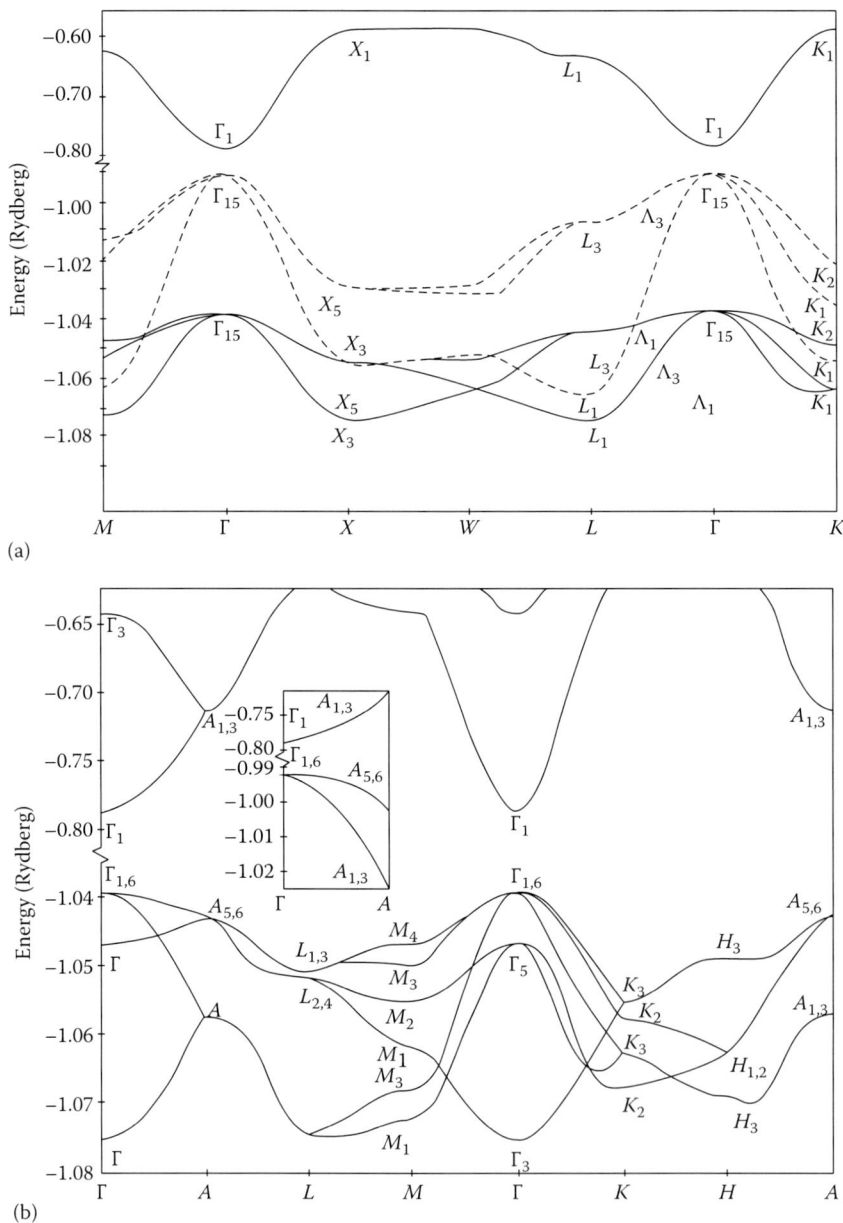

FIGURE 7.19 (a) s- and p-bands of β AgI. Dashed curves: Ag 4d interactions included, full curves: Ag 4d electrons omitted. (b) TBA band structure of wurtzite β AgI, neglecting Ag 4d electrons. Inset: the conduction band Γ_1-$A_{1,3}$, and the valence band Γ_6-$A_{5,6}$ Γ_1-$A_{1,3}$ energy levels arising when the silver 4d electrons are included.

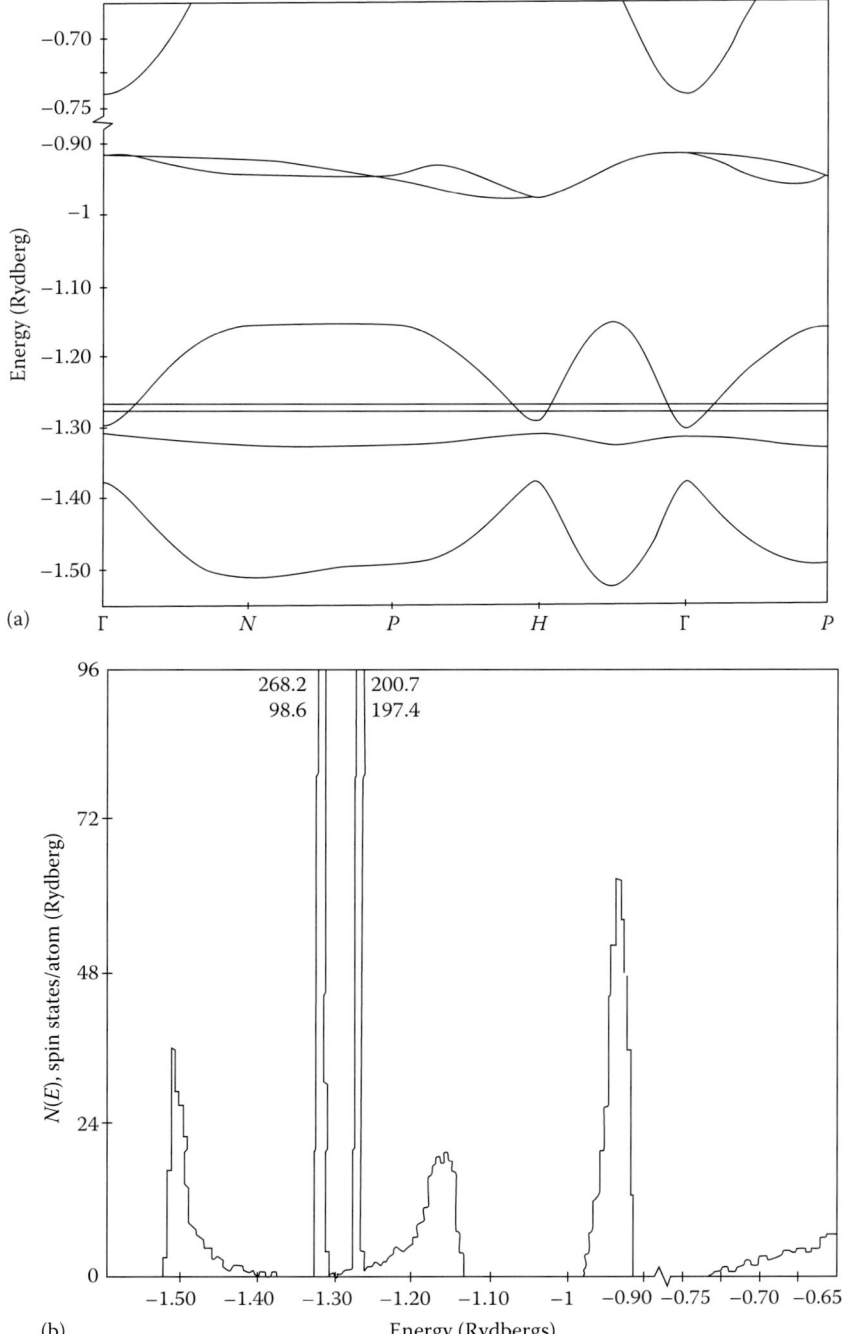

FIGURE 7.20 (a) TBA band structure of α-AgI. (b) Density of states of α-AgI corresponding to 2500 wave vectors in the 1/24th section of the bcc Brillouin zone.

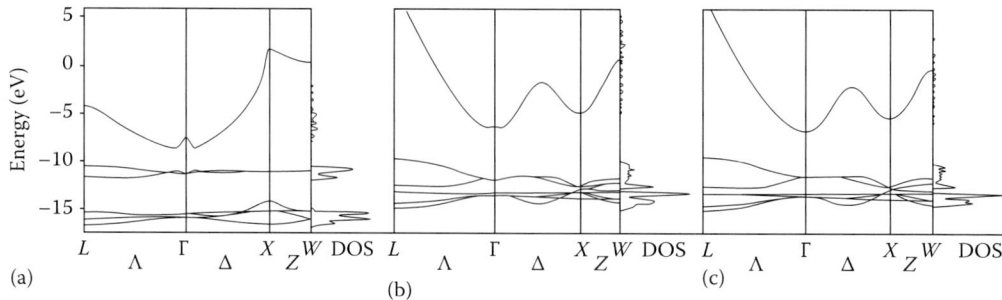

FIGURE 7.21   Band structures and density of states (DOS) of (a) AgF, (b) AgCl, and (c) AgBr. Note the differences between M-X distances in the molecule and crystal to be compared with the band gap (eV) in that order: AgF 198 pm, 247 pm, 2.8 eV; AgCl 228, 277, 3.25; AgBr 239, 289, 2.69. (From Glaus, S. and Calzaferri, G., *Photochem. Photobiol. Sci.*, 2, 398, 2003.)

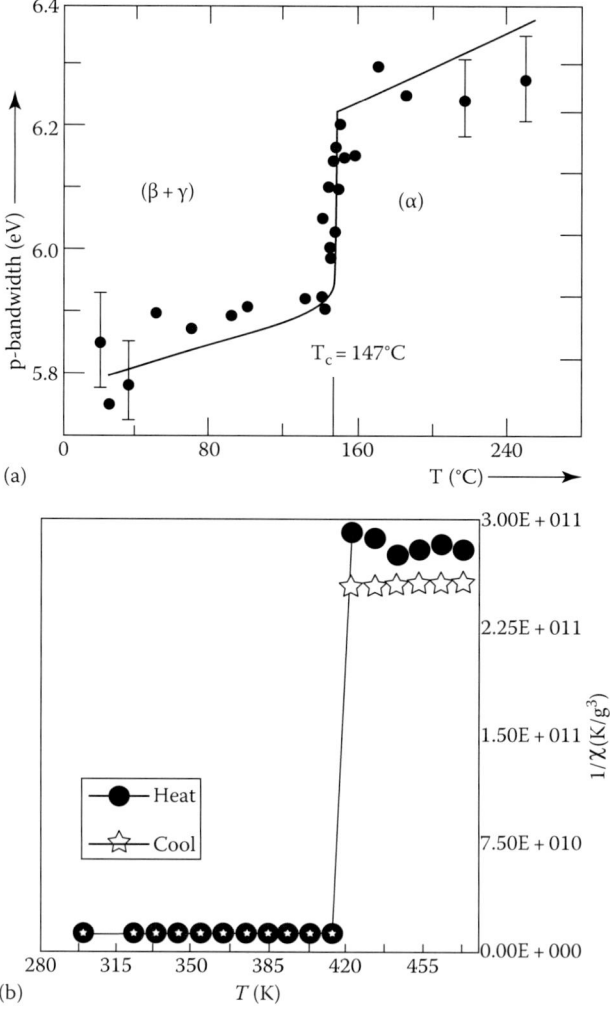

FIGURE 7.22   (a) A spectacular display of the ZB/W to bcc structural phase transition in AgI basically triggered by p–d hybridization. (From Akopyan, I.Kh. et al., *Phys. Stat. Solidi* (a), 119, 363, 1990.) (b) Temperature dependence of reciprocal susceptibility ($1/\chi$) obtained from EPR of mechanochemical reaction synthesized AgI powder depicting the sharp transition at ~420 K. (From Bharathi Mohan, D. and Sunandanam, C.S., *J. Phys. Chem. B*, 110, 4569, 2006.).

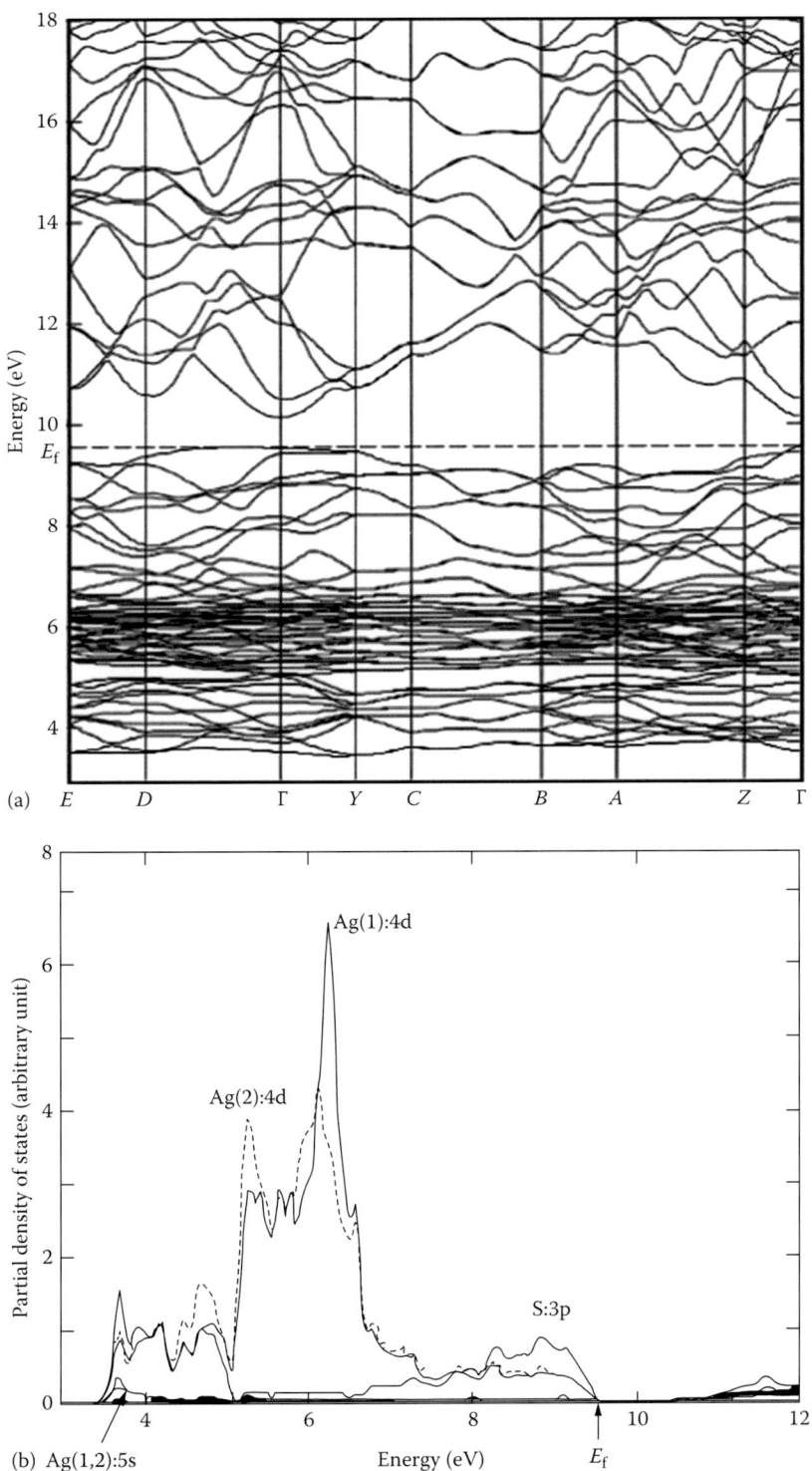

FIGURE 7.23 (a) Band structure of beta Ag$_2$S and (b) density of states. (From Kashida, S. et al., *SSI*, 158, 167, 2003.)

Kikuchi et al. [7.33] have studied the electronic properties of silver and copper tellurides Ag_2Te and Cu_2Te with antifluorite structure using first-principles DFT calculations adopting the linearized augmented plane wave method. In order to highlight the strength of the coupling for the d-bands and p-bands, they selectively shifted downward the d-band from Ag states in Ag_2Te or the p-band from the Te p states in Cu_2Te. The conclusion is that the two tellurides have remarkably different degree of p–d hybridization, with the d-bands of Ag ions much more weakly coupled with the p-bands of Te ions (Figure 7.24).

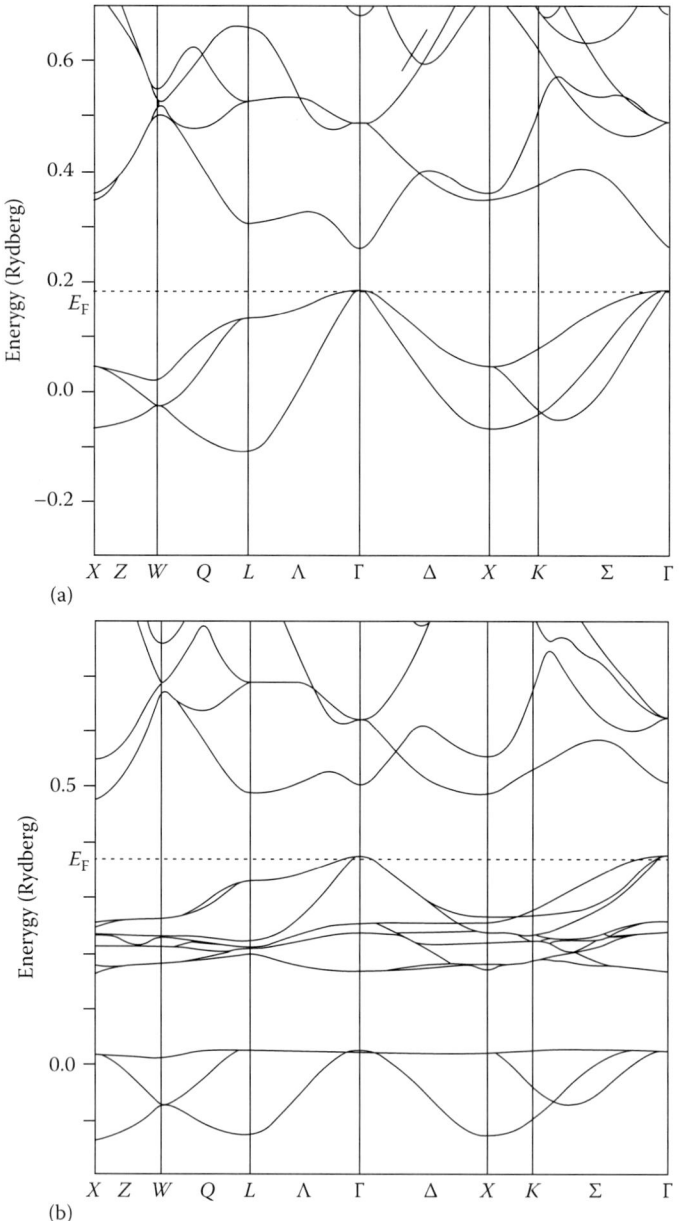

FIGURE 7.24 Band structures of (a) Ag_2Te and (b) Cu_2Te calculated at the equilibrium lattice constants. (From Kikuchi, H. et al., *JPCM*, 10, 11439, 1998.)

Ono et al. [7.34a,b,c] performed band calculations of AgC, AgBr, γ-AgI, and γ-CuX (X = Cl, Br, I) using the linear combination of atomic orbitals (LCAO) model outlined earlier. The general observations are as follows: the upper region of the valence band of AgX is mainly occupied by the p-band of X$^-$ while that of CuX is mainly occupied by the d-band of X$^-$! A selective downward shift of the Ag$^+$ d-bands for AgX and p-band of X$^-$ for CuX (as done before) highlights the strength of the d- and p-band coupling in CuX. Let λ be the average value of d-band in AgI for example. In another band calculation, the energy value of d-band was downshifted by δ and the average value of new d-band is taken as λ'. The actual shift is given by $\lambda - \lambda'$. In AgX, the two bands are easily separable that the d-band of Ag$^+$ preserved its localized nature from hybridization. On the other hand, CuX gives rise to strong admixture of the d-band of Cu$^+$ and the p-band of X$^-$.

These results agree with those of Kikuchi et al. δ is related to δ_0 as

$$\delta = \left(k_{p\text{-}d}\right)^{-1} \delta_0 \tag{7.66}$$

where k_{p-d} is the strength of p–d hybridization. k_{p-d} is maximum for CuI and decrease in the order CuBr, CuCl, AgCl, AgBr, and AgI. Tracing it in the reverse order, it corresponds to the order of magnitude of observed ionic conductivity in the low temperature phase (γ):

$$\sigma\left(AgI\right) > \sigma\left(AgBr\right) > \sigma\left(AgCl\right) > \sigma\left(CuCl\right) > \sigma\left(CuBr\right) > \sigma\left(CuI\right)$$

The most interesting conclusion is that the activation energy for ionic conduction in this series of compounds has an inverse correlation with k_{p-d} (Figure 7.25). Smaller degree of hybridization leads to a lower activation energy for ion migration. Thus, there is a fundamental connection between electronic structure and ionic conductivity.

In the next section, we discuss the formalism and application of DFT, which has proved to be a very versatile tool for both materials and process evaluation in solid state ionics.

7.5 Density Functional Theory of Electronic Structure of Superionic Conductors

The electron density provides, in principle, *all the information* of a many-electron wavefunction of a multi-electron atom either free or present in a molecule or cluster or a nanocrystal or a bulk crystal. This was the realization of Hohenberg and Kohn who along with Sham formulated the first-principles or *ab initio* approach to the band structure problem using *density functionals* instead of single-electron wavefunctions as in Schrodinger's approach. What is a density functional? It is defined as

$$n(r) = \left\langle \Psi \left| \Sigma_{i=1}^{N} \delta\left(r - R_1\right) \right| \Psi \right\rangle \tag{7.67}$$

$$= N \int dr_1 \ldots dr_N \Psi^*\left(r_1, r_2, \ldots, r_N\right) \delta\left(r - r_i\right) \Psi\left(r_1, r_2, \ldots, r_N\right) \tag{7.68}$$

Suppose we know the density of the ground state of a many-electron system. Then, we can deduce from it the *external* potential in which the electrons reside up to an overall constant.

In what ways *two* many-electron problems *differ*? (1) external potential U and (2) number of electrons that reside in the potentials. *Both* are determined by electron density. Thus, the density *completely* determines the many-body problem. This is the essence of the DFT. Indeed DFT focuses on quantities in the real, 3D coordinate space principally on the electron density of the ground state. Other quantities of great interest are as follows: (1) the exchange correlation hole density $n_{xc}(r, r')$ describes how the presence of an

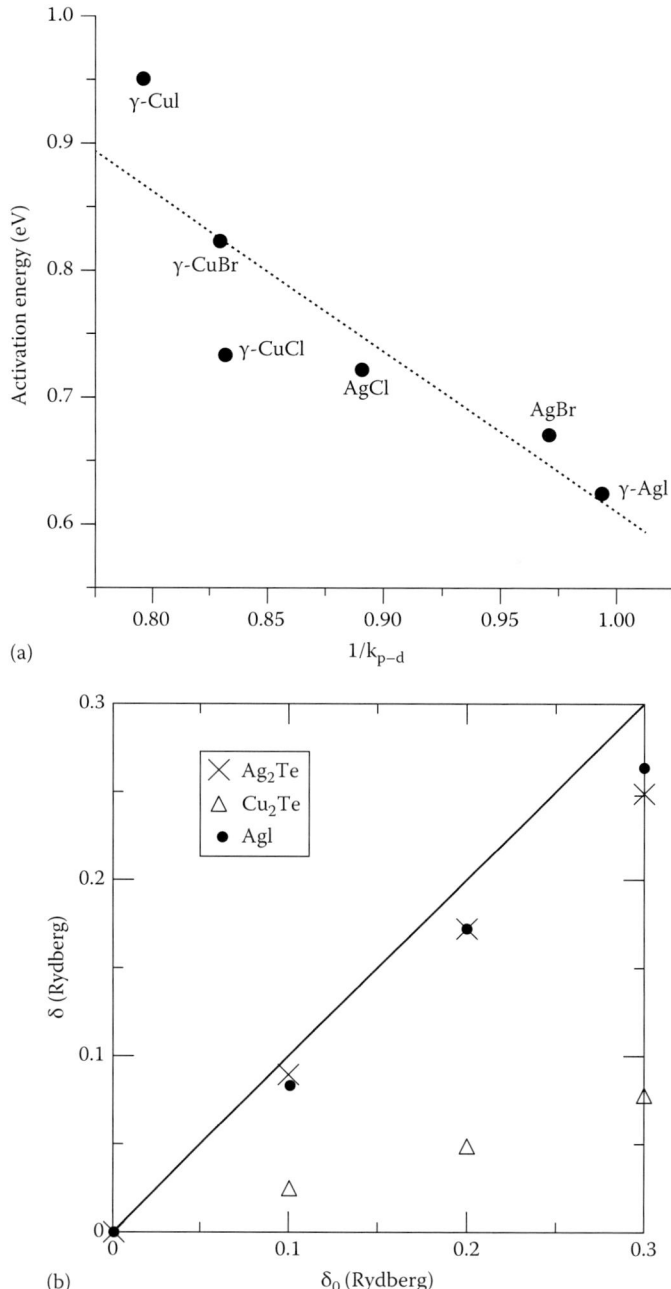

FIGURE 7.25   (a) Activation energy for ion conduction versus $1/k_{p-d}$; (b) Actual shift δ vs. δ_0 defined before. (From Ono, S., et al., *Solid State Ionics*, 139, 249, 2001.)

electron at the point r depletes the total density of the other electrons at point r'. (2) The linear response function $\chi(r, r', \omega)$ describes the change of total density at the point r due to a perturbing potential at the point r' with frequency ω. Quantities 1 and 2 are physical, independent of representation and easily visualizable even for very large systems. For periodic solids, it is referred to as the standard model [7.35a]. Beginners are advised to refer [7.35b].

What about the complete quantum mechanical wavefunction? It needs N variables for its description! See Equation 7.68 given earlier. The claim of the DFT is proved by *Reductio ad absurdum*, a form of logic that starts by assuming the opposite and goes on to show that it is absurd.

So how does the DFT arise? Suppose we have many different solid systems, say, AgI, CuBr, and CuI among other solid state ionic materials. They all consist of electrons (s, p, and d) interacting with one another through Coulomb potential, moving in potential U and obeying Schrodinger's equation.

If in each case we are given the charge density as a function of space, then in principle, one can deduce U; solve for all properties of the system.

Let us consider ground state energy E and kinetic energy T as functionals of the density. So, we can write

$$E\left[n\right] = T\left[n\right] + U\left[n\right] + U_{ee}\left[n\right] \tag{7.69}$$

where n means $n(r)$, a function of space, and T, U, and U_{ee} are Coulomb interaction between electrons—an important component of $E[n]$—which are all functions of n. In mathematical language, a functional is just a function of a function. If we can find the functional $E[n]$, then the true ground state density $n(r)$ minimizes it. Subject only to the constraint

$$\int dr\, n\left(r\right) = N \tag{7.70}$$

One can write the energy functional E as

$$E\left[n\right] = \int dr\, n\left(r\right)U\left(r\right) + F_{HK}\left[n\right] \tag{7.71}$$

where

$$F_{HK} = T\left[n\right] + U_{ee}\left[n\right] \tag{7.72}$$

(H and K stand for Hohenberg and Kohn).

Note that F_{HK} *does not* depend upon potential $U(r)$. Therefore, it is a *universal* functional for *all* systems of N particles.

Let us now focus on $F[n]$. Define a functional $F[n]$ as the minimum over all wavefunctions that produce density $n(r)$ of F:

$$F[n] \equiv \min_{\Psi \to n}\langle\Psi|T + U_{ee}|\Psi\rangle \tag{7.73}$$

This F can be defined even if U does not exist! That would produce density n for *some* quantum mechanical ground state.

How then does one find the ground state E_0 of a many-body system?

$$E_0 = \min_{\Psi}\langle\Psi|T + U + U_{ee}|\Psi\rangle \tag{7.74}$$

which means to minimize all wave functions Ψ that produce n and then to minimize all densities.

$$E_0 = \min_n \left[\min_{\Psi \grave{a} n} \langle \Psi | T + U + U_{ee} | \Psi \rangle \right] \tag{7.75}$$

$$= \min_n \left[\min_{\Psi \grave{a} n} \langle \Psi | T + U_{ee} | \Psi \rangle + \grave{o} U(r) n(r) dr \right] \tag{7.76}$$

Because U depends on the density

$$E_0 = \min_n \left[F[n] + \int U(r) n(r) dr \right] \tag{7.77}$$

$$\equiv \min_n E[n] \tag{7.78}$$

Observe that defining E and accommodating degenerate ground states are done simultaneously. This formalism has been used to compute all properties of solids including thermodynamic, structural, mechanical, and optical properties. An important quantity of relevance to solids state ionics—the dielectric functions—has been computed as part of band structure calculations. We shall illustrate the DFT method as applied to $LiFePO_4$, an important cathode material for Li-ion batteries.

Cathode materials are the most critical challenge for the large-scale application of Li-ion batteries in electric vehicles and for the storages of electricity (as would be discussed in the next chapter). The first-principles calculations play an important role in the development and optimization of novel cathode materials. First-principles calculations of energy, volume change, band-gap, phase diagram, and Li-ion transport mechanism of cathode materials focus on the design of such materials.

In Chapter 1, we have shown the 3D crystal structure of $LiFePO_4$. It belongs to the orthorhombic system. With a space group *Pnma*, it has four formula units per unit cell. The four Fe atoms in this structure are of particular interest. From the position of $Fe_1(x, y, z)$, the other three Fe atoms are located: Fe_2 is at $(-x + 1/2, -y, z - 1/2)$, while Fe_3 and Fe_4 have coordinates $(-x, y + 1/2, -z + 1/2)$ and $(x + 1/2, -y + 1/2, -z + 1/2)$. Cut (010) oriented $LiFePO_4$ through (000) and (0½0). Through this operation, Li and O atoms are exposed at the surface. However if we cut the same (010) through (0¼0) and (0¾0), all the three atoms Fe, P, and O will be exposed! Why are we focusing on the (010) surface? The electronic structure of the (010) surface of $LiFePO_4$ is quite similar to that of the bulk or the macroscopic crystal. Furthermore, a crystal with (010) surface is most easily produced. The system stability depends on the exposed atoms. Adsorption occurs only at the most stable cleaving surface. Relaxation of all these possible cut surfaces determines surfaces with the lowest surface energy that enable carbon adsorption calculations to be done. The importance of carbon-coated $LiFePO_4$ materials comes from its good rate capability and stability as a Li-battery material (batteries are discussed in Chapter 8). The crucial question is what is the relation between the coating C layer and the interface of the $LiFePO_4$ material?

A computer calculation method of the first-principles pseudopotentials is used to explore the important properties of the C-coated $LiFePO_4$ material [7.36]. Geometric structure and electronic characteristics of the (010) surface enable band gap, conductive properties, and formation energy between the adsorption surface and bulk material to be obtained. Apart from revealing the mechanism of adsorption, one can explain how the change in the microscopic structure can improve the electrochemical performance of $LiFePO_4$ (Figures 7.26 through 7.28)

This is to be compared with the DOS of subsurface and of bulk. Observe the significant differences. To give a flavor of the sensitivity of the DFT technique to local variations in DOS, look at Figure 7.29.

The calculations were performed using the CASTEP program developed by Segall et al. [7.37]. The band gap of the surface is 1.23 eV less than the 2.38 eV obtained for the bulk. Three new bands occur at the Fermi level. The DOS are obtained for three cases of bulk, sub-surface, and surface. A large difference arises between bulk and surface, while that between surface and subsurface is little. The important

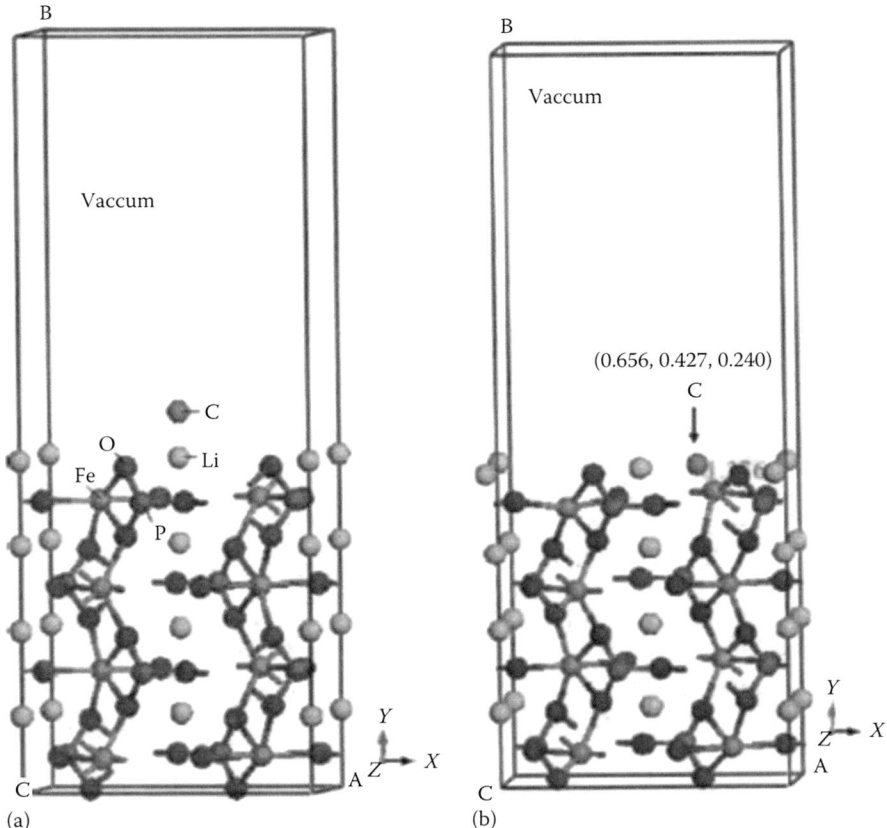

FIGURE 7.26 Slab model of (a) the initial C-adsorbed (010) surface and (b) relaxed structure. (From Zhang, P. et al., *J. Alloys Comp.*, 540, 121, 2012.)

results on PDOS or partial density of states reveal that Fe_3 atoms bond with six O atoms, resulting in FeO_6 octahedron found in the bulk sample. Significantly, Fe4 terminates at the surface with five atoms to form FeO_5 "octahedron" because of the broken $Fe–O_3$ bond.

C-adsorbed (010) surface reveals a band gap of 1.55 eV as against the value 1.23 eV for the unadsorbed (010) surface. New bands created at the Fermi level arise from the effects on the d orbitals which exist only in the Fe atom. Surface (010) forms only five Fe–O bonds while Fe in bulk form six Fe–O bonds. Significantly, the breakage of Fe–O bonds at the surface increases the activity of Fe and leads to (1) a dramatic *increase* in the number of electrons in the Fe_3 d orbital and (2) formation of new bonds at the Fermi level. A nanoeffect is inferred: smaller the particles, the more Li atoms exposed on the terminal surface. A transfer of Li between particles is thus favored.

An important result is that PO_4 tetrahedron is preserved if the surface is cut through (000).

Fe atoms in $LiFePO_4$ are of high spin (weak field). Crystal field theory predicts five d orbitals from different degenerate orbitals in the non-spherically symmetric ligand field. The spin configuration is $3d^6$ $4t_{2g}2e_g$. In t_{2g} configuration, spins of three electrons are ↑↑↑ and one is ↓. These form bonds with p-orbitals of O_3 alone. e_g has two electrons with spins ↑↑, which bond with p-orbitals of O_1 and O_2.

Lowered band gap and newly formed bands favor the transfer of electron from valence band through forbidden band to the conduction band.

Therefore, the use of very small particles improves electrochemical properties through better transfer of electrons.

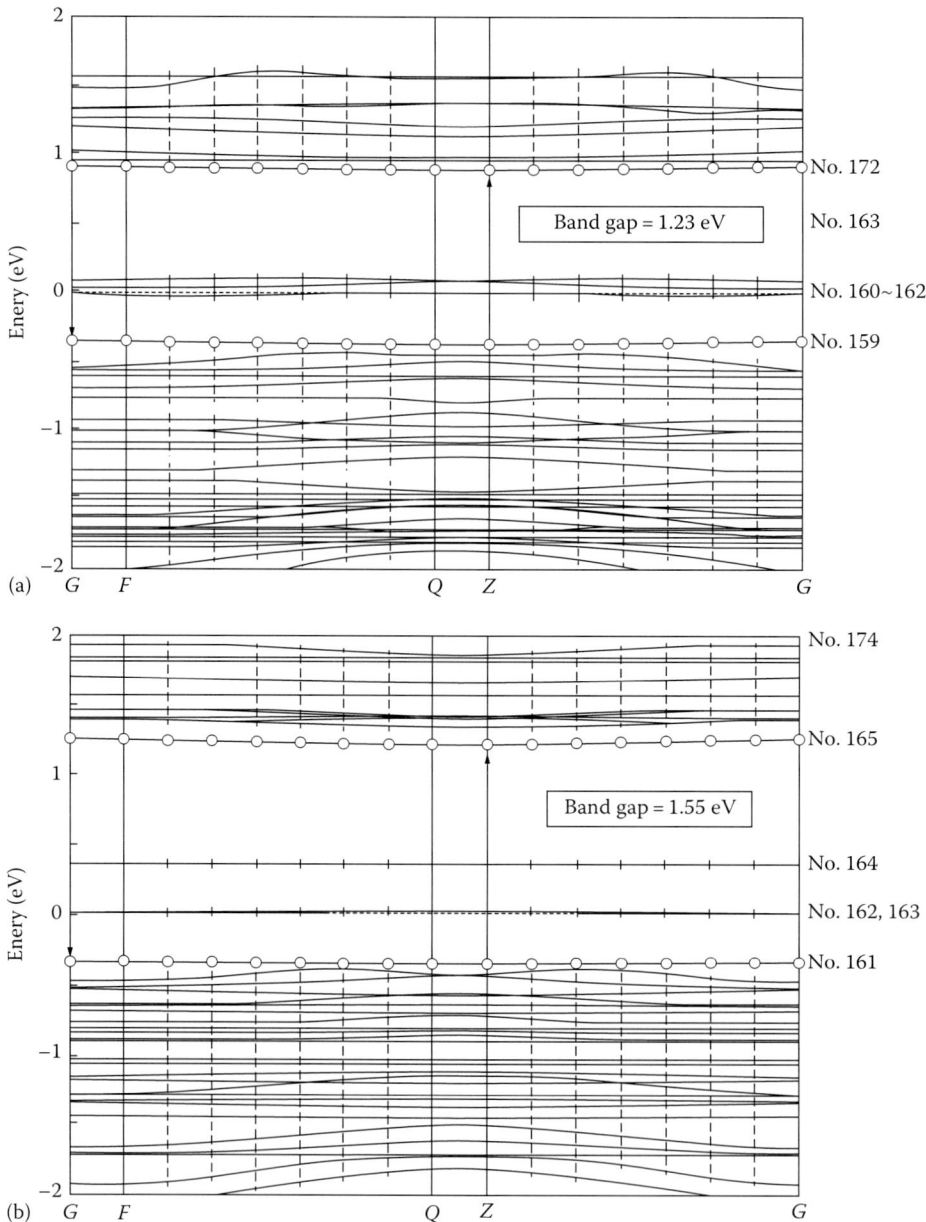

FIGURE 7.27 The band structure at the Fermi level for (a) uncoated and (b) C-coated LiFePO$_4$ surfaces. (From Zhang, P. et al., *J. Alloys Comp.*, 540, 121, 2012.)

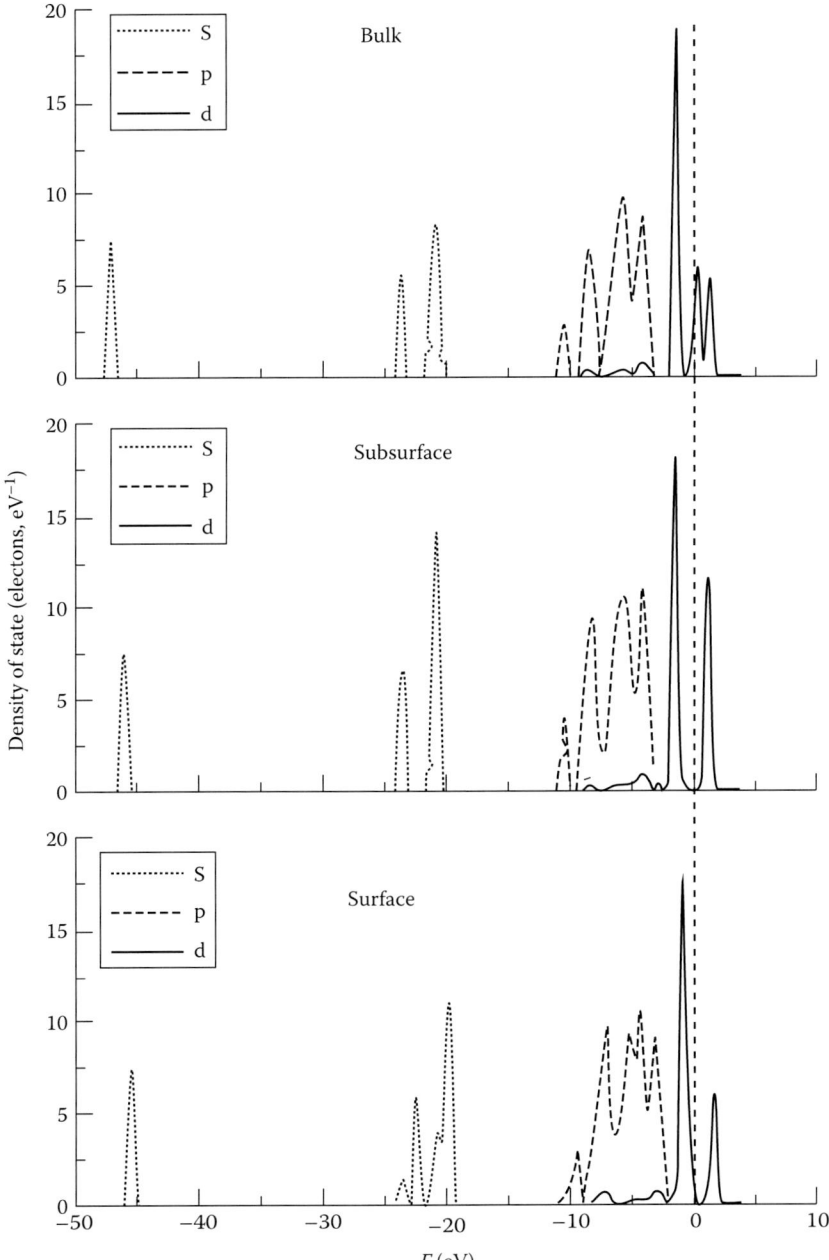

FIGURE 7.28 The density of states (DOS) of the LiFePO$_4$ surface (010). (From Zhang, P. et al., *J. Alloys Comp.*, 540, 121, 2012.)

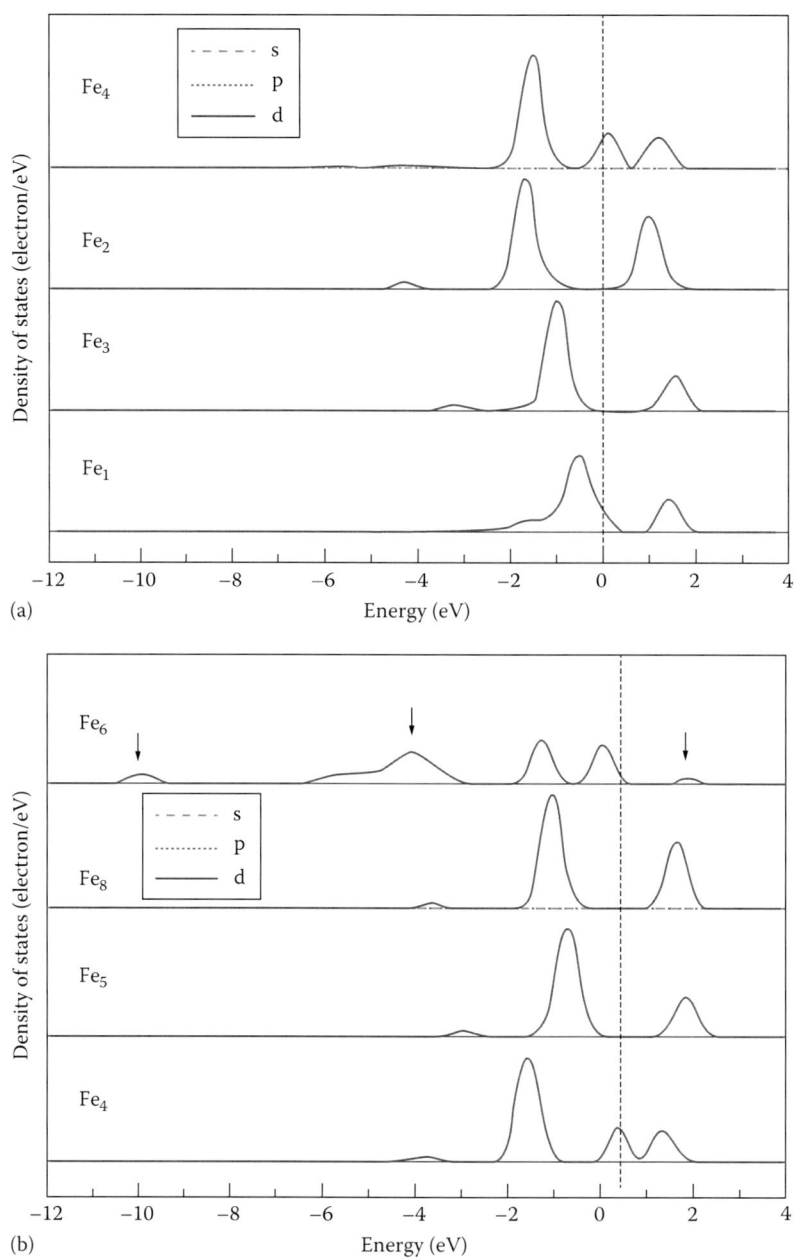

FIGURE 7.29   DOS of d-states in LiFePO$_4$ corresponding to different positions in (010) LiFePO$_4$. (a) Uncoated sample and (b) carbon-coated sample. (From Zhang, P. et al., *J. Alloys Comp.*, 540, 121, 2012.)

7.6 Electron Relaxation in LiFePO$_4$

LiFePO$_4$ is an unusual example of a practical solid state ionic material that exhibits, in addition to useful electrochemical activity of Li$^+$ ions, a non-negligible electronic conductivity and very interesting type of magnetism due to the existence of Fe ions only in the Fe^{++} state.

Electron relaxation in fast ion conductors is intimately linked to the conductivity of mixed ionic–electronic conductors. In these materials, electronic conductivity makes a nontrivial contribution to both physics and chemistry as well as to electrochemical activity, which is the basis of applications. An example is $LiFePO_4$. Electronic transport in olivine $LiFePO_4$ and its Mn, Co, and Ni analogues is mostly dominated by polaron hopping.

Let us ask at this stage: what is a polaron? The electron–phonon interaction in a semiconductor or ionic crystal causes an apparent increase in electron mass because the electron drags the heavy ion cores with it. In such materials, the combination of the electron and its strain field is known as a polaron. The effect is large in ionic crystals because of the strong Coulomb interaction between ions and electrons.

The basic precursor is $FePO_4$ into which Li can be "intercalated" to give Li_xFePO_4 and "deintercalated." In transition metal compounds including oxides and phosphates, the current carriers are accompanied by a self-induced lattice distortion that may cause self trapping. Conceptually speaking, in order to generate a hole polaron in $LiFePO_4$, an electron is removed from one of the Fe ions in a hypothetical "supercell" so that atomic relaxation can happen [7.38].

The transport of a hole polaron occurs by hopping between two fully relaxed equilibrium configurations. The polaronic conductivity σ_p in $LiFePO_4$ is described by the general formula of Mott:

$$\sigma_p = c_a (c_a - 1) e^2 v_e / k_B T \exp(-2\alpha R)/R \exp(-E_a/2K_B T) \tag{7.79}$$

In polaronic materials such as $LiFePO_4$, $LiCoO_2$, and $Li Mn_2O_4$, conductivity depends on both the availability of carriers and their mobility. Indeed most intercalation compounds are likely polaronic. For instance, lithium removal from $LiCoO_2$ causes an insulator to metal transition due to overlap of hole states created by Li^+ removal. Charge compensation on transition metal ions that accompany Li^+ insertion or removal creates mixed valence compounds with a large number of potential carriers. Two aspects are of interest: (1) migration energy of polarons ($E_{polaron}$) and (2) polaron binding to Li^+ or vacancies ($E_{pol-binding}$). In $LiFePO_4$, for example, the two energies are ~0.2 and ~0.37–0.5 eV [7.39].

Another aspect of electron relaxation arising in $LiFePO_4$ has to do with the very interesting magnetism and magnetic phase transitions in that compound whose origin is to be found in the Fe^{2+} ions in the structure. Orthorhombic $LiFePO_4$ (space group P_{nma}) undergoes a magnetic phase transition from paramagnetic to antiferromagnetic phase at 50 K. At and below this Neel temperature (T_N), the Fe^{++} spins are oriented along the crystallographic b-axis (Figure 7.30).

The Fe^{++} ($S = 2$) moments exist in the highly temperature-sensitive microscopic crystalline environments in which they exist. The closely packed oxygen framework of $LiFePO_4$ accommodates Fe^{++} in distorted octahedral surroundings where they share the oxygens with PO_4 tetrahedra. This circumstance leads to the formation of Fe^{++}-based corrugated layers that are stacked along the [100] crystallographic axis. Nearest neighbors in the bc plane are coupled magnetically by a relatively strong exchange interaction J_1 through a Fe–O–Fe oxygen bond. The in-plane nearest neighbors are coupled (J_2) via Fe–O–O bonds [7.41]. Interlayer magnetic coupling is mediated by a phosphate ion via Fe–O–P–O–Fe bonding [7.42]. Thus arises the highly anisotropic of $LiFePO_4$ that are intermediate between two- and three-dimensional systems [7.43]. In particular, the magnetic properties are determined by the electronic states reflecting the potential advantage of $LiFePO_4$ among the members of the Olivine family. The dynamics of the magnetic phase transitions is of interest and Mossbauer spectroscopy (discussed in Chapter 4) is a powerful tool to investigate this aspect. Elucidation of the electronic level scheme of Fe^{++} in the crystal field approach and the determination of the electronic ground state at low temperatures are conveniently done using this technique. More importantly, the time scale of Mossbauer spectroscopic transitions falls in the nanosecond regime and therefore the technique allows determination of a range of relaxation times not easily measureable by other conventional techniques. By the Mossbauer method, the electron spin relaxation frequency is estimated through the line broadening of the absorption peaks [7.44a,b]. Let us briefly describe and discuss the relaxation phenomena observed in $LiFePO_4$ in the temperature range 20–300 K through ^{57}Fe Mossbauer spectroscopy. Figure 7.31 shows the Mossbauer spectra of $LiFePO_4$ at selected temperatures in (1) paramagnetic regime and (2) the antiferromagnetic regime [7.45]. The Mossbauer parameters extracted from an analysis of spectrum at 300 K are the isomer shift $\delta = 1.222$ mm/s and quadrupole splitting $\Delta E_0 = 3.000$ mm/s typical of Fe^{++} in ferrous compounds. They exhibit lower electron density due to the six 3d electrons on high spin Fe^{++}. The asymmetry of

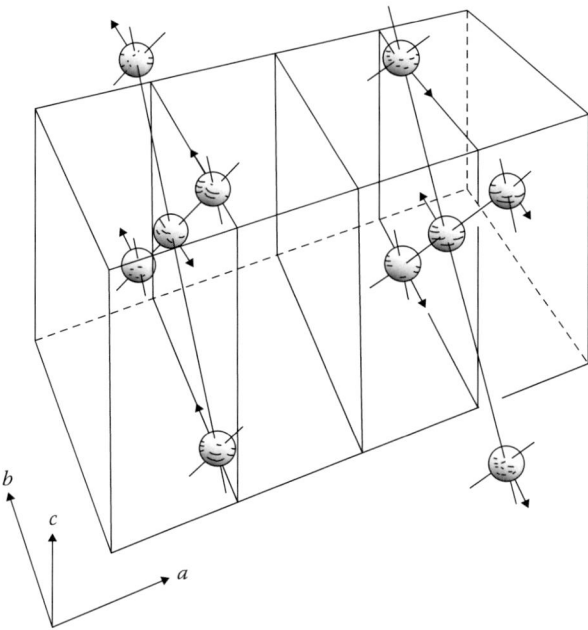

FIGURE 7.30 Magnetic structure of $LiFePO_4$ showing the spin orientations of Fe^{++} ions. (From Santoro, R.P. and Newnham, R.E., *Acta Crystallogr.*, 22, 144, 1967.)

the doublet is ascribed to Fe^{++} in distorted FeO_6 octahedra. Notice that the asymmetry increases as one approaches the Neel temperature. The origin of this asymmetry must be found in the slow electron-spin relaxation that occurs when the electron spin flips at a rate comparable to the nuclear precessional frequency [7.46]. The spectra below T_N are characteristic Fe^{++} present in a magnetically ordered single-phase $LiFePO_4$. The well-resolved octuplet has been analyzed by the diagonalization of the magnetic hyperfine and quadrupole interaction Hamiltonian to obtain the magnetic hyperfine field (B_{hf}) and the quadrupole interaction ΔE_Q. They reveal that the Fe^{++} magnetic moments are aligned along the b-axis as shown in Figure 7.31. B_{hf} values at 48 and 20 K are 26.2 and 122.9 kOe, respectively. The temperature dependence of B_{hf} is found to be

$$B_{hf}(T) = B_{hf}(0)\left\{1 - (T/T_N)^p\right\}^q \qquad (7.80)$$

where $p = 16.43$, $q = 2.1$, $T_N \sim 50$ K, and the saturation value of the hyperfine field at $T = 0$ K $\sim$125 kOe. The nonspherical ferrous ions are strongly coupled to lattice vibrations resulting in spin–lattice relaxation leading to asymmetric broadening of Mossbauer peaks. The relaxation frequency (τ^{-1}) estimated from the width of the absorption line on the higher velocity side shows a temperature dependence as shown in Figure 7.32 in paramagnetic and antiferromagnetic regimes.

Thus, the electrons in Fe^{++} ion ($3d^6$, ^5D) in $LiFePO_4$ undergoes a relaxation process of the spin–lattice type from 300 K down to 20 K. the relaxation time at the latter temperature being 1.199×10^7 Hz.

Partial removal of Li from $LiFePO_4$ to get $Li_{0.59}FePO_4$ leads to a two-phase asymmetric eight-line pattern due to different electric quadrupole interactions in tryphylite and heterosite phases of $LiFePo_4$, due to $Fe^{2+}(3d^6)$ and $Fe^{3+}(3d^5)$ [7.47].

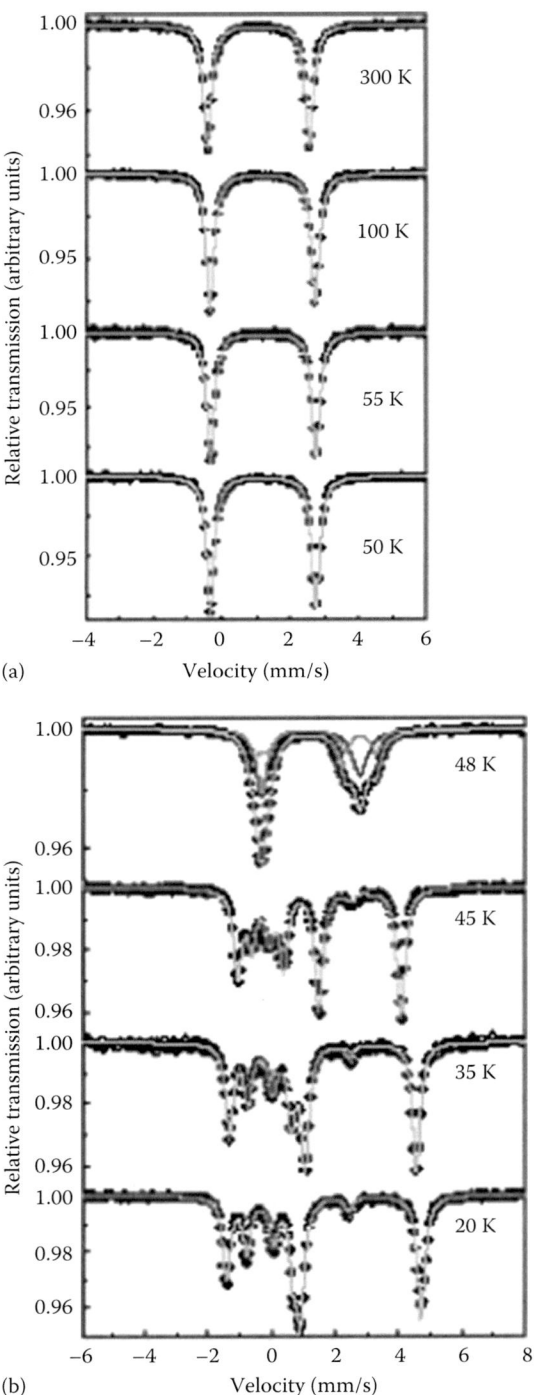

FIGURE 7.31 Mossbauer spectra of LiFePO$_4$ (a) at 300, 100, 55, and 50 K in the paramagnetic regime all of which show a quadrupole-split doublet and (b) at 48, 54, 35, and 20 K in the antiferromagnetic regime where the electron relaxation effects described in the text are seen. These are analyzed by a diagonalization of the full static Hamiltonian to obtain full lines. The well-resolved sextet is typical of a magnetically ordered ferrous ion in monophasic LiFePO$_4$. (From Sundarayya, Y. et al., *Phys. Stat. Sol. B*, 250, 1599, 2003.)

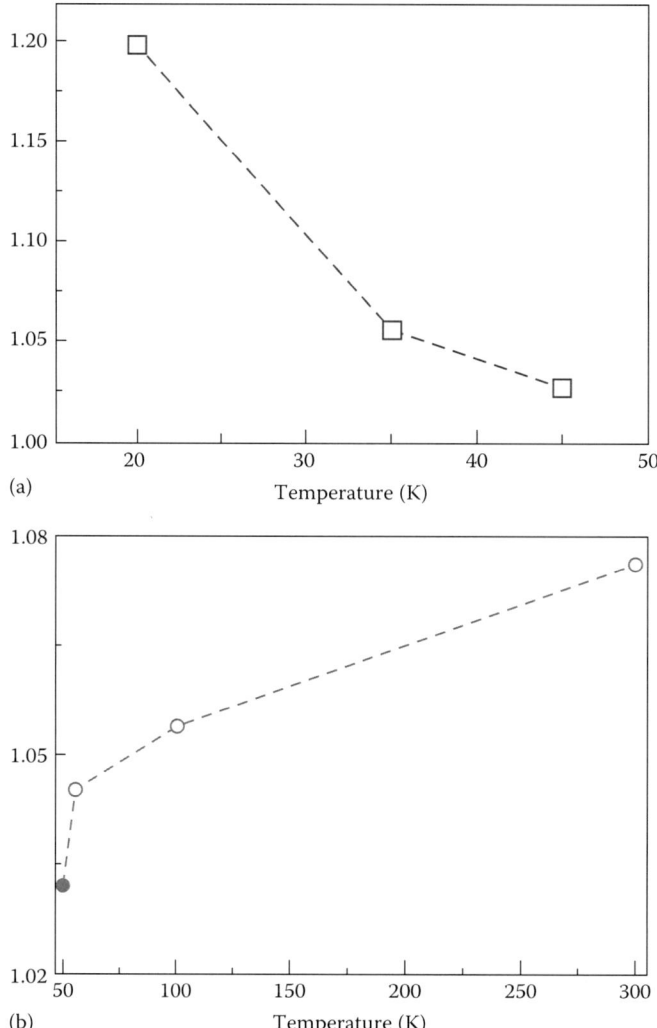

(a)

(b)

FIGURE 7.32   Electron relaxation times in (a) paramagnetic and (b) antiferromagnetic regimes of LiFePO$_4$. (From Sundarayya, Y. et al., *Phys. Stat. Sol. B*, 250, 1599, 2003.)

The electronic energy band diagram of LiCoO$_2$ and its partially and fully deintercalated phases are shown in Figure 7.33a. The main feature resulting from the complete removal of Li from LiCoO$_2$ is the merging of the Co(3d) band with the O(sp) band [7.48].

Coelectronic spin state transitions in nanometric LiCoO$_2$ upon delithiation have been observed through NMR (Figures 7.34 and 7.35) [7.49].

Nanosize effect on the magnetic susceptibility of LiCoO$_2$ is shown in Figure 7.36.

These transitions occur on the surfaces of stoichiometric LiCoO$_2$ where Co^{+++} adjacent to the surface adopt an intermediate spin state if their coordination is square pyramidal. On the other hand, if they are pseudo-tetrahedrally coordinated, a high spin state of Co^{+++} is realized. Note that the crystal field splitting of Co orbitals is modified at the surface due to missing Co–O bonds. The electronic spin transition has a significant impact on the Co^{3+}/Co^{4+} redox potential, which is an important input in the design of Li ion batteries to be discussed in the next chapter.

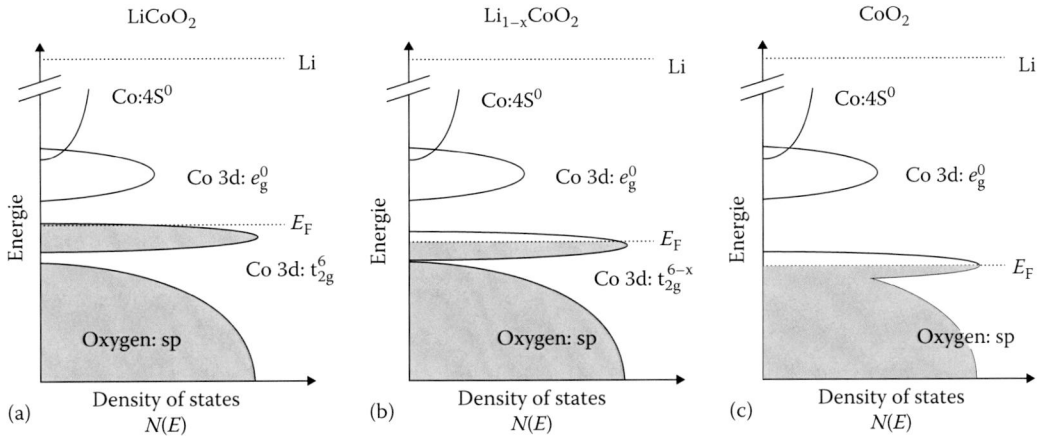

FIGURE 7.33 Band structure evolution of LiCoO$_2$ as Li is removed from the material. Note the merger of Co 3d orbitals and oxygen sp orbitals.

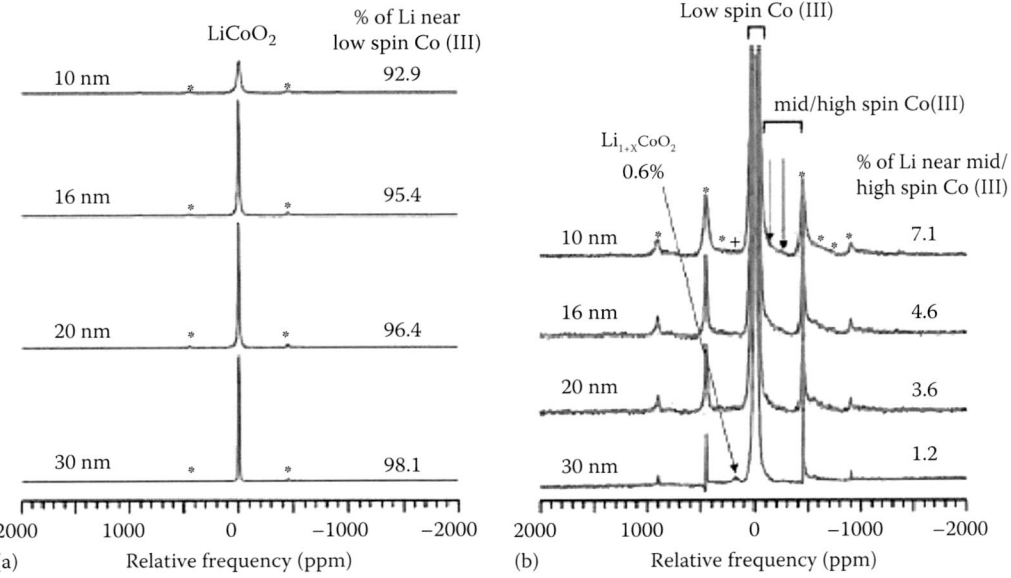

FIGURE 7.34 ^{7}Li NMR spectral observation of a change in the electronic spin state of the LiCoO$_2$ surfaces as a function of particle size in nanoscale LiCoO$_2$. (a) Percentages of Li nearby low-spin Co(III) in LiCoO$_2$ are mentioned on the right corresponding to particle sizes noted on the left. (b) Peak assignments of low-spin, mid-/high-spin Co(III). * denotes modulation side bands introduced by nagic angle spinning. (From Qian, D. et al., *J. Am. Chem. Soc.*, 134, 6096, 2012.)

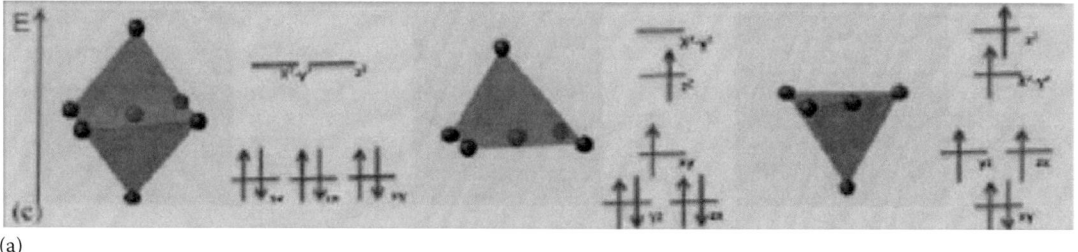

(a)

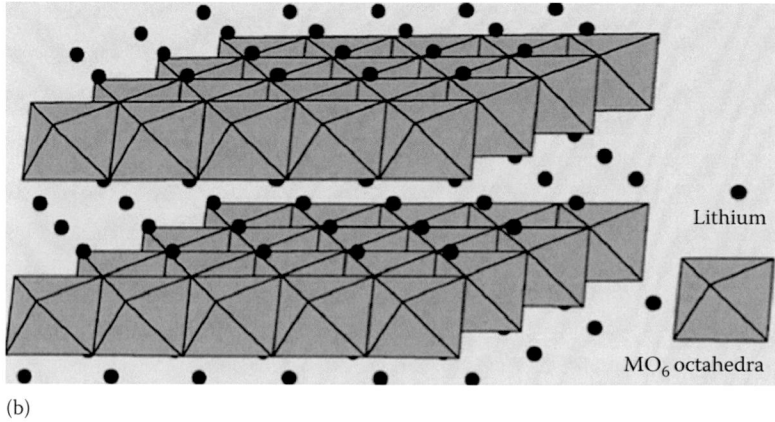

(b)

FIGURE 7.35 (a) Co^{+++} coordinations and the coresponding spin states. Octahedral, square-pyramidal, and pseudo-tetrahedral oxygen coordinations of Co^{+++} (the central ion), and the corresponding spin configurations are shown. (b) 3D structure of $LiMO_2$ (M = Mn, Co, Ni). (From Qian, D. et al., *J. Am. Chem. Soc.*, 134, 6096, 2012.)

7.7 Role of Electronic Structure in 2D and 3D Li and Na Insertion Compounds

Figure 7.37a shows l-Li_xMO_2 (layered) and s-LiM_2O_4 (spinel) structures, where M = 3d transition metal. M occupies octahedral sites in *both* structures. In l-Li_xMO_2, M and Li (and/or vacancies) alternately occupy (111) planes of the ccp oxygen sublattice. The (111) plane parallel to the M layers is shown by the black line between the layered and spinel structures. The [111] direction is also shown. In s-$Li_{1/2}MnO_2$, (111) planes with three-fourths of the Mn alternate with (111) planes with one-fourth of the Mn. Li ions occupy tetrahedral sites in the planes with one-fourth of the Mn. *The planes with three-fourths of the Mn are free of Li*. In fully lithiated spinel-like s-$Li_2Mn_2O_4$, Li moves into octahedral sites. Three-fourths of the Li are in the (111) plane with one-fourth of the Mn, and one-fourth of the Li are in the plane with three-fourths of the Mn.

The ability to resist phase transition impacts the overall performance of a Li insertion compound used as an electrode in a Li rechargeable battery [7.50]. Two questions that arise are as follows:

1. What role does electronic structure play in determining the site preference and mobility of 3d transition metal ion in an oxide?
2. How do these factors in turn affect the resistance of metastable 3d metal oxides against transformation?

Consider the main prototype Li–Mn–O system. Its advantages, namely high capacity and good safety, stem from the relative stability of the fully charged MnO_2. Note that Mn is far less expensive than Co.

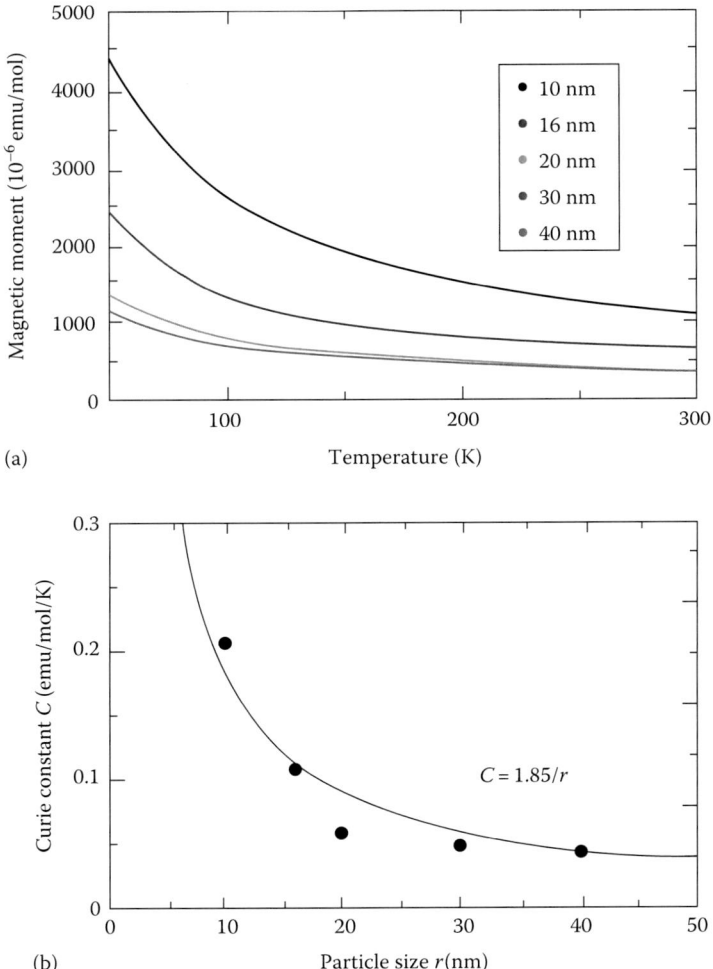

FIGURE 7.36 (a) Temperature dependence of molar magnetic susceptibilities of nanoscale $LiCoO_2$ at different particle sizes (10, 16, 20) and (b) Curies constants as s function of particle sizes. (From Reed, J. and Ceder, G., *Chem. Rev.*, 104, 4513, 2004.)

Three important crystal structures in the Li–Mn–O system are as follows:

1. Spinel $LiMn_2O_4$
2. $NaFeO_2$ type layered
3. Orthorhombic $LiMnO_2$ (Pmna)

There is structural transformation among 2 and 3 upon electrochemical cycling. The transformation kinetics bears on electronic structure.

Now refer to Figure 7.37b. For phase transformations involving rearrangement of transition-metal cations over octahedral sites within a fixed cubic close-packed (ccp) oxide framework, such as the transformation of layered Li_xMnO_2 to spinel, the results of the work of study indicate that the low-energy pathway for transition-metal migration between octahedral sites is through a shared nearest neighbor tetrahedral site (i.e., $O_h \rightarrow T_d \rightarrow Oh$) rather than directly between octahedral sites ($O_h \rightarrow O_h$). This suggests that the smaller the energy change is for transition-metal ion movement between octahedral and

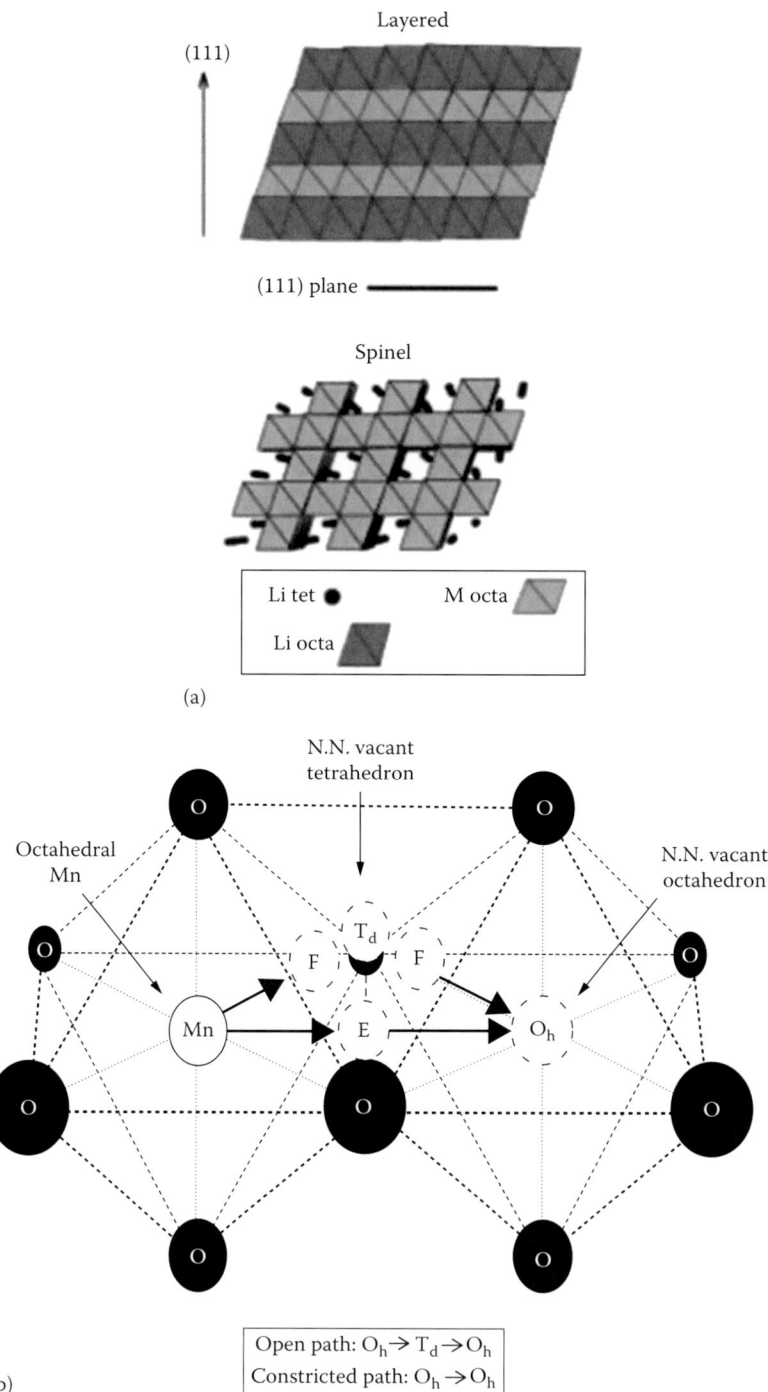

FIGURE 7.37 (a) Layered (top) and spinel structures (bottom) in the Li–Mn–O system. (b) Most open and constricted paths. (From Reed, J. and Ceder, G., *Chem. Rev.*, 104, 4513, 2004.)

tetrahedral coordination, the more easily the transition metal ions should be able to rearrange in a (ccp) oxide when transforming from a metastable structure to a stable one. As a result, the resistance against transformation of metastable transition-metal oxides with a ccp oxygen sublattice (e.g., Li_xMnO_2 with R-$NaFeO_2$ structure) will depend on the relative stability of the transition metal in octahedral coordination compared to tetrahedral coordination. The availability of empty tetrahedra without any cations occupying nearest neighbor face-sharing octahedra is also an important factor in the ability of TM ions to migrate through a ccp oxide. Such tetrahedra provide a relatively low-energy pathway by avoiding the highly repulsive interaction between face-sharing cations.

7.8 Summary

We have discussed in this chapter the chemical bonding and electronic structure of solid state ionic materials as they connect to ion transport in the ionic and superionic phases. Beginning with a nearly free ionic approximation–based model for an isotropic semiconductor, a DFT inspired empirical discussion of the superionic bond, the question of critical iconicity has been discussed. This is followed by a development of the tight binding approximation and a fairly extensive discussion of the DFT and its application to the band structure of $LiFePO_4$. The most interesting case of electronic relaxation in $LiFePO_4$, which has a bearing on its performance as a battery electrode has been discussed in the context of Mossbauer spectroscopy of magnetic phase transitions that also connects to Jahn–Teller effect and polarons that are not discussed. The spin state transitions of Co in nanoscale $LiCoO_2$ are discussed briefly. The chapter closes with a discussion on the role of electronic structure on the (de)intercalation behavior of $LiCoO_2$. These results have a natural bearing on Li ion battery electrodes, which are discussed in the next chapter.

Problems

7.1 *How does the energy gap arise?* Represent the time-independent state of an electron wave in a crystal by a standing wave. Form two different standing waves from two traveling waves $\exp(\pm i\pi x/a)$ where a is the lattice parameter of the 1D lattice. These two waves accumulate electrons at different regions. So they have different potential energy. What are the charge densities ρ^+ and ρ^- for the two waves? Sketch them. An energy gap E_g arises if ρ^+ and ρ^- differ by E_g. Assuming a potential energy $U(x) = U\cos 2\pi x/a$, show that $E_g = U$. [*Hint*: See Ref. [7.21]].

7.2 *Nearly free electron (NFE) model.* Figure 7.1 illustrates the use of a fundamental model called the nearly free electron (NFE) or weak binding approximation. This is the opposite of tight binding approximation discussed in this chapter. Formulate the NFE approximation and solve for $E(k)$. The term "weak binding" implies that the periodic potential of the crystal is so weak that the electrons in the crystal are nearly free. Essentially, it involves the solution of the Schrodinger equation in the limit of a very weak periodic potential. See http://web.mit.edu/course/6/6.732/www/6.732-pt1.pdf.

7.3 *E(k) vs k curves.* (a) Obtain an expression for the effective mass m^* in the tight binding approximation assuming that (i) the electron moves in the x-direction (i.e., $k_y = k_z = 0$) and (ii) $k_x \ll \pi/a$.

(b) Work out the $E(k)$ relation for a fcc structure in the tight-binding approximation. Plot the curve. Plot also the $E(k)$ relation for the bcc structure worked out and compare the two.

7.4 *CuBr and AgBr as I–VII semiconductors.* Look at the energy band model of the semiconductor in the following. A filled lower band is separated from an empty upper band by a band gap. CuBr like GaAs has a direct band gap of 2.91 eV while AgBr has an indirect gap of 2.5 eV.

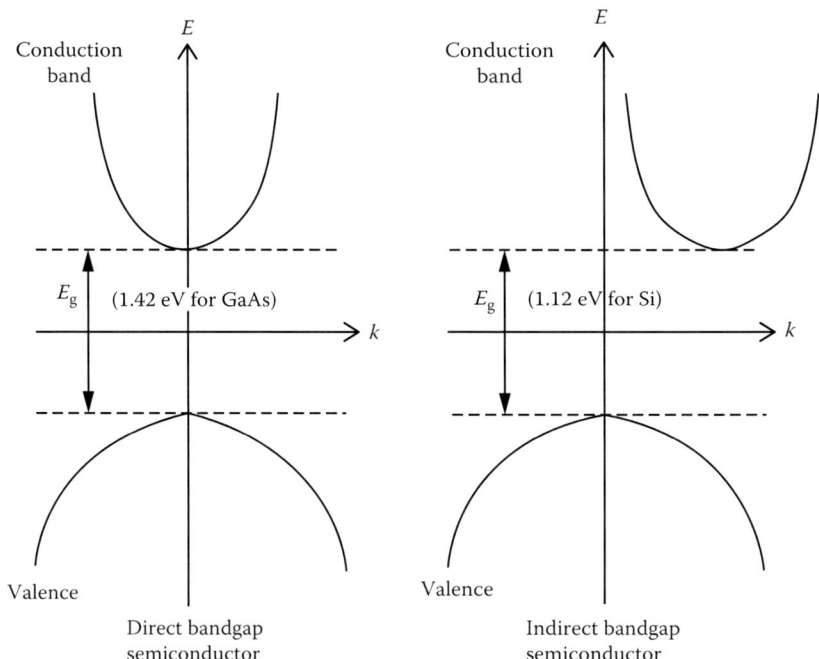

Now CuBr is a superionic conductor and AgBr is not. From the nature of the metal–halogen band and the corresponding crystal structure, reason out why this is so. Also note that both CuBr and AgBr have cationic Frenkel defects but the concentration of these in AgBr until melting point is not enough to make it a superionic conductor. What makes CuBr a superionic conductor? Locate the "defect band" in the CuBr band structure.

7.5 *Band structure development in AgI and CuI. The plasmon–exciton transition.* Plasmons are quantized as collective oscillations of conduction electrons in a metal such as Ag or Cu. The optical absorption spectrum in the UV–visible region recorded on a thin film sample of Ag or Cu (coated on a glass substrate) shows the existence of such plasmons. When such a film is exposed to a stream of iodine vapors, the conduction electrons of plasmons are gradually captured by iodine to form a thin film of AgI (or CuI) on the metal "substrate." These electrons form bound states with the holes of the valence band of AgI or CuI called excitons which form a spectrum akin to hydrogen atom spectra. Discuss the basic physics of plasmons and excitons. What information on AgI or CuI band structure may be obtained from the plasmons and excitons? Sketch the expected response. Also, discuss a plausible mechanism of the plasmon–exciton transition. Consult Ref. [7.51]. Plasmon–exciton transition is discussed in Ref. [7.52].

7.6 *Density functional theory* has been applied to compute the electronic energy band structure and optical properties of γ-Ag_3SI. Access and follow the work of Erdinc and Akkus [7.53]. They have calculated the band gap of γ-Ag_3SI to be 0.46 eV. Look for the experimental band gap and discuss the difference.

7.7 *Li-intercalation and electronic structure*—the case of LMN electrode. Intercalation means partial or complete insertion of Li into an electrode during the charging process of a Li ion battery. The following figure shows the energy band diagrams for $Li_xNi_yMn_yCo_{1-2y}O_2$ (LMN) with different Li contents in three states as mentioned.

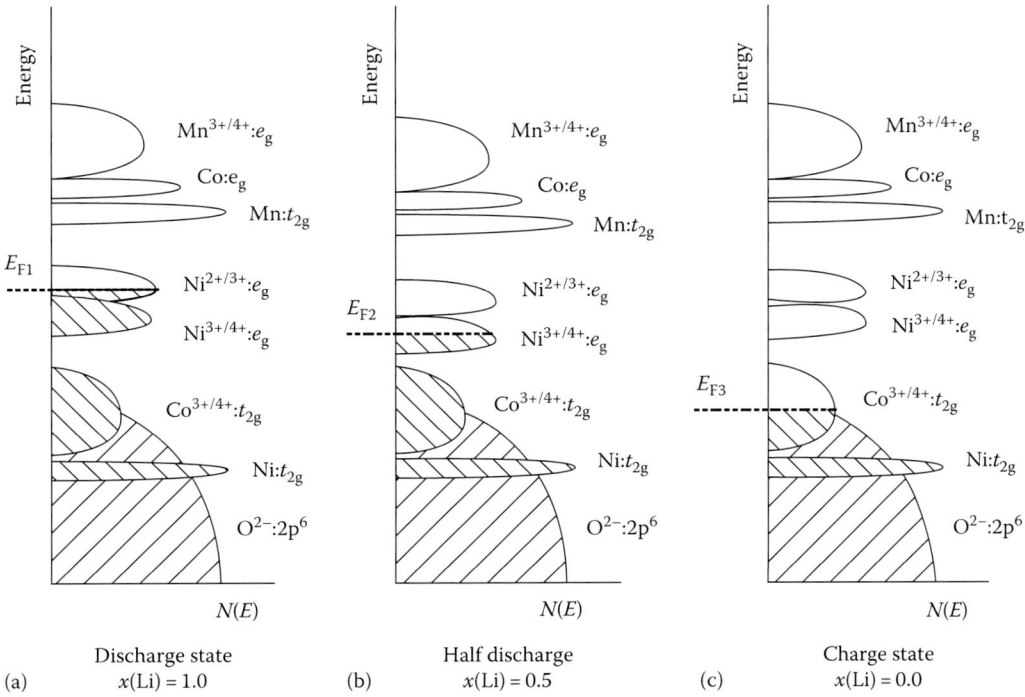

	Discharge state		Half discharge		Charge state
(a)	x(Li) = 1.0	(b)	x(Li) = 0.5	(c)	x(Li) = 0.0

Note the discrete states of Mn, Co, and Ni and their relative positions with respect to the Fermi level (E_F). Observe how the Fermi level shifts as the Li content changes in the three states: charge, half discharge, and full discharge. Discuss the origin of these changes. See Ref. [7.54]

7.8 *Polaronic conduction* (a) Electrical conductivity measurements on $LaCoO_3$ have shown that the polarons are indeed responsible for carrying current in this compound. See the following figure.

Carefully examine the plots (a) and (b). Note that the total conductivity in (a) consists of a DC part plus a "relaxation" part which come from polarons. From plot (b), deduce dc conductivity at three temperatures and make a log (conductivity) vs $1/T$ plot and find the activation energy for transport in $LaCoO_3$. Think of fitting the entire impedance plot of (a) shown by "o" to an equation of the type Equation 7.79 to deduce polaron coupling constant and activation energy for polaron hopping [7.55].

(b) Dielectric relaxation in $LiMn_{1.95}Al_{0.05}O_4$ has been measured to deduce polaron conduction parameters. The origin of polarons has been discussed in terms of Jahn–Teller interaction of the Mn^{3+} ion. See Iguchi et al. [7.56] for a fundamental discussion on dielectric measurements and origin of polaron transport.

7.9 *Kohn–Sham version of DFT* and new developments. The spirit of Kohn–Sham DFT (KS DFT) is this: the **complex** many-body problem of interacting electrons in a static external potential is reduced to a *solvable* problem of *noninteracting electrons* moving in *an effective potential*. The effective potential includes the external potential and the effects of the Coulomb interactions between the electrons, for example, the exchange and correlation interactions. Modeling the latter two interactions becomes the task of KS DFT. The simplest approximation is the local-density approximation (LDA). LDA is based upon exact exchange energy for a uniform electron gas, which can be obtained from the Thomas–Fermi model, and from fits to the correlation energy for a uniform electron gas. Noninteracting systems are relatively easy to solve as the wavefunction can be represented as a Slater determinant of orbitals. Further, the kinetic energy functional of such a system is known exactly. The exchange-correlation part of the total-energy functional remains unknown and must be approximated.

Another approach is orbital-free density functional theory (OFDFT). Here approximate functionals are also used for the kinetic energy of the non-interacting system. A fundamentally different way to construct the DFT is as a Legendre transformation from external potential to electron density (see Ref. [7.57]). Electronic structure problem in finite periodic systems can be mathematically handled in DFT. Follow and understand the method by seeing the original references cited in this chapter.

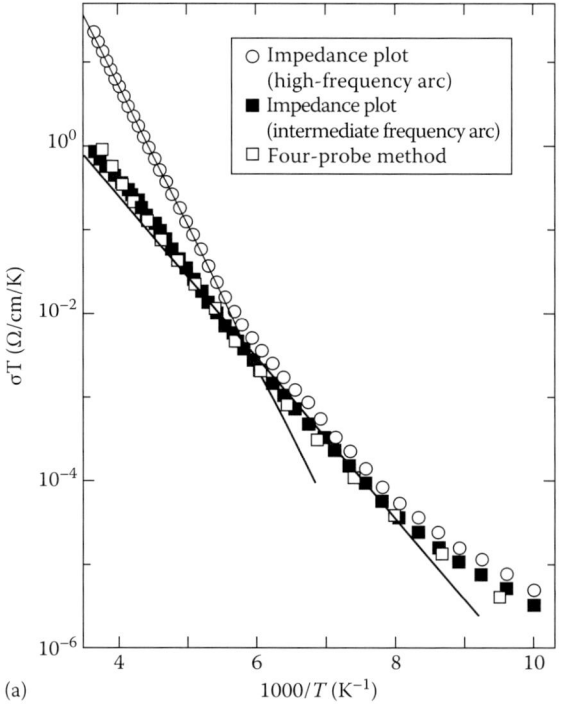

(a)

FIGURE P.8   (a) Arrhenius plots for $LaCoO_3$.   *(Continued)*

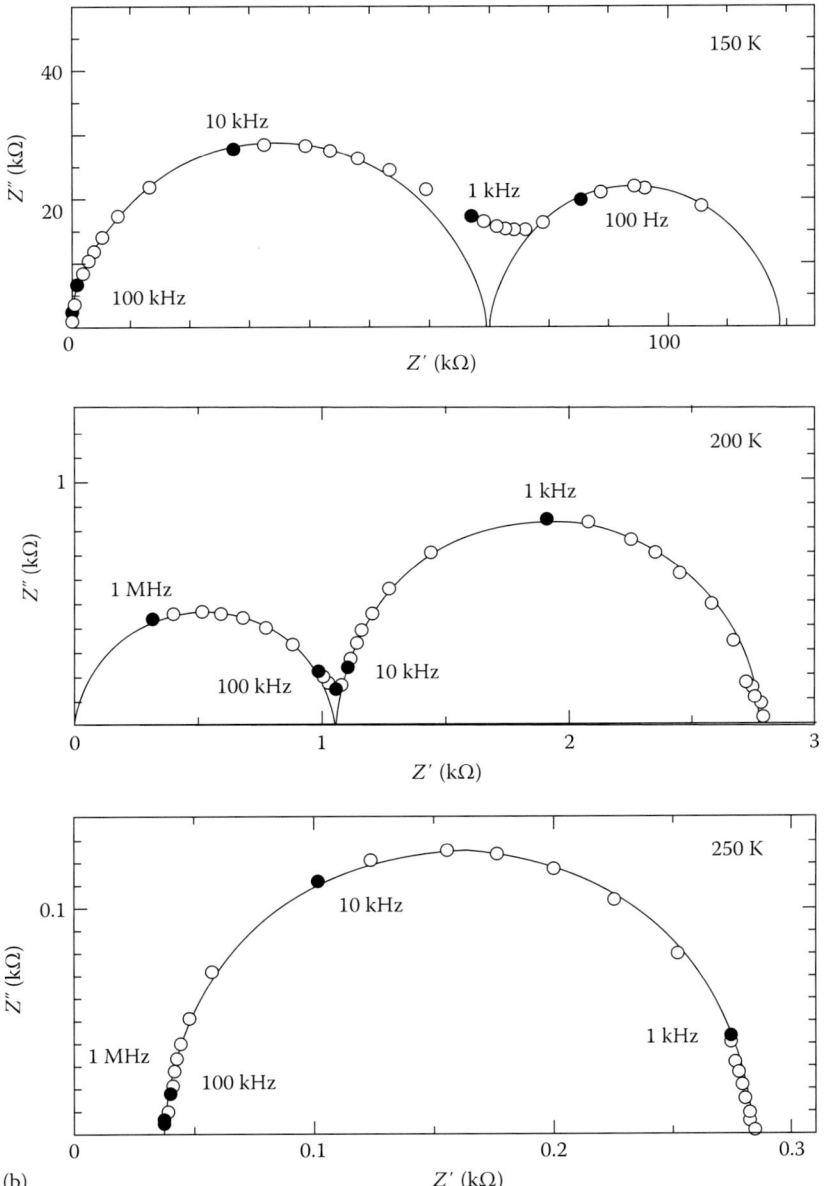

FIGURE P.8 (*Continued*)   (b) Complex-plane impedance (Z' vs. Z") analyses for LaCoO at 150 K, 200 K, and 250 K. Z' and Z" are the real and imaginary parts of the total impedance at each applied frequency. The *highest-frequency arc* crosses the origin at *each temperature*. The highest resistance of *this arc* corresponding to the lowest resistance of the *intermediate arc* gives the *bulk resistance*. The highest resistance of the intermediate arc is the *total resistance in grains and grain boundaries*.

REFERENCES

7.1a.   V. K. Rashenov, M. G. Fiogel, R. A. Alarashi, *Phys. Status Solidi (b)* 54(1972) 355.

7.1b.   R. A. Alarashi et al., *Phys. Status Solidi b* 54(1972) K5.

7.2.   E. Schrodinger, *What is Life?* Cambridge University Press, Cambridge, U.K., 1993, Chap. 5, p. 58.

7.3.   J. C. Phillips, *Rev. Mod. Phys.* 42(1970) 317.

7.4.   K. Wakamura, *SSC* 86(1993) 503.

7.5.   C. R. A. Catlow, A. M. Stoneham, *J. Phys. C* 16(1983) 4321.

7.6.   M. Aniya, *J. Phys. Soc. Jpn.* 61(1992) 4474.

7.7.   F. Billi, H. E. Roman, W. Dietrich, *Solid State Ionics* 28–30(1988) 58.

7.8.   F. Seitz, *Modern Theory of Solids*, Dover, New York, 1987, p. 547.

7.9.   W. Hayes, A. M. Stoneham, *Defects and Defect Processes in Nonmetallic Solids*, Dover, Mineola, NY, 2004.

7.10.   L. Pauling, *Nature of the Chemical Bond*, Oxford & IBH, New Delhi, India, 1975, Chap. 3.

7.11.   C. Goycoolea et al., *Int. J. Quant. Chem.* 36(1989) 455.

7.12.   C. S. Sunandana, *J. Phys. Chem. Solids* 58(1997) 1359.

7.13.   C. S. Sunandana, *Bull. Mater. Sci.* 21(1998) 1.

7.14.   S. Chandra, *Superionic Solids*, North Holland, Amsterdam, the Netherlands, 1981.

7.15.   S. Geller, B. B. Owens, *J. Phys. Chem. Solids* 33(1972) 1241.

7.16.   D. Mazumdar, PA Govindacharyulu, DN Bose, *J. Phys. Chem. Solids* 43(1982) 933.

7.17.   R. G. Pearson, *J. Amer. Chem. Soc.* 85(1963) 3533; 107(1985) 6801.

7.18.   M. Aniya, Physics of solid state ionics, in *Chemical Bonding in Superionic Conductors*, T. Sakuma, H. Takahashi (Eds.), Research Signpost, Trivandrum, India, 2006, pp. 45–64; *Integrated Ferroelectrics* 115(2010) 81.

7.19.   J. A. Van Vechten, *Phys. Rev.* 182(1969) 891; 187(1969) 1007.

7.20.   M. Kobayashi et al., Physical studies of electronic structure and ion dynamics in superionic conductors, *Superionic Conductor Physics: Proceedings of the First International Discussion Meeting*, Kyoto, Japan, September 10–14, 2003, pp. 15–20.

7.21.   C. Kittel, *Introduction to Solid State Physics*, 7th edn., Wiley, New York, 1996.

7.22.   J. C. Phillips, *Bonds and Bands in Semiconductors*, Academic Press, New York, 1973.

7.23.   Y. W. Tsang, M. L. Cohen, *Phys. Rev. B* 9(1974) 3541.

7.24.   M. Kobayashi, Electronic structure of superionic conductors, in *Chemical Bonding in Superionic Conductors*, T. Sakuma, H. Takahashi (Eds.), Research Signpost, Trivandrum, India, 2006, pp. 1–15.

7.25.   J. Jhu, R. Pandey, M. Gu, *J. Phys. Condens. Matter* 24(2012) 475503.

7.26.   Y. Kowada, H. Adachi, M. Tatsumisago, T. Minami, *J. Non-Cryst. Solids* 232–234(1998) 497.

7.27.   M. S. Dresselhaus, Solid state physics, Fall, 2001, http://web.mit.edu/course/6/6.732/www/6.732-pt1.pdf (accessed August 26, 2015).

7.28.   P. V. Smith, *J. Phys. Chem. Solids* 37(1976) 581, 589.

7.29.   S. Glaus, G. Calzaferri, *Photochem. Photobiol. Sci.* 2(2003) 398.

7.30.   I. Kh. Akopyan et al., *Phys. Status Solidi (a)* 119(1990) 363.

7.31.   D. Bharathi Mohan, C. S. Sunandana, *J. Phys. Chem. B* 110(2006) 4569.

7.32.   S. Kashida et al., *SSI* 158(2003) 167.

7.33.   H. Kikuchi, H. Iyetomi, A. Hasegawa, *JPCM* 10(1998) 11439; 9(1997) 6031.

7.34a. S. Ono et al., *Solid State Ionics* 139(2001) 249.

7.34b. M. Kobayashi et al., *Int. J. Modern Phys. B* 15(2001) 678.

7.34c. M. Kobayashi et al., *Solid State Ionics*, 154–155(2002), 209.

7.35a. W. Kohn, *Rev. Modern Phys.*, 71(1999) 1253; http://www.nobelprize.org/nobel_prizes/chemistry/laureates/1998/kohn-lecture.pdf.

7.35b. P. Koksinen, V. Makinen, *Comp. Mater. Sci.* 47(2009) 237; http://users.jyu.fi/~pekkosk/resources/pdf/koskinen_CMS_09.pdf.

7.36.   P. Zhang, D. Zhang, L. Huang, Q. Wei, M. Lin, X. Ren, *J. Alloy Comp.*, 540(2012) 121.

7.37.   M. D. Segall, P. J. D. Lindan, M. J. Probert, C. J. Pickard, P. J. Hasnip, S. J. Clark, M. C. Payne, *J. Phys. Condens. Matter* 14(2002) 2717.

7.38.   J. Lee et al., *Appl. Phys. Lett.* 101(2011) 033901.

7.39.   G. Ceder et al., *MRS Bull.* 36(2011) 185.

7.40.   R. P. Santoro, R. E. Newnham, *Acta Crystallogr.* 22(1967) 144.

7.41.   D. Dai et al., *Inorg. Chem.* 44(2005) 2407.

7.42.   J. M. Maye, *Phys. Rev.* 131(1963) 38.

7.43.   D. Vaknin et al., *Phys. Rev. B* 65(2002) 224414.

7.44a. H. N. Oak, J. G. Mullen, *Phys. Rev.* 168(1968) 563.

7.44b. K. K. P. Srivastava, S. N Mishra, *Phys. Status Solidi B* 118(1982) 93.

7.45.   Y. Sundarayya et al., *Phys. Status Solidi B* 250(2013) 1599.

7.46.   M. Blume, J. A. Tjon, *Phys Rev.* 165(1968) 446.

7.47.   I. K. Lee et al., *J. Appl. Phys.* 107(2010) 09A522.

7.48.   J. M. Tarascon et al., *Dalton Trans.* (2004) 2988–2994.

7.49.   D. Qian et al., *J. Am. Chem. Soc.* 134(2012) 6096.

7.50.   J. Reed, G. Ceder, *Chem. Rev.* 104(2004) 4513.

7.51.   C. Kittel, *Introduction to Solid State Physics*, 7th edn., Wiley, Hoboken, NJ, 2004.

7.52.   M. Gnanavel, C. S. Sunandana, *Physics of Iodized Silver & Silver-Copper Thin-Film Nanostructures*, Omniscriptum Publishing Group, Saarbruken, Germany, 2013.

7.53.   B. Erdinc, H. Akkus, *J. Korean Phys. Soc.* 56(2012) 796. www.kps.or.kr/jkps/downloadPdf. asp?articleuid=%7B15A0353A.

7.54.   C. M. Julien et al., *Inorganics* 2(2014) 132.

7.55.   E. Iguchi et al., *Phys. Rev. B* 54(1996) 17431.

7.56.   E. Iguchi et al., *J. Appl. Phys.* 91(2002) 2149.

7.57.   G. Kotter et al., *Rev. Mod. Phys.* 78(2006) 865.

7.58.   P. K. Chattaraj, *Curr. Sci.*, 61(1991) 391.

8

The All Solid State Battery

8.1 Perspective

The uniqueness of this chapter is that it deals with the prime mover—the electrochemical energy power convertor used in laptops and mobile telephones—whose daily charging and discharging processes play out many of the earlier chapters of this book! Indeed, the Li-ion battery (LIB) is the paradigm of the energy- and mobility-hungry twenty-first century!! Building a better battery is the current story of reinventing rechargeable cells to (1) cut costs and (2) boost capacity. A cheap long life for a secondary battery is apparently the goal of applied solid state ionics.

This chapter provides a unified discussion of the most important application of solid state ionics, namely the (futuristic) rechargeable all-solid state battery in the background of the currently prevalent battery systems that use a liquid electrolyte. Following up on Sections 1.1.2 and 1.1.5, battery basics are discussed. The Li^+ ion physics of battery components is considered followed by a discussion of specific battery types including thin film battery, Na^+ ion battery, polymer battery, and highly flexible printed batteries. The flowing sections are devoted to advanced Li battery technologies and aspects of system architecture and integration, the electric vehicle challenge, and hybrid batteries. Special purpose applications are briefly mentioned.

To underline the importance of the developing LIB, note that in 2014, commercial batteries were made by Amprius Corp. with a volumetric energy density of 650 wH/L (20% higher than before), using a silicon anode, for smartphone manufacturers (http://en.wikipedia.org/wiki/Lithium-ion_battery). For specific details on Si anode refer Reference 8.1.

A motivating article aptly titled "Recharging lithium battery research with first-principles methods" by Ceder et al. [8.2] underscores the importance of computational modeling in battery development. Another one emphasizing experimental aspects is the paper by Van Noorden [8.3].

An ion-based battery—LIB for instance—is a storehouse of "ionic electricity." It is an electrochemical generator of direct current electricity made possible by electrochemical reactions at the cathode–electrolyte and the electrolyte–anode "half-cells" of an electrochemical cell. This chapter discusses Li ion and other cation batteries, which are the foremost industrial and low-power applications of solid state ionic materials.

The all-solid state battery comprises bulk solid/thin film solid electrolytes besides anode and cathode and the charge separators and is the most desirable and challenging goal in solid state ionics today. Lithium-ion- and sodium-ion-based components of such a battery are currently in an active state of research and development.

For a start, let us focus on the practical primary versus the secondary battery.

Look at the well-sealed inverter pack consisting of liquid electrolyte-based battery (What is an anode? What is a cathode?) and at the cellphone/laptop battery. How is it constructed?

The first one needs just a topping with double distilled water once in a few months. The second one? Well it needs almost daily "charging."

A battery designed and fabricated has to be made ready for use. It has to be charged. A charged battery is an electrochemical source of energy. Ions wait to be initiated into a dynamic mode of electrochemical to electrical energy conversion. This is simply done by connecting it to a load. That is when "discharge" begins to happen. Enabling your cellphone or laptop or camera to function is an example.

A fully discharged battery is a primary battery such as the hearing aid battery that has to be discarded after use.

A fully discharged laptop battery has to be charged again, used, and then charged again. It is an apparently limitless cycle. The charge–discharge "drama" challenges us to explain the process in a dynamic and fundamental way. This "explanation" enables the development of better batteries, which includes (1) materials, (2) components, and (3) concept development and innovation.

To put the discussion in perspective, Figure 8.1 compares the button cell with a liquid electrolyte (1), and an all-solid state battery (2) employing a Li^+-ion glassy electrolyte whose structural unit is shown in Figure 8.1c. The cell of Figure 8.1b was designed for electrochemical measurements [8.4]. Figure 8.1d

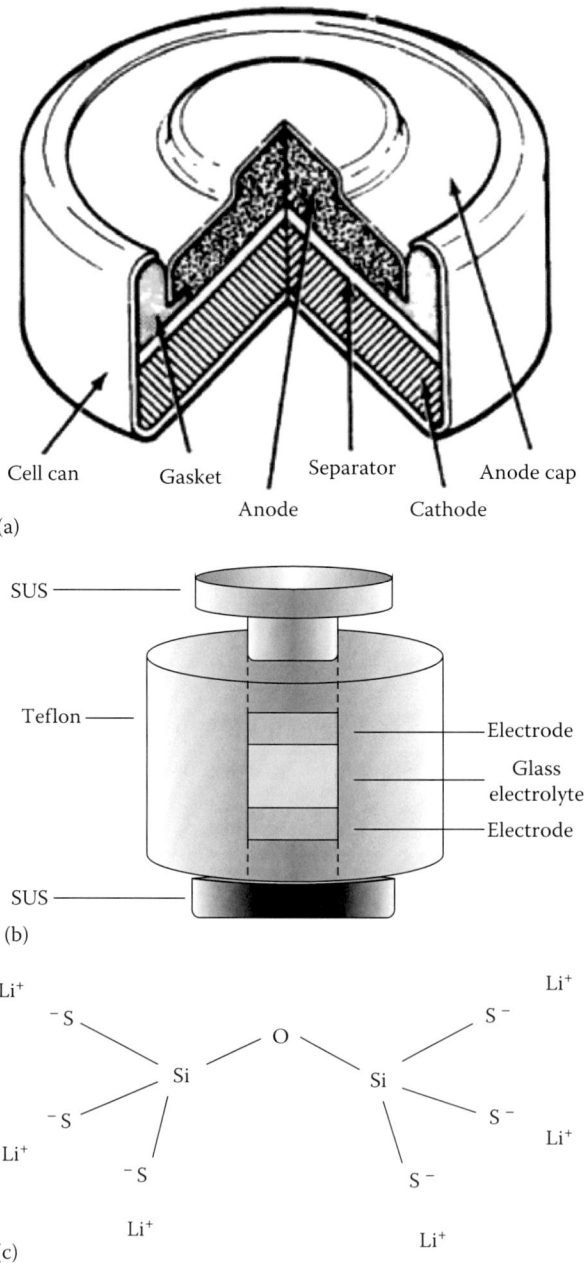

FIGURE 8.1 (a) An alkali-manganese-type button cell, (b) all-solid state cell, (c) glass electrolyte structural unit. (From Bard, A.J. et al. (eds.), *Electrochemistry Dictionary*, Springer-Verlag, Berlin, Germany, 2012.)   *(Continued)*

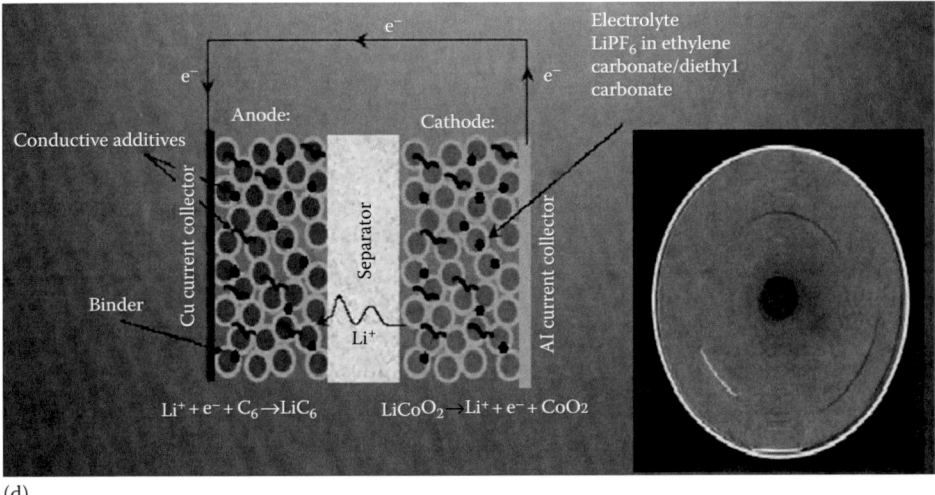

(d)

FIGURE 8.1 (*Continued*) (d) a Li ion battery with all parts labeled. (From Minami, T. (Ed. in chief), *Solid State Ionics for Batteries*, Springer-Verlag, Tokyo, Japan, 2005 p. 126.)

shows for comparison a schematic of a modern Li battery with all components labeled. The main issue involved in comparison is the expected absence of leakage and associated issues that are the bugbear of liquid electrolyte-based cells.

A battery is characterized by three parameters: (1) open-circuit voltage (abbreviated as OCV or VOC) is the difference of electrical potential between two terminals of a device when disconnected from any circuit. (2) Energy density expressed in two ways: (a) gravimetric energy density that is a measure of how much energy a battery contains in comparison to its weight and is typically expressed in Watt-hours/kilogram (Wh/kg) and (b) volumetric energy density, a measure of how much energy a battery contains in comparison to its volume and is typically expressed in Watt-hours/liter (Wh/L). For a LIB, the typical numbers are 90 and 210 (Wh/kg). (3) Batteries gradually self-discharge even if not connected and delivering current. Li^+ rechargeable batteries have a self-discharge rate typically stated by manufacturers to be 1.5%–2% per month.

8.2 Nature's Own Battery and the Car Battery

8.2.1 Nature's Own Battery

In nature, the energy created (usually through enzymes) is used immediately. To store that electrical potential for later use, one has to find a way to keep the enzyme in a charge-separated state for a longer period of time. It is thus possible to extend the length of time a battery-like enzyme can store energy from seconds to hours [8.5].

In its natural configuration, the enzyme is perfectly embedded in the cell's outer layer, known as the lipid membrane. The enzyme's structure allows it to quickly recombine the charges and recover from a charge-separated state.

However, when different lipid molecules make up the membrane, as in Kálmán's experiments [8.6], there is a mismatch between the shape of the membrane and the enzyme embedded within it. Both the enzyme and the membrane end up changing their shapes to find a good fit. The changes make it more difficult for the enzyme to recombine the charges, thereby allowing the electrical potential to last much longer.

These are biocompatible batteries with no toxic metals in them. Imagine batteries made of enzymes and other biological molecules. These could help monitor a patient from the inside postsurgery, and biocompatible batteries could be left inside the body without causing harm.

As a general comparison of solid state ionics to solid state electronics (Chapter 1) observe that the progress on improving battery capacity is slow compared to increases in the computer processing capacity. But then electrons do not take up space in a processor so their size does not limit processing capacity but lithographic constraints do. Ions in a battery are different. They do take up space and potentials are dictated by the thermodynamics of the relevant chemical reactions. Therefore, significant improvements in battery capacity can come only by changing to a different chemistry.

Let us start the discussion through a simple view of the battery.

A battery generally consists of three major components: (1) the anode, (2) the cathode, and (3) the electrolyte/separator (Figure 8.2).

The anode is the negative electrode, the cathode is the positive electrode, and the electrolyte/separator *physically separates*, but *electrochemically connects*, the two electrodes inside the battery. When a battery is connected to a circuit and is *discharging*, electrons leave the anode, flow through the external circuit, and return to the battery at the cathode.

When a battery is *charging,* an *external voltage source* is now part of the circuit.

When a sufficient voltage is applied to drive the electrons, they leave the cathode, flow through the external circuit, and return to the battery at the anode.

A typical LIB based on the layered oxide $LiCoO_2$ in which Li^+ can move freely "in and out" is a fairly small (~2 cm × ~7 cm), light-weight (~50 gm) device. It has a capacity of ~2500 mAh (0.2 C, 3.8 V) possessing an average voltage of 3.7 V, charge voltage of 4.2 V taking 2½ h to charge, an energy density of 530 Wh/dm^3, gravimetric energy density of 205 Wh/kg, and a cyclability of 80% at 300th charge–discharge cycle.

The temperature ranges of operation are as follows: (1) for charge 0°C–450°C, (2) for discharge from −200°C to +600°C, and (3) for storage from −20°C to +450°C. A polymer electrolyte battery is very thin (~0.5 × 3.5 × 0.7 cm^3) compared to a crystalline electrolyte battery weighing less than 20 g while all the other aspects are comparable. A schematic of an LIB illustrating its principle (a), the performance characteristics (b), [8.7] and the events during first charge (c) is shown in Figure 8.3.

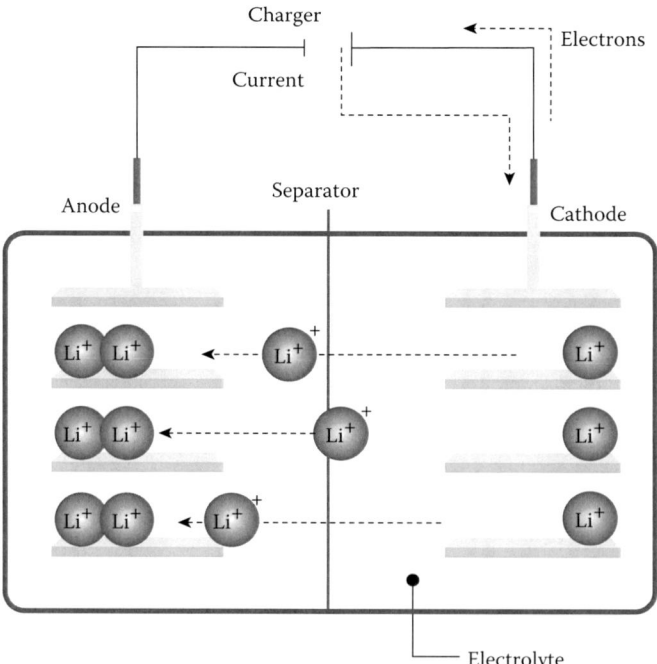

FIGURE 8.2 A typical battery with a charger current flowing in the external circuit. Note the specific function of the separator. (http://www.lgchem.com.) Note that Li-ion cells are charged using constant voltage (*C–V*) chargers. They can accept only 1 h recharges taking typically 1.5 h for fast charging.

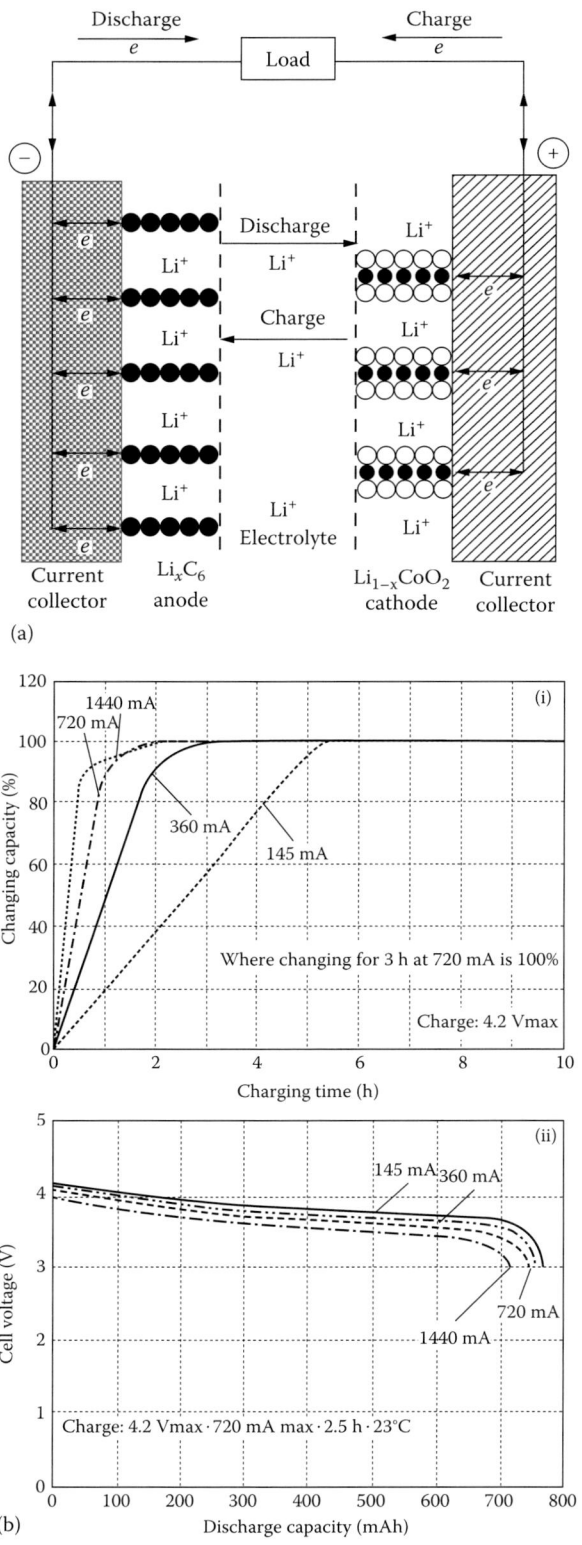

FIGURE 8.3 (a) Illustrating the action of an LIB. (b) Battery characteristics: (i) charging and (ii) discharging through load. (From Sony, *Lithium Ion Battery Technical Handbook*, www.sony.com.) (*Continued*)

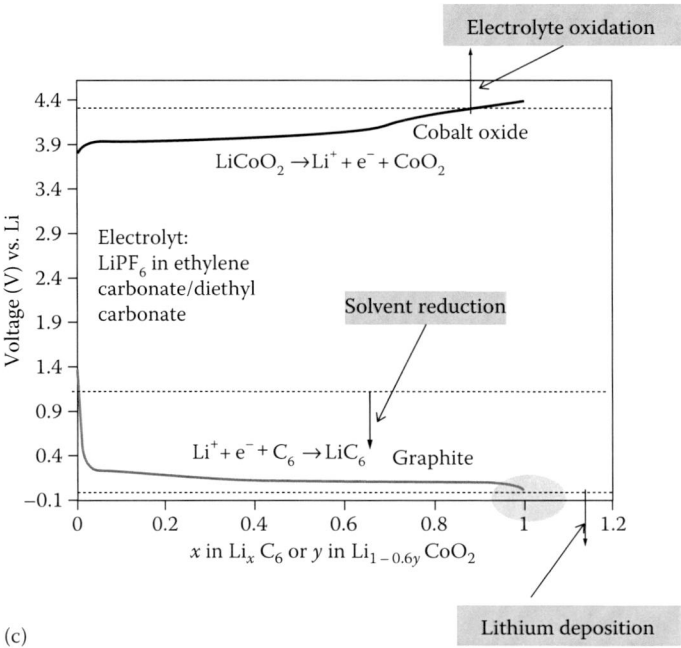

(c)

FIGURE 8.3 (*Continued*) (c) Illustrating events during the first charge. (Dunning, J. and Forbes, D., The inside story of the Li ion battery, http://www.calpoly.edu/~research/documents/The_Inside_Story_of_the_Lithium_Ion_Battery_0511.pdf, modified December 3, 2014.)

Essentially, the LIB possesses an initial charge: during manufacture, Li-ion migrates from Li compound of the cathode ($LiCoO_2$) to carbon material of the anode:

$$LiCoO_2 + C \rightarrow Li_{1-x}CoO_2 + Li_xC \qquad (8.1)$$

Subsequent discharge reactions occur through the migration of Li ions from anode back to the cathode:

$$Li_{1-x}CoO_2 + Li_xC \underset{charging}{\overset{discharging}{\rightleftarrows}} Li_{1-x+dx} + Li_{x-dx}C \qquad (8.2)$$

The state-of-the-art LIBs are considered to be the best rechargeable batteries on the market. They consist of a graphitic-carbon anode, a liquid electrolyte comprising lithium salts dissolved in organic solvents, a microporous polymer separator, and a lithium-intercalated transition-metal-oxide positive electrode. Their theoretical specific energy is about 400–500 Wh/g.

Let us now discuss briefly the car battery.

8.2.2 Car Battery

The good old lead–acid battery or the car battery helps in the understanding of the basic processes at work in an electrochemical device. More importantly, it allows us to develop the electrochemical principles behind its construction and operation (Figure 8.4).

Electrolyte contains aqueous ions H^+ and SO_4^-. The conduction mechanism is through migration of ions via drift and diffurion.

The car battery is essentially an SLI or starter-lighting-ignition battery designed to provide short burst of high current of ~500 A to crank the engine. It has a typical lifetime of 500 cycles at 25% depth of edischarge.

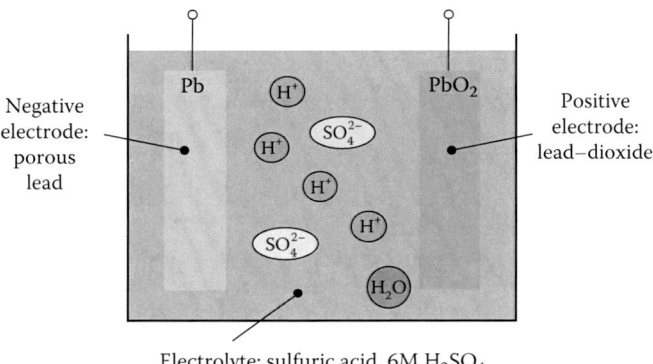

Electrolyte: sulfuric acid, 6M H_2SO_4

FIGURE 8.4 The anode (porous Pb), the liquid electrolyte (6M H_2SO_4), and the cathode (PbO_2) are the main components of the lead–acid battery.

Figure 8.5 compares the charge–discharge characteristics and self-dischrge of a sealed lead–acid battery with others including LIB.

Pollet et al. [8.8] consider in detail relations of the lead–acid battery with other types of batteries.

With the advent of battery-aided vehicles for ground transportation, lead–acid batteries are again in contention. Of course, large uninterruptible power applications use flooded lead–acid batteries as a back-up solution, with a service life of over 20 years. Negative plates made of lead or lead alloy are sandwiched between positive plates made of lead or lead alloy with calcium or antimony as additive. The insulator or

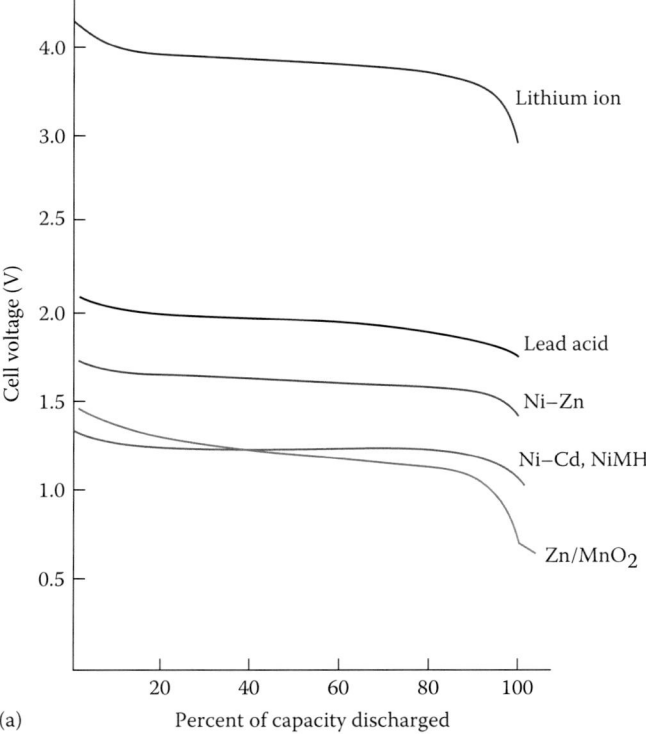

(a)

FIGURE 8.5 (a) Discharge characteristics of lead–acid battery compared with Li-ion, Ni–Zn, Ni–Cd (NiMH), and Zn/MnO₂ batteries. *(Continued)*

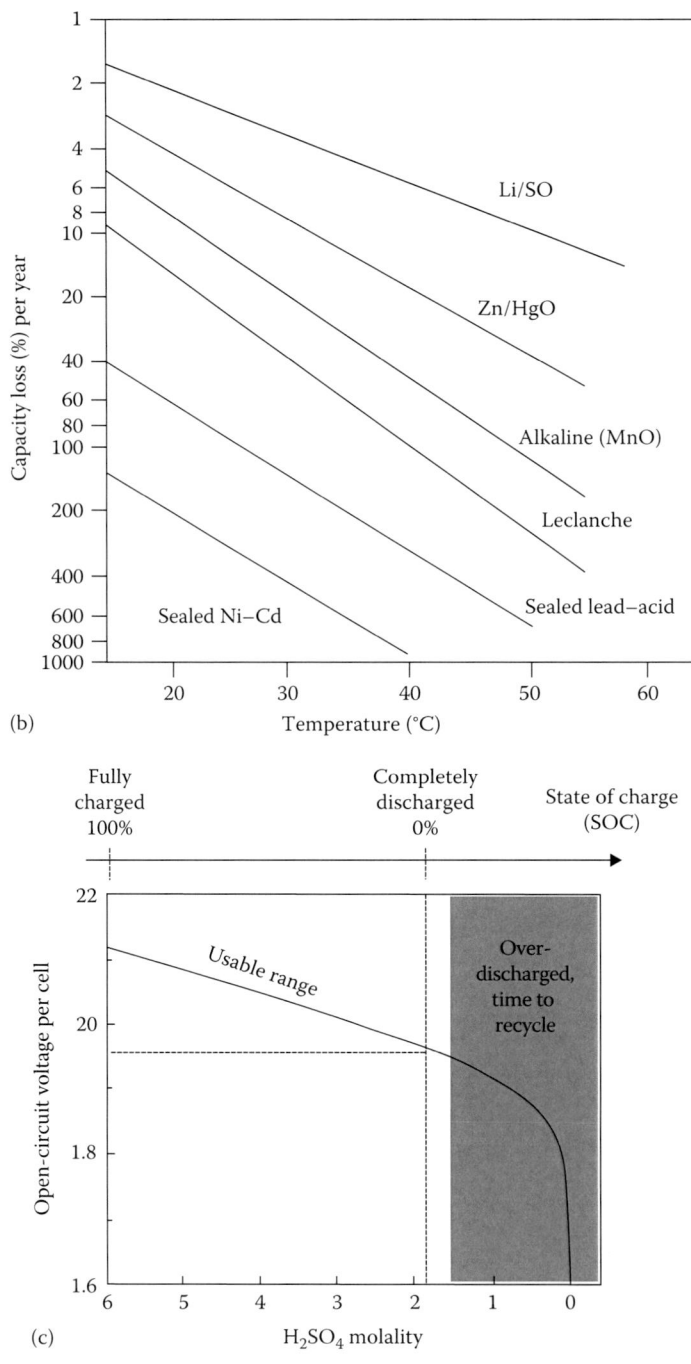

FIGURE 8.5 (*Continued*)   (a) Discharge characteristics of lead–acid battery compared with Li-ion, Ni–Zn, Ni–Cd (NiMH), and Zn/MnO$_2$ batteries. (b) Self-discharge of sealed lead–acid battery compared with other batteries. (c) A "theoretical" open-circuit voltage vs. the state of charge/molality of H$_2$SO$_4$ curve for the lead–acid battery. (From York, R.A., Lead-acid batteries, Lecture notes, University of California, Santa Barbara. http://my.ece.ucsb.edu/York/Bobsclass/194/default.htm.)

separator is a microporous material that allows chemical reactions to take place while preventing the electrodes from shorting owing to contact. The negative and positive plates are pasted with an active material lead dioxide or lead sulfate. The active material provides a large surface area for storing electrochemical energy. Each positive plate welded together is attached to a "+" terminal post. The same welding is used to weld each negative plate and is attached to the "−" terminal post. Plate assembly is housed in a polypropylene casing with a cover provided with a vent cap. The container assembly and the cover plate are glued to form a leak-proof seal. The container is filled with an electrolyte solution—a combination of sulfuric acid and distilled water—made up to a specific gravity of 1.215.

Upon charging or application of an electric current, electrochemical reactions are initiated in the battery creating the cell's potential or voltage. Interestingly, the Nernst–Einstein equation relates, in a useful approximation, the chemical reaction energy E at a given concentration to electrolyte energy E^0 at standard 1 molar concentration: $E = E^0 + k_B T/q \ln Q$, where Q is the molar concentration and $k_B T/q = 26$ mV at 298 K. See Reference 8.9.

A "theoretical" discharge curve of the lead–acid battery is shown in Figure 8.5c.

8.3 Li$^+$-Ion Physics and Battery Components

Let us now delineate the elements of the LIB. To put the discussion in perspective, Figure 8.6 gives a bird's eye view of the functional "positive" and "negative" materials with Li metal as the reference. The three functional elements of the LIB, namely the anode, the cathode, and the electrolyte have specific needs as detailed in Table 8.1.

Let us look at the elements one by one.

8.3.1 Cathode

What are the conduction processes in a cathode particle, say, of LiFePO$_4$? Refer Figure 8.7.

Consider a single composite cathode particle—in contact with conductive additive and binder—during charge. When a Li$^+$ ion diffuses out of the cathode giving rise to ionic conduction during the

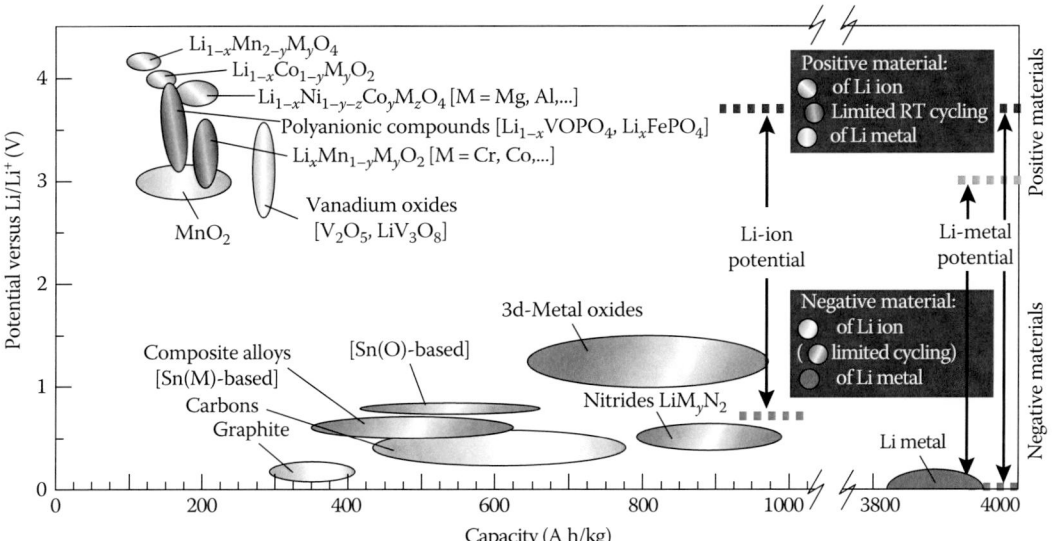

FIGURE 8.6 Voltage versus capacity plots for various positive and negative materials. The tremendous difference in capacity between Li metal and other negative electrodes makes the problem of dendrite growth a serious one. (From Tarascon, J.-M. and Armand, M., *Nature*, 414, 395, 2001.)

TABLE 8.1

Li Ion Battery Elements

Element	Specific Need	Examples
Battery anode	Electropositive material; no reactions or solubility in electrolyte; high σ_{el}	Li and other metals
Battery cathode	Electronegative material; high σ_{el}	NiS, FeS$_2$, Li$_x$CoO$_3$, PbI$_2$, I$_2^+$P$_2$VP
Battery electrolyte	Low leakage current; chemical inertness, Low σ_{el}	β-alumina for the Na/S battery; Al$_2$O$_3$:Li for pacemaker, LiI for pocket calculator, also Li$_3$N, LiPON

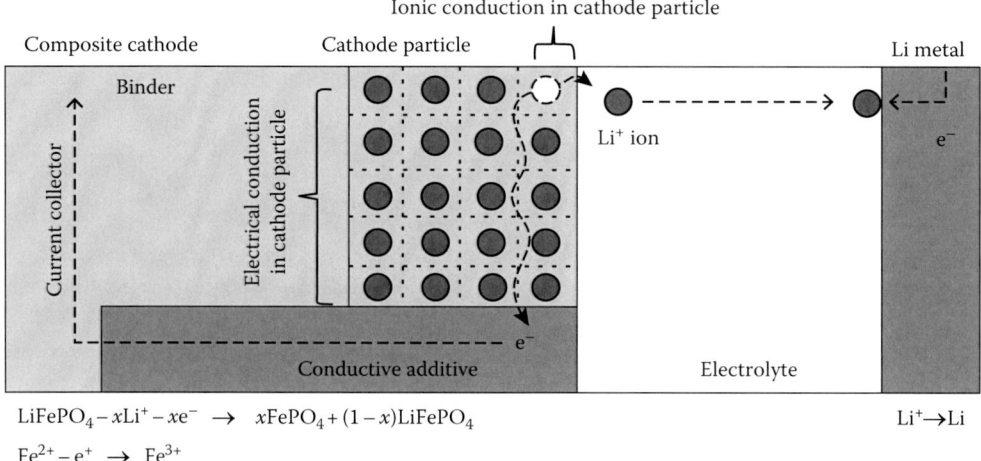

$$LiFePO_4 - xLi^+ - xe^- \;\rightarrow\; xFePO_4 + (1-x)LiFePO_4 \qquad\qquad Li^+ \rightarrow Li$$

$$Fe^{2+} - e^+ \;\rightarrow\; Fe^{3+}$$

FIGURE 8.7 Conduction processes in a LiFePO$_4$ cathode particle. The relevant chemical reactions are shown at the bottom. (From Park, M. et al., *J. Power Sources*, 2010.)

charge cycle, the valence state of the transition metal ion changes giving rise to electronic conduction. We say Fe^{++} ion is oxidized to Fe^{+++}. The simultaneously occurring electronic (or electrical) and ionic conductivities must be in principle optimized in cathode materials. This is essentially because either of these properties can dictate the overall cell properties mainly capacity and cycle life. However, diffusion generally governs charge/discharge rates so that ionic conductivity becomes more important. The Li$^+$ ion diffusivity and electrical conductivity of 2D layered structure LiCoO$_2$ are from 1.5×10^{-10} to 8.0×10^{-8} (single particle) and 0.2 S/cm (single crystal). Such high values (compared to other cathodes LiMn$_2$O$_4$ and LiFePO$_4$) point to its higher realized theoretical capacity.

The fundamental process of diffusion of a Li$^+$ ion in a cathode particle depends on the Li ion–host material (structure) interaction. The Huggins model describes the diffusion path in terms of relevant energies as follows.

Total potential energy W_T = Coulomb interaction W_C + van der Waals interaction W_P + overlap repulsion W_R between closed shell ions. The 2D layered structure of LiCoO$_2$ ensures highest Li ion diffusivity. First-principles calculations on Li$_x$CoO$_2$ ($0 \leq x \leq 1$) have helped find reasons for wide variations in diffusivity based on activation barrier change and divacancy mechanism.

LiFePO$_4$ is currently under active study as a potential cathode for LIBs. Thin film form of LiFePO$_4$ seems to be attractive. Figure 8.8a shows microstructure and impedance characteristics of thin film cathode deposited on Al foil. The anode, the cathode, and the electrolyte are all seen performing their role in transport. Figure 8.8b shows the impedance (Z) of the composite cathode deposited on Al foil serving as a current collector (top left) and silver paste substrate (top right). The effect of substrate is quite considerable. Al foil substrate gives rise to a high frequency arc in the Cole–Cole plot, which has

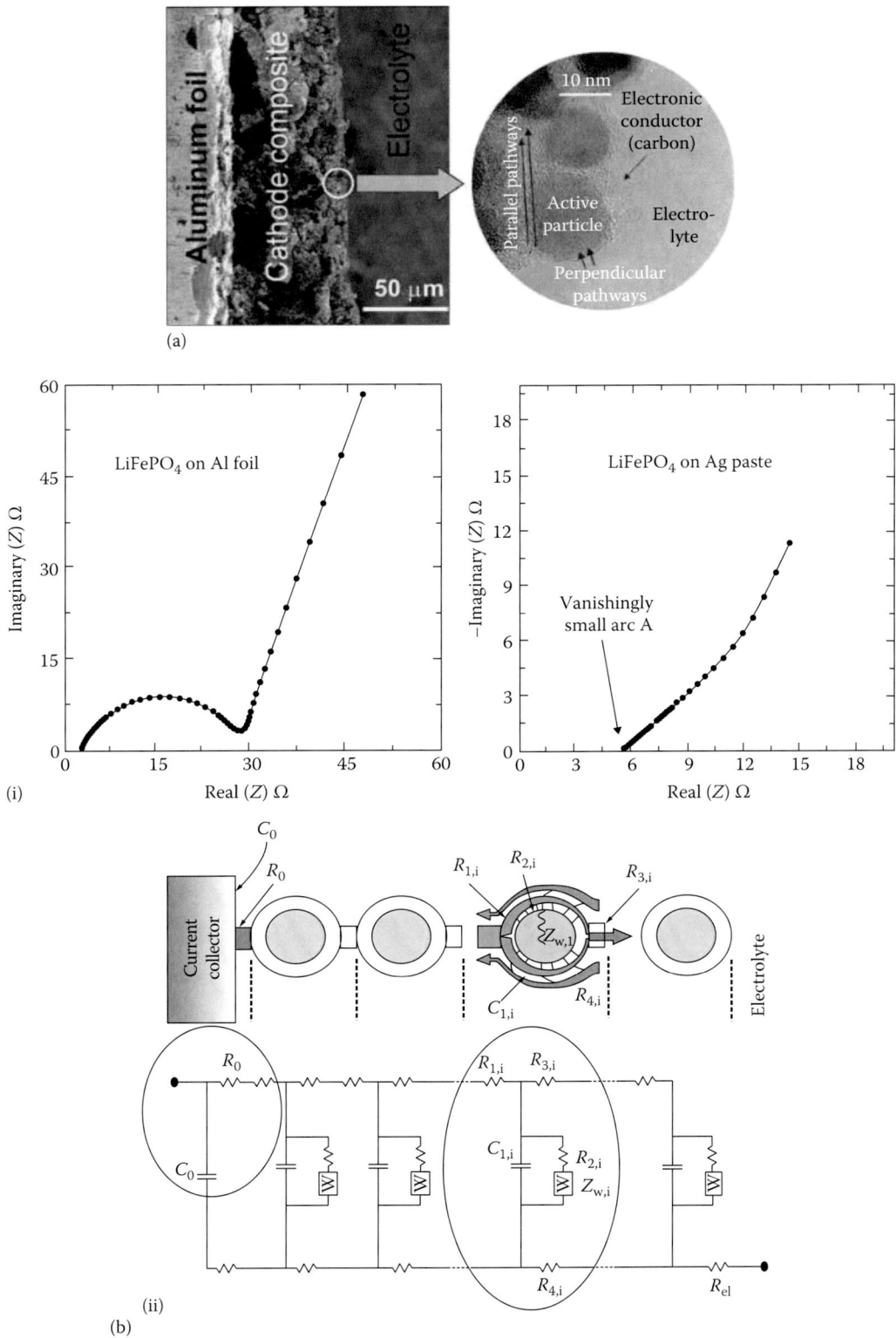

FIGURE 8.8 LiFePO$_4$ thin film/Al foil as a Li ion battery cathode. (a) Microstructure with active particle. (b) Cole–Cole plots (i) and circuit model (ii). (From Jamnik, J. and Gaberscek, M., *MRS Bull.*, 34, 942, 2009.)

been routinely attributed to the Li incorporation resistance into Li_xFePO_4 together with the $LiFePO_4$/ electrolyte interfacial capacitance forming an RC circuit. But it actually corresponds to the poor interfacial contact between Al foil and cathode composite, which may be improved by using a silver paste. Typical contacts found in a Li battery and their approximate equivalent circuits are shown at the bottom of Figure 8.8b. This is an ingenious attempt to map the electrode microstructure and the resulting impedance spectrum [8.12]. It is instructive to understand the meanings of the R's and C's that occur in the equivalent circuit. R_0 is the resistance for electron transfer from the metallic substrate to the first layer of active particles. $R_{1,i}$ is the electronic resistance carbon coating while $R_{2,i}$ is charge incorporation resistance between the active particles and the electrolyte. $R_{3,i}$ is the inter-particle contact resistance while $R_{4,I}$ is the resistance of the ionic pathways between the particles. C_0 is the double-layer capacitance at the surface of the current collector while $C_{1,I}$ is the double-layer capacitance at the active particle/electrolyte interface. Finally, $Z_{w,i}$ represents the finite-length chemical diffusion in the solid active particles while R_{el} is the bulk resistance. The perpendicular pathways indicated in Figure 8.8a refer to the effect of incorporation of charge from external phases (electrolyte or carbon black) to the active insertion particle namely Li^+. This analysis demonstrates the role of electronic transfer from the current collector to the composite cathode material.

8.3.2 Electrolyte

The role of the electrolyte is to provide an ion conducting path between the anode and the cathode. Enhancement of ionic conductivity is therefore the major goal in battery development. A room temperature conductivity of at least 1 mS/cm is the basic requirement for a functional electrolyte in a consumer battery. Electrolytes could be liquid or solid and the latter category includes micro-/nanocrystalline forms besides glassy/glass ceramic/polymer types. Let us consider these in brief.

8.3.2.1 Liquid Electrolytes

Liquid electrolytes for most part have a higher Li ionic conductivity than their solid counterparts. They naturally allow free passage of mobile ions from anode to cathode and vice versa. Further, they ensure a good liquid to solid contact between the electrolyte and an electrode. Thus, they remain the most common electrolyte in today's rechargeable Li batteries. Solid $LiPF_6$ in a nonaqeous solvent (usually ethyl carbonate) has been the ubiquitous choice. A porous separator prevents touching of electrodes.

Suppose $LiPF_4$ is dissolved in an organic solvent. Then Li^+ and PF_4^- are produced by dissociation. Dissociation of a salt depends on the dielectric constant of the solvent implying the strong solvating power. As the Li^+ ions are surrounded by solvent molecules, the influence of the anion is reduced. Solvents with larger anions distribute their charge better preventing ion pairing and thus optimizing conductivity and solubility. Furthermore, a lower viscosity facilitates better ion movement.

8.3.2.2 Solid Electrolytes

Although one always speaks of "solid electrolytes," most practical battery systems employ liquid electrolytes that despite being better than solid electrolytes in terms of overall conductivity cause the twin problems of leakage and occasional explosion. These drawbacks arise from high temperature and overcharge. Thus, the development of solid state ionic materials that can effectively replace liquid electrolytes is important to achieve a truly all-solid state battery. The solid electrolyte enables batteries enable the use of higher energy electrode materials.

The most important parameter of electrolytes used in electrochemical cells is ionic conductivity. The use of solid state electrolytes has been limited due to low ionic conductivity caused by their immobile matrix regardless of their own merits such as no leak, nonvolatility, mechanical strength, and processing flexibility. Another important parameter we should consider is transference of the number of ions. Electrolytes are characterized by their ionic conductivity. It is desirable that overall ionic conductivity results from the dominant contribution of the ions of interest. However, high values of the cationic transference number achieved by solid or gel electrolytes have resulted in low ionic conductivity, leading to inferior cell performances. In a recent development, organogel electrolyte for LIB has been discovered with ionic conductivity close to liquid electrolytes but with better thermal stability. Organogel networks can form via polymerization. This mechanism converts a precursor solution of monomers with various reactive sites into polymeric chains that grow into a single covalently linked network. At a critical concentration (the gel point), the polymeric network becomes large enough so that on the macroscopic scale, the solution starts to exhibit gel-like physical properties: an extensive continuous solid network, no steady-state flow, and solid-like rheological properties The two novel characteristics of this organogel are (1) an irreversible thermal gelation and (2) a high value of the Li^+ transference number [8.13].

Another type of solid electrolytes under development are the solid composite polymer electrolytes consisting of high-molecular-weight polyethylene oxide (PEO) with KI and $NaClO_4$ as electrolyte salts reinforced with additives such as CeO_2 and ZrO_2 nanoparticles. Room temperature conductivity of ~2 mS/cm has been realized in PEO_{15}–KI doped with 20% CeO_2. Addition of nanoscale ceramic fillers helps improve the mechanical and potential stability of the electrolyte. A Lewis acid–base reaction between the ceramic filler and the polymer makes it happen. Lewis acid sites existing on the surface of these nanoparticles and Lewis base centers of ether oxygen found on the polymer act as the prime movers of the reaction [8.14].

8.3.2.2.1 Li-Based Solid Electrolytes

Three examples of oxide-based electrolytes are of interest.

1. $Li_{3x}La_{(3/2)x}TiO_3$. Possessing a perovskite structure with $\sigma \sim 1$ mS/cm but grain boundary resistance lowers conductivity by two orders.
2. Li phosphate with NASICON structure: $Li_{1.3}Al_{0.3}Ti_{1.7}(PO_4)_3$ with $\sigma_{total} \sim 10$ mS/cm which is much better than (1), although Ti(IV) is common to both. Note that Ti is easily reduced around ~1.8 V versus Li/Li^+ for (a) and ~2.5 V versus Li/Li^+ for (b).
3. Lithium sulfide solid electrolytes
   a. Li_2S–GeS_2–P_2S_5 crystal stable against oxidation is suitable as cathode.
   b. LiI–Li_2S–P_2S_5 glass stable against reduction is suitable as anode [8.15].
4. The layered Li–Ni–Mn oxide with or without Co is a promising candidate material. $Li_{1/3}Mn_{1/3}Co_{1/3}$ is interesting from the point of view of (a) rate capability, (b) cycle life, and (c) safety.

8.3.2.2.2 Garnet-Type $Li_7La_3Zr_2O_{12}$.

Recent research has shown that certain Li-oxide garnets with high mechanical, thermal, chemical, and electrochemical stability are excellent fast Li-ion conductors. However, the detailed crystal chemistry of Li-oxide garnets is not well understood, the relationship between crystal chemistry and conduction behavior is not clear either. An investigation undertaken by Geiger et al. [8.15] has provided an understanding of the crystal chemical and structural properties, as well as the stability relations, of Li(7) La(3)Zr(2)O(12) garnet, which is the best conducting Li-oxide garnet discovered to date. Two different sintering methods produced Li-oxide garnet but with slightly different compositions and different grain sizes: (1) the first method involves ceramic crucibles in initial synthesis steps and later sealed Pt capsules, which produced single crystals up to roughly 100 μm in size. Electron microprobe and laser ablation inductively coupled plasma mass spectrometry (ICP-MS) measurements show small amounts of Al in the garnet, probably originating from the crucibles. The crystal structure of this phase was determined using x-ray single-crystal diffraction every 100 K from 100 to 500 K. The crystals are cubic (Figure 8.9a) with space group $Ia\overline{3}d$ *at all temperatures*. The atomic displacement parameters and Li-site occupancies were measured. Li atoms could be located on at least two structural sites that are partially occupied, while other Li atoms in the structure appear to be delocalized. ^{27}Al NMR spectra show two main resonances that are interpreted as indicating that minor Al occurs on the two different Li sites. Li NMR spectra show a single narrow resonance at 1.2–1.3 ppm indicating fast Li-ion diffusion at

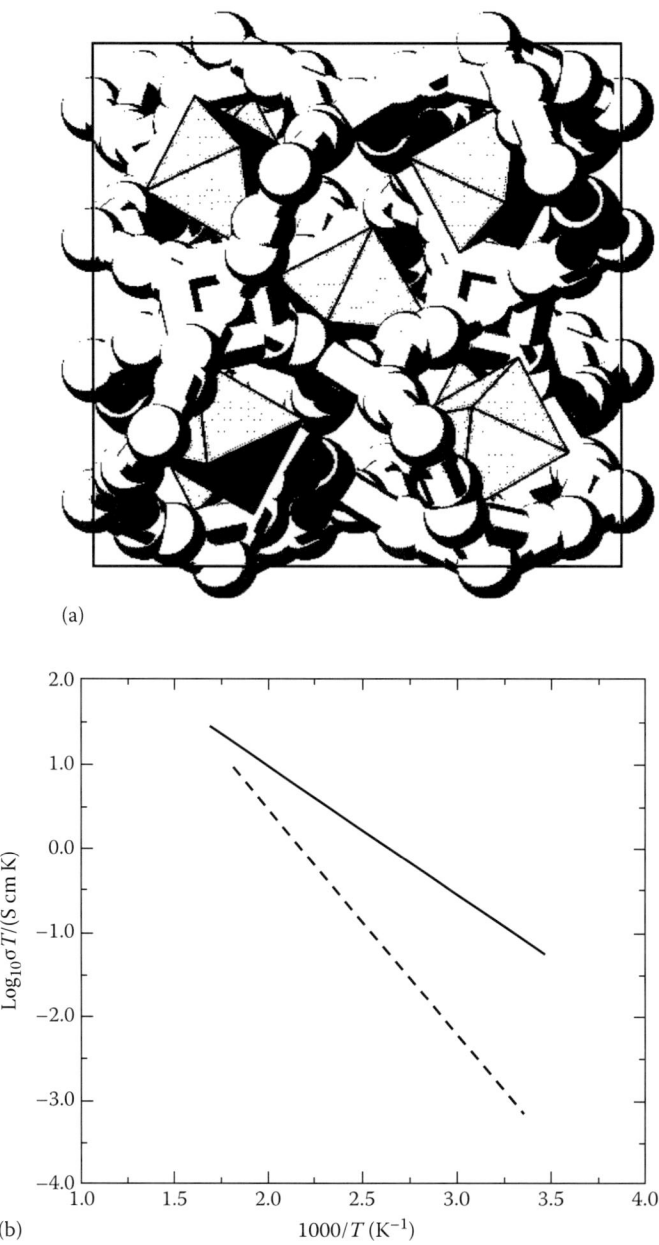

(a)

(b)

FIGURE 8.9 (a) Crystal structure of the cubic fast lithium ion conductor: Garnet-type $Li_7La_3Zr_2O_{12}$ showing ZrO_6 octahedra and the mobile Li^+ ions. The unit cell edge measures 1.29682 nm. (b) Arrhenius plots of conductivity for cubic (——) and tetragonal (------) phases. (From Geiger, C.A. et al., *Inorg. Chem.*, 50, 1089, 2011.)

room temperature. The chemical shift value indicates that the Li atoms spend most of their time at the tetrahedrally coordinated C (24d) site. (2) The second synthesis method, using solely Pt crucibles during sintering, produced fine-grained Li(7)La(3)Zr(2)O(12) crystals. This material was studied by x-ray powder diffraction at different temperatures between 25°C and 200°C. This phase is tetragonal at room temperature and undergoes a phase transition to a cubic phase between 100°C and 150°C. Cubic "Li(7) La(3)Zr(2)O(12)" may be stabilized at ambient conditions relative to its slightly less conducting tetragonal modification via small amounts of Al^{3+}. Several crystal chemical properties appear to promote the high Li-ion conductivity in cubic Al-containing Li(7)La(3)Zr(2)O(12) (Figure 8.9b). They are (1) isotropic three-dimensional Li-diffusion pathways, (2) closely spaced Li sites and Li delocalization that allow for easy and fast Li diffusion, and (3) low occupancies at the Li sites, which may also be enhanced by the heterovalent substitution Al(3+) ⇔ 3Li.

8.3.2.2.3 PVA/Ionic Liquid-Based Polymer Electrolytes

Ionic liquids (ILs) are room-temperature molten salts that possess unique properties, such as negligible vapor pressure, good thermal stability, and nonflammability, together with high ionic conductivity and a wide window of electrochemical stability. They also act as an ion source and a plasticizer for solid polymer electrolyte films. Combining ILs with polymer electrolytes offers the prospect of new applications in batteries and fuel cells, where they surpass the performance of conventional media such as organic solvents (in batteries) or water (in polymer electrolyte membrane fuel cells), resulting in improved safety and a higher operating temperature range. However, the most important challenge is how to immobilize ILs in polymer matrices (generally based polyvinyl alcohol (PVA)–LiClO$_4$ composites) while retaining their sought-after properties [8.16]. The use of an IL such as 1-ethyl-3 methyl imidazolium ethylsulfate or EMIM EtSO$_4$ was found to significantly improve the Li$^+$-ion conductivity besides allowing a comprehensive investigation of dielectric relaxation behavior [8.17].

8.3.3 Anode

The anode is an electrode where oxidation occurs and electrons flow from electrolyte to electrode. (At the cathode electrons flow from electrode to electrolyte.) Thus, an anode is the negative electrode in a battery (but a positive electrode in electrolysis!). The redox species that gives electrons to the anode is not necessarily an anion.

Anode materials: Although Li and graphite are currently being employed as anode materials, high capacity materials such as Si are currently being explored. Si, however, causes deterioration of electrochemical performance due to large volumetric changes in charge–discharge cycles due to its open crystal structure. Nanotechnology could help solve this problem.

Figure 8.10 shows a LIB in operation during charge and discharge cycles bringing into full play the functions of the anode and the cathode.

Let us discuss a couple of anode materials. Spinel-structured Li$_4$Ti$_5$O$_{12}$ is a reliable anode for LIB because it (1) has good Li insertion/deinsertion or intercalation/deintercalation reversiblity and (2)

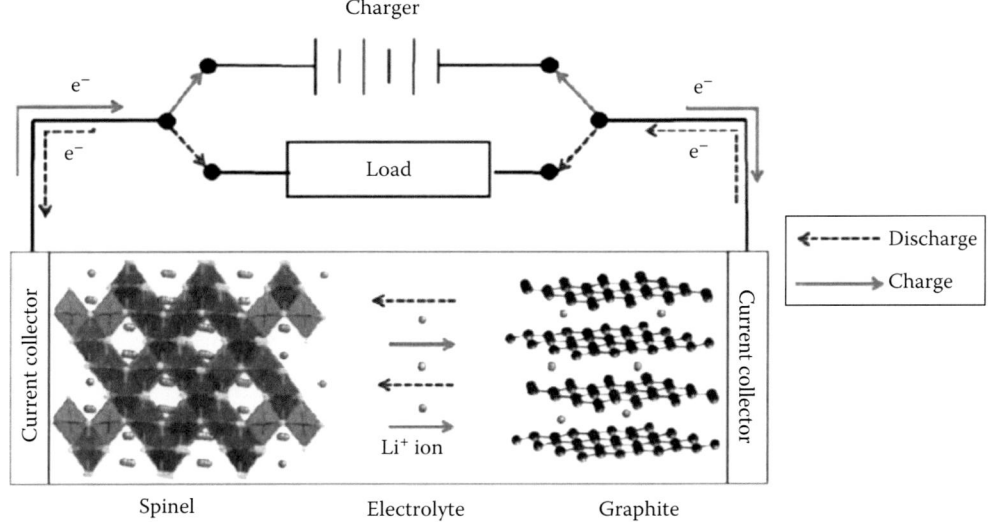

FIGURE 8.10   A Li$^+$ ion battery in operation. Li spinel cathode and graphite anode are used.

exhibits no structural change during charge–discharge cycling. Ink-jet processed (printed) $Li_4Ti_5O_{12}$ anode thin films have very desirable electrochemical characteristics [8.18] as depicted in Figure 8.11.

The utility of a graphene paper (GP) as an anode material has been demonstrated by Liu et al. [8.19]. Figure 8.12 shows the SEM and digital photos of GP and the charge–discharge profiles of Li/GP half cell. Cycle life experiments show that facile Li intercalation takes place after 15 cycles.

Titania (TiO_2) with anatase structure has been explored as a viable Li battery anode material [8.20a]. Heterogeneous nanostructured electrodes using carbon nanosheets (CNS) and TiO_2 exhibit high electronic and ionic conductivities. CNS with an optimum amount of TiO_2 coating is proposed as a promising approach for the fabrication of electrodes for chip compatible thin-film LIBs [8.20b].

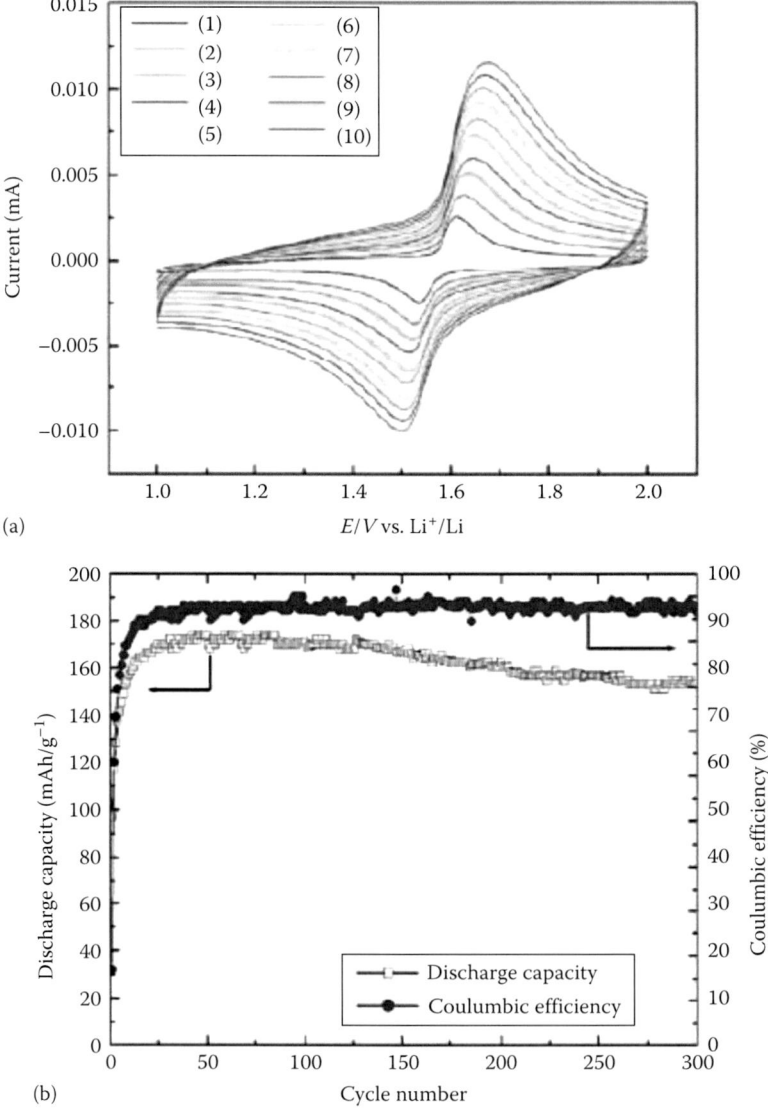

FIGURE 8.11 (a) Cyclic voltammograms of as-prepared $Li_4Ti_5O_{12}$ thin film electrode for scan rates of 0.1–1 mV/s in 10 equal steps 0.1 displayed as 1–10. (b) Discharge capacity and Coulombic efficiency of as-printed $Li_4Ti_5O_{12}$ thin film electrode as functions of cycle number during first 300 cycles at a constant discharge current density of 10.4 µA/cm² in potential range 1.0–2.0 V. (From Zhao, Y. et al., *J. Solid State Electrochem.*, 13, 705, 2009.)

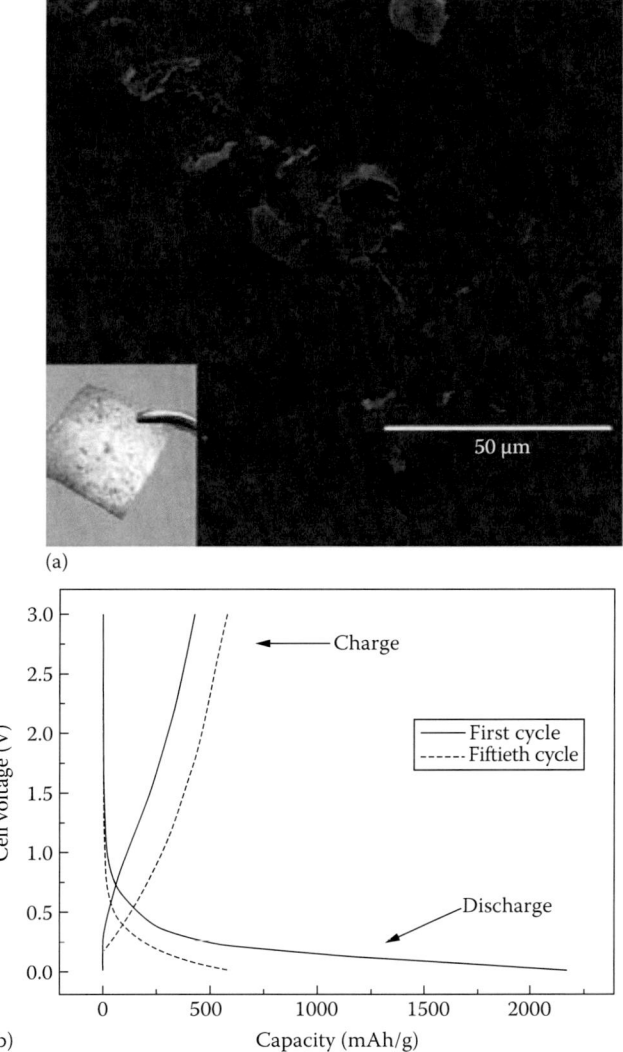

(a)

(b)

FIGURE 8.12 SEM and digital photo (inset) of graphene paper (GP) anode material. (b) Charge–discharge profiles of Li/GP half cell under 50 mA/g current rate. (From Liu, X. et al., *Eur. Phys. J. Appl. Phys.*, 66, 30301, 2014.)

High-energy-capacity LIBs using Si film anodes are found to have a long cyclic life by successful relaxation of the stress induced during lithiation and delithiation. A soft elastomer substrate in the Si anode plays a beneficial role for improved performance of the battery, which suggests a general way to solve the bottleneck problem for Si anodes [8.20c].

The separator, a vital component of the battery, is discussed next.

8.3.4 Separator

A separator is a porous membrane placed between electrodes of opposite polarity permeable to ion flow but forbidding electrical contact of electrodes. Separators have been manufactured from cellulose in papers and cellophane to nanowoven fabrics, foams, ion-exchange membranes, and microporous flatsheets made from polymeric materials. Separators play a key role in all batteries. Their main function is to keep positive and negative electrodes apart to prevent electrical short circuit. At the same time, rapid transport of ionic charge carriers is essential to complete the circuit during the passage of current

TABLE 8.2

Two Celgard Separators

Separator/Properties	Celgrad 2730	Celgrad 2400
Structure	Single layer	Single layer
Composition	PE	PP
Thickness (μm)	20	25
Gurley (s)	22	24
Ionic resistivity (Ω cm^2)	2.23	2.55
Porosity (%)	43	40
Melt. temp. (°C)	135	65

Source: Arora, P. and Zhang, Z.J., *Chem. Rev.*, 104, 4419, 2004.
Note: Gurley measured in seconds is a measure of mechanical strength of the separator.

in an electrochemical cell. They should be good electronic conductors and capable of conducting ions by either intrinsic ionic conductor or by soaking electrolyte. Properties of two commercial separators are given in Table 8.2.

Typically, a separator used in an LIB is <25 μm thick and has a resistance of <2 Ω cm^2, a pore size of <1 μm, a porosity of 40%, and a high + temperature integrity of >150°C.

A Na ion conducting gel polymer electrolyte successfully competes with Celgard 2730 with nearly four times higher conductivity than Celgard at RT (0.60 mS/cm) and nearly double the transport number of 0.30.

Next, we will present a brief discussion on Li–air battery.

8.4 Li–Air (Li–Oxygen) Battery

Electrochemical power sources based on metal/oxygen chemical couples are unique because oxygen, the cathode active material, does not have to be stored in the battery, but rather it can be accessed from the environment.

Li–air batteries with air as cathode, based on their high theoretical specific energy (about 3458 Wh/kg), are an attractive technology for electrical energy storage [8.22]. They could make long-range electric vehicles widely affordable. The key to this high energy density is the presence of "air," since the batteries capture atmospheric oxygen to use in the cathode reaction instead of storing their own oxidizing agent. In nonaqueous Li/air cell, reversible chemistry with a high current efficiency needs to be established. The deposition of an electrically resistive reaction product apparently limits the capacity. In nonaqueous Li/air batteries, the two reactions of interest are:

$$2Li + \tfrac{1}{2}\,O_2 \leftrightarrow Li_2O \tag{8.3}$$

$$2Li + O_2 \leftrightarrow Li_2O_2 \tag{8.4}$$

Full reduction of O_2 to Li_2O is desirable because of its high specific energy and energy density but Li_2O_2 forms more readily than Li_2O.

A typical experimental nonaqueous Li/air cell has a piece of Li metal, a nonaqueous solvent, a separator (glass fiber or Celgard) and an air electrode (typically carbon black), a polymer bider (PVDF or PTFE or cellulose), and an organic solvent (NMP, acetone). Slurry is coated onto a metal grid or metal foam by a film coating process. Commonly used positive electrode supports are Al grid or Ni foam. Air cathode should have (a) high surface area at reasonable pore volume, (b) good electronic (>1 S/cm) and ionic(>10 mS/cm) conductivities, and (c) a design that supports fast gas transport to the reaction center during discharge.

Figure 8.13a shows a typical Li–air gas diffusion cathode.

Figure 8.14 shows a Li–air cell designed by Zhang et al. [8.23] for ambient operation.

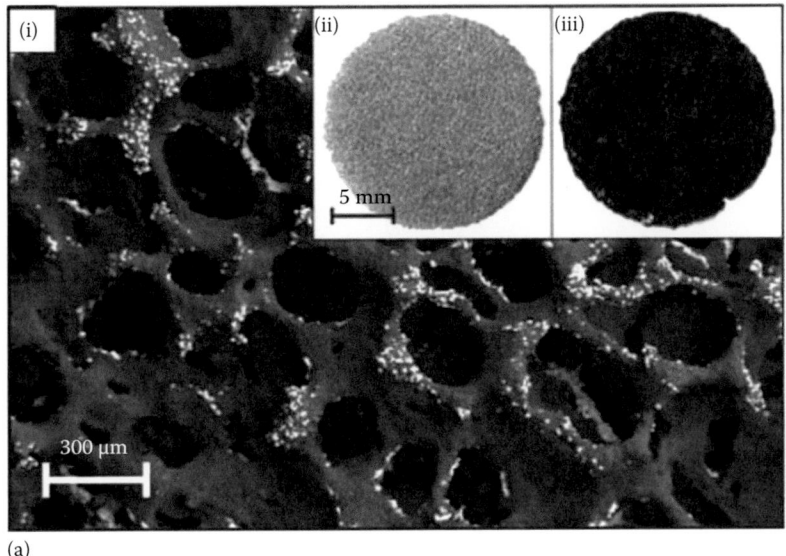

(a)

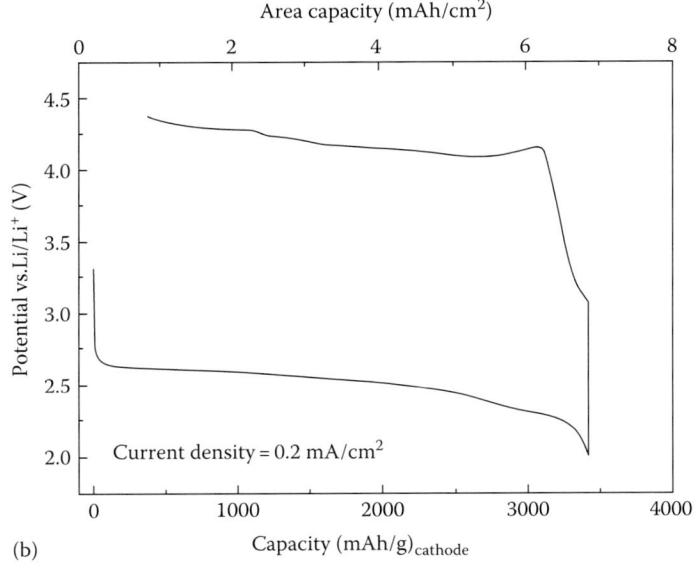

(b)

FIGURE 8.13   (a): (i) Ni-foam-based Li–air cathode coated with 5 wt.% Super P carbon black, 40% PVDF binder, and 10% MnO_2 additive. Inset (ii): SEM image of coated Ni foam and (iii) optical image of uncoated Ni foam. (b) Sample discharge and charge curve for a nonaqueous Li/air battery with a hydrophobized gas diffusion cathode. Cycling done at a current density of 0.2 mA/cm². The electrolyte contains 1 M $LiPF_6$ in TEGDME. The cathode as described in (a). Loading in each case is 2 mg/cm². Electrode thickness is 50 μm. The capacity is normalized to the total mass of the cathode excluding the current collector. Top *x*-axis shows the capacity normalized to electrode surface for comparison. (From Christensen, J. et al., *J. Electrochem. Soc.*, 159, R1, 2012.)

Ou et al. have developed an oxide-on-metal inverted design of oxygen electrocatalysts for nonaqueous Li/air batteries [8.24]. Apparently Li/air batteries are on a fairly long path of research and development.

Replacing the lithium anode with a sodium anode may offer an unexpected path toward making metal-air batteries rechargeable while still offering a relatively high energy density (1605 Wh/kg). Sodium and oxygen form NaO_2 (sodium superoxide), a more stable compound. Since NaO_2 does not decompose, the reaction can be reversed during charging. A Na/air fuel-cell-type battery has been developed and tested (Figure 8.15).

For more information on Na/air batteries, see Reference 8.26.

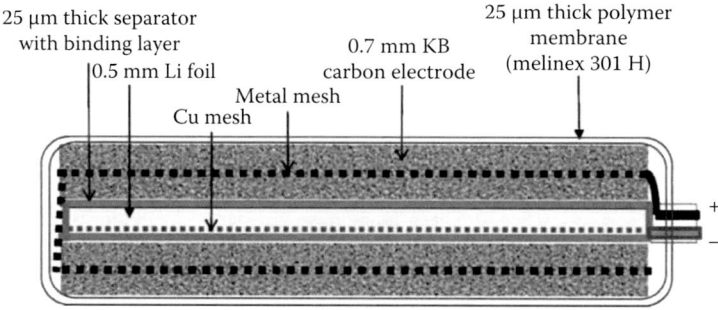

FIGURE 8.14 A sealed Li–air test cell for ambient operation. (From Zhang, J. et al., *J. Power Sources*, 195, 4332, 2010.)

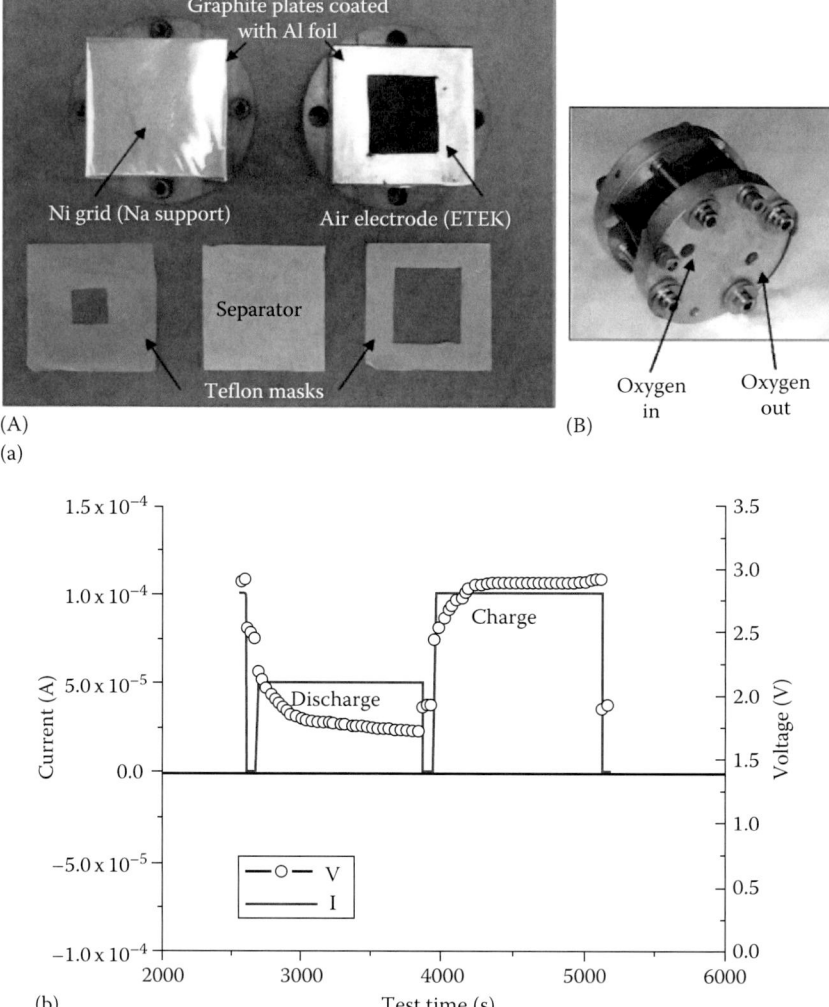

FIGURE 8.15 A, Components of a fuel-cell type Na-air battery; B, Fuel-cell hardware, with electrode area 1 cm² using ETEK cathode. The electrolyte consists of 0.1 M CP, 1 M $NaClO_4$, and 1% (w/w) Al_2O_3 nanopowder dispersed in PEGDME 2000 and propylene carbonate (PC) (90:10). (b) Discharge/charge curves of a Na/O_2 cell at 105°C, 1.5–3.0 V (or 20-min operating time), discharge and charge currents are 50 μA and 100 μA, respectively. (From Saravanan, K. et al., *Energy Environ. Sci.*, 3, 457, 2010; Moitzheim, S. et al., *Nanotechnol.*, 25, 504008, 2014; Yu, C. et al., *Adv. Energy Mater.*, 2, 68, 2012.)

8.5 Thin Film Battery

All-solid state high-voltage Li-ion secondary batteries are leak proof, mechanically robust, and may be used over a wide range of temperatures. Solid Li electrolytes generally have a transference number $t_{\mathrm{Li+}}$ of $\mathrm{Li^+}$ ions of unity. This is in contrast to common liquid and polymer electrolytes where both cations and anions are mobile. Often $t_{\mathrm{Li^+}} \ll t_{\mathrm{anions}}$. The high mobility of ions other than $\mathrm{Li^+}$ ions may thus lead more readily to the formation of solid electrolyte interfacial layer causing deterioration and limiting battery lifetime. The negligible mobility of ions other than $\mathrm{Li^+}$ ions *in the solid state* would lead to superior chemical stability.

Although solid Li-ion electrolytes have a generally lower $\mathrm{Li^+}$ ionic conductivity than liquid/gel electrolytes, application of thin film technology offers a solution to this problem. Physical vapor deposition techniques discussed in Chapter 3 may be used to prepare thin film nano-to-micro metal and metal oxide layers.

Thin films show much lower resistance than bulk samples. Therefore the low electrical conductivity of solid electrolytes will not be a problem. Thin film electrodes (apart from thin film electrolytes) are advantageous because of lower diffusion lengths (and thus faster equilibration) compared to thick films. These circumstances allow the development of all solid state thin film Li batteries. These show high power densities and high energy densities. Also, the all-solid state batteries have lower gravimetric densities compared to conventional rocking chair batteries.

We briefly discuss the all-solid state battery based on Al anode [8.27]. Besides substituting Li anode by Al (note that Al forms alloys with Li via electrochemical reactions), the cathode is the 3D spinel-structured $\mathrm{Li_2MMn_3O_8}$ (M = Fe, Co) to increase the operating voltage (with respect to 2D $\mathrm{LiCoO_2}$).

The Al anode may be made by sputtering of a Al target, the electrolyte Li-phosphorous oxy nitride (LiPON) may be made by reactive sputtering of $\mathrm{Li_3PO_4}$ in nitrogen atmosphere, and the cathode may be fabricated by e-beam evaporation of precursor powders made by glycine combustion of high-purity $\mathrm{LiNO_3}$, $\mathrm{Co(NO_3)_3}$ $6\mathrm{H_2O}$, $\mathrm{Fe(NO_3)_2}$ $9\mathrm{H_2O}$, $\mathrm{Mn(NO_3)_2}$ $4\mathrm{H_2O}$, and glycine ($\mathrm{H_2NCH_2COOH}$). Approximately one mole of glycine per two moles of nitrates is to be mixed in the required molar ratio and dissolved in a minimum amount of deionized water. The solutions are heated to 300°C using a hot plate. Complete evaporation of water followed by sudden ignition leads to black powders. These are collected and annealed at 700°C for 2 h in air using alumina crucibles. Powders pressed into pellets with 20% (by weight) excess $\mathrm{LiNO_3}$ may be used as targets for e-beam evaporation. Let us focus on the electrochemical characterization of the thin film battery. Figure 8.16 shows the experimental setup.

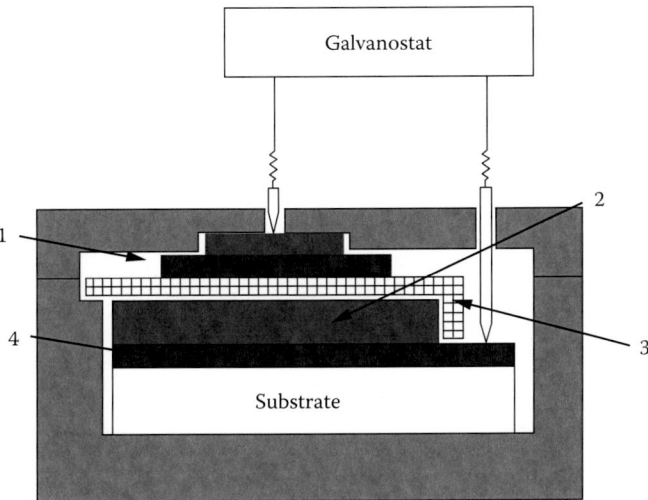

FIGURE 8.16 Setup for characterization of individual thin film electrode and electrolyte using auxiliary liquid electrolyte: (1) Li, (2) Electrode (500–1000 nm), (3) electrolyte-microporous flat sheet membrane soaked in 1 mol/L $\mathrm{LiPF_6}$ in EC-DEC (2:1), and (4) Pt-10% Rh (100 nm) current collector. Electrode performance tested using galvanostat by constant current discharge and coulometric titration vs. Li metal. (From Schwenzel, J. et al., *J. Power Sources*, 154, 232, 2006.)

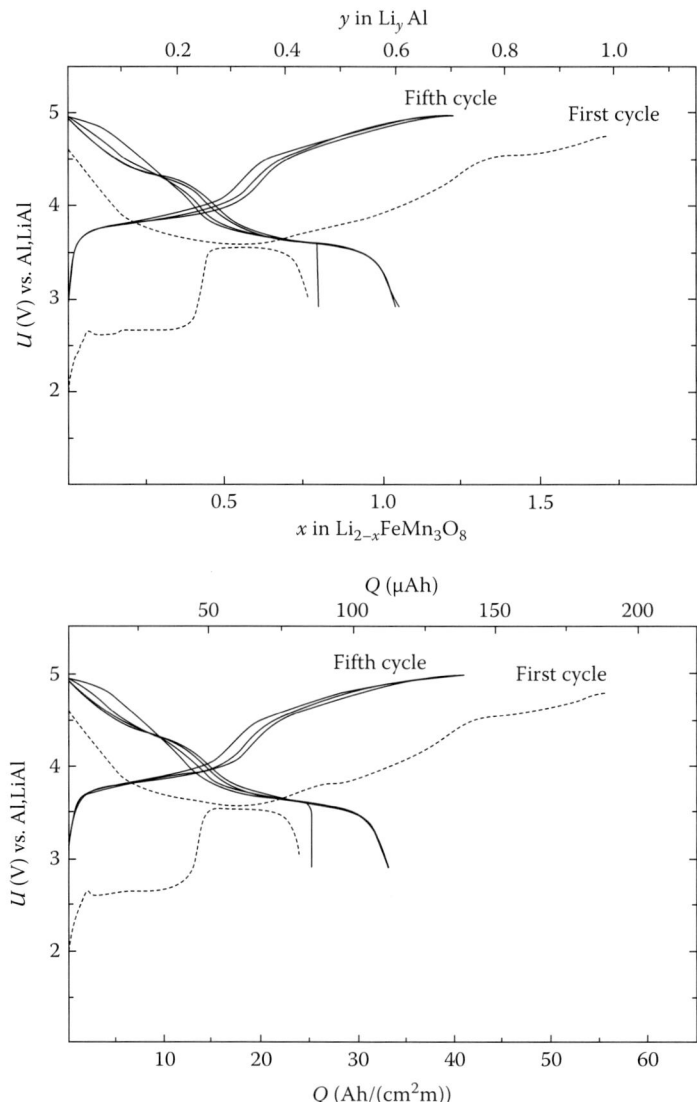

FIGURE 8.17   Galvanostatics of the cell Al│Li$_2$Fe Mn$_3$O$_8$│Pt-10%Rh. Applied current: 5 µA and electrode area: 6.7 cm^2. (From Schwenzel, J. et al., *J. Power Sources*, 154, 232, 2006.)

Figure 8.17 shows the galvanostatic charge and discharge measurements of the cell set-up Al│Li$_2$Fe Mn$_3$O$_8$│Pt-10%Rh. The first cycle sees a charging process until a potential of 4.8 V is reached. A maximum of 5 V is allowed during subsequent cycles. A chemical diffusion coefficient may be determined from a long time relaxation behavior when a linear time dependence of cell voltage is seen.

8.6 The Na$^+$-Ion Battery

Gel-based sodium sulfate battery of Faraday described earlier is perhaps the first report to show that the Na$^+$-ion battery is feasible.

Due to the high cost of production and relatively less abundance of Li, Na is being explored as a desirable alternative to Li in the development of rechargeable battery systems. Let us consider the

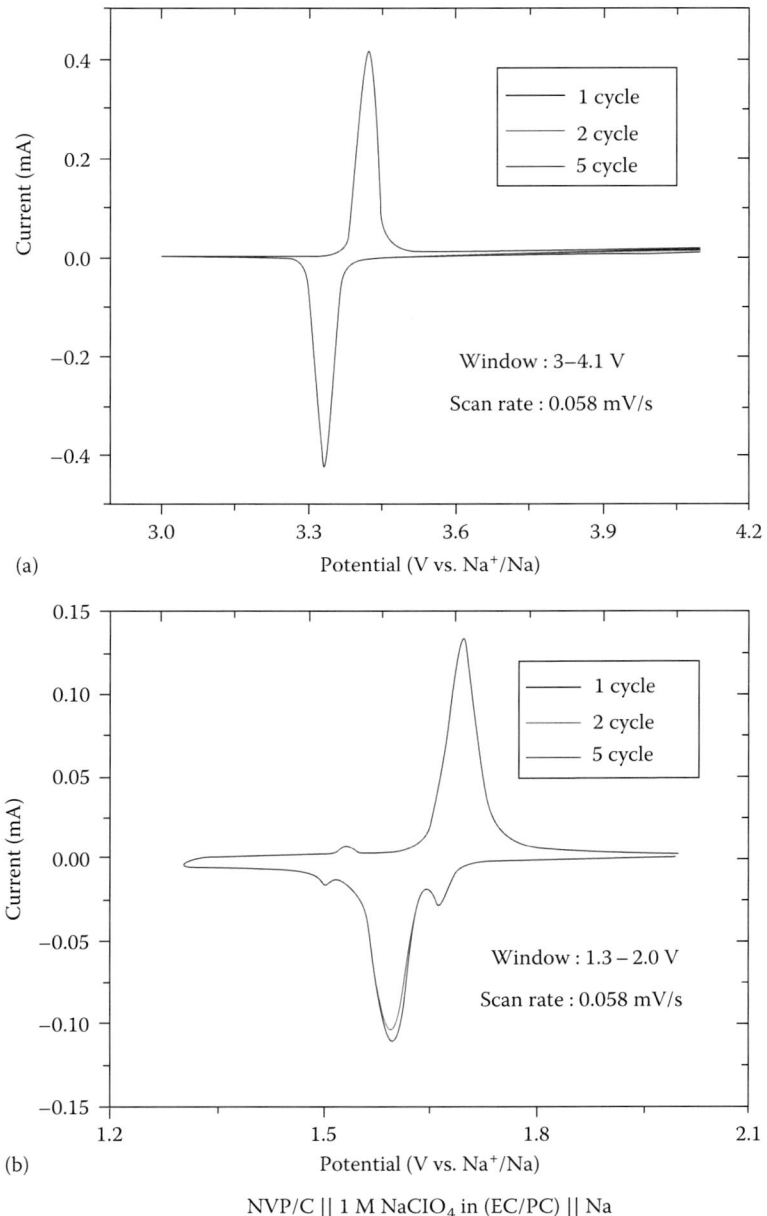

FIGURE 8.18 Cyclic voltammograms of Na^+ ion battery using $Na_3V_2(PO_4)_3$ (NVP) as Na^+ insertion host (a) Window 3 – 4.1 V, (b) Window 1.3 – 2.0 V. (From Saravanan, K. et al., *Adv. Energy Mater.*, 3, 444, 2013.)

example of $Na_3V_2(PO_4)_3$ as a potential Na^+ insertion host. This compound has a rhombohedral structure [8.28] and possesses a theoretical capacity: 117 mAh/g for extraction of 2 Na^+ ions. The expected energy density is ~400 Wh/kg (117 mAh/g × 3.4 V) vs. Na metal. The cyclic voltammograms are shown in Figure 8.18 for five cycles with respect to $NaClO_4$ in ethylene carbonate/propylene carbonate mixed nonaqueous solvent. Two sharp redox peaks are observed in the low-voltage and high-voltage regions. Cell potential at 1.64 V vs. Na^+/Na in the low-voltage region is attributed to the V^{3+}/V^{2+} redox couple and high-voltage peak at 3.37 V vs. Na^+/Na corresponds to the V^{4+}/V^{3+} redox couple (Figure 8.19).

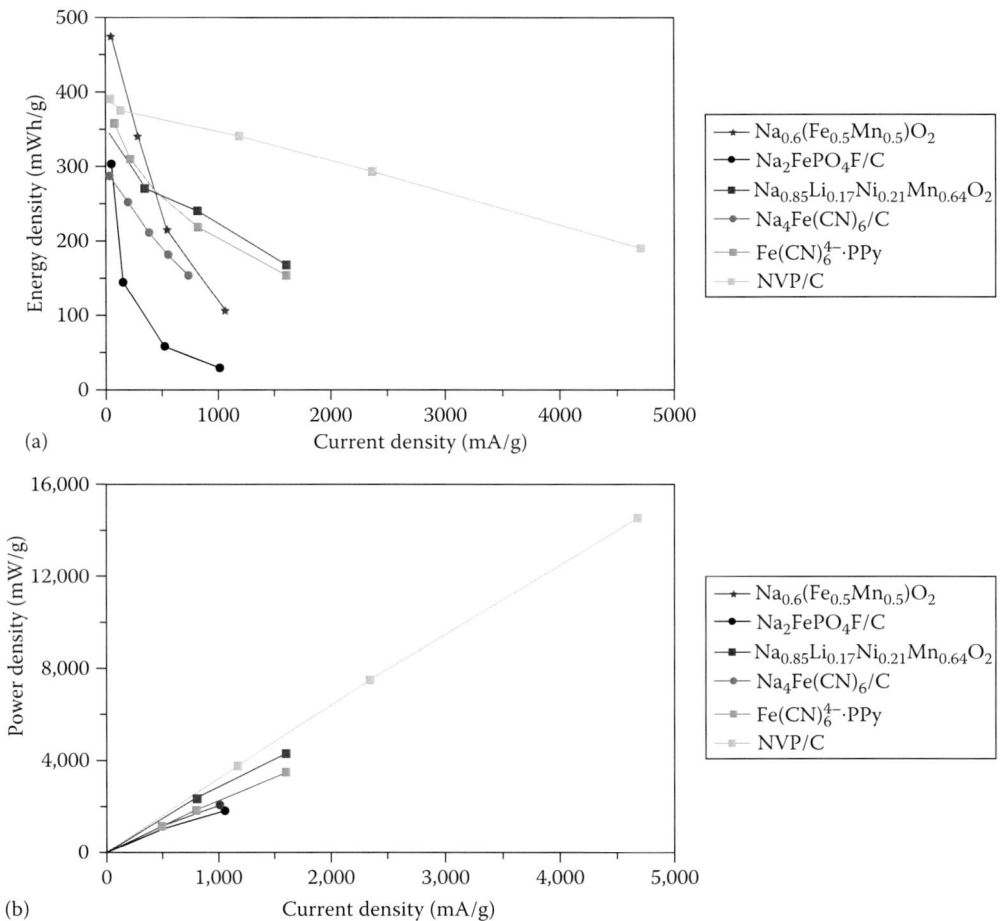

FIGURE 8.19 Energy density and power density of NVP in comparison with other Na$^+$ insertion hosts. NVP performs better than the five other sodium compounds. (From Saravanan, K. et al., *Adv. Energy Mater.*, 3, 444, 2013.)

8.7 Li$^+$ Polymer Battery

Following up on the Li–air battery, we now consider a polymer-electrolyte based Li–air battery. It is a nonaqueous thin film battery [8.29], fabricated from a thin Li metal foil anode, a thin solid polymer electrolyte membrane that conducts Li ions, and a thin carbon composite electrode sheet made up of high surface area carbon. On the carbon-composite electrode oxygen, the electroactive cathode material, accessed from the environment, is reduced during battery discharge to generate electric power. The organic polymer electrolyte membrane serves as (1) a separator that electronically insulates the cathode from the anode and (2) a medium through which Li ions are transported from the Li anode to the oxygen cathode during discharge. This Li/oxygen cell appears to be rechargeable due to the use of nonaqueous electrolyte. Note that, in traditional polymer electrolyte-based Li batteries, the cathode comprises Li-intercalating solid state materials such as TiS_2, V_6O_{11}, $LiMn_2O_4$, and $LiCoO_2$.

Figure 8.20 shows a schematic of this battery. The Li/O_2 polymer battery is fabricated with the oxygen-permeable membrane on the cathode side of the plastic envelope sealed with a peelable tape. Prior to use, the tape can be removed to let oxygen into the carbon electrode and activate the battery. The battery can be resealed after each use or it can be recharged and used. The materials used for the polymer electrolyte and other components of the battery are nontoxic so that it is possible to fabricate a completely environmentally friendly battery (Figure 8.21).

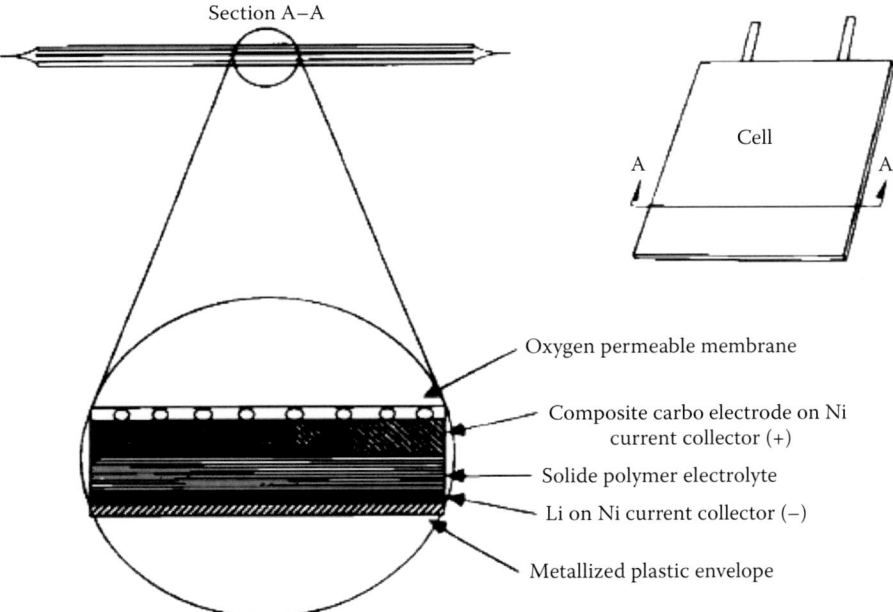

Section A–A

Cell

A   A

Oxygen permeable membrane

Composite carbo electrode on Ni
current collector (+)

Solide polymer electrolyte

Li on Ni current collector (−)

Metallized plastic envelope

FIGURE 8.20 Schematic of a polymer-electrolyte-based Li-oxygen battery. The calculated open-circuit voltage of this battery for the idealized reaction $4Li + O_2 \rightarrow 2Li_2O$ is 2.01 V. Theoretical specific energies including and excluding O_2 are 5.200 and 22.140 Wh/kg, respectively. (From Abraham, K.M. and Jiang, Z., *J. Electrochem. Soc.*, 143, 1, 1996.)

8.8 Advanced Lithium-Ion Battery Technology

Advanced materials such as graphene and carbon nanotubes have provided an impetus for the development of advanced LIB technologies. We illustrate one technology based on a graphene ink anode and a $LiFePO_4$ cathode [8.30]. A careful balance of cell composition and suppression of the initial irreversible capacity of the anode over a few cycles, an optimal battery performance has been demonstrated. A specific capacity of 165 mAh/g and an estimated energy density of ~190 Wh/kg have been achieved. A stable

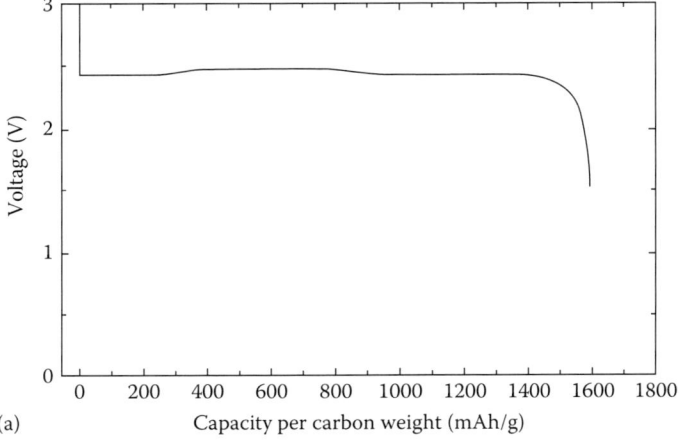

(a)

FIGURE 8.21 (a) The discharge curve of a Li/PAN-based polymer electrolyte oxygen cell at a current density of 0.1 mA/cm² at room temperature. The cathode contained Chevron acetylene black carbon. The cell was packaged in metallized plastic envelope and discharged by exposing the carbon electrode to laboratory air. *(Continued)*

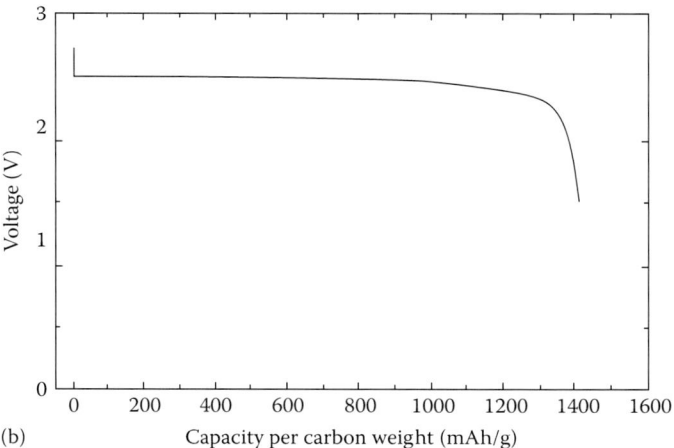

(b)

FIGURE 8.21 (Continued) (b) The discharge curve of a Li/PAN-based polymer electrolyte/oxygen cell at a current density of 0.1 mA/cm² at room temperature in an atmosphere of oxygen. The cathode contained Chevron acetylene block carbon. The cell was packaged in a cell can having the shape of letter "D," and O₂ from a tank was used to maintain a flowing O₂ atmosphere. (From Abraham, K.M. and Jiang, Z., *J. Electrochem. Soc.*, 143, 1, 1996.)

operation for over 80 charge–discharge cycles has been realized. What makes this technology attractive is the low cost of components of the battery and potential upscalability. The microstructure of the anode and cathode and the performance curve are shown in Figure 8.22.

As a prelude to Chapter 10, the following observations on advanced technologies are in order. A number of advances have been made in the LIB field *by controlling particle size* in addition *to composition, structure, and morphology* in order to design better electrodes and electrolyte components. Decreasing electrochemically active materials to *sub-micrometer and smaller sizes* combined with *carbon-coating* approaches to achieve core–shell morphologies has led to new directions in electrode materials. Reaction mechanisms and materials systems once discarded are now sensibly being reconsidered for the next generation of LIBs. Moving from bulk materials to nanosize particles has enabled the following: (1) the use of new Li-reaction mechanisms, in which conversion–reaction electrodes show enormous capacity gains;

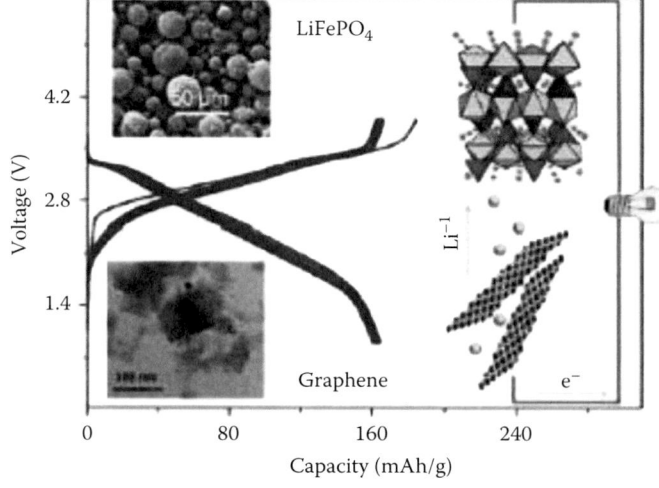

FIGURE 8.22 Microstructure (right), structure (left), and performance (voltage vs. capacity) of a battery based on graphene ink anode and a LiFePO₄ cathode. (From Hassoun, J. et al., *Nano Lett.*, 14, 4901, 2014.)

(2) the use of negative electrodes based on alloy reactions—Tin (Sn)-based LIB technologies are here, and Si-based ones are emerging; (3) the identification of poorly conducting polyanionic compounds or fluorine-based compounds that exhibit excellent electrochemical performance; and (4) the transformation of the poorly conducting lithium iron phosphate ($LiFePO_4$) insertion electrode into a valuable electrode material for electric vehicle applications. LIBs based on $LiFePO_4$ are extremely attractive because of safety and cost. Safety because the operating voltage of the $LiFePO_4$ system is compatible with the thermodynamic stability of the electrolyte, cost because of the use of abundant and low-cost constituents. In addition to being an attractive LIB for the electric vehicle market, $LiFePO_4$-based batteries are also being screened for stationary energy storage.

Life cycle costs represent another important consideration. A foreseeable strategy for battery processing will involve the use of electroactive organic electrode materials synthesized from "green chemistry" concepts through low-cost processes free of toxic solvents; this will also enlist the use of natural organic sources [carbon dioxide (CO_2)–harvesting entities] as precursors, which will be biodegradable and easily destroyed by combustion (providing CO_2) so that the battery assembly/recovery processes will have a minimal CO_2 footprint. At the research level, there is interest in rechargeable LIB systems that have significantly higher energy densities. Although the $Li–O_2$ system has been around as a primary battery, the prospect of developing it into a reversible (secondary) battery has become tantalizing because of a projected three- to four-fold increase in gravimetric energy density as compared with the current Li-ion technology.

8.9 Batteries for Electrical Vehicles

Electric cars use the energy stored in a battery (or a series of batteries) for vehicle propulsion. Electric motors provide a clean and safe alternative to the polluting internal combustion engine. There are many pros and cons about electric cars. The electric vehicle is known to have faster acceleration but shorter distance range than conventional engines. They produce no exhaust but require long charging times.

Note that EV batteries are quite different from those used in laptops and cellphones. They should handle power up to 100 kW and energy capacity up to tens of kilowatts and operate within the constraints of limited space and weight, besides being affordable. The current two major battery technologies used in electrical vehicles (EVs) are the more complete and predominant nickel metal hydride or NiMH and the upcoming lithium ion (Li ion) [8.31]. The latter technology would be relevant for plug-in hybrid electric vehicles (PHEV) and battery electric vehicles (BEV).

Argonne National Laboratory's patented composite NMC (Li–(Ni–Mn–Co)O_2) cathode materials technology for LIBs is expected to yield longer lasting, safer batteries for hybrid-electric vehicles, cell phones, laptop computers, and other applications. This material is designed at the molecular level enabling batteries to store more energy. http://phys.org/news124626282.html#jCp.

It is curious to learn that in the early 1900s, the number of EVs was almost double that of petrol power cars. By 1920, internal combustion engines dominated the cars completely [8.32].

Performance of an EV battery changes as the operating conditions such as temperature, charging or discharging current, state of charge, and service time vary.

Issues specific to EVs are battery power management and reuse of second-hand EV battery for stationary power grid applications.

Following Young et al. [8.32], let us focus on the basics of power and energy of electrical propulsion which is rooted in basic physics.

Part or all of the propulsion power and energy of an EV is supplied by the battery located inside the vehicle. Consider a pure battery EV. As in a regular petrol/diesel fueled vehicle, the power train in an EV needs to provide power for the vehicle under *all kinds of road conditions and driving modes*.

Additionally, an EV needs to handle regenerative braking so that the kinetic energy of the moving vehicle can be captured and stored in the battery for future use.

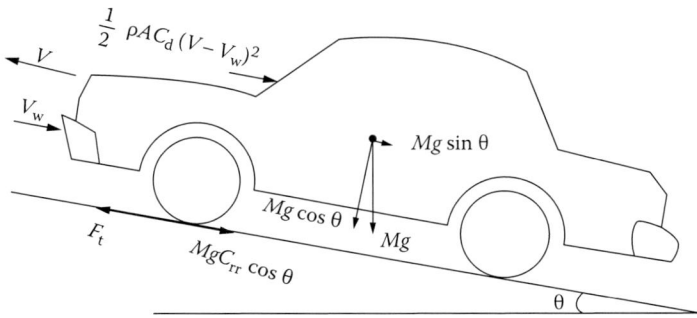

FIGURE 8.23   Forces applied on a vehicle. (From Young, K. et al., in Garcia-Valle, R. and Pecas Lopes, J.A. (eds.), *Electric Vehicle Integration into Modern Power Networks, Power Electronics and Power Systems*, Springer Science, New York, 2013, Chapter 2.)

Acceleration of a vehicle is determined by *all* the forces applied on it (Figure 8.23). Applying Newton's second law to the problem,

$$f_m Ma = F_t - \Sigma F_r \tag{8.5}$$

where
   M is the overall mass of vehicle
   a is the vehicle acceleration
   f_m is the mass factor that converts the rotational inertias of rotating components into equivalent translational mass
   F_t is the total traction force to the vehicle
   ΣF_r is the total resistive force

F_rs are the rolling resistance between the tyre and road surface, aerodynamic drag and uphill grading resistance.

$$\Sigma F_r = MgC_{rr}\cos\theta + \frac{1}{2}\rho AC_d(v - v_w)^2 + Mg\sin\theta \tag{8.6}$$

where
   g is the acceleration due to gravity
   C_{rr} is the coefficient of rolling resistance between the tyre and road surface
   ρ is the density of the ambient air
   A is the vehicle frontal area
   C_d is the aerodynamic drag coefficient
   v is the vehicle speed
   v_w is the wind speed in the vehicle in the direction of its motion
   θ is the slope angle

For a downhill slope, θ is negative.
   Now the total propulsion force

$$F_t = f_m Ma + MgC_{rr}\cos\theta + \frac{1}{2}\rho AC_d(v - v_w)^2 + Mg\sin\theta \tag{8.7}$$

The power to drive the vehicle at speed v is

$$P = F_t v = f_m Mav + MgC_{rr}v\cos\theta + \frac{1}{2}\rho AC_d v(v - v_w)^2 + Mgv\sin\theta \tag{8.8}$$

For a vehicle on a flat road, $\theta = 0$. At the early state of acceleration, the propulsion power is mainly used to accelerate the vehicle and also to overcome the rolling resistance. When the speed is reached, the power is used to keep the speed by overcoming the rolling resistance and aerodynamic drag force. For an EV, the battery power has to be enough to meet acceleration needs. To accelerate a vehicle with the typical parameters $M = 1360$ kg, $f_m = 1.05$, $a = (0\text{--}96.6$ km/h in 10 s$) = 2.68$ m^2/s, $C_{rr} = 0.02$, $\rho = 1.225$ kg/m^3, $A = 2$ m^2, $C_d = 0.5$, w = 0 m/s, $\theta = 0°$, an average power to accelerate the vehicle to 96.6 km/h is 61 kW.

In regenerative braking, the electric propulsion motor of the EV works as a generator to convert the kinetic energy of vehicle motion into electrical energy and charge battery. The braking power P_b is

$$P_b = F_b v = f_m m \mu v - M g C_{rr} v \cos\theta - \frac{1}{2}\rho A C_d v(v - v_w)^2 - M g v \sin\theta \qquad (8.9)$$

where
  F_b is the braking force
  μ is the deceleration of the vehicle

For the same vehicle with the aforementioned parameters, the peak braking power for bringing the vehicle moving at 96.6 km/h to a stop in 5 s could be as high as 186 kW. This is because of the shorter time for deceleration than for acceleration. The power train battery should therefore be capable of both supplying and absorbing the high power.

What is the energy capability of the battery? The key factor is the capacity of the energy storage device. Note that although petrol gives a theoretical specific energy of 13,000 Wh/kg as against 120 Wh/kg of typical LIBs, the battery scores over the petrol-fueled ICE (internal combustion engine) because the efficiency of the battery for EV propulsion is 80% while ICE has an efficiency of only 20%. So, a battery pack of ~30 kWh in an EV can let the EV achieve 100 km range.

How is charging and discharging of the EV battery done? There are two methods for charging: (1) constant voltage and (2) constant current. A combination of both methods is used in practice. Figure 8.24 gives the LIB charging profile. Initially, the battery is precharged at a low, constant current if not done before. Then, it is switched to charge the battery with constant current at a higher value. When the battery voltage reaches a certain threshold, the charging mode is changed to constant voltage charge. Constant voltage charge can be used to maintain the battery voltage later if the DC charging supply is still available. Batteries for EVs should handle issues such as random charging and when to stop charging.

An EV battery is composed of a positive (+) electrode at a higher potential and a negative electrode (−) at a lower potential. An ion conductive but electrically insulating electrolyte is present between the two electrodes. During charging, the + is the anode with the reduction reaction and the − is the cathode with the oxidation reaction. During discharge, however, the reaction is reversed and the + and − electrodes become cathode and anode, respectively. (Conventionally positive and negative electrode materials are referred to as cathode and anode.) In a sealed cell, the liquid electrolyte is held in a separator. The separator (1) prevents direct short between the two electrodes and acts as (2) a reservoir for extra electrolyte, (3) a space saver allowing for electrode expansion, (4) an ammonia trap in Ni metal hydride (NiMH) battery, and (5) a safety device for preventing shortage due to Li dendrite formation in LIB.

NiMH batteries are predominantly used in EVs, although LIBs have also entered the EV though not in a big way currently. We close this chapter by briefly describing NiMH batteries. We have already discussed LIB.

Figure 8.25 gives the schematic of the NiMH battery.

The active material in the negative electrode is metal hydride (MH), a special type of intermetallic alloy capable of chemically absorbing and desorbing hydrogen. It is an AB$_5$ composition with a CaCu$_5$ crystal structure. A is a mixture of La, Ce, Pr, and Nd (all lanthanides) while B is made from Ni, Co, Mn, and Al (three transition metals plus Al). Active material in the positive cathode is Ni(OH)$_2$. The poor intrinsic conductivity of Ni hydride is improved by coprecipitation of other atoms and formation of conductive

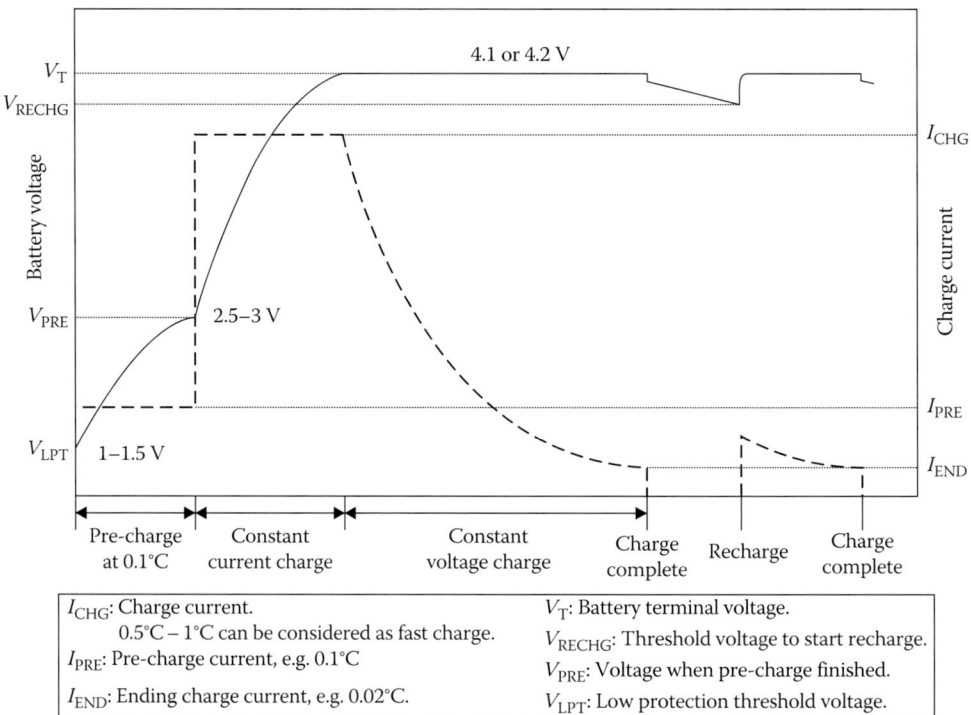

FIGURE 8.24   Charge profile of a Li-ion cell.

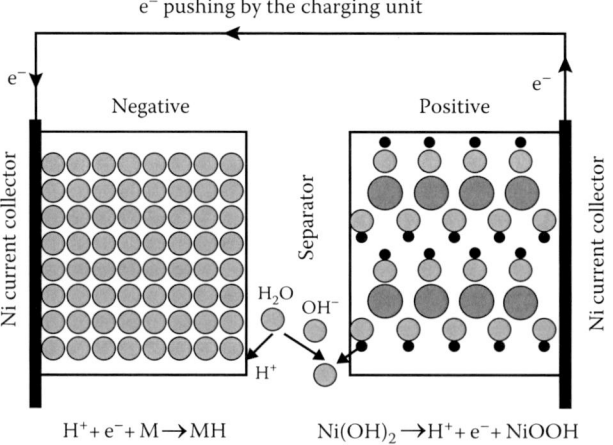

FIGURE 8.25   The NiMH rechargeable battery schematic. Reactions at the − and + electrodes are indicated.

networks outside the particle or multilayer coating structure. The separator is made from grafted polyethylene (PE)-/polypropylene (PP)-woven fabric and 30 wt.% KOH with a pH value of about 14.3.

During charge, water is split into protons and hydroxyl ions by the voltage supplied from the charger. The proton enters the negative electrode, neutralizes the electron supplied by the charger through the current collector, and hops between adjacent storage sites via quantum mechanical tunneling. The voltage is equivalent to the applied hydrogen pressure in a gas phase reaction. This voltage will remain at a near constant value before protons occupy all of the available sites. OH⁻ generated by charging will add to the

OH⁻ already present in the KOH electrolyte. On the surface of the positive electrode, some OH⁻ will recombine with the protons coming from $Ni(OH)_2$ and form water molecules. The overall charging reaction is

$$M + Ni(OH)_2 \rightarrow MH + NiOOH \tag{8.10}$$

Note that neither water nor OH⁻ is consumed, so no change in pH occurs during charge/discharge. As protons are consumed at the surface of the positive electrode, more protons are driven out of the bulk from both the voltage and the concentration gradients. Loss of one proton increases the oxidation state of Ni to 3+ in NiOOH. Electrons are collected by Ni-form or perforated Ni-plate and moved back to the charging unit to complete the circuit.

The whole process is reversed during discharge. In the negative electrode, protons are sent to the electrolyte and recombine with the OH⁻ as electrons are pushed to the outside load. The electrons reenter the positive electrode side of the battery through the outside load and neutralize the protons generated from the water split on the surface of the positive electrode.

8.10 Summary

In this chapter, we have discussed the most important application of solid state ionics namely the all-solid state LIB. Starting from the lead–acid car battery used for starting lighting and ignition, the Li^+ ion physics and battery components are discussed. This is followed by a description of Li–air or Li–oxygen battery concepts and construction, thin film battery, the Na^+-ion battery, and the polymer battery. The polymer battery along with thin film battery are true all-solid state batteries. An advanced battery technology involving graphene and $LiFePO_4$ as anode and cathode shows the shape of things to come in this cutting edge topic in energy conversion. Finally, the car comes full circle—this time propelled by a battery—the hybrid/plug-in hybrid/battery electric vehicle (HEV or PHEV or BEV) technology that would revolutionize the car industry and thus ensuring more and more freedom from dependence on fossil fuels. The next chapter would focus on two other applications—fuel cells and electrochemical sensors.

Problems

8.1 An electrochemical cell converts free energy liberated in a chemical reaction or from a change in concentration of a species into electrical energy. This chemical reaction is essentially an oxidation–reduction reaction (also called "redox" reaction) whereby the reactants taking part in the reaction are chemically reduced/oxidized at an electrode. Oxidation at a negatively charged anode (say made of metal A): $A \rightarrow A^{v+} + ve^-$. The electrons so released travel around an external electrical circuit to the positively charged cathode made of metal B, say, where reduction occurs: $Bv+ + ve^- \rightarrow B$. What is the overall reaction? The change in the Gibbs free energy ΔG of this reaction relates to the electromotive force (EMF) of the cell, E, and the electric potential difference between anode and cathode: $\Delta G = -vFE$, where F is the Faraday (the charge of a mole of electrons). Also, Gibbs free energy or chemical potential of each reacting species i relates to its activity a_i by mu $\mu_A = \mu_{A0}(T, p) + k_B \ln(a_i)$ and thus to the overall reaction. Show that the free energy change is also given by

$$\Delta G = \Delta G^{std} + RT \ln \frac{[a(A^{v+})a(B)]}{[a(A)a(B^{v+})]}$$

where ΔG^{std} is the standard free energy (the value when all species have unit activity).

Show finally that the EMF of the cell is given by the Nernst equation

$$\varepsilon = \varepsilon^{std} - \left[\frac{RT}{vF}\right]\ln\frac{[a(A^{v+})a(B)]}{[a(A)a(B^{v+})]}$$

where ε^{std} is the standard EMF of the cell, for all species of unit activity.

8.2 What factors influence the performance of solid state batteries quantified as (1) the open-circuit voltage, (2) the energy density, and (3) the self-discharge lifetime? Note that energy density can be (a) gravimetric and (b) volumetric.

8.3 In a Li$^+$-ion battery, Li ions reversibly intercalate and deintercalate into/from the anode and cathode materials on operation. The materials consist of a host material with Li$^+$ ion accessible to interatomic sites. The intercalation/deintercalation process causes a change in the charge distribution inside the host material skeleton and *an overall change* in the material charge. This in turn causes electron flow in the external circuit. In the absence of the corrosion reactions, the thermodynamic value of the Li ion cell voltage is equal to the open circuit potential (V_{OC}). For the operating battery, this thermodynamic value is determined by the difference in the chemical potentials of Li into the cathode μ^{Li+}_{cath} and the anode μ^{Li+}_{an}: $V_{cell} = \left|(\mu^{Li+}_{cath} - \mu^{Li+}_{an})/F\right|$ where F is the Faraday constant. Using this relation find V_{cell} for Li, Mg, Na, Cu, F, and O-based cells.

8.4 Redox potential of a battery is an important parameter for its design and development. This has been studied by using density functional theory outlined in Chapter 7. For motivation and possible adaptation to a new project and first-principles prediction of redox potentials in transition-metal compounds with LDA+, see Reference 8.33.

8.5 GITT and PITT for Li battery testing. (a) Galvanostatic intermittent titration technique (GITT) measurement: the cell is charged and discharged for a few cycles (say 10 cycles) to make the electrode/electrolyte interface reach stable state. Charging is done at a small current (say 0.2 mA) for a few 100 s (say 200 s). This is followed by open circuit relaxation. The procedure is continued until the

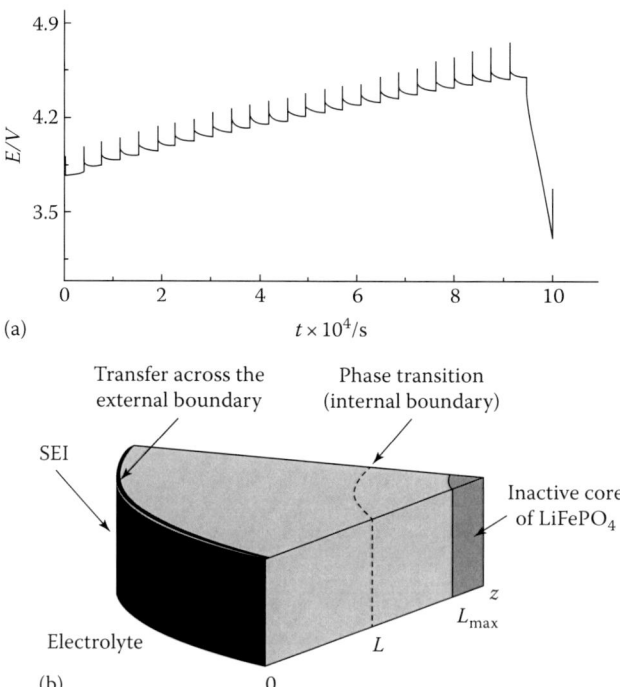

FIGURE P8.5 (a) GITT plot for NCM cathode. (b) Structure of a single LiFePO$_4$ particle after lithiation/delithiation cycles. (After Zheng, W. et al., *Bull. Mat. Sci.*, 36, 495, 2013.)

voltage reaches the required value. A typical GITT curve for $LiNi_{1/3}Co_{1/3}Mn_{1/3}O_2$ (NCM) cathode. From the plot shown in Figure P8.5a estimate the required voltage and the number of cycles needed. (b) PITT is potentiometric intermittent titration technique. Both GITT and PITT have been used to determine the Li^+ diffusion coefficient in $LiFePO_4$ electrode by Churikov et al. [8.35]. Following this work, solve 1D Fick's second-order equation

$$\frac{\partial c}{\partial t} = D\frac{\partial^2 c}{\partial t^2}$$

where, c is the ion concentration, D is the diffusion coefficient, $0 < z < L$, $t \geq 0$. Consider the three cases: (1) current on, (2) current off, and (3) potential on (see Figure P8.5b). Note that the D for Li^+ varies between $10^{-11} - 10^{-16}$ cm^2/s depending on the Li content in the solid solution Li_xFePO_4–$Li_{1-x}FePO_4$ ($x < 0.02$) or the $LiFePO_4/FePO_4$ phase ratio.

8.6 The fundamental basis for battery design and development is the so-called electrochemical series. According to this series, the standard potential of Li for the half reaction $Li^+ + e^- = Li$ is -3.04 V, while that standard potential of Na for the half reaction $Na^+ + e^- = Na$ is -2.713 V. Interpret these numbers in the contest of Li-ion and Na-ion batteries. Also find such data for other battery elements of interest (see http://www.smauro.it/chimica_fisica/Dati%20Termodinamici/Electrochemical% 20Series%202.pdf).

A partial electrochemical series is given in the following as a guide to plan and design battery configurations. Based on this data, design an Mg^{++}-ion battery.

The electrochemical series is shown here.

Elements	Electrode Reaction	E°_{red} (volts)
	Oxidised Form + ne⁻ ⟶ Reduced Form	
Li	$Li^+(aq) + e^- \longrightarrow Li(s)$	– 3.05
K	$K^+(aq) + e^- \longrightarrow K(s)$	– 2.93
Ba	$Ba^{2+}(aq) + 2e^- \longrightarrow Ba(s)$	– 2.90
Ca	$Ca^{2+}(aq) + 2e^- \longrightarrow Ca(s)$	– 2.87
Na	$Na^+(aq) + e^- \longrightarrow Na(s)$	– 2.71
Mg	$Mg^{2+}(aq) + 2e^- \longrightarrow Mg(s)$	– 2.37
Al	$Al^{3+}(aq) + 3e^- \longrightarrow Al(s)$	– 1.66
Zn	$Zn^{2+}(aq) + 2e^- \longrightarrow Zn(s)$	– 0.76
Cr	$Cr^{3+}(aq) + 3e^- \longrightarrow Cr(s)$	– 0.74
Fe	$Fe^{2+}(aq) + 2e^- \longrightarrow Fe(s)$	– 0.44
	$H_2O(l) + e^- \longrightarrow \frac{1}{2}H_2(g) + OH^-(aq)$	– 0.41
Cd	$Cd^{2+}(aq) + 2e^- \longrightarrow Cd(s)$	– 0.40
Pb	$PbSO_4(s) + 2e^- \longrightarrow Pb(s) + SO_4^{2-}(aq)$	– 0.31
Co	$Co^{2+}(aq) + 2e^- \longrightarrow Co(s)$	– 0.28
Ni	$Ni^{2+}(aq) + 2e^- \longrightarrow Ni(s)$	– 0.25
Sn	$Sn^{2+}(aq) + 2e^- \longrightarrow Sn(s)$	– 0.14
Pb	$Pb^{2+}(aq) + 2e^- \longrightarrow Pb(s)$	– 0.13
H₂	$2H^+ + 2e^- \longrightarrow H_2(g)$ (standard electrode)	0.00
Cu	$Cu^{2+}(aq) + 2e^- \longrightarrow Cu(s)$	+ 0.34
I₂	$I_2(s) + 2e^- \longrightarrow 2I^-(aq)$	+ 0.54
Fe	$Fe^{3+}(aq) + e^- \longrightarrow Fe^{2+}(aq)$	+ 0.77
Hg	$Hg_2^{2+}(aq) + 2e^- \longrightarrow 2Hg(l)$	+ 0.79
Ag	$Ag^+(aq) + e^- \longrightarrow Ag(s)$	+ 0.80
Hg	$Hg^{2+}(aq) + 2e^- \longrightarrow Hg(l)$	+ 0.85
N₂	$NO_3^- + 4H^+ + 3e^- \longrightarrow NO(g) + 2H_2O$	+ 0.97
Br₂	$Br_2(aq) + 2e^- \longrightarrow 2Br^-(aq)$	+ 1.08
O₂	$O_2(g) + 2H_3O^+(aq) + 2e^- \longrightarrow 3H_2O$	+ 1.23
Cr	$Cr_2O_7^{2-} + 14H^+ + e^- \longrightarrow 2Cr^{3+} + 7H_2O$	+ 1.33
Cl₂	$Cl_2(g) + 2e^- \longrightarrow 2Cl^-(aq)$	+ 1.36
Au	$Au^{3+}(aq) + 3e^- \longrightarrow Au(s)$	+ 1.42
Mn	$MnO_4^-(aq) + 8H_3O^+(aq) + 5e^- \longrightarrow Mn^{2+}(aq) + 12H_2O(l)$	+ 1.51
F₂	$F_2(g) + 2e^- \longrightarrow 2F^-(aq)$	+ 2.87

(a) Tendency for oxidation to occur (b) Power as reducing agent — Increase

(a) Tendency for reduction to occur (b) Power as oxidising agent

Increase

REFERENCES

8.1 X. Li et al., *Nat. Commun.* 5(2014).

8.2 G. Ceder et al., *MRS Bull.* 36(3)(2011) 185.

8.3 R. Van Noorden, *Nature* 507(2014) 26.

8.4 A. J. Bard et al. (Eds.), *Electrochemistry Dictionary*, Springer-Verlag, Berlin, Germany, 2012.

8.5 T. Minami (Ed. in chief), *Solid State Ionics for Batteries*, Springer-Verlag, Tokyo, Japan, 2005, p. 126.

8.6 S. S. Deshmukh, K. Tang, L. Kálmán, *J. Am. Chem. Soc.* 133(2011) 16309, doi:110.1021/ja207750z.

8.7 *Lithium Ion Battery Technical Handbook*, www. sony.com.

8.7a J. Dunning and D. Forbes, The inside story of the Li ion battery, http://www.calpoly.edu/~research/documents/The_Inside_Story_of_the_Lithium_Ion_Battery_0511.pdf, modified December 3, 2014.

8.8 B. G. Pollet et al., *Electrochim. Acta* 84(2012) 235.

8.9 R. S. Treptow, *J. Chem. Educ.* 79(3)(2002).

8.10 J.-M. Tarascon, M. Armand, *Nature* 414(2001) 359.

8.11 M. Park et al., *J. Power Sources* 195(2011) 7855.

8.12 J. Jamnik, M. Gaberscek, *MRS Bull.* 34(2009) 942.

8.13 Y.-S. Kim, Y.-G. Cho, D. O. N. Park, H.-K. Song, *Sci. Rep.* 3(2013) 1917.

8.14 A. Dey, S. Karan, S. K. Dey, *Indian J. Pure Appl. Phys.* 51(2013) 281.

8.15 C. A. Geiger et al., *Inorg. Chem.* 50(2011) 1089.

8.16 Y.-S. Ye, J. Rick, B.-J. Huang, *J. Mater. Chem. A* 1(2013) 2719.

8.17 A. L. Saroj, R. K. Singh, *J. Phys. Chem. Solids* 73(2012) 162.

8.18 Y. Zhao et al., *J. Solid State Electrochem.* 13(2009) 705.

8.19 X. Liu et al., *Eur. Phys. J. Appl. Phys.* 66(2014) 30301.

8.20a K. Saravanan et al., *Energy Environ. Sci.* 3(4)(2010) 457–464.

8.20b S. Moitzheim et al., *Nanotechnology* 25(2014) 504008.

8.20c C. Yu et al., *Adv. Energy Mater.* 2(2012) 68.

8.21 P. Arora, Z. J. Zhang, *Chem. Rev.* 104(2004) 4419.

8.22 J. Christensen et al., *J. Electrochem. Soc.* 159(2012) R1.

8.23 J. Zhang et al., *J. Power Sources* 195(2010) 4332.

8.24 J. Ou et al., *Nanoscale* (2014) 12324.

8.25 B. Dunn et al., *Science* 334(6058) (2011) 928.

8.26 P. Hartmann et al., *Nat. Mater.* 12(2013) 228.

8.27 J. Schwenzel et al., *J. Power Sources* 154(2006) 232.

8.28 K. Saravanan, C. W. Mason, A. Rudola, K. H. Wong, P. Balaya, *Adv. Energy Mater.* 3(2013) 444.

8.29 K. M. Abraham, Z. Jiang, *J. Electrochem. Soc.* 143(1996) 1.

8.30 J. Hassoun et al., *Nano Lett.* 14(2014) 4901.

8.31 U. Kohler, J. Kumpers, M. Ullrich, *J. Power Sources* 105(2002) 139–144.

8.32 K. Young et al., Electric vehicle battery technologies, in R. Garcia-Valle, J. A. Pecas Lopes (Eds.), *Electric Vehicle Integration into Modern Power Networks, Power Electronics and Power Systems*, Springer Science, New York, 2013, Chap. 2.

8.33 U. Zhou, M. Cococcioni, C. A. Marianetti, D. Morgan, G. Ceder, *Phys. Rev. B* 70(December 2004) 235121.

8.34 W. Zheng et al., *Bull. Mater. Sci.* 36(2013) 495.

8.35 A. V. Churikov et al., *Electrochim. Acta* 55(2010) 2939.

9

Fuel Cells and Sensors

9.1 Perspective

Batteries discussed in the last chapter and fuel cells [9.1] and supercapacitors (besides sensors) to be discussed in this chapter belong to the same family of energy conversion devices. They are all based on the fundamentals of electrochemical thermodynamics and kinetics. All the three are needed to serve the wide energy requirements of various devices and systems. A battery is a closed system while a fuel cell is an open system in a thermodynamic sense. Winter and Brodd [9.2] have thoroughly answered the question: "What are Batteries, Fuel Cells and Supercapacitors?"

9.2 Fuel Cells

Fuel cells have emerged as an answer to the adverse environmental impact of internal combustion engines. A fuel cell, like a combustion engine, uses a sort of chemical fuel as its energy input, but like a battery, the chemical energy is *directly* converted to electrical energy, without the polluting and relatively inefficient combustion step. Besides high efficiency and low emissions, fuel cells are unique because of their modular and distributed nature and zero noise pollution. They are also naturally linked to hydrogen fuel economy where hydrogen is visualized as the fuel of the future. It is interesting to note that the first fuel cell employing solid electrolytes is due to Haber [9.3] who took a patent in 1905 (Figure 9.1).

9.2.1 Chemical to Electrical Energy Converter

The principle of electric power generation in fuel cells is entirely identical to that of ordinary batteries such as dry batteries or primary batteries in general. But in the case of batteries, the reactants involved in the electrochemical reactions are stored within the battery itself. Also, electric power generation stops the moment these reactants are consumed. The secondary battery such as the Li ion battery *regenerates* the chemical reactants by application of electrical energy from outside.

In contrast, the chemical reactants of a fuel cell are supplied from outside. Furthermore, the reaction products are continuously *discharged* outside the fuel cell. Therefore, a fuel cell can continuously supply electrical power as long as an external supply of reactants is maintained.

Fuel cells and batteries *both* use electrochemical reactions to convert the chemical (bond) energy of the fuel into electrical energy. Fuel cell consists of two types of electron-conducting electrodes—the anode and the cathode—and an ion-conducting electrolyte.

At the interface between the electrodes and the electrolyte, electrochemical reaction generates charge carriers such as electron and ion. What happens at the anode and the cathode? At the *anode,* an *oxidation* reaction occurs that *generates electrons.* At the *cathode,* a *reduction* reaction occurs that *consumes electrons.* Note that *ordinary* oxidation and reduction reactions occur at the *same place,* as in combustion flames that release the chemical energy of the fuel as heat. But in *electrochemical* reactions, the oxidation and the reduction reactions occur at *different electrodes separated by the electrolyte.* Electrical current can be drawn when *these electrodes* are *connected* in *an external circuit.*

Let us discuss an example. Electrical power production via fuel cell is by inducing a reaction *to oxidize hydrogen* by reforming natural gas or other fuels and a reaction *to reduce oxygen* in the air. Each of these

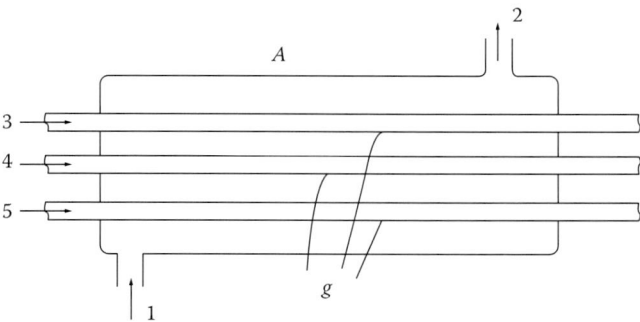

FIGURE 9.1   Haber's fuel cell (1905).

two reactions occurs at separate electrodes connected to an external circuit. The basic fuel cell reaction is the opposite of the electrolysis of water. Why? *Hydrogen* and *oxygen* undergo an electrochemical reaction to produce *electrical power* and *water*. Fuel cells used in space vehicles produce water in electricity generation useful for daily routine within the space vehicle.

9.2.2 Basics of Conversion Process

A basic motivation to look at the fuel-cell-based conversion of chemical energy of the fuel to electrical energy comes from the biological fuel cells. For instance, these mimic human blood circulation in that a fluidic system could be designed based on enzyme-based biofuel cell [9.2]. Biofuel cells converting chemical energy into electricity upon biocatalyzed chemical reactions are promising alternative sources of sustainable electrical energy. Implantable biofuel cells as micropower sources operating in living organisms offer a design challenge within bioelectronics systems. Implantable medical devices, such as cardiac defibrillators/pacemakers, cochlear implants, and insulin pumps among others, powered by implanted biofuel cells extracting electrical energy from a human body are possible resulting in bionic human hybrids.

We shall consider the bio-inspired single chamber fuel cell at the end of the next section.

The primary components of a fuel cell are (1) an ion conducting electrolyte, (2) a cathode, and (3) an anode (Figure 9.2) collectively referred to as membrane electrode assembly or a single-cell fuel cell. In an intuitive example, a fuel such as hydrogen is brought into the anode compartment and an oxidant typically oxygen is brought into the cathode compartment. There is an overall chemical driving force for hydrogen and oxygen to react and produce water. Direct chemical combustion is prevented by the electrolyte that separates the fuel from the oxidant. The electrolyte serves as a barrier to gas diffusion but lets the ions migrate across it.

Half-cell reactions occur at the anode and the cathode producing ions that can traverse the electrolyte. Suppose the electrolyte conducts oxide ions. Then, oxygen will be electro-reduced at the cathode to produce O^{2-} ions and consume electrons. Oxide ions after migrating across the electrolyte react at the anode with hydrogen and release electrons:

Cathode:

$$\tfrac{1}{2}O_2 + 2e^- \rightarrow O^2 \tag{9.1}$$

Anode:

$$H_2 + O^= \rightarrow H_2O + 2e^- \tag{9.2}$$

Overall:

$$\tfrac{1}{2}O_2 + H_2 \rightarrow H_2O \tag{9.3}$$

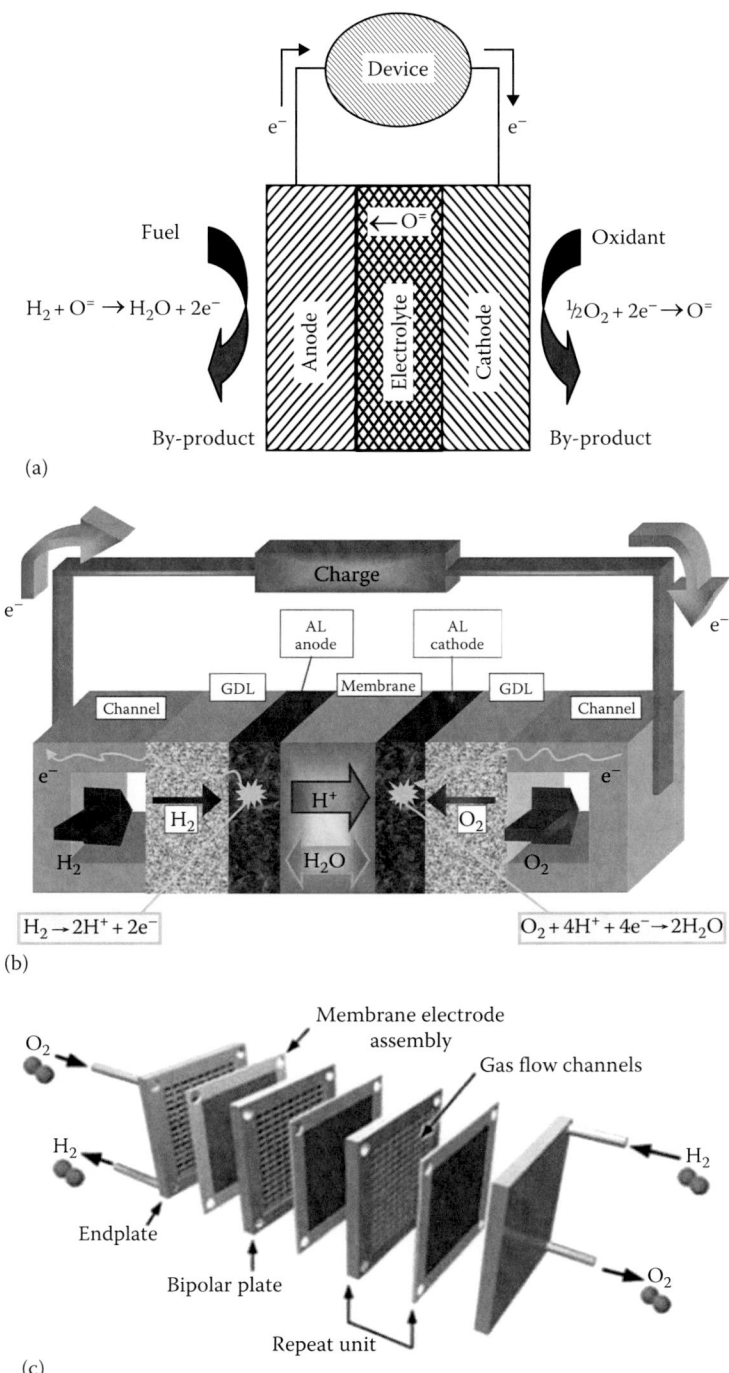

FIGURE 9.2 (a) Schematic of a fuel cell circuit with a device in the external circuit. Half cell reactions involving fuel (H_2) and oxidant (O_2) are shown. (From Haile, S.M., *Acta Materialia*, 51, 5981, 2003.) (b) Structure and function of a polymer electrolyte membrane fuel cell. (c) An exploded view of the membrane electrode assembly and gas flow channels also showing input and output to/from the fuel cell. (Parts (b) and (c) from Nano-CAT Project, http://nanocat-project.eu/Presentation-PEMFC. © European Union, 2014.)

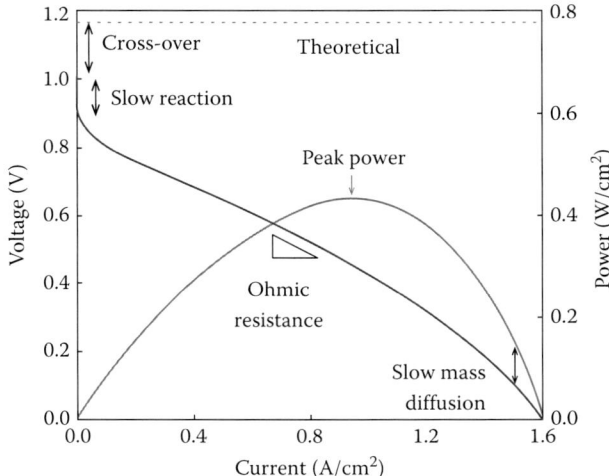

FIGURE 9.3   Fuel cell polarization and current density curves (schematic). (From Haile, S.M., *Acta Mater.*, 51, 5981, 2003.)

What are the analogous reactions for a proton conducting electrolyte?

The flow of ionic charge through the electrolyte must be balanced by the flow of electronic charge through the cathode circuit. *This balance produces the electrical power.*

The performance of a fuel cell is the plot of the voltage output as function of electrical current density shown in Figure 9.3.

The measured voltage E may be written as

$$E = E_{eq} - E_L - \eta_{act} - \eta_{iR} - \eta_{diff} \tag{9.4}$$

where

   E_{eq} is the equilibrium or expected Nernstian voltage
   E_L is the loss in voltage due to leaks across the electrolyte
   η_{act} is the activation overpotential due to slow electrode reactions (at low temperatures)
   η_{iR} is the overpotential due to ohmic resistances in the cell
   η_{diff} is the overpotential due to mass diffusion limitations

E_{eq} depends on the reaction in question. Focus on reaction Equation 9.4. Note that $E_{eq} = -\Delta G/nF$, where ΔG is the Gibbs free energy change, n is the number of electrons transferred in the reaction, and F is the Faraday constant. Now

$$\Delta G = \Delta G°(T) + RT \ln \frac{P_{H_2} P_{O_2}^{1/2}}{P_{H_2O}} \tag{9.5}$$

The first term on the right (tabulated as in Chapter 2) defines the standard potential

$$E°(T) = \frac{-\Delta G°(T)}{nF} \tag{9.6}$$

For reaction Equation 9.4, it is −242 to +45.8 kJ/mol * T.

In low-temperature fuel cells, the reaction kinetics is slow. In this case, the open circuit potential E_0 is given by

$$E_0 - E_{eq} = b \log I_0 \tag{9.7}$$

where b is the so-called Tafel slope in the empirical Tafel equation which is a plot of the overpotential versus the log of current density:

$$\eta_{act} = b \log I_0 - b \log I \tag{9.8}$$

and I and I_0 are the current density and the exchange current density. In hydrogen/oxygen polymer electrolyte membranes cells to be discussed, this offset voltage is ~400 mV.

Fuel cell technologies are primarily based on the nature of electrolytes employed. We shall focus on two types: (1) solid oxide fuel cells (SOFC) and (2) polymer electrolyte membrane fuel cells (PEMFC).

9.2.3 Solid Oxide Fuel Cells

SOFC employ solid, nonporous metal-oxide-based electrolytes. Ion conduction occurs through the migration of oxide ions in the crystal lattice. Stabilized zirconia is the typical electrolyte and the cell operates at temperatures in the range 900°C–1100°C. The schematic of SOFC along with the operating principle is shown in Figure 9.4.

SOFCs are designed in two modes: (1) tubular mode and (2) planar mode. Figure 9.5 shows these in schematic form.

The cell designs of Figure 9.5 are quite demanding from processing point of view. Figure 9.6 shows an easily fabricated (by a sol–gel type of technique) electrolyte or electrode-supported structure.

A critical part of fuel cells is the so-called triple phase boundary (TPB). The actual electrochemical reactions take place at the boundary between reactant gas, electrode, and electrolyte. The porous nature of the

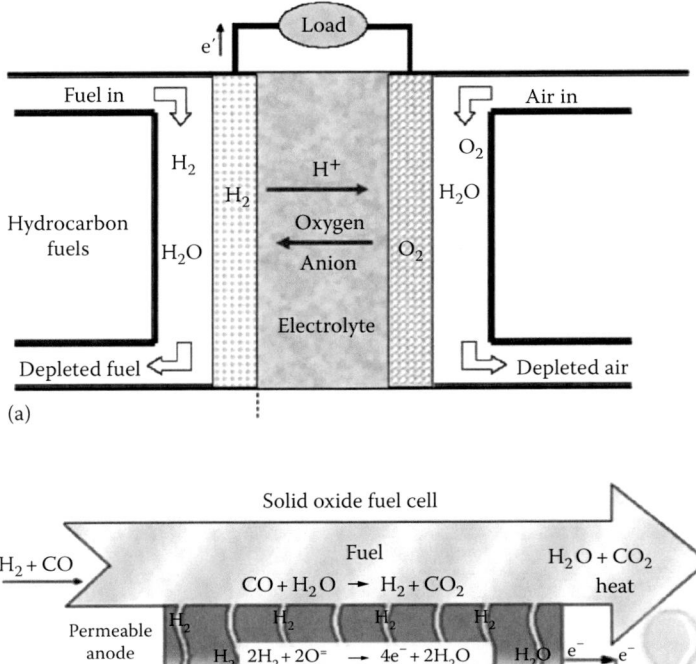

(a)

(b)

FIGURE 9.4 (a) Schematic of SOFC. (b) Working principle of SOFC. Note that the anode and cathode are permeable while electrolyte is impermeable. The electrical power produced by the fuel cell is shown to light a bulb. (From Zou, C. et al., in Aparicio, M. et al. (eds.), *Sol-Gel Processing for Conventional and Alternative Energy*, Springer, New York, 2012, Chapter 2.)

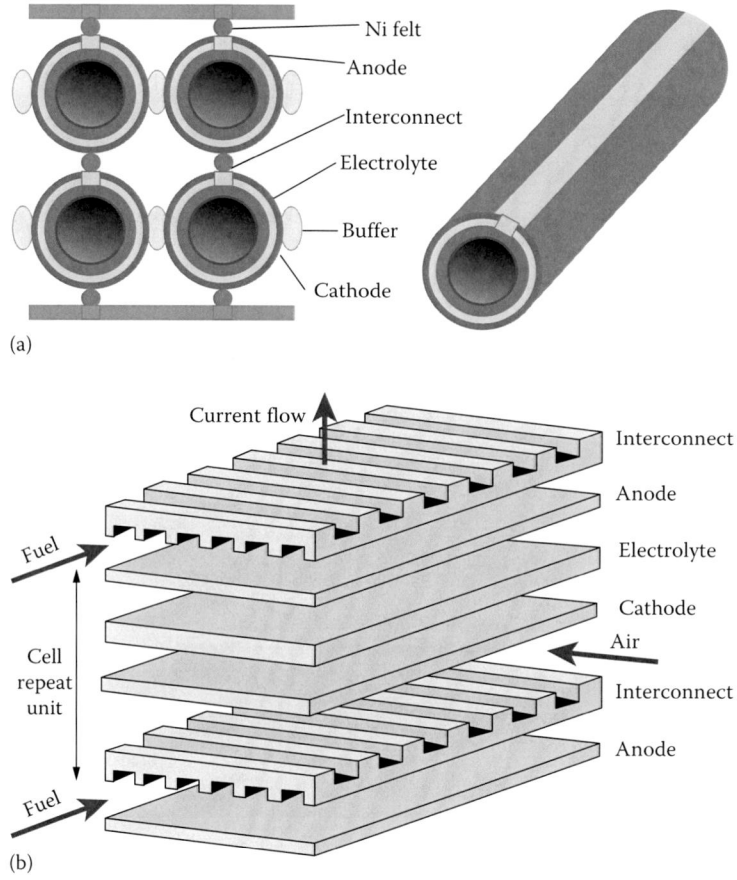

(a)

(b)

FIGURE 9.5 Two typical SOFC designs. (a) Tubular design. Either cathode or anode is made into a long tube with a porous wall. (b) Planar serial design. Each cell is made into a flat (square or rectangular). The cell repeat unit has anode (porous), electrolyte (dense), cathode (porous), airspace, and interconnect. Cells are put in series and connected by inter-connect plates. Fuel enters between interconnect and the anode. (From Zou, C. et al., in Aparicio, M. et al. (eds.), *Sol-Gel Processing for Conventional and Alternative Energy*, Springer, New York, 2012, Chapter 2.)

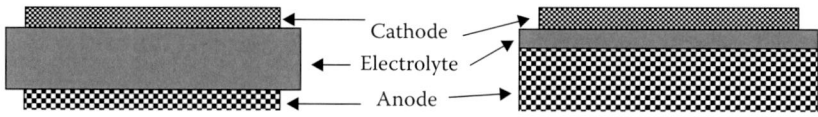

FIGURE 9.6 Electrode supported SOFC designs. Left: electrolyte supported structure, right: anode-supported structure. (From Zou, C. et al., in Aparicio, M. et al. (eds.), *Sol-Gel Processing for Conventional and Alternative Energy*, Springer, New York, 2012, Chapter 2.)

anode and the cathode achieves high surface area which significantly increases the length of the TPB. Note the relatively thin electrolyte in Figure 9.6 which reduces the resistive losses in the cell, thereby reducing the operating temperature. Apart from 8YSZ or 8% yttria-doped zirconia that can operate at $T > 900°C$, Gd-doped ceria or $Ce_{0.9}Gd_{0.1}O_2$ has high ionic conductivity between 500°C and 700°C that favors the fuel cell operation at lower temperatures. Figure 9.7 depicts three processes that occur at the TPB.

Fabrication of anode and electrolytes for SOFCs is more of a fine art than of a routine materials synthesis. This is emphasized in Figure 9.8 that displays the sol–gel method along with flowcharts for making a cathode and an anode.

Table 9.1 lists a few SOFC materials and their performance [9.3].

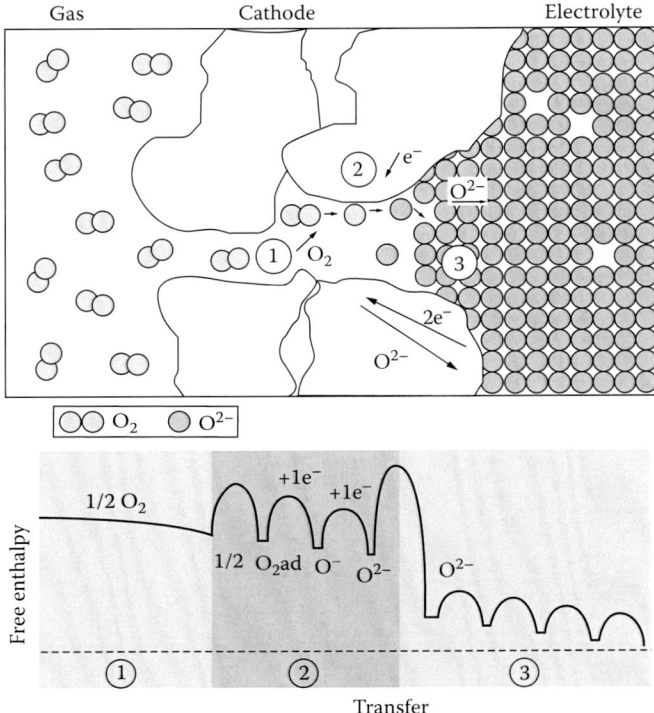

FIGURE 9.7 Three important physicochemical processes visualized to occur at the three-phase fuel cell system interfaces: (1) Gas diffusion, (2) adsorption and dissociation or ionization, and (3) oxygen ion conduction. For simplicity, it has been assumed in constructing this picture that oxygen is completely ionized while entering the electrolyte. (From Zou, C. et al., in: *Sol-Gel Processing for Conventional and Alternative Energy*, Aparicio, M. et al. (Eds.), Springer, New York, 2012, Chap 2.)

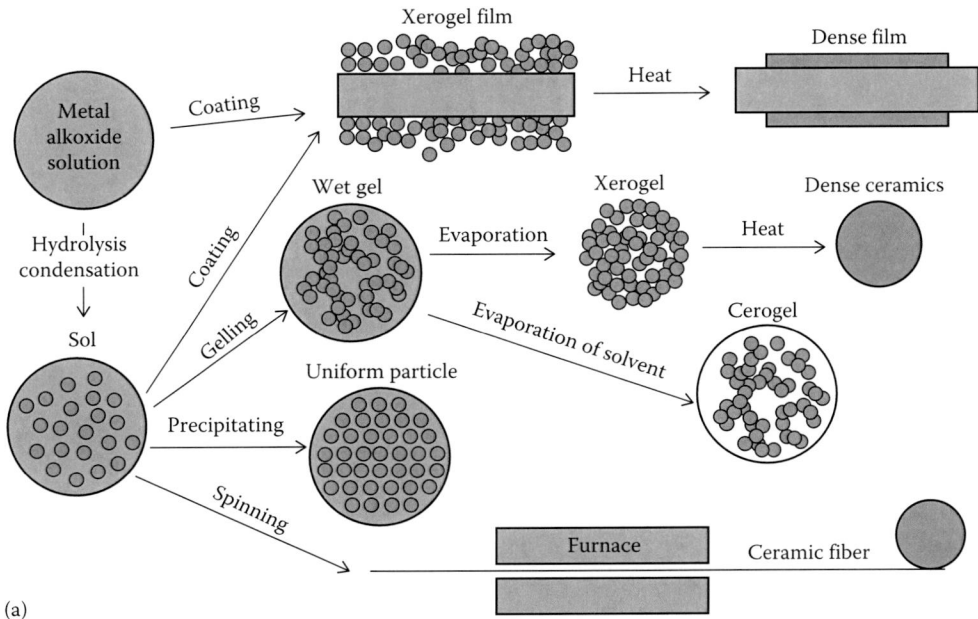

FIGURE 9.8 (a) Sol–gel processes relevant to SOFC materials synthesis in the form of dense film, dense ceramic, aerogel, and ceramic fiber.   (*Continued*)

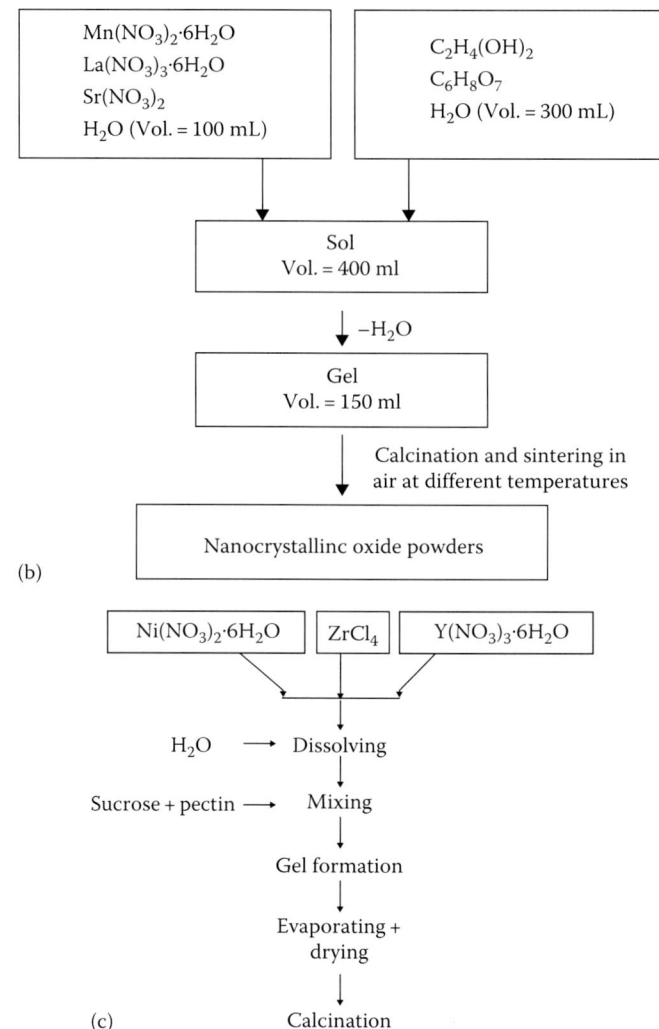

FIGURE 9.8 (Continued)   (b) Flowchart for sol–gel synthesis of LSM La$_{0.8}$Sr$_{0.2}$MnO$_{3-\delta x}$ cathode powder. Particles with cubic structure and 40 nm size are obtained. (c) Synthesis of Ni–YSZ anode by sol–gel method.

TABLE 9.1

Materials and Performance of SOFCs Operated at 650°C with Hydrogen/Air as Input Gases

Electrolyte	Thickness (µm)	Anode	Cathode	Peak Power Density
LSGMC	205	Ni–SDC	SSC	240–410
Ce$_{0.9}$Gd$_{0.1}$O$_{1.95}$	–40	Ni–Ru–GDC	Sm$_{0.5}$Sr$_{0.5}$CoO$_{3-\delta}$	770[a]
Ce$_{0.9}$Gd$_{0.1}$O$_{1.95}$	5–10	Ni–YSZ	La$_{0.6}$Sr$_{0.4}$Co$_{0.2}$Fe$_{0.8}$O$_{3-\delta}$	150
Ce$_{0.9}$Gd$_{0.1}$O$_{1.95}$	150	Ni–GDC	La$_{0.6}$Sr$_{0.4}$Co$_{0.2}$Fe$_{0.8}$O$_{3-\delta}$	110
YSZ	~30	Ni–YSZ	LSM + SDC	190
YSZ	~10	Ni–YSZ	La$_{0.8}$Sr$_{0.2}$FeO$_{3-\delta}$	400

Note: LSGMC, La$_{0.8}$Sr$_{0.2}$Ga$_{0.8}$Mg$_{0.15}$C0$_{0.05}$O$_{3-\delta}$.

[a] Measured at 600°C. An even higher value possible for 650°C.

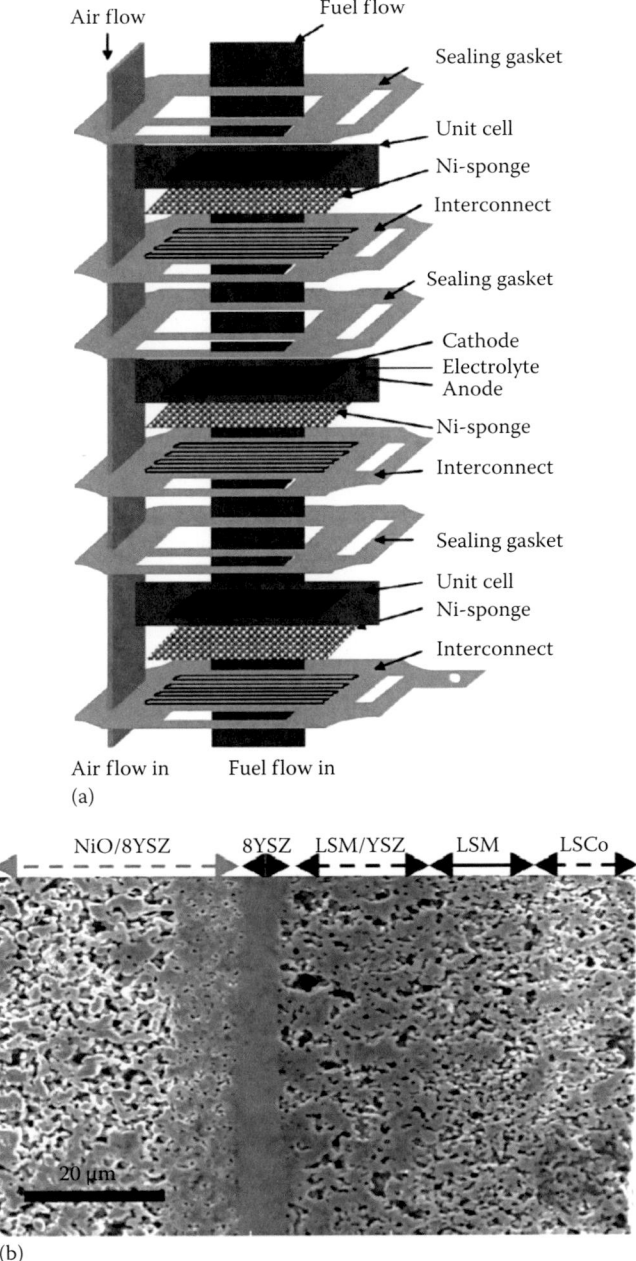

FIGURE 9.9 (a) Planar single-columned three-cell SOFC stack based on anode-supported 10 cm × 10 cm unit cells. (b) A cross-sectional view of the unit cell of the 10 × 10 SOFC stack. (After Jung, H.Y. et al., *J. Power Sources*, 159, 478, 2006.)

To appreciate the design and dimensions of a practical planar fuel cell, see Figure 9.9 [9.6].

Now, we will get back to the discussion of the bio-inspired fuel cell. A single-chamber fuel cell has been demonstrated (Figure 9.10). Such a cell utilizes fuel/oxidant mixtures in a single gas inlet. Carefully selected anode and cathode catalysts produce well-controlled half-cell reactions at the anode and the cathode:

Anode:

$$CH_4 + \frac{1}{2}O_2 \rightarrow CO + 2H_2 \text{ (chemical)} \tag{9.9}$$

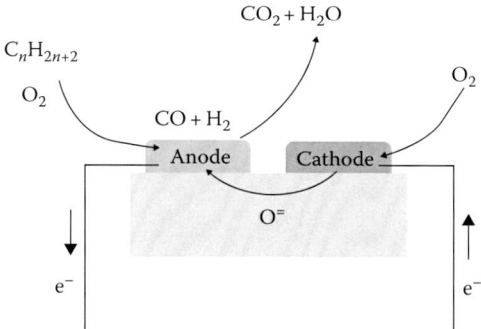

FIGURE 9.10 Mimicking biology. A single chamber fuel cell. Here a hydrocarbon fuel is partially oxidized at the anode. Co and H_2 are produced consuming O_2. The oxygen partial pressure thus developed drives the electrochemical reaction of the fuel cell.

$$H_2 + O^- \rightarrow H_2O + 2e^- \text{ (electrochemical)} \quad (9.10)$$

$$CO + O^- \rightarrow CO_2 + 2e^- \text{ (electrochemical)} \quad (9.11)$$

Cathode:

$$\tfrac{1}{2}O_2 + 2e^- \rightarrow O_2 \text{ (electrochemical)} \quad (9.12)$$

Note that the simple chemical oxidation of the hydrocarbon that would yield CO_2 and H_2O does not take place. Instead, only partial oxidation occurs at the anode and the products of this intermediate reaction are consumed electrochemically, while oxygen is consumed electrochemically at the cathode. Note that there are no complications due to sealing, greatly simplifying design and enhancing thermal and mechanical shock resistance. The chemical precision envisaged in this design is routinely utilized in nature. Biofuel cells operating on aqueous glucose not separated from ambient oxygen demonstrate essentially perfect electrode selectivity [9.7].

For more information, see Takizawa [9.8].

Next, we will discuss polymer electrolyte membrane fuel cells.

9.2.4 Polymer Electrolyte Membrane Fuel Cells

Unlike SOFCs, PEMFCs use an ion-exchange membrane such as Nafion as an electrolyte. Nafion is the trade name for hydrated perfluoro sulfonic acid which is impermeable to gases. Figure 9.11a shows the schematic of a typical PEFMC and Figure 9.11b depicts a residential high-temperature PEFMC. Typical advantages of a PEMFC are (1) a nonvolatile electrolyte below 100°C, (2) negligible electrolyte leaching, and (3) a thin electrolyte layer. The excellent oxygen solubility of polymer electrolyte coupled with favorable diffusivity characteristics enables PEFMCs to deliver better performance than SOFCs.

The operating temperature of a LT-PEMFC (not shown) is less than 100°C, so it is necessary to burn part of the fuel and the off-gas to produce steam for the reformation process. On the other hand, a HT-PEMFC operates in the temperature region of 150°C–170°C. So steam can be produced by the exhaust heat from the cell stack (Figure 9.11b). Therefore the fuel utilization of a HT-PEMFC is higher than that of a LT-PEMFC. In the case of an SOFC, operating at a temperature of about 700°C, there is no material that is suitable for use as a gas seal. The electrical efficiency of the system is generally determined as roughly the product of the generation efficiency (cell voltage) and the system energy efficiency (fuel utilization). The electrical efficiency of a HT-PEMFC, which can achieve a fuel utilization of 85%–90%, is higher than that of a LT-PEMFC or an SOFC, which can achieve only 70%–75% fuel utilization.

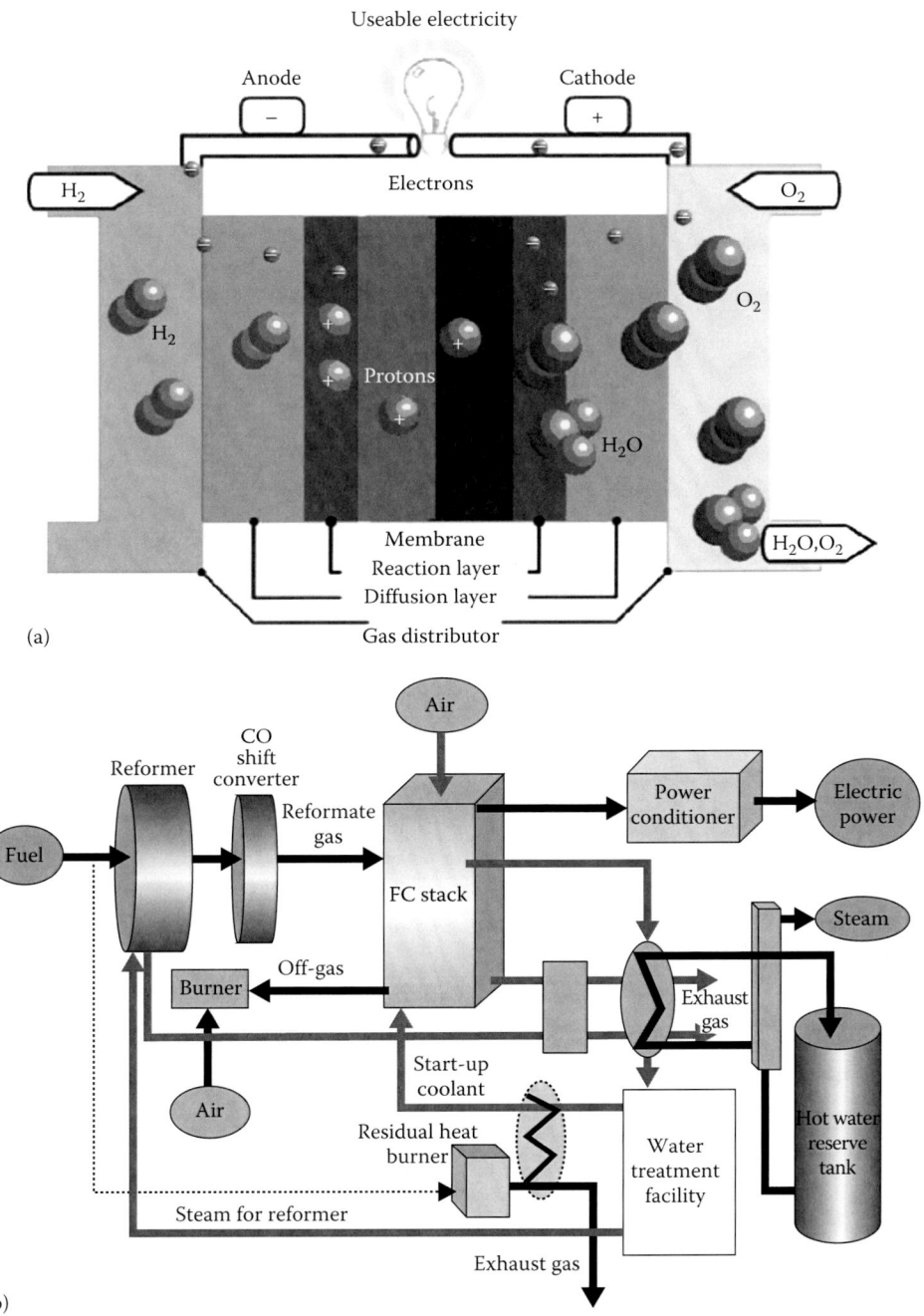

FIGURE 9.11 (a) Schematic of a polymer electrolyte membrane fuel cell (PEMFC). (From DLR, Stuttgart, Germany.) (b) Schematic diagram of residential high temperature PEMFC. *(Continued)*

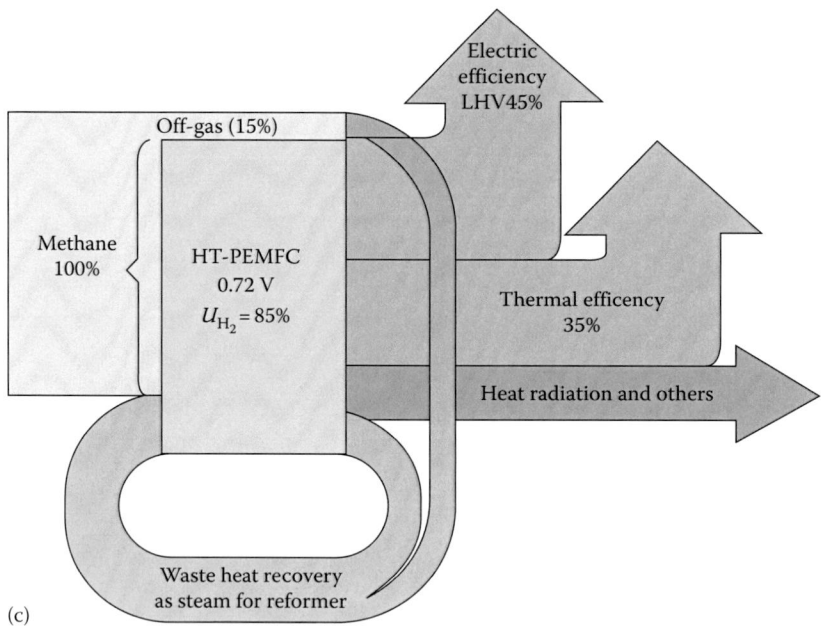

FIGURE 9.11 (*Continued*)   (c) Heat balance schematic. (From www.daido-it.ac.jp.)

9.3 Electrochemical Supercapacitors

Electrochemical supercapacitors possess tens to hundreds of times larger capacitance than conventional "parallel-plate" capacitor that consists of two metal plates separated by a gas or vacuum as dielectric. When a metal or an electronic conductor is brought into contact with a solid/liquid ionic conductor, a charge accumulation is achieved electrostatically on either side of the interface. This charge accumulation leads to the formation of an electrical double layer which is essentially a molecular dielectric. Note that no charge transfer takes place across the interface so that the current generated is essentially a displacement current via a rearrangement of charges. This is an example of a non-Faradaic process [9.9].

Charge storage is also achieved through a charge transfer via an electron transfer that produces changes in the oxidation states in electrostatic materials according to Faraday's laws in relation to the electrode potaential. This is a typical example of a Faradaic process that helps develop supercapacitors of which there are two types:

1. Based on the charging and discharging of the interfacial electric double layer.
2. Based on the charge–discahrge mechnism involving the electric charge transfer between the phases but no bulk phase transformation, resulting in pseudo capacitors or redox capacitors.

The important difference between non-Faradaic and Faradaic processes is the following. The electrons involved in the first process–charging of the electrical double layer—are the itinerant conduction band electrons of metal or carbon electrode. The electrons in the second Faradaic process are those transferred to or from valence electron state (or orbitals) of the redox component-anode or cathode.The electrons could come into or go away from the conduction band states of the electronically conducting support material. This depends on whether the Fermi level (the highest occupied energy level in a material typically metal) of the electronically conducting support lies below the highest occupied molecular orbital of the reductant or above the lowest unoccupied molecular orbital states of the oxidant.

In the aforementioned second type, namely the pseudocapacitors, the double-layer charging process (non-Faradaic) is accompanied by a Faradaic charge transfer. Thus, the capacitance of a supercapacitor is given by

$$C = C_{double\ layer} + C_{pseudo} \tag{9.13}$$

C_{pseudo} arises from a surface redox reaction of the type

$$O_{ad} + ne \rightarrow R_{ad} \tag{9.14}$$

where O_{ad} and R_{ad} are the adsorbed oxidant and reductant, respectively.

A supercapacitor schematic is given in Figure 9.12a. Like a battery, it consists of a negative and a positive electrode and an electrolyte separator.

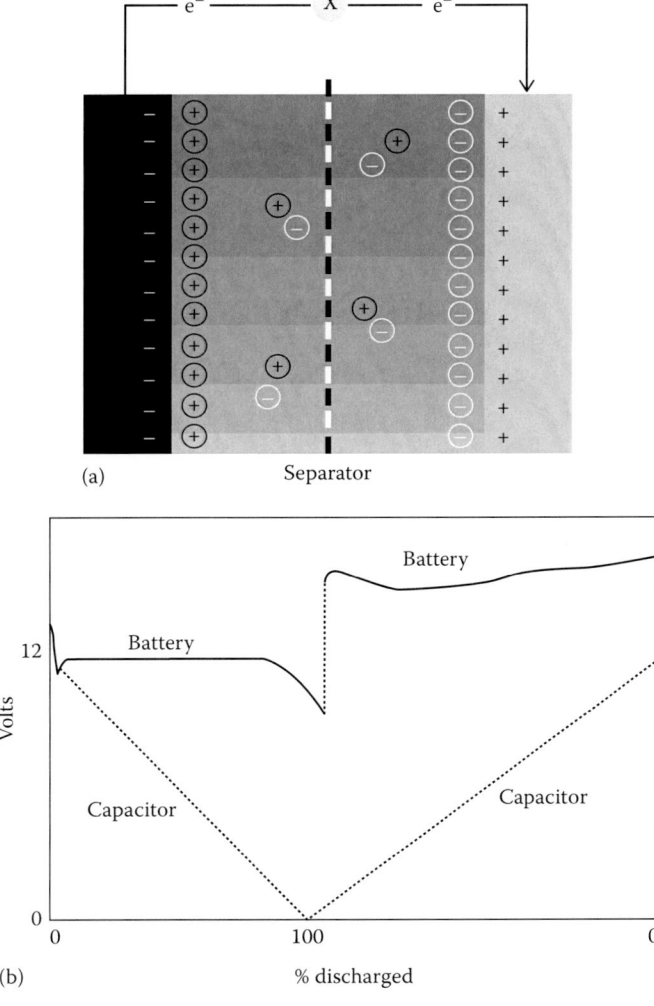

FIGURE 9.12 (a) Supercapacitor schematic. (b) Difference between a battery and a supercapacitor. (From Shukla, A.K., *Resonance*, 6, 72, 2001.)

Before considering an example of a supercapacitor, let us clarify the functional difference between a battery and a supercapacitor (Figure 9.12b). Note that a capacitor unlike the battery can be rapidly charged and discharged 100%.

In Equation 9.12, the charges assocaited with $C_{double\ layers}$ and C_{pseudo} are typically 20 and 220 μC, respectively. A more direct comparison is based on energy $E = \frac{1}{2}CV^2$. As an exercise, compare the energy stored in (1) 1 V electrical double-layer capacitor made of a porous carbon electrode with surface area 1000 m^2/g and (b) 1 V supercapacitor with a pseudocapacitance held between two carbon electrodes of similar area.

A supercapacitor based on a composite material made of precipitaion-synthesized amorphous manganese dioxide MnO_2 and commercial single-walled carbon nanotube has been demonstrated by Subramanian et al. [9.10]. It has a long cycle performance at a high charge–discharge curent of 2 A/g.

Now, we will move on to a discussion on electrochemical sensors.

9.4 Electrochemical Sensors

Electrochemical sensors are small, self-powered micro fuel cells that produce an output signal as a result of a chemical reaction. They are lightweight and power efficient which make them ideally suited for personal monitoring devices that are small enough to fit into a shirt pocket and that can operate continuously for as long as four months without the need to replace the battery.

In automobile emissions control systems, electrochemical sensors positioned in the exhaust pipe feed data to the engine management computer. The aim of this sensor is to help the engine run efficiently by detecting oxygen levels in the exhaust flow. The computer uses the sensor readings to ensure that the engine is being given the right amount of fuel—too much or too little fuel can generate nitrogen oxide pollutants and, in some cases, lead to poor performance and even engine damage.

In hospitals, oxygen electrochemical sensors are designed as area monitors that have upper and lower alarm levels so that oxygen enriched atmospheres may also be monitored. Sensor output can be recorded on a strip chart recorder so that a permanent record of the oxygen in the atmosphere may be maintained.

In industrial applications such as oil drilling, oil refining, and wastewater treatment, hydrogen sulfide (H_2S) is a common hazard. Sensors are used in these industries in portable and fixed systems to give an early and reliable warning of the presence of H_2S.

A useful electrochemical sensor must meet a number of experimental design criteria [9.11].

Many of these are linked to its potential benefits. Among the most important are the following:

1. For amperometric and voltammetric sensors, the species to be determined is electroactive within the sensor's potential range, and whether there is the addition of an inert, supporting electrolyte to carry the current perturbs the equilibria in solution.
2. For potentiometric sensors, there is an adequate electrode material, free from interferences.
3. The concentration of electroactive species can be determined with sufficient accuracy and precision.
4. The measurements are sufficiently reliable and repeatable.
5. The response time of the sensor is sufficiently fast.
6. The drift or diminution of sensor response with time owing to electrode degradation or surface fouling is sufficiently small.
7. Calibration is simple and easy to perform, or not necessary.
8. The detection limit is sufficiently low for the purpose envisaged.

The relative importance of these factors depends on the monitoring necessities as well as on the technique employed and the electrode and cell configuration.

Figure 9.13 shows a typical electrochemical sensor of the amperometric type.

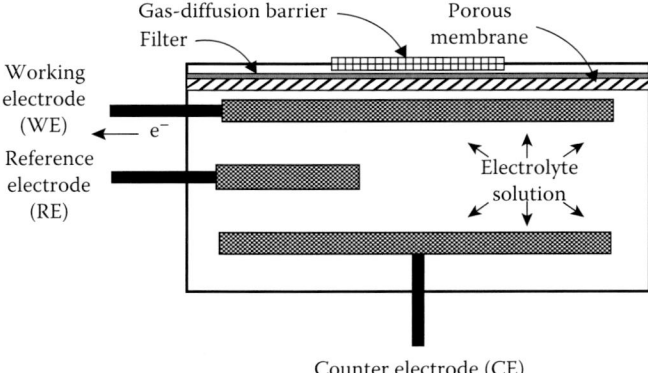

FIGURE 9.13 A basic electrochemical sensor with filter. A porous membrane to control the amount of gas molecules that diffuses into a cell containing liquid electrolyte and three electrodes. The three electrodes are made of noble metal and stacked parallel to each other. (From electronicdesign.com.)

TABLE 9.2

Commonly Detected Gases and the Corresponding Electrodes

Gas	Measuring Electrode	Counter Electrode
H_2S	$H_2S + 4H_2O \rightarrow H_2SO_4 + 7H^+ + 8e^-$	$O_2 + 4H + 4e^- \rightarrow 2H_2O$
HCN	$2HCN \rightarrow 2H + 2CN^-$	$O_2 + 2H + 2e^- \rightarrow H_2O$
CO	$CO + H_2O \rightarrow CO_2 + 2H+ + 2e^-$	$O_2 + 4H + 4e^- \rightarrow 2H_2O$
Cl_2	$2H_2O \rightarrow O_2 + 4H^+ + 4e^-$	$2H^+ + Cl_2 + 2e^- \rightarrow 2HCl$
SO_2	$SO_2 + 2H_2O \rightarrow H_2SO_4 + 2H^+ + 2e^-$	$O_2 + 4H + 4e^- \rightarrow 2H_2O$
H_2	$H_2 \rightarrow 2H^+ + 2e^-$	$O_2 + 4H + 2e^- \rightarrow H_2O$
NO	$NO + 2H_2O \rightarrow HNO_3 + 3H^+ + 3e^-$	$O_2 + 4H + 4e^- \rightarrow 2H_2O$
NO_2	$2H_2O \rightarrow O_2 + 4H^+ + 4e^-$	$NO_2 + 2H + 2e^- \rightarrow NO + H_2O$

The electrodes, namely the measuring and counter electrode (CE), depend upon the gas to be sensed or detected. Table 9.2 gives a list of commonly detected gases and the corresponding electrodes.

Depending on the design of the sensor, all three electrodes can be made of different material to complete the cell reaction. The thin layer of electrolyte facilitates the cell reaction and carries the ionic charge across the electrodes efficiently. This chemical interaction generates a small current proportional to the concentration of the gas. Because of the current generated in this process, the electrochemical sensor is referred to as an amperometric gas sensor (or a micro fuel cell). A scrubber filter installed in front of the sensor filters out unwanted gases. Activated charcoal is a common filter material.

When gas comes in contact with the sensor, it passes through the membrane to reach and react at the surface of the working electrode (WE). The working electrode is where the potential is controlled and a current flow is generated that is proportional to the gas concentration.

The performance of the sensor deteriorates over time because of the continuous electrochemical reaction of the changes in potential occurring on the electrode. The deterioration must be reduced while maintaining a constant sensitivity with a good linearity. For this purpose, a reference electrode (RE) is placed close to the working electrode. RE will anchor the working electrode at the correct potential. RE must maintain a constant potential, so no current should flow through it.

The CE is a conductor that completes the cell circuit. Current that flows into the solution via the working electrode leaves the solution via the CE.

When the sensor is exposed to the target gas, such as carbon monoxide, the reaction at the working electrode oxidizes the carbon monoxide to become carbon dioxide, which diffuses out of the sensor.

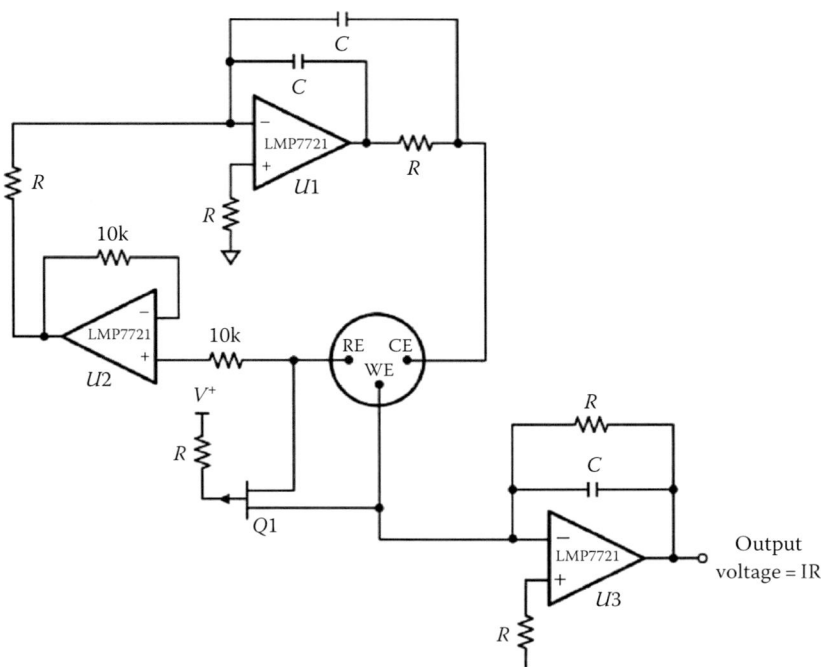

FIGURE 9.14 An amperometric sensor connected in an op amp circuit.

Hydrogen ions and electrons are generated. The hydrogen ions migrate through the electrolyte toward the CE. This process leaves a negative charge deposited on the working electrode.

The electronic circuit of an amperometric sensor is shown in Figure 9.14 which comprises three amplifiers and one JFET transistor.

The control loop made up of amplifiers $U1$ and $U2$ provides the current to the CE to balance the current required by the working electrode. Amplifier $U1$ provides the current to maintain the working electrode at the same potential as the RE.

The voltage follower ($U2$) is connected to the reference electrode and cannot draw any current from the reference electrode.

The input bias current of the amplifier is very critical. LMP7721 ultra low input bias current amplifier is used to ensure that the RE will maintain constant potential by having less than 3 fA of bias current. The control amplifier ($U1$) provides the current to the CE to balance the current required by the working electrode.

The current-to-voltage converter ($U3$) is configured as a trans-impedance amplifier. It converts the signal current from the working electrode into a voltage proportional to the applied gas concentration. *Output voltage is the sensor current multiplied by the feedback resistor value.* For more details, see http://electronicdesign.com/site-files/electronicdesign.com/files/archive/electronicdesign.com/content/14978/59899_fig_03.jpg.

The typical specifications of the sensor of Figure 9.15 are given in Table 9.3.

In a sensor, either a chemical reaction takes place or the charge transport is modulated by the reaction. Electrochemical sensing always requires a closed circuit. Current must flow to make a measurement. Since we need a closed loop, we need at least two electrodes. Three parameters characterize the use of sensor [9.12]: (1) sensitivity, (2) selectivity, and (3) response time.

Sensitivity is the ability of the sensor to quantitatively measure the test gas under given conditions. It is governed by the inherent physical and chemical properties of the materials used. Selectivity of a sensor is its ability to sense a particular gas free from interference. Response time is a measure of how quickly the maximum signal change is achieved with gas concentration changes. In addition, reversibility, long-term stability, size, and power consumption are other factors influencing the overall performance of the sensor.

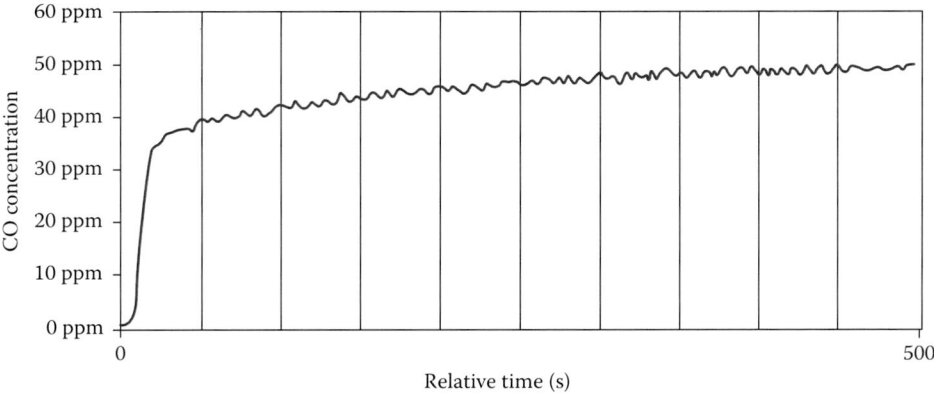

FIGURE 9.15   Typical response of CO sensor. (From http://forum.eepw.com.cn/thread/228170/.)

TABLE 9.3

CO Sensor Specifications

Parameter	Value
Sensitivity	55–100 nA/ppm (65 nA/ppm typ)
Response time (t_{90} from 0 to 400 ppm CO)	<30 s
Range (ppm CO, guaranteed performance)	0–2000 ppm
Overrange limit (specifications not guaranteed)	4000 ppm

To illustrate the characteristics of a typical sensor, an ammonia gas sensor based on a thin film of poly-electrolyte templated polyaniline is presented [9.13]. Poly(4-styrenesulfonate-co-maleic acid) or PSSM acts as both the template and counter ion. The apparatus used for sensing is shown in Figure 9.16a. The characteristics of the sensor are displayed in Figures 9.16b and 9.17.

This sensor evaluation demonstrates the following: The PANI–PSSM sensor exhibits stable, reproducible, and reversible resistance changes in the presence of ammonia in the 5–250 ppm range. The response and recovery times are quite short, in the range 50–1000 s depending on the ammonia concentration. Response at 150 s is shown to vary linearly with ammonia concentrations in this range.

Apart from amperometric sensors which use the principle of polarographic oxygen pumping, there are potentiometric (voltage measuring) sensors based on Nernst principle.

Let us now illustrate a potentiometric sensor.

Yttria-stablized zirconia is an important ion-conducting membrane for potentiometric oxygen concentration cells. Let different partial pressures of oxygen p_1 and p_2 be maintained across the membrane. Then, an emf E is developed at temperature T given by the Nernst equation

$$E = \left(\frac{RT}{4F}\right) \ln\left(\frac{p_1}{p_2}\right) \tag{9.15}$$

where
  F is the Faraday
  R is the gas constant

Bulk zirconia cells based on Equation 9.15 are an important control element in automotive fuel ignition systems. The maximum electrical resistance of zirconia allowed by electronic determines the temperature of operation of such cells. For a mechanically strong ceramic, this means a 1-mm-thick zirconia membrane and a greater than 600°C operating temperature.

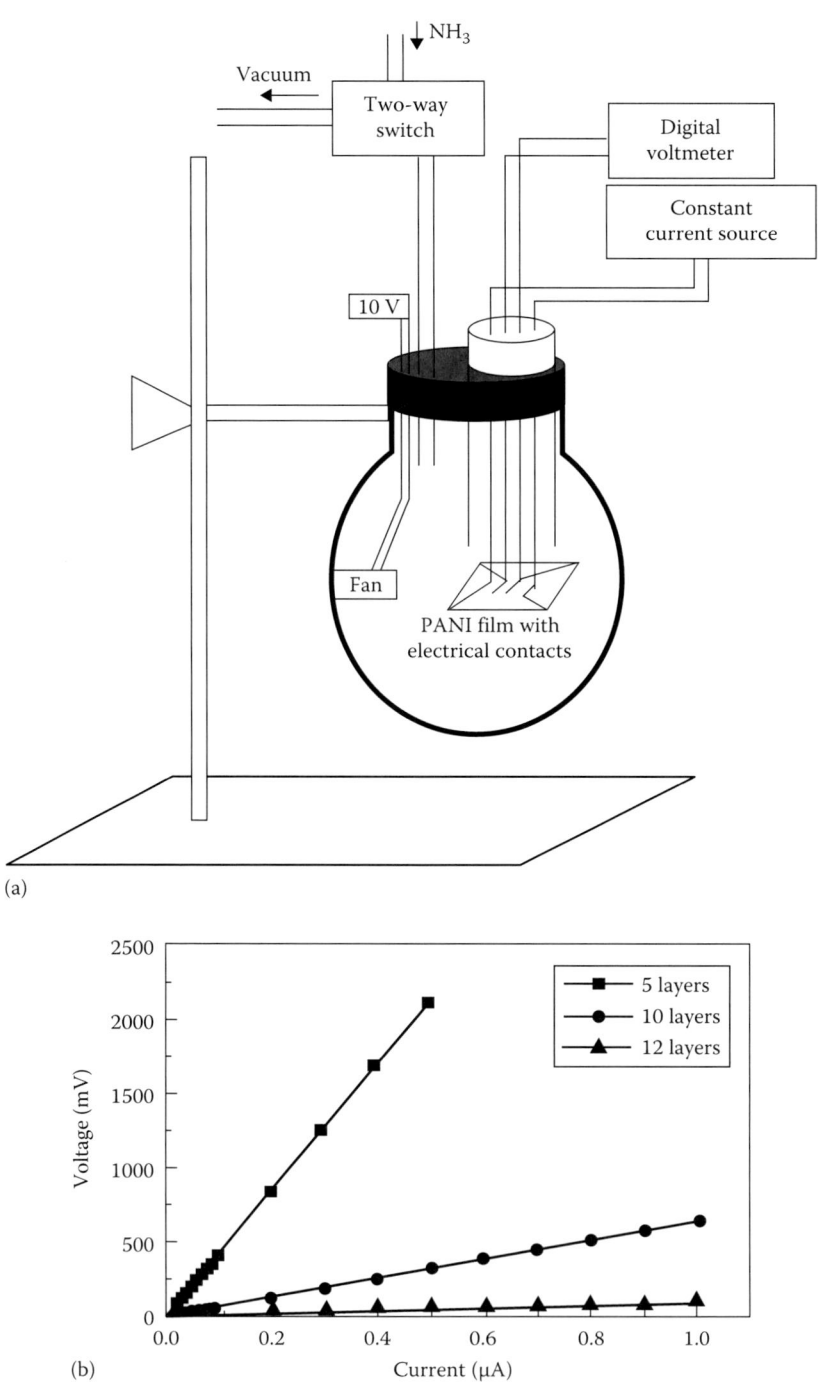

FIGURE 9.16 (a) A simple apparatus for template-based PANI thin film ammonia sensing at room temperature. (b) *I–V* characteristics of PANI–PSSM thin films of varying layer thickness fabricated by spin casting. (From Prasad, G.K. et al., *Sens. Actuators*, 106, 626, 2005.)

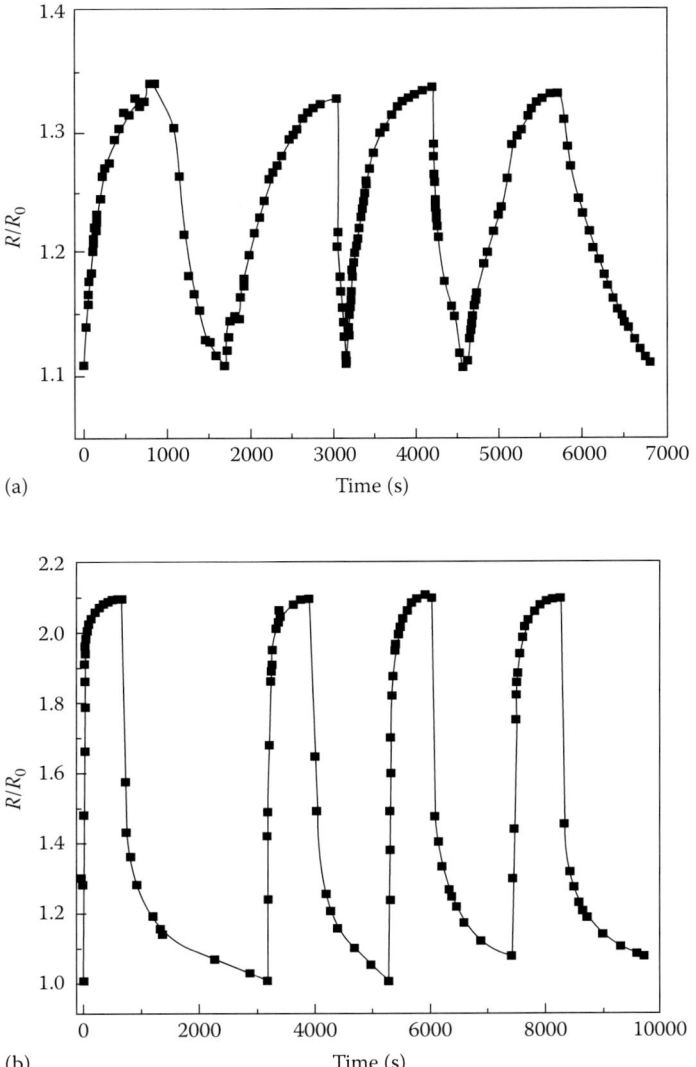

FIGURE 9.17 Time dependence of the resistance ratio R/R_0 of PANI–PSSM thin film (five layers) on repeated exposure and removal of ammonia; concentration of ammonia in the air–ammonia mixture: (a) 5 ppm and (b) 250 ppm. Here R_0 is the resistance of the sensor film in vacuum before introduction of ammonia gas. R is the resistance as a function of time of exposure to ammonia gas. (From Prasad, G.K. et al., *Sens. Actuators*, 106, 626, 2005.)

How could one reduce the temperature of operation? By changing the design itself to a thin film configuration. A typical potentiometric thin film gas sensor based on a sol–gel zirconia membrane employs a ceramic film a few microns thick in either a planar or a coaxial geometry. This uses the oxygen equilibrium pressure within a Ni/NiO layer to establish a reference oxygen partial pressure. Sol–gel thin films technology could provide gas sensors integrated onto silicon or in the form of inline coaxial sensors operating at or close to room temperature [9.12].

As a potentiometric sensor example, we present LaF$_3$ thin film oxygen sensor [9.14]. The design and the response time characteristics of this sensor are shown in Figure 9.18.

Figure 9.19 displays the linear response of the electrode response and the log of the oxygen partial pressure. Such an electrode reaction follows single electron Nernst equation.

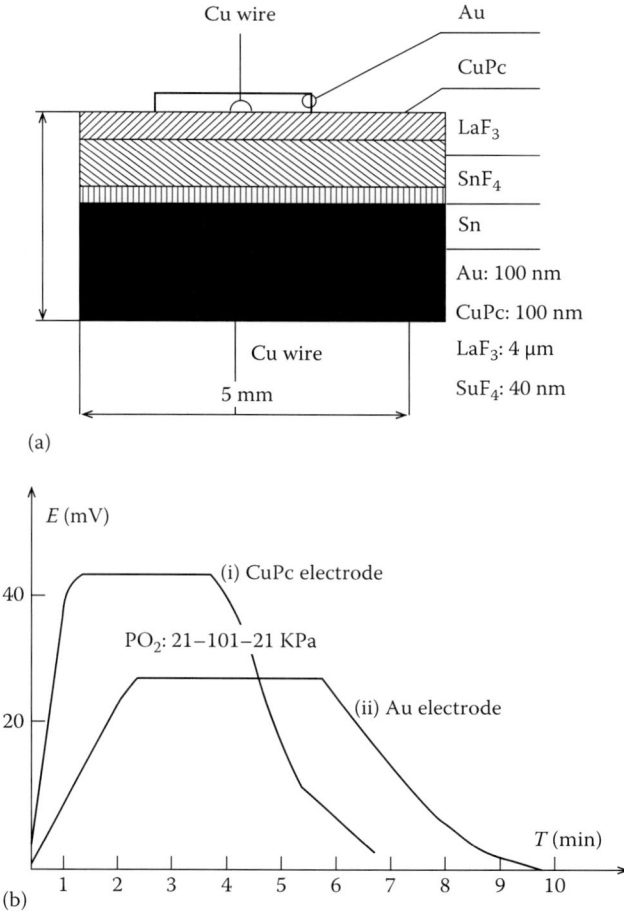

FIGURE 9.18 LaF$_3$ thin film oxygen sensor CuPc is copper phthalocyanine. Other symbols are self explanatory. The thin film LaF$_3$ electrolyte together with the sensitive CuPc was fabricated by RF sputtering. (a) Design and (b) time response characteristics. (From Tan, G.L. et al. *Sens. Actuators, B* 34, 417, 1996.)

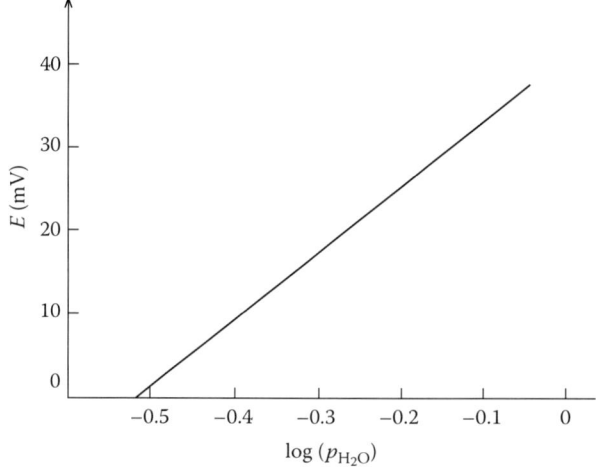

FIGURE 9.19 Electrode response E (mV) vs. log(p_{H_2O}) for the LaF$_3$ electrolyte-based thin film (<4 μm thick) sensor.

$$E = E_0 + \frac{RT}{nF} \ln P_{O_2} \tag{9.16}$$

which could be written as

$$O_2 + e^{-O} = O_2^-, \quad O_2 + H_2O = OH^- + HO_2^-$$

so that the overall reaction is

$$O_2 + H_2O + e^- = OH^- + HO_2^- \tag{9.17}$$

Apart from these standard sensors of auto exhaust type, a recent development of sensors and biosensors based on nanomaterials and nanostructures must be mentioned [9.15]. Another significant development is the advent of electrospun nanofibers as sensor materials [9.16].

Finally, Guth et al. [9.17] have discussed the developments in electrode materials for electrochemical sensors.

9.5 Outlook

Solid state ionic materials as fuel cell electrolytes, supercapacitors, and electrochemical sensors are in an active state of research and development with goals such as miniaturization and feature size reduction to sub-micron size. Entirely new types of materials development such as electrospun fibers have integrated these materials into the frontier area of advanced functional materials. Future years would see a development in both materials variety and device capability making solid state ionics a cutting edge technology.

9.6 Summary

In this chapter, we have provided brief accounts of three specific applications of solid state ionics namely fuel cells, supercapacitors, and sensors all of which have a symbiotic relationship with the batteries discussed in the last chapter. As the feature size of the principal elements of these devices reaches sub-micron dimensions, these devices could be made more efficient and could be integrated into solid state electronics. The next chapter would focus on the emerging area of nanoionics, an important adjunct to nanoelectronics.

Problems

9.1 Typical specifications for a home-located air-cooled PEM fuel cell unit are as follows: power output = 600 W and output voltage 24 V DC. It is to be fueled by compressed hydrogen. Design a fuel cell stack to deliver required power.

9.2 Pollution monitoring sensors for the industry need to be comprehensive. What is a total NO_x sensor in this context? How is a laminated-type sensor designed and fabricated? (See Reference 9.18).

9.3 What is the difference between an amperometric and a potentiometric sensor? Design using a polymeric membranes of stable β-AgI, a potentiometric sensor to monitor iodine content in table salt (see Reference 9.19).

REFERENCES

9.1 S. C. Singhal, K. Kendall, *High Temperature Solid Oxide Fuel Cells*, Elsevier, New York, 2004.

9.2 M. Winter, R. J. Brodd, *Chem. Rev.* 104(2004) 4245.

9.3 K. MacVittie et al., *Energy Environ. Sci.* 6(2013) 81.

9.4 S. M. Haile, *Acta Mater.* 51(2003) 5981.

9.5 C. Zou et al., Solid oxide fuel cells, in *Sol-Gel Processing for Conventional and Alternative Energy*, M. Aparicio et al. (Eds.), Springer, New York, 2012, Chap. 2.

9.6 H. Y. Jung et al., *J. Power Sources* 159(2006) 478.

9.7 N. Mano et al., *J. Am. Chem. Soc.* 124(2002) 12962.

9.8 K. Takizawa, Electrochemistry of fuel cell, in *Energy Carriers and Conversion Systems*, Vol. II.

9.9 A. K. Shukla, *Resonance* 6(2001) 72.

9.10 V. Subramanian et al., *Electrochem. Commun.* 8(2006) 827.

9.11 C. M. A. Brett, *Pure Appl. Chem.* 73(2001) 1969–1977.

9.12 R. Ramamoorthy, P. K. Dutta, S. A. Akbar, *J. Mater. Sci.* 38(2003) 4271.

9.13 G. K. Prasad et al., *Sens. Actuators* 106(2005) 626.

9.14 G. L. Tan et al., *Sens. Actuators B* 34(1996) 417.

9.15 G. Zhou et al., *Anal. Chem.* 87(2015) 230.

9.16 J. Huang, in *Nanotechnology and Nanomaterials Advances in Nanofibers*, R. Macguire (Ed.), 2013.

9.17 U. Guth et al., New developments in electrode materials for electrochemical sensors, in *Advancement in Sensing Technology SSM1*, S. C. Mukhopadhyay et al. (Eds.), Springer-Verlag, Berlin, Germany, 2013, pp. 181–189.

9.18 S. Zhaikov, N. Muria, *Sens. Actuators* 121(2007) 639.

9.19 D. James, T. Prasada Rao, *Electrochim. Acta* 66(2012) 340.

10

Nanoionics

10.1 Motivation

Perhaps the best motivation to this chapter dealing with nanostructures comes from Kittel's book [10.1]:

> The term nanostructure denotes a condensed matter structure having a minimum dimension approximately between 1 nm (10 A) and 10 nm (100 A). These structures may be fine particles, fine wires, or thin films. Fine particles typically contain between 10 and 1000 atoms. Semiconductor technology has made it possible to fabricate small pools of electrons called in various ways: single-electron transistors, quantum dots, artificial atoms, Coulomb islands, or quantum corrals. The unusual physical properties of nanostructures compared with bulk solids are attributed to several factors:
>
> - The ratio of the number of atoms on the surface to the number of atoms in the interior may be of the order of unity.
> - The ratio of surface energy to total energy may be of the order of unity.
> - The conduction or valence electrons are confined to a small length or volume so that the quantum wavelength of the lowest electronic state is constricted and consequently the minimum wavelength is shorter than in the bulk solid.
> - A wavelength or boundary condition shift will change the optical absorption spectrum.
> - Assemblies of nanoclusters may have great harness and yield strength because it is difficult to move dislocations in spatially confined regions.

The confinement referred to in the third point above is very interesting and it is applicable to semiconductor-based nanostructures such as AgI. Indeed, strain-induced confinement of quasi-free excitons has been observed in the optical spectra of vapor-quenched metastable Ag−Cu thin films progressively iodized under ambient conditions. Delayed evolution and inhomogeneous broadening of the exciton absorption at 420 nm of the spatially confined γ-AgI nanoparticles were clearly seen by the development of $Z_{1,2}$ exciton peak—iodization being controlled by the as-quenched Ag−Cu clusters. Cu addition not only restricts the particle size but also favors the island-type growth mode [10.2].

Let us clarify how physics changes when one goes from a macrosolid to a nanosolid by using the concept of bandgap. The electronic energy band gap of a macrosolid is the minimal energy required to excite an electron from the valence band to the conduction band, leaving an electron hole in the valence band. These electronic charge carriers may form an exciton (via Coulomb interaction) with an energy slightly lower than the band gap energy. The distance between the electron and the electron hole defines the Bohr radius of the exciton ($a_{exciton}$) relative to the usual Bohr radius a_0:

$$a_{exciton} = \left(\frac{4\pi\varepsilon_0}{m_0 e^2} \right) \varepsilon_\infty \left[\frac{1}{m_0^*} + \frac{1}{m_h^*} \right] = a_0 \, \varepsilon_\infty \left[\frac{1}{m_0^*} + \frac{1}{m_h^*} \right] \tag{10.1}$$

As the particle size approaches the Bohr radius of the exciton, the electron–hole pair gets spatially confined. It assumes a state of higher kinetic energy.

When the system is approached by a spherical particle with an infinitely high potential energy outside the sphere, the energy of the lowest excited state varies with the particle size R as [10.3]:

$$E = E_g + \frac{h^2}{8m_0 R^2}\left(\frac{1}{m_e^*} + \frac{1}{m_h^*}\right) - \frac{1.8e^2}{4\pi\varepsilon_0\varepsilon_\infty R} \tag{10.2}$$

In Equation 10.2 valid in the effective mass (m_e^*, m_h^*) model of a semiconductor, the second term on the right-hand side is a confinement term increasing as R^{-2} while the third term is the Coulomb attraction increasing as R^{-1}. As $R \to \infty$, the value for E approaches that of E_g.

Thin films (~0.70 nm thick) of the oxide ion conductor zirconia-16% yttria show a rapid increase of band gap when the grain size of the microstructure decreases below 30 nm down to 10 nm. The expression for the band gap is [10.4].

$$E_g\left(eV\right) = 8.68(\text{nm}^2)(eV)R^{-2} + 1.24(\text{nm})(eV)R^{-1} \tag{10.3}$$

With this perspective, let us try to answer a question in order to set the pace for this chapter that deals with the emerging area of nanoionics. In Chapter 1, we have briefly compared nanoionics with nanoelectronics.

10.2 What Is Nanoionics?

To start from an application perspective, an ionic nanosolid offers relative to its bulk counterpart (1) better electrolyte contact and (2) shorter transport length for, say, Li^+ ions in $LiFePO_4$. There is only one diffusion channel for Li^+, namely [010]; point (2) is very helpful in exploring better cathodes for Li-ion batteries by focusing on the synthesis of size- and shape-controlled nanocrystals.

Nanoscale condensed matter is the scale on which macroscopic and microscopic concepts meet as in the transition between cluster science and solid state science.

Nanoionics is a branch of nanoscience and nanotechnology devoted to the study of phenomenology and applications of nanoscale ionic systems in the solid state. It covers the properties, effects, and mechanisms of processes connected with fast ion transport in these systems. Fundamental properties of oxide and halide ceramics at nanometer length scales and fast ion conductor (advanced superionic conductor)/electronic conductor heterostructures are typical areas of interest, while potential applications are in electrochemical devices (electrical double layer devices) for conversion and storage of energy, charge, and information. Two classes of solid state ionic nanosystems are (1) nanosystems based on solids with low ionic conductivity and (2) nanosystems based on advanced superionic conductors (e.g., α-AgI and rubidium silver iodide family), leading to two specific areas of nanoionics differing from each other in the design of interfaces. In the first case, the focus is on the role boundaries play in the creation of conditions for high concentrations of charged defects (vacancies and interstitials) in a disordered space-charge layer. In the second case, what matters is the conservation of the original highly ionic conductive crystal structures of advanced superionic conductors at ordered (lattice-matched) hetero boundaries.

Nanoionics phenomena can be defined as the *indirect effects* caused by the *local migration* of mobile ions at *a heterogeneous interface*. A space charge layer builds up in mixed or ion conducting materials upon heterogeneous contact with materials of higher carrier density, such as metals and semiconductors, because of the difference in work function of the two materials. The potential profile is like the Schottky barrier commonly invoked to explain the electronic band structure of the heterogeneous junction at the metal/semiconductor interface to understand various electronic functions of the barrier, such as rectification.

In the case of mixed/ionic conductors, mobile ions also play an important role in *relaxation at the interface* to form the space charge layer, analogous to the *electrochemical double layer* in liquid electrolyte systems. This is because the modulated concentration of both ionic and electronic defects can contribute to (1) the enhanced ionic and electronic conductivity at the space charge layer and (2) the charge transfer reaction kinetics at the interface region. For nanoionics phenomena at the heterogeneous interface, the reduced thickness of the layer could lead to the increased surface charge density, thereby modulating the surface adsorption behavior of surrounding gases.

The International Technology Roadmap for Semiconductors (ITRS) recognizes nanoionics-based resistive switching memories as emerging research devices based on ionic memory. The gray area intersecting nanoelectronics and nanoionics leads to nanoelionics. Like nanoelectronics, the field of nanoionics relies on the metastability of the phase distribution and is characterized by a high information content on a mesoscopic scale.

Examples of nanoionic devices include (1) all-solid state supercapacitors with fast ion transport at the functional heterojunctions (nanoionic supercapacitors), (2) lithium batteries and fuel cells with nanostructured electrodes, and (3) nano-switches with quantized conductivity on the basis of fast ion conductors. These are well compatible with sub-voltage and deep-sub-voltage nanoelectronics. Autonomous micro power sources, microelectro mechanical systems (MEMS), and reconfigurable memory cell arrays are some areas that nanoionic devices could find wide applications.

An important case of fast ionic conduction in the solid state is that in surface space-charge layer of ionic crystals. Such conduction was first predicted by Kurt Lehovec. A significant role of boundary conditions with respect to ionic conductivity was first experimentally discovered by C.C. Liang who found an anomalously high conduction in the $LiI–Al_2O_3$ two-phase system. Because a space-charge layer with specific properties has nanometer thickness, the effect is directly related to nanoionics. The Lehovec effect has become the basis for the creation of a multitude of nanostructured fast ion conductors which are used in modern portable lithium batteries and fuel cells. Recently, a 1D structure-dynamic approach was developed in nanoionics for detailed description of the space charge formation and relaxation processes in irregular potential relief (direct problem) and interpretation of characteristics of nanosystems with fast ion transport (inverse problem), as example, for the description of a collective phenomenon: coupled ion transport and dielectric-polarization processes which lead to Jonsher's "universal" dynamic response.

At the easily perceptible level, size effects in ionic systems are extremely important for ion transport as well as mass storage. The impact of size can be manifold due to the interfacial symmetry breaking and the consequent carrier redistribution. The contribution goes beyond the mesoscopic case, where any contribution of unperturbed bulk defect structure has disappeared. Indeed, it extrapolates into the regime of atomistic sizes! One might imagine the transition of heterolayered systems to layered crystals, of composites to mixed crystals as well as of nanocrystalline state to the amorphous state. This brings in a true unification of all solid state and materials physical pheneomenology.

Writing with sagacity, Schoonman concludes his solid state ionics review [10.3] as follows:

> 5. Concluding Remarks
> The field of Nanoionics is attracting increased attention. For materials for all-solid state rechargeable lithium-ion batteries, the reduction to the nanoscale has been quite beneficial. The same holds true for varistor, solid oxide fuel cell, and gas sensor materials. While the quantum confinement regime has been explored for some ionic materials, it is evident that thus far only optical properties have been related to the quantum confinement regime. Electrical properties of solid electrolytes or mixed ionic–electronic conducting materials in the quantum confinement regime have not been described yet and represent a challenge to this field. However, it is evident that lowering the length scale may lead to enhanced defect densities and, therefore, to enhanced ionic conductivities in the space charge regions, irrespective of yet unknown mobility effects. With regard to the safe storage of (sustainable) hydrogen, new exciting avenues are being opened for storage using nano-sized and nanostructured materials. Consolidation of nanosized powders by dynamic compaction techniques opens up the possibility to manufacture nanostructured microstructures, which is beneficial for electrical properties. In addition, the interfaces between the components of (rechargeable) solid state Li-ion batteries are improved by dynamic compaction, leading to reduced battery dc-resistivities.

Coming to specifics, let us focus on Schoonman [10.3].

10.3 Structure Defects and Conductivity

The structure of an ideal bulk solid at $T = 0$ K temperature is defined by Kittel:

$$\text{Lattice} + \text{Basis} = \text{Ideal crystal structure.} \tag{10.4}$$

Defects arise in a crystal at $T > 0$ K so that the Maier definition is more appropriate:

$$\text{Real structure} = \text{Perfect (ground) invariant structure} + \text{Defect (chemically excited) structure.} \tag{10.5}$$

The second one depends on temperature, dopant concentration, and component potential (e.g., oxygen partial pressure).

In the infinite bulk, electrical neutrality of charge applies, and global and local equilibrium at constant T and total pressure follow once Gibbs free energy is minimized with respect to particle number.

Solid state ionic materials by and large possess Frenkel disorder induced by cation Frenkel defects as in AgCl. The defect chemistry is well described at low defect concentrations at not too high temperatures. At high temperatures, when native defect concentrations are of the order of mobile ion population, long-range Coulomb interactions arise, which lower defect formation energy and super-Arrhenius conductivity behavior and lead to incipient phase transition as in AgBr or a real phase transition in β-AgI.

When interfaces become important, the real structure of (1021) is modified as

$$\text{Real structure} = \text{Perfect bulk structure} + \text{Perfect core structure} + \text{Inhomogeneous defect structure.} \tag{10.6}$$

Focusing on the defect structure, the carrier concentrations are now *smeared out around the interface.*

As the size of a bulk crystal is decreased, the distance between neighboring interfaces decreases likewise (Figure 10.1a). This leads to a significant change in overall transport properties due to increased

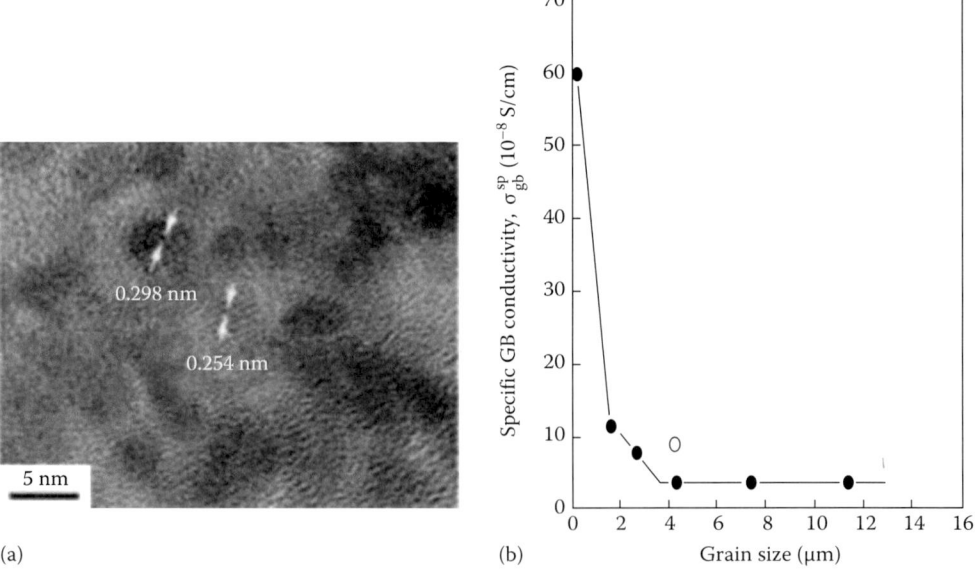

(a)   (b)

FIGURE 10.1 (a) High-resolution TEM of 8 mol% yttria-stabilized cubic zirconia nanocrystals showing spherical structures (dark features) separated by grain boundaries (white features) of *comparable* dimensions synthesized by a vapor phase hydrolysis process. (From Shi, F. et al., *Prog. Nat. Sci. Mater. Int.*, 22, 15, 2012.) (b) The clearest correlation between grain size and ionic conductivity. Specific grain boundary conductivity in 15 mol% calcia-stabilized zirconia at 773 K vs. grain size. The specific grain boundary conductivity drops sharply initially increasing grain size and ultimately saturates above ~4 μm. (From Aoki, M. et al., *J. Am. Ceram. Soc.*, 79, 1169, 1996.)

interface to volume ratio (Figure 10.1b). Local properties are still invariant. Transport along core regions may dominate in nanocrystalline samples though not in microcrystalline samples. Space charge transport too may follow this trend. We can visualize a mesoscopic regime in which the influence of the interface is perceived throughout the sample (Figure 10.2).

In practical terms, suppose the thickness of the ionic conductor sample is not very large compared to the Debye length, λ, of a solid electrolyte which describes the shielding of an electric field by the charge carriers. The electrostatic potential drops to $1/e$ of its value within λ. This very small length (compared to

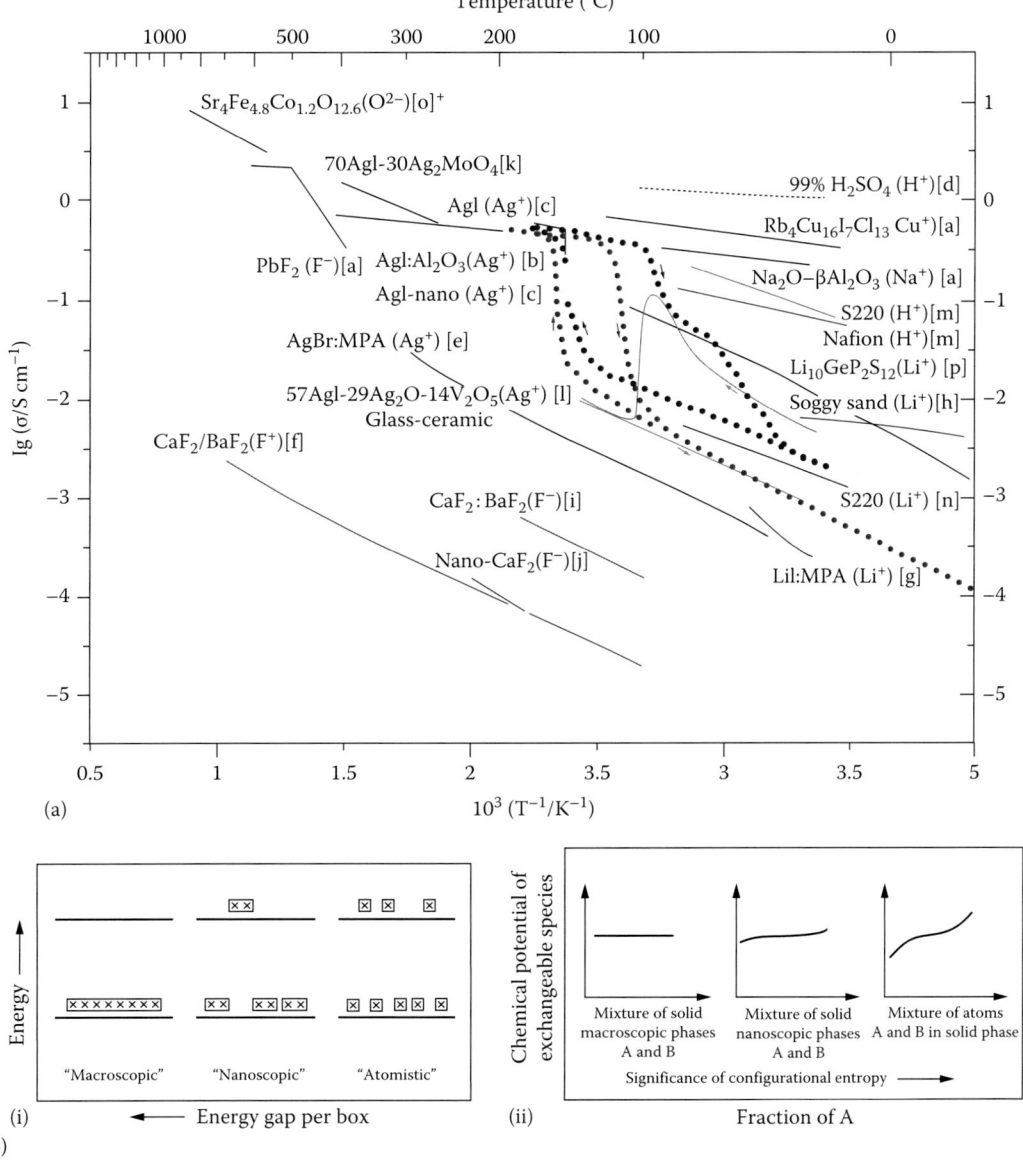

FIGURE 10.2 (a) Ionic conductivity profiles of interfacially dominated solid state ionic materials. (b) Nanoscopic system in (i) energy and (ii) configurational entropy perspectives. In the energy perspective while macroscopic systems are "localized" through a well-defined energy gap, the predominant configurational entropy of atomistic systems allows only a distributed ground state and excited states at $T > 0$ K. Nanoscopics take an intermediate position. A first-order transition in a macroscopic sample gets smeared out in a nanoscopic sample. (From Maier, J., *Chem. Mater.*, 26, 348, 2014.)

that in liquid electrolytes) is in the region of one atomic layer. This implies that the formation of an ionically blocking reaction product at the interface would make a solid state battery behave like a capacitor. But the formation of a good ionic or mixed conducting product at the interface (for example, formation of Ag_3SI between AgI [electrolyte] and Ag_2S [electrode] in the solid state galvanic cell), which is stable with both the electrode and the electrolyte may improve the contact between the two phases.

Maier has visualized the effect of size in terms of the sample size relative to λ (Figure 10.3).

Let us now discuss a realistic model for conductivity enhancement in an ionic nanocomposite in which the space charge layer interaction is specifically considered.

Modeling space charge layer interaction and conductivity enhancement in nanoionic composites [10.8].

Goodyear [10.8] has computed numerically potential conductivity enhancement due to formation of space charge layers in nanoionic composites for several structures of non-conducting nanoparticles in a bulk ionic conductor. Optimum loading fractions extracted from simulation results are found to depend strongly on the thickness of the space charge layer relative to the size of the particles. The model is also applied to a space charge layer depletion scenario, which is in good agreement with results from an experimental study. Figure 10.4 shows the schematic of the model.

Nanoparticles in the ordered structures are treated as uniform hard spheres of radius R. These are surrounded by a space charge layer in which ionic conductivity decreases exponentially with position from a maximum at the surface σ_0. It approaches asymptotically the bulk value of the ionic conducting

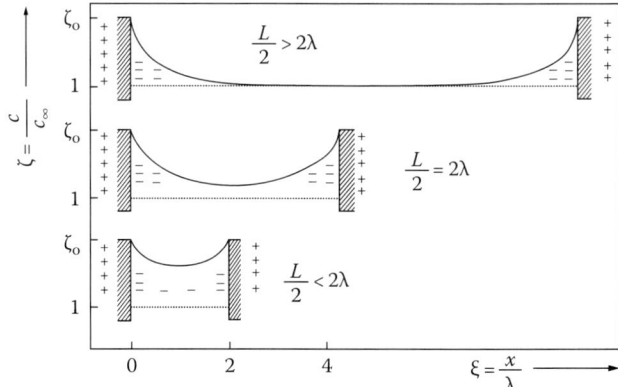

FIGURE 10.3 In very thin films and nanocrystalline materials whose sizes (L) are not too great compared to Debye length λ, the mesoscopic situation arises with anomalous effects due to space charge interaction when $(L/2) < 2\lambda$. These effects may be approximated to a size factor $g \sim 4\lambda/L$. (From Maier, J., *Solid State Ionics*, 70/71, 43, 1994.)

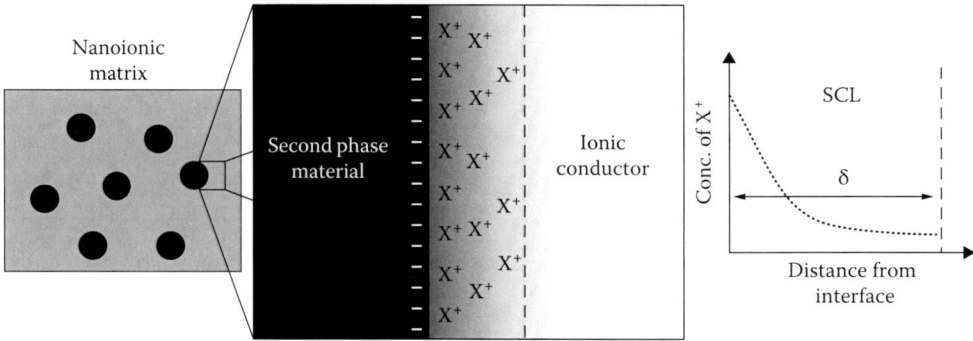

FIGURE 10.4 Nanoionic composite and space charge layer created at the inter. (From Goodyear, C.E. et al., *Electrochimica Acta*, 56, 929, 2011.)

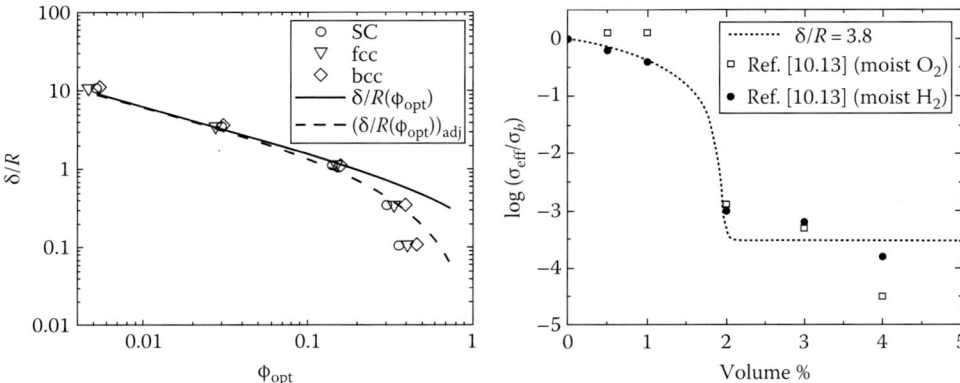

FIGURE 10.5 A typical plot of conductivity enhancement. (After Goodyear, C.E. et al., *Electrochim. Acta*, 56, 9295, 2011.)

bulk phase, σ_b, far from the surface. Specifically, in the absence of other particles, the spatially varying conductivity for particle j (σ_j) varies with radial position r_j from the center of particle j as

$$\sigma_i = \sigma_b + (\sigma_0 - \sigma_b)\frac{R}{r_i}\exp\left[-\frac{r_i - R}{\delta}\right] \tag{10.7}$$

where δ defines the characteristic length of the spatial decay of conductivity enhancement in the space charge layer. Δ related but not equivalent to the Debye length of the material is of the order of nanometers. The additional radial factor R/r_j outside the exponential arises from the spherical geometry of the problem. Equation 10.7 used to model the variation of ionic conductivity in the space charge region follows from the standard double layer theory and applies for relatively small conductivity perturbations. The focus here is on accumulating ionic defects/charge carriers which lead to conductivity enhancement. The depletion case is also considered briefly and connects to the experiment (Figure 10.5). For further details, see [10.8].

Next, we discuss basics of thermodynamics and electrostatics.

10.4 Thermodynamics and Electrostatics

The thermodynamic stability of materials depends on particle size due to the competition between surface and bulk energies [10.9]. Thermodynamics of nanocrystals is conveniently discussed in relation to thermodynamics of bulk materials. The latter is governed by the Gibbs free energy difference (ΔG) necessary for the formation of the bulk solid. In the case of the formation of nanosolid, a substantial component of the total energy is the surface energy. This surface energy when added to the bulk energy determines the ΔG form necessary to form the nanophase. Surface energy depends on the size of the nanoparticle so that particle size is a crucial independent variable. Indeed, the particle size $d = (V_0)^{1/3}$, where V_0 is the volume of the particle. The case of NaO_2 nanoscale thermodynamic relation between oxygen partial pressure Po_2 (atm) and particle size (nm) is illustrated in Figure 10.6a. Due to low surface energies, NaO_2 particles are stable over NaO_2 at small particle sizes. For particle sizes bigger than 6 nm, the bulk phase Na_2O_2 is preferred over NaO_2. Figure 10.6b describes the phase diagram of NaO_2–Na_2O_2 as a oxygen partial pressure (Po_2) vs. temperature (T) plot. Note that the Pnnm NaO_2 structure transforms to the Fm3m NaO_2 structure at 230–240 K when $Po_2 = 1$ atm. Na_2O_2 is in equilibrium with Fm3m NaO_2 at 8.5 atm when $T = 300$ K. The Fm3m NaO_2, the Pnnm NaO_2 structure, and the Na_2O_2 structures are shown in Figure 10.6c in relation to those of metallic Na and Na_2O.

Let us briefly discuss "electrostatics" of the space charge.

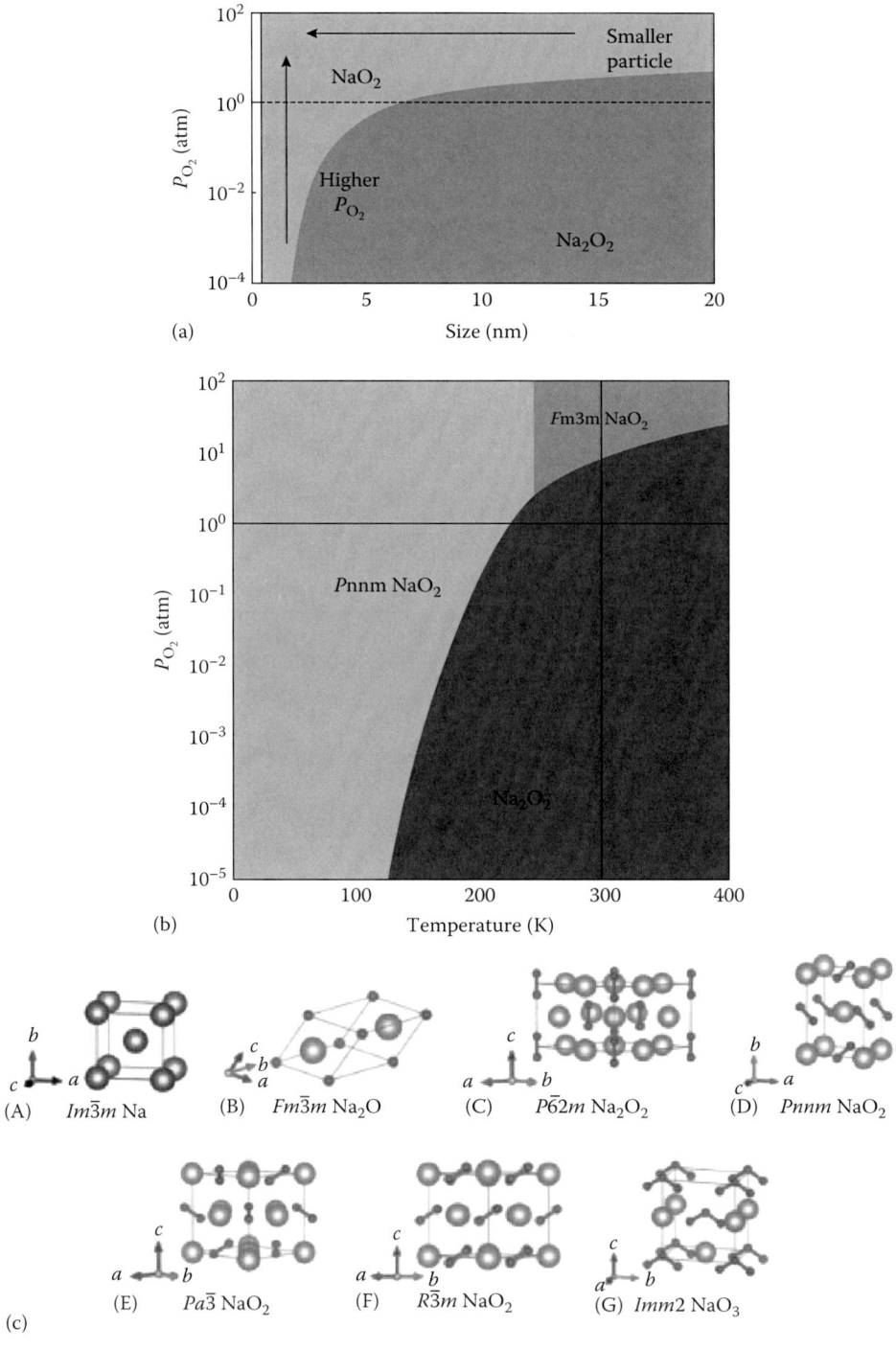

FIGURE 10.6 Illustrating the thermodynamics of (a) nanoparticles with reference to that of (b) bulk using NaO_2 phase diagram as example. The horizontal dashed line in (a) represents $P_{O_2} = 1$ atm. This phase diagram P_{O2} versus particle size is at 300 K at the O_2 gas limit. The dark gray domain in (b) is where Na_2O_2 is stable. The light gray domain is for Pnnm NaO_2 and the medium gray domain is for Fm3m NaO_2. The horizontal dashed line denotes $P_{O_2} = 1$ atm and the vertical dashed line denotes $T = 300$ K. (c) Crystal structures of Na_2O (B), Na_2O_2 (C), and three polymorphs of NaO_2 (D–F) and NaO_3 (G) in relation to that of metallic Na (A). Big spheres are Na, small ones are O·O–O bonds are indicated in (C–G). (From Kang, S.Y. et al., *Nano. Lett.*, 14, 1016, 2014.)

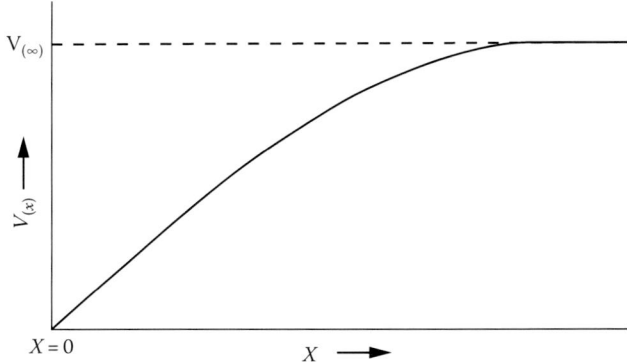

FIGURE 10.7 Potential distribution $V(x)$ normal to the surface of an ionic crystal. (From Lehovec, K., *J. Chem. Phys.*, 21, 1123, 1953.)

The influence of atomic structure on the ion-transport and polarization processes in the space charge region of a heterojunction is apparently the challenge of nanoionics. A beginning was made by Lehovec [10.10] to discuss the space charge layer and distribution of interstitial ions at the surface of an ionic crystal. The condition of zero space charge invoked in the calculation of the concentrations of lattice defects assuming that the energies of creation of anion and cation interstitials and vacancies are known is not fulfilled in general for the surface zone of the crystal comprising many atomic layers. *The space charge in the surface zone causes the electrostatic potential between the bulk and the surface of the crystal.*

Figure 10.7 shows the potential distribution. How does this arise? Consider a perfect crystal with zero defects. Allow Frenkel defects (interstitials) to form starting at the surface. Let the energy of creating cation interstitials be less than that of creating anion interstitials. Then, initially more cation interstitials will form and migrate into the interior than anion interstitials. This will lead to a *positive* space charge layer and leave a corresponding negative charge at the surface of the crystal. The corresponding potential will *obstruct* the migration of cations from the surface to the interior but *enhance* migration of anions, which is necessary to obtain a stationary state.

Thus, in thermal equilibrium, symmetry demands that the bulk of a macrocrystal must be at a constant potential. Therefore, space charge is zero for the bulk of the crystal and the potential difference $V(\infty)$ between the bulk of the crystal and the surface can be calculated from the zero space charge condition. Poisson's equation $d^2V/dx^2 = -\rho/\varepsilon\varepsilon_0$, where ρ is the space charge, ε is material dielectric constant, and ε_0 is the free space dielectric constant.

Of course, the goal is to calculate the concentration of lattice defects as a function of distance from the surface. Lehovec's paper calculates the surface charge density for NaCl. An important point of relevance to nanoionics is that the implications of the surface space charge layer on the ionic conduction of a pure crystal may be tested by measurement of conductivity of samples of varying ratio of surface to volume. The difference between the concentration of lattice defects near the surface and that in the bulk leads to a "surface conduction."

Let us now discuss briefly about phase transitions.

10.5 Phase Transitions in Nanoionics

Striking qualitative effects are met if new phenomena arise at the nanoscale. Examples are the appearance of new phase such as the stabilization of the hexagonal LiI phase in the form of thin films on sapphire (Figure 10.8).

Another example is that of AgI nanoplates that are fully penetrated by stacking faults leading to the highest room temperature ionic conductivity for a binary solid material (Figure 10.9).

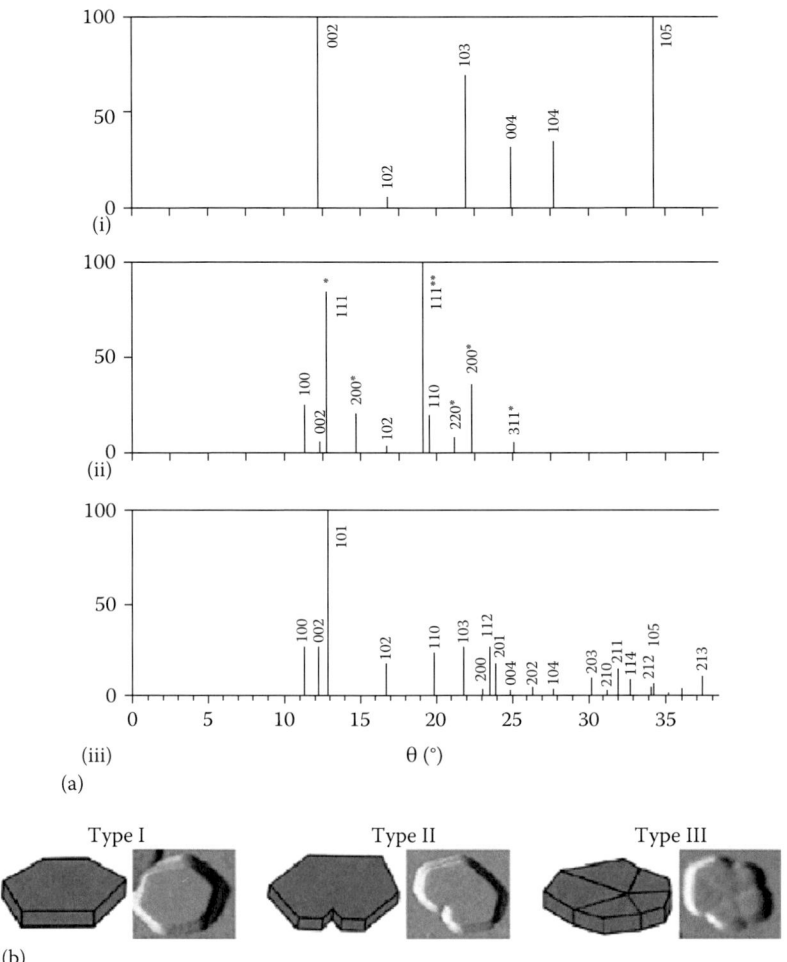

FIGURE 10.8 (a) XRD (relative intensity vs. Bragg theta plots) of hexagonal LiI. (i) Reflection experiment, (ii) transmission/reflection experiment, and (iii) calculated pattern. (b) In the metastable hexagonal phase (radius ratio of $Li^+/I^- =$ 0.252) at room temperature ($a = 0.4541$ nm, $c = 0.7305$ nm), the Li ions occupy small tetrahedral sites in the close-packed structure. The rattling motion of Li^+ ions in the cubic structure arises from the occupation by Li ions of large volume octahedral sites. Further characterization of hexagonal phase comes from infrared transmission minimum at 268 cm^{-1} which is the restrahl frequency. (From Wassermann, B. and Martin, T.P., *Solid State Commun.*, 65, 561, 1988.)

Phase transition in a nanocrystalline solid is closely associated with the total energy (internal energy plus surface energy) and the energy barrier for a structural transition [10.12–10.17].

Silver chalcogenides synthesized at nanoscale have proved to be excellent materials for investigation of phase transitions because of their narrow-band gap semiconducting nature at room temperature and semiconductor–superionic phase transition at not too high temperatures. Thus, while nano Ag_2S transforms at 454 K, nano Ag_2Se transforms at 408 K from orthorhombic phase at room temperature to the superionic cubic phase. What is the nature of this phase transition? A nano alloy phase Ag_4SeS undergoes a transition at 355 K [10.18]. Let us consider the example of Ag_2Se.

Figure 10.10 illustrates the phase transition in nano Ag_2Se detected thermally (DSC) and structurally (x-ray diffraction, XRD). The first-order transformation with a considerable thermal hysteresis and a peak in heat capacity (not shown) are striking. Intriguingly, the latter feature along with a dip in the thermal diffusivity suggests a mixed phase in the hysteresis zone. This aspect has been examined by Wang et al. in an integrated study of the orthorhombic–tetragonal–cubic phase transition in Ag_2Se.

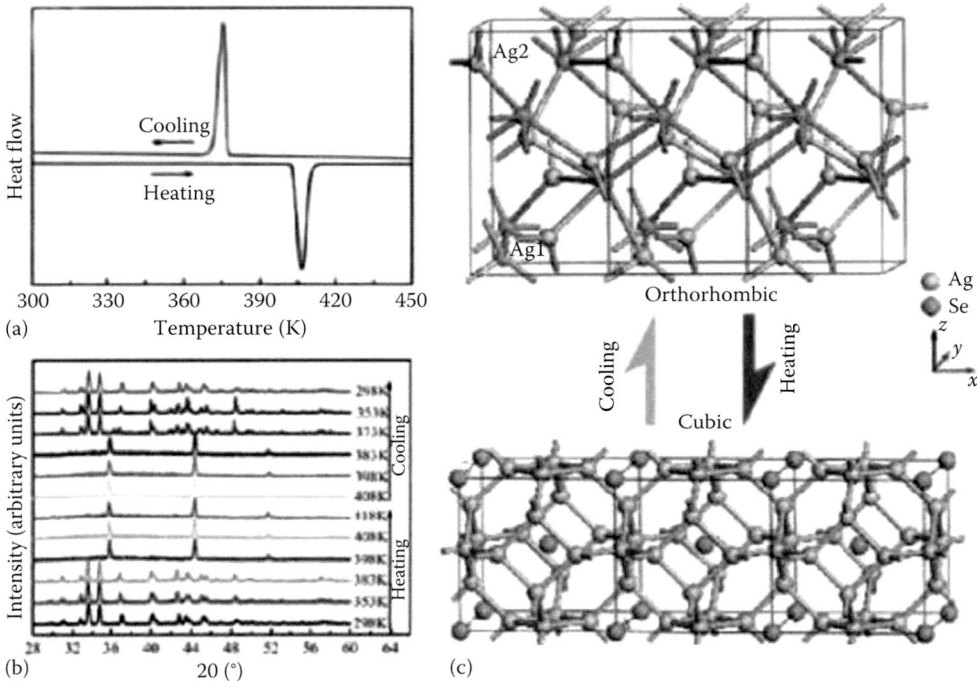

FIGURE 10.9 (a–c) Three types of nanoplates of AgI including unusual polytypes 7R (type II) and 9 R (type III) with mesoscopic superionic conductivity at room temperature. Prepared by solution-based method using polyelectrolytes, the nanoplates exhibit extremely high conductivity, which is enhanced by four orders of magnitude compared to the macroscopic AgI phase (β-AgI) and surprising conductivity isotropy. Mesoscopic ionic-conductivity effects in polytype heterostructures explain these observations. (From Guo, Y.G. et al., *Adv. Mater.*, 17, 2815, 2005.)

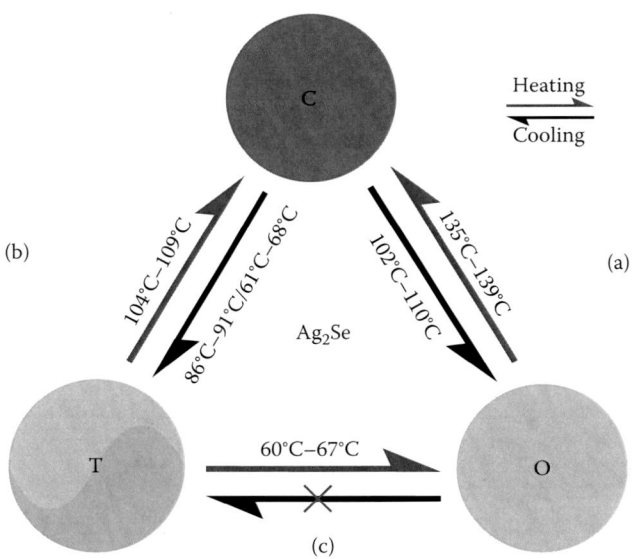

FIGURE 10.10 First-order phase transition (PT) in Ag_2Se nanocrystals from room-temperature orthorhombic phase to high-temperature cubic phase. (a) DSC. Note the characteristic thermal hysteresis of 29° in the heating (endothermic peak, 408 K) and cooling (exothermic peak, 379 K) cycle. (b) XRD. (c) Structural schematic of orthorhombic to cubic and vice versa PT. Spectacular effects at the phase transition temperature include a peak in the heat capacity and a dip in the thermal diffusivity pointing to the involvement of phonons. (From Xiao, C. et al., *JACS*, 134, 4287, 2012.)

FIGURE 10.11   Irreversible and reversible phase transitions in nanocrystalline Ag_2Se. The stability of the tetragonal phase and the reversibility of the tetragonal to cubic phase transition depend on the method of synthesis. (From Wang, J. et al., *Chem. Mater.*, 26, 5647.)

Figure 10.9 gives a glimpse of the nature of the phase transitions between tetragonal (t), orthorhombic (β), and cubic (α) phases. The stability and phase transition behavior of t-Ag_2Se nanocrystals depends on the method of preparation. t-Ag2Se made by oleylamine-mediated synthesis shows a higher temperature and time-sensitive metastability and undergoes a $t \rightarrow \beta \rightarrow \alpha \rightarrow \beta$ phase transition sequence during thermal cycling in which $t \rightarrow \beta$ transition is exothermic and irreversible while the $\beta \rightarrow \alpha$ transition is reversible. In contrast, t-phase Ag_2Se ($a = b = 0.706$ nm, $c = 0.498$ nm) nanocrystals synthesized by poly vinyl pyrrolidone (PVP) assisted solvothermal method are more stable and exhibit a direct, reversible $t \rightarrow \alpha$ phase transition *without undergoing* the β-phase. Sintering at $T > {\sim}250°C$ destroys the stability of the t-phase and the irreversibility of the $t \rightarrow \alpha$ transition due to size increase of the sample.

Figure 10.11 illustrates the irreversible phase transitions in nanocrystalline Ag_2Se.

Let us look at the physical basis for nanoionic devices.

10.6 Physics for Devices

Nanoionic hard drives use nanoionic technology allowing for smaller devices while doing away with moving parts and the mechanical failures that are associated with previous hard disk drives. Nanoionic hard drives are currently the most state-of-the-art drives on the market and nanoionics was not utilized in hard drives until February 2014. Nanoionic devices were first proposed in 1992: "The results obtained show that it is possible to form arrays of electrochemical devices with single elements ~10 nm in size in the films" [10.2]. The basis of design of nanoionic devices is the creation of nanostructures with nanoionic parameter $\lambda/L \sim 1$, where L is the size of device structure and λ is the characteristic size of specific region, where the property of fast ionic transport is realized. "Possibilities to influence on these specific regions $\langle\lambda\rangle$ in a controllable manner may appear in short sized devices." Ion–electronic hybrid devices should be considered as a step on a way to the future nanoelectronics–nanoionics (nanoelionics) that was first proposed in 1996 [10.20].

10.7 Nanobatteries

A TEM cross-sectional observation of an all-solid state battery is shown in Figure 10.12.

First-order phase transitions in nanoscopic systems have been considered in the framework of nanothermodynamics of Hill (see Appendix 10.A). Let us now look an example of a nanoionic device—nanobattery. The equilibrium conditions and a generalized version of the Clapeyron–Clausius equation for a two-phase nanoscopic system that contains two phases have been obtained. The equilibrium conditions obtained are the same as the ones that result from the equivalence between Tsallis thermodynamics and Hill's nanothermodynamics [10.26].

Understanding and improving the behavior of interfaces is essential to the development of safer and high-performance Li-based batteries, regardless of their range of applications. Indirect methods, such as impedance spectroscopy, or direct methods, such as the live in situ observation of batteries cycled within a scanning electron microscope (in situ SEM), are helpful in looking at the interface microstructure/composition evolution upon cycling. These methods directly link interface properties and battery performance. Very significantly, they also enable us to spot local interface defects that are crucial to the development of say a 2D solid state microbattery. Indeed, this technology could power the new generation of microelectromechanical systems (MEMS). A very interesting ex situ TEM observation of "nanobatteries" is obtained by cross-sectioning a microbattery [10.27] using focus ion beam (FIB) in a dual beam SEM (Figure 10.12). Then, TEM analyses between pristine, cycled, and faulted all solid state $LiCoO_2$/solid electrolyte/SnO Li-ion batteries have revealed drastic changes such as the presence, depending on the battery fabrication process, of both cavities within the solid electrolyte layers and low wetting points between the electrolyte and the negative electrode. Moreover, postmortem TEM observations of cycled microbatteries have revealed a rapid deterioration of the interface upon cycling because of the migration of the chemical elements between stacked layers. Such findings are involved both in the improvement of the reliability of the 2D all solid state battery assembling process and in the enhancement of their cycling performances. A future target is the development of live in situ TEM observation of "nanobatteries" cycled *within the microscope.*

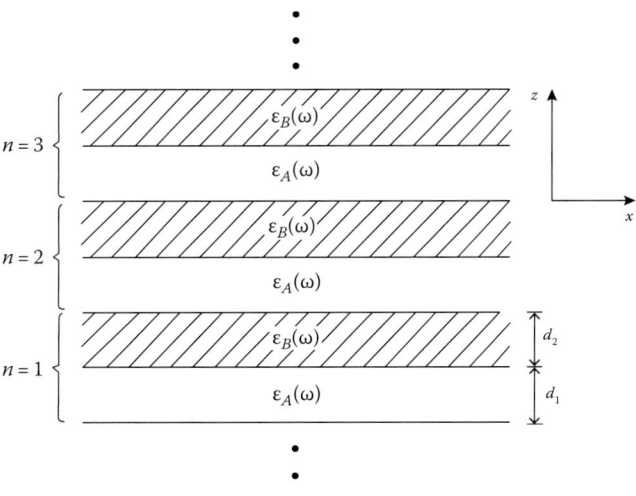

FIGURE 10.12 First cross-sectional observation of an all-solid state lithium-ion "nanobattery" by transmission electron microscopy. (From Brazier, A. et al., *Chem. Mater.*, 20, 2352, 2008.)

10.8 Superionic Superlattices

Interesting collective excitations arise in superionic superlattices as examined within the framework of electromagnetic theory.

Aniya and Kobayashi [10.22] have considered in the framework of electromagnetic theory the collective modes in the superlattice system composed of superionic conductors and ionic conductors (Figure 10.13). In a hydrodynamical model, the anion cage is immersed in the cation liquid. The behavior of the modes is analyzed in terms of the coupling strength between excitations pertaining to different layers. The coupling strength is controlled by varying the slab thicknesses. An interesting behavior is that the diffusion mode transforms to relaxation mode when the coupling strength is varied from strong to weak. The coupling strength has an effect on the acoustical and optical modes. The dispersion results are displayed in Figure 10.14.

Figure 10.15 illustrates structural disorder at a heterophase boundary and the space charge region of an oxygen ion conductor.

We now consider experimental systems involving heterostructures and briefly discuss the conductivity results (Figures 10.16 and 10.17).

1. Ceria-stabilized zirconia (CSZ)/Al_2O_3: In multilayer systems consisting of an ionic conductor and an electrical insulator, the ionic current can flow both across the bulk and the heterophase boundaries. In the CSZ/Al_2O_3 multilayer system, the total conductivity has been measured parallel to the interfaces as a function of temperature with thickness of CSZ and alumina layers varied systematically. It increases by two orders of magnitude when thickness of individual CSZ layers decreased from 0.78 μm to 40 nm, depending linearly on the reciprocal thickness of the individual layers. That is on the number of CSZ/alumina layers. This indicates a parallel connection between individual conduction paths and the interfacial regions. Figures 10.16 through 10.18 show TEM, XRD, and conductivity results.

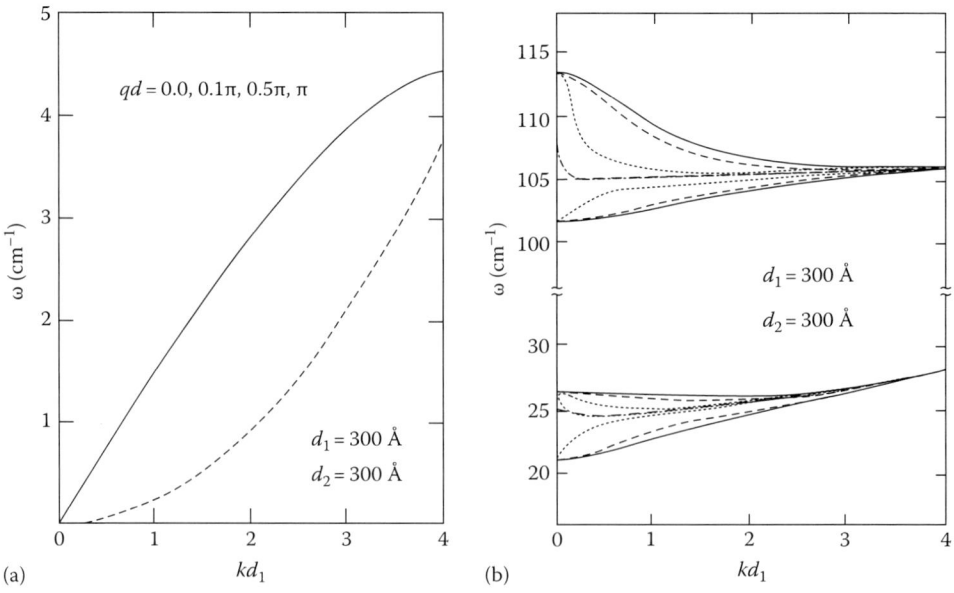

FIGURE 10.13 A superlattice structure composed of superionic conductor (a) and ionic crystal (b). (From Aniya, M. and Kobayashi, M., *Appl. Phys.*, A 49, 641, 1989.)

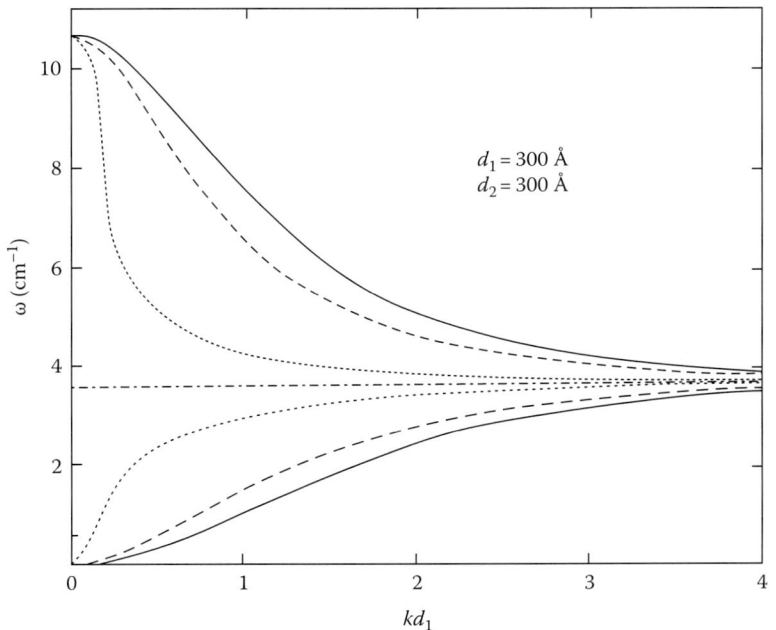

FIGURE 10.14 Different types of predicted mode dispersions (angular frequency vs. superlattice separation) in super-ionic superlattices. (a) Acoustic modes, (b) optic modes, (c) relaxation modes (top) and, diffusive (small kd_1) and relaxational (large kd_1) modes (bottom). qd values are 0,0 (–•–), 0–1π (•••), 0–5π (—), π (continuous curve). (From Aniya, M. and Kobayashi, M., *Appl. Phys., A* 49, 641, 1989.)

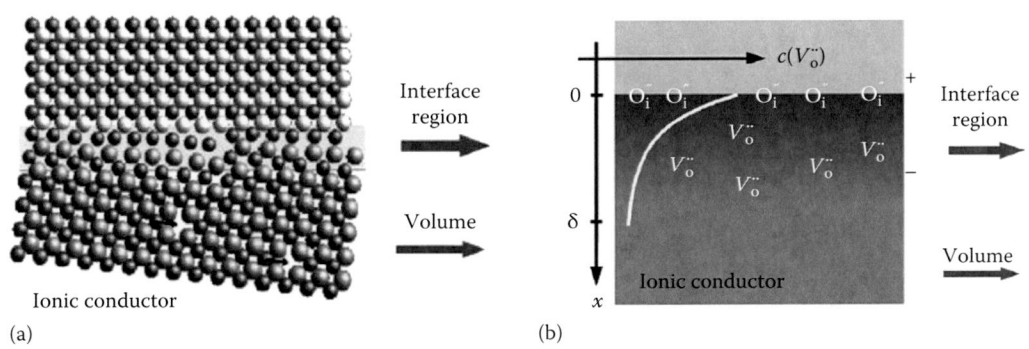

FIGURE 10.15 (a) Structurally disordered region at a heterophase boundary. (b) Space charge region of an oxygen ion conductor with an enhanced density of mobile ionic charge carriers. (From Peters, A. et al., *Solid State Ionics*, 178, 67, 2007.)

2. The system CaF_2/BaF_2 involves heterojunctions in two-phase systems that are unusually interesting because they (1) the help improve ionic conduction and (2) are expected to show qualitatively different behavior when the interface spacing is comparable to or smaller than the width of the space charge regions in relatively large crystals. Figure 10.19 shows the structure of the heterojunctions and the corresponding conductivity data.

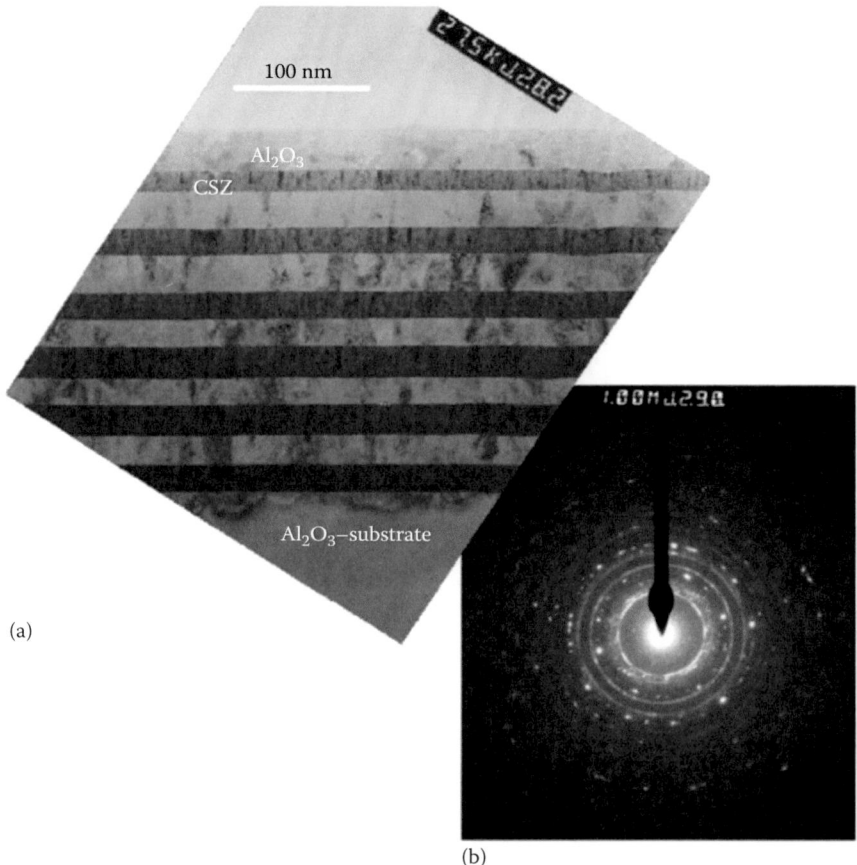

(a)

(b)

FIGURE 10.16 (a) Bright-field TEM picture and (b) selected area electron diffraction (SAED) pattern of 6-layer CSZ/6-layer Al_2O_3 multilayer system heat-treated at 600°C for ~100 h. (From Peters, A. et al., *Solid State Ionics*, 178, 67, 2007.)

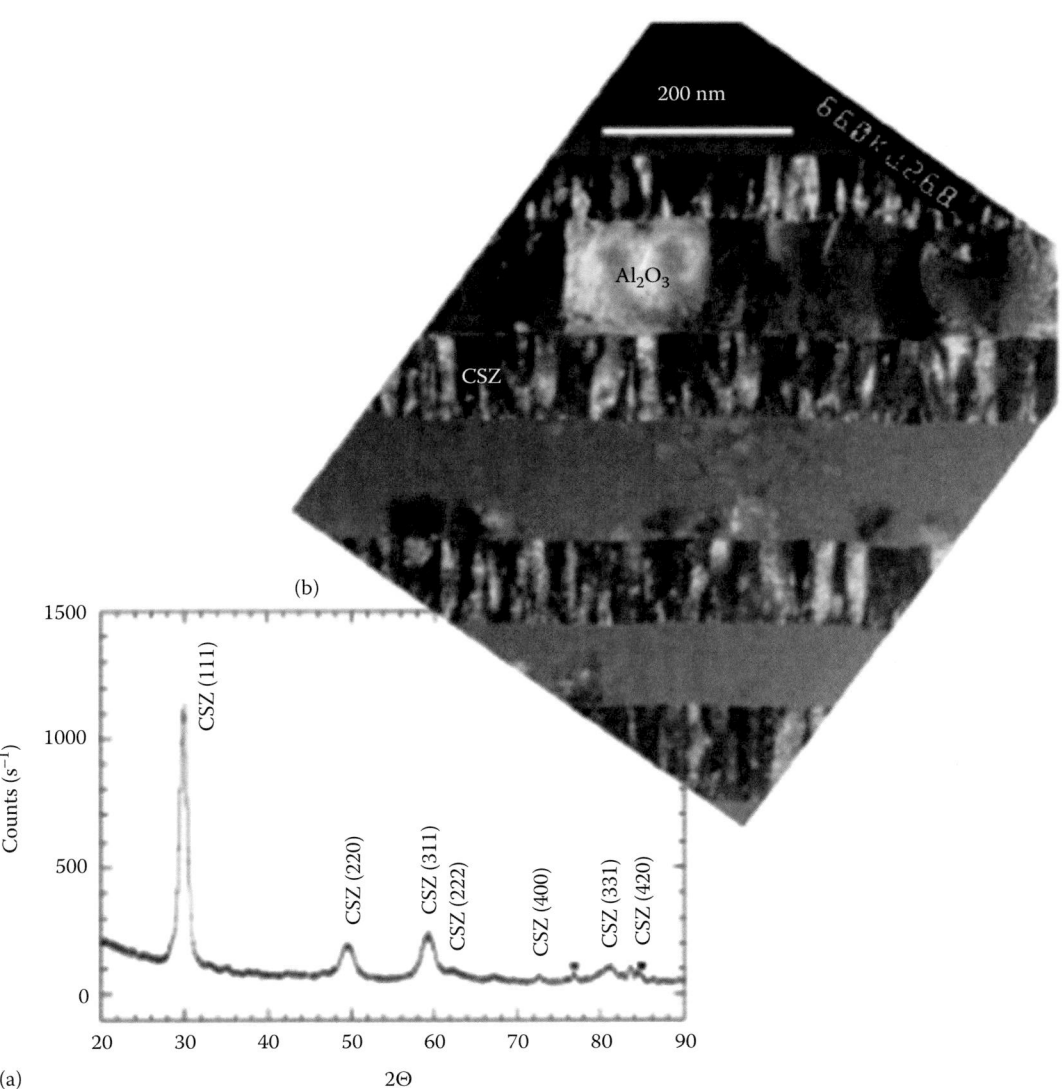

FIGURE 10.17 (a) XRD pattern (Bragg–Brentano geometry, Cu-K$_\alpha$ radiation) and (b) bright-field TEM of just made 4 CSZ–3 alumina multilayer by pulsed laser deposition.

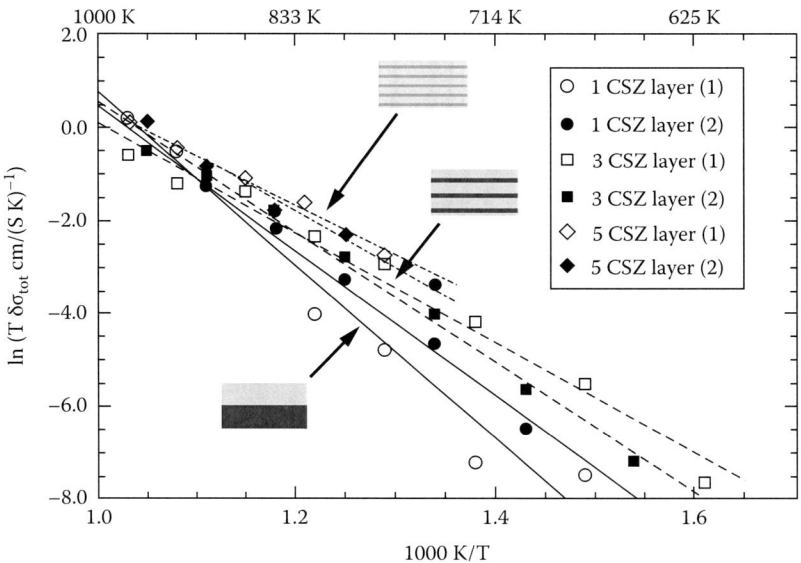

FIGURE 10.18 Arrhenius plot of total conductivity for 1, 3, and 5 CSZ-alumina layers. (From Peters, A. et al., *Solid State Ionics*, 178, 67, 2007.)

3. Alumina/zirconia (Al_2O_3/ZrO_2) multilayer thin films were deposited on Si (100) substrates at an optimized oxygen partial pressure of 3 Pa at room temperature by pulsed laser deposition. The Al_2O_3/ZrO_2 multilayers of 10:10, 5:10, 5:5, and 4:4 nm with 40 bilayers were deposited alternately in order to stabilize a high-temperature phase of zirconia at room temperature. All these films were characterized by XRD, cross-sectional transmission electron microscopy (XTEM), and atomic force microscopy. The XRD studies of all the multilayer films showed only a tetragonal structure of zirconia and amorphous alumina. The high-temperature XRD (HTXRD) studies of a typical 5:5 nm film indicated the formation of tetragonal zirconia at room temperature and high thermal stability. It was found that the critical layer thickness of zirconia is ≤10 nm, below which tetragonal zirconia is formed at room temperature. The XTEM studies on the as-deposited (Al_2O_3/ZrO_2) 5:10 nm multilayer film showed distinct formation of multilayers with sharp interface and consists of mainly tetragonal phase and amorphous alumina, whereas the annealed film (5:10 nm) showed the inter-diffusion of layers at the interface (Figure 10.19).

10.9 Outlook

Ever since Despotuli and Nikolaichic took the first step toward nanoionics [10.26]. There has been an interesting progress in nanoionics most significantly in multilayers. Nanoionics continues to innovate the most significant of which has been the atomic switch [10.27]. The future years will see progress in supercapacitors [10.28] and electrochemical storage in confined systems [10.29]. Nanoionics has a bright future [10.30].

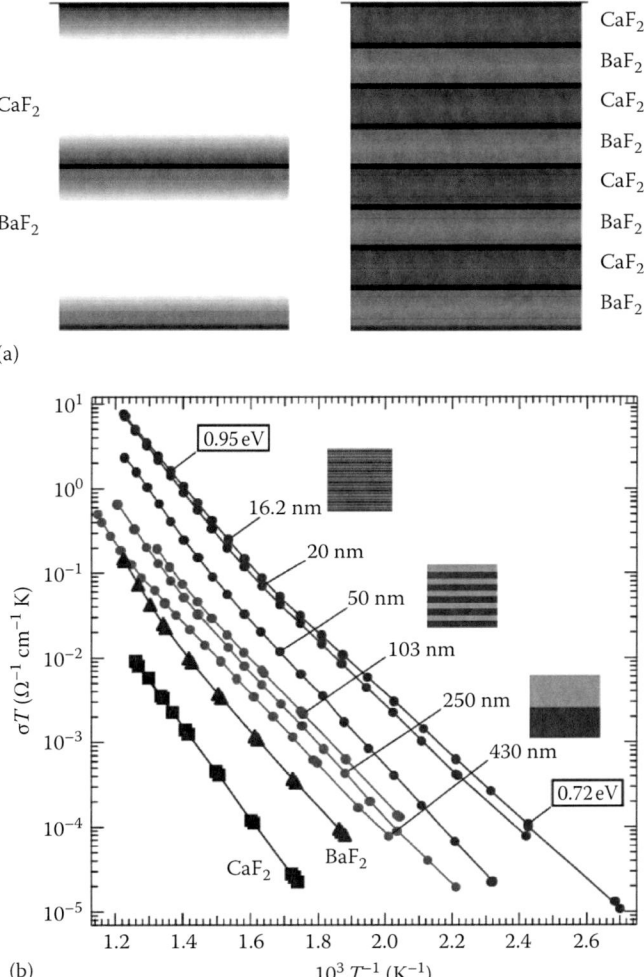

FIGURE 10.19 (a) Unusual artificial ionic conductors. Conductivity profiles in the semi-infinite space-charge and mesoscale situations. The concentration or (parallel) conductivity profiles are sketched for the semi-infinite space-charge situation (period $>8\lambda$, left), and for the mesoscale situation (period $<8\lambda$, right), in which the space-charge regions overlap and bulk values are exceeded even in the centers of the individual layers. (b) Parallel-to-the-interface ionic conductivity data of the CaF_2, BaF_2, and CaF_2–BaF_2 heterostructure films. MBE grown heterostructure films with various periods and interfacial densities in the 430 ± 16 nm range possess microstructures displayed alongside Arrhenius plots. The overall thickness is approximately the same in all cases (500 nm). σ, conductivity; T, temperature. The different shades refer to different site regimes (gray: semi-infinite space charge zones; dark gray: finite space charge zones). If the period of the heterostructure is greater than 100 nm, the conductivity increases linearly with N/L, approximately linearly with the number of heterojunctions N. Heterolayers with mixed conductors in which both ionic and electronic conductivities are present are more interesting to explore for mesoscopic behavior. (From Sata, N. et al., *Nature*, 408, 946, 2000.)

10A.1 Appendix: Nanothermodynamics or Nonextensive Thermodynamics

Hill's nanothermodynamics is a contribution to the route toward the "nanorevolution" started by Feynman's famous 1959 lecture "There's plenty of room at the bottom."

An extension of the classic thermodynamics theory to nanometer scale has generated a new interdisciplinary theory—nanothermodynamics. It serves as a bridge between macroscopic and nanoscopic systems. Nanothermodynamics theories play a critical role in investigating the size-dependent physicochemical properties of nanomaterials and are thus a cutting-edge topic. The focus and emphasis are on the utilization of nanothermodynamics models to investigate the size-dependent thermal stability, magnetic properties, photoelectric behaviors, thermoelectric phenomena, mechanical properties, electrical properties, etc., of nanomaterials—the last one being relevant to solid state ionics.

The effects of size, dimensionality, and composition when investigated through a quantitative nanothermodynamics model reveal that (a) the size dependence of these properties can be universally reconciled to the effect of severe bond dangling; (b) for the same material size, the sequence of size effects on the properties, from strong to weak, is nanoparticles, nanowires, and thin films; and (c) the composition effects on the properties of nanoalloys are substantial, having a nonlinear relationship. The model also reveals that vacancy formation determined by the cohesive energy variation is one of the intrinsic factors that dominate the size-dependent physicochemical properties of nanomaterials.

The relevance of nanothermodynamics to materials synthesis has been established recently [10.35]. Hill anticipated that nanothermodynamics could be used in two ways: (1) An aid to analyze, classify, and correlate equilibrium experimental data on "small systems" including crystallites, macromolecules, polymers, and polyelectrolytes; (2) To verify, stimulate, and provide a framework for statistical thermodynamical analysis of models of finite systems. His vision of experimental small systems is as condensed phases such as a small solid/liquid particle or a macromolecule among others.

Consider an ensemble of identical nanoscopic systems that represent a macroscopic system. Focus on a macroscopic system that contains two phases in thermodynamical equilibrium. Divide each phase into identical nanoscopic systems. What are the differentials of internal energy for the considered phases?

The statistical thermodynamics models of systems with a finite number of particles N often predict nonextensive thermodynamic potentials. By taking the thermodynamic limit ($N \to \infty$, $V \to \infty$, with N/V fixed), these potentials become extensive, as required in macroscopic thermodynamics. But, we need not take the thermodynamic limit if we are indeed interested in the finite-size system. In contrast to macroscopic thermodynamics, in nanothermodynamics, we are interested in thermodynamic functions and relations for a *single small system,* including, in general, variations in the size of the small system. Considering the size of the system as a variable is central to nanothermodynamics.

Nanothermodynamics aims at obtaining the thermodynamic properties of a small system. The thermodynamic concepts are applied not to a single small system *directly* but rather to *a large sample of small systems, a Gibbs "ensemble."* The use of an ensemble provides a macroscopic framework and within this framework the average values of fluctuating extensive magnitudes (e.g., the average energy, $U = \langle E \rangle$) are evaluated.. Since experiments are, in general, carried out on "ensembles" of small systems, the ensemble average is just what is required operationally.

Consider a system composed of N identical noninteractive subsystems (or small systems), and hence referred to as the composite system. Each subsystem contains N particles of a single component and the composite system contains $N_t = NN$ particles. The thermodynamic magnitudes of the composite system are identified by a subscript t. The Gibbs equation for the composite system is

$$dU_t = TdS_t - pdV_t + mdN_t \tag{10A.1}$$

Since the composite system is macroscopic, its internal energy U_t is a first-order homogeneous function of the extensive state variables (S_t, V_t, N_t). Application of Euler's theorem for homogenous functions leads to the Euler equation

$$U_t = TS_t - pV_t + mN_t \tag{10A.2}$$

Differentiation of Equation 10.5 and comparison with Equation 10A.1, yields the Gibbs–Duhem equation

$$-S_t dT + V_t dp - N_t dm = 0 \qquad (10A.3)$$

This equation implies that (T, p, m) cannot be used as independent (intensive) state variables to characterize the state of the composite system.

The extensive state quantities of one of the N identical subsystems are

$$U = U_t/N, \; S = S_t/N, \; V = V_t/N, \; N = N_t/N \qquad (10A.4)$$

Dividing Equations 10.4–10.6 by N, we deduce the corresponding equations for a subsystem as

$$dU = TdS - pdV + \mu dN$$

$$U = TS - pV + \mu N$$

$$-SdT + Vdp - Nd\mu = 0$$

They are similar to Equations 10A.1–10A.3 because classical thermodynamics implicitly assumes that the subsystems are also macroscopic, and hence extensive. But, *what if the subsystems are not macroscopic?* The answer is that the thermodynamic equations for small systems need to be *modified to account for finite-size effects.*

Hill's nanothermodynamics is a generalization of macroscopic thermodynamics to account for finite-size effects via the introduction of a *new* thermodynamic potential called *the subdivision potential.* To understand its meaning, compare two composite systems (or ensembles) with the same extensive variables S_t, V_t, and N_t and differing in the number of subsystems (or small systems). One of them is composed of N_1 subsystems, while each of them characterized by the extensive variables $S_1 = S_t/N_1$, $V_1 = V_t/N_1$, and $N_1 = N_t/N_1$ and with internal energy U_1. Analogously, the other composite system is composed of N_2 subsystems, each of them characterized by the extensive variables $S_2 = S_t/N_2$, $V_2 = V_t/N_2$, and $N_2 = N_t/N_2$ and with internal energy U_2. In classical thermodynamics, the Euler equations $U_1 = TS_1 - pV_1 + \mu N_1$ and $U_2 = TS_2 - pV_2 + \mu N_2$ let us conclude that $N_1 U_1 = N_2 U_2$.

Equivalently, the two composite systems have the same energy:

$$U_{t1}(S_t, V_t, N_t, N_1) = U_{t2}(S_t, V_t, N_t, N_2) \qquad (10A.5)$$

In other words, in macroscopic thermodynamics U_t is considered to be a function of (S_t, V_t, N_t) and N is *not needed as a state variable.* Experimentally, if the equality is not satisfied, then the Euler equation needs corrections. The subdivision potential is defined as

$$E \approx U_t(S_t, V_t, N_t, N + 1) - U_t(S_t, V_t, N_t, N) \qquad (10A.6)$$

It is the energy required to increase in one unit the number of subdivisions of the system, while keeping constant the extensive state variables (S_t, V_t, N_t). More exactly, the definition of E is

$$E \equiv (\partial U_t/\partial N)_{S_t, V_t, N_t} \qquad (10A.7)$$

E can be positive or negative, depending on the nature of the small systems. This potential vanishes in classical thermodynamics (negligible in a macroscopic system). Note that the Gibbs equation, $dU_t = TdS_t - pdV_t + \mu dN_t$, forbids the variation of U_t while keeping constant (S_t, V_t, N_t). Note also that E is not the same as

$$U \equiv (\partial U_t/\partial N)_{S, V, N}$$

which is the energy required to add another identical subsystem (characterized by the state variables S, V, and N) to the composite system.

The idea here is that the Gibbs, Euler, and Gibbs–Duhem equations may cease to be valid for one subsystem if it is so small that it does not satisfy the *macroscopic limit*. But a system composed of a large number N of small systems is *macroscopic* and satisfies the Gibbs, Euler, and Gibbs–Duhem equations of macroscopic thermodynamics. But, U_t must be considered a function of (S_t, V_t, N_t) and the number N of subsystems and, therefore, the Gibbs equation of the composite system is

$$dU_t = TdS_t - pdV_t + \mu dN_t + EdN \qquad (10A.8)$$

Thus, (E, N) becomes *a new pair of conjugate quantities* similar to (T, S_t), (p, V_t), and (μ, N_t). The composite system is macroscopic and the small systems are *noninteracting*, their separation being larger than the range of their interactions.

So the energy U_t can be assumed to be a first-order homogeneous of its extensive variables, N included. Euler and Gibbs–Duhem equations now take the form

$$U_t = TS_t - pV_t + \mu N_t + EN \qquad (10A.9)$$

$$NdE = -S_t dT + V_t dp - N_t d\mu \qquad (10A.10)$$

Divide Equations 10A.9 and 10A.10 by N, and use Equation A10.4, to obtain the Euler and Gibbs–Duhem equations of a small system:

$$U = TS - pV + \mu N + E \text{ (small system)} \qquad (10A.11)$$

$$dE = -SdT + Vdp - Nd\mu \text{ (small system)} \qquad (10A.12)$$

From these two, the Gibbs equation of a small system is

$$dU = TdS - pdV + \mu dN \text{ (small and macroscopic system)} \qquad (10A.13)$$

which is the same as in classical thermodynamics! On the contrary, the Gibbs—Duhem equation (Equation 10A.12), differs from the macroscopic Gibbs--Duhem equation

$$Nd\mu^\infty = -SdT + Vdp \text{ (macroscopic system)}.$$

In a macroscopic system, fixing T and p also implies that μ^∞ is fixed. Equation 10A.12 reflects one of the key features of nanothermodynamics. *It is that the intensive parameters* T, p, *and* μ *can now be varied independently due to the additional degree of freedom brought by the variable size of the small system.* Thus, from Equation 10A.12

$$(d\mu)_{T,p} = -dE/N \qquad (10A.14)$$

Alternatively, if the environment fixes T and μ of the small system, its pressure can still be varied because the size can be varied.

This discussion lays the foundations of nanothermodynamics. For further details of theory and applications, seminal references [10.36] and [10.37] may be consulted.

To conclude, nanothermodynamics allows nanosystem size to be taken into account while being general enough to retain the macroscopic nature of the overall system.

Problems

10.1 Particle size determination. Use the Scherrer formula to determine the particle sizes from the x-ray diffraction patterns shown below for yttria-stabilized zirconia nanocrystals.

Also, do a Voigt function-based Rietveld fit to estimate the size of the nanocrystals [10.31] (Figure P10.1).

10.2 The mixed phase (orthorhombic–tetragonal) region in nano Ag_2Se exhibits a λ-shaped specific heat anomaly peaking at 408 K while thermal diffusivity shows a dip at the same temperature (see Figure P10.2). This is also the temperature at which semiconducting Ag_2Se becomes a fast Ag^+-ion conductor. Explain this behavior in terms of phonon scattering in Ag_2Se.

10.3 Furusawa model: This model deals with the thickness dependence of ionic conductivity on basic classical electrodynamics:

$$\log \delta\sigma = \log \{(\sigma\lambda - \sigma_\infty)\lambda\} - \log d$$

A straight line with slope $(\sigma\lambda - \sigma_\infty)\lambda$ passes through origin. $\log \delta\sigma$ versus $\log d$ plot is a straight line with slope -1 for sample thickness $d \gg \lambda$. For $0 < d < \lambda$ one gets a straight line with slope zero. From the crossover of the two straight lines, estimate the thickness λ of the high

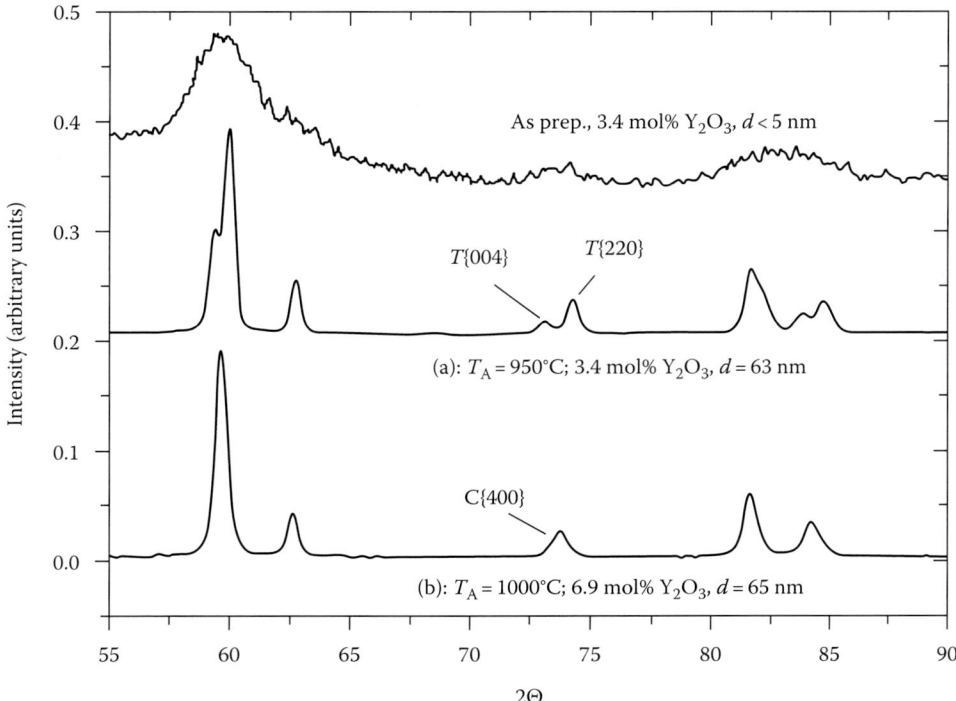

FIGURE P10.1 XRD patterns of nanocrystalline yttria-stabilized zirconia samples. Line broadening and qualitative analysis of the crystal structure help estimate crystallite size. (From Knoner, G. et al., *Proc. Natl. Acad. Sci. USA*, 100, 3870, 2003.)

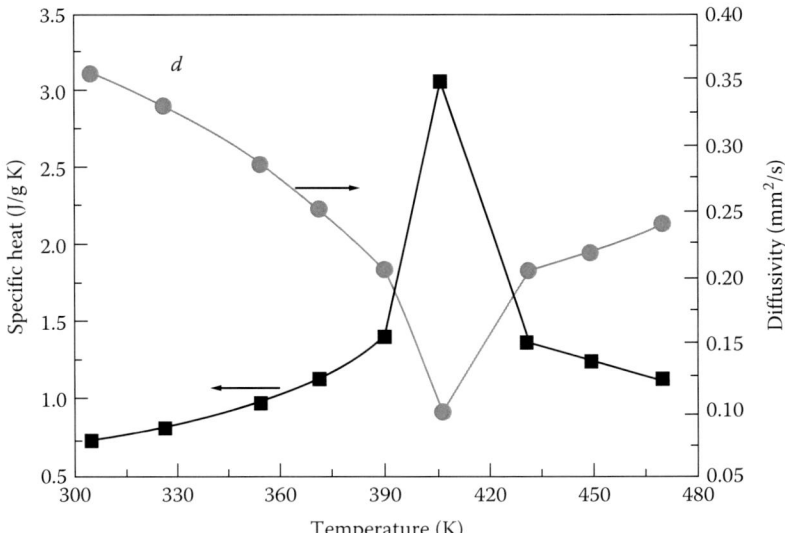

FIGURE P10.2   Specific heat maximum and thermal diffusivity minimum in Ag_2Se nanocrystals. (From Xiao C. et al., *J. Am. Chem. Soc.*, 134, 4287, 2012.)

ionic conduction region. Using estimated λ and slope of straight line $(\sigma\lambda - \sigma_\infty)\lambda$, $\sigma\lambda$ would be estimated. This procedure is effective for thin film samples of thickness $d > \lambda$. For further details see [10.35].

REFERENCES

10.1   C. Kitttel, *Introduction to Solid State Physics*, 7th edn., John Wiley & Sons, 2004, p. 168.

10.2   P. Senthil Kumar, C. S. Sunandana, *Nano Lett.*, 2(2002) 431.

10.3   C. Nan et al., *NanoResearch*, 6(7)(2013) 469.

10.4   J. Maier, *Solid State Ionics* 131(2000) 13; F. Shi et al., *Prog. Nat. Sci. Mater. Int.*, 22(2012) 15.

10.5   M. Aoki et al., *J. Am. Ceram. Soc.*, 79(1996) 1169.

10.6   J. Maier, *Chem. Mater.*, 26(2014) 348.

10.7   J. Maier, *Solid State Ionics*, 70/71(1994) 43.

10.8   Goodyear, C. E., *Electrochim. Acta,* 56(2011) 9295.

10.9   S. Y. Kang et al., *Nano Lett.*, 14(2014) 1016.

10.10   K. Lehovec, *J. Chem. Phys.*, 21(1953) 1123.

10.11   Y. G. Guo et al., *Adv. Mater.*, 17(2005) 2815.

10.12   B. Wassermann, T. P. Martin, *Solid State Commun.*, 65(1988) 561.

10.13   H. Zheng et al., *Science*, 333(2011) 20.

10.14   J. M. McHale et al., *Science*, 277(1997) 788.

10.15   C. C. Chen et al., *Science*, 276(1997) 398.

10.16   J. B. Rivest et al., *J. Phys. Chem. Lett.*, 2(2011) 2402.

10.17   K. Jacobs et al., *Science*, 293(2001) 1803.

10.18   C. Xiao et al., *JACS*, 134(2012) 428.

10.19   J. Wang et al., *Chem. Mater.*, 26(2014) 5647.

10.20   http://en.wikipedia.org/wiki/Nanoionic_device.

10.21   A. Brazier et al., *Chem. Mater.*, 20(2008) 2352–2359.

10.22   M. Aniya, M. Kobayashi, *Appl. Phys. A*, 49(1989) 641–646.

10.23   A. Peters et al., *Solid State Ionics*, 178(2007) 67.

10.24   N. Sata et al., *Nature*, 408(2000) 946.

10.25   G. Balakrishnan et al., *Nanoscale Res. Lett.*, 8(2013) 82.

10.26   A. L. Despotuli, V. I. Nikolaichic, *Solid State Ionics*, 60(4)(1993) 275–278.

10.27   R. Waser, M. Aono, *Nat. Mater.*, 6(11)(2007) 833.

10.28   A. L. Despotuli, A. V. Andreeva, *Mod. Electron.*, 7(2007) 24–29. Russian: [1], English translation: [2].

10.29   J. Maier, *Nat. Mater.*, 4(11)(2005) 805.

10.30   S. Yamaguchi, *Sci. Technol. Adv. Mater.*, 8(6)(2007) 503.

10.31   K. Reimann, R. Wurschum, *J. Appl. Phys.*, 81(1997) 7186.

10.32   G. Knoner et al., *Proc. Natl. Acad. Sci. USA*, 100(2003) 3870.

10.33   A van Dijken, Optical properties and quantum confinement of nanocrystalline II–VI semiconductor particles, PhD Thesis, Utrecht University (1999).

10.34   I. Kosacki et al. *Appl. Phys. Lett.* 74(1999).

10.35   S. Furusawa, *Journal of the Physical Society of Japan*, 70(2001) 3585.

10.35   J. N. Lalena et al., *Inorganic Materials Synthesis and Fabrication,* Wiley, 2008.

10.36   T. L. Hill *Nano Letters*, 2001, 1 (5), pp 273–275.

10.37   T. L. Hill, *Thermodynamics of Small Systems*, Dover, New York, 1994.

11

In Lieu of an Epilogue

Ions' Meet*

(Ions rarely meet like planets. Here is a lyrical view of what happened at the "in camera" event).

When ions met
At the periodic table
Of the cosmic theatre,
Emeritus Professor Proton of
Helios island was the Chair.
At the opening Cat session
Dr Lithium who became professor
On fast track declared
"I'm the prime mover of the 21st century"
"Note this"-pointing to the
Listless nobleman silver
Matronly Ms Iodine sighed in sympathy
Doctors Natter and Pot-ash
Didn't make any bones about
Their bioactivity-Chairman nodded
Adding "Of course-I superwise"
The Afternoon Aunt session
Featured Dr Ms Florina in her
Stunning mink acid-proof wear
(Even Faraday would blush!)
I collude with carbon-you would
Realize trifle lately-allowing
The prime mover function
Not to be outdone, our life
Saver Prof Oxy fuelled
His arguments to conclude
"I'm the ultimate solution to pollution"
Doctor Chlora gently reminded-
"Don't be too sure and let your
Car battery down-not yet"
Carbon, zirconium, yttrium
And other senior engineers
Added cautiously
"Dare you forget us
Not any skeletons in your cupboard"
Even as Prof. Proton
Readied to conclude

---

* This unpublished poem was presented at the *Fourth National Conference on Solid State Ionics* held at IIT Bombay, India, during March 3–5, 2000, and sums up Solid State Ionics in an informal way.

Index

A

Absolute rate theory (ART), 359–361
AFM, *see* Atomic force microscope (AFM)
Alumina/zirconia (Al_2O_3/ZrO_2) multilayer thin films, 506
Anharmonicity
    anisotropic Debye–Waller factor, 348
    Bragg reflections, 348
    deviation of system, 346
    energy and entropy, 347
    equilibrium energy, 347
    fourth-order anharmonicity, 348
    LO phonon, 348
    nonzero thermal expansion, solid, 348
    oscillating system, 346
    phonon frequencies, 348
    physical properties of Li^+, 348
    Raman pressure measurements, 348
    SHO, 346
    temperature dependence, Li_2S single crystal, 348–349
    thermal expansion of solids, 346
    thermodynamics, 347
    x-ray diffraction experiments, 348
Anisotropic Debye–Waller factor, 348
Anode
    carbon nanosheets, 448
    graphene paper, 448–449
    LIB, 447
    materials, 447
    titania, 448
Arrhenius plot, 268
ART, *see* Absolute rate theory (ART)
Athermal phase transitions
    Arrhenius slopes, 296
    description, 292
    Landau theory, 295
    Li-7 and Na-23 NMR study, 296
    SO_{4-} groups, 293
    strain components, 293
    thermal and athermal strains, 294
Atomic force microscope (AFM), 82–83, 180, 183

B

Batterivity
    car battery, 14, 16
    cathode–electrolyte interface, 16
    crystal structure, 14
    electrons, 14
    lattice, 17
    lead–acid batteries, 14
    mobile/laptop/EV battery, 16
    1s orbital, 16
Beer–Lambert law, 229
Bloembergen–Purcell–Pound (BPP) formula, 247

Bond fluctuations and phase transitions
    Aniya's model, 381
    chemical bonds, 380
    equilibrium condition, 382
    iconicity concepts, 381
    polar covalently bond, 380–381
    stimuli and responses, 383
    tetrahedral bonds, 381
Bond overlap population (BOP) distributions, 396
Born–von Karman (B–vK) theory, 321–322
Bose–Einstein statistics, 201
Bouguer–Lambert–Beer law, 227
BPP formula, *see* Bloembergen–Purcell–Pound (BPP) formula
Bragg–Brentano focus geometry, 211
Bragg scattering angle, 159
Brillouin zone (BZ)
    boundary, 319
    Ewald sphere, 29
    fcc crystal, 315, 317
    geometrical construction, 29
    Jones zone, 385
    primitive unit cells, 31
    reciprocal lattice, 30, 33
    soft mode, 362
    solid state and materials, 29
    TBA, 401–402
    Wigner–Seitz cell, 29

C

Car battery
    anode, 439
    battery-aided vehicles, 441
    charge–discharge characteristics, 439
    electrochemical reactions, 441
    electrolyte, 439
    lead–acid battery, 439–440
Cathode
    Al foil, thin film, 443–444
    conduction processes, $LiFePO_4$, 442
Catlow–Stoneham model, 375
CE, *see* Counter electrode (CE)
Ceria-stabilized zirconia (CSZ)/Al_2O_3, 502
Coin battery, 152
Common salt
    face-centered cubic NaI structure, 4
    fluorite structure, 3
    ionic conductivity, 5
    lead–acid battery, 2
    microionic and nanoionic power devices, 3
    NaI and AgI, 4
    XRD, 4
Computer calculation method, 412

For Product Safety Concerns and Information please contact our EU representative GPSR@taylorandfrancis.com Taylor & Francis Verlag GmbH, Kaufingerstraße 24, 80331 München, Germany

Printed and bound by CPI Group (UK) Ltd, Croydon, CR0 4YY

10/10/2025

01973586-0001